Discriminant Analysis and Statistical Pattern Recognition

Probability and Mathematical Statistics (Continued)

SERFLING · Approximation Theorems of Mathematical Statistics
SHORACK and WELLNER · Empirical Processes with Applications to Statistics
STAUDTE and SHEATHER · Robust Estimation and Testing
STOYANOV · Counterexamples in Probability
STYAN · The Collected Papers of T. W. Anderson: 1943–1985
WHITTAKER · Graphical Models in Applied Multivariate Statistics
YANG · The Construction Theory of Denumerable Markov Processes

Applied Probability and Statistics

ABRAHAM and LEDOLTER · Statistical Methods for Forecasting
AGRESTI · Analysis of Ordinal Categorical Data
AGRESTI · Categorical Data Analysis
ANDERSON and LOYNES · The Teaching of Practical Statistics
ANDERSON, AUQUIER, HAUCK, OAKES, VANDAELE, and WEISBERG · Statistical Methods for Comparative Studies
ASMUSSEN · Applied Probability and Queues
*BAILEY · The Elements of Stochastic Processes with Applications to the Natural Sciences
BARNETT · Interpreting Multivariate Data
BARNETT and LEWIS · Outliers in Statistical Data, *Second Edition*
BARTHOLOMEW, FORBES, and McLEAN · Statistical Techniques for Manpower Planning, *Second Edition*
BATES and WATTS · Nonlinear Regression Analysis and Its Applications
BELSLEY · Conditioning Diagnostics: Collinearity and Weak Data in Regression
BELSLEY, KUH, and WELSCH · Regression Diagnostics: Identifying Influential Data and Sources of Collinearity
BHAT · Elements of Applied Stochastic Processes, *Second Edition*
BHATTACHARYA and WAYMIRE · Stochastic Processes with Applications
BIEMER, GROVES, LYBERG, MATHIOWETZ, and SUDMAN · Measurement Errors in Surveys
BLOOMFIELD · Fourier Analysis of Time Series: An Introduction
BOLLEN · Structural Equations with Latent Variables
BOX · R. A. Fisher, the Life of a Scientist
BOX and DRAPER · Empirical Model-Building and Response Surfaces
BOX and DRAPER · Evolutionary Operation: A Statistical Method for Process Improvement
BOX, HUNTER, and HUNTER · Statistics for Experimenters: An Introduction to Design, Data Analysis, and Model Building
BROWN and HOLLANDER · Statistics: A Biomedical Introduction
BUCKLEW · Large Deviation Techniques in Decision, Simulation, and Estimation
BUNKE and BUNKE · Nonlinear Regression, Functional Relations and Robust Methods: Statistical Methods of Model Building
CHATTERJEE and HADI · Sensitivity Analysis in Linear Regression
CHATTERJEE and PRICE · Regression Analysis by Example, *Second Edition*
CLARKE and DISNEY · Probability and Random Processes: A First Course with Applications, *Second Edition*
COCHRAN · Sampling Techniques, *Third Edition*
*COCHRAN and COX · Experimental Designs, *Second Edition*
CONOVER · Practical Nonparametric Statistics, *Second Edition*
CONOVER and IMAN · Introduction to Modern Business Statistics
CORNELL · Experiments with Mixtures, Designs, Models, and the Analysis of Mixture Data, *Second Edition*
COX · A Handbook of Introductory Statistical Methods
*COX · Planning of Experiments
CRESSIE · Statistics for Spatial Data
DANIEL · Applications of Statistics to Industrial Experimentation
DANIEL · Biostatistics: A Foundation for Analysis in the Health Sciences, *Fifth Edition*
DAVID · Order Statistics, *Second Edition*
DEGROOT, FIENBERG, and KADANE · Statistics and the Law
*DEMING · Sample Design in Business Research
DILLON and GOLDSTEIN · Multivariate Analysis: Methods and Applications
DODGE and ROMIG · Sampling Inspection Tables, *Second Edition*
DOWDY and WEARDEN · Statistics for Research, *Second Edition*

*Now available in a lower priced paperback edition in the Wiley Classics Library.

Applied Probability and Statistics (Continued)

DRAPER and SMITH · Applied Regression Analysis, *Second Edition*
DUNN · Basic Statistics: A Primer for the Biomedical Sciences, *Second Edition*
DUNN and CLARK · Applied Statistics: Analysis of Variance and Regression, *Second Edition*
ELANDT-JOHNSON and JOHNSON · Survival Models and Data Analysis
FLEISS · The Design and Analysis of Clinical Experiments
FLEISS · Statistical Methods for Rates and Proportions, *Second Edition*
FLEMING and HARRINGTON · Counting Processes and Survival Analysis
FLURY · Common Principal Components and Related Multivariate Models
GALLANT · Nonlinear Statistical Models
GROSS and HARRIS · Fundamentals of Queueing Theory, *Second Edition*
GROVES · Survey Errors and Survey Costs
GROVES, BIEMER, LYBERG, MASSEY, NICHOLLS, and WAKSBERG · Telephone Survey Methodology
HAHN and MEEKER · Statistical Intervals: A Guide for Practitioners
HAND · Discrimination and Classification
HEIBERGER · Computation for the Analysis of Designed Experiments
HELLER · MACSYMA for Statisticians
HOAGLIN, MOSTELLER, and TUKEY · Exploratory Approach to Analysis of Variance
HOAGLIN, MOSTELLER, and TUKEY · Exploring Data Tables, Trends and Shapes
HOAGLIN, MOSTELLER, and TUKEY · Understanding Robust and Exploratory Data Analysis
HOCHBERG and TAMHANE · Multiple Comparison Procedures
HOEL · Elementary Statistics, *Fifth Edition*
HOGG and KLUGMAN · Loss Distributions
HOLLANDER and WOLFE · Nonparametric Statistical Methods
HOSMER and LEMESHOW · Applied Logistic Regression
IMAN and CONOVER · Modern Business Statistics
JACKSON · A User's Guide to Principle Components
JOHN · Statistical Methods in Engineering and Quality Assurance
JOHNSON · Multivariate Statistical Simulation
JOHNSON and KOTZ · Distributions in Statistics
 Discrete Distributions
 Continuous Univariate Distributions—1
 Continuous Univariate Distributions—2
 Continuous Multivariate Distributions
JUDGE, GRIFFITHS, HILL, LÜTKEPOHL, and LEE · The Theory and Practice of Econometrics, *Second Edition*
JUDGE, HILL, GRIFFITHS, LÜTKEPOHL, and LEE · Introduction to the Theory and Practice of Econometrics, *Second Edition*
KALBFLEISCH and PRENTICE · The Statistical Analysis of Failure Time Data
KASPRZYK, DUNCAN, KALTON, and SINGH · Panel Surveys
KISH · Statistical Design for Research
KISH · Survey Sampling
LAWLESS · Statistical Models and Methods for Lifetime Data
LEBART, MORINEAU, and WARWICK · Multivariate Descriptive Statistical Analysis: Correspondence Analysis and Related Techniques for Large Matrices
LEVY and LEMESHOW · Sampling of Populations: Methods and Applications
LINHART and ZUCCHINI · Model Selection
LITTLE and RUBIN · Statistical Analysis with Missing Data
MAGNUS and NEUDECKER · Matrix Differential Calculus with Applications in Statistics and Econometrics
MAINDONALD · Statistical Computation
MALLOWS · Design, Data, and Analysis by Some Friends of Cuthbert Daniel
MANN, SCHAFER, and SINGPURWALLA · Methods for Statistical Analysis of Reliability and Life Data
MASON, GUNST, and HESS · Statistical Design and Analysis of Experiments with Applications to Engineering and Science
McLACHLAN · Discriminant Analysis and Statistical Pattern Recognition
MILLER · Survival Analysis
MONTGOMERY and PECK · Introduction to Linear Regression Analysis, *Second Edition*

*Now available in a lower priced paperback edition in the Wiley Classics Library.

Continued on back end papers

Discriminant Analysis and Statistical Pattern Recognition

GEOFFREY J. McLACHLAN
Department of Mathematics
The University of Queensland
St. Lucia, Queensland, Australia

A Wiley-Interscience Publication
JOHN WILEY & SONS, INC.
New York • Chichester • Brisbane • Toronto • Singapore

Library of Congress Cataloging in Publication Data:

McLachlan, Geoffrey J., 1946–
Discriminant Analysis and Statistical Pattern Recognition/
Geoffrey J. McLachlan
p. cm.—(Wiley series in probability and mathematical statistics. Applied probability and statistics)
"A Wiley-Interscience publication."
Includes bibliographical references and index.
ISBN 0-471-61531-5
1. Discriminant analysis. 2. Pattern perception. I. Title.
II. Series.
QA278.65.M38 1992
519.5′35–dc20 91-29342
CIP

Printed and bound in the United States of America

10 9 8 7 6 5 4 3

To
Beryl, Jonathan, and Robbie

Contents

Preface xiii

1. General Introduction 1

1.1. Introduction, 1
1.2. Basic Notation, 4
1.3. Allocation Rules, 6
1.4. Decision-Theoretic Approach, 7
1.5. Unavailability of Group-Prior Probabilities, 9
1.6. Training Data, 11
1.7. Sample-Based Allocation Rules, 12
1.8. Parametric Allocation Rules, 13
1.9. Assessment of Model Fit, 16
1.10. Error Rates of Allocation Rules, 17
1.11. Posterior Probabilities of Group Membership, 21
1.12. Distances Between Groups, 22

2. Likelihood-Based Approaches to Discrimination 27

2.1. Maximum Likelihood Estimation of Group Parameters, 27
2.2. A Bayesian Approach, 29
2.3. Estimation of Group Proportions, 31
2.4. Estimating Disease Prevalence, 33
2.5. Misclassified Training Data, 35
2.6. Partially Classified Training Data, 37
2.7. Maximum Likelihood Estimation for Partial Classification, 39

2.8. Maximum Likelihood Estimation for Partial Nonrandom Classification, 43
2.9. Classification Likelihood Approach, 45
2.10. Absence of Classified Data, 46
2.11. Group-Conditional Mixture Densities, 50

3. Discrimination via Normal Models **52**

3.1. Introduction, 52
3.2. Heteroscedastic Normal Model, 52
3.3. Homoscedastic Normal Model, 59
3.4. Some Other Normal-Theory Based Rules, 65
3.5. Predictive Discrimination, 67
3.6. Covariance-Adjusted Discrimination, 74
3.7. Discrimination with Repeated Measurements, 78
3.8. Partially Classified Data, 86
3.9. Linear Projections of Homoscedastic Feature Data, 87
3.10. Linear Projections of Heteroscedastic Feature Data, 96

4. Distributional Results for Discrimination via Normal Models **101**

4.1. Introduction, 101
4.2. Distribution of Sample NLDF (W-Statistic), 101
4.3. Moments of Conditional Error Rates of Sample NLDR, 107
4.4. Distributions of Conditional Error Rates of Sample NLDR, 112
4.5. Constrained Allocation with the Sample NLDR, 118
4.6. Distributional Results for Quadratic Discrimination, 122

5. Some Practical Aspects and Variants of Normal Theory-Based Discriminant Rules **129**

5.1. Introduction, 129
5.2. Regularization in Quadratic Discrimination, 130
5.3. Linear Versus Quadratic Normal-Based Discriminant Analysis, 132
5.4. Some Models for Variants of the Sample NQDR, 137
5.5. Regularized Discriminant Analysis (RDA), 144
5.6. Robustness of NLDR and NQDR, 152
5.7. Robust Estimation of Group Parameters, 161

6. Data Analytic Considerations with Normal Theory-Based Discriminant Analysis 168

6.1. Introduction, 168
6.2. Assessment of Normality and Homoscedasticity, 169
6.3. Data-Based Transformations of Feature Data, 178
6.4. Typicality of a Feature Vector, 181
6.5. Sample Canonical Variates, 185
6.6. Some Other Methods of Dimension Reduction to Reveal Group Structure, 196
6.7. Example: Detection of Hemophilia A Carriers, 201
6.8. Example: Statistical Diagnosis of Diabetes, 206
6.9. Example: Testing for Existence of Subspecies in Fisher's *Iris* Data, 211

7. Parametric Discrimination via Nonnormal Models 216

7.1. Introduction, 216
7.2. Discrete Feature Data, 216
7.3. Parametric Formulation for Discrete Feature Data, 218
7.4. Location Model for Mixed Features, 220
7.5. Error Rates of Location Model-Based Rules, 229
7.6. Adjustments to Sample NLDR for Mixed Feature Data, 232
7.7. Some Nonnormal Models for Continuous Feature Data, 238
7.8. Case Study of Renal Venous Renin in Hypertension, 243
7.9. Example: Discrimination Between Depositional Environments, 249

8. Logistic Discrimination 255

8.1. Introduction, 255
8.2. Maximum Likelihood Estimation of Logistic Regression Coefficients, 259
8.3. Bias Correction of MLE for $g = 2$ Groups, 266
8.4. Assessing the Fit and Performance of Logistic Model, 270
8.5. Logistic Versus Normal-Based Linear Discriminant Analysis, 276
8.6. Example: Differential Diagnosis of Some Liver Diseases, 279

9. Nonparametric Discrimination **283**

9.1. Introduction, 283
9.2. Multinomial-Based Discrimination, 284
9.3. Nonparametric Estimation of Group-Conditional Densities, 291
9.4. Selection of Smoothing Parameters in Kernel Estimates of Group-Conditional Densities, 300
9.5. Alternatives to Fixed Kernel Density Estimates, 308
9.6. Comparative Performance of Kernel-Based Discriminant Rules, 312
9.7. Nearest Neighbor Rules, 319
9.8. Tree-Structured Allocation Rules, 323
9.9. Some Other Nonparametric Discriminant Procedures, 332

10. Estimation of Error Rates **337**

10.1. Introduction, 337
10.2. Some Nonparametric Error-Rate Estimators, 339
10.3. The Bootstrap, 346
10.4. Variants of the Bootstrap, 353
10.5. Smoothing of the Apparent Error Rate, 360
10.6. Parametric Error-Rate Estimators, 366
10.7. Confidence Intervals, 370
10.8. Some Other Topics in Error-Rate Estimation, 373

11. Assessing the Reliability of the Estimated Posterior Probabilities of Group Membership **378**

11.1. Introduction, 378
11.2. Distribution of Sample Posterior Probabilities, 379
11.3. Further Approaches to Interval Estimation of Posterior Probabilities of Group Membership, 384

12. Selection of Feature Variables in Discriminant Analysis **389**

12.1. Introduction, 389
12.2. Test for No Additional Information, 392
12.3. Some Selection Procedures, 396
12.4. Error-Rate-Based Procedures, 400
12.5. The F-Test and Error-Rate-Based Variable Selections, 406

12.6. Assessment of the Allocatory Capacity of the Selected Feature Variables, 410

13. Statistical Image Analysis **413**

13.1. Introduction, 413
13.2. Markov Random Fields, 417
13.3. Noncontextual Methods of Segmentation, 421
13.4. Smoothing Methods, 422
13.5. Individual Contextual Allocation of Pixels, 425
13.6. ICM Algorithm, 428
13.7. Global Maximization of the Posterior Distribution of the Image, 435
13.8. Incomplete-Data Formulation of Image Segmentation, 438
13.9. Correlated Training Data, 443

References **447**

Author Index **507**

Subject Index **519**

Preface

Over the years a not inconsiderable body of literature has accumulated on discriminant analysis, with its usefulness demonstrated over many diverse fields, including the physical, biological and social sciences, engineering, and medicine. The purpose of this book is to provide a modern, comprehensive, and systematic account of discriminant analysis, with the focus on the more recent advances in the field. Discriminant analysis or (statistical) discrimination is used here to include problems associated with the statistical separation between distinct classes or groups and with the allocation of entities to groups (finite in number), where the existence of the groups is known *a priori* and where typically there are feature data on entities of known origin available from the underlying groups. It thus includes a wide range of problems in statistical pattern recognition, where a pattern is considered as a single entity and is represented by a finite dimensional vector of features of the pattern.

In recent times, there have been many new advances made in discriminant analysis. Most of them, for example those based on the powerful but computer-intensive bootstrap methodology, are now computationally feasible with the relatively easy access to high-speed computers. The new advances are reported against the background of the extensive literature already existing in the field. Both theoretical and practical issues are addressed in some depth, although the overall presentation is biased toward practical considerations.

Some of the new advances that are highlighted are regularized discriminant analysis and bootstrap-based assessment of the performance of a sample-based discriminant rule. In the exposition of regularized discriminant analysis, it is noted how some of the sample-based discriminant rules that have been proposed over the years may be viewed as regularized versions of the normal-based quadratic discriminant rule. Recently, there has been proposed a more sophisticated regularized version, known as regularized discriminant analysis. This approach, which is a sample-based compromise between normal-based linear and quadratic discriminant analyses, is considered in some detail, given the highly encouraging results that have been reported for its performance in such difficult circumstances, as when the group-sample sizes are small relative to the number of feature variables. On the role of the bootstrap in estimation problems in discriminant analysis, particular attention is given to its usefulness in providing improved nonparametric estimates of the error rates of sample-based discriminant rules in their applications to unclassified entities.

With the computer revolution, data are increasingly being collected in the form of images, as in remote sensing. As part of the heavy emphasis on recent advances in the literature, an account is provided of extensions of discriminant analysis motivated by problems in statistical image analysis.

The book is a monograph, not a textbook. It should appeal to both applied and theoretical statisticians, as well as to investigators working in the many diverse areas in which relevant use can be made of discriminant techniques. It is assumed that the reader has a fair mathematical or statistical background.

The book can be used as a source of reference on work of either a practical or theoretical nature on discriminant analysis and statistical pattern recognition. To this end, an attempt has been made to provide a broad coverage of the results in these fields. Over 1200 references are given.

Concerning the coverage of the individual chapters, Chapter 1 provides a general introduction of discriminant analysis. In Chapter 2, likelihood-based approaches to discrimination are considered in a general context. This chapter also provides an account of the use of the EM algorithm in those situations where maximum likelihood estimation of the group-conditional distributions is to be carried out using unclassified feature data in conjunction with the training feature data of known group origin.

As with other multivariate statistical techniques, the assumption of multivariate normality provides a convenient way of specifying a parametric group structure. Chapter 3 concentrates on discrimination via normal theory-based models. In the latter part of this chapter, consideration is given also to reducing the dimension of the feature vector by appropriate linear projections. This process is referred to in the pattern recognition literature as linear feature selection. Chapter 4 reports available distributional results for normal-based discriminant rules. Readers interested primarily in practical applications of discriminant analysis may wish to proceed directly to Chapter 5, which discusses practical aspects and variants of normal-based discriminant rules. The aforementioned approach of regularized discriminant analysis is emphasized there.

Chapter 6 is concerned primarily with data analytic considerations with normal-based discriminant analysis. With a parametric formulation of problems in discriminant analysis, there is a number of preliminary items to be addressed. They include the detection of apparent outliers among the training sample, the question of model fit for the group-conditional distributions, the use of data-based transformations to achieve approximate normality, the assessment of typicality of the feature vector on an unclassified entity to be allocated to one of the specifed groups, and low-dimensional graphical representations of the feature data for highlighting and/or revealing the underlying group structure. Chapter 7 is devoted to parametric discrimination via nonnormal models for feature variables that are either all discrete, all continuous, or that are mixed in that they consist of both types of variables. A semiparametric approach is adopted in Chapter 8 with a study of the widely used logistic model for discrimination. Nonparametric approaches to discrimination are presented in Chapter 9. Particular attention in this chapter is given to kernel discriminant analysis, where the nonparametric kernel method is used to estimate the group-conditional densities in the formation of the posterior probabilities of group membership and the consequent discriminant rule.

Chapter 10 is devoted fully to the important but difficult problem of assessing the various error rates of a sample-based discriminant rule on the basis of the same data used in its construction. The error rates are useful in summarizing the global performance of a discriminant rule. Of course, for a specific case as, for example, in medical diagnosis, it is more appropriate to concentrate on the estimation of the posterior probabilities of group membership. Accordingly, a separate chapter (Chapter 11) is devoted to this problem.

Chapter 12 is on the selection of suitable feature variables using a variety of criteria. This is a fundamental problem in discriminant analysis, as there are many practical and theoretical reasons for not using all of the available feature variables. Finally, Chapter 13 is devoted to the statistical analysis of image data. Here the focus is on how to form contextual allocation rules that offer improved performance over the classical noncontextual rules, which ignore the spatial dependence between neighboring images.

Thanks are due to the authors and owners of copyrighted material for permission to reproduce published tables and figures. The author also wishes to thank Tuyet-Trinh Do for her assistance with the preparation of the typescript.

GEOFFREY J. MCLACHLAN

Brisbane, Queensland
January, 1991

CHAPTER 1

General Introduction

1.1 INTRODUCTION

Discriminant analysis as a whole is concerned with the relationship between a categorical variable and a set of interrelated variables. More precisely, suppose there is a finite number, say, g, of distinct populations, categories, classes, or groups, which we shall denote here by $G_1, \ldots, G_g$. We will henceforth refer to the G_i as groups. Note that in discriminant analysis, the existence of the groups is known *a priori*. An entity of interest is assumed to belong to one (and only one) of the groups. We let the categorical variable z denote the group membership of the entity, where $z = i$ implies that it belongs to group G_i $(i = 1, \ldots, g)$. Also, we let the p-dimensional vector $\mathbf{x} = (x_1, \ldots, x_p)'$ contain the measurements on p available features of the entity.

In this framework, the topic of discriminant analysis is concerned with the relationship between the group-membership label z and the feature vector $\mathbf{x}$. Within this broad topic there is a spectrum of problems, which corresponds to the inference-decision spectrum in statistical methodology. At the decision end of the scale, the group membership of the entity is unknown and the intent is to make an outright assignment of the entity to one of the g possible groups on the basis of its associated measurements. That is, in terms of our present notation, the problem is to estimate z solely on the basis of $\mathbf{x}$. In this situation, the general framework of decision theory can be invoked. An example in which an outright assignment is required concerns the selection of students for a special course, where the final decision to admit students is based on their answers to a questionnaire. For this decision problem, there are two groups with, say, G_1, referring to students who complete the course successfully, and G_2 to those who do not. The feature vector $\mathbf{x}$ for a student contains his/her answers to the questionnaire. A rule based on $\mathbf{x}$ for allocating a student to either G_1 or G_2 (that is, either accepting or rejecting the student into the course) can be formed from an analysis of the feature vectors of past students from each of the two groups. The construction of suitable allocation rules is to be pursued in the subsequent sections of this chapter.

At the other extreme end of the spectrum, no assignment or allocation of the entity to one of the possible groups is intended. Rather the problem is to draw inferences about the relationship between z and the feature variables in $\mathbf{x}$. An experiment might be designed with the specific aim to provide insight and understanding into the predictive structure of the feature variables. For example, a political scientist may wish to determine the socio-economic factors that have the most influence on the voting patterns of a population of voters.

Between these extremes lie most of the everyday situations in which discriminant analysis is applied. Typically, the problem is to make a prediction or tentative allocation for an unclassified entity. For example, concerning prediction, an economist may wish to forecast on the basis of his or her most recent accounting information, those members of the corporate sector that might be expected to suffer financial losses leading to failure. For this purpose, a discriminant rule may be formed from accounting data collected on failed and surviving companies over many past years. An example where allocation, tentative or otherwise, is required is with the discrimination between an earthquake and an underground nuclear explosion on the basis of signals recorded at a seismological station (Elvers, 1977). An allocation rule is formed from signals recorded on past seismic events of known classification.

Examples where prediction or tentative allocation is to be made for an unclassified entity occur frequently in medical prognosis and diagnosis. A source for applications of discriminant analysis to medical diagnosis is the bibliography of Wagner, Tautu and Wolbler (1978) on problems in medical diagnosis. In medical diagnosis, the definitive classification of a patient often can be made only after exhaustive physical and clinical assessments or perhaps even surgery. In some instances, the true classification can be made only on evidence that emerges after the passage of time, for instance, an autopsy. Hence, frequent use is made of diagnostic tests. Where possible, the tests are based on clinical and laboratory-type observations that can be made without too much inconvenience to the patient. The financial cost of the test is also sometimes another consideration, particularly in mass screening programs. Suppose that the feature vector $\mathbf{x}$ contains the observations taken on a patient for his or her diagnosis with respect to one of g possible disease groups $G_1, \ldots, G_g$. Then the relative plausibilities of these groups for a patient with feature vector $\mathbf{x}$ as provided by a discriminant analysis may be of assistance to the clinician in making a diagnosis. This is particularly so with the diagnosis of Conn's syndrome in patients with high blood pressure, as reported in Aitchison and Dunsmore (1975, Chapter 1). The two possible groups represent the cause, which is either a benign tumor in one adrenal gland, curable by surgical removal (G_1), or a more diffuse condition affecting both adrenal glands, with the possibility of control of blood pressure by drugs (G_2). The actual cause can be confirmed only by microscopic examination of adrenal tissue removed at an operation. However, because surgery is inadvisable for patients in G_2, a clinician is faced with a difficult treatment decision. Thus, a realistic preoperative assessment that a patient with a particular feature vector belongs to either G_1 or G_2 would be most valuable to the clinician. The available feature variables on a patient relate to age, plasma concentrations of sodium, potassium, bicarbonate, renin, and aldosterone, and systolic and diastolic blood pressures.

The relative plausibilities of group membership for a patient with an associated feature vector are also useful in medical prognosis. Here the vector is measured after the onset of some medical condition, say, an injury, and the groups represent the possible outcomes of the injury. There are several reasons why an initial prediction of the eventual outcome of the injury may be needed. For instance, in situations where the management of the patient is closely linked to the outcome, it provides a guide to the clinician as to whether his or her particular course of management is appropriate. It also provides a firmer basis for advice to relatives of an injured patient on the chances of recovery. These and other reasons are discussed by Titterington et al. (1981) and Murray et al. (1986) in the context of the prognosis for patients with severe head injuries. For these patients, the three possible outcomes were dead or vegetative, severe disability, and moderate or good recovery. The feature vector for a patient included background information such as age and cause of injury and four clinical variables (eye opening, motor response, motor response pattern, and pupil reaction).

Situations in medical diagnosis where outright rather than tentative allocations to groups are made occur in mass screening programs. Suppose that in the detection of a disease, G_1 consists of those individuals without the disease and G_2 of those with the disease. Then in a screening program for this disease, a patient is assigned outright to either G_1 or G_2, according to whether the diagnostic test is negative or positive. Usually, with a positive result, further testing is done before a final assignment is made. For example, with the enzyme-linked immunosorbent assay (ELISA) test used to screen donated blood for antibodies to the AIDS virus, a positive test would result in a more definitive test such as the Western blot being performed (Gastwirth, 1987). J. A. Anderson (1982) has given an example on patient care where an irrevocable outright assignment has to be made. It concerns the decision on whether to administer a preoperative anticoagulant therapy to a patient to reduce the risk of postoperative deep vein thrombosis.

Discriminant analysis is widely used also in the field of pattern recognition, which is concerned mainly with images. The aim of pattern recognition is to automate processes performed by humans. For example, automatic analysis and recognition of photomicrographs of tissue cells can be used in blood tests, cancer tests, and brain-tissue studies. Another example of much current interest concerns the automatic recognition of images remotely sensed from earth satellites. It is considered in Chapter 13.

The branch of pattern recognition known as statistical pattern recognition has close ties with statistical decision theory and areas of multivariate analysis, in particular discriminant analysis. In statistical pattern recognition, each pattern is considered as a single entity and is represented by a finite dimensional vector of features of the pattern. Hence, the recognition of patterns with respect to a finite number of predefined groups of patterns can be formulated within the framework of discriminant analysis. The number of features required for recognition of a pattern may become very large if the patterns under study are very complex or if, as in fingerprint identification, the number of possible pattern groups is very large. Consequently, the above approach may have to be modified; see, for example, Fu (1986) and Mantas (1987).

By now, there is an enormous literature on discriminant analysis, and so it is not possible to provide an exhaustive bibliography here. However, we have attempted to cover the main results, in particular the more recent developments. Additional references on the earlier work may be found in the books devoted to the topic as a whole or in part by Lachenbruch (1975a), Goldstein and Dillon (1978), Klecka (1980), and Hand (1981a, 1982). They have been supplemented recently by the volume edited by S. C. Choi (1986), the notes of Hjort (1986a), and the report by a panel of the Committee on Applied and Theoretical Statistics of the Board on Mathematical Sciences of the National Research Council, chaired by Professor R. Gnanadesikan (Panel on Discriminant Analysis, Classification and Clustering, 1989). Further references may be found in the symposium proceedings edited by Cacoullos (1973) and Van Ryzin (1977), the review article by Lachenbruch and Goldstein (1979), and in the encyclopedia entry by Lachenbruch (1982). There are also the relevant chapters in the rapidly growing list of textbooks on multivariate analysis. Another source of references is the pattern recognition literature. Fukunaga (1972, 1990), Patrick (1972), Duda and Hart (1973), Young and Calvert (1974), and Devijver and Kittler (1982) are examples of texts on statistical pattern recognition. A single source of references in discriminant and cluster analyses and in pattern recognition is the book edited by Krishnaiah and Kanal (1982).

1.2 BASIC NOTATION

We let $\mathbf{X}$ denote the p-dimensional random feature vector corresponding to the realization $\mathbf{x}$ as measured on the entity under consideration. The associated variable z denoting the group of origin of the entity is henceforth replaced by a g-dimensional vector $\mathbf{z}$ of zero-one indicator variables. The ith component of $\mathbf{z}$ is defined to be one or zero according as $\mathbf{x}$ (really the entity) belongs or does not belong to the ith group G_i $(i = 1,\ldots,g)$; that is,

$$\begin{aligned} z_i &= 1, \qquad \mathbf{x} \in G_i, \\ &= 0, \qquad \mathbf{x} \notin G_i, \end{aligned}$$

for $i = 1,\ldots,g$. Where possible, random variables are distinguished from their realizations by the use of the corresponding uppercase letters.

The probability density function (p.d.f.) of $\mathbf{X}$ in group G_i is denoted by $f_i(\mathbf{x})$ for $i = 1,\ldots,g$. These group-conditional densities are with respect to arbitrary measure on $\mathbb{R}^p$, so that $f_i(\mathbf{x})$ can be a mass function by the adoption of counting measure. Under the mixture model approach to discriminant analysis, it is assumed that the entity has been drawn from a mixture G of the g groups $G_1,\ldots,G_g$ in proportions $\pi_1,\ldots,\pi_g$, respectively, where

$$\sum_{i=1}^{g}\pi_i = 1 \qquad \text{and} \qquad \pi_i \geq 0 \qquad (i = 1,\ldots,g).$$

The p.d.f. of $\mathbf{X}$ in G can therefore be represented in the finite mixture form

$$f_X(\mathbf{x}) = \sum_{i=1}^{g}\pi_i f_i(\mathbf{x}). \tag{1.2.1}$$

An equivalent assumption is that the random vector $\mathbf{Z}$ of zero-one group indicator variables with $\mathbf{z}$ as its realization is distributed according to a multinomial distribution consisting of one draw on g categories with probabilities $\pi_1,\ldots,\pi_g$, respectively; that is,

$$\text{pr}\{\mathbf{Z} = \mathbf{z}\} = \pi_1^{z_1}\pi_2^{z_2}\cdots\pi_g^{z_g}. \tag{1.2.2}$$

We write

$$\mathbf{Z} \sim \text{Mult}_g(1, \boldsymbol{\pi}), \tag{1.2.3}$$

where $\boldsymbol{\pi} = (\pi_1,\ldots,\pi_g)'$. Note that with a deterministic approach to the problem, $\mathbf{z}$ is taken to be a parameter rather than a random variable as here. The distribution function of $\mathbf{Y} = (\mathbf{X}', \mathbf{Z}')'$ is denoted by $F(\mathbf{y})$, where the prime denotes vector transpose. We let $F_i(\mathbf{x})$ and $F_X(\mathbf{x})$ denote the distribution functions corresponding to the densities $f_i(\mathbf{x})$ and $f_X(\mathbf{x})$, respectively.

The ith mixing proportion π_i can be viewed as the prior probability that the entity belongs to G_i ($i = 1,\ldots,g$). With $\mathbf{X}$ having been observed as $\mathbf{x}$, the posterior probability that the entity belongs to G_i is given by

$$\begin{aligned}\tau_i(\mathbf{x}) &= \text{pr}\{\text{entity} \in G_i \mid \mathbf{x}\} \\ &= \text{pr}\{Z_i = 1 \mid \mathbf{x}\} \\ &= \pi_i f_i(\mathbf{x})/f_X(\mathbf{x}) \qquad (i = 1,\ldots,g).\end{aligned} \tag{1.2.4}$$

In the next section, we consider the formation of an optimal rule of allocation in terms of these posterior probabilities of group membership $\tau_i(\mathbf{x})$.

The term "classification" is used broadly in the literature on discriminant and cluster analyses. To avoid any possible confusion, throughout this monograph, we reserve the use of classification to describe the original definition of the underlying groups. Hence, by a classified entity, we mean an entity whose group of origin is known. A rule for the assignment of an unclassified entity to one of the groups will be referred to as a discriminant or allocation rule. In the situation where the intention is limited to making an outright assignment of the entity to one of the possible groups, it is perhaps more appropriate to use the term allocation rather than discriminant to describe the rule. However, we will use either nomenclature regardless of the underlying situation. In the pattern recognition jargon, such a rule is referred to as a classifier.

1.3 ALLOCATION RULES

At this preliminary stage of formulating discriminant analysis, we consider the pure decision case, where the intent is to make an outright assignment of an entity with feature vector $\mathbf{x}$ to one of the g possible groups. Let $r(\mathbf{x})$ denote an allocation rule formed for this purpose, where $r(\mathbf{x}) = i$ implies that an entity with feature vector $\mathbf{x}$ is to be assigned to the ith group G_i $(i = 1,\ldots,g)$. In effect, the rule divides the feature space into g mutually exclusive and exhaustive regions $R_1,\ldots,R_g$, where, if $\mathbf{x}$ falls in R_i, then the entity is allocated to group G_i $(i = 1,\ldots,g)$.

The allocation rates associated with this rule $r(\mathbf{x})$ are denoted by $e_{ij}(r)$, where

$$e_{ij}(r) = \mathrm{pr}\{r(\mathbf{X}) = j \mid \mathbf{X} \in G_i\}$$

is the probability that a randomly chosen entity from G_i is allocated to G_j $(i,j = 1,\ldots,g)$. It can be expressed as

$$e_{ij}(r) = \int_{R_j} f_i(\mathbf{x})\,d\nu,$$

where ν denotes the underlying measure on $\mathbb{R}^p$ appropriate for $f_X(\mathbf{x})$. The probability that a randomly chosen member of G_i is misallocated can be expressed as

$$\begin{aligned} e_i(r) &= \sum_{j \neq i}^{g} e_{ij}(r) \\ &= \int_{\overline{R}_i} f_i(\mathbf{x})\,d\nu, \end{aligned}$$

where $\overline{R}_i$ denotes the complement of R_i $(i = 1,\ldots,g)$.

For a diagnostic test using the rule $r(\mathbf{x})$ in the context where G_1 denotes the absence of a disease and G_2 its presence, the error rate $e_{12}(r)$ corresponds to the probability of a false positive, and $e_{21}(r)$ is the probability of a false negative. The correct allocation rates

$$e_{22}(r) = 1 - e_{21}(r) \qquad \text{and} \qquad e_{11}(r) = 1 - e_{12}(r)$$

are known as the sensitivity and specificity, respectively, of the diagnostic test.

1.4 DECISION-THEORETIC APPROACH

Decision theory provides a convenient framework for the construction of discriminant rules in the situation where an outright allocation of an unclassified entity is required. The present situation where the prior probabilities of the groups and the group-conditional densities are taken to be known is relatively straightforward.

Let c_{ij} denote the cost of allocation when an entity from G_i is allocated to group G_j, where $c_{ij} = 0$ for $i = j = 1,\ldots,g$; that is, there is zero cost for a correct allocation. We assume for the present that the costs of misallocation are all the same. We can then take the common value of the c_{ij} $(i \neq j)$ to be unity, because it is only their ratios that are important.

For given $\mathbf{x}$, the loss for allocation performed on the basis of the rule $r(\mathbf{x})$ is

$$l\{\mathbf{z}, r(\mathbf{x})\} = \sum_{i=1}^{g} z_i Q[i, r(\mathbf{x})], \tag{1.4.1}$$

where, for any u and v, $Q[u,v] = 0$ for $u = v$ and 1 for $u \neq v$. The expected loss or risk, conditional on $\mathbf{x}$, is given by

$$E[l\{\mathbf{Z}, r(\mathbf{x})\} \mid \mathbf{x}] = \sum_{i=1}^{g} \tau_i(\mathbf{x}) Q[i, r(\mathbf{x})], \tag{1.4.2}$$

since from (1.2.4),

$$E(Z_i \mid \mathbf{x}) = \tau_i(\mathbf{x}).$$

An optimal rule of allocation can be defined by taking it to be the one that minimizes the conditional risk (1.4.2) at each value $\mathbf{x}$ of the feature vector. In decision-theory language, any rule that so minimizes (1.4.2) for some $\pi_1,\ldots,\pi_g$ is said to be a Bayes rule. It can be seen from (1.4.2) that the conditional risk is a linear combination of the posterior probabilities, where all coefficients are zero except for one, which is unity. Hence, it is minimized by taking $r(\mathbf{x})$ to be the label of the group to which the entity has the highest posterior probability of belonging. Note that this is the "intuitive solution" to the allocation problem.

If we let $r_o(\mathbf{x})$ denote this optimal rule of allocation, then

$$r_o(\mathbf{x}) = i \quad \text{if} \quad \tau_i(\mathbf{x}) \geq \tau_j(\mathbf{x}) \qquad (j = 1,\ldots,g;\ j \neq i). \tag{1.4.3}$$

The rule $r_o(\mathbf{x})$ is not uniquely defined at $\mathbf{x}$ if the maximum of the posterior probabilities of group membership is achieved with respect to more than one group. In this case the entity can be assigned arbitrarily to one of the groups for which the corresponding posterior probabilities are equal to the maximum value. If

$$\text{pr}\{\tau_i(\mathbf{X}) = \tau_j(\mathbf{X})\} = 0 \qquad (i \neq j = 1,\ldots,g),$$

then the optimal rule is unique for almost all $\mathbf{x}$ relative to the underlying measure ν on R^p appropriate for $f_X(\mathbf{x})$.

As the posterior probabilities of group membership $\tau_i(\mathbf{x})$ have the same common denominator $f_X(\mathbf{x})$, $r_o(\mathbf{x})$ can be defined in terms of the relative sizes of the group-conditional densities weighted according to the group-prior probabilities; that is,

$$r_o(\mathbf{x}) = i \qquad \text{if} \quad \pi_i f_i(\mathbf{x}) \geq \pi_j f_j(\mathbf{x}) \qquad (j = 1,\ldots,g;\ j \neq i). \tag{1.4.4}$$

Note that as the optimal or Bayes rule of allocation minimizes the conditional risk (1.4.2) over all rules r, it also minimizes the unconditional risk

$$\begin{aligned} e(r) &= \sum_{i=1}^{g} E\{\tau_i(\mathbf{X})Q[i, r(\mathbf{X})]\} \\ &= \sum_{i=1}^{g} \pi_i \int_{R_i} f_i(\mathbf{x})\, d\nu \\ &= \sum_{i=1}^{g} \pi_i e_i(r), \end{aligned}$$

which is the overall error rate associated with r.

Discriminant analysis in its modern guise was founded by Fisher (1936). His pioneering paper, which did not take the group-conditional distributions to be known, is to be discussed later in this chapter in the context of sample-based allocation rules. Concerning initial work in the case of known group-conditional distributions, Welch (1939) showed for $g = 2$ groups that a rule of the form (1.4.4) is deducible either from Bayes theorem if prior probabilities are specified for the groups or by the use of the Neyman–Pearson lemma if the two group-specific errors of allocation are to be minimized in any given ratio. Wald (1939, 1949) developed a general theory of decision functions, and von Mises (1945) obtained the solution to the problem of minimizing the maximum of the errors of allocation for a finite number of groups, which was in the general theme of Wald's work. Rao (1948) discussed explicit solutions of the form (1.4.4) and also the use of a doubtful region of allocation. In a subsequent series of papers, he pursued related problems and extensions; see Rao (1952, 1954) for an account. There is an extensive literature on the development of allocation rules. The reader is referred to Das Gupta (1973) for a comprehensive review.

Up to now, we have taken the costs of misallocation to be the same. For unequal costs of misallocation c_{ij}, the conditional risk of the rule $r(\mathbf{x})$ is

$$\sum_{i \neq r(\mathbf{x})}^{g} \tau_i(\mathbf{x}) c_{i,r(\mathbf{x})}. \tag{1.4.5}$$

Let $r_o(\mathbf{x})$ be the optimal or Bayes rule that minimizes (1.4.5). Then it follows that $r_o(\mathbf{x}) = i$ if

$$\sum_{h \neq i}^{g} \tau_h(\mathbf{x}) c_{hi} \leq \sum_{h \neq j}^{g} \tau_h(\mathbf{x}) c_{hj} \qquad (j = 1, \ldots, g;\ j \neq i). \tag{1.4.6}$$

For $g = 2$ groups, (1.4.6) reduces to the definition (1.4.3) or (1.4.4) for $r_o(\mathbf{x})$ in the case of equal costs of misallocation, except that π_1 is replaced now by $\pi_1 c_{12}$ and π_2 by $\pi_2 c_{21}$. As it is only the ratio of c_{12} and c_{21} that is relevant to the definition of the Bayes rule, these costs can be scaled so that

$$\pi_1 c_{12} + \pi_2 c_{21} = 1.$$

Hence, we can assume without loss of generality that $c_{12} = c_{21} = 1$, provided π_1 and π_2 are now interpreted as the group-prior probabilities adjusted by the relative importance of the costs of misallocation. Due to the rather arbitrary nature of assigning costs of misallocation in practice, they are often taken to be the same in real problems. Further, the group-prior probabilities are often specified as equal. This is not as arbitrary as it may appear at first sight. For example, consider the two-group situation, where G_1 denotes a group of individuals with a rare disease and G_2 those without it. Then, although π_1 and π_2 are disparate, the cost of misallocating an individual with this rare disease may well be much greater than the cost of misallocating a healthy individual. If this is so, then $\pi_1 c_{12}$ and $\pi_2 c_{21}$ may be comparable in magnitude and, as a consequence, the assumption of equal group-prior probabilities with unit costs of misallocation in the formation of the Bayes rule $r_o(\mathbf{x})$ is apt. Also, it would avoid in this example the occurrence of highly unbalanced group-specific error rates. The latter are obtained if $r_o(\mathbf{x})$ is formed with extremely disparate prior probabilities π_i and equal costs of misallocation. This imbalance between the group-specific error rates is a consequence of $r_o(\mathbf{x})$ being the rule that minimizes the overall error rate. In the next section, we consider the construction of rules that are optimal with respect to other criteria. In particular, it will be seen that by specifying the prior probabilities π_i in $r_o(\mathbf{x})$ so that its consequent error rates are equal, we obtain the rule that minimizes the maximum of the group-specific error rates.

1.5 UNAVAILABILITY OF GROUP-PRIOR PROBABILITIES

In some instances in practice, the prior probabilities π_i of the groups G_i are able to be assigned or reasonable estimates are available. For example, in the context of medical diagnosis where the groups represent the possible disease categories to which an individual is to be allocated, the prior probabilities can be taken to be the prevalence rates of these diseases in the population from which the individual has come. However, as to be discussed further in Section

2.3, in some instances, the very purpose of forming an allocation rule for the basis of a screening test is to estimate the prevalence rates of diseases. Also, with a deterministic approach to the construction of an allocation rule, prior probabilities are not relevant to the formulation of the problem.

We now consider the selection of suitable allocation rules where the prior probabilities of the groups are not available. We will only give a brief coverage of available results. For further details, the reader is referred to T. W. Anderson (1984, Chapter 6), who has given a comprehensive account of the decision-theoretic approach to discriminant analysis.

In the absence of prior probabilities of the groups, we cannot define the risk either unconditional or conditional on the feature vector $\mathbf{x}$. Hence, some other criterion must be used. Various other criteria have been discussed by Raiffa (1961). One approach is to focus on the group-specific unconditional losses and to look for the class of admissible rules; that is, the set of rules that cannot be improved upon. For an entity from G_i, the unconditional loss for a rule $r(\mathbf{x})$ is

$$l_i(r) = \sum_{j \neq i}^{g} c_{ij} \text{pr}\{r(\mathbf{X}) = j \mid \mathbf{X} \in G_i\}$$

$$= \sum_{j \neq i}^{g} c_{ij} e_{ij}(r) \qquad (i = 1, \dots, g).$$

A rule $r^*(\mathbf{x})$ is at least as good as $r(\mathbf{x})$ if

$$l_i(r^*) \leq l_i(r) \qquad (i = 1, \dots, g). \tag{1.5.1}$$

If at least one inequality in (1.5.1) is strict, then $r^*(\mathbf{x})$ is better than $r(\mathbf{x})$. The rule $r(\mathbf{x})$ is said to be admissible if there is no other rule $r^*(\mathbf{x})$ that is better.

It can be shown that if $\pi_i > 0$ $(i = 1, \dots, g)$, then a Bayes rule is admissible. Also, if $c_{ij} = 1$ $(i \neq j)$, and

$$\text{pr}\{f_i(\mathbf{X}) = 0 \mid \mathbf{X} \in G_j\} = 0 \qquad (i, j = 1, \dots, g),$$

then a Bayes rule is admissible. The converse is true without conditions (except that the parameter space is finite). The proofs of these and other related results can be found in T. W. Anderson (1984, Chapter 6) and in the references therein.

A principle that usually leads to the selection of a unique rule is the minimax principle. A rule is minimax if the maximum unconditional loss is a minimum. In the present context, the rule $r(\mathbf{x})$ is minimax if the maximum of $l_i(r)$ over $i = 1, \dots, g$ is a minimum over all allocation rules. The minimax rule is the Bayes procedure for which the unconditional losses are equal (von Mises, 1945).

1.6 TRAINING DATA

We have seen in the previous section that the absence of prior probabilities for the groups introduces a complication into the process of obtaining a suitable allocation rule. A much more serious issue arises when the group-conditional densities are either partially or completely unknown.

A basic assumption in discriminant analysis is that in order to estimate the unknown group-conditional densities, there are entities of known origin on which the feature vector $\mathbf{X}$ has been recorded for each. We let $\mathbf{x}_1,\ldots,\mathbf{x}_n$ denote these recorded feature vectors and $\mathbf{z}_1,\ldots,\mathbf{z}_n$ the corresponding vectors of zero-one indicator variables defining the known group of origin of each. We let

$$\mathbf{y}_j = (\mathbf{x}_j', \mathbf{z}_j')' \qquad (j = 1,\ldots,n).$$

The collection of data in the matrix $\mathbf{t}$ defined by

$$\mathbf{t}' = (\mathbf{y}_1,\ldots,\mathbf{y}_n) \tag{1.6.1}$$

is referred to in the literature as either the initial, reference, design, training, or learning data. The last two have arisen from their extensive use in the context of pattern recognition. Also in the latter field, the formation of an allocation rule from training data of known origin is referred to as supervised learning.

There are two major sampling designs under which the training data $\mathbf{T}$ may be realized, joint or mixture sampling and $\mathbf{z}$-conditional or separate sampling. They correspond, respectively, to sampling from the joint distribution of $\mathbf{Y} = (\mathbf{X}', \mathbf{Z}')'$ and to sampling from the distribution of $\mathbf{X}$ conditional on $\mathbf{z}$. The first design applies to the situation where the feature vector and group of origin are recorded on each of n entities drawn from a mixture of the possible groups. Mixture sampling is common in prospective studies and diagnostic situations. In a prospective study design, a sample of individuals is followed and their responses recorded.

With most applications in discriminant analysis, it is assumed that the training data are independently distributed. For a mixture sampling scheme with this assumption, $\mathbf{x}_1,\ldots,\mathbf{x}_n$ are the realized values of n independent and identically distributed (i.i.d.) random variables $\mathbf{X}_1,\ldots,\mathbf{X}_n$ with common distribution function $F_X(\mathbf{x})$. We write

$$\mathbf{X}_1,\ldots,\mathbf{X}_n \overset{iid}{\sim} F_X.$$

The associated group indicator vectors $\mathbf{z}_1,\ldots,\mathbf{z}_n$ are the realized values of the random variables $\mathbf{Z}_1,\ldots,\mathbf{Z}_n$ distributed unconditionally as

$$\mathbf{Z}_1,\ldots,\mathbf{Z}_n \overset{iid}{\sim} \mathrm{Mult}_g(1, \boldsymbol{\pi}). \tag{1.6.2}$$

The assumption of independence of the training data is to be relaxed in Chapter 13. Examples in remote sensing are given there where the assumption of independence is not valid.

With separate sampling in practice, the feature vectors are observed for a sample of n_i entities taken separately from each group G_i ($i = 1,\ldots,g$). Hence, it is appropriate to retrospective studies, which are common in epidemiological investigations. For example, with the simplest retrospective case-control study of a disease, one sample is taken from the cases that occurred during the study period and the other sample is taken from the group of individuals who remained free of the disease. As many diseases are rare and even a large prospective study may produce few diseased individuals, retrospective sampling can result in important economies in cost and study duration. Note that as separate sampling corresponds to sampling from the distribution of $\mathbf{X}$ conditional on $\mathbf{z}$, it does not provide estimates of the prior probabilities π_i for the groups.

1.7 SAMPLE-BASED ALLOCATION RULES

We now consider the construction of an allocation rule from available training data $\mathbf{t}$ in the situation where the group-conditional densities and perhaps also the group-prior probabilities are unknown. The initial approach to this problem, and indeed to discriminant analysis in its modern guise as remarked earlier, was by Fisher (1936). In the context of $g = 2$ groups, he proposed that an entity with feature vector $\mathbf{x}$ be assigned on the basis of the linear discriminant function $\mathbf{a}'\mathbf{x}$, where $\mathbf{a}$ maximizes an index of separation between the two groups. The index was defined to be the magnitude of the difference between the group-sample means of $\mathbf{a}'\mathbf{x}$ normalized by the pooled sample estimate of its assumed common variance within a group. The derivation of Fisher's (1936) linear discriminant function is to be discussed further in Section 3.3, where it is contrasted with normal theory-based discriminant rules.

The early development of discriminant analysis before Fisher (1936) dealt primarily with measures of differences between groups based on sample moments or frequency tables, and ignored correlations among different variates in the feature vector (Pearson, 1916; Mahalanobis, 1927, 1928). One of Fisher's first contacts with discriminant problems was in connection with M. M. Barnard's (1935) work on the secular variation of Egyptian skull characteristics. By 1940, Fisher had published four papers on discriminant analysis, including Fisher (1938) in which he reviewed his 1936 work and related it to the contributions by Hotelling (1931) on his now famous T^2 statistic and by Mahalanobis (1936) on his D^2 statistic and earlier measures of distance. Das Gupta (1980) has given an account of Fisher's research in discriminant analysis.

With the development of discriminant analysis through to the decision-theoretic stage (Wald, 1944; Rao, 1948, 1952, 1954; Hoel and Peterson, 1949), an obvious way of forming a sample-based allocation rule $r(\mathbf{x};\mathbf{t})$ is to take it to be an estimated version of the Bayes rule $r_o(\mathbf{x})$ where, in (1.4.3), the posterior probabilities of group membership $\tau_i(\mathbf{x})$ are replaced by some estimates

$\hat{\tau}_i(\mathbf{x};\mathbf{t})$ formed from the training data $\mathbf{t}$. One approach to the estimation of the posterior probabilities of group membership is to model the $\tau_i(\mathbf{x})$ directly, as with the logistic model to be presented in Chapter 8. Dawid (1976) calls this approach the diagnostic paradigm.

A more common approach, called the sampling approach by Dawid (1976), is to use the Bayes formula (1.2.4) to formulate the $\tau_i(\mathbf{x})$ through the group-conditional densities $f_i(\mathbf{x})$. With this approach, the Bayes rule is estimated by the so-called plug-in rule,

$$r(\mathbf{x};\mathbf{t}) = r_o(\mathbf{x};\hat{F}), \tag{1.7.1}$$

where we now write the optimal or Bayes rule as $r_o(\mathbf{x};F)$ to explicitly denote its dependence on the distribution function $F(\mathbf{y})$ of $\mathbf{Y} = (\mathbf{X}',\mathbf{Z}')'$. As before, $\mathbf{X}$ is the feature observation and $\mathbf{Z}$ defines its group of origin. In (1.7.1), $\hat{F}$ denotes an estimate of F that can be obtained by estimating separately each group-conditional distribution from the training data $\mathbf{t}$.

The group-prior probabilities can be estimated by the proportion of entities from each group at least under mixture sampling. Their estimation under separate sampling is considered in the next chapter, commencing in Section 2.3. Concerning the estimates of the group-conditional distribution functions, a nonparametric approach may be adopted using, say, kernel or nearest-neighbor methods. These along with other nonparametric methods are to be discussed in Chapter 9. A commonly used approach is the parametric, which is introduced in the next section in a general context. It is to be considered further in Chapter 3 for the specific choice of normal models and in Chapter 7 for nonnormal models. There is also the work, in the spirit of the empirical Bayes approach of Robbins (1951, 1964), on the allocation of a sequence of unclassified entities whose group-indicator vectors and features are independently distributed. Results under various assumptions on the available information on the underlying distributions have been obtained by Johns (1961), Samuel (1963a, 1963b), Hudimoto (1968), K. Choi (1969), Wojciechowski (1985), and Stirling and Swindlehurst (1987), among others.

1.8 PARAMETRIC ALLOCATION RULES

Under the parametric approach to the estimation of the group-conditional distributions, and hence of the Bayes rule, the group-conditional distributions are taken to be known up to a manageable number of parameters. More specifically, the ith group-conditional density is assumed to belong to a family of densities

$$\{f_i(\mathbf{x};\boldsymbol{\theta}_i) : \boldsymbol{\theta}_i \in \Theta_i\}, \tag{1.8.1}$$

where $\boldsymbol{\theta}_i$ is an unknown parameter vector belonging to some parameter space Θ_i $(i = 1,\ldots,g)$. Often the group-conditional densities are taken to belong to the same parametric family, for example, the normal.

The density functions of $\mathbf{X}$ and $\mathbf{Y} = (\mathbf{X}', \mathbf{Z}')'$ are written now as $f_X(\mathbf{x}; \boldsymbol{\Psi})$ and $f(\mathbf{y}; \boldsymbol{\Psi})$, respectively, where

$$\boldsymbol{\Psi} = (\boldsymbol{\pi}', \boldsymbol{\theta}')' \tag{1.8.2}$$

and $\boldsymbol{\theta}$ is the vector consisting of the elements of $\boldsymbol{\theta}_1, \ldots, \boldsymbol{\theta}_g$ known *a priori* to be distinct. For example, if the group-conditional distributions are assumed to be multivariate normal with means $\boldsymbol{\mu}_1, \ldots, \boldsymbol{\mu}_g$ and common covariance matrix $\boldsymbol{\Sigma}$, then $\boldsymbol{\theta}_i$ consists of the elements of $\boldsymbol{\mu}_i$ and of the distinct elements of $\boldsymbol{\Sigma}$, and $\boldsymbol{\theta}$ consists of the elements of $\boldsymbol{\mu}_1, \ldots, \boldsymbol{\mu}_g$ and of the distinct elements of $\boldsymbol{\Sigma}$. Note that since the elements of the vector $\boldsymbol{\pi}$ of the mixing proportions π_i sum to one, one of them is redundant in $\boldsymbol{\Psi}$, but we will not modify $\boldsymbol{\Psi}$ accordingly, at least explicitly. However, in our statements about the distribution of any estimator of $\boldsymbol{\Psi}$, it will be implicitly assumed that one of the mixing proportions has been deleted from $\boldsymbol{\Psi}$.

With the so-called estimative approach to the choice of a sample-based discriminant rule, unknown parameters in the adopted parametric forms for the group-conditional distributions are replaced by appropriate estimates obtained from the training data $\mathbf{t}$. Hence, if $r_o(\mathbf{x}; \boldsymbol{\Psi})$ now denotes the optimal rule, then with this approach,

$$r(\mathbf{x}; \mathbf{t}) = r_o(\mathbf{x}; \hat{\boldsymbol{\Psi}}),$$

where $\hat{\boldsymbol{\Psi}}$ is an estimate of $\boldsymbol{\Psi}$ formed from $\mathbf{t}$. Provided $\hat{\boldsymbol{\theta}}_i$ is a consistent estimator of $\boldsymbol{\theta}_i$ and $f_i(\mathbf{x}; \boldsymbol{\theta}_i)$ is continuous in $\boldsymbol{\theta}_i$ $(i = 1, \ldots, g)$, then $r_o(\mathbf{x}; \hat{\boldsymbol{\Psi}})$ is a Bayes risk consistent rule in the sense that its risk, conditional on $\hat{\boldsymbol{\Psi}}$, converges in probability to that of the Bayes rule, as n approaches infinity. This is assuming that the postulated model (1.8.1) is indeed valid and that the group-prior probabilities are estimated consistently as possible, for instance, with mixture sampling of the training data. If the conditional risk for $r_o(\mathbf{x}; \hat{\boldsymbol{\Psi}})$ converges almost surely to that of the Bayes rule as n approaches infinity, then it is said to be Bayes risk strongly consistent. Consistency results for sample-based allocation rules have been obtained by Van Ryzin (1966) and Glick (1972, 1976). Initial references on the notion of consistency for sample-based allocation rules include Hoel and Peterson (1949) and Fix and Hodges (1951). The latter technical report, which also introduced several important nonparametric concepts in a discriminant analysis context, has been reprinted in full recently at the end of a commentary on it by Silverman and Jones (1989).

Given the widespread use of maximum likelihood as a statistical estimation technique, the plug-in rule $r_o(\mathbf{x}; \hat{\boldsymbol{\Psi}})$ is usually formed with $\hat{\boldsymbol{\Psi}}$, or at least $\hat{\boldsymbol{\theta}}$, taken to be the maximum likelihood estimate. This method of estimation in the context of discriminant analysis is to be considered further in the next section. Since their initial use by Wald (1944), Rao (1948, 1954), and T. W. Anderson (1951), among others, plug-in rules formed by maximum likelihood estimation under the assumption of normality have been extensively applied in practice. The estimation of $r_o(\mathbf{x};\boldsymbol{\Psi})$ by $r_o(\mathbf{x};\hat{\boldsymbol{\Psi}})$, where $\hat{\boldsymbol{\Psi}}$ is the maximum likelihood estimate of $\boldsymbol{\Psi}$, preserves the invariance of an allocation rule under monotone transformations.

Concerning some other parametric approaches to constructing a sample-based rule, there is the likelihood ratio criterion. The unknown vector $\mathbf{z}$ of zero-one indicator variables defining the group of origin of the unclassified entity is treated as a parameter to be estimated, along with $\boldsymbol{\Psi}$, on the basis of $\mathbf{t}$ and also $\mathbf{x}$. It differs from the estimative approach in that it includes the unclassified observation $\mathbf{x}$ in the estimation process. Hence, in principle, there is little difference between the two approaches although, in practice, the difference may be of some consequence, in particular for disparate group-sample sizes.

Another way of proceeding with the estimation of the group-conditional densities, and, hence, of $r_o(\mathbf{x};\boldsymbol{\Psi})$, is to adopt a Bayesian approach, which is considered in Section 2.2. Among other criteria proposed for constructing allocation rules is minimum distance. With this criterion, an entity with feature vector $\mathbf{x}$ is allocated to the group whose classified data in the training set $\mathbf{t}$ is closest to $\mathbf{x}$ in some sense. Although minimum-distance rules are often advocated in the spirit of distribution-free approaches to allocation, they are predicated on some underlying assumption for the group-conditional distributions. For example, the use of Euclidean distance as a metric corresponds to multivariate normal group-conditional distributions with a common spherical covariance matrix, and Mahalanobis distance corresponds to multivariate normal distributions with a common covariance matrix. The aforementioned parametric allocation rules are discussed in more detail with others in Chapter 3 in the context of normal theory-based discrimination.

Often, in practice, the total sample size is too small relative to the number p of feature variables in $\mathbf{x}$ for a reliable estimate of $\boldsymbol{\theta}$ to be obtained from the full set $\mathbf{t}$ of training data. This is referred to as "the curse of dimensionality," a phrase due to Bellman (1961). Consideration then has to be given to which variables in $\mathbf{x}$ should be deleted in the estimation of $\boldsymbol{\theta}$ and the consequent allocation rule. Even if a satisfactory discriminant rule can be formed using all the available feature variables, consideration may still be given to the deletion of some of the variables in $\mathbf{x}$. This is because the performance of a rule fails to keep improving and starts to fall away once the number of feature variables has reached a certain threshold. This so-called peaking phenomenon of a rule is discussed further in Chapter 12, where the variable-selection problem is to be addressed. It is an important problem in its own right in discriminant analysis, as with many applications, the primary or sole aim is not one of allocation, but rather to infer which feature variables of an entity are most useful in explaining the differences between the groups. If some or all of the group-sample sizes n_i of the classified data are very small, then consideration may have to be given to using unclassified data in the estimation of $\boldsymbol{\theta}$, as discussed in Section 2.7.

1.9 ASSESSMENT OF MODEL FIT

If the postulated group-conditional densities provide a good fit and the group-prior probabilities are known or able to be estimated with some precision, then the plug-in rule $r_o(\mathbf{x};\hat{F})$ should be a good approximation to the Bayes rule $r_o(\mathbf{x};F)$. However, even if $\hat{F}$ is a poor estimate of F, $r_o(\mathbf{x};\hat{F})$ may still be a reasonable allocation rule. It can be seen from the definition (1.4.4) of $r_o(\mathbf{x};F)$ that for $r_o(\mathbf{x};\hat{F})$ to be a good approximation to $r_o(\mathbf{x};F)$, it is only necessary that the boundaries defining the allocation regions,

$$\{\mathbf{x} : \pi_i f_i(\mathbf{x}) = \pi_j f_j(\mathbf{x}), \qquad i < j = 1,\ldots,g\}, \tag{1.9.1}$$

be estimated precisely. This implies at least for well-separated groups that in consideration of the estimated group-conditional densities, it is the fit in the tails rather than in the main body of the distributions that is crucial. This is what one would expect. Any reasonable allocation rule should be able to allocate correctly an entity whose group of origin is obvious from its feature vector. Its accuracy is really determined by how well it can handle entities of doubtful origin. Their feature vectors tend to occur in the tails of the distributions.

If reliable estimates of the posterior probabilities of group membership $\tau_i(\mathbf{x})$ are sought in their own right and not just for the purposes of making an outright assignment, then the fit of the estimated density ratios $\hat{f}_i(\mathbf{x})/\hat{f}_j(\mathbf{x})$ is important for all values of $\mathbf{x}$ and not just on the boundaries (1.9.1). It can be seen in discriminant analysis that the estimates of the group-conditional densities are not of interest as an end in themselves, but rather how useful their ratios are in providing estimates of the posterior probabilities of group membership or at least an estimate of the Bayes rule. However, for convenience, the question of model fit in practice is usually approached by consideration of the individual fit of each estimated density $\hat{f}_i(\mathbf{x})$.

Many different families of distributions may be postulated for the group-conditional densities, although some may be difficult to deal with analytically or computationally. The normal assumption is commonly adopted in practice. In some cases for this to be reasonable, a suitable transformation of the feature variables is required. In many practical situations, some variables in the feature vector $\mathbf{x}$ may be discrete. Often treating the discrete variables, in particular binary variables, as if they were normal in the formation of the discriminant rule is satisfactory. However, care needs to be exercised if several of the feature variables are discrete. The use of nonnormal models, including for mixed feature vectors where some of the variables are continuous and some are discrete, is discussed in Chapter 7, and Chapter 3 is devoted entirely to discrimination via normal models. Practical aspects such as robust methods of estimating group-conditional parameters, use of transformations to achieve approximate normality, testing for normality, and detection of atypical entities are discussed in Chapters 5 and 6.

1.10 ERROR RATES OF ALLOCATION RULES

1.10.1 Types of Error Rates

The allocation rates associated with the optimal or Bayes rule are given by

$$eo_{ij}(F) = \text{pr}\{r_o(\mathbf{X};F) = j \mid \mathbf{X} \in G_i\} \qquad (i,j = 1,\ldots,g), \quad (1.10.1)$$

where $eo_{ij}(F)$ is the probability that a randomly chosen entity from G_i is allocated to G_j on the basis of $r_o(\mathbf{x};F)$. The error rate specific to the ith group G_i is

$$eo_i(F) = \sum_{j \neq i}^{g} eo_{ij}(F) \qquad (i = 1,\ldots,g),$$

and the overall error rate is

$$eo(F) = \sum_{i=1}^{g} \pi_i eo_i(F).$$

As seen in Section 1.4, $r_o(\mathbf{x};F)$ is the rule that minimizes the overall error rate in the case of unit costs of misallocation. Consequently, $eo(F)$ is referred to as the optimal (overall) error rate. The optimal overall error rate can be used as a measure of the degree of separation between the groups, as to be considered in Section 1.12.

We proceed now to define the error rates of a sample-based rule. Let $r(\mathbf{x};\mathbf{t})$ denote an allocation rule formed from the training data $\mathbf{t}$. Then the allocation rates of $r(\mathbf{x};\mathbf{t})$, conditional on $\mathbf{t}$, are defined by

$$ec_{ij}(F_i;\mathbf{t}) = \text{pr}\{r(\mathbf{X};\mathbf{t}) = j \mid \mathbf{X} \in G_i, \mathbf{t}\}, \qquad (1.10.2)$$

which is the probability, conditional on $\mathbf{t}$, that a randomly chosen entity from G_i is allocated to G_j $(i,j = 1,\ldots,g)$. The group-specific conditional error rates are given by

$$ec_i(F_i;\mathbf{t}) = \sum_{j \neq i}^{g} ec_{ij}(F_i;\mathbf{t}) \qquad (i = 1,\ldots,g),$$

and the overall conditional error rate by

$$ec(F;\mathbf{t}) = \sum_{i=1}^{g} \pi_i ec_i(F_i;\mathbf{t}).$$

For equal costs of misallocation, the rule $r(\mathbf{x};\mathbf{t})$ is Bayes risk consistent (strongly consistent) if $ec(F;\mathbf{t})$ converges in probability (almost surely) to $eo(F)$, as n approaches infinity.

On averaging the conditional allocation rates over the distribution of the training data, we obtain the expected or unconditional rates defined as

$$\begin{aligned} eu_{ij}(F) &= E\{ec_{ij}(F_i;\mathbf{T})\} \\ &= \text{pr}\{r(\mathbf{X};\mathbf{T}) = j \mid \mathbf{X} \in G_i\} \qquad (i,j = 1,\ldots,g). \quad (1.10.3) \end{aligned}$$

In the case of separate sampling where $\mathbf{t}$ is based on a fixed number of entities from each group, we should, strictly speaking, denote these unconditional rates as $eu_{ij}(F_1,\ldots,F_g)$, rather than $eu_{ij}(F)$. The group-specific unconditional error rates are given by

$$eu_i(F) = \sum_{j \neq i}^{g} eu_{ij}(F),$$

and the overall unconditional error rate by

$$eu(F) = \sum_{i=1}^{g} \pi_i eu_i(F).$$

We are following Hills (1966) here in referring to the $ec_i(F_i;\mathbf{t})$ and $ec(F;\mathbf{t})$ as the conditional or actual error rates, and to the $eu_i(F)$ and $eu(F)$ as the unconditional or expected error rates. Before the introduction of his careful terminology for the various types of error rates, there had been a good deal of confusion in the literature; see the comment of Cochran (1966).

1.10.2 Relevance of Error Rates

Concerning the relevance of error rates in discriminant analysis, for allocation problems, they play a major role in providing a measure of the global performance of a discriminant rule. It has been suggested (Lindley, 1966) that more attention should be paid to the unconditional losses. However, as remarked earlier, the specification of costs in practice is often arbitrary.

On the use of error rates to measure the performance of a sample-based allocation rule, it is the conditional error rates that are of primary concern once the rule has been formed from the training data $\mathbf{t}$. If $\mathbf{t}$ denotes all the available data of known origin, then one is stuck with this training set in forming a rule. An example where these error rates enter naturally into an analysis is when the rule $r(\mathbf{x};\mathbf{t})$ forms the basis of a diagnostic test for estimating the prevalence rates of a disease, as covered in Section 2.3.

The average performance of the rule over all possible realizations of $\mathbf{t}$ is of limited interest in applications of $r(\mathbf{x};\mathbf{t})$. However, the unconditional error rates are obviously relevant in the design of a rule. They relate the average performance of the rule to the size of the training set and to the group-conditional distributions as specified. For example, consider the case of two groups G_1 and G_2 in which the feature vector $\mathbf{X}$ is taken to have a multivariate normal distribution with means $\boldsymbol{\mu}_1$ and $\boldsymbol{\mu}_2$, respectively, and common covariance matrix $\boldsymbol{\Sigma}$. For separate sampling with equal sample sizes $n/2$ from each of the two groups, the sample-based analogue of the Bayes rule with equal group-prior probabilities has equal unconditional error rates. Their common value, equal to the overall error rate for equal priors, is given by

$$eu(F) \approx eo(F) + n^{-1}\{\phi(\tfrac{1}{2}\Delta)/4\}\{p\Delta + 4(p-1)\Delta^{-1}\}, \qquad (1.10.4)$$

where

$$eo(F) = \Phi(-\tfrac{1}{2}\Delta)$$

and

$$\Delta = \{(\boldsymbol{\mu}_1 - \boldsymbol{\mu}_2)'\boldsymbol{\Sigma}^{-1}(\boldsymbol{\mu}_1 - \boldsymbol{\mu}_2)\}^{1/2} \qquad (1.10.5)$$

is the Mahalanobis (1936) distance between G_1 and G_2. In this and subsequent work, Φ and ϕ denote the standard normal distribution and density, respectively. The error of the approximation (1.10.4) is of order $O(n^{-2})$ (Okamoto, 1963). The derivation of (1.10.4) is discussed in Section 4.2.

From (1.10.4), we can determine approximately how large n must be for a specified Δ and p in order for the unconditional error rate not to exceed too far the best obtainable, as given by the optimal rate $eo(F)$. For instance, for $\Delta = 1$ representing two groups that are close together, n on the basis of (1.10.4) has be to at least 40 with $p = 3$ for the rate to be less than 1/3 on average; that is, not more than 0.0248 in excess of the optimal rate of 0.3085. The latter value shows that it is not possible to design an accurate allocation rule in this case. Indeed, if n is small, then for $p > 1$, the error rate is not far short of 1/2, which is the error rate for a randomized rule that ignores the feature vector and makes a choice of groups according to the toss of a fair coin.

It can be seen from (1.10.2) and (1.10.3) that the conditional and unconditional allocation rates of a sample-based rule depend on the unknown group-conditional distributions and so must be estimated. In the absence of any further data of known origin, these rates must be estimated from the same data $\mathbf{t}$ from which the rule has been formed. Hence, there are difficulties in obtaining unbiased estimates of the error rates of a sample-based rule in its application to data of unknown origin, distributed independently of the training sample. Estimation of the error rates of allocation rules is thus a difficult but important problem in discriminant analysis. It is taken up in Chapter 10, which is devoted fully to it.

We have seen that the situation where some of the errors of allocation are more serious than others can be handled through the specification of unequal costs of misallocation in the definition (1.4.6) of the Bayes rule. Another approach would be to introduce regions of doubt in the feature space where no allocation is made. This approach was adopted by J. A. Anderson (1969) in his design of a rule with upper bounds specified on the errors of allocation. It was used also by Habbema, Hermans, and van der Burgt (1974b) in their development of a decision-theoretic model for allocation. Previously, Marshall and Olkin (1968) had considered situations where direct assessment of the group of origin is possible, but expensive. In these situations, after the feature vector has been observed, there is a choice between allocation and extensive group assessment. Another approach where there is an alternative to an outright allocation of the entity after its feature vector has been observed was given by Quesenberry and Gessaman (1968). Their nonparametric procedure constructs tolerance regions for each group, and an entity is allocated to the set of those groups whose tolerance regions contain the feature vector $\mathbf{x}$. If $\mathbf{x}$ falls within all or outside all the tolerance regions, then the entity is not allocated; see also Gessaman and Gessaman (1972). Broffitt, Randles, and Hogg (1976) introduced a rank method for partial allocation with constraints imposed on the unconditional error rataes. This nonparametric approach to partial discrimination in the presence of constraints is discussed in Section 9.9.

A parametric approach to constrained discrimination with unknown group-conditional densities has been investigated by T. W. Anderson (1973a, 1973b) and McLachlan (1977b) for the sample normal-based linear discriminant rule. Their work is described in Section 4.5. Also, Gupta and Govindarajulu (1973) considered constrained discrimination in the special case of univariate normal group-conditional distributions with multiple independent measurements available on the entity to be allocated.

The error rates are not the only measure of the global accuracy of an allocation rule. Breiman et al. (1984, Section 4.6) have proposed a global measure in terms of estimates of the posterior probabilities of group membership for a rule $r(\mathbf{x};\mathbf{t})$ defined analogously to the Bayes rule $r_o(\mathbf{x};F)$. That is, $r(\mathbf{x};\mathbf{t})$ is equal to i if the estimated posterior probabilities satisfy

$$\hat{\tau}_i(\mathbf{x};\mathbf{t}) \geq \hat{\tau}_j(\mathbf{x};t) \qquad (j = 1,\ldots,g;\ j \neq i). \tag{1.10.6}$$

Their proposed measure of the accuracy (conditional here on the training data $\mathbf{t}$) of the rule $r(\mathbf{x};\mathbf{t})$ is

$$E\left[\sum_{i=1}^{g}\{\hat{\tau}_i(\mathbf{X};\mathbf{t}) - \tau_i(\mathbf{X})\}^2 \mid \mathbf{t}\right]. \tag{1.10.7}$$

They noted that if the mean-squared error (conditional on $\mathbf{t}$) of the rule $r(\mathbf{x};\mathbf{t})$ is defined as

$$\text{MSE}(r) = E\left[\sum_{i=1}^{g}\{\hat{\tau}_i(\mathbf{X};\mathbf{t}) - Z_i\}^2 \mid \mathbf{t}\right], \tag{1.10.8}$$

then it can be decomposed into the two terms,

$$\text{MSE}(r) = \text{MSE}(r_o) + E\left[\sum_{i=1}^{g}\{\hat{\tau}_i(\mathbf{X};\mathbf{t}) - \tau_i(\mathbf{X})\}^2 \mid \mathbf{t}\right],$$

where

$$\text{MSE}(r_o) = E\left[\sum_{i=1}^{g}\{\tau_i(\mathbf{X}) - Z_i\}^2\right]$$

is the mean-squared error of the Bayes rule $r_o(\mathbf{x})$. Hence, a comparison in terms of the accuracy (1.10.7) of different rules of the form (1.10.6) can be made in terms of their conditional mean-squared errors. This provides a significant advantage as, unlike (1.10.7), MSE(r) can be estimated directly from $\mathbf{t}$ as

$$\frac{1}{n}\sum_{j=1}^{n}\sum_{i=1}^{g}\{\hat{\tau}_i(\mathbf{x}_j;\mathbf{t}) - z_{ij}\}^2,$$

where $z_{ij} = (\mathbf{z}_j)_i$, and $\mathbf{z}_j$ is the vector of zero-one indicator variables defining the known group of origin of the jth feature vector $\mathbf{x}_j$ in the training data $\mathbf{t}$ $(j = 1,\ldots,n)$.

Note that by virtue of their definition, error rates are concerned only with the allocatory performance of a rule. Hence, for rules of the form (1.10.6), they are concerned only with the relative sizes of the estimated posterior probabilities of group membership. By contrast, the criterion (1.10.7) attempts to measure the accuracy of a rule of the form (1.10.6) by assessing the absolute fit of the posterior probabilities of group membership.

Other ways of assessing the discriminatory performance of a fitted model have been considered by Habbema, Hilden, and Bjerregaard (1978b, 1981); Hilden, Habbema, and Bjerregaard (1978a, 1978b); and Habbema and Hilden (1981).

1.11 POSTERIOR PROBABILITIES OF GROUP MEMBERSHIP

It was shown in Section 1.8 that the posterior probabilities of group membership $\tau_i(\mathbf{x})$ or their estimates may play no role in the formation of some allocation rules in the pure decision context. On the other hand with the Bayes rule or a sample version, the relative sizes of the posterior probabilities of group membership $\tau_i(\mathbf{x})$ form the basis of the subsequent outright allocation to be made. In many real problems, only a tentative allocation is contemplated before consideration is to be given to taking an irrevocable decision as to the group of origin of an unclassified entity. For these problems, the probabilistic allocation rule implied by the $\tau_i(\mathbf{x})$ or their estimates provides a concise way of expressing the uncertainty about the group membership of an unclassified entity with an observed feature vector $\mathbf{x}$.

It has been argued (Spiegelhalter, 1986) that the provision of accurate and useful probabilistic assessments of future events should be a fundamental task for biostatisticians collaborating in clinical or experimental medicine. To this end, the posterior probabilities of group membership play a major role in patient management and clinical trials. For example, in the former context with the groups corresponding to the possible treatment decisions, the uncertainty over which decision to make is conveniently formulated in terms of the posterior probabilities of group membership. Moreover, the management of the patient may be only at a preliminary stage where an outright assignment may be premature particularly, say, if the suggested treatment decision is not clearcut and involves major surgery on the patient. The reliability of these estimates is obviously an important question to be considered, especially in applications where doubtful cases of group membership arise.

If the posterior probabilities of group membership have been estimated for the express purpose of forming an allocation rule, then their overall reliability can be assessed through the global performance of this rule as measured by its associated error rates. However, as emphasized by Critchley and Ford (1985), even if all its error rates are low, there may still be entities about which there is great uncertainty as to their group of origin. Conversely, these global measures may be high, yet it may still be possible to allocate some entities with great certainty. Thus, in some situations, it may not be appropriate to consider an assessment in terms of the error rates. Indeed, as pointed out by Aitchison and Kay (1975), in clinical medicine, the Hippocratic oath precludes any criterion of average results over individual patients (such as error rates), so that conditioning on the feature vector **x** is an apt way to proceed. In Chapter 11, we consider methods for assessing the reliability of the estimates of the posterior probabilities of group membership from the same training data used to form these estimates in the first instance.

1.12 DISTANCES BETWEEN GROUPS

Over the years, there have been proposed many different measures of distance, divergence, or discriminatory information between two groups. Krzanowski (1983a) has put them broadly into two categories: (a) measures based on ideas from information theory and (b) measures related to Bhattacharyya's (1943) measure of affinity.

Some members of category (a) are considered first. There is the Kullback–Leibler (1951) measure of discriminatory information between two groups with distribution functions F_1 and F_2, admitting densities $f_1(\mathbf{x})$ and $f_2(\mathbf{x})$, respectively, with respect to some measure ν. This measure is given by

$$\delta_{KL}(F_1,F_2) = \int f_1(\mathbf{x})\log\{f_1(\mathbf{x})/f_2(\mathbf{x})\}\,d\nu.$$

It is a directed divergence in that it also has a directional component, since generally, $\delta_{KL}(F_1,F_2) \neq \delta_{KL}(F_2,F_1)$; that is, it is not a metric. Jeffreys' (1948) measure is a symmetric combination of the Kullback–Leibler information,

$$\delta_J(F_1,F_2) = \delta_{KL}(F_1,F_2) + \delta_{KL}(F_2,F_1).$$

A third measure in category (a) is

$$\delta_S(F_1,F_2) = \tfrac{1}{2}[\delta_{KL}\{F_1,\tfrac{1}{2}(F_1+F_2)\} + \delta_{KL}\{F_2,\tfrac{1}{2}(F_1+F_2)\}],$$

which is Sibson's (1969) information radius given in its simple form.

Rényi (1961) generalized both Shannon (1948) entropy and the Jeffreys–Kullback–Leibler information by introducing a scalar parameter. Recently, Burbea and Rao (1982), Burbea (1984), and Taneja (1983) have proposed various alternative ways to generalize $\delta_J(F_1,F_2)$. The proposed measures of Burbea and Rao (1982) and Burbea (1984) involve one parameter, and the measures proposed by Taneja (1983) involve two parameters. The definitions of these generalized measures may be found in Taneja (1987). Another measure that has been proposed is the power divergence corresponding to the power-divergence family of goodness-of-fit statistics introduced by Cressie and Read (1984); see also Read and Cressie (1988, Section 7.4).

Concerning members of category (b), Bhattacharyya's original measure of affinity is

$$\rho = \int \{f_1(\mathbf{x})f_2(\mathbf{x})\}^{1/2}\,d\nu. \tag{1.12.1}$$

Bhattacharyya (1946) subsequently proposed

$$\delta_B(F_1,F_2) = \cos^{-1}(\rho),$$

and, in unpublished notes, A. N. Kolmogorov used

$$\delta_{KO}(F_1,F_2) = 1-\rho.$$

Chernoff (1952) introduced the more general distance measure

$$\delta_C(F_1,F_2) = -\log \int \{f_1(\mathbf{x})\}^{\alpha}\{f_2(\mathbf{x})\}^{1-\alpha}\,d\nu,$$

where $\alpha \in [0,1]$. It reduces to $-\log\rho$ in the special case of $\alpha = 0.5$. If $f_1(\mathbf{x})$ and $f_2(\mathbf{x})$ are multivariate normal densities with means $\boldsymbol{\mu}_1$ and $\boldsymbol{\mu}_2$ and covariance matrices $\boldsymbol{\Sigma}_1$ and $\boldsymbol{\Sigma}_2$, respectively, then

$$\begin{aligned}-\log\rho &= \tfrac{1}{8}(\boldsymbol{\mu}_1-\boldsymbol{\mu}_2)'\boldsymbol{\Sigma}^{-1}(\boldsymbol{\mu}_1-\boldsymbol{\mu}_2)\\ &\quad+\tfrac{1}{2}\log[|\boldsymbol{\Sigma}|/\{|\boldsymbol{\Sigma}_1||\boldsymbol{\Sigma}_2|\}^{1/2}],\end{aligned}$$

where

$$\boldsymbol{\Sigma} = \tfrac{1}{2}(\boldsymbol{\Sigma}_1+\boldsymbol{\Sigma}_2);$$

see, for example, Kailath (1967).

Matusita (1956) subsequently defined the distance measure

$$\begin{aligned}\delta_M(F_1,F_2) &= \left[\int \{\sqrt{f_1(\mathbf{x})}-\sqrt{f_2(\mathbf{x})}\}^2\,d\nu\right]^{1/2}\\ &= (2-2\rho)^{1/2}\\ &= \{2\delta_{KO}(F_1,F_2)\}^{1/2}.\end{aligned} \tag{1.12.2}$$

The distance (1.12.2) is sometimes referred to as Hellinger's distance; see, for example, Le Cam (1970) and Beran (1977). There is little practical difference between these functionally related measures in category (b).

Additional distance measures to these defined above may be found in Ben-Bassat (1982), who summarized findings on the current measures, including their relationships with lower and upper bounds on the overall optimal error rate. A recent paper on the latter topic is Ray (1989a), who considered the maximum of the span s between the upper and lower bounds on the overall error rate $eo(F)$ of the Bayes rule, as provided by ρ. Hudimoto (1956–1957, 1957–1958) had shown that

$$\tfrac{1}{2}-\tfrac{1}{2}(1-4\pi_1\pi_2\rho^2)^{1/2} \le eo(F) \le (\pi_1\pi_2)^{1/2}\rho,$$

with span

$$s = (\pi_1\pi_2)^{1/2}\rho - \tfrac{1}{2} + \tfrac{1}{2}(1 - 4\pi_1\pi_2\rho^2)^{1/2}.$$

Ray (1989a) showed that the maximum value of s is $\frac{1}{2}(\sqrt{2}-1)$, which can be attained for values of π_1, and hence values of π_2, lying inside the interval

$$\{(2-\sqrt{2})/4,\ (2+\sqrt{2})/4\}.$$

In another related paper, Ray (1989b) considered the maximum of the span between the upper and lower bounds on $eo(F)$ as provided by the generalized distance measure of Lissack and Fu (1976). This measure is

$$\delta_{LF}(F_1, F_2) = \int |\tau_1(\mathbf{x}) - \tau_2(\mathbf{x})|^\alpha f_X(\mathbf{x})\, d\nu \qquad 0 < \alpha < \infty,$$

where $\tau_i(\mathbf{x})$ is the posterior probability of membership of group G_i $(i = 1, 2)$, as defined by (1.2.4).

It is a generalization of the Kolmogorov variational distance defined by

$$\int |\pi_1 f_1(\mathbf{x}) - \pi_2 f_2(\mathbf{x})|\, d\nu.$$

For $\alpha = 1$, $\delta_{LF}(F_1, F_2)$ reduces to this distance with

$$\begin{aligned} \delta_{LF}(F_1, F_2) &= \int |\pi_1 f_1(\mathbf{x}) - \pi_2 f_2(\mathbf{x})|\, d\nu \\ &= 2\int \max\{\pi_1 f_1(\mathbf{x}), \pi_2 f_2(\mathbf{x})\}\, d\nu - 1 \\ &= 1 - 2eo(F); \end{aligned} \qquad (1.12.3)$$

see Rao (1948, 1977) and Glick (1973b).

For $0 < \alpha \leq 1$, Lissack and Fu (1976) showed that

$$\tfrac{1}{2}\{1 - \delta_{LF}(F_1, F_2)\} \leq eo(F) \leq \tfrac{1}{2}[1 - \{\delta_{LF}(F_1, F_2)\}^{1/\alpha}],$$

and for $1 \leq \alpha < \infty$,

$$\tfrac{1}{2}[1 - \{\delta_{LF}(F_1, F_2)\}^{1/\alpha}] \leq eo(F) \leq \tfrac{1}{2}\{1 - \delta_{LF}(F_1, F_2)\}.$$

The lower and upper bounds coincide for $\alpha = 1$ in accordance with the result (1.12.3). For $\alpha > 1$, Ray (1989b) showed that the maximum of their difference is

$$\tfrac{1}{2}\{\alpha^{-1/(\alpha-1)} - \alpha^{-\alpha/(\alpha-1)}\},$$

and that it increases from 0 to 0.5 as α increases from 1 to infinity. He also established that the maximum difference between the upper and lower bounds increases from 0 to 0.5 as α decreases from 1 to 0.

An early study of distance measures was by Adhikari and Joshi (1956). A general class of coefficients of divergence of one distribution from another was considered by Ali and Silvey (1966), who demonstrated that the measures above are members of this class. Furthermore, if $f_1(\mathbf{x})$ and $f_2(\mathbf{x})$ are multivariate normal densities with means $\boldsymbol{\mu}_1$ and $\boldsymbol{\mu}_2$, respectively, and common covariance matrix $\boldsymbol{\Sigma}$, then every coefficient in their class is an increasing function of the Mahalanobis distance Δ defined by (1.10.5). For example, the affinity ρ then is given by

$$\rho = \exp(-\Delta^2/8), \tag{1.12.4}$$

as calculated previously by Matusita (1973). The Mahalanobis distance Δ has become the standard measure of distance between two groups when the feature variables are continuous.

Atkinson and Mitchell (1981) have shown how Δ arises naturally from Rao's (1945) procedure for determining distances between members of a well-behaved parametric family of distributions. The relevant family in this case is the multivariate normal with common shape but varying location. Concerning alternative models for multivariate data, a rich source of models is provided by the class of elliptic distributions whose densities have elliptical contours and which include the multivariate normal, multivariate Student's t, and Cauchy. The p-dimensional random variable $\mathbf{X}$ is said to have an elliptic distribution with parameters $\boldsymbol{\mu}$ ($p \times 1$ vector) and $\boldsymbol{\Sigma}$ ($p \times p$ positive-definite matrix) if its density is of the form

$$|\boldsymbol{\Sigma}|^{-1/2} f_S[\{\delta(\mathbf{x}, \boldsymbol{\mu}; \boldsymbol{\Sigma})\}^{1/2}], \tag{1.12.5}$$

where $f_S(\cdot)$ is any function such that $f_S(\|\mathbf{x}\|)$ is a density on $\mathbb{R}^p$ and

$$\delta(\mathbf{x}, \boldsymbol{\mu}; \boldsymbol{\Sigma}) = (\mathbf{x} - \boldsymbol{\mu})'\boldsymbol{\Sigma}^{-1}(\mathbf{x} - \boldsymbol{\mu}).$$

The class of elliptic densities defined by (1.12.5) can be generated by a nonsingular transformation of $\mathbf{x}$ from the class of spherically symmetric densities $f_S(\|\mathbf{x}\|)$, where $\|\mathbf{x}\| = (\mathbf{x}'\mathbf{x})^{1/2}$ denotes the Euclidean norm of $\mathbf{x}$. It follows in (1.12.5) that $\boldsymbol{\mu}$ is the mean of $\mathbf{X}$ and $\boldsymbol{\Sigma}$ is a scalar multiple of the covariance matrix of $\mathbf{X}$. Mitchell and Krzanowski (1985) have shown that the Mahalanobis distance Δ remains appropriate when the family of distributions under consideration is one of the elliptic class having fixed shape but varying location. One implication of this result noted by Mitchell and Krzanowski (1985) is that the sample analogue of Δ is the appropriate measure of the distance between the estimates of $f_1(\mathbf{x})$ and $f_2(\mathbf{x})$ fitted using either the estimative or the predictive approaches under the assumption of multivariate normal densities with a common covariance matrix. As discussed in the next chapter, the fitted densities are multivariate normal in the estimative case and multivariate Student's t in the predictive case.

Bhattacharyya's (1946) measure $\delta_B(F_1, F_2)$, which has become known as the angular separation between the groups, was the first measure proposed for discrete feature data. An alternative approach was adopted by Balakrishnan and Sanghvi (1968) and Kurezynski (1970), who attempted to create Mahalanobis-like distance measures for discrete feature data. These subsumed some earlier measures based on chi-squared statistics (Sanghvi, 1953). Bhattacharyya's

(1946) measure in the context of multinomial data has received strong support from Edwards (1971), who provided a further measure through the stereographic projection as an approximation to the angular separation.

The affinity-based measure (1.12.1) has been developed by Matusita (1964, 1967a, 1967b, 1971, 1973). For $g > 2$ groups, the affinity of the group-conditional distributions $F_1,\ldots,F_g$ is defined as

$$\rho_g(F_1,\ldots,F_g) = \int \{f_1(\mathbf{x})\cdots f_g(\mathbf{x})\}^{1/g}\,d\nu.$$

The affinity ρ_g is connected with the distance

$$\left|\int [\{f_i(\mathbf{x})\}^{1/g} - \{f_j(\mathbf{x})\}^{1/g}]^g\,d\nu\right|^{1/g}$$

between any pair of distributions F_i and F_j, $i \neq j = 1,\ldots,g$. In the case of two groups, there is a complete duality between the distance measure and the affinity measure of Matusita (1956). However, this is not clear where there are more than two groups (Ahmad, 1982).

Toussaint (1974b) has extended the definition of $\rho_g(F_1,\ldots,F_g)$ to

$$\int [\{f_1(\mathbf{x})\}^{c_1}\cdots\{f_g(\mathbf{x})\}^{c_g}]\,d\nu,$$

where

$$\sum_{i=1}^{g} c_i = 1,$$

and $c_i \geq 0\,(i = 1,\ldots,g)$. It reduces to $\rho_g(F_1,\ldots,F_g)$ when $c_i = 1/g$ for $i = 1, \ldots,g$. As noted by Glick (1973b), a measure of separation between $F_1,\ldots,F_g$ is provided by

$$1 - \rho_g^{g/2}(F_1,\ldots,F_g). \tag{1.12.6}$$

Since it can be shown that

$$\rho_g^g(F_1,\ldots,F_g) \leq \min_{i\neq j} \rho_2^2(F_i,F_j),$$

it implies that the separation among the g groups according to (1.12.6) is not less than the separation between any two of them. This measure also has the other desirable properties of a separation measure in that it is symmetric in its arguments and has a minimum value of zero at $F_1 = F_2 = \cdots = F_g$. Glick (1973b) also generalized the two-group result (1.12.3) by showing that $1 - 2eo(F)$ can be viewed as a separation measure for an arbitrary number g of groups when in equal proportions; see also Cleveland and Lachenbruch (1974).

In a series of papers, Krzanowski (1983a, 1984a, 1987a) has studied the use of Matusita's (1956) measure of distance in situations where the feature vector consists of continuous and discrete variables. His work is discussed in Section 7.4, where discriminant analysis in the case of mixed features is presented.

CHAPTER 2

Likelihood-Based Approaches to Discrimination

2.1 MAXIMUM LIKELIHOOD ESTIMATION OF GROUP PARAMETERS

We consider maximum likelihood estimation of the vector $\boldsymbol{\theta}$ containing all the unknown parameters in the parametric families (1.8.1) postulated for the g group-conditional densities of the feature vector $\mathbf{X}$. As in the previous work, we let $\mathbf{t}' = (\mathbf{y}_1, \ldots, \mathbf{y}_n)$ denote the observed training data, where $\mathbf{y}_j = (\mathbf{x}_j', \mathbf{z}_j')'$ for $j = 1, \ldots, n$. For training data $\mathbf{t}$ obtained by a separate sampling scheme, the likelihood function $L(\boldsymbol{\theta})$ for $\boldsymbol{\theta}$ is formed by evaluating at their observed values $\mathbf{x}_1, \ldots, \mathbf{x}_n$ the joint density of the feature vectors conditional on their known group-indicator vectors $\mathbf{z}_1, \ldots, \mathbf{z}_n$. We proceed here under the assumption that $\mathbf{y}_1, \ldots, \mathbf{y}_n$ denote the realized values of n independent training observations. Then the log likelihood function for $\boldsymbol{\theta}$ is given under (1.8.1) by

$$\log L(\boldsymbol{\theta}) = \sum_{i=1}^{g} \sum_{j=1}^{n} z_{ij} \log f_i(\mathbf{x}_j; \boldsymbol{\theta}_i), \tag{2.1.1}$$

where log denotes the natural logarithm. An estimate $\hat{\boldsymbol{\theta}}$ of $\boldsymbol{\theta}$ can be obtained as a solution of the likelihood equation

$$\partial L(\boldsymbol{\theta})/\partial \boldsymbol{\theta} = \mathbf{0},$$

or, equivalently,

$$\partial \log L(\boldsymbol{\theta})/\partial \boldsymbol{\theta} = \mathbf{0}. \tag{2.1.2}$$

Briefly, the aim of maximum likelihood estimation (Lehmann, 1980, 1983) is to determine an estimate for each n ($\hat{\boldsymbol{\theta}}$ in the present context) so that it defines a sequence of roots of the likelihood equation that is consistent and asymptotically efficient. Such a sequence is known to exist under suitable regularity conditions (Cramér, 1946). With probability tending to one, these roots correspond to local maxima in the interior of the parameter space. For estimation models in general, the likelihood usually has a global maximum in the interior of the parameter space. Then typically a sequence of roots of the likelihood equation with the desired asymptotic properties is provided by taking $\hat{\boldsymbol{\theta}}$ for each n to be the root that globally maximizes the likelihood; that is, $\hat{\boldsymbol{\theta}}$ is the maximum likelihood estimate. We will henceforth refer to $\hat{\boldsymbol{\theta}}$ as the maximum likelihood estimate, even though it may not globally maximize the likelihood. Indeed, in some of the examples on mixture models to be presented, the likelihood is unbounded. However, for these models, there may still exist under the usual regularity conditions a sequence of roots of the likelihood equation with the properties of consistency, efficiency, and asymptotic normality; see McLachlan and Basford (1988, Chapter 1).

For $\mathbf{t}$ obtained under mixture sampling, the log likelihood function for $\boldsymbol{\Psi} = (\boldsymbol{\pi}', \boldsymbol{\theta}')'$ is given by

$$\log L(\boldsymbol{\theta}) + \sum_{i=1}^{g} \sum_{j=1}^{n} z_{ij} \log \pi_i.$$

It follows from consideration of the likelihood equation for $\boldsymbol{\Psi}$ that it is estimated by

$$\hat{\boldsymbol{\Psi}} = (\hat{\boldsymbol{\pi}}', \hat{\boldsymbol{\theta}}')',$$

where $\hat{\boldsymbol{\theta}}$ is defined as before and $\hat{\boldsymbol{\pi}} = (\hat{\pi}_1, \ldots, \hat{\pi}_g)'$, and where

$$\begin{aligned} \hat{\pi}_i &= \sum_{j=1}^{n} z_{ij}/n \\ &= n_i/n \qquad (i = 1, \ldots, g). \end{aligned}$$

Given that a statistical model is at best an approximation to reality, it is worth considering here the behavior of the maximum likelihood estimate $\hat{\boldsymbol{\theta}}$ if the postulated parametric structure for the group-conditional densities is not valid. Suppose now that the group-conditional densities $f_i(\mathbf{x})$ do not belong to the parametric families postulated by (1.8.1). Working with a mixture sampling scheme, the true mixture density of $\mathbf{X}$ can be expressed as

$$f_X(\mathbf{x}) = \sum_{i=1}^{g} \pi_{i,o} f_i(\mathbf{x}),$$

where $\pi_{i,o}$ denotes the true value of π_i $(i = 1, \ldots, g)$.

As seen previously, regardless of whether a mixture or separate sampling scheme applies, $\hat{\boldsymbol{\theta}}$ is obtained by consideration of the same function $\log L(\boldsymbol{\theta})$

given by (2.1.1). Following Hjort (1986a, 1986b), we have that as $n \to \infty$, $1/n$ times $\log L(\boldsymbol{\theta})$ tends almost surely to

$$\sum_{i=1}^{g} \pi_{i,o} \left\{ \int f_i(\mathbf{x}) \log f_i(\mathbf{x}; \boldsymbol{\theta}_i) \right\} d\nu. \tag{2.1.3}$$

Suppose there is a unique value of $\boldsymbol{\theta}, \boldsymbol{\theta}_o$, that maximizes (2.1.3) with respect to $\boldsymbol{\theta}$. Then this value also minimizes the quantity

$$\sum_{i=1}^{g} \pi_{i,o} \left[\int f_i(\mathbf{x}) \log\{f_i(\mathbf{x})/f_i(\mathbf{x}; \boldsymbol{\theta}_i\} \, d\nu \right],$$

which is a mixture in the true proportions $\pi_{1,o}, \ldots, \pi_{g,o}$ of the Kullback–Leibler distances between the true and the postulated group-conditional densities of $\mathbf{X}$.

Under mild regularity conditions, it follows that if $\hat{\boldsymbol{\theta}}$ is chosen by maximization of $\log L(\boldsymbol{\theta})$, it tends almost surely to $\boldsymbol{\theta}_o$. Hence, the maximum likelihood estimator $\hat{\boldsymbol{\theta}}$ under the invalid model (1.8.1) is still a meaningful estimator in that it is a consistent estimator of $\boldsymbol{\theta}_o$, the value of $\boldsymbol{\theta}$ that minimizes the Kullback–Leibler distances between the actual group-conditional densities of $\mathbf{X}$ and the postulated parametric families, mixed in the proportions in which the groups truly occur.

2.2 A BAYESIAN APPROACH

A review of the Bayesian approach to discriminant analysis has been given by Geisser (1966, 1982). This approach is based on the concept of the predictive density of the feature vector $\mathbf{X}$. The predictive density of $\mathbf{X}$ within group G_i is defined by

$$\hat{f}_i^{(P)}(\mathbf{x}; \mathbf{t}) = \int f_i(\mathbf{x}; \boldsymbol{\theta}_i) p(\boldsymbol{\theta} \mid \mathbf{t}) \, d\boldsymbol{\theta} \qquad (i = 1, \ldots, g), \tag{2.2.1}$$

where $p(\boldsymbol{\theta} \mid \mathbf{t})$ can be regarded either as some weighting function based on $\mathbf{t}$ or as a full Bayesian posterior density function for $\boldsymbol{\theta}$ based on $\mathbf{t}$ and a prior density $p(\boldsymbol{\theta})$ for $\boldsymbol{\theta}$. In the latter case,

$$p(\boldsymbol{\theta} \mid \mathbf{t}) \propto p(\boldsymbol{\theta}) L(\boldsymbol{\theta}; \mathbf{t}),$$

where $L(\boldsymbol{\theta}; \mathbf{t})$ denotes the likelihood function for $\boldsymbol{\theta}$ formed from the training data $\mathbf{t}$. Note that for economy of notation, we are using $p(\cdot)$ here as a generic symbol for a density function. In the subsequent discussion, the vector $\boldsymbol{\pi} = (\pi_1, \ldots, \pi_g)'$ defining the group-prior probabilities is taken to be specified, so that the vector $\boldsymbol{\Psi}$ of parameters to be estimated is reduced to $\boldsymbol{\theta}$. We therefore write the posterior probability of membership of the ith group $\tau_i(\mathbf{x}; \boldsymbol{\Psi})$ as $\tau_i(\mathbf{x}; \boldsymbol{\pi}, \boldsymbol{\theta})$.

The predictive estimate of $\tau_i(\mathbf{x};\boldsymbol{\pi},\boldsymbol{\theta})$ is obtained by substituting the predictive estimates of the group-conditional densities in its defining expression (1.2.4) for $\tau_i(\mathbf{x};\boldsymbol{\pi},\boldsymbol{\theta})$ to give

$$\hat{\tau}_i^{(P)}(\mathbf{x};\mathbf{t}) = \pi_i \hat{f}_i^{(P)}(\mathbf{x};\mathbf{t})/\hat{f}_X^{(P)}(\mathbf{x};\mathbf{t}) \qquad (i = 1,\ldots,g), \tag{2.2.2}$$

where

$$\hat{f}_X^{(P)}(\mathbf{x};\mathbf{t}) = \sum_{j=1}^{g} \pi_j \hat{f}_j^{(P)}(\mathbf{x};\mathbf{t}).$$

According to Aitchison and Dunsmore (1975, Chapter 11), the predictive approach was first presented explicitly by Geisser (1964) for multivariate normal group-conditional distributions, and in Dunsmore (1966). For moderately large or large-size training samples, the predictive and estimative approaches give similar results for the assessment of the posterior probabilities of group membership. However, for small sample sizes, there can be dramatic differences. This appears to have been first noted by Aitchison and Kay (1975). These two approaches are to be compared under normal models in Section 3.8. It will be seen there that if the estimates produced by the estimative approach are corrected for bias, then the differences between the two approaches are considerably reduced (Moran and Murphy, 1979). For further discussion of the estimation of $\tau_i(\mathbf{x};\boldsymbol{\pi},\boldsymbol{\theta})$ in a Bayesian framework, the reader is referred to Critchley, Ford, and Rijal (1987).

We now consider the semi-Bayesian approach as adopted by Geisser (1967) and Enis and Geisser (1970) in the estimation of the log odds under normal models for $g = 2$ groups. With the semi-Bayesian approach to the estimation of $\tau_i(\mathbf{x};\boldsymbol{\pi},\boldsymbol{\theta})$, its posterior distribution is calculated on the basis of the information in $\mathbf{t}$ but not $\mathbf{x}$. The mean of this posterior distribution so calculated is given by

$$\int \tau_i(\mathbf{x};\boldsymbol{\pi},\boldsymbol{\theta}) p(\boldsymbol{\theta} \mid \mathbf{t})\, d\boldsymbol{\theta} \qquad (i = 1,\ldots,g). \tag{2.2.3}$$

Corresponding to a squared-error loss function, (2.2.3) can be used as an estimate of $\tau_i(\mathbf{x};\boldsymbol{\pi},\boldsymbol{\theta})$. By using different loss functions, other estimates of $\tau_i(\mathbf{x};\boldsymbol{\pi},\boldsymbol{\theta})$ can be obtained, for example, the median or the mode of the posterior distribution of the probability of membership of G_i.

It is of interest to contrast the estimate (2.2.3) of $\tau_i(\mathbf{x};\boldsymbol{\pi},\boldsymbol{\theta})$ with the predictive estimate (2.2.2). Following Rigby (1982), we have from (2.2.1) and (2.2.2) that

$$\begin{aligned}\hat{\tau}_i^{(P)}(\mathbf{x};\mathbf{t}) = \frac{\pi_i \hat{f}_i^{(P)}(\mathbf{x};\mathbf{t})}{\hat{f}_X^{(P)}(\mathbf{x};\mathbf{t})} &= \int \frac{\pi_i f_i(\mathbf{x};\boldsymbol{\theta}_i) p(\boldsymbol{\theta} \mid \mathbf{t})}{p(\mathbf{x} \mid \mathbf{t})} d\boldsymbol{\theta} \\ &= \int \frac{\pi_i f_i(\mathbf{x};\boldsymbol{\theta}_i) p(\mathbf{x},\boldsymbol{\theta} \mid \mathbf{t})}{p(\mathbf{x} \mid \boldsymbol{\theta},\mathbf{t}) p(\mathbf{x} \mid \mathbf{t})} d\boldsymbol{\theta} = \int \tau_i(\mathbf{x};\boldsymbol{\pi},\boldsymbol{\theta}) p(\boldsymbol{\theta} \mid \mathbf{x};\mathbf{t})\, d\boldsymbol{\theta}.\end{aligned} \tag{2.2.4}$$

It can be seen from (2.2.4) that the predictive estimate of $\tau_i(\mathbf{x},\boldsymbol{\pi};\boldsymbol{\theta})$ corresponds to a fully Bayesian approach, as it averages $\tau_i(\mathbf{x};\boldsymbol{\pi},\boldsymbol{\theta})$ over the posterior distribution of $\boldsymbol{\theta}$ given both $\mathbf{t}$ and $\mathbf{x}$. On comparing it with the semi-Bayesian estimate of $\tau_i(\mathbf{x};\boldsymbol{\pi},\boldsymbol{\theta})$ given by (2.2.3), it follows that these two estimates will be practically the same if the information provided by $\mathbf{x}$ about $\boldsymbol{\theta}$ is negligible compared to that provided by $\mathbf{t}$.

2.3 ESTIMATION OF GROUP PROPORTIONS

We consider now the problem where the aim is to estimate the proportions $\pi_1,\ldots,\pi_g$ in which a mixture G of g distinct groups $G_1,\ldots,G_g$ occur. McLachlan and Basford (1988, Chapter 4) have given several examples where this problem arises. One example concerns crop-acreage estimation on the basis of remotely sensed observations on a mixture of several crops; see, for instance, Chhikara (1986) and the references therein. The problem is to estimate the acreage of a particular crop as a proportion of the total acreage. Training data are available on each of the crops to provide estimates of the unknown parameters in the distribution for an individual crop. Another example concerns the case study of Do and McLachlan (1984), where the aim was to assess the rat diet of owls through the estimation of the proportion of each of seven species of rats consumed.

If the training data $\mathbf{t}$ have been obtained by sampling from the mixture of interest G, then the proportion π_i can be estimated simply by its maximum likelihood estimate n_i/n, where n_i denotes the number of entities known to belong to G_i $(i = 1,\ldots,g)$. Therefore, we consider the problem in the context where the n_i provide no information on the proportions π_i. This would be the case if the training data were obtained by sampling separately from each of the groups or from some other mixture of these groups. In order to obtain information on the desired proportions π_i, it is supposed that there is available a random sample of size m, albeit unclassified, from the relevant mixture G. We let $\mathbf{x}_j$ $(j = n+1,\ldots,n+m)$ denote the observed feature vectors on these m unclassified entities having unknown group-indicator vectors $\mathbf{z}_j$ $(j = n+1,\ldots,n+m)$.

An obvious and computationally straightforward way of proceeding is to form a discriminant rule $r(\mathbf{x};\mathbf{t})$ from the classified training data $\mathbf{t}$, and then to apply it to the m unclassified entities with feature vectors $\mathbf{x}_j$ $(j = n+1,\ldots,n+m)$ to find the proportion assigned to the ith group G_i $(i = 1,\ldots,g)$. That is, if m_i denotes the number of the m unclassified entities assigned to G_i, then a rough estimate of π_i is provided by m_i/m $(i = 1,\ldots,g)$. Unless $r(\mathbf{x};\mathbf{t})$ is an infallible rule, m_i/m will be a biased estimator of π_i. For $g = 2$, it can be easily seen that, conditional on $\mathbf{t}$,

$$E(m_1/m) = \pi_1 ec_{11} + \pi_2 ec_{21} \tag{2.3.1}$$

and

$$E(m_2/m) = \pi_1 ec_{12} + \pi_2 ec_{22}, \tag{2.3.2}$$

with either equation giving the so-called discriminant analysis estimator of π_1,

$$\hat{\pi}_{1D} = (m_1/m - ec_{21})/(ec_{11} - ec_{21}),$$

as an unbiased estimator of π_1. In the equations above, the conditional allocation rates of $r(\mathbf{x};\mathbf{t})$, $ec_{ij}(F_i;\mathbf{t})$, are written simply as ec_{ij} for convenience $(i,j = 1,2)$. If $\hat{\pi}_D$ is outside [0,1], then it is assigned the appropriate value zero or one.

On considering (2.3.1) and (2.3.2) simultaneously, $\hat{\pi}_{1D}$ and $\hat{\pi}_{2D} = 1 - \hat{\pi}_{1D}$ can be expressed equivalently as

$$\hat{\boldsymbol{\pi}}_D = \mathbf{J}^{-1}(m_1/m,\ m_2/m)', \tag{2.3.3}$$

where $\hat{\boldsymbol{\pi}}_D = (\hat{\pi}_{1D}, \hat{\pi}_{2D})'$ and

$$\mathbf{J} = \begin{pmatrix} ec_{11} & ec_{21} \\ ec_{12} & ec_{22} \end{pmatrix}.$$

The result (2.3.3) can be generalized to $g > 2$ groups to give

$$\hat{\boldsymbol{\pi}}_D = \mathbf{J}^{-1}(m_1/m,\ldots,m_g/m)', \tag{2.3.4}$$

where the (i,j)th element of the confusion matrix $\mathbf{J}$ is equal to

$$(\mathbf{J})_{ij} = ec_{ji} \qquad (i,j = 1,\ldots,g).$$

According to Macdonald (1975), $\hat{\boldsymbol{\pi}}_D$ seems to have been first suggested by Worlund and Fredin (1962). For known conditional allocation rates $ec_{ij}(F_i;\mathbf{t})$, $\hat{\boldsymbol{\pi}}_D$ is the maximum likelihood estimate of $\boldsymbol{\pi} = (\pi_1,\ldots,\pi_g)'$ based on the proportions m_i/m $(i = 1,\ldots,g)$, and is unbiased. However, in practice, these conditional allocation rates are unknown and must be estimated before $\hat{\boldsymbol{\pi}}_D$ can be calculated from (2.3.4).

It can be seen that if a nonparametric rule $r(\mathbf{x};\mathbf{t})$ is used and the conditional allocation rates are estimated nonparametrically, then the discriminant analysis estimator $\hat{\boldsymbol{\pi}}_D$ can be made distribution-free. In this sense, it should be more robust than a parametric estimator of $\boldsymbol{\pi}$, for example, the maximum likelihood estimator $\hat{\boldsymbol{\pi}}$ whose computation is to be described in Section 2.7. The latter, which is based on the classified training data $\mathbf{t}$ in conjunction with the unclassified data $\mathbf{x}_j$ $(j = n+1,\ldots,n+m)$, is of course fully efficient if the assumed parametric structure holds. Ganesalingam and McLachlan (1981) have investigated the efficiency of $\hat{\boldsymbol{\pi}}_D$ relative to $\hat{\boldsymbol{\pi}}$ in the case of two groups in which the feature vector has a multivariate normal distribution with a common covariance matrix. They concluded that the relative efficiency of $\hat{\boldsymbol{\pi}}_D$ can be quite high provided the mixing proportions are not too disparate and n is not too small relative to m. More recently, for the same normal model, Lawoko and McLachlan (1989) have studied the bias of $\hat{\pi}_{1D}$ as a consequence of using estimates of the conditional allocation rates in its formation. They also considered the case where the classified training observations are correlated.

There are other methods of estimation of the mixing proportions. For a mixture of $g = 2$ groups, π_1 can be estimated also by

$$\hat{\pi}_{1M} = \{(\bar{\mathbf{x}}_1 - \bar{\mathbf{x}}_2)'\mathbf{S}_u^{-1}(\bar{\mathbf{x}}_u - \bar{\mathbf{x}}_2)\}/\{(\bar{\mathbf{x}}_1 - \bar{\mathbf{x}}_2)'\mathbf{S}_u^{-1}(\bar{\mathbf{x}}_1 - \bar{\mathbf{x}}_2)\},$$

where

$$\bar{\mathbf{x}}_i = \sum_{j=1}^{n} z_{ij}\mathbf{x}_j/n_i \qquad (i = 1,2),$$

$$\bar{\mathbf{x}}_u = \sum_{j=n+1}^{n+m} \mathbf{x}_j/m,$$

and

$$\mathbf{S}_u = \sum_{j=n+1}^{n+m} (\mathbf{x}_j - \bar{\mathbf{x}}_u)(\mathbf{x}_j - \bar{\mathbf{x}}_u)'/(m-1).$$

The estimator $\hat{\pi}_{1M}$ can be viewed as the moment estimator of π_1 after transformation of the original feature data $\mathbf{x}_j$ from $\mathbb{R}^p$ to $\mathbb{R}$ by

$$\mathbf{x}_j'\mathbf{S}_u^{-1}(\bar{\mathbf{x}}_1 - \bar{\mathbf{x}}_2) \qquad (j = 1,\ldots,n+m).$$

The asymptotic relative efficiency of $\hat{\pi}_{1M}$ has been derived by McLachlan (1982) for a mixture of two multivariate normal distributions with a common covariance matrix.

There are also minimum-distance estimators. The discriminant analysis, maximum likelihood, and moment estimators of the mixing proportions can all be obtained by using the method of minimum distance through an appropriate choice of the distance measure. A review of these various estimators of the mixing proportions can be found in McLachlan and Basford (1988, Chapter 4).

2.4 ESTIMATING DISEASE PREVALENCE

A situation where the model in the previous section is appropriate occurs in epidemiological studies, where an important aim is to estimate disease prevalence within a population. The groups represent the possible disease categories. Without some idea of prevalence, it is very difficult to plan prospective studies, to interpret retrospective studies, or to make rational health planning decisions; see Rogan and Gladen (1978). It is often impracticable to examine the entire population and so a random sample is taken. Further, it is usually too expensive and perhaps too arduous an experience for the individual being tested for a definitive classification of the disease to be made. Also, even if exhaustive physical and clinical tests were carried out, a true diagnosis may still not be possible. Hence, typically, the sample drawn from the population is allocated to the various groups on the basis of some straightforward but fallible diagnostic test, whose error rates are assessed by applying it to patients with known disease classification. Screening programs often use tests in this way. Even if the prime purpose of the program is finding cases of disease rather than estimating prevalence, the performance of the test is still of interest.

The performance of a screening test designed to detect the presence of a single disease can be evaluated in terms of the sensitivity and specificity of the diagnostic rule $r(\mathbf{x};\mathbf{t})$, which are given conditional on $\mathbf{t}$ by $ec_{22}(F_2;\mathbf{t})$ and $ec_{11}(F_1;\mathbf{t})$, respectively. Here the two groups G_1 and G_2 refer to the absence or presence of the disease. An individual is assigned to G_1 or G_2 according as to whether the test is negative or positive; that is, according as $r(\mathbf{x};\mathbf{t})$ equals 1 or 2. We can write the sensitivity of the test (conditional on $\mathbf{t}$) as

$$\text{pr}\{r(\mathbf{X};\mathbf{t}) = 2 \mid Z_2 = 1\}, \tag{2.4.1}$$

and the specificity as

$$\text{pr}\{r(\mathbf{X};\mathbf{t}) = 1 \mid Z_1 = 1\},$$

where Z_i is one or zero according as the individual with feature vector $\mathbf{x}$ belongs to G_i or not $(i = 1, 2)$.

It is sometimes mistakenly assumed in the confusion over conditional probabilities that, because a test has a high sensitivity as given by the conditional probability (2.4.1), the reverse conditional probability

$$\text{pr}\{Z_2 = 1 \mid r(\mathbf{x};\mathbf{t}) = 2\} \tag{2.4.2}$$

must also be high. This is what Diaconis and Freedman (1981) have referred to in a general context as the "fallacy of the transposed conditional." Although a test may have high sensitivity, the conditional probability (2.4.2), called the predictive value of a positive test (PVP), may be small. Hence, the usefulness of a test is often evaluated in terms of the PVP and the PVN, where the latter denotes the predictive value of a negative test, defined by

$$\text{pr}\{Z_1 = 1 \mid r(\mathbf{x};\mathbf{t}) = 1\}.$$

By Bayes theorem, the PVP can be expressed as

$$\text{pr}\{Z_2 = 1 \mid r(\mathbf{x};\mathbf{t}) = 2\} = \pi_2 ec_{22}/\{\pi_2 ec_{22} + (1-\pi_2)(1-ec_{11})\}, \tag{2.4.3}$$

with a similar expression for the PVN. In (2.4.3), the sensitivity and specificity of the test are abbreviated to ec_{22} and ec_{11}, respectively. It can be seen from (2.4.3) that in order to evaluate the PVP and PVN of a test, the disease prevalence rate π_2, as well as the sensitivity and specificity, must be assessed.

Recently, Gastwirth (1987) established the asymptotic normality of the estimator of the PVP as given by (2.4.3), where π_2 is estimated according to (2.3.3) and where the sensitivity and specificity are replaced by independent binomial estimators formed by applying the test to subsequent data of known origin. A nonmedical example where the abovementioned methodology is applicable is with the use of lie detectors and the associated issues of veracity and admissibility of polygraph evidence in judicial and preemployment screening of applicants.

Where not explicitly stated, it is implicitly assumed in the above that the results are conditional on the training data **t**. In some screening applications, the diagnostic test may be based on a rather sophisticated discriminant rule $r(\mathbf{x};\mathbf{t})$ formed from **t**. For example, for the screening of keratoconjunctivitis sicca in rheumatoid arthritic patients, J. A. Anderson (1972) proposed a diagnostic test based on ten symptoms. However, with many screening tests, in particular presymptomatic ones, the diagnostic test is based on some ad hoc rule, generally using only one feature variable. In the latter situation, the variable is usually measured on a continuous scale and a threshold is imposed to define a positive test. The training data **t** are used then solely to assess the performance of the test for a given threshold. In an example considered by Boys and Dunsmore (1987), patients were designated as either malnourished or nonmalnourished according as their plasma cholesterol levels were less or greater than some threshold.

The choice of threshold in such tests depends on the role in which they are applied. With the ELISA test applied in the context of routine screening of blood donations for AIDS, the threshold for declaring the ELISA assay to be positive is set so that the test is highly sensitive at the expense of having rather low specificity. A high specificity is achieved subsequently by following a positive ELISA with a confirmatory Western blot test (Weiss et al., 1985).

Hand (1986c, 1987a) cautioned that, as the aims of screening and estimation of disease prevalence are somewhat different, the threshold should be chosen with the particular aim in mind. For prevalence estimation, the aim is to minimize the variance of $\hat{\pi}_2$, whereas with screening, the aim is to maximize the accuracy of the rule, that is, some function of the sensitivity and specificity, such as their sum. He investigated the choice of threshold separately for each aim, but using a different sampling scheme to that taken above. In his scheme, the entities of known origin were part of the unclassified data. They were obtained by sampling from each lot of m_i entities assigned to G_i by the fallible diagnostic test and then using an infallible rule to classify them correctly.

2.5 MISCLASSIFIED TRAINING DATA

It is usually assumed in applications of discriminant analysis that the training entities are correctly classified. The classification of a training set is often expensive and difficult, as noted in the previously presented examples on discriminant analysis in the context of medical diagnosis. Another applicable example concerns the classification of remotely sensed imagery data. An important consideration besides the expense and difficulty in procuring classified training observations is that the actual classification may well be subject to error. Indeed, the concept of a true diagnosis is probably inappropriate in some medical fields, and certainly in some such as psychiatry. In the remote-sensing example, say, of crop patterns, the classification of the training pixels may be undertaken visually and hence be prone to error.

In considering now the possibility of misclassification in the training set, we let $\alpha_i(\mathbf{x})$ denote the probability that an entity from group G_i and with feature vector $\mathbf{x}$ is misclassified in the formation of the training set $(i = 1,\ldots,g)$. The misclassification is said to be nonrandom or random depending on whether the $\alpha_i(\mathbf{x})$ do or do not depend on the feature vector $\mathbf{x}$. The simple error structure of random misclassification,

$$\alpha_i(\mathbf{x}) \equiv \alpha_i \qquad (i = 1,\ldots,g), \tag{2.5.1}$$

may rise, for example, where the classification of the training data is made on the basis of machine output (for example, X-ray interpretation or blood test results) and either the output or interpretation for each entity is inaccurate in a way that is independent of its feature vector $\mathbf{x}$. In this same context of medical diagnosis, nonrandom misclassification should be more applicable than random if the classification of the patients in the training set was carried out by clinicians using symptoms closely related to the feature variables.

In his initial work on this problem, Lachenbruch (1966) considered the effect of random misclassification of training entities on the error rates of the sample linear discriminant rule obtained by plugging in the usual maximum likelihood estimates of the parameters in the case of two multivariate normal group-conditional distributions with a common covariance matrix. McLachlan (1972a) derived the asymptotic theory for this model under the additional assumption that one group is not misclassified. For instance, it is often reasonable to assume in the formation of a diagnostic rule that the training sample of healthy patients are all correctly classified. Lachenbruch (1979) subsequently showed that whereas random misclassification is not a serious issue for the sample normal-based linear discriminant rule if α_1 and α_2 are similar, it is for its quadratic counterpart. Lachenbruch (1974) and Chhikara and McKeon (1984) considered the problem under more general misclassification models to allow for nonrandom misclassification. The general conclusion is that ignoring errors in the classification of the training set can be quite harmful for random misclassification. For nonrandom misclassification, the error rates of the sample discriminant rule appear to be only slightly affected, although the optimism of its apparent error rates is considerably increased.

Random misclassification of training entities has been investigated further by Chittineni (1980, 1981) and by Michalek and Tripathi (1980), who also considered the effect of measurement error in the feature variables. More recently, Grayson (1987) has considered nonrandom misclassification of training entities in the context of two groups G_1 and G_2, representing healthy and unhealthy patients, respectively. He supposed that the health status of a patient as specified by the group-indicator vector $\mathbf{z}$ can only be ascertained unreliably. Recall that $z_i = (\mathbf{z})_i$ is one or zero, according as the entity belongs or does not belong to G_i $(i = 1,2)$. We let $\tilde{\mathbf{z}}$ denote the value of $\mathbf{z}$ under the uncertain (noisy) classification. Corresponding to the posterior probability that a patient with feature vector $\mathbf{x}$ belongs to G_i, we let $\tilde{\tau}_i(\mathbf{x})$ be the probability that the ith element $\tilde{Z}_i$ of $\tilde{\mathbf{Z}}$ is one for a patient with feature vector $\mathbf{x}$ $(i = 1,2)$. Under very general conditions on the misclassification errors,

$$\alpha_1(\mathbf{x}) = \text{pr}\{\tilde{Z}_1 = 0 \mid Z_1 = 1, \mathbf{x}\}$$

and

$$\alpha_2(\mathbf{x}) = \text{pr}\{\tilde{Z}_2 = 0 \mid Z_2 = 1, \mathbf{x}\},$$

Grayson (1987) showed that the likelihood ratio for arbitrary group-conditional densities is not ordinally affected. That is, logit$\{\tau_i(\mathbf{x})\}$ is a monotonic function of logit$\{\tilde{\tau}_i(\mathbf{x})\}$. Thus, the class of admissible decision rules is unaffected by this error structure in the classification of the training data. As an illustration of a consequence of the monotonicity of the scales, Grayson (1987) gave an example where there is a need to select the 40 most ill patients (only 40 hospital beds being available). The same patients would be chosen regardless of whether the $\tau_i(\mathbf{x})$ or the $\tilde{\tau}_i(\mathbf{x})$ were used.

2.6 PARTIALLY CLASSIFIED TRAINING DATA

In this section, we consider the problem of forming a discriminant rule, using classified data in conjunction with data on entities unclassified with respect to the g underlying groups $G_1, \ldots, G_g$. We will see there is a number of separate problems in discriminant analysis that fall within this context.

Consistent with our previous notation, we let $\mathbf{t}' = (\mathbf{y}_1, \ldots, \mathbf{y}_n)$ contain the information on the classified entities, where $\mathbf{y}_j = (\mathbf{x}_j', \mathbf{z}_j')'$, and $\mathbf{z}_j$ denotes the known group-indicator vector for the jth entity with observed feature vector $\mathbf{x}_j$ $(j = 1, \ldots, n)$. It is supposed that in addition to the classified training data $\mathbf{t}$, there are available the feature vectors $\mathbf{x}_j$ $(j = n+1, \ldots, n+m)$ observed on m entities of unknown origin. The latter are assumed to have been drawn from a mixture G of $G_1, \ldots, G_g$ in some unknown proportions $\pi_1, \ldots, \pi_g$. We let

$$\mathbf{t}_u' = (\mathbf{x}_{n+1}, \ldots, \mathbf{x}_{n+m}).$$

The unknown group-indicator vector associated with the unclassified feature vector $\mathbf{x}_j$ is denoted by $\mathbf{z}_j$ $(j = n+1, \ldots, n+m)$. If the aim is solely to make a tentative or outright allocation of the m unclassified entities to $G_1, \ldots, G_g$, then we can proceed as discussed in the previous sections. A discriminant rule $r(\mathbf{x}; \mathbf{t})$ can be formed from the classified training data $\mathbf{t}$ and then applied in turn to each of the m unclassified entities with feature vector $\mathbf{x}_j$ $(j = n+1, \ldots, n+m)$ to produce an assessment of its posterior probabilities of group membership and, if required, an outright allocation.

However, in some situations, it is desired to construct a discriminant rule using the unclassified data $\mathbf{t}_u$ in conjunction with the classified training set $\mathbf{t}$. The updating problem falls within this framework. The observing of feature vectors and the subsequent allocation of the corresponding unclassified entities is an ongoing process and, after a certain number of unclassified observations has been obtained, the discriminant rule is updated on the basis of all the observed data. In most updating problems, it is the allocation of unclassified entities subsequent to those whose features have been observed that is of prime concern. Those unclassified entities whose features have been observed may be reallocated as part of the updating process, but their new allocations are generally of no practical consequence in their own right. The latter would be the case if irrevocable decisions had to be made on the basis of the original allocations. For instance, in medical diagnosis, one does not always have the luxury of being able to wait until further information becomes available before making a decision.

If there are sufficiently many classified observations available from each of the groups, then updating may not be a worthwhile exercise. However, often in practice, there are impediments to the procurement of entities of known origin, which limit the number of classified entities available. We have seen in the examples on medical diagnosis given in the previous sections that it may be physically inconvenient to the patient, as well as very expensive, to attempt to make a true diagnosis of the diseased status. In such situations where there is an adequate number of unclassified entities whose features can be measured easily, updating provides a way of improving the performance of the discriminant rule formed solely from the limited classified training data. Of course, as an unclassified observation contains less information than a classified one, many unclassified entities may be needed to achieve an improvement of practical consequence; see Ganesalingam and McLachlan (1978, 1979), O'Neill (1978), and McLachlan and Ganesalingam (1982). Their work was obtained under the assumption of multivariate normality with a common covariance matrix for the two group-conditional distributions. Amoh (1985) later considered the case of inverse normal group-conditional distributions.

Updating procedures appropriate for nonnormal group-conditional densities have been suggested by Murray and Titterington (1978), who expounded various approaches using nonparametric kernel methods, and J. A. Anderson (1979), who gave a method for the logistic discriminant rule. A Bayesian approach to the problem was considered by Titterington (1976), who also considered sequential updating. The more recent work of Smith and Makov (1978) and their other papers on the Bayesian approach to the finite mixture problem, where the observations are obtained sequentially, are covered in Titterington, Smith, and Makov (1985, Chapter 6).

Another problem where estimation on the basis of both classified and unclassified data arises is in the assessment of the proportions in which g groups occur in a mixture G. This problem was discussed in Section 2.3. It is supposed that the classified data have been obtained by a scheme, such as separate sampling, under which they provide no information on the mixing proportions. For the purposes of estimation of the latter, a random but unclassified sample is available from the mixture G. The mixing proportions can be estimated then by the discriminant analysis estimator (2.3.4). However, this estimator is not fully efficient in cases where a parametric family is postulated for the group-conditional distributions. Hence, in such cases, maximum likelihood estimation of the mixing proportions on the basis of both $\mathbf{t}$ and $\mathbf{t}_u$ might be considered.

A third type of problem where estimation on the basis of both classified and unclassified data arises is in situations where the classified data from each group do not represent an observed random sample. This is pursued further in Section 2.8, where it is shown that maximum likelihood estimation is effected by consideration of the same likelihood as in the case of a randomly classified sample.

2.7 MAXIMUM LIKELIHOOD ESTIMATION FOR PARTIAL CLASSIFICATION

Whatever the reason for wishing to carry out estimation on the basis of both classified and unclassified data, it can be undertaken in a straightforward manner, at least in principle, by using maximum likelihood to fit a finite mixture model via the EM algorithm of Dempster, Laird, and Rubin (1977).

With this parametric approach to discrimination, the problem is to fit the mixture model

$$f_X(\mathbf{x};\boldsymbol{\Psi}) = \sum_{i=1}^{g} \pi_i f_i(\mathbf{x};\boldsymbol{\theta}_i) \tag{2.7.1}$$

on the basis of the classified training data $\mathbf{t}$ and the unclassified feature vectors $\mathbf{x}_j$ $(j = n+1,\ldots,n+m)$ in $\mathbf{t}_u$. As before, $\boldsymbol{\Psi} = (\boldsymbol{\pi}',\boldsymbol{\theta}')'$ denotes the vector of all unknown parameters, where $\boldsymbol{\theta}$ consists of those elements of $\boldsymbol{\theta}_1,\ldots,\boldsymbol{\theta}_g$ that are known *a priori* to be distinct.

For classified training data $\mathbf{t}$ obtained by mixture sampling, the log likelihood for $\boldsymbol{\Psi}$ formed from $\mathbf{t}$ and $\mathbf{t}_u$ is given by

$$\begin{aligned}\log L(\boldsymbol{\Psi}) &= \sum_{i=1}^{g}\sum_{j=1}^{n} z_{ij}\log\{\pi_i f_i(\mathbf{x}_j;\boldsymbol{\theta}_i)\}\\ &\quad + \sum_{j=n+1}^{n+m} \log f_X(\mathbf{x}_j;\boldsymbol{\Psi}).\end{aligned} \tag{2.7.2}$$

As discussed in McLachlan and Basford (1988, Chapter 2), an estimate $\hat{\boldsymbol{\Psi}}$ of $\boldsymbol{\Psi}$ is provided by an appropriate root of the likelihood equation

$$\partial \log L(\boldsymbol{\Psi})/\partial\boldsymbol{\Psi} = \mathbf{0}.$$

Since they have described in some detail the application of the EM algorithm to this problem, only a brief outline is given here. The complete data are taken to be $\mathbf{t}, \mathbf{t}_u$, and the unobserved $\mathbf{z}_j$ $(j = n+1,\ldots,n+m)$. For this specification, the complete-data log likelihood is

$$\log L_C(\boldsymbol{\Psi}) = \sum_{i=1}^{g}\sum_{j=1}^{n+m} z_{ij}\log f_i(\mathbf{x}_j;\boldsymbol{\theta}_i) + \sum_{i=1}^{g}\sum_{j=1}^{n+m} z_{ij}\log\pi_i.$$

The EM algorithm is applied to this model by treating $\mathbf{z}_j$ ($j = n+1,\ldots,n+m$) as missing data. It is easy to program and proceeds iteratively in two steps, E (for expectation) and M (for maximization). Using some initial value for Ψ, say, $\Psi^{(0)}$, the E-step requires the calculation of

$$H(\Psi,\Psi^{(0)}) = E\{\log L_C(\Psi) \mid \mathbf{t},\mathbf{t}_u;\Psi^{(0)}\},$$

the expectation of the complete-data log likelihood $\log L_C(\Psi)$, conditional on the observed data and the initial fit $\Psi^{(0)}$ for Ψ. This step is effected here simply by replacing each unobserved indicator variable z_{ij} by its expectation conditional on $\mathbf{x}_j$, given by

$$E(Z_{ij} \mid \mathbf{x}_j;\Psi^{(0)}) = \tau_i(\mathbf{x}_j;\Psi^{(0)}) \qquad (i = 1,\ldots,g).$$

That is, z_{ij} is replaced by the initial estimate of the posterior probability that the jth entity with feature vector $\mathbf{x}_j$ belongs to G_i ($i = 1,\ldots,g$; $j = n+1,\ldots,n+m$).

On the M-step first time through, the intent is to choose the value of Ψ, say, $\Psi^{(1)}$, that maximizes $H(\Psi,\Psi^{(0)})$, which, from the E-step, is equal here to $\log L_C(\Psi)$ with each z_{ij} replaced by $\tau_i(\mathbf{x}_j;\Psi^{(0)})$ for $j = n+1,\ldots,n+m$. It leads, therefore, to solving the equations

$$\hat{\pi}_i^{(1)} = \left\{ n_i + \sum_{j=n+1}^{n+m} \tau_i(\mathbf{x}_j;\Psi^{(0)}) \right\} \Big/ (n+m) \qquad (i = 1,\ldots,g) \tag{2.7.3}$$

and

$$\sum_{i=1}^{g}\sum_{j=1}^{n} z_{ij}\partial \log f_i(\mathbf{x}_j;\boldsymbol{\theta}_i^{(1)})/\partial\boldsymbol{\theta}$$
$$+\sum_{i=1}^{g}\sum_{j=n+1}^{n+m} \tau_i(\mathbf{x}_j;\Psi^{(0)})\partial \log f_i(\mathbf{x}_j;\boldsymbol{\theta}_i^{(1)})/\partial\boldsymbol{\theta} = \mathbf{0}, \tag{2.7.4}$$

where $\partial \log f_i(\mathbf{x}_j;\boldsymbol{\theta}_i^{(1)})/\partial\boldsymbol{\theta}$ denotes $\partial \log f_i(\mathbf{x}_j;\boldsymbol{\theta}_i)/\partial\boldsymbol{\theta}$ evaluated at the point $\boldsymbol{\theta}_i = \boldsymbol{\theta}_i^{(1)}$. In the case where the classified entities have been sampled separately from each group, the equation

$$\hat{\pi}_i^{(1)} = \sum_{j=n+1}^{n+m} \tau_i(\mathbf{x}_j;\Psi^{(0)})/m$$

and not (2.7.3), is appropriate for the estimate of π_i ($i = 1,\ldots,g$).

One nice feature of the EM algorithm is that the solution to the M-step often exists in closed form, as is to be seen for mixtures of normals in Section 3.8, where the actual form of (2.7.4) is to be given.

The E- and M-steps are alternated repeatedly, where in their subsequent executions, the initial fit $\Psi^{(0)}$ is replaced by the current fit for Ψ, say, $\Psi^{(k-1)}$, on the kth cycle. Another nice feature of the EM algorithm is that the likelihood for the incomplete-data specification can never be decreased after an EM sequence. Hence,

$$L(\Psi^{(k+1)}) \geq L(\Psi^{(k)}),$$

which implies that $L(\Psi^{(k)})$ converges to some L^* for a sequence bounded above. Dempster et al. (1977) showed that if the very weak condition that $H(\phi, \Psi)$ is continuous in both ϕ and Ψ holds, then L^* will be a local maximum of $L(\Psi)$, provided the sequence is not trapped at some saddle point. A detailed account of the convergence properties of the EM algorithm in a general setting has been given by Wu (1983), who addressed, in particular, the problem that the convergence of $L(\Psi^{(k)})$ to L^* does not automatically imply the convergence of $\Psi^{(k)}$ to a point Ψ^*. Louis (1982) has devised a procedure for extracting the observed information matrix when using the EM algorithm, as well as developing a method for speeding up its convergence. Meilijson (1989) has since provided a unification of EM methodology and Newton-type methods; see also McLachlan and Basford (1988, Section 1.9), and Jones and McLachlan (1990b).

Let $\hat{\Psi}$ be the chosen solution of the likelihood equation. It is common in practice to estimate the inverse of the covariance matrix of a maximum likelihood solution by the observed information matrix, rather than the expected information matrix evaluated at the solution; see Efron and Hinkley (1978). The observed information $\mathbf{I}(\hat{\Psi})$ for the classified and unclassified data combined is equal to

$$\mathbf{I}(\hat{\Psi}) = \mathbf{I}_c(\hat{\Psi}) + \mathbf{I}_u(\hat{\Psi}),$$

where

$$\mathbf{I}_c(\Psi) = -\sum_{i=1}^{g}\sum_{j=1}^{n} z_{ij}\,\partial^2 \log\{\pi_i f_i(\mathbf{x}_j; \boldsymbol{\theta}_i)\}/\partial\Psi\,\partial\Psi'$$

evaluated at the point $\Psi = \hat{\Psi}$ is the observed information matrix for the classified data and

$$\mathbf{I}_u(\Psi) = -\sum_{j=n+1}^{n+m} \partial^2 \log f_X(\mathbf{x}_j; \Psi)/\partial\Psi\,\partial\Psi'$$

at $\Psi = \hat{\Psi}$ is the observed information matrix for the unclassified data.

It is computationally attractive to compute $\mathbf{I}(\hat{\Psi})$ using an approximation that requires only the gradient vector of the complete-data log likelihood. This approximation is

$$\mathbf{I}(\hat{\Psi}) \simeq \sum_{j=1}^{n}\{\partial \log L_C(\hat{\Psi}; \mathbf{x}_j, \mathbf{z}_j)/\partial\Psi\}\{\partial \log L_C(\hat{\Psi}; \mathbf{x}_j, \mathbf{z}_j)/\partial\Psi\}'$$

$$+ \sum_{j=n+1}^{n+m}\{\partial \log L_C(\hat{\Psi}; \mathbf{x}_j, \tilde{\mathbf{z}}_j)/\partial\Psi\}\{\partial \log L_C(\hat{\Psi}; \mathbf{x}_j, \tilde{\mathbf{z}}_j)/\partial\Psi\}',$$

where the unknown group-indicator vector $\mathbf{z}_j$ for the unclassified feature vector $\mathbf{x}_j$ $(j = n+1,\ldots,n+m)$ is replaced by the vector of its estimated posterior probabilities of group membership. That is, $\partial \log L_C(\hat{\Psi};\mathbf{x}_j,\tilde{\mathbf{z}}_j)/\partial\Psi$ is equal to $\partial \log L_C(\hat{\Psi};\mathbf{x}_j,\mathbf{z}_j)/\partial\Psi$ evaluated at

$$\mathbf{z}_j = (\tau_1(\mathbf{x}_j;\hat{\Psi}),\ldots,\tau_g(\mathbf{x}_j;\hat{\Psi}))'$$

for $j = n+1,\ldots,n+m$. Here $L_C(\Psi;\mathbf{x}_j,\mathbf{z}_j)$ denotes the complete-data likelihood for Ψ when formed from just the jth observation $\mathbf{y}_j = (\mathbf{x}_j',\mathbf{z}_j')'$, $j = 1,\ldots,n+m$. Expressions for the elements of the gradient vector of $\log L_C(\Psi;\mathbf{x}_j,\mathbf{z}_j)$ are given in McLachlan and Basford (1988, Section 2.4) in the case of multivariate normal group-conditional distributions.

Maximum likelihood estimation is facilitated by the presence of data of known origin with respect to each group in the mixture. For example, there may be singularities in the likelihood on the edge of the parameter space in the case of group-conditional densities that are multivariate normal with unequal covariance matrices. However, no singularities will occur if there are more than p classified observations available from each group. Also, there can be difficulties with the choice of suitable starting values for the EM algorithm. But in the presence of classified data, an obvious choice of a starting point is the maximum likelihood estimate of Ψ based solely on the classified training data $\mathbf{t}$, assuming there are enough classified entities from each group for this purpose. In the case where $\mathbf{t}$ provides no information on $\pi_1,\ldots,\pi_g$, an initial estimate of π_i is provided by the estimator (2.3.3).

In fitting the mixture model (2.7.1) to the partially classified data, it would be usual to compare the three fits, $\hat{\Psi}^{(0)}$, $\hat{\Psi}_u$, and $\hat{\Psi}$, where $\hat{\Psi}^{(0)}$ and $\hat{\Psi}_u$ denote the maximum likelihood estimates of Ψ computed from $\mathbf{t}$ and $\mathbf{t}_u$, respectively. Also, if the group memberships of the m unclassified entities are still of interest in their own right at the updating stage, then particular consideration would be given to a comparison of the fits $\tau_i(\mathbf{x}_j;\Psi^{(0)})$, $\tau_i(\mathbf{x}_j;\hat{\Psi}_u)$, and $\tau_i(\mathbf{x}_j;\hat{\Psi})$ for the posterior probabiities of group membership $(i = 1,\ldots,g;\ j = n+1,\ldots,n+m)$. In the case where $\mathbf{t}$ provides no information on the mixing proportions, we can form an estimate of only $\boldsymbol{\theta}$ on the basis of $\mathbf{t}$. In any event, it is the updating of the estimate of $\boldsymbol{\theta}$ containing the parameters of the group-conditional densities that is of primary concern. In Section 6.7, we give an example on updating in which each unclassified entity has its own, but known, prior probabilities of group membership.

A comparison of the three abovementioned estimates of Ψ, or at least of $\boldsymbol{\theta}$, is of much interest in those situations where it is felt that the groups from which the classified entities have been drawn have undergone some change before or during the sampling of the unclassified entities. This change might be able to be explained by the fact there was quite a lengthy period between the sampling of the classified and unclassified entities. For example, the problem may be to identify the sex of fulmar petrel birds on the basis of their bill length and bill depth. These measurements on birds of known sex may be available only from old museum records for a population of birds captured many years ago.

The question of whether a given unclassified entity is atypical of each of the g specified groups is discussed in Section 6.4. However, the unclassified entities may well be typical with respect to at least one of the groups, yet they may well be from a mixture of groups that are different from those that are the source of the classified entities. This last possibility is much more difficult to address.

We note here in passing that the EM algorithm also provides a way of carrying out maximum likelihood estimation where some of the entities in the training set have observations missing on their feature variables; see the monograph of Little and Rubin (1987) for an account of this approach in a general statistical context. Also, Little (1988) has considered robust estimation of the mean and covariance matrix from data with missing values. Studies on the missing data problem in the context of discriminant analysis include those by Chan and Dunn (1972, 1974); Srivastava and Zaatar (1972); Chan, Gilman, and Dunn (1976); Little (1978); and, more recently, Hufnagel (1988). Blackhurst and Schluchter (1989) and Fung and Wrobel (1989) have considered this problem for the logistic model, which is introduced in Chapter 8.

2.8 MAXIMUM LIKELIHOOD ESTIMATION FOR PARTIAL NONRANDOM CLASSIFICATION

As mentioned previously in Section 2.6, in some situations in practice with a partially classified training set, the classified data may not represent an observed random sample from the sample space of the feature vector. For example, in medical screening, patients are often initially diagnosed on the basis of some simple rule, for instance, whether one feature variable is above or below a certain threshold, corresponding to a positive or negative test. Patients with a positive test are investigated further from which their true condition may be ascertained. However, patients with a negative test may be regarded as apparently healthy and so may not be investigated further. This would be the case if a true diagnosis can be made, say, by only an invasive technique whose application would not be ethical in apparently healthy patients. In these circumstances, if only the data of known origin were used in the estimation of the unknown parameters, then it would generally bias the results, unless appropriate steps were taken, such as fitting truncated densities or using logistic regression. The latter approach is considered in Chapter 8. Another approach that avoids this bias problem is to perform the estimation on the basis of all the data collected, including the data of unknown origin. This approach was adopted by McLachlan and Gordon (1989) in their development of a probabilistic allocation rule as an aid in the diagnosis of renal artery stenosis (RAS), which is potentially curable by surgery. Their case study is described in Section 7.8.

We now proceed to show for this last approach that the maximum likelihood estimate of Ψ in the mixture model (2.7.1) is obtained by consideration of the same likelihood (2.7.2) as in the case where the classified data are randomly classified. It is supposed that the classification of an entity is only undertaken if its feature vector falls in some region, say, R, of the feature space. A random sample of size M is taken, where m denotes the number of entities with feature vectors falling in the complement of R, and $n = M - m$ denotes the number with feature vectors in R. The latter n entities are subsequently classified with n_i found to come from group G_i ($i = 1,2$), and $n = n_1 + n_2$. The feature vectors are relabeled so that $\mathbf{x}_j$ ($j = 1,\ldots,n$) denote the n classified feature vectors and $\mathbf{x}_j$ ($j = n+1,\ldots,n+m$) the m unclassified feature vectors.

Under this sampling scheme, the probability that n entities have feature vectors falling in R is given by

$$P_R(n) = \binom{M}{n} c_R^n (1 - c_R)^m,$$

where

$$c_R = \int_R f_X(\mathbf{x};\Psi)\,d\nu.$$

Given n, the probability that n_1 of these entities are from G_1 and n_2 from G_2 is

$$P(n_1 \mid n) = \binom{n}{n_1} \pi_{1R}^{n_1} \pi_{2R}^{n_2},$$

where

$$\begin{aligned}\pi_{iR} &= \mathrm{pr}\{Z_i = 1 \mid \mathbf{X} \in R\} \\ &= \pi_i c_{iR}/c_R,\end{aligned}$$

and

$$c_{iR} = \int_R f_i(\mathbf{x};\boldsymbol{\theta}_i)\,d\nu \qquad (i = 1,2).$$

Let $L(\Psi;\mathbf{t},\mathbf{t}_u,n,n_1)$ denote the likelihood function for Ψ formed on the basis of the classified data $\mathbf{t}$, the unclassified feature vectors in $\mathbf{t}_u$, and also n and n_1. Then

$$\begin{aligned}\log L(\Psi;\mathbf{t},\mathbf{t}_u,n,n_1) &= \log\{P_R(n)P(n_1 \mid n)\} \\ &\quad + \sum_{i=1}^{2}\sum_{j=1}^{n} z_{ij}\log\{f_i(\mathbf{x}_j;\boldsymbol{\theta}_i)/c_{iR}\} \\ &\quad + \sum_{j=n+1}^{n+m} \log\{f_X(\mathbf{x}_j;\Psi)/(1-c_R)\},\end{aligned}$$

which, apart from a combinatorial additive term, reduces to $\log L(\Psi)$ as given by (2.7.2).

In some instances, the classification of an entity with feature vector falling in the region R is not always available at the time of the analysis. The classification may have been mislaid or the patient for some reason may not have been initially classified. In any event, provided the classified entities may be viewed as a random sample from those entities with feature vectors falling in R, we can form the likelihood in the same manner as (2.7.2).

2.9 CLASSIFICATION LIKELIHOOD APPROACH

Another likelihood-based approach to the allocation of the m unclassified entities in the partial classification scheme introduced in Section 2.6 is what is sometimes called the classification likelihood approach. With this approach, $\boldsymbol{\Psi}$ and the unknown $\mathbf{z}_j$ are chosen to maximize $L_C(\boldsymbol{\Psi})$, the likelihood for the complete-data specification adopted in the application of the EM algorithm to the fitting of the mixture model (2.7.1) in Section 2.7. That is, the unknown $\mathbf{z}_j$ are treated as parameters to be estimated along with $\boldsymbol{\Psi}$. Accordingly, the maximization of $L_C(\boldsymbol{\Psi})$ is over the set of zero-one values of the elements of the unknown $\mathbf{z}_j$, corresponding to all possible assignments of the m entities to the g groups, as well as over all admissible values of $\boldsymbol{\Psi}$. A recent reference on this approach is found in McLeish and Small (1986), and additional references can be found in McLachlan and Basford (1988, Chapter 1). It can be seen that the classification likelihood approach is equivalent to using the likelihood ratio criterion for discrimination, whose implementation for $m = 1$ was given in Section 1.8.

In principle, the maximization process for the classification likelihood approach can be carried out for arbitrary m, since it is just a matter of computing the maximum value of $L_C(\boldsymbol{\Psi})$ over all possible partitions of the m entities to the g groups $G_1,\ldots,G_g$. In some situations, for example, with multivariate normal group-conditional distributions having unequal covariance matrices, the restriction that at least $p+1$ observations belong to each G_i is needed to avoid the degenerate case of infinite likelihood. Unless m is small, however, searching over all possible partitions is prohibitive. If $\hat{\mathbf{z}}_j$ $(j = n+1,\ldots,n+m)$ denotes the optimal partition of the m unclassified entities, then $\hat{z}_{ij} = 1$ or 0 according as to whether

$$\hat{\pi}_i f_i(\mathbf{x}_j;\hat{\boldsymbol{\theta}}_i) \geq \hat{\pi}_h f_h(\mathbf{x}_j;\hat{\boldsymbol{\theta}}_h) \qquad (h = 1,\ldots,g;\ h \neq i)$$

holds or not, where $\hat{\boldsymbol{\theta}}_i$ and $\hat{\pi}_i$ are the maximum likelihood estimates of $\boldsymbol{\theta}_i$ and π_i, respectively, for the m entities partitioned according to $\hat{\mathbf{z}}_{n+1},\ldots,\hat{\mathbf{z}}_{n+m}$. Hence a solution corresponding to a local maximum can be computed iteratively by alternating a modified version of the E-step but the same M-step, as described in Section 2.7 for the application of the EM algorithm in fitting the mixture model (2.7.1). In the E-step on the $(k+1)$th cycle of the iterative process, z_{ij} is replaced not by the current estimate of the posterior probability that the jth entity belongs to G_i, but by one or zero according as to whether

$$\pi_i^{(k)} f_i(\mathbf{x}_j;\boldsymbol{\theta}_i^{(k)}) \geq \pi_h^{(k)} f_h(\mathbf{x}_j;\boldsymbol{\theta}_h^{(k)}) \qquad (h = 1,\ldots,g;\ h \neq i)$$

holds or not ($i = 1,\ldots,g;\ j = n+1,\ldots,n+m$).

The iterative updating of a discriminant rule proposed by McLachlan (1975a, 1977a) can be viewed as applying the classification likelihood approach from a starting point equal to the estimate of $\boldsymbol{\Psi}$ based solely on the classified data $\mathbf{t}$. For $g = 2$ groups with equal prior probabilities and with multivariate normal distributions having a common covariance matrix, McLachlan (1975a) showed that it leads asymptotically to an optimal partition of the unclassified entities. O'Neill (1976) subsequently showed how this process can be easily modified to give an asymptotically optimal solution in the case of unknown prior probabilities.

2.10 ABSENCE OF CLASSIFIED DATA

We now discuss the fitting of the mixture model (2.7.1) in the absence of feature data on entities that have been classified with respect to the components of the mixture model to be fitted. This is usually referred to as unsupervised learning in the pattern recognition literature. Corresponding to this situation of no data on classified training entities, we set $n = 0$ in the partial classification scheme defined in Section 2.6. The feature vectors on the m unclassified entities are now labeled $\mathbf{x}_1,\ldots,\mathbf{x}_m$, so that

$$\mathbf{t}_u' = (\mathbf{x}_1,\ldots,\mathbf{x}_m).$$

By fitting a mixture model with g components to the data $\mathbf{t}_u$, we obtain a probabilistic clustering of the m unclassified entities in terms of their estimated posterior probabilities of component membership $\tau_i(\mathbf{x}_j;\hat{\boldsymbol{\Psi}})$, where $\hat{\boldsymbol{\Psi}}$ now denotes the maximum likelihood estimate computed on the basis of $\mathbf{t}_u$ ($i = 1,\ldots,g;\ j = 1,\ldots,m$). A partition of the m entities into g nonoverlapping clusters is given by $\hat{\mathbf{z}}_1,\ldots,\hat{\mathbf{z}}_m$, where $\hat{z}_{ij} = (\hat{\mathbf{z}}_j)_i$ is one or zero, according as

$$\tau_i(\mathbf{x}_j;\hat{\boldsymbol{\Psi}}) \geq \tau_h(\mathbf{x}_j;\hat{\boldsymbol{\Psi}}) \qquad (h = 1,\ldots,g;\ h \neq i)$$

holds or not ($i = 1,\ldots,g;\ j = 1,\ldots,m$). If the maximum of $\tau_i(\mathbf{x}_j;\hat{\boldsymbol{\Psi}})$ over $i = 1,\ldots,g$ is near to one for most of the observations $\mathbf{x}_j$, then it suggests that the mixture likelihood approach can put the m entities into g distinct clusters with a high degree of certainty. Conversely, if the maximum is generally well below one, it indicates that the components of the fitted mixture model are too close together for the m entities to be clustered with any certainty. Hence, these estimated posterior probabilities can be used to provide a measure of the strength of the clustering. To this end, Ganesalingam and McLachlan (1980) and, more recently, Basford and McLachlan (1985) have investigated the use of the estimated posterior probabilities of component membership in forming estimates of the overall and individual allocation rates of the clustering with respect to the components of the fitted mixture model. In those cases where the clusters do not correspond to *a priori* defined groups, the estimated allocation rates can be interpreted as estimates of the allocation rates that would exist if the clustering were assumed to reflect an externally existing partition of the data. This is considered further in Chapter 10.

The recent monograph of McLachlan and Basford (1988) provides an in-depth account of the fitting of finite mixture models. Briefly, with mixture models in the absence of classified data, the likelihood typically will have multiple maxima; that is, the likelihood equation will have multiple roots. If so, the selection of suitable starting values for the EM algorithm is crucial. There are obvious difficulties with this selection in the typical cluster analysis setting, where there is no *a priori* knowledge of any formal group structure on the underlying population. McLachlan (1988) has considered a systematic approach to the choice of initial values for Ψ, or, equivalently, for the posterior probabilities of component membership of the mixture to be fitted. He proposed that two-dimensional scatter plots of the data after appropriate transformation, using mainly principal component analysis, be used to explore the data initially for the presence of clusters. Any visual clustering so obtained can be reflected in the initial values specified for the posterior probabilities of component membership of the mixture. This approach, along with methods that use three-dimensional plots, is considered further in Section 6.6.

Fortunately, with some applications of finite mixture models, in particular in medicine, there is available some *a priori* information concerning a possible group structure for the population. In some instances, this information may extend to a provisional grouping of the m unclassified entities. Then Ψ can be estimated as if this provisional grouping were valid to yield an initial value for use in the EM algorithm. In medical applications, a provisional grouping may correspond to a clinical diagnosis of the patients using one or perhaps some of the feature variables. Often a clinical diagnosis may focus on only one of the feature variables. If all features are used, then it is generally in a limited and rather ad hoc way. For example, a patient might be diagnosed as unhealthy if any of the measured feature variables falls in a range conventionally regarded as abnormal in healthy patients. It is of interest, therefore, to compute the statistical diagnosis as given by $\hat{\mathbf{z}}_1, \ldots, \hat{\mathbf{z}}_m$, and to compare it with the clinical diagnosis. The statistical diagnosis is attractive in that it provides an objective grouping of the patients that takes into account all of their feature variables simultaneously. Also, it is not dependent upon arbitrary clinical decisions. An example using the diabetes data originally analyzed by Reaven and Miller (1979) is given in Section 6.8.

With some applications, the *a priori* information on the group structure extends to knowing the number g of underlying groups in the population. For example, in the screening for a disease, there is an obvious dichotomy of the population into disease-free and diseased groups. In some situations where the possible groups represent the various causes of a disease, the assumed number of groups g may not be the actual number, as there may be further, as of yet undetected, causes.

Assessing the true value of g is an important but very difficult problem. In the present framework of finite mixture models, an obvious approach is to use the likelihood ratio statistic λ to test for the smallest value of the number g of components in the mixture compatible with the data. Unfortunately, with mixture models, regularity conditions do not hold for $-2\log\lambda$ to have its usual asymptotic null distribution of chi-squared with degrees of freedom equal to the difference between the number of parameters under the null and alternative hypotheses; see McLachlan and Basford (1988, Chapter 1) and the references therein, including McLachlan (1987a) and Quinn, McLachlan, and Hjort (1987).

One way of assessing the null distribution is to use a resampling method, which can be viewed as a particular application of the general bootstrap approach of Efron (1979, 1982). More specifically, for the test of the null hypothesis of $g = g_0$ versus the alternative of $g = g_1$, the log likelihood ratio statistic can be bootstrapped as follows. Proceeding under the null hypothesis, a so-called bootstrap sample is generated from the mixture density $f_X(\mathbf{x};\hat{\boldsymbol{\Psi}})$, where $\hat{\boldsymbol{\Psi}}$ is taken to be the maximum likelihood estimate of $\boldsymbol{\Psi}$ formed under the null hypothesis from the original observed sample $\mathbf{t}_u$. The value of $-2\log\lambda$ is calculated for the bootstrap sample after fitting mixture models for $g = g_0$ and $g = g_1$ in turn to it. This process is repeated independently a number of times K, and the replicated values of $-2\log\lambda$ evaluated from the successive bootstrap samples can be used to assess the bootstrap, and hence the true, null distribution of $-2\log\lambda$. In particular, it enables an approximation to be made to the achieved level of significance P corresponding to the value of $-2\log\lambda$ evaluated from the original sample. The value of the jth-order statistic of the K replications can be taken as an estimate of the quantile of order $j/(K+1)$, and the P-value can be assessed by reference with respect to the ordered bootstrap replications of $-2\log\lambda$. Actually, the value of the jth-order statistic of the K replications is a better approximation to the quantile of order $(3j-1)/(3K+1)$ (Hoaglin, 1985). This bootstrap approach to the testing of g is applied in Section 6.9.

For some hypotheses, the null distribution of λ will not depend on any unknown parameters. In this case, there will be no difference between the bootstrap and true null distributions of $-2\log\lambda$. An example is the case of multivariate normal components with all parameters unknown, where $g_0 = 1$ under the null hypothesis. In this situation, where it is the actual statistic $-2\log\lambda$ and not its bootstrap analogue that is being resampled, the resampling may be viewed as an application of the Monte Carlo approach to the construction of a hypothesis test having an exact level of desired significance. This approach was proposed originally by Barnard (1963); see Hope (1968) and Hall and Titterington (1989). Aitkin, Anderson, and Hinde (1981) appear to have been the first to apply the resampling approach in the context of finite mixture models.

We return now to the question of misclassified training data as first raised in Section 2.5. One way of proceeding if there is some doubt as to the veracity of the given classification of some or all of the training entities is to leave their group-indicator vectors unspecified and to fit the mixture model (2.7.1) to the consequent partially or completely unclassified sample. As reported in Section 2.5 on the effects of misclassified training data, it appears that ignoring any misclassification does not lead to much harm provided it is those entities whose feature vectors fall in the doubtful region of the feature space that are more prone to error in classification. However, if the training entities are misclassified at random, then ignoring the misclassification can distort appreciably the learning process.

As a consequence of this, Krishnan and Nandy (1987) proposed a model to take random misclassification of the training entities into account in the design of the sample discriminant rule. Working with respect to $g = 2$ groups, they proposed that in place of the vector $\mathbf{z}_j$ defining the group of origin of the jth entity in the training set, there is a number v_j between zero and one ($j = 1,\ldots,m$). This number represents the supervisor's assessment of the chance that the jth entity belongs to group G_1 rather than to G_2. They assumed that in group G_i, this number is distributed, independently of $\mathbf{x}_j$, with density $q_i(v_j)$, where $q_i(\cdot)$ and $q_2(\cdot)$ denote the beta(η_1,η_2) and beta(η_2,η_1) densities, respectively, where η_1 and η_2 are unknown. The group-conditional distributions were taken to be multivariate normal with a common covariance matrix. As this is an incomplete-data problem in the absence of values of $\mathbf{z}_1,\ldots,\mathbf{z}_m$, maximum likelihood estimation of η_1 and η_2 and the other parameters can be approached using the EM algorithm. For this model, Krishnan and Nandy (1990a) have derived the efficiency of the sample linear discriminant rule so formed relative to its version based on a perfectly classified training set.

Unfortunately, with the abovementioned model, it is impossible to carry out the M-step exactly, so far as the parameters η_1 and η_2 are concerned. Titterington (1989) consequently suggested an alternative model incorporating a stochastic supervisor for which the EM algorithm can be easily implemented to compute the maximum likelihood estimates. Following the way of Aitchison and Begg (1976) for indicating uncertain diagnosis, Titterington (1989) proposed the use of

$$u_j = \log\{v_j/(1 - v_j)\}$$

instead of v_j as a basic indicator of the supervisor's assessment that the jth entity arises from G_1. As the sample space for U_j is the real line, Titterington (1989) adopted the natural assumption of a normal distribution for U_j within each group. In order to maintain symmetry, as in the beta-based formulation of Krishnan and Nandy (1987), the normal group-conditional distributions for U_j were taken to have a common variance and means ω_1 and ω_2 with $\omega_2 = -\omega_1$. For a knowledgeable supervisor, one should expect $\omega_1 > 0$, just as one should expect $\eta_1 > \eta_2$ in the beta-based model. Titterington (1989) noted that his model is just as flexible as the beta-based model, as Aitchison and Begg (1976) have described how the same variety of shapes can be produced for the density of U_j starting from a normal distribution, as can be produced by the family of beta distributions.

The relative efficiency of the sample linear discriminant rule formed under this model has been considered by Krishnan and Nandy (1990b). In related work, Katre and Krishnan (1989) have considered maximum likelihood estimation via the EM algorithm in the case where an unequivocal classification of the training data is supplied, but incorrectly with unknown random errors under the model (2.5.1). Krishnan (1988) has investigated the relative efficiency of the sample linear discriminant function formed under this model.

2.11 GROUP-CONDITIONAL MIXTURE DENSITIES

The fitting of finite mixture models in discriminant analysis also arises if some or all of the group-conditional densities are modeled as finite mixtures. In medical diagnosis in the context of healthy or unhealthy patients, the distribution of the features for the latter group is often appropriately modeled as a mixture, as discussed by Lachenbruch and Kupper (1973) and Lachenbruch and Broffitt (1980). An example can be found in Emery and Carpenter (1974). In the course of their study of sudden infant death syndrome, they concluded tentatively from estimates of the density of the degranulated mast cell count in sudden infant death cases that it was a mixture of the control density with a smaller proportion of a contaminating density of higher mean. The control group consisted of infants who died from known causes that would not affect the degranulated mast cell count.

Rawlings et al. (1984) have given an example that arose in a study of alcohol-related diseases where both the groups under consideration, corresponding to two diseased states, are each a mixture of two subgroups. Another example where both the groups are modeled as mixtures of subgroups has been given by McLachlan and Gordon (1989) on the diagnosis of renal artery stenosis. It is discussed in Section 7.8.

A further situation where the group-conditional densities are taken to be mixtures is with a latent class approach, to discrimination with discrete feature data. With this approach, considered further in Section 7.3.3, the group-conditional densities are modeled as mixtures of the same component densities.

A finite mixture model for, say, the ith group-conditional density can be fitted to the training data from G_i by using equations (2.7.3) and (2.7.4) with $n = 0$. This assumes that there are no parameters in this mixture density in common with the other group-conditional parameters, although the equations can be easily modified if this is not so.

Group-conditional mixture densities are used also in the modeling of digitized images of cervical cells from the PAP smear slide. On the basis that a cervical cytology specimen consists of 13 cell types, the problem is to discriminate between a mixture of 5 normal squamous and nonsquamous cell types and a mixture of 8 abnormal and malignant tumor cell types (Oliver et al., 1979). In this example, the fitting of the group-conditional mixture densities is facilitated by the presence of training data for each of the 13 cell types.

CHAPTER 3

Discrimination via Normal Models

3.1 INTRODUCTION

We have seen in Section 1.4 that discriminant analysis is relatively straightforward for known group-conditional densities. Generally, in practice, the latter are either partially or completely unknown, and so there is the problem of their estimation from data **t** on training entities, as defined by (1.6.1). As with other multivariate statistical techniques, the assumption of multivariate normality provides a convenient way of specifying a parametric structure. Hence, normal models for the group-conditional densities provide the basis for a good deal of the theoretical results and practical applications in discriminant analysis. In this chapter, we therefore focus on discrimination via normal-based models. In the latter part, we consider the reduction of the dimension of the feature vector through linear projections that optimize certain separatory and allocatory measures for normal models.

3.2 HETEROSCEDASTIC NORMAL MODEL

3.2.1 Optimal Rule

Under a heteroscedastic normal model for the group-conditional distributions of the feature vector **X** on an entity, it is assumed that

$$\mathbf{X} \sim N(\boldsymbol{\mu}_i, \boldsymbol{\Sigma}_i) \quad \text{in} \quad G_i \qquad (i = 1,\ldots,g), \tag{3.2.1}$$

where $\boldsymbol{\mu}_1,\ldots,\boldsymbol{\mu}_g$ denote the group means, and $\boldsymbol{\Sigma}_1,\ldots,\boldsymbol{\Sigma}_g$ the group-covariance matrices. Corresponding to (3.2.1), the ith group-conditional density $f_i(\mathbf{x};\boldsymbol{\theta}_i)$

is given by

$$\begin{aligned} f_i(\mathbf{x};\boldsymbol{\theta}_i) &= \phi(\mathbf{x};\boldsymbol{\mu}_i,\boldsymbol{\Sigma}_i) \\ &= (2\pi)^{-p/2}|\boldsymbol{\Sigma}_i|^{-1/2}\exp\{-\tfrac{1}{2}(\mathbf{x}-\boldsymbol{\mu}_i)'\boldsymbol{\Sigma}_i^{-1}(\mathbf{x}-\boldsymbol{\mu}_i)\}, \end{aligned}$$

where $\boldsymbol{\theta}_i$ consists of the elements of $\boldsymbol{\mu}_i$, and the $\frac{1}{2}p(p+1)$ distinct elements of $\boldsymbol{\Sigma}_i$ $(i = 1,\dots,g)$. It is assumed that each $\boldsymbol{\Sigma}_i$ is nonsingular. There is no loss of generality in so doing, since singular group-covariance matrices can always be made nonsingular by an appropriate reduction of dimension.

If $\pi_1,\dots,\pi_g$ denote the prior probabilities for the groups $G_i,\dots,G_g$, then we let

$$\begin{aligned} \boldsymbol{\Psi}_U &= (\pi_1,\dots,\pi_g,\boldsymbol{\theta}_U')' \\ &= (\boldsymbol{\pi}',\boldsymbol{\theta}_U')', \end{aligned}$$

where $\boldsymbol{\theta}_U$ consists of the elements of $\boldsymbol{\mu}_1,\dots,\boldsymbol{\mu}_g$ and the distinct elements of $\boldsymbol{\Sigma}_1,\dots,\boldsymbol{\Sigma}_g$. The subscript "$U$" emphasizes that the group-covariance matrices are allowed to be unequal in the specification of the normal model (3.2.1).

The posterior probability that an entity with feature vector $\mathbf{x}$ belongs to group G_i is denoted by $\tau_i(\mathbf{x};\boldsymbol{\Psi}_U)$ for $i = 1,\dots,g$. In estimating these posterior probabilities, it is more convenient to work in terms of their log ratios. Accordingly, we let

$$\begin{aligned} \eta_{ig}(\mathbf{x};\boldsymbol{\Psi}_U) &= \log\{\tau_i(\mathbf{x};\boldsymbol{\Psi}_U)/\tau_g(\mathbf{x};\boldsymbol{\Psi}_U)\} \\ &= \log(\pi_i/\pi_g) + \xi_{ig}(\mathbf{x};\boldsymbol{\theta}_U), \end{aligned} \tag{3.2.2}$$

where

$$\xi_{ig}(\mathbf{x};\boldsymbol{\theta}_U) = \log\{f_i(\mathbf{x};\boldsymbol{\theta}_i)/f_g(\mathbf{x};\boldsymbol{\theta}_g)\} \qquad (i = 1,\dots,g-1).$$

The definition (3.2.2) corresponds to the arbitrary choice of G_g as the base group.

Under the heteroscedastic normal model (3.2.1),

$$\begin{aligned} \xi_{ig}(\mathbf{x};\boldsymbol{\theta}_U) = &-\tfrac{1}{2}\{\delta(\mathbf{x},\boldsymbol{\mu}_i;\boldsymbol{\Sigma}_i) - \delta(\mathbf{x},\boldsymbol{\mu}_g;\boldsymbol{\Sigma}_g)\} \\ &- \tfrac{1}{2}\{\log|\boldsymbol{\Sigma}_i|/|\boldsymbol{\Sigma}_g|\} \qquad (i = 1,\dots,g-1), \end{aligned} \tag{3.2.3}$$

where

$$\delta(\mathbf{x},\boldsymbol{\mu}_i;\boldsymbol{\Sigma}_i) = (\mathbf{x}-\boldsymbol{\mu}_i)'\boldsymbol{\Sigma}_i^{-1}(\mathbf{x}-\boldsymbol{\mu}_i)$$

is the squared Mahalanobis distance between $\mathbf{x}$ and $\boldsymbol{\mu}_i$ with respect to $\boldsymbol{\Sigma}_i$ $(i = 1,\dots,g)$. The notation $\delta(\mathbf{a},\mathbf{b};\mathbf{C})$ for the squared Mahalanobis distance

$$(\mathbf{a}-\mathbf{b})'\mathbf{C}^{-1}(\mathbf{a}-\mathbf{b})$$

between two vectors $\mathbf{a}$ and $\mathbf{b}$ with respect to some positive definite symmetric matrix $\mathbf{C}$ applies throughout this book. For brevity, we henceforth abbreviate $\delta(\mathbf{x},\boldsymbol{\mu}_i;\boldsymbol{\Sigma}_i)$ to $\delta_i(\mathbf{x})$ for $i = 1,\dots,g$.

In this setting, the optimal or Bayes rule $r_o(\mathbf{x};\mathbf{\Psi}_U)$ assigns an entity with feature vector $\mathbf{x}$ to G_g if

$$\eta_{ig}(\mathbf{x};\mathbf{\Psi}_U) \leq 0 \qquad (i = 1,\ldots,g-1)$$

is satisfied. Otherwise, the entity is assigned to G_h if

$$\eta_{ig}(\mathbf{x};\mathbf{\Psi}_U) \leq \eta_{hg}(\mathbf{x};\mathbf{\Psi}_U) \qquad (i = 1,\ldots,g-1;\ i \neq h)$$

holds. In defining the Bayes rule above and in the sequel, we will take the costs of misallocation to be the same. However, unequal costs of misallocation can be incorporated into the formulation of the Bayes rule, as considered in a general context in Section 1.4. In the subsequent work, we will refer to $r_o(\mathbf{x};\mathbf{\Psi}_U)$ as the normal-based quadratic discriminant rule (NQDR).

3.2.2 Plug-In Sample NQDR

In practice, $\boldsymbol{\theta}_U$ is generally taken to be unknown and so must be estimated from the available training data $\mathbf{t}$, as given by (1.6.1). With the estimative approach to discriminant analysis, the posterior probabilities of group membership $\tau_i(\mathbf{x};\mathbf{\Psi}_U)$ and the consequent Bayes rule $r_o(\mathbf{x};\mathbf{\Psi}_U)$ are estimated simply by plugging in some estimate $\hat{\boldsymbol{\theta}}_U$, such as the maximum likelihood estimate, for $\boldsymbol{\theta}_U$ in the group-conditional densities.

The maximum likelihood estimates of $\boldsymbol{\mu}_i$ and $\mathbf{\Sigma}_i$ computed under (3.2.1) from the training data $\mathbf{t}$ are given by the sample mean $\bar{\mathbf{x}}_i$ and the sample covariance matrix $\hat{\mathbf{\Sigma}}_i$, respectively, where

$$\bar{\mathbf{x}}_i = \sum_{j=1}^{n} z_{ij}\mathbf{x}_j/n_i$$

and

$$\hat{\mathbf{\Sigma}}_i = \sum_{j=1}^{n} z_{ij}(\mathbf{x}_j - \bar{\mathbf{x}}_i)(\mathbf{x}_j - \bar{\mathbf{x}}_i)'/n_i$$

for $i = 1,\ldots,g$. Consistent with our previous notation,

$$n_i = \sum_{j=1}^{n} z_{ij}$$

denotes the number of entities from group G_i in the training data $\mathbf{t}$ ($i = 1,\ldots,g$). It is assumed here that $n_i > p$, so that $\hat{\mathbf{\Sigma}}_i$ is nonsingular ($i = 1,\ldots,g$). In the subsequent work, we follow the usual practice of estimating $\mathbf{\Sigma}_i$ by the unbiased estimator

$$\mathbf{S}_i = n_i\hat{\mathbf{\Sigma}}_i/(n_i - 1) \qquad (i = 1,\ldots,g).$$

With $\boldsymbol{\theta}_U$ estimated as above,

$$\xi_{ig}(\mathbf{x};\hat{\boldsymbol{\theta}}_U) = -\tfrac{1}{2}\{\hat{\delta}_i(\mathbf{x}) - \hat{\delta}_g(\mathbf{x})\} - \tfrac{1}{2}\log\{|\mathbf{S}_i|/|\mathbf{S}_g|\} \qquad (i = 1,\ldots,g-1),$$

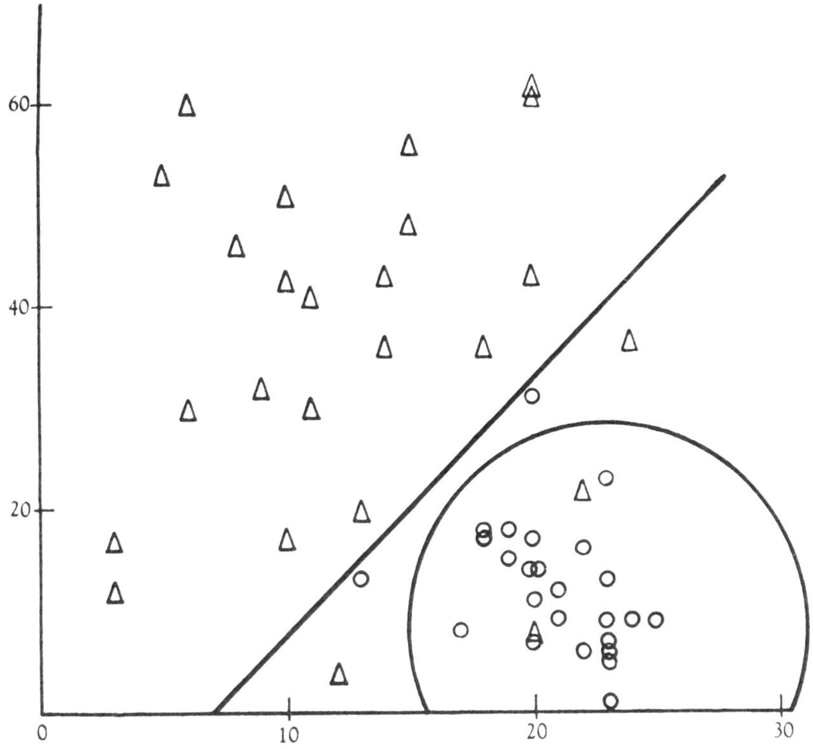

FIGURE 3.1. Boundaries for the sample normal-based quadratic and linear discriminant functions with a zero cutoff point, as computed from psychological data with circles and triangles representing normals and psychotics, respectively. From Smith (1947).

where

$$\hat{\delta}_i(\mathbf{x}) = \delta(\mathbf{x}, \overline{\mathbf{x}}_i; \mathbf{S}_i)$$

$$= (\mathbf{x} - \overline{\mathbf{x}}_i)' \mathbf{S}_i^{-1} (\mathbf{x} - \overline{\mathbf{x}}_i) \qquad (i = 1, \ldots, g).$$

One of the earliest applications of the sample normal-based quadratic discriminant function $\xi_{ig}(\mathbf{x}; \hat{\boldsymbol{\theta}}_U)$ was given by Smith (1947) in the case of $g = 2$ groups. It concerned discriminating between normal persons and psychotics on the basis of bivariate data on $n_1 = 25$ normals and $n_2 = 25$ psychotics. For this data set, the boundary for the sample normal-based discriminant function with a zero cutoff point is plotted in Figure 3.1, along with the straight-line boundary under the constraint of equal group-covariance matrices.

In some situations in practice, individual training observations on a feature variable are not available. For example, the classified feature data may have been collected in the form of frequencies of observations falling in fixed class intervals. A further problem that is often encountered is truncation of the data; observations below and above certain readings are often not available. In such situations, the maximum likelihood estimates of the group means and variances can be computed using the algorithm given in Jones and McLachlan (1990a). It uses maximum likelihood to fit either univariate normal distributions or mixtures of a specified number of them to categorized data that may also be in truncated form.

3.2.3 Bias Correction of Plug-In Sample NQDR

In forming an estimate of $\xi_{ig}(\mathbf{x};\boldsymbol{\theta}_U)$, there are no compelling reasons for using the unbiased $\mathbf{S}_i$ in place of the biased $\hat{\boldsymbol{\Sigma}}_i$ $(i = 1,\ldots,g)$. If unbiasedness is of prime concern, then the focus should be on the provision of unbiased estimates of the group-conditional densities or, more appropriately, of the ratios of the group-conditional densities because the latter directly determine the Bayes rule. An unbiased estimate of $f_i(\mathbf{x};\boldsymbol{\theta}_i)$ is available from the results of Ghurye and Olkin (1969) on unbiased estimation of multivariate densities. The unbiased minimum variance estimator of $f_i(\mathbf{x};\boldsymbol{\theta}_i)$ is given by

$$(2\pi)^{-(1/2)p}k_i|(1-n_i^{-1})\mathbf{S}_i|^{-1/2}$$
$$\times |\mathbf{I}_p - n_i(n_i-1)^{-2}\mathbf{S}_i^{-1}(\mathbf{x}-\bar{\mathbf{x}}_i)(\mathbf{x}-\bar{\mathbf{x}}_i)'|^{(1/2)(n_i-p-3)},$$

where

$$k_i = (n_i-1)^{-(1/2)p}\{c(p,n_i-2)/c(p,n_i-1)\},$$

$$c(p,n_i) = [2^{(1/2)pn_i}\pi^{(1/4)p(p-1)}\prod_{j=1}^{p}\Gamma\{\tfrac{1}{2}(n_i-j+1\}]^{-1},$$

and $\mathbf{I}_p$ is the $p \times p$ identity matrix.

It is more appropriate, however, in the present context of estimation of the Bayes rule to consider unbiased estimates of the ratios of the group-conditional densities $f_i(\mathbf{x};\boldsymbol{\theta}_i)$. By using certain standard results on the expectations of quadratic forms and Wishart distributions, the minimum variance unbiased estimator of $\xi_{ig}(\mathbf{x};\boldsymbol{\theta}_U)$ can be obtained as follows (Moran and Murphy, 1979). It is straightforward to calculate the expectation of $\hat{\delta}_i(\mathbf{x})$, conditional on $\mathbf{x}$, using the result in Das Gupta (1968) that

$$E(\mathbf{S}_i^{-1}) = c_1(n_i-1)\boldsymbol{\Sigma}_i^{-1} \qquad (i = 1,\ldots,g), \tag{3.2.4}$$

where $c_1(\cdot)$ is defined by

$$c_1(n_i) = n_i/(n_i-p-1). \tag{3.2.5}$$

With this result, we have that

$$\begin{aligned}E\{\hat{\delta}_i(\mathbf{x})\} &= E[\mathrm{tr}\{(\mathbf{x}-\bar{\mathbf{x}}_i)'\mathbf{S}_i^{-1}(\mathbf{x}-\bar{\mathbf{x}}_i)\}] \\ &= E[\mathrm{tr}\{\mathbf{x}-\bar{\mathbf{x}}_i)(\mathbf{x}-\bar{\mathbf{x}}_i)'\mathbf{S}_i^{-1}\}] \\ &= \mathrm{tr}[E\{(\mathbf{x}-\bar{\mathbf{x}}_i)(\mathbf{x}-\bar{\mathbf{x}}_i)'\}E(\mathbf{S}_i^{-1})] \\ &= c_1(n_i-1)\mathrm{tr}[\{\boldsymbol{\Sigma}_i/n_i + (\mathbf{x}-\boldsymbol{\mu}_i)(\mathbf{x}-\boldsymbol{\mu}_i)'\}\boldsymbol{\Sigma}_i^{-1}] \\ &= c_1(n_i-1)\{(p/n_i)+\delta_i(\mathbf{x})\} \qquad (i=1,\ldots,g). \end{aligned} \tag{3.2.6}$$

This implies that

$$\tilde{\delta}_i(\mathbf{x}) = \{\hat{\delta}_i(\mathbf{x})/c_1(n_i-1)\} - p/n_i \tag{3.2.7}$$

is an unbiased estimator of $\delta_i(\mathbf{x})$ for $i = 1,\ldots,g$.

We also need the result that

$$E\{\log|\mathbf{S}_i|\} = \log|\boldsymbol{\Sigma}_i| - p\log(n_i-1) + c_2(n_i), \tag{3.2.8}$$

where

$$c_2(n_i) = \sum_{k=1}^{p}\psi\{\tfrac{1}{2}(n_i-k)\} \qquad (i=1,\ldots,g), \tag{3.2.9}$$

and where $\psi(y) = d\log\Gamma(y)/dy$ is the digamma function as discussed in Abramowitz and Stegun (1965, page 258). As noted by Critchley and Ford (1985), it may be efficiently evaluated using the algorithm of Bernardo (1976).

It follows directly from (3.2.7) and (3.2.8) that an unbiased estimator of $\xi_{ig}(\mathbf{x};\boldsymbol{\theta}_U)$ is given by

$$\begin{aligned}\hat{\xi}_{ig,U}(\mathbf{x}) = &-\tfrac{1}{2}\hat{\delta}_i(\mathbf{x})/c_1(n_i-1) + \tfrac{1}{2}\hat{\delta}_g(\mathbf{x})/c_1(n_g-1) \\ &+ \tfrac{1}{2}p(n_i^{-1}-n_g^{-1}) - \tfrac{1}{2}\log\{|\mathbf{S}_i|/|\mathbf{S}_g|\} \\ &- \tfrac{1}{2}p\log\{(n_i-1)/(n_g-1)\} + \tfrac{1}{2}\{c_2(n_i)-c_2(n_g)\}\end{aligned} \tag{3.2.10}$$

for $i = 1,\ldots,g-1$. As $\hat{\xi}_{ig,U}(\mathbf{x})$ is a function of the complete sufficient statistics $\bar{\mathbf{x}}_i$ and $\mathbf{S}_i$ $(i = 1,\ldots,g)$, it is the uniform minimum variance unbiased estimator of $\xi_{ig}(\mathbf{x};\boldsymbol{\theta}_U)$.

In this chapter, only point estimates of the log likelihood ratios $\xi_{ig}(\mathbf{x};\boldsymbol{\theta}_U)$ and the posterior probabilities of group membership $\tau_i(\mathbf{x};\boldsymbol{\Psi}_U)$ are considered. The sampling distributions of these quantities and, in particular, their standard errors for the provision of confidence intervals are considered separately in Chapter 11 in the context of assessing their reliability on the basis of the same training data from which they were formed.

3.2.4 Equal-Mean Discrimination

There are some applications of discriminant analysis in practice in which the groups may be assumed to have the same mean vectors. In this case for multivariate normal group-conditional distributions, the optimal discriminant rule

$r_o(\mathbf{x};\boldsymbol{\theta}_U)$ has to be based on the differences between the group-covariance matrices. Okamoto (1961) studied this problem for $g = 2$ groups. Besides developing the theory, he presented an application in which the aim is to label twin pairs of like sex as monozygotic or dizygotic on the basis of a set of measurements of physical characteristics. Bartlett and Please (1963) considered the same type of application, using the now well-known twins data of Stocks (1933) to study the usefulness of the quadratic discriminant rule in discriminating between monozygotic and dizygotic twins. They adopted the uniform covariance structure

$$\boldsymbol{\Sigma}_i = \sigma_i^2\{(1-\rho_i)\mathbf{I}_p + \rho_i \mathbf{1}_p\mathbf{1}_p'\} \qquad (i = 1,2), \tag{3.2.11}$$

where $\mathbf{1}_p = (1,1,\ldots,1)'$ is the $p \times 1$ vector of ones. It is not uncommon in biological work for the correlations to be more or less the same in magnitude within a group.

From Bartlett (1951a), the inverse of $\boldsymbol{\Sigma}_i$ as given by (3.2.11) is

$$\boldsymbol{\Sigma}_i^{-1} = c_{1i}\mathbf{I}_p - c_{2i}\mathbf{1}_p\mathbf{1}_p', \tag{3.2.12}$$

where

$$c_{1i} = 1/\{\sigma_i^2 + (1-\rho_i)\},$$

and

$$c_{2i} = \rho_i/[\sigma_i^2(1-\rho_i)\{(1+(p-1)\rho_i\}]$$

for $i = 1,2$. By using (3.2.12) in (3.2.3), it follows that the optimal discriminant function $\xi(\mathbf{x},\boldsymbol{\theta}_U)$ is given, apart from an additive constant, as

$$\xi(\mathbf{x},\boldsymbol{\theta}_U) = -\tfrac{1}{2}(c_{11}-c_{12})Q_1 + \tfrac{1}{2}(c_{21}-c_{22})Q_2,$$

where $Q_1 = \mathbf{x}'\mathbf{x}$, and $Q_2 = (\mathbf{x}'\mathbf{1}_p)^2$. Because $g = 2$ only, we have written $\xi_{12}(\mathbf{x};\boldsymbol{\theta}_U)$ as $\xi(\mathbf{x};\boldsymbol{\theta}_U)$. In Penrose's (1946–1947) conception of "size" and "shape" components, Q_2 is the square of the "size" component. The "shape" component does not arise here as the group means are all the same.

Kshirsagar and Musket (1972) have considered a method for choosing the cutoff point with the use of $\xi(\mathbf{x};\boldsymbol{\theta}_U)$ under the model (3.2.11) so that the two error rates are the same. More recently, Marco, Young, and Turner (1987b) have derived the asymptotic expectation of what can be regarded as a plug-in-type estimate of the overall error rate of the rule based on a sample version of $\xi(\mathbf{x};\boldsymbol{\theta}_U)$ formed under (3.2.11) with $\rho_1 = \rho_2$. Earlier, Schwemer and Mickey (1980) derived the expected error rates of the linear discriminant rule applied to data from two groups with the same means but proportional covariance matrices. Lachenbruch (1975b) noted that a quadratic discriminant rule can be avoided by working with the absolute value of each variate in the feature vector. In its application to the twins data of Stocks (1933), Lachenbruch (1975b) found that his so-called absolute linear discriminant rule performed slightly worse than the quadratic discriminant rule as used earlier by Bartlett and Please (1963) and Desu and Geisser (1973). However, when he induced contamination into this data set, his absolute linear rule performed reasonably well, but the performance of the quadratic discriminant rule deteriorated considerably. Geisser (1964), Geisser and Desu (1968), Enis and Geisser (1970), Desu and Geisser (1973), and Lee (1975) have investigated the equal-mean discrimination problem via Bayesian approaches. Note that the case of unequal group-means under the uniform covariance structure (3.2.11) has been considered by Han (1968).

3.3 HOMOSCEDASTIC NORMAL MODEL

3.3.1 Optimal Rule

It can be seen from (3.2.3) that a substantial simplification occurs in the form for the posterior probabilities of group membership and the consequent Bayes rule if the group-covariance matrices $\boldsymbol{\Sigma}_1, \ldots, \boldsymbol{\Sigma}_g$ are all the same. This is because the quadratic term in $\mathbf{x}, \mathbf{x}'\boldsymbol{\Sigma}_i^{-1}\mathbf{x}$, in the exponent of the ith group-conditional density is now the same for all groups, and so vanishes in the pairwise ratios of these densities as specified by (3.2.3).

We let

$$\boldsymbol{\Psi}_E = (\boldsymbol{\pi}', \boldsymbol{\theta}_E')',$$

where $\boldsymbol{\theta}_E$ denotes $\boldsymbol{\theta}_U$ under the constraint that

$$\boldsymbol{\Sigma}_i = \boldsymbol{\Sigma} \qquad (i = 1, \ldots, g).$$

The subscript "E" emphasizes the specification of equality for the group-covariance matrices. This definition of $\boldsymbol{\theta}_E$ does mean that it contains $\frac{1}{2}p(p+1)(g-1)$ redundant parameters. However, it conveniently allows the posterior probabilities of group membership to be written as $\tau_i(\mathbf{x}; \boldsymbol{\Psi}_E)$ under the homoscedastic normal model

$$\mathbf{X} \sim N(\boldsymbol{\mu}_i, \boldsymbol{\Sigma}) \quad \text{in} \quad G_i \qquad (i = 1, \ldots, g). \tag{3.3.1}$$

As the notation $\tau_i(\mathbf{x}; \boldsymbol{\Psi}_E)$ implies, the posterior probabilities of group membership under the homoscedastic normal model (3.3.1) are obtained by replacing $\boldsymbol{\Psi}_U$ with $\boldsymbol{\Psi}_E$ in the expressions for the posterior probabilities $\tau_i(\mathbf{x}; \boldsymbol{\Psi}_U)$ under the heteroscedastic version (3.2.1). Effectively, we replace $\boldsymbol{\theta}_U$ by $\boldsymbol{\theta}_E$ in the log likelihood ratios $\xi_{ig}(\mathbf{x}; \boldsymbol{\theta}_U)$.

Corresponding to (3.2.2), we have under the homoscedastic normal model (3.3.1) that

$$\begin{aligned}\eta_{ig}(\mathbf{x}; \boldsymbol{\Psi}_E) &= \log\{\tau_i(\mathbf{x}; \boldsymbol{\Psi}_E)/\tau_g(\mathbf{x}; \boldsymbol{\Psi}_E)\} \\ &= \log(\pi_i/\pi_g) + \xi_{ig}(\mathbf{x}; \boldsymbol{\theta}_E),\end{aligned} \tag{3.3.2}$$

where

$$\begin{aligned}\xi_{ig}(\mathbf{x}; \boldsymbol{\theta}_E) &= -\tfrac{1}{2}\{\delta_{i,E}(\mathbf{x}) - \delta_{g,E}(\mathbf{x})\} \\ &= \{\mathbf{x} - \tfrac{1}{2}(\boldsymbol{\mu}_i + \boldsymbol{\mu}_g)\}'\boldsymbol{\Sigma}^{-1}(\boldsymbol{\mu}_i - \boldsymbol{\mu}_g) \\ &\qquad\qquad (i = 1, \ldots, g-1),\end{aligned} \tag{3.3.3}$$

and where

$$\delta_{i,E}(\mathbf{x}) = \delta(\mathbf{x}, \boldsymbol{\mu}_i; \boldsymbol{\Sigma}) \qquad (i = 1, \ldots, g).$$

The optimal or Bayes rule $r_o(\mathbf{x}; \boldsymbol{\Psi}_E)$ assigns an entity with feature vector $\mathbf{x}$ to G_g if

$$\eta_{ig}(\mathbf{x}; \boldsymbol{\Psi}_E) \leq 0 \qquad (i = 1, \ldots, g-1)$$

is satisfied. Otherwise, the entity is assigned to G_h if

$$\eta_{ig}(\mathbf{x}; \boldsymbol{\Psi}_E) \leq \eta_{hg}(\mathbf{x}; \boldsymbol{\Psi}_E) \qquad (i = 1, \ldots, g-1;\ i \neq h)$$

holds. We will refer to $r_o(\mathbf{x}; \boldsymbol{\Psi}_E)$ as the normal-based linear discriminant rule (NLDR).

In the case of $g = 2$ groups, we write $\eta_{12}(\mathbf{x}; \boldsymbol{\Psi}_E)$ and $\xi_{12}(\mathbf{x}; \boldsymbol{\theta}_E)$ as $\eta(\mathbf{x}; \boldsymbol{\Psi}_E)$ and $\xi(\mathbf{x}; \boldsymbol{\theta}_E)$, respectively. For future reference, we express the normal-based linear discriminant function (NLDF) $\xi(\mathbf{x}; \boldsymbol{\theta}_E)$ in the form

$$\xi(\mathbf{x}; \boldsymbol{\theta}_E) = \beta_{0E}^o + \boldsymbol{\beta}_E' \mathbf{x}, \tag{3.3.4}$$

where

$$\beta_{0E}^o = -\tfrac{1}{2}(\boldsymbol{\mu}_1 + \boldsymbol{\mu}_2)' \boldsymbol{\Sigma}^{-1} (\boldsymbol{\mu}_1 - \boldsymbol{\mu}_2),$$

and

$$\boldsymbol{\beta}_E = \boldsymbol{\Sigma}^{-1} (\boldsymbol{\mu}_1 - \boldsymbol{\mu}_2).$$

Thus, we can write $\eta(\mathbf{x}; \boldsymbol{\Psi}_E)$ as

$$\begin{aligned} \eta(\mathbf{x}; \boldsymbol{\Psi}_E) &= \log(\pi_1/\pi_2) + \beta_{0E}^o + \boldsymbol{\beta}_E' \mathbf{x} \\ &= \beta_{0E} + \boldsymbol{\beta}_E' \mathbf{x} \\ &= \boldsymbol{\alpha}_E' (1, \mathbf{x}')', \end{aligned} \tag{3.3.5}$$

where

$$\beta_{0E} = \log(\pi_1/\pi_2) + \beta_{0E}^o,$$

and

$$\boldsymbol{\alpha}_E = (\beta_{0E}, \boldsymbol{\beta}_E')'.$$

3.3.2 Optimal Error Rates

It can be seen from (3.3.3) that the NLDR $r_o(\mathbf{x}; \boldsymbol{\Psi}_E)$ is linear in $\mathbf{x}$. One consequence of this is that it is straightforward to obtain closed expressions for the optimal error rates at least in the case of $g = 2$ groups. For it can be seen from (3.3.2) that in this case, $r_o(\mathbf{x}; \boldsymbol{\Psi}_E)$ is based on the single linear discriminant function $\xi(\mathbf{x}; \boldsymbol{\theta}_E)$ with cutoff point

$$k = \log(\pi_2/\pi_1).$$

That is, an entity with feature vector $\mathbf{x}$ is assigned to G_1 or G_2 according as to whether $\xi(\mathbf{x}; \boldsymbol{\theta}_E)$ is greater or less than k.

The optimal error rate specific to G_1 is therefore given by

$$\begin{aligned} eo_{12}(\Psi_E) &= \text{pr}\{\xi(\mathbf{X};\boldsymbol{\theta}_E) < k \mid \mathbf{X} \in G_1\} \\ &= \Phi\{(k - \tfrac{1}{2}\Delta^2)/\Delta\}, \end{aligned}$$

where

$$\Delta^2 = (\boldsymbol{\mu}_1 - \boldsymbol{\mu}_2)'\boldsymbol{\Sigma}^{-1}(\boldsymbol{\mu}_1 - \boldsymbol{\mu}_2)$$

is the squared Mahalanobis distance between G_1 and G_2, and $\Phi(\cdot)$ denotes the standard normal distribution function. Similarly,

$$eo_{21}(\Psi_E) = \Phi\{-(k + \tfrac{1}{2}\Delta^2)/\Delta\}.$$

In the above, we have followed the notation for the allocation rates as introduced in Section 1.10, so that $eo_{ij}(\Psi_E)$ denotes the probability that a randomly chosen entity from group G_i is allocated to G_j on the basis of the Bayes rule $(i,j = 1,\ldots,g)$.

For a zero cutoff point $(k = 0)$, the group-specific error rates $eo_{12}(\Psi_E)$ and $eo_{21}(\Psi_E)$ are equal with the common value of $\Phi(-\frac{1}{2}\Delta)$. Often, in practice, k is taken to be zero. Besides corresponding to the case of equal prior probabilities, it also yields the minimax rule, as defined in Section 1.5.

3.3.3 Plug-In Sample NLDR

For unknown $\boldsymbol{\theta}_E$ in the case of an arbitrary number g of groups, the maximum likelihood estimate of $\boldsymbol{\mu}_i$ is given as under heteroscedasticity by the sample mean $\bar{\mathbf{x}}_i$ of the feature observations from G_i in the training data $\mathbf{t}$ $(i = 1,\ldots,g)$. The maximum likelihood estimate $\hat{\boldsymbol{\Sigma}}$ of the common group-covariance matrix $\boldsymbol{\Sigma}$ is the pooled (within-group) sample covariance matrix. That is,

$$\begin{aligned} \hat{\boldsymbol{\Sigma}} &= \sum_{i=1}^{g}(n_i/n)\hat{\boldsymbol{\Sigma}}_i \\ &= \sum_{i=1}^{g}\sum_{j=1}^{n} z_{ij}(\mathbf{x}_j - \bar{\mathbf{x}}_i)(\mathbf{x}_j - \bar{\mathbf{x}}_i)'/n. \end{aligned}$$

In using these estimates to form the plug-in sample versions $\xi_{ig}(\mathbf{x};\hat{\boldsymbol{\theta}}_E)$ and $\tau_i(\mathbf{x};\hat{\boldsymbol{\Psi}}_E)$ of the log likelihood ratios and the group-posterior probabilities, we follow here the usual practice of first correcting $\hat{\boldsymbol{\Sigma}}$ for bias, so that

$$\mathbf{S} = n\hat{\boldsymbol{\Sigma}}/(n-g)$$

is used instead of $\hat{\boldsymbol{\Sigma}}$. With $\boldsymbol{\theta}_E$ estimated as above,

$$\xi_{ig}(\mathbf{x};\hat{\boldsymbol{\theta}}_E) = \{\mathbf{x} - \tfrac{1}{2}(\bar{\mathbf{x}}_i + \bar{\mathbf{x}}_g)\}'\mathbf{S}^{-1}(\bar{\mathbf{x}}_i - \bar{\mathbf{x}}_g) \qquad (i = 1,\ldots,g-1). \tag{3.3.6}$$

It can be seen from (3.3.6) that for $g = 2$ groups, the plug-in sample version $r_o(\mathbf{x};\hat{\Psi}_E)$ of the normal-based linear discriminant rule (NLDR) $r_o(\mathbf{x};\Psi_E)$ is based on the sample version

$$\xi(\mathbf{x};\hat{\theta}_E) = \{\mathbf{x} - \tfrac{1}{2}(\bar{\mathbf{x}}_1 + \bar{\mathbf{x}}_2)\}\mathbf{S}^{-1}(\bar{\mathbf{x}}_1 - \bar{\mathbf{x}}_2) \qquad (3.3.7)$$

of the NLDF $\xi(\mathbf{x};\theta_E)$, as defined by (3.3.4). The sample NLDF $\xi(\mathbf{x};\hat{\theta}_E)$ is often referred to in the literature as the W classification statistic (Wald, 1944).

It is essentially the same as Fisher's (1936) linear discriminant function derived without the explicit adoption of a normal model. As noted in Section 1.7, Fisher's linear discriminant function is given by $\mathbf{a}'\mathbf{x}$, where $\mathbf{a}$ maximizes the quantity

$$\{\mathbf{a}'(\bar{\mathbf{x}}_1 - \bar{\mathbf{x}}_2)\}^2/\mathbf{a}'\mathbf{S}\mathbf{a}, \qquad (3.3.8)$$

which is the square of the difference between the group-sample means of $\mathbf{a}'\mathbf{x}$ scaled by the (bias-corrected) pooled sample variance of $\mathbf{a}'\mathbf{x}$. The maximization of (3.3.8) is achieved by taking $\mathbf{a}$ proportional to $\hat{\beta}_E$, where

$$\hat{\beta}_E = \mathbf{S}^{-1}(\bar{\mathbf{x}}_1 - \bar{\mathbf{x}}_2).$$

This leads to the sample linear discriminant function

$$\mathbf{x}'\hat{\beta}_E = \mathbf{x}'\mathbf{S}^{-1}(\bar{\mathbf{x}}_1 - \bar{\mathbf{x}}_2). \qquad (3.3.9)$$

Because

$$(\bar{\mathbf{x}}_1 - \bar{\mathbf{x}}_2)'\mathbf{S}^{-1}(\bar{\mathbf{x}}_1 - \bar{\mathbf{x}}_2) \geq 0,$$

the sample mean of $\mathbf{x}'\hat{\beta}_E$ in group G_1 is not less than what it is in G_2. Hence, an entity with feature vector $\mathbf{x}$ can be assigned to G_1 for large values of $\mathbf{x}'\hat{\beta}_E$ and to G_2 for small values. If the cutoff point for (3.3.9) is taken to be equidistant between its group-sample means, then it is equivalent to the sample NLDF $\xi(\mathbf{x};\hat{\theta}_E)$ applied with a zero cutoff point.

Normal theory-based discriminant analysis is now widely available through computing packages such as BMDP (1988), GENSTAT (1988), IMSL (1984), P–STAT (1984), S (Becker, Chambers, and Wilks, 1988), SAS (1990), SPSS–X (1986), and SYSTAT (1988). There are also those packages with regression programs that can be applied indirectly to discriminant analysis; for example, Minitab (Ryan et al., 1982) and GLIM (1986). The reader is referred to Lachenbruch (1982) and to Section 4.2 in the report by Panel on Discriminant Analysis, Classification, and Clustering (1989) for an account of some of these packages with respect to their facilities for discriminant analysis. In addition to the programs for discriminant analysis that exist as part of a general-purpose package of statistical programs, there are those algorithms that have been designed specifically for discriminant analysis. For example, James (1985) has provided some algorithms written in BASIC for discriminant analysis. Other algorithms for performing some procedure or procedures in discriminant analysis are to be referenced in the subsequent text where the related topic is considered.

3.3.4 Fisher's Linear Regression Approach

Fisher (1936) also derived the sample linear discriminant function (3.3.9) using a linear regression approach. The discrimination problem can be viewed as a special case of regression, where the regressor variables are given by the feature vector $\mathbf{x}$ and the dependent variables by the vector $\mathbf{z}$ of group-indicator variables. In the case of $g = 2$ groups, the dependent variable associated with the entity having feature vector $\mathbf{x}_j$ can be taken to be $z_{1j} = (\mathbf{z}_j)_1$ where, as defined previously, z_{1j} is one or zero according as $\mathbf{x}_j$ belongs to G_1 or G_2. Then for a linear relationship between the dependent and regressor variables, we have

$$z_{1j} = a_0 + \mathbf{a}'\mathbf{x}_j + \epsilon_j \qquad (j = 1,\dots,n), \tag{3.3.10}$$

where $\epsilon_1,\dots,\epsilon_n$ are the errors. The two values taken by the dependent variable in (3.3.10) are irrelevant, provided they are distinct for each group. Fisher (1936) actually took

$$z_{1j} = (-1)^i n_i/n \qquad \text{if} \quad \mathbf{x}_j \in G_i \qquad (i = 1,2).$$

The least-squares estimate of $\mathbf{a}$ satisfies

$$\mathbf{V}\mathbf{a} = \sum_{j=1}^{n}(z_{1j} - \bar{z}_1)(\mathbf{x}_j - \bar{\mathbf{x}}), \tag{3.3.11}$$

where

$$\mathbf{V} = \sum_{j=1}^{n}(\mathbf{x}_j - \bar{\mathbf{x}})(\mathbf{x}_j - \bar{\mathbf{x}})'$$

denotes the total sums of squares and products matrix. It can be decomposed into the within-group matrix, given here by $(n-2)\mathbf{S}$, and the between-group matrix of sums of squares and products. The latter is given by

$$\sum_{i=1}^{2} n_i(\bar{\mathbf{x}}_i - \bar{\mathbf{x}})(\bar{\mathbf{x}}_i - \bar{\mathbf{x}})',$$

which can be expressed as

$$(n_1 n_2/n)(\bar{\mathbf{x}}_1 - \bar{\mathbf{x}}_2)(\bar{\mathbf{x}}_1 - \bar{\mathbf{x}}_2)'.$$

Concerning the right-hand side of (3.3.11), it equals

$$\begin{aligned}\sum_{j=1}^{n}(z_{1j} - \bar{z}_1)(\mathbf{x}_j - \bar{\mathbf{x}}) &= \sum_{j=1}^{n} z_{1j}(\mathbf{x}_j - \bar{\mathbf{x}}) \\ &= n_1(\bar{\mathbf{x}}_1 - \bar{\mathbf{x}}) \\ &= (n_1 n_2/n)(\bar{\mathbf{x}}_1 - \bar{\mathbf{x}}_2).\end{aligned} \tag{3.3.12}$$

On using (3.3.12) and substituting

$$\mathbf{V} = (n-2)\mathbf{S} + (n_1 n_2/n)(\bar{\mathbf{x}}_1 - \bar{\mathbf{x}}_2)(\bar{\mathbf{x}}_1 - \bar{\mathbf{x}}_2)'$$

into the left-hand side of (3.3.11), it follows that

$$(n-2)\mathbf{Sa} = (\bar{\mathbf{x}}_1 - \bar{\mathbf{x}}_2)[(n_1 n_2/n)\{1-(\bar{\mathbf{x}}_1 - \bar{\mathbf{x}}_2)'\mathbf{a}\}],$$

which shows that the least-squares estimate of **a** satisfies

$$\mathbf{a} \propto \mathbf{S}^{-1}(\bar{\mathbf{x}}_1 - \bar{\mathbf{x}}_2).$$

Further discussion of this may be found in T. W. Anderson (1984, Section 6.5) and Siotani, Hayakawa, and Fujikoshi (1985, Section 9.4).

3.3.5 Bias Correction of Sample NLDR

In practice, estimation of the posterior probabilities of group membership and outright allocation are usually based on the $\xi_{ig}(\mathbf{x};\hat{\boldsymbol{\theta}}_E)$ uncorrected for bias. However, it is a simple exercise to correct each $\xi_{ig}(\mathbf{x};\hat{\boldsymbol{\theta}}_E)$ for bias. From (3.3.3), we can express $\xi_{ig}(\mathbf{x};\hat{\boldsymbol{\theta}}_E)$ as

$$\xi_{ig}(\mathbf{x};\hat{\boldsymbol{\theta}}_E) = -\tfrac{1}{2}\{\hat{\delta}_{i,E}(\mathbf{x}) - \hat{\delta}_{g,E}(\mathbf{x})\} \qquad (i = 1,\ldots,g-1), \tag{3.3.13}$$

where

$$\hat{\delta}_{i,E}(\mathbf{x}) = \delta(\mathbf{x},\bar{\mathbf{x}}_i;\mathbf{S}) \qquad (i = 1,\ldots,g).$$

On noting that $(n-g)\mathbf{S}$ has a Wishart distribution with expectation $(n-g)\boldsymbol{\Sigma}$ and degrees of freedom $n-g$, the result (3.2.4) implies that

$$E(\mathbf{S}^{-1}) = c_1(n-g)\boldsymbol{\Sigma}^{-1}.$$

Hence, corresponding to the derivation of (3.2.6) in the heteroscedastic case, we have that the expectation of $\hat{\delta}_{i,E}(\mathbf{x})$, conditional on **x**, is

$$E\{\hat{\delta}_{i,E}(\mathbf{x})\} = c_1(n-g)\{\delta_{i,E}(\mathbf{x}) + p/n_i\} \tag{3.3.14}$$

for $i = 1,\ldots,g$. From (3.3.13) and (3.3.14), the expectation of $\xi_{ig}(\mathbf{x};\hat{\boldsymbol{\theta}}_E)$, conditional on **x**, is given by

$$\begin{aligned} E\{\xi_{ig}(\mathbf{x};\hat{\boldsymbol{\theta}}_E)\} &= -\tfrac{1}{2}c_1(n-g)\{\delta_{i,E}(\mathbf{x}) - \delta_{g,E}(\mathbf{x}) + p(n_i^{-1} - n_g^{-1})\} \\ &= c_1(n-g)\{\xi_{ig}(\mathbf{x};\boldsymbol{\theta}_E) - \tfrac{1}{2}p(n_i^{-1} - n_g^{-1})\} \end{aligned} \tag{3.3.15}$$

for $i = 1,\ldots,g-1$. The uniform minimum variance unbiased estimator of $\xi_{ig}(\mathbf{x};\boldsymbol{\theta}_E)$ is therefore simply

$$\hat{\xi}_{ig,E}(\mathbf{x}) = \xi_{ig}(\mathbf{x};\hat{\boldsymbol{\theta}}_E)/c_1(n-g) + \tfrac{1}{2}p(n_i^{-1} - n_g^{-1}) \qquad (i = 1,\ldots,g-1). \tag{3.3.16}$$

It can be seen from (3.3.16) that if the group-sample sizes are equal, then the estimated posterior probabilities of group membership formed with equal prior probabilities imply the same outright allocation no matter whether each $\xi_{ig}(\mathbf{x};\boldsymbol{\theta}_E)$ is estimated by $\xi_{ig}(\mathbf{x};\hat{\boldsymbol{\theta}}_E)$ or its bias-corrected version $\hat{\xi}_{ig,E}(\mathbf{x})$.

3.4 SOME OTHER NORMAL THEORY-BASED RULES

3.4.1 Likelihood Ratio Rule: Heteroscedastic Normal Model

We give here the derivation of the allocation rule based on the likelihood ratio criterion under normality, firstly for the heteroscedastic model (3.2.1). Let $\mathbf{z}$ be the unknown indicator vector defining the group of origin of an unclassified entity with feature vector $\mathbf{x}$. That is, $z_i = (\mathbf{z})_i = 1$ if the entity comes from G_i and zero otherwise $(i = 1,\ldots,g)$. The likelihood ratio rule for the allocation of this entity with feature vector $\mathbf{x}$ is constructed by treating $\mathbf{z}$ as an unknown parameter, along with Ψ_U, to be estimated by maximum likelihood on the basis of the training data $\mathbf{t}$ and also $\mathbf{x}$. The log likelihood to be maximized is

$$\log L(\Psi_U, \mathbf{z}; \mathbf{t}, \mathbf{x}) = \sum_{i=1}^{g} \left\{ \sum_{j=1}^{n} z_{ij} \log f_i(\mathbf{x}_j; \boldsymbol{\theta}_i) + z_i f_i(\mathbf{x}; \boldsymbol{\theta}_i) \right\} + \sum_{i=1}^{g} (n_i + z_i) \log \pi_i, \tag{3.4.1}$$

where $f_i(\mathbf{x};\boldsymbol{\theta}_i) = \phi(\mathbf{x};\boldsymbol{\mu}_i,\boldsymbol{\Sigma}_i)$, the p-variate normal density with mean $\boldsymbol{\mu}_i$ and covariance matrix $\boldsymbol{\Sigma}_i$ $(i = 1,\ldots,g)$. The maximization of (3.4.1) is over the set of zero-one elements of $\mathbf{z}$ corresponding to the separate assignment of $\mathbf{x}$ to each of the g groups, as well as over all admissible values of Ψ_U.

The log likelihood (3.4.1) has been formed in the case where the training data $\mathbf{t}$ have been obtained by mixture sampling and where the entity has been drawn from the same mixture of the g underlying groups. The case of separate sampling can be handled simply by taking the estimates of the group-prior probabilities π_i to be the same in the subsequent computations (T. W. Anderson, 1984, Chapter 6).

The values of π_i, $\boldsymbol{\mu}_i$, $\boldsymbol{\Sigma}_i$, and $\mathbf{z}$ that maximize the log likelihood (3.4.1) satisfy

$$\hat{\pi}_i = (n_i + \hat{z}_i)/(n + 1), \tag{3.4.2}$$

$$\hat{\boldsymbol{\mu}}_i = (n_i \bar{\mathbf{x}}_i + \hat{z}_i \mathbf{x})/(n_i + \hat{z}_i), \tag{3.4.3}$$

and

$$\hat{\boldsymbol{\Sigma}}_i = \left\{ \sum_{j=1}^{n_i} \hat{z}_{ij} (\mathbf{x}_j - \hat{\boldsymbol{\mu}}_i)(\mathbf{x}_j - \hat{\boldsymbol{\mu}}_i)' + \hat{z}_i (\mathbf{x} - \hat{\boldsymbol{\mu}}_i)(\mathbf{x} - \hat{\boldsymbol{\mu}}_i)' \right\} \Big/ (n_i + \hat{z}_i) \tag{3.4.4}$$

for $i = 1,\ldots,g$. It can be shown that $\hat{\boldsymbol{\Sigma}}_i$ may be expressed as

$$\hat{\boldsymbol{\Sigma}}_i = \left\{ (n_i - 1)\mathbf{S}_i + \frac{\hat{z}_i n_i}{n_i + 1} (\mathbf{x} - \bar{\mathbf{x}}_i)(\mathbf{x} - \bar{\mathbf{x}}_i)' \right\} \Big/ (n_i + \hat{z}_i).$$

For given i ($i = 1,\ldots,g$), let $\mathbf{z}^{(i)}$ denote the vector whose ith element is one and whose other elements are zero. Also, let $\hat{\boldsymbol{\theta}}_U^{(i)}$ denote the value of $\hat{\boldsymbol{\theta}}_U$ computed from (3.4.2) to (3.4.4) for $\hat{\mathbf{z}} = \mathbf{z}^{(i)}$. Then $\hat{\boldsymbol{\theta}}_U$ and $\hat{\mathbf{z}}$ are computed by first calculating $\hat{\boldsymbol{\theta}}_U^{(i)}$ and then evaluating

$$L(\hat{\boldsymbol{\theta}}_U^{(i)}, \mathbf{z}^{(i)}; \mathbf{t}, \mathbf{x}) \tag{3.4.5}$$

for each i in turn from $i = 1,\ldots,g$. Then $\hat{\mathbf{z}}$ is equal to $\mathbf{z}^{(k)}$ if (3.4.5) is maximized over $i = 1,\ldots,g$ at $i = k$.

3.4.2 Likelihood Ratio Rule: Homoscedastic Normal Model

For the homoscedastic normal model (3.3.1), the solution is given by (3.4.2) and (3.4.3), and with the common group-covariance matrix $\boldsymbol{\Sigma}$ estimated as

$$\hat{\boldsymbol{\Sigma}} = \sum_{i=1}^{g} \left\{ \sum_{j=1}^{n} \hat{z}_{ij}(\mathbf{x}_j - \hat{\boldsymbol{\mu}}_i)(\mathbf{x}_j - \hat{\boldsymbol{\mu}}_i)' + \hat{z}_i(\mathbf{x} - \hat{\boldsymbol{\mu}}_i)(\mathbf{x} - \hat{\boldsymbol{\mu}}_i)' \right\} \Big/ (n+1),$$

which can be expressed as

$$\hat{\boldsymbol{\Sigma}} = \left\{ (n-g)\mathbf{S} + \sum_{i=1}^{g} \frac{\hat{z}_i n_i}{n_i + 1}(\mathbf{x} - \bar{\mathbf{x}}_i)(\mathbf{x} - \bar{\mathbf{x}}_i)' \right\} \Big/ (n+1).$$

Often, the estimated group-prior probabilities $\hat{\pi}_i$ are taken to be the same, and so then they play no role in the construction of the likelihood ratio. If this is done in the homoscedastic case, then the log likelihood ratio

$$\log\{L(\hat{\boldsymbol{\theta}}_E^{(i)}, \mathbf{z}^{(i)}; \mathbf{t}, \mathbf{x}) / L(\hat{\boldsymbol{\theta}}_E^{(g)}, \mathbf{z}^{(g)}; \mathbf{t}, \mathbf{x})\}$$

simplifies to

$$-\tfrac{1}{2}(n+1)\log \frac{1 + n_i(n_i+1)^{-1}(n-g)^{-1}\hat{\delta}_{i,E}(\mathbf{x})}{1 + n_g(n_g+1)^{-1}(n-g)^{-1}\hat{\delta}_{g,E}(\mathbf{x})}. \tag{3.4.6}$$

This result was first given by T. W. Anderson (1958, Chapter 6) in the case of $g = 2$ groups. In this case, the consequent allocation rule is based on the negative of the statistic

$$Z = \frac{n_1}{n_1 + 1}\hat{\delta}_{1,E}(\mathbf{x}) - \frac{n_2}{n_2 + 1}\hat{\delta}_{2,E}(\mathbf{x}), \tag{3.4.7}$$

which has become known in the discriminant analysis literature as the Z-statistic. It reduces to minus twice the classical sample linear discriminant function or W-statistic for equal group-sample sizes. The univariate version of (3.4.6) for arbitrary g was proposed initially by Rao (1954), who considered the standard t-test for the compatibility of the feature variable x with each of g groups in which X was taken to have a univariate normal distribution with a common variance. This test and its multivariate version based on Hotelling's

T^2 are to be discussed in Section 6.4 in the context of assessing the typicality of $\mathbf{x}$. Further work on the Z-statistic was undertaken by Kudo (1959, 1960) and John (1960a, 1963). Das Gupta (1965, 1982) has shown that the sample rule based on the Z-statistic is minimax and unbiased in the sense that its expected error rate for each group does not exceed 0.5. It thus avoids the severe imbalance that can occur between the group-specific error rates of the normal-based linear discriminant rule with unequal group-sample sizes (Moran, 1975).

3.4.3 Miscellaneous Results

Besides the results presented in the preceding sections, various other work has been done over the years on the provision of suitable allocation rules under the assumption of normality for the group-conditional distributions. For example, there is the combinatoric approach developed by Dunn and Smith (1980, 1982) and Dunn (1982). In another example, Adegboye (1987) adopted the two-stage sampling approach of Stein (1945) to construct a sample discriminant rule whose error rates do not depend on the unknown but common variance for univariate normal group-conditional distributions in the case of $g = 2$ groups.

Another approach to allocation where the feature vector has unequal group-covariance matrices is that proposed recently by Dudewicz and Taneja (1989), using the so-called heteroscedastic method of Dudewicz and Bishop (1979). It applies in situations where there is control over the number of training entities to be sampled from the specified groups. This approach is being incorporated into ESS^{TM}, the Expert Statistical System, and has been illustrated on data collected by magnetic resonance imaging of the human brain; see Martin et al. (1989) and Dudewicz et al. (1989) and the references therein.

On examples of rules constructed under variations of the basic assumptions, Ellison (1965) extended the discrimination problem to the situation where the groups represent g linear manifolds, in one of which the mean of a p-dimensional normally distributed random variable lies. Burnaby (1966), Rao (1966a), and Srivastava (1967a) considered discrimination between multivariate normal group-conditional distributions in the presence of some structure on the means.

3.5 PREDICTIVE DISCRIMINATION

3.5.1 Predictive Sample Rule

We consider in this section discrimination based on the predictive estimates of the group-conditional densities, as obtained under the assumption of normality. The predictive approach to discriminant analysis has been defined in Section 2.2. Under the heteroscedastic normal model (3.2.1), the ith-group conditional density $f_i(\mathbf{x};\boldsymbol{\theta}_i)$ is taken to be $\phi(\mathbf{x};\boldsymbol{\mu}_i,\boldsymbol{\Sigma}_i)$, the p-variate normal density with mean $\boldsymbol{\mu}_i$ and covariance matrix $\boldsymbol{\Sigma}_i$ $(i = 1,\ldots,g)$.

In the absence of any prior objective knowledge of $\boldsymbol{\mu}_i$ and $\boldsymbol{\Sigma}_i$, it is convenient to adopt the conventional improper or vague prior

$$p(\boldsymbol{\mu}_1,\ldots,\boldsymbol{\mu}_g,\boldsymbol{\Sigma}_1^{-1},\ldots,\boldsymbol{\Sigma}_g^{-1}) \propto \prod_{i=1}^{g} |\boldsymbol{\Sigma}_i|^{(1/2)(p+1)} \tag{3.5.1}$$

for $\boldsymbol{\mu}_i$ and $\boldsymbol{\Sigma}_i^{-1}$ $(i = 1,\ldots,g)$. The predictive estimate of $f_i(\mathbf{x};\boldsymbol{\theta}_i)$ is given then by

$$\hat{f}_i^{(P)}(\mathbf{x}) = \int \phi(\mathbf{x};\boldsymbol{\mu}_i,\boldsymbol{\Sigma}_i)p(\boldsymbol{\mu}_i,\boldsymbol{\Sigma}_i \mid \overline{\mathbf{x}}_i,\mathbf{S}_i)d\boldsymbol{\mu}_i d\boldsymbol{\Sigma}_i, \tag{3.5.2}$$

where $p(\boldsymbol{\mu}_i,\boldsymbol{\Sigma}_i \mid \overline{\mathbf{x}}_i,\mathbf{S}_i)$ is the posterior density of $(\boldsymbol{\mu}_i,\boldsymbol{\Sigma}_i)$ given the sample statistics $(\overline{\mathbf{x}}_i,\mathbf{S}_i)$ and assuming either a normal-Wishart prior distribution or the limiting case of vague prior knowledge (Aitchison and Dunsmore, 1975, Chapter 2). After integration, (3.5.2) yields a p-dimensional Student-type density

$$\hat{f}_{i,U}^{(P)}(\mathbf{x}) = f_{St}(\mathbf{x};n_i - p,p,\overline{\mathbf{x}}_i,(1+n_i^{-1})M_i\mathbf{S}_i) \qquad (i = 1,\ldots,g), \tag{3.5.3}$$

where $M_i = (n_i - 1)/(n_i - p)$. Here $f_{St}(\mathbf{x};m,p,\mathbf{a},\mathbf{B})$ denotes the classical p-dimensional t-density with m degrees of freedom,

$$f_{St}(\mathbf{x};m,p,\mathbf{a},\mathbf{B}) = c(m,p)|m\mathbf{B}|^{-1/2}\{1+\delta(\mathbf{x},\mathbf{a};m\mathbf{B})\}^{-(1/2)(m+p)}, \tag{3.5.4}$$

where

$$c(m,p) = \Gamma\{\tfrac{1}{2}(m+p)\}/[\pi^{(1/2)p}\Gamma(\tfrac{1}{2}m)];$$

see, for example, Johnson and Kotz (1972, page 134). Similarly, under the homoscedastic normal model (3.3.1),

$$\hat{f}_{i,E}^{(P)}(\mathbf{x}) = f_{St}(\mathbf{x};n-g-p+1,p,\overline{\mathbf{x}}_i,(1+n_i^{-1})M\mathbf{S}) \qquad (i = 1,\ldots,g), \tag{3.5.5}$$

where $M = (n-g)/(n-g-p+1)$.

The predictive estimates of the posterior probabilities of group membership are formed with the group-conditional densities replaced by their predictive estimates as given above. As part of this process, the log likelihood ratio $\xi_{ig}(\mathbf{x};\boldsymbol{\theta}_U)$ is estimated by

$$\begin{aligned}\hat{\xi}_{ig,U}^{(P)}(\mathbf{x}) = &-\tfrac{1}{2}n_i\log\{1+n_i(n_i^2-1)^{-1}\hat{\delta}_i(\mathbf{x})\} + \tfrac{1}{2}n_g\log\{1+n_g(n_g^2-1)^{-1}\hat{\delta}_g(\mathbf{x})\}\\ &-\tfrac{1}{2}\log\{|\mathbf{S}_i|/|\mathbf{S}_g|\} + \tfrac{1}{2}p\log[\{n_i(n_i^2-1)^{-1}\}/\{n_g(n_g^2-1)^{-1}\}]\\ &+\log\{c(n_i-1)/c(n_g-1)\} \qquad (i = 1,\ldots,g-1),\end{aligned} \tag{3.5.6}$$

where $\hat{\delta}_i(\mathbf{x}) = \delta(\mathbf{x}, \bar{\mathbf{x}}_i; \mathbf{S}_i)$ for $i = 1,\ldots,g$. The corresponding predictive estimate under homoscedasticity is

$$\hat{\xi}^{(P)}_{ig,E}(\mathbf{x}) = -\tfrac{1}{2}(n-g+1)\log\frac{1+n_i(n_i+1)^{-1}(n-g)^{-1}\hat{\delta}_{i,E}(\mathbf{x})}{1+n_g(n_g+1)^{-1}(n-g)^{-1}\hat{\delta}_{g,E}(\mathbf{x})} + \tfrac{1}{2}p\log[\{n_i(n_g+1)\}/\{n_g(n_i+1)\}] \quad (i = 1,\ldots,g-1), \tag{3.5.7}$$

where $\hat{\delta}_{i,E}(\mathbf{x}) = \delta(\mathbf{x}, \bar{\mathbf{x}}_i; \mathbf{S})$ for $i = 1,\ldots,g$.

3.5.2 Semi-Bayesian Approach

We consider now the semi-Bayesian approach where the posterior distributions of the quantities to be estimated are formed on the basis of the information in $\mathbf{t}$ but not also in $\mathbf{x}$.

For $g = 2$ homoscedastic normal groups, Geisser (1967) showed that the mean of the posterior distribution of the log odds, or equivalently $\xi(\mathbf{x};\boldsymbol{\theta}_E)$, formed in this way is

$$\hat{\xi}^{(S)}_E(\mathbf{x}) = -\tfrac{1}{2}\{\hat{\delta}_{1,E}(\mathbf{x}) - \hat{\delta}_{2,E}(\mathbf{x})\} - \tfrac{1}{2}p(1/n_1 - 1/n_2).$$

For the heteroscedastic model, Enis and Geisser (1970) showed that the posterior mean of $\xi(\mathbf{x};\boldsymbol{\theta}_U)$ equals

$$\hat{\xi}^{(S)}_U(\mathbf{x}) = \tfrac{1}{2}\sum_{i=1}^{2}(-1)^i\{\hat{\delta}_i(\mathbf{x}) + \log|\mathbf{S}_i| + p/n_i + p\log(n_i - 1) - c_2(n_i)\},$$

where $c_2(\cdot)$ is defined by (3.2.9). A comparison of these posterior means with the bias-corrected versions (3.2.10) and (3.3.16) of the corresponding estimates given by the estimative approach shows that they are quite similar.

It can be seen from (3.5.7) that the form of the predictive estimate of the posterior log odds $\xi(\mathbf{x};\boldsymbol{\theta}_E)$ can be markedly different from that of the estimate produced by either the semi-Bayesian or estimative approach. For example, for unequal group-sample sizes, it is a quadratic function in $\mathbf{x}$. This difference in form arises because the predictive approach, in contrast to the semi-Bayesian or estimative approach, uses the information in $\mathbf{x}$ as well as in $\mathbf{t}$.

In related work, Enis and Geisser (1974) noted that, as $\xi(\mathbf{x};\boldsymbol{\theta}_E)$ is linear $\mathbf{x}$, it may be for many intuitively compelling to use a linear estimate. They consequently derived the sample linear estimate of $\xi(\mathbf{x};\boldsymbol{\theta}_E)$ that minimizes the implied overall predictive error rate in the case of the prior distribution

$$p(\boldsymbol{\mu}_1, \boldsymbol{\mu}_2, \boldsymbol{\Sigma}^{-1}) \propto |\boldsymbol{\Sigma}|^{(1/2)(p+1)}$$

for $\boldsymbol{\mu}_1$, $\boldsymbol{\mu}_2$, and $\boldsymbol{\Sigma}^{-1}$; see also Geisser (1977).

We now proceed to explore the differences between the assessments produced by the estimative and predictive methods.

3.5.3 Comparison of Predictive and Estimative Methods

On contrasting the predictive estimate (3.5.3) with the estimative assessment

$$f_i(\mathbf{x};\hat{\boldsymbol{\theta}}_i) = \phi(\mathbf{x};\bar{\mathbf{x}}_i,\mathbf{S}_i)$$

for the ith group-conditional density $f_i(\mathbf{x};\boldsymbol{\theta}_i)$ under the heteroscedastic normal model (3.2.1), it can be seen that they are both centered on the same vector $\bar{\mathbf{x}}_i$ and with the same class of ellipsoids of concentration, but with $\hat{f}_{i,U}^{(P)}(\mathbf{x})$ less concentrated than $f_i(\mathbf{x};\hat{\boldsymbol{\theta}}_i)$ about $\bar{\mathbf{x}}_i$. Aitchison (1975) used a measure based on the Kullback–Leibler divergence measure to compare the density estimates produced by these two methods. In terms of the present notation, the measure for the estimation of the ith group-conditional density $f_i(\mathbf{x};\boldsymbol{\theta}_i)$ was defined as

$$E\left[\int f_i(\mathbf{x};\boldsymbol{\theta}_i)\log\{\hat{f}_{i,U}^{(P)}(\mathbf{x})/f_i(\mathbf{x};\hat{\boldsymbol{\theta}}_i)\}\,d\mathbf{x}\right], \tag{3.5.8}$$

where E refers to expectation over the distribution of the training data. It may be viewed as the average of the log difference between the estimative and predictive estimates, averaged firstly over values of $\mathbf{x}$ in the ith group G_i and then over the training data. With $f_i(\mathbf{x};\boldsymbol{\theta}_i)$ taken to be $\phi(\mathbf{x};\boldsymbol{\mu}_i,\boldsymbol{\Sigma}_i)$, Aitchison (1975) showed that (3.5.8) does not depend on the parameters $\boldsymbol{\mu}_i$ and $\boldsymbol{\Sigma}_i$ and is positive. The latter indicates greater overall closeness of the predictive estimate to the true density. Recently, El-Sayyad, Samiuddin, and Al-Harbey (1989) have provided a critical assessment of the use of (3.5.8) as a criterion for density estimation.

Murray (1977a) subsequently showed that in terms of the Kullback–Leibler measure, $\hat{f}_{i,U}^{(P)}(\mathbf{x})$ is preferable to all other estimates of $f_i(\mathbf{x};\boldsymbol{\theta}_i)$ that are invariant under translation and nonsingular linear transformations. As emphasized by Han (1979), the predictive density estimate $\hat{f}_{i,U}^{(P)}(\mathbf{x})$ therefore has an interpretation in a frequentist framework.

As evident from their definitions above, the practical differences between the estimates of the group-conditional densities produced by the estimative and predictive methods will be small for large sample sizes. This was demonstrated in the medical examples analyzed in Hermans and Habbema (1975). However, for small sample sizes, Aitchison et al. (1977) illustrated in the context of the differential diagnosis of Conn's syndrome the large differences that can be obtained between the estimative and predictive estimates of the posterior probabilities of group membership. Their example demonstrated how the estimative approach gives an exaggerated view of the group-posterior probabilities, whereas the predictive moderates this view. Aitchison et al. (1977) also conducted a simulation study in the case of two groups with equal (known) prior probabilities, but not always equal covariance matrices in their multivariate normal distributions. They concluded overall that the predictive method was superior to the estimative in terms of the mean absolute deviation from the true posterior log odds. To highlight the differences that can occur using the two methods, they took n_1 and n_2 to be very small relative to p, and so the reliability of either method in such situations is questionable. It should be noted too that the differences between the estimative and predictive approaches are not nearly as pronounced in terms of the allocation rules they produce as they are in terms of the estimates they give for the group-posterior probabilities. For example, for equal group-sample sizes under the homoscedastic normal model, it can be seen from (3.3.13) and (3.5.7) that the estimative and predictive methods both give the same outright allocation for any feature vector **x** if the group-prior probabilities are specified to be the same.

3.5.4 Predictive Versus Estimative Assessment of Posterior Log Odds

In the remainder of this section, we focus on the case of $g = 2$ homoscedastic groups for which the posterior log odds are given by

$$\begin{aligned} \eta(\mathbf{x};\boldsymbol{\Psi}_E) &= \log\{\tau_1(\mathbf{x};\boldsymbol{\Psi}_E)/\tau_2(\mathbf{x};\boldsymbol{\Psi}_E)\} \\ &= \log(\pi_1/\pi_2) + \xi(\mathbf{x};\boldsymbol{\theta}_E), \end{aligned} \tag{3.5.9}$$

where $\eta(\mathbf{x};\boldsymbol{\Psi}_E)$ and $\xi(\mathbf{x};\boldsymbol{\theta}_E)$ correspond to the previously defined quantities $\eta_{ig}(\mathbf{x};\boldsymbol{\Psi}_E)$ and $\xi_{ig}(\mathbf{x};\boldsymbol{\theta}_E)$, respectively, with their subscripts suppressed since $g = 2$ only. Also, as we are taking the prior probabilities to be specified, $\boldsymbol{\theta}_E$ and not $\boldsymbol{\Psi}_E = (\boldsymbol{\pi}', \boldsymbol{\theta}_E')'$ is the vector of all unknown parameters.

McLachlan (1979) provided an asymptotic account of the relative performance of the estimates $\tau_1(\mathbf{x};\boldsymbol{\pi},\hat{\boldsymbol{\theta}}_E)$ and $\hat{\tau}_{1,E}^{(P)}(\mathbf{x};\boldsymbol{\pi},\mathbf{t})$ of the posterior probability $\tau_1(\mathbf{x};\boldsymbol{\Psi}_E)$, as produced by the estimative and predictive methods, respectively, under the homoscedastic normal model (3.3.1). It supported the previous empirical findings that the estimative approach gives a more extreme assessment than the predictive of the posterior probabilities of group membership.

Geisser (1979) subsequently pointed out that for equal group-sample sizes, $n_1 = n_2 = \frac{1}{2}n$, it could be proved exactly that the predictive posterior odds are always closer to the prior odds than the estimative odds. The derivation of this result relies on the inequalities

$$1 \leq \{(1 + a_1/m)/(1 + a_2/m)\}^{-m} \leq \{1 + (a_2 - a_1)/m\} \leq \exp(a_2 - a_1) \tag{3.5.10}$$

for $a_2 \geq a_1 \geq 0$. These inequalities are reversed for $a_1 \geq a_2 \geq 0$. A somewhat similar result was used in Desu and Geisser (1973) in the case of equal group-means and differing group-covariance matrices.

To show how these inequalities (3.5.10) can be used to establish the result stated above on the sizes of the estimative and predictive posterior odds relative to the prior odds, we first note from (3.3.13) that the estimative assessment of the log likelihood ratio $\xi(\mathbf{x};\boldsymbol{\theta}_E)$ can be written as

$$\xi(\mathbf{x};\hat{\boldsymbol{\theta}}_E) = -\tfrac{1}{2}\{\hat{\delta}_{1,E}(\mathbf{x}) - \hat{\delta}_{2,E}(\mathbf{x})\}.$$

From (3.5.7), we have that the corresponding predictive estimate for $n_1 = n_2 = \frac{1}{2}n$ can be expressed as

$$\hat{\xi}_E^{(P)}(\mathbf{x}) = -\tfrac{1}{2}(n-1)\log\left[\{1+(n-1)^{-1}c_n\hat{\delta}_{1,E}(\mathbf{x})\}/\{1+(n-1)^{-1}c_n\hat{\delta}_{2,E}(\mathbf{x})\}\right], \tag{3.5.11}$$

where

$$c_n = n(n-1)/(n^2-4).$$

In order for $\mathbf{S}$ to be nonsingular, we need $n > p+1$, and so in the case of equal group-sample sizes, n cannot be less than 4 for all p. Hence, we can proceed on the basis that $c_n \leq 1$.

By applying (3.5.10) to (3.5.11), it follows that if $\xi(\mathbf{x};\hat{\boldsymbol{\theta}}_E) \geq 0$, then

$$0 \leq \hat{\xi}_E^{(P)}(\mathbf{x}) \leq c_n\xi(\mathbf{x};\hat{\boldsymbol{\theta}}_E) \leq \xi(\mathbf{x};\hat{\boldsymbol{\theta}}_E), \tag{3.5.12}$$

and that if $\xi(\mathbf{x};\hat{\boldsymbol{\theta}}_E) \leq 0$, then

$$\xi(\mathbf{x};\hat{\boldsymbol{\theta}}_E) \leq c_n\xi(\mathbf{x};\hat{\boldsymbol{\theta}}_E) \leq \hat{\xi}_E^{(P)}(\mathbf{x}) \leq 0. \tag{3.5.13}$$

Using (3.5.12) and (3.5.13) in conjunction with (3.5.9), we have that

$$\pi_1/\pi_2 \leq \hat{\tau}_{1,E}^{(P)}(\mathbf{x};\boldsymbol{\pi},\mathbf{t})/\hat{\tau}_{2,E}^{(P)}(\mathbf{x};\boldsymbol{\pi},\mathbf{t}) \leq \tau_1(\mathbf{x};\boldsymbol{\pi},\hat{\boldsymbol{\theta}}_E)/\tau_2(\mathbf{x};\boldsymbol{\pi},\hat{\boldsymbol{\theta}}_E)$$

for $\pi_1/\pi_2 \geq 1$, and that

$$\tau_1(\mathbf{x};\boldsymbol{\pi},\hat{\boldsymbol{\theta}}_E)/\tau_2(\mathbf{x};\boldsymbol{\pi},\hat{\boldsymbol{\theta}}_E) \leq \hat{\tau}_{1,E}^{(P)}(\mathbf{x};\boldsymbol{\pi},\mathbf{t})/\hat{\tau}_{2,E}^{(P)}(\mathbf{x};\boldsymbol{\pi},\mathbf{t}) \leq \pi_1/\pi_2$$

for $\pi_1/\pi_2 \leq 1$. Hence, the predictive posterior odds are always closer to the prior odds than the estimative posterior odds, thus confirming the less extreme assessment provided by the former approach. This result can be extended to the case of $g > 2$ groups.

3.5.5 Biases of Predictive and Estimative Assessments of Posterior Log Odds

Concerning the biases of the estimative and predictive estimates of the log likelihood ratio $\xi(\mathbf{x};\boldsymbol{\theta}_E)$, we have from (3.3.15) that the bias of $\xi(\mathbf{x};\hat{\boldsymbol{\theta}}_E)$, conditional on $\mathbf{x}$, is

$$\text{bias}\{\xi(\mathbf{x};\hat{\boldsymbol{\theta}}_E)\} = c_1(n-2)\{\xi(\mathbf{x};\boldsymbol{\theta}_E) - \tfrac{1}{2}p(1/n_1 - 1/n_2)\} - \xi(\mathbf{x};\boldsymbol{\theta}_E),$$

where $c_1(n-2) = (n-2)/(n-p-3)$. On taking the expectation of this bias with respect to the distribution of $\mathbf{X}$ in the first group, we obtain

$$E[\text{bias}\{\xi(\mathbf{x};\hat{\boldsymbol{\theta}}_E)\} \mid \mathbf{X} \in G_1] = c_1(n-2)\{\tfrac{1}{2}\Delta^2 - \tfrac{1}{2}p(1/n_1 - 1/n_2)\} - \tfrac{1}{2}\Delta^2,$$

TABLE 3.1 Mean Biases of the Estimative and Predictive Estimates of the Posterior Log Odds for Equal Group-Covariance Matrices and Known Equal Group-Prior Probabilities when $p = 4$ and $n_1 = n_2$

		Approach			
		Estimative		Predictive	
Δ	$\Phi(-\frac{1}{2}\Delta)$	$n_i = 8$	$n_i = 16$	$n_i = 8$	$n_i = 16$
1.049	.3	.31	.11	−.07	−.04
1.683	.2	.79	.28	−.24	−.13
2.563	.1	1.83	.66	−.78	−.45

Source: Adapted from Moran and Murphy (1979).

where Δ is the Mahalanobis distance between G_1 and G_2. For $n_1 = n_2$, it can be seen that

$$\mathrm{E}[\mathrm{bias}\{\xi(\mathbf{x};\hat{\boldsymbol{\theta}}_E)\} \mid \mathbf{X} \in G_1] = \tfrac{1}{2}\Delta^2\{c_1(n-2)-1\} > 0.$$

Hence, the unconditional bias of $\xi(\mathbf{x};\hat{\boldsymbol{\theta}}_E)$ is always positive for entities from G_1. This bias will be in the opposite direction for entities from G_2.

Moran and Murphy (1979) have shown that for the predictive method

$$\mathrm{E}[\mathrm{bias}\{\xi_E^{(P)}(\mathbf{x})\} \mid \mathbf{X} \in G_1] = \tfrac{1}{2}p\log[\{n_1(n_2+1)\}/\{n_2(n_1+1)\}] + \sum_{k=0}^{\infty}\frac{n-1}{n-1+2k}\mathrm{pr}\left\{\chi^2_{2+2k} \le \frac{n_2}{n_2+1}\Delta^2\right\} - \frac{1}{2}\Delta^2.$$

It can be seen that the mean bias of the predictive estimate of $\xi(\mathbf{x};\boldsymbol{\theta}_E)$ does not depend on p for $n_1 = n_2$. A comparison of these mean biases of the estimative and predictive approaches can be found in Moran and Murphy (1979). It shows that $\xi(\mathbf{x};\hat{\boldsymbol{\theta}}_E)$ gives a considerably more biased estimate of $\xi(\mathbf{x};\boldsymbol{\theta}_E)$ than $\xi_E^{(P)}(\mathbf{x})$, as illustrated in Table 3.1.

Moran and Murphy (1979) also noted from their simulation study that it is the bias of the estimative method that accounts for a large part of the poor performance of $\xi(\mathbf{x};\hat{\boldsymbol{\theta}}_E)$ and $\xi(\mathbf{x};\hat{\boldsymbol{\theta}}_U)$ in their estimation of $\xi(\mathbf{x};\boldsymbol{\theta}_E)$ and $\xi(\mathbf{x};\boldsymbol{\theta}_U)$, the log likelihood ratio under homoscedasticity and heteroscedasticity, respectively. Consistent with this, they found in their study that the bias-corrected versions $\hat{\xi}_E(\mathbf{x})$ and $\hat{\xi}_U(\mathbf{x})$ of the estimative assessments of $\xi(\mathbf{x};\boldsymbol{\theta}_E)$ and $\xi(\mathbf{x};\boldsymbol{\theta}_U)$ were more comparable in performance with the corresponding predictive estimates. More specifically, their simulations suggest that, in terms of the criterion of mean absolute deviation, the predictive approach is still preferable to the estimative after bias correction for the estimation of the log likelihood ratio, or, equivalently, the posterior log odds since the prior probabilities were specified. On the relative performances of the allocation rules based on these estimates of the posterior odds, the estimative method with correction for bias appears to be comparable to the predictive.

3.5.6 Some Additional Comments on Predictive Versus Estimative Methods

It has been noted for the homoscedastic normal model (3.3.1) that the estimative and predictive estimates of the log likelihood ratio $\xi_{ig}(\mathbf{x};\boldsymbol{\theta}_E)$ lead to the same outright allocation in the case of equal group-sample sizes. However, even with disparate group-sample sizes, there is a frequentist approach that gives an allocation rule very similar to that obtained with the predictive method. It can be seen from (3.4.6) that the rule obtained by the likelihood ratio criterion is very similar to the predictive rule based on (3.5.7). This similarity, which was noted by Aitchison and Dunsmore (1975, page 235), is not surprising. We have seen that the predictive method forms estimates by effectively using the posterior distributions of the parameters formed on the basis of the information in both $\mathbf{t}$ and $\mathbf{x}$. Likewise both $\mathbf{t}$ and $\mathbf{x}$, albeit in a frequentist framework, are used in forming estimates of the parameters with the likelihood ratio criterion.

As to be considered further in Section 6.4, huge discrepancies can occur between the predictive and estimative assessments of the typicality index of a feature vector with respect to a group, as proposed by Aitchison and Dunsmore (1975, Chapter 11). However, as pointed out by Moran and Murphy (1979), the predictive assessment of this typicality index is the same as the P-value of the standard frequentist test for compatibility of $\mathbf{x}$ with respect to a given group.

It is not surprising that, where discrimination is feasible, the estimative assessment of the posterior probabilities of group membership, the Bayes rule, and the typicality index can in each instance be modified or replaced by another frequentist assessment that yields the same or practically the same assessment as the predictive approach. For if a data set is informative, then an effective frequentist analysis should lead to essentially the same conclusions from a practical point of view as those drawn from a Bayesian analysis; see Durbin (1987).

3.6 COVARIANCE-ADJUSTED DISCRIMINATION

3.6.1 Formulation of Problem

Suppose that the p-dimensional feature vector $\mathbf{X}$ can be partitioned into two subvectors, $\mathbf{X}^{(1)}$ and $\mathbf{X}^{(2)}$, where the latter subvector has the same distribution in each group. That is, the distribution of $\mathbf{X}^{(2)}$ is not group-dependent. Hence, it is of no discriminatory value in its own right but, in conjunction with $\mathbf{X}^{(1)}$, may still be of use in the role of a vector of covariates to $\mathbf{X}^{(1)}$. The variables in $\mathbf{X}^{(2)}$ are referred to also as concomitant or ancillary variables (Rao, 1966b).

In this situation, there are two obvious ways of proceeding in forming a discriminant rule. One can use the full observed feature vector $\mathbf{x}$, thereby ignoring the available knowledge that $\mathbf{x}^{(2)}$ is really only a subvector of covariates.

Alternatively, in the light of this knowledge, one might use $\mathbf{x}^{(1)}$ adjusted for $\mathbf{x}^{(2)}$, where the former is not independent of the latter.

A third option would be to form a discriminant rule using just $\mathbf{x}^{(1)}$. For known group-conditional distributions (or an infinitely sized training set), the use of $\mathbf{x}^{(1)}$ instead of the full feature vector $\mathbf{x}$ cannot decrease the overall error rate. However, for training sets of finite size, a reduction in the error rate may be achieved in some situations. This is to be considered further in Chapter 12, where the variable-selection problem in discriminant analysis is to be addressed.

The development of discriminant rules employing covariates started with Cochran and Bliss (1948) and continued with papers by Rao (1949, 1966b), Cochran (1964b), Subrahmaniam and Subrahmaniam (1973), and Subrahmaniam and Subrahmaniam (1976). In these papers, the emphasis has been on the testing of whether there is a significant difference between the group means of $\mathbf{X}^{(1)}$ with and without the use of the observed value $\mathbf{x}^{(2)}$ of $\mathbf{X}^{(2)}$. That is, they are more concerned with the variable-selection problem in the presence of covariates rather than with the adjustment of the discriminant rule to allow for covariates in the feature vector.

To examine more closely the formation of a discriminant rule adjusted for the covariate vector $\mathbf{x}^{(2)}$, we note that

$$f_i(\mathbf{x}) = f_i(\mathbf{x}^{(2)})f_i(\mathbf{x}^{(1)} \mid \mathbf{x}^{(2)}), \tag{3.6.1}$$

where, for economy of notation in (3.6.1), we have used the one symbol f_i to denote the joint, marginal, and conditional density of $\mathbf{X}^{(1)}$ and $\mathbf{X}^{(2)}$, of $\mathbf{X}^{(2)}$, and of $\mathbf{X}^{(1)}$ conditional on $\mathbf{x}^{(2)}$, respectively, in G_i $(i = 1,2)$. If the subvector $\mathbf{X}^{(2)}$ has the same distribution in both groups, so that

$$f_1(\mathbf{x}^{(2)}) \equiv f_2(\mathbf{x}^{(2)}),$$

then it can be seen from (3.6.1) that the Bayes rule is the same no matter whether it is formed using the group-conditional densities of the full feature vector $\mathbf{X}$ or of the subvector $\mathbf{X}^{(1)}$ conditional on $\mathbf{X}^{(2)} = \mathbf{x}^{(2)}$. Hence, in the situation where $\mathbf{X}$ has known group-conditional densities, there is nothing to be gained by working in terms of the group-conditional densities of $\mathbf{X}^{(1)}$ given $\mathbf{x}^{(2)}$, which is equivalent, at least under a normal model, to taking $\mathbf{x}^{(1)}$ as the feature vector after adjusting it for the subvector of covariates $\mathbf{x}^{(2)}$. However, in the situation where the group-conditional densities of $\mathbf{X}$ have to be estimated, there will be a loss of efficiency in using $\mathbf{x}$ as the feature vector without incorporating the information that $X^{(2)}$ has a common distribution within each group. Further, if the group-conditional distributions of $\mathbf{X}^{(1)}$ given $\mathbf{x}^{(2)}$ are taken to be multivariate normal, then only the mean and covariance matrix of the subvector $\mathbf{X}^{(2)}$ of covariates need to be estimated instead of its complete density as with the use of $\mathbf{x}$.

Without loss of generality, we can relabel the variables in $\mathbf{x}$ so that

$$\mathbf{x} = (\mathbf{x}^{(1)\prime}, \mathbf{x}^{(2)\prime})'.$$

Corresponding to this partition of $\mathbf{x}$, we let

$$\boldsymbol{\mu}_i = (\boldsymbol{\mu}_i^{(1)\prime}, \boldsymbol{\mu}^{(2)\prime})'$$

and

$$\boldsymbol{\Sigma}_i = \begin{pmatrix} \boldsymbol{\Sigma}_{11i} & \boldsymbol{\Sigma}_{12i} \\ \boldsymbol{\Sigma}_{21i} & \boldsymbol{\Sigma}_{22} \end{pmatrix}$$

be the corresponding partitions of $\boldsymbol{\mu}_i$ and $\boldsymbol{\Sigma}_i$, respectively, for $i = 1,\ldots,g$. Consistent with the assumption that $\mathbf{X}^{(2)}$ has the same distribution in each group, $\boldsymbol{\mu}^{(2)}$ and $\boldsymbol{\Sigma}_{22}$ are not group-dependent. A common assumption is to take the distribution of $\mathbf{X}^{(1)}$ conditional on $\mathbf{x}^{(2)}$ within G_i to be multivariate normal with mean

$$\boldsymbol{\mu}_i^{(1)} + \boldsymbol{\Sigma}_{12i}\boldsymbol{\Sigma}_{22}^{-1}(\mathbf{x}^{(2)} - \boldsymbol{\mu}^{(2)}) \tag{3.6.3}$$

and covariance matrix

$$\boldsymbol{\Sigma}_{11i} - \boldsymbol{\Sigma}_{12i}\boldsymbol{\Sigma}_{22}^{-1}\boldsymbol{\Sigma}_{21i} \tag{3.6.4}$$

for $i = 1,\ldots,g$. This assumption will be valid if $\mathbf{X}^{(1)}$ and $\mathbf{X}^{(2)}$ have a joint multivariate normal distribution, but of course it is not a necessary condition. A discussion of other possible models for the distribution of $\mathbf{X}^{(1)}$ given $\mathbf{x}^{(2)}$ can be found in Lachenbruch (1977).

Corresponding to the partition (3.6.2) of $\mathbf{x}$, let

$$\bar{\mathbf{x}}_i = (\bar{\mathbf{x}}_i^{(1)\prime}, \bar{\mathbf{x}}_i^{(2)\prime})', \qquad \bar{\mathbf{x}} = (\bar{\mathbf{x}}^{(1)\prime}, \bar{\mathbf{x}}^{\prime(2)})',$$

and

$$\mathbf{S}_i = \begin{pmatrix} \mathbf{S}_{11i} & \mathbf{S}_{12i} \\ \mathbf{S}_{21i} & \mathbf{S}_{22i} \end{pmatrix}, \qquad \mathbf{S} = \begin{pmatrix} \mathbf{S}_{11} & \mathbf{S}_{12} \\ \mathbf{S}_{21} & \mathbf{S}_{22} \end{pmatrix}$$

be the corresponding partitions of $\bar{\mathbf{x}}_i$, $\bar{\mathbf{x}}$, $\mathbf{S}_i$, and $\mathbf{S}$, respectively, where

$$\bar{\mathbf{x}} = (n_1\bar{\mathbf{x}}_1 + n_2\bar{\mathbf{x}}_2)/n.$$

Then the ith group mean (3.6.3) and covariance matrix (3.6.4) of $\mathbf{X}^{(1)}$ given $\mathbf{x}^{(2)}$ can be estimated by

$$\bar{\mathbf{x}}_i^{(1)} + \mathbf{S}_{12i}\mathbf{S}_{22}^{-1}(\mathbf{x}^{(2)} - \bar{\mathbf{x}}^{(2)})$$

and

$$\mathbf{S}_{11i} - \mathbf{S}_{12i}\mathbf{S}_{22}^{-1}\mathbf{S}_{21i},$$

respectively. There are more efficient estimators of $\boldsymbol{\Sigma}_{12i}$ and $\boldsymbol{\Sigma}_{22}$ available than $\mathbf{S}_{12i}$ and $\mathbf{S}_{22}$, because the latter are computed using the sample mean of $\mathbf{X}^{(2)}$ specific to each group rather than the overall sample mean $\bar{\mathbf{x}}^{(2)}$.

The model above was adopted by Rawlings et al. (1986), who considered discrimination between a group of outpatients in an alcoholism treatment program and a group of nonalcoholic outpatients. In their example, the subvector $\mathbf{x}^{(1)}$ consisted of measurements on clinical laboratory tests for chloride, carbon dioxide, potassium, sodium, and glucose. The subvector $\mathbf{x}^{(2)}$ consisted of age as the single covariate.

3.6.2 Covariance-Adjusted Sample NLDF

Under the additional assumption of homoscedasticity, the ith group mean and covariance matrix of $\mathbf{X}^{(1)}$ given $\mathbf{x}^{(2)}$ can be estimated by

$$\bar{\mathbf{x}}_i^{(1)} + \mathbf{S}_{12}\mathbf{S}_{22}^{-1}(\mathbf{x}^{(2)} - \bar{\mathbf{x}}^{(2)}) \qquad (i = 1,\ldots,g) \tag{3.6.5}$$

and

$$\mathbf{S}_{11.2} = \mathbf{S}_{11} - \mathbf{S}_{12}\mathbf{S}_{22}^{-1}\mathbf{S}_{21}. \tag{3.6.6}$$

For $g = 2$ groups under the homoscedastic normal model, we let $\xi_{1.2}(\mathbf{x};\hat{\boldsymbol{\theta}}_E)$ be the sample covariance-adjusted NLDF formed by plugging these estimates into the group-conditional densities of $\mathbf{X}^{(1)}$ given $\mathbf{x}^{(2)}$. It is equal to $\xi(\mathbf{x};\hat{\boldsymbol{\theta}}_E)$ with $\mathbf{x}$, $\bar{\mathbf{x}}_i$, and $\mathbf{S}$ replaced by $\mathbf{x}^{(1)}$, $\bar{\mathbf{x}}_i^{(1)} + \mathbf{S}_{12}\mathbf{S}_{22}^{-1}(\mathbf{x}^{(2)} - \bar{\mathbf{x}}^{(2)})$, and $\mathbf{S}_{11.2}$, respectively. An alternative way of forming a covariance-adjusted LDF is to work in terms of $\mathbf{x}^{(1)}$ adjusted for $\mathbf{x}^{(2)}$, that is,

$$\mathbf{x}_{1.2} = \mathbf{x}^{(1)} - \boldsymbol{\Sigma}_{12}\boldsymbol{\Sigma}_{22}^{-1}\mathbf{x}^{(2)},$$

and to form the rule in terms of the group-conditional densities of $\mathbf{X}_{1.2}$. This rule is the same as obtained by considering the group densities of $\mathbf{X}^{(1)}$ conditional on $\mathbf{x}^{(2)}$. However, their sample analogues differ slightly. The sample linear discriminant function that is obtained by using the group-conditional densities of $\mathbf{X}_{1.2}$, and then replacing the parameters with their sample analogues in terms of $\bar{\mathbf{x}}_i$ and $\mathbf{S}$, is given by

$$\left[\mathbf{x}_{1.2} - \tfrac{1}{2}\{\bar{\mathbf{x}}_1^{(1)} + \bar{\mathbf{x}}_2^{(1)} - \mathbf{S}_{12}\mathbf{S}_{22}^{-1}(\bar{\mathbf{x}}_1^{(2)} + \bar{\mathbf{x}}_2^{(2)})\}\right]'$$
$$\times\, \mathbf{S}_{11.2}^{-1}\{\bar{\mathbf{x}}_1^{(1)} - \bar{\mathbf{x}}_2^{(1)} - \mathbf{S}_{12}\mathbf{S}_{22}^{-1}(\bar{\mathbf{x}}_1^{(2)} - \bar{\mathbf{x}}_2^{(2)})\}. \tag{3.6.7}$$

If $\bar{\mathbf{x}}_i^{(2)}$ is set equal to $\bar{\mathbf{x}}^{(2)}$ for $i = 1,2$ in (3.6.7), then it reduces to $\xi_{1.2}(\mathbf{x};\hat{\boldsymbol{\theta}}_E)$. Hence, the latter uses the more efficient pooled sample mean $\bar{\mathbf{x}}^{(2)}$ to estimate the common mean of $\mathbf{X}^{(2)}$ in each group.

3.6.3 Asymptotic Unconditional Error Rates

Let

$$\Delta = \{(\boldsymbol{\mu}_1 - \boldsymbol{\mu}_2)'\boldsymbol{\Sigma}^{-1}(\boldsymbol{\mu}_1 - \boldsymbol{\mu}_2)\}^{1/2}$$

and

$$\Delta_{1.2} = \{(\boldsymbol{\mu}_1^{(1)} - \boldsymbol{\mu}_2^{(1)})'(\boldsymbol{\Sigma}_{11} - \boldsymbol{\Sigma}_{12}\boldsymbol{\Sigma}_{22}^{-1}\boldsymbol{\Sigma}_{21})^{-1}(\boldsymbol{\mu}_1^{(1)} - \boldsymbol{\mu}_2^{(1)})\}^{1/2}.$$

As the expectation of $\mathbf{X}^{(2)}$ is the same in both groups, it follows that $\Delta = \Delta_{1.2}$. Analogous to the expansion of the ith group-conditional distribution of

$$\{\xi(\mathbf{x};\hat{\boldsymbol{\theta}}_E) + (-1)^i\tfrac{1}{2}\Delta^2\}/\Delta$$

by Okamoto (1963), Memon and Okamoto (1970) have expanded the ith group-conditional distribution of

$$\{\xi_{1.2}(\mathbf{x};\hat{\boldsymbol{\theta}}_E) + (-1)^i\tfrac{1}{2}\Delta^2\}/\Delta$$

up to terms of order $O(N^{-2})$ under the homoscedastic normal model (3.3.1), where $N = n_1 + n_2 - 2$. McGee (1976) has since obtained the third-order expansions. The asymptotic unconditional error rates of the sample normal-based linear discriminant rule based on $\xi_{1.2}(\mathbf{x};\hat{\boldsymbol{\theta}}_E)$ are available from these expansions.

The comparative performance of the sample rules based on $\xi_{1.2}(\mathbf{x};\hat{\boldsymbol{\theta}}_E)$ and $\xi(\mathbf{x};\hat{\boldsymbol{\theta}}_E)$ have been considered by Cochran (1964b) and Memon and Okamoto (1970). More recently, Leung and Srivastava (1983b) have compared their asymptotic overall unconditional error rates in the case of separate sampling with equal group-sample sizes and with a zero cutoff point. In the latter case, we let $eu_{1.2}(\Delta, p_1, p_2)$ and $eu(\Delta, p)$ denote the overall unconditional error rates of the sample rules based on $\xi_{1.2}(\mathbf{x};\hat{\boldsymbol{\theta}}_E)$ and $\xi(\mathbf{x};\hat{\boldsymbol{\theta}}_E)$, respectively, where p_i is the dimension of $\mathbf{x}^{(i)}$ $(i = 1,2)$. Leung and Srivastava (1983b) have established that

$$eu_{1.2}(\Delta, p_1, p_2) < eu(\Delta, p)$$

up to terms of the first order if and only if

$$\frac{\Delta}{4}\left(p_1 + \frac{p_2}{N_o - 1}\right) + \frac{p-1}{\Delta}\left(1 + \frac{p_2}{N_o - 1}\right) < \frac{p\Delta}{4} + \frac{p-1}{\Delta}, \qquad (3.6.8)$$

where $N_o = N - p_2$. It is obvious that (3.6.8) holds if $N_o \gg p_2$, which is usually the case in most practical situations.

In another study on the use of the sample covariance-adjusted NLDF $\xi_{1.2}(\mathbf{x};\hat{\boldsymbol{\theta}}_E)$, Leung (1988b) has investigated the increase of order $O(N^{-1})$ in the error rate as a consequence of having to estimate the unknown group-prior probabilities in the formation of the plug-in sample version of the Bayes rule based on $\xi_{1.2}(\mathbf{x};\hat{\boldsymbol{\theta}}_E)$ with cutoff point $\log(\pi_2/\pi_1)$.

For this covariance problem, Fujikoshi and Kanazawa (1976) have considered the rule obtained by the likelihood ratio criterion. They have given expansions of the group-conditional distributions of the associated statistic and modified versions, including those obtained by Studentization; see Siotani (1982).

3.7 DISCRIMINATION WITH REPEATED MEASUREMENTS

3.7.1 Optimal Rule

In some practical situations, repeated measurements are made on the entity to be allocated. In a medical context, a patient may be recalled for further repeated measurement of the same clinical variable(s) with the intent of providing a firmer basis for diagnosis or prognosis of his or her condition. For example, in diagnosing hypertension, a patient's systolic and diastolic blood pressures (inter alia) may be measured repeatedly over weeks, months, or even years before any decision on specific therapy is taken. In the example considered by S. C. Choi (1972), an individual is diagnosed on the basis of multiple measurements of cell-membrane thickness. A muscle capillary is examined at

20 sites around its perimeter. In addition to using the averages of the measurements across these sites, the variability between sites can be exploited in the formation of a sample discriminant rule for discriminating between diseased and nondiseased individuals.

We first focus on the case where the repeated measurements have the same mean and are equicorrelated. These assumptions are relaxed in the last two subsections, where consideration is given to the fitting of time series and growth curves to the repeated measurements.

Suppose that on the entity to be classified, K repeated measurements are made on each of its p features so that its feature vector $\mathbf{x}$ is now of dimension pK and is given by

$$\mathbf{x} = (\mathbf{x}'_{m_1},\ldots,\mathbf{x}'_{m_K})',$$

where $\mathbf{x}_{m_k}$ denotes the kth measurement on $\mathbf{X}$ $(k = 1,\ldots,K)$. In the initial study on this problem by S. C. Choi (1972) and the subsequent studies by Gupta (1980, 1986), $\mathbf{x}_{m_k}$ was taken to be the realization of a random vector modeled in group G_i $(i = 1,\ldots,g)$ as

$$\mathbf{X}_{m_k} = \boldsymbol{\mu}_i + \mathbf{V}_i + \boldsymbol{\epsilon}_{ik} \qquad (k = 1,\ldots,K). \tag{3.7.1}$$

In this model, $\mathbf{V}_i$ is distributed $N(\mathbf{0},\boldsymbol{\Sigma}_{ei})$ independently of the $\boldsymbol{\epsilon}_{ik}$, which are distributed as

$$\boldsymbol{\epsilon}_{i1},\ldots,\boldsymbol{\epsilon}_{iK} \overset{iid}{\sim} N(\mathbf{0},\boldsymbol{\Sigma}_{mi}) \qquad (i = 1,\ldots,g).$$

Within G_i, the covariance matrix $\boldsymbol{\Sigma}_{ei}$ can be viewed as arising from the variation in the features over entities, whereas $\boldsymbol{\Sigma}_{mi}$ reflects the variation between the repeated measurements on the features of a given entity.

Under (3.7.1), the ith group-conditional density $f_{im}(\mathbf{x};\boldsymbol{\theta}_U)$ of the full feature vector $\mathbf{X}$ is multivariate normal with mean

$$(\boldsymbol{\mu}'_i,\ldots,\boldsymbol{\mu}'_i)'$$

and covariance matrix

$$\boldsymbol{\Sigma}_{ei} \otimes \mathbf{1}_K\mathbf{1}'_K + \boldsymbol{\Sigma}_{mi} \otimes \mathbf{I}_K, \tag{3.7.2}$$

where $\mathbf{1}_K$ denotes the K-dimensional vector whose elements are all unity and $\otimes$ denotes the Kronecker product. The expression (3.7.2) displays the equicorrelated structure of the covariance matrix of the feature vector under this model. Here now the vector $\boldsymbol{\theta}_i$ of parameters contains the elements of $\boldsymbol{\mu}_i$ and the distinct elements of $\boldsymbol{\Sigma}_{ei}$ and $\boldsymbol{\Sigma}_{mi}$ $(i = 1,\ldots,g)$, and $\boldsymbol{\theta}_U$ contains the elements of $\boldsymbol{\theta}_1,\ldots,\boldsymbol{\theta}_g$.

S. C. Choi (1972) and Gupta (1980) noted that $f_{im}(\mathbf{x};\boldsymbol{\theta}_U)$ can be expressed as

$$\begin{aligned} f_{im}(\mathbf{x};\boldsymbol{\theta}_i) =& (2\pi)^{-(1/2)pK}K^{-(1/2)p}|\boldsymbol{\Sigma}_{mi}|^{-(1/2)(K-1)} \\ &\times |\boldsymbol{\Sigma}_{ei} + K^{-1}\boldsymbol{\Sigma}_{mi}|^{-1/2}\exp(-\tfrac{1}{2}Q_i), \end{aligned} \tag{3.7.3}$$

where

$$Q_i = (K-1)\mathrm{tr}(\mathbf{S}_x \boldsymbol{\Sigma}_{mi}^{-1}) + (\overline{\mathbf{x}}_m - \boldsymbol{\mu}_i)'(\boldsymbol{\Sigma}_{ei} + K^{-1}\boldsymbol{\Sigma}_{mi})^{-1}(\overline{\mathbf{x}}_m - \boldsymbol{\mu}_i),$$

$$\mathbf{S}_x = \sum_{k=1}^{K}(\mathbf{x}_{m_k} - \overline{\mathbf{x}})(\mathbf{x}_{m_k} - \overline{\mathbf{x}})',$$

and

$$\overline{\mathbf{x}}_m = \sum_{k=1}^{K}\mathbf{x}_{m_k}/K.$$

It follows from (3.7.3) that in the case of $g = 2$ groups, the optimal or Bayes rule is based on the discriminant function

$$\begin{aligned}\xi_m(\mathbf{x};\boldsymbol{\theta}_U) &= \log\{f_{1m}(\mathbf{x};\boldsymbol{\theta}_U)/f_{2m}(\mathbf{x};\boldsymbol{\theta}_U)\}\\ &= \sum_{i=1}^{2}\frac{1}{2}(-1)^i\{(\overline{\mathbf{x}} - \boldsymbol{\mu}_i)'(\boldsymbol{\Sigma}_{ei} + K^{-1}\boldsymbol{\Sigma}_{mi})^{-1}(\overline{\mathbf{x}} - \boldsymbol{\mu}_i)\\ &\quad + (K-1)\mathrm{tr}(\mathbf{S}_x\boldsymbol{\Sigma}_{mi}^{-1}) + (K-1)\log|\boldsymbol{\Sigma}_{mi}|\\ &\quad + \log|\boldsymbol{\Sigma}_{ei} + K^{-1}\boldsymbol{\Sigma}_{mi}|\}. \qquad (3.7.4)\end{aligned}$$

Gupta (1986) has obtained expressions for the error rates of the Bayes rule in the case of a single feature variable ($p = 1$). The case $p > 1$ has been studied recently by Gupta and Logan (1990).

If

$$\boldsymbol{\Sigma}_{e1} = \boldsymbol{\Sigma}_{e2} = \boldsymbol{\Sigma}_e \qquad (3.7.5)$$

and

$$\boldsymbol{\Sigma}_{m1} = \boldsymbol{\Sigma}_{m2} = \boldsymbol{\Sigma}_m, \qquad (3.7.6)$$

then (3.7.4) reduces to

$$\xi_m(\mathbf{x};\boldsymbol{\theta}_E) = \{\overline{\mathbf{x}}_m - \tfrac{1}{2}(\boldsymbol{\mu}_1 + \boldsymbol{\mu}_2)\}'(\boldsymbol{\Sigma}_e + K^{-1}\boldsymbol{\Sigma}_m)^{-1}(\boldsymbol{\mu}_1 - \boldsymbol{\mu}_2), \qquad (3.7.7)$$

the linear discriminant function based on the sample mean $\overline{\mathbf{x}}_m$ of the K measurements made on the entity to be allocated. Here $\boldsymbol{\theta}_E$ denotes $\boldsymbol{\theta}_U$ under the constraints (3.7.5) and (3.7.6) of equal group-covariance matrices. It is clear that for $K = 1$, it coincides with the familiar linear discriminant function based on a single measurement of each feature variable.

3.7.2 Posterior Log Odds: Univariate Case

In the univariate case of $p = 1$ under the homoscedastic constraints (3.7.5) and (3.7.6), Andrews, Brant, and Percy (1986) have considered the effect of repeated measurements on the posterior log odds

$$\eta_m(\mathbf{x};\boldsymbol{\theta}_E) = \log(\pi_1/\pi_2) + \xi_m(\mathbf{x};\boldsymbol{\theta}_E).$$

In the case of $p = 1$, we write Σ_e and Σ_m as σ_e^2 and σ_m^2, respectively, and we put

$$\rho = \sigma_e^2/\sigma^2,$$

where $\sigma^2 = \sigma_e^2 + \sigma_m^2$. Then for $p = 1$,

$$(\Sigma_e + K^{-1}\Sigma_m)^{-1} = \sigma^{-2}K/\{1 + (K-1)\rho\}.$$

By using this result in (3.7.7), it follows for $p = 1$ that

$$\log \eta_m(\mathbf{x};\boldsymbol{\theta}_E) = \log(\pi_1/\pi_2) + c\{\bar{x}_m - \tfrac{1}{2}(\mu_1 + \mu_2)\}(\mu_1 - \mu_2)\sigma^{-2}, \qquad (3.7.8)$$

where

$$c = K/\{1 + (K-1)\rho\}.$$

It can be seen from (3.7.8) that for $\rho = 0$, for which repeated measurements bring new information, the prior odds are modified by the factor

$$\exp\left[K\left\{\bar{x}_m - \tfrac{1}{2}(\mu_1 + \mu_2)\right\}(\mu_1 - \mu_2)\sigma^{-2}\right].$$

As ρ increases, that is, repeated measurements became less informative, the factor c decreases from K at $\rho = 0$ to 1 at $\rho = 1$ in the case of no new information. The case $\rho = 0$ had been considered previously by Lachenbruch (1980).

3.7.3 Plug-In Sample Rule

Returning now to the multivariate heteroscedastic model in which the group-conditional densities are given by (3.7.3), we can form a sample rule by plugging in estimates of the $\boldsymbol{\mu}_i, \Sigma_{ei}$, and the Σ_{mi} in the case of unknown $\boldsymbol{\theta}_U$. The training data $\mathbf{t}$ on the classified entities are denoted as previously by

$$\mathbf{t}' = (\mathbf{y}_1, \ldots, \mathbf{y}_n),$$

where $\mathbf{y}_j = (\mathbf{x}_j', \mathbf{z}_j')'$, and where now

$$\mathbf{x}_j = (\mathbf{x}_{jm_1}', \ldots, \mathbf{x}_{jm_K}')'$$

contains the K repeated measurements of the feature vector on the jth classified entity ($j = 1, \ldots, n$).

Let

$$\bar{\mathbf{x}}_i = \sum_{j=1}^{n}\sum_{k=1}^{K} z_{ij}\mathbf{x}_{jm_k}/(n_i K),$$

$$\bar{\mathbf{x}}_{ij} = \sum_{k=1}^{K} z_{ij}\mathbf{x}_{jm_k}/K \qquad (j = 1, \ldots, n_i),$$

$$\mathbf{S}_{mi} = \sum_{j=1}^{n}\sum_{k=1}^{K} z_{ij}(\mathbf{x}_{jm_k} - \bar{\mathbf{x}}_{ij})(\mathbf{x}_{jm_k} - \bar{\mathbf{x}}_{ij})'/\{n_i(K-1)\},$$

and

$$\mathbf{S}_i = K \sum_{j=1}^{n_i} (\overline{\mathbf{x}}_{ij} - \overline{\mathbf{x}}_i)(\overline{\mathbf{x}}_{ij} - \overline{\mathbf{x}}_i)'/(n_i - 1)$$

for $i = 1,\ldots,g$. Then following S. C. Choi (1972), $\boldsymbol{\mu}_i$, $\boldsymbol{\Sigma}_{mi}$, and $\boldsymbol{\Sigma}_{ei}$ in (3.7.3) can be estimated as

$$\hat{\boldsymbol{\mu}}_i = \overline{\mathbf{x}}_i,$$

$$\hat{\boldsymbol{\Sigma}}_{mi} = \mathbf{S}_{mi},$$

and

$$\hat{\boldsymbol{\Sigma}}_{e_i} = (\mathbf{S}_i - \mathbf{S}_{m_i})/K$$

for $i = 1,\ldots,g$.

This problem has been considered also by Ellison (1962), Bertolino (1988), and by Logan (1990), who presented the predictive approach.

3.7.4 Discrimination Between Time Series

Often the multiple measurements on an entity are recorded over time, and so it is appropriate to adopt a time-series approach to the modeling of the group-conditional distributions of the feature vector. Note that this situation is distinct from that in which the training replicates of the feature vector in a given group are viewed collectively as the realization of a time series. The latter is to be considered in Section 13.9, which is on discriminant analysis with dependent training data.

Schumway (1982) has provided an extensive list of references and applications of discriminant analysis for time series. Applications listed there include discriminating between presumed earthquakes and underground nuclear explosions, the detection of a signal imbedded in a noise series, discriminating between different classes of brain wave recordings, and discriminating between various speakers or speech patterns on the basis of recorded speech data. Additional references may be found in the work by Browdy and Chang (1982) and Krzyśko (1983) on discrimination between time series.

Suppose that for a single ($p = 1$) feature variable x of the entity, the multiple measurements $x_{m_1},\ldots,x_{m_K}$ have been recorded at K points in time $s_1,\ldots,s_K$. One may also observe multivariate series, but we focus on the univariate case in the interest of simplifying the notation. On writing

$$X_{m_k} = \mu_{ik} + \epsilon_{ik} \qquad (k = 1,\ldots,K)$$

in group G_i ($i = 1,\ldots,g$), we put

$$\mathbf{X} = (X_{m_1},\ldots,X_{m_k})',$$

$$\boldsymbol{\mu}_i = (\mu_{i1},\ldots,\mu_{iK})',$$

and

$$\boldsymbol{\epsilon}_i = (\epsilon_{i1},\ldots,\epsilon_{iK})',$$

where ϵ_i is taken to have a multivariate normal distribution with mean zero and covariance matrix Σ_i $(i = 1,\ldots,g)$. Thus,

$$\mathbf{X} \sim N(\boldsymbol{\mu}_i, \boldsymbol{\Sigma}_i) \quad \text{in} \quad G_i \qquad (i = 1,\ldots,g). \tag{3.7.9}$$

The simple case of detecting a fixed signal embedded in Gaussian noise is represented by taking $g = 2$, $\boldsymbol{\mu}_1 = \mathbf{s}$, and $\boldsymbol{\mu}_2 = \mathbf{0}$ in (3.7.9). This also covers the case of the detection of a stochastic signal, provided an additive model is assumed for the signal and noise, so that $\boldsymbol{\Sigma}_i$ represents the covariance matrix of the signal plus noise within group G_i $(i = 1,2)$.

A common assumption in practice is that ϵ_i arises from a zero-mean stationary discrete time parameter process so that $(\boldsymbol{\Sigma}_i)_{u,v}$ is a function of $|u - v|$ for $u, v = 1,\ldots,K$. A special case is a first-order autoregressive process. Then

$$(\boldsymbol{\Sigma}_i)_{u,v} = \sigma_i^2 \rho_i^{|u-v|},$$

where σ_i^2 is the ith group-conditional variance of X_{m_k}, and $|\rho_i| < 1$ $(i = 1, \ldots,g)$.

The rather cumbersome matrix calculations with the time-domain approach can be avoided through consideration of the more easily computed spectral approximations. Schumway (1982) has presented some of the spectral approximations that make discriminant analysis in the frequency domain such an attractive procedure.

Broemeling and Son (1987) and Marco, Young, and Turner (1988) have considered the Bayesian approach to discriminant analysis of time series in the case of univariate autoregressive processes.

3.7.5 Discrimination Between Measurements Made at Different Points in Time

We consider here the case of discrimination within a time series, corresponding to multiple multivariate measurements recorded over time on an entity. For an observation $\mathbf{x}$ recorded on the entity at an unknown point in time, the aim is to determine its time of measurement from a choice of two specified time points, say, s_1 and s_2. Under the model introduced initially by Das Gupta and Bandyopadhyay (1977), $\mathbf{x}$ is taken to be the realization of a random vector distributed normally with means $\boldsymbol{\mu}_1$ and $\boldsymbol{\mu}_2$, and common covariance matrix $\boldsymbol{\Sigma}$, at the time points s_1 and s_2, respectively. Further, in order to form a sample discriminant rule in situations where the group parameters are not all known, it is supposed that there are available $n_i = \frac{1}{2}n$ feature observations $\mathbf{x}_{ij}$ $(j = 1,\ldots,n_i)$ known to be measured at time s_i $(i = 1,2)$. A stationary Gaussian process is assumed for the errors in the time series, so that each $(\mathbf{x}_{1j}', \mathbf{x}_{2j}')'$ denotes a realization of a $2p$-dimensional random vector, having a multivariate normal distribution with mean $(\boldsymbol{\mu}_1', \boldsymbol{\mu}_2')'$ and covariance matrix

$$\begin{pmatrix} \boldsymbol{\Sigma} & \boldsymbol{\Sigma}_{12} \\ \boldsymbol{\Sigma}_{21} & \boldsymbol{\Sigma} \end{pmatrix}.$$

In the special case of a first-order autoregressive process,

$$\boldsymbol{\Sigma}_{12} = \rho\boldsymbol{\Sigma}, \tag{3.7.10}$$

where $|\rho| < 1$. Thus, unless $\boldsymbol{\Sigma}_{12} = \mathbf{0}$, the classified training observations are not independent across the two groups, representing the two specified points in time. But as usual, they are independent within a group; that is, $\mathbf{x}_{ij}$ for $j = 1, \ldots, \frac{1}{2}n$ denote $\frac{1}{2}n$ independent realizations ($i = 1, 2$).

Das Gupta and Bandyopadhyay (1977) have derived the asymptotic expansions of the sample plug-in version of the optimal discriminant rule and its unconditional error rates for various forms of $\boldsymbol{\Sigma}_{12}$, including (3.7.10). In the latter special case, the problem has been investigated further in a series of papers by Bandyopadhyay (1977, 1978, 1979). Bandyopadhyay (1982, 1983) and Leung and Srivastava (1983a) have studied the covariance-adjusted sample discriminant rule for this problem.

3.7.6 Allocation of Growth Curves

We consider here the allocation of an entity on the basis of a growth curve fitted to the repeated measurements made on it. Allocation of growth curves was first considered by Lee (1977) from a Bayesian viewpoint and later extended by Nagel and de Waal (1979); see Lee (1982) for a review. The growth-curve discrimination problem can be formulated in the just considered context of repeated measurements on the features of the entity to be allocated to one of the g underlying groups. Suppose that for a single ($p = 1$) feature variable x of the entity, the multiple measurements $x_{m_1}, \ldots, x_{m_K}$ have been recorded at K points in time $s_1, \ldots, s_K$. If the growth curve is taken to be a polynomial of degree $q-1$, we can write in group G_i

$$x_{m_k} = \mu_{i_k} + \epsilon_{i_k} \qquad (k = 1, \ldots, K),$$

where

$$\mu_{i_k} = \omega_{i0} + \omega_{i1}s_k + \cdots + \omega_{i,q-1}s_k^{q-1} \qquad (i = 1, \ldots, g).$$

The usual assumption about the vector of residuals

$$\boldsymbol{\epsilon}_i = (\epsilon_{i1}, \ldots, \epsilon_{iK})'$$

is that it is multivariate normal with mean zero and with an unknown covariance matrix $\boldsymbol{\Sigma}_i$. This covariance matrix is not specified to be diagonal as the observations in time on the same entity are not independently distributed in general.

The unknown coefficients ω_{ij} ($j = 0, \ldots, q-1$) and $\boldsymbol{\Sigma}_i$ ($i = 1, \ldots, g$) can be estimated from the training data $\mathbf{t}$ containing growth curves observed on n classified entities. Here $\mathbf{x}_j = (x_{jm_1}, \ldots, x_{jm_K})'$ contains the K measurements in time on the single feature of the jth classified entity ($j = 1, \ldots, n$). To represent

t in the form of the growth model proposed by Potthoff and Roy (1964) under the assumption of homoscedasticity $\boldsymbol{\Sigma}_i = \boldsymbol{\Sigma}$ $(i = 1,\ldots,g)$, we first let

$$\mathbf{t}'_x = (\mathbf{x}_1,\ldots,\mathbf{x}_n)$$

and

$$\mathbf{t}'_z = (\mathbf{z}_1,\ldots,\mathbf{z}_n),$$

where it is assumed now that the data have been relabeled so that the first n_1 feature vectors contain the growth curves on the entities from G_1, the next n_2 contain the growth curves on the entities from G_2, and so on, until the last n_g feature vectors contain the growth curves on the entities from G_g. Consistent with our previous notation,

$$n_i = \sum_{j=1}^{n} z_{ij}$$

denotes the number of entities from G_i, where $z_{ij} = (\mathbf{z}_j)_i$ is one or zero according as to whether the jth entity belongs to G_i or not $(i = 1,\ldots,g)$.

With this notation, the training data **t** can be expressed in the form of the growth model of Potthoff and Roy (1964) as

$$\mathbf{t}'_x = \mathbf{C}\boldsymbol{\omega}\mathbf{t}'_z + \boldsymbol{\epsilon}, \tag{3.7.11}$$

where **C** is the $K \times q$ matrix with

$$(\mathbf{C})_{uv} = s_u^{v-1} \qquad (u = 1,\ldots,K;\ v = 1,\ldots,q)$$

and where $\boldsymbol{\omega}$ is the $q \times g$ matrix with

$$(\boldsymbol{\omega})_{uv} = \omega_{v,u-1} \qquad (u = 1,\ldots,q;\ v = 1,\ldots,g).$$

The $p \times n$ matrix $\boldsymbol{\epsilon}$ of residuals is given by

$$\boldsymbol{\epsilon} = (\boldsymbol{\epsilon}_1,\ldots,\boldsymbol{\epsilon}_n),$$

where $\boldsymbol{\epsilon}_j$ denotes the vector of residuals for $\mathbf{x}_j$ $(j = 1,\ldots,n)$. In accordance with our previous assumptions, the $\boldsymbol{\epsilon}_j$ are distributed $N(\mathbf{0},\boldsymbol{\Sigma})$, independently of each other.

For further details on the fitting of the growth-curve model (3.7.11) to the training data **t**, the reader is referred to Lee (1982) and the references therein.

3.7.7 Sequential Discrimination

When the repeated measurements on an entity are able to be observed sequentially, consideration may be given to the use of sequential allocation rules, particularly if it is possible to treat the repeated observations as being independent. References related to the sequential approach to discrimination include Wald (1947, 1950), Armitage (1950), Kendall (1966), Freedman (1967), Simons (1967), Fu (1968), Meilijson (1969), Roberts and Mullis (1970), Yarborough (1971), Srivastava (1973a), Smith and Makov (1978), Geisser (1982), and Titterington, Smith, and Makov (1985, Chapter 6).

Mallows (1953) studied the sequential approach in the situation where there is only a single measurement on the feature variables of an entity, but the latter are observed sequentially.

3.8 PARTIALLY CLASSIFIED DATA

Up to now, in this chapter, we have concentrated on the typical situation in discriminant analysis, where the training data are completely classified. We now consider the computation of the maximum likelihood estimate of $\mathbf{\Psi}_U$ in the case of partially classified feature data, as considered in Section 2.7 for arbitrary parametric forms for the group-conditional distributions. The same notation as introduced there is adopted here, where the available data consist of the classified training data $\mathbf{t}$ containing $\mathbf{y}_j = (\mathbf{x}_j', \mathbf{z}_j')'$ for $j = 1, \ldots, n$ and the unclassified data $\mathbf{t}_u$ containing the m unclassified feature observations $\mathbf{x}_j$ ($j = n+1, \ldots, n+m$). The maximum likelihod estimate of $\mathbf{\Psi}_U$ from $\mathbf{t}$ and $\mathbf{t}_u$ can be computed iteratively via the EM algorithm, using equations (2.7.3) and (2.7.4) as described in Section 2.7.

In the present situation of multivariate normal group-conditional distributions with unequal covariance matrices, the solution of (2.7.4) exists in closed form. It follows that on the M-step of the $(k+1)$th cycle, the current fit for the group means and covariance matrices is given explicitly by

$$\boldsymbol{\mu}_i^{(k+1)} = \left\{ \sum_{j=1}^{n} z_{ij}\mathbf{x}_j + \sum_{j=n+1}^{n+m} \tau_i(\mathbf{x}_j; \mathbf{\Psi}_U^{(k)})\mathbf{x}_j \right\} \Big/ M_i(\mathbf{\Psi}_U^{(k)}) \tag{3.8.1}$$

and

$$\begin{aligned}\mathbf{\Sigma}_i^{(k+1)} = \Bigg\{ &\sum_{j=1}^{n} z_{ij}(\mathbf{x}_j - \boldsymbol{\mu}_i^{(k+1)})(\mathbf{x}_j - \boldsymbol{\mu}_i^{(k+1)})' \\ &+ \sum_{j=n+1}^{n+m} \tau_i(\mathbf{x}_j; \mathbf{\Psi}_U^{(k)})(\mathbf{x}_j - \boldsymbol{\mu}_i^{(k+1)})(\mathbf{x}_j - \boldsymbol{\mu}_i^{(k+1)})' \Bigg\} \Big/ M_i(\mathbf{\Psi}_U^{(k)}),\end{aligned} \tag{3.8.2}$$

where

$$M_i(\mathbf{\Psi}_U^{(k)}) = \left\{ n_i + \sum_{j=n+1}^{n+m} \tau_i(\mathbf{x}_j; \mathbf{\Psi}_U^{(k)}) \right\} \qquad (i = 1, \ldots, g).$$

The posterior probability $\tau_i(\mathbf{x}_j;\boldsymbol{\Psi}_U)$ that the jth entity with feature vector $\mathbf{x}_j$ belongs to G_i can be computed from

$$\tau_i(\mathbf{x};\boldsymbol{\Psi}_U) = \frac{(\pi_i/\pi_g)\exp\{\xi_{ig}(\mathbf{x};\boldsymbol{\theta}_U)\}}{1+\sum_{h=1}^{g-1}(\pi_h/\pi_g)\exp\{\xi_{hg}(\mathbf{x};\boldsymbol{\theta}_U)\}} \qquad (i = 1,\ldots,g-1),$$

where $\xi_{ig}(\mathbf{x};\boldsymbol{\theta}_U)$ is defined by (3.2.3).

Under the homoscedastic version of the normal model where $\boldsymbol{\Sigma}$ is the common group-covariance matrix, $\boldsymbol{\Psi}_U^{(k)}$ is replaced by $\boldsymbol{\Psi}_E^{(k)}$ in (3.8.1) and the group-specific estimate (3.8.2) is replaced by

$$\boldsymbol{\Sigma}^{(k+1)} = (n+m)^{-1}\left\{\sum_{i=1}^{g}\sum_{j=1}^{n} z_{ij}(\mathbf{x}_j - \boldsymbol{\mu}_i^{(k+1)})(\mathbf{x}_j - \boldsymbol{\mu}_i^{(k+1)})' \right.$$
$$\left. + \sum_{j=n+1}^{n+m} \tau_i(\mathbf{x}_j;\boldsymbol{\Psi}_E^{(k+1)})(\mathbf{x}_j - \boldsymbol{\mu}_i^{(k+1)})(\mathbf{x}_j - \boldsymbol{\mu}_i^{(k+1)})'\right\}.$$

McLachlan and Basford (1988) have provided a FORTRAN listing of computer programs for fitting heteroscedastic and homoscedastic normal mixture models to partially classified data with either separate or mixture sampling of the classified observations and also to completely unclassified data.

3.9 LINEAR PROJECTIONS OF HOMOSCEDASTIC FEATURE DATA

3.9.1 Introduction

In dealing with multivariate feature observations, it often facilitates visualization and understanding to represent them in a lower-dimensional space. In particular, two- and three-dimensional scatter plots are often helpful in exploring relationships between the groups, assessing the group-conditional distributions, and identifying atypical feature observations. However, if the dimension p of the data is greater than about 7 or 10, then considerable patience and concentration are needed for a careful scrutiny of all $\binom{p}{2}$ and $\binom{p}{3}$ scatter plots of pairs and triples of the feature variables. One approach to reducing the effort involved in such an exercise is first to transform linearly the p original feature variables into a smaller number q of variables. This process is referred to in the pattern recognition literature as linear feature selection.

For the linear projection $\mathbf{C}_q$, where $\mathbf{C}_q$ is a $q \times p$ matrix of rank q $(q \leq p)$, there is the problem of how to choose $\mathbf{C}_q$ so as to best preserve the distinction between the groups, where q may or may not be specified. Often q will be specified to be at most 2 or 3 for convenience of the subsequent analysis, in particular the graphical representations of the transformed feature data. In some situations, there is interest in finding the single linear combination that best distinguishes the g groups, and so q is specified to be one.

We proceed here under the homoscedastic normal model (3.3.1) for the group-conditional distributions. For this model, Geisser (1977) has provided a comprehensive account of the topic. There is also the paper of Hudlet and Johnson (1977), which appeared in the same proceedings as the former reference. More recent papers include Schervish (1984) and McCulloch (1986). Previously, Guseman, Peters, and Walker (1975) had considered the problem in the more general situation of unequal group-covariance matrices. Results for this heteroscedastic case, which has been considered recently by Young, Marco, and Odell (1987), are discussed in Section 3.10.

3.9.2 Canonical Variate Analysis

A starting point in the consideration of linear projections of $\mathbf{x}$ is a canonical variate analysis, which expresses the differences between the means $\boldsymbol{\mu}_1, \ldots, \boldsymbol{\mu}_g$ in $d = \min(p, b_o)$ dimensions, where b_o is the rank of the matrix $\mathbf{B}_o$ defined in what follows. A canonical variate analysis does not depend on the assumption of normality, as only knowledge of the first two moments of the group-conditional distributions is required. Let

$$\mathbf{B}_o = \frac{1}{g-1}\sum_{i=1}^{g}(\boldsymbol{\mu}_i - \overline{\boldsymbol{\mu}})(\boldsymbol{\mu}_i - \overline{\boldsymbol{\mu}})',$$

where

$$\overline{\boldsymbol{\mu}} = \sum_{i=1}^{g}\boldsymbol{\mu}_i/g.$$

We will reserve the notation $\mathbf{B}$ for the between-group sums of squares and products matrix on its degrees of freedom,

$$\mathbf{B} = \frac{1}{g-1}\sum_{i=1}^{g} n_i(\overline{\mathbf{x}}_i - \overline{\mathbf{x}})(\overline{\mathbf{x}}_i - \overline{\mathbf{x}})'.$$

For mixture sampling in equal proportions from the groups, $\mathbf{B}/n$ converges in probability to $\mathbf{B}_o/g$, as $n \to \infty$. The matrix $\mathbf{B}_o$ is of rank $b_o \leq g-1$, where $b_o = g-1$ if $\boldsymbol{\mu}_1, \ldots, \boldsymbol{\mu}_g$ are linearly independent.

The canonical variates of $\mathbf{x}$ are defined by

$$\mathbf{v} = \boldsymbol{\Gamma}_d \mathbf{x}, \tag{3.9.1}$$

where

$$\boldsymbol{\Gamma}_d = (\boldsymbol{\gamma}_1, \ldots, \boldsymbol{\gamma}_d)', \tag{3.9.2}$$

and where $\boldsymbol{\gamma}_1$ maximizes the ratio

$$\boldsymbol{\gamma}'\mathbf{B}_o\boldsymbol{\gamma}/\boldsymbol{\gamma}'\boldsymbol{\Sigma}\boldsymbol{\gamma}. \tag{3.9.3}$$

For $k = 2, \ldots, d$, $\boldsymbol{\gamma}_k$ maximizes the ratio (3.9.1) subject to

$$\boldsymbol{\gamma}_k'\boldsymbol{\Sigma}\boldsymbol{\gamma}_h = 0 \qquad (h = 1, \ldots, k-1). \tag{3.9.4}$$

Hence, the correlation between $\gamma_h'\mathbf{X}$ and $\gamma_k'\mathbf{X}$ is zero for $h \neq k = 1,\ldots,d$. The usual normalization of γ_k is

$$\gamma_k'\mathbf{\Sigma}\gamma_k = 1 \qquad (k = 1,\ldots,d), \tag{3.9.5}$$

which implies that $\gamma_k'\mathbf{X}$ has unit variance. This normalization along with the constraint (3.9.4) implies that

$$\mathbf{\Gamma}_d\mathbf{\Sigma}\mathbf{\Gamma}_d' = \mathbf{I}_d,$$

where $\mathbf{I}_d$ is the $d \times d$ identity matrix.

It will be seen from both allocatory and separatory aspects, the new set of coordinates $v_1,\ldots,v_d$ is the complete set of multiple linear discriminant functions. Hence, sometimes in the literature they are referred to as discriminant coordinates rather than by the more usual name of canonical variates.

The computation of $\mathbf{\Gamma}_d$ is well-covered in standard textbooks on multivariate analysis; see, for example, Seber (1984, Chapter 5). The transpose of its kth row, γ_k, is the eigenvector corresponding to the kth largest (nonzero) eigenvalue of $\mathbf{\Sigma}^{-1}\mathbf{B}_o$. Hence, γ_k satisfies

$$(\mathbf{B}_o - \lambda_{o,k}\mathbf{\Sigma})\gamma_k = 0 \qquad (k = 1,\ldots,d), \tag{3.9.6}$$

where $\lambda_{o,1},\ldots,\lambda_{o,k}$ denote the nonzero eigenvalues of $\mathbf{\Sigma}^{-1}\mathbf{B}_o$ ordered in decreasing size. From (3.9.5) and (3.9.6),

$$\gamma_k'\mathbf{B}_o\gamma_k = \lambda_{o,k} \qquad (k = 1,\ldots,d).$$

The eigenvalues and eigenvectors of $\mathbf{\Sigma}^{-1}\mathbf{B}_o$ can be found using a singular-value decomposition algorithm; see Seber (1984, Section 10.1). Their computation is usually considered in the literature for the typical situation in practice where $\mathbf{B}_o$ and $\mathbf{\Sigma}$ are unknown and must be replaced by estimates formed from the available training data. The sample version of a canonical variate analysis, which is taken up later in Section 6.7, is just the multiple-group generalization of Fisher's (1936) approach to discriminant analysis in the case of $g = 2$ groups for which, as seen in Section 3.3.3,

$$\gamma_1 \propto \mathbf{S}^{-1}(\overline{\mathbf{x}}_1 - \overline{\mathbf{x}}_2).$$

By building on the initial work by Fisher (1936) and Hotelling (1935, 1936), the technique of canonical variate analysis was developed during the 1940s by Rao (1948) and others.

For $p > d$, we let $\gamma_{d+1},\ldots,\gamma_p$ denote the eigenvectors of $\mathbf{\Sigma}^{-1}\mathbf{B}_o$ corresponding to its $p-d$ zero eigenvalues, normalized as

$$\gamma_k'\mathbf{\Sigma}\gamma_k = 1 \qquad (k = d+1,\ldots,p).$$

We put

$$\mathbf{\Gamma} = (\mathbf{\Gamma}_d', \mathbf{\Gamma}_{p-d}')', \tag{3.9.7}$$

where

$$\mathbf{\Gamma}_{p-d} = (\gamma_{d+1},\ldots,\gamma_p)'.$$

It follows then that

$$\mathbf{\Gamma X} \sim N(\mathbf{\Gamma}\boldsymbol{\mu}_i, \mathbf{I}_p) \quad \text{in} \quad G_i \quad (i = 1,\ldots,g), \tag{3.9.8}$$

where, corresponding to the partition (3.9.7) of $\mathbf{\Gamma}$,

$$\mathbf{\Gamma}\boldsymbol{\mu}_i = \begin{pmatrix} \mathbf{\Gamma}_d\boldsymbol{\mu}_i \\ \mathbf{\Gamma}_{p-d}\boldsymbol{\mu}_i \end{pmatrix},$$

and

$$\mathbf{\Gamma}_{p-d}\boldsymbol{\mu}_i = \mathbf{\Gamma}_{p-d}\overline{\boldsymbol{\mu}} \quad (i = 1,\ldots,g). \tag{3.9.9}$$

3.9.3 Discrimination in Terms of Canonical Variates

It is clear from (3.9.8) and (3.9.9) that for the purposes of allocation, the last $p-d$ canonical variates can be discarded without an increase in any of the group-specific error rates. This is because $\mathbf{\Gamma}_{p-d}\mathbf{X}$ is distributed independently of $\mathbf{\Gamma}_d\mathbf{X}$, with the same distribution in each group. A concise reference on this is Kshirsagar and Arvensen (1975).

We now verify this result that the optimal rule $r_o(\mathbf{x};\mathbf{\Psi}_E)$ formed from the vector $\mathbf{x}$ of p original feature variables is the same as the optimal rule $r_o(\mathbf{v};\mathbf{\Psi}_E(\mathbf{\Gamma}_d))$ formed from the vector $\mathbf{v}$ containing the d canonical variates of $\mathbf{x}$. Here $\mathbf{\Psi}_E(\mathbf{\Gamma}_d)$ is the analogue of $\mathbf{\Psi}_E$ for the canonical variates. First, note that for multivariate normal group-conditional distributions with a common covariance matrix, the optimal rule $r_o(\mathbf{x};\mathbf{\Psi}_E)$ makes an allocation on the basis of the minimum of the discriminant scores,

$$\min_i\{(\mathbf{x}-\boldsymbol{\mu}_i)'\mathbf{\Sigma}^{-1}(\mathbf{x}-\boldsymbol{\mu}_i) - 2\log\pi_i\}. \tag{3.9.10}$$

Further, it is invariant under a nonsingular transformation, so that it is the same as the optimal rule $r_o(\mathbf{\Gamma x},\mathbf{\Psi}_E(\mathbf{\Gamma}))$ using the full transformed feature vector $\mathbf{\Gamma x}$. Hence, on noting that $\mathbf{\Gamma\Sigma\Gamma}' = \mathbf{I}_p$, it follows that $r_o(\mathbf{x};\mathbf{\Psi}_E)$ is the same as the rule based on

$$\min_i\{(\mathbf{\Gamma x}-\mathbf{\Gamma}\boldsymbol{\mu}_i)'(\mathbf{\Gamma x}-\mathbf{\Gamma}\boldsymbol{\mu}_i) - 2\log\pi_i\}. \tag{3.9.11}$$

This result can be established directly by noting that (3.9.10) can be written as (3.9.11) since $\mathbf{\Gamma\Gamma}' = \mathbf{\Sigma}^{-1}$.

Considering now (3.9.11),

$$\{\mathbf{\Gamma}(\mathbf{x}-\boldsymbol{\mu}_i)\}'\{\mathbf{\Gamma}(\mathbf{x}-\boldsymbol{\mu}_i)\} = \{\mathbf{\Gamma}_d(\mathbf{x}-\boldsymbol{\mu}_i)\}'\{\mathbf{\Gamma}_d(\mathbf{x}-\boldsymbol{\mu}_i)\} + (\mathbf{\Gamma}_{p-d}\mathbf{x}-\mathbf{\Gamma}_{p-d}\overline{\boldsymbol{\mu}})'(\mathbf{\Gamma}_{p-d}\mathbf{x}-\mathbf{\Gamma}_{p-d}\overline{\boldsymbol{\mu}}),$$

and so it is equivalent to

$$\min_i\{\mathbf{v}-\mathbf{\Gamma}_d\boldsymbol{\mu}_i)'(\mathbf{v}-\mathbf{\Gamma}_d\boldsymbol{\mu}_i) - 2\log\pi_i\}. \tag{3.9.12}$$

This last result (3.9.12) defines $r_o(\mathbf{v};\mathbf{\Psi}_E(\mathbf{\Gamma}_d))$, because the covariance matrix of the vector of canonical variates is the identity matrix. Hence, $r_o(\mathbf{x};\mathbf{\Psi}_E)$ is the same rule as $r_o(\mathbf{v};\Psi_E(\mathbf{\Gamma}_d))$. It can be seen from (3.9.11) that the latter assigns an entity to the group whose mean in the canonical variate space is closest in Euclidean distance to $\mathbf{v}$, after adjustment for any disparity in the prior probabilities π_i of the groups.

Of course, the overall error rate of the Bayes rule will be increased if it is based on a linear projection of $\mathbf{x}$, $\mathbf{C}_q\mathbf{x}$, where the rank q of $\mathbf{C}_q$ is less than d. We now proceed to consider the allocatory and separatory aspects of discrimination on the basis of $\mathbf{C}_q\mathbf{x}$. It will be seen for a given $q < d$ in the case of $g > 2$ groups that, although an intuitively desirable projection $\mathbf{C}_q$ would be one that minimizes the error rate while maximizing the separation between the groups, the optimal choice of $\mathbf{C}_q$ depends on whether the error rate or a separatory measure is to be optimized. The latter aspect of the choice of $\mathbf{C}_q$ is considered first.

3.9.4 Separatory Measures

One approach to assessing the effectiveness of a linear map of the feature vector $\mathbf{x}$ from R^p down to $R^q (q \leq p)$ is to use a measure of spread of the g groups. For this purpose, Geisser (1977) and McCulloch (1986) have considered a class of separatory criteria, which are intuitively motivated functions of the group means and their common covariance matrix. This class does not require any distributional assumptions, only the first and second moments.

For the linear projection $\mathbf{C}_q$, they defined their class of measures of spread of the g groups to consist of measures of the form

$$h\{(\mathbf{C}_q\mathbf{\Sigma}\mathbf{C}_q')^{-1}(\mathbf{C}_q\mathbf{B}_o\mathbf{C}_q')\} = h_{q_o}(\kappa_1,\ldots,\kappa_{q_o}),$$

where $h(\cdot)$ is any scalar function so that $h_{q_o}(\kappa_1,\ldots,\kappa_{q_o})$ is increasing in the nonzero eigenvalues $\kappa_1,\ldots,\kappa_{q_o}$ of

$$(\mathbf{C}_q\mathbf{\Sigma}\mathbf{C}_q')^{-1}(\mathbf{C}_q\mathbf{B}_o\mathbf{C}_q'),$$

where $q_o = \min(q, b_o)$. This class contains well-known measures such as those of Hotelling and Wilks. The former measure is based on the sum of the eigenvalues, and the latter is based on their product.

For given q, let s_q be the measure of spread defined by

$$s_q = \max_{\mathbf{C}_q} h\{(\mathbf{C}_q\mathbf{\Sigma}\mathbf{C}_q')^{-1}(\mathbf{C}_q\mathbf{B}_o\mathbf{C}_q')\}.$$

The maximum value of s_q over q is attained at $q = d = \min(p, b_o)$. A choice of $\mathbf{C}_d$ that achieves this is $\mathbf{\Gamma}_d$ defined by (3.9.2). Hence, there is no interest in the following work in considering $q \geq d$.

Geisser (1977) showed that for a given q $(q < d)$,

$$s_q = h_q(\lambda_{o,1},\ldots,\lambda_{o,q}),$$

which can be achieved by taking $\mathbf{C}_q = \mathbf{\Gamma}_q$, where

$$\mathbf{\Gamma}_q = (\gamma_1,\ldots,\gamma_q)'.$$

An alternative proof was given recently by McCulloch (1986), using the singular-value decomposition of $(\boldsymbol{\mu}_1, \ldots, \boldsymbol{\mu}_g)$ after the feature vector has been transformed so that $\boldsymbol{\Sigma} = \mathbf{I}_p$. Hence, for this class of measures of spread, the first q canonical variates, $v_1 = \boldsymbol{\gamma}_1'\mathbf{x}, \ldots, v_q = \boldsymbol{\gamma}_q'\mathbf{x}$, provide the optimal projection of $\mathbf{x}$ from $\mathbb{R}^p$ to $\mathbb{R}^q$.

3.9.5 Allocatory Aspects

Another way of defining the effectiveness of a linear projection $\mathbf{C}_q$ of the feature vector is to use the overall error rate $eo(\boldsymbol{\Psi}_E(\mathbf{C}_q))$ of the optimal rule $r_o(\mathbf{C}_q\mathbf{x}; \boldsymbol{\Psi}_E(\mathbf{C}_q))$ formed from the q feature variables in $\mathbf{C}_q\mathbf{x}$. Here $\boldsymbol{\Psi}_E(\mathbf{C}_q)$ denotes the elements of $\mathbf{C}_q\boldsymbol{\mu}_1, \ldots, \mathbf{C}_q\boldsymbol{\mu}_g$ and the distinct elements of $\mathbf{C}_q\boldsymbol{\Sigma}\mathbf{C}_q'$. As seen in Section 3.9.3, there is no increase in the error rate $eo(\boldsymbol{\Psi}_E(\mathbf{C}_q))$ over $eo(\boldsymbol{\Psi}_E)$ for the full feature vector $\mathbf{x}$, if we take $\mathbf{C}_q = \boldsymbol{\Gamma}_d$ and so use the vector $\mathbf{v} = \boldsymbol{\Gamma}_d\mathbf{x}$ containing the d canonical variates, where $d = \min(p, b_0)$ and b_0 is the rank of $\mathbf{B}_o$. However, if we use a projection $\mathbf{C}_q\mathbf{x}$ with $q < d$, then there will be an increase in the error rate $eo(\boldsymbol{\Psi}_E(\mathbf{C}_q))$. So far as the spread of the groups is concerned, it was seen in the last section that the choice $\mathbf{C}_q = \boldsymbol{\Gamma}_q$, corresponding to the first q canonical variates, is the optimal linear projection from $\mathbb{R}^p$ to $\mathbb{R}^q$ for the class of measures of spread

$$h\{\mathbf{C}_q\boldsymbol{\Sigma}\mathbf{C}_q')^{-1}(\mathbf{C}_q\mathbf{B}_o\mathbf{C}_q')\}, \tag{3.9.13}$$

where $h(\cdot)$ is any scalar function that is increasing in the nonzero eigenvalues of

$$(\mathbf{C}_q\boldsymbol{\Sigma}\mathbf{C}_q')^{-1}(\mathbf{C}_q\mathbf{B}_o\mathbf{C}_q').$$

However, it does not follow that $eo(\boldsymbol{\Psi}_E(\mathbf{C}_q))$ attains its minimum over $\mathbf{C}_q$ for $\mathbf{C}_q = \boldsymbol{\Gamma}_q$. But, obviously, this choice can be regarded as an approximation to the optimal solution for the error rate.

We now examine this in more detail for linear combinations $\mathbf{C}_1\mathbf{x}$ of the original p feature variables. For convenience of notation, we henceforth write $\mathbf{C}_1$ as $\mathbf{a}$.

3.9.6 Best Linear Combination in Terms of Error Rate

Geisser (1977) has given an excellent account of linear discrimination for the homoscedastic normal model (3.3.1) as assumed now. He showed that for g groups with equal prior probabilities, the overall error rate for the Bayes rule based on the linear combination $\mathbf{a}'\mathbf{x}$ is

$$eo(\boldsymbol{\Psi}_E(\mathbf{a})) = (2/g)\sum_{i=1}^{g-1} \Phi\{\tfrac{1}{2}(\nu_{(i+1)} - \nu_{(i)})\}, \tag{3.9.14}$$

where $\nu_{(1)} \geq \nu_{(2)} \geq \cdots \geq \nu_{(g)}$ are the ordered values of $\nu_1, \ldots, \nu_g$, and

$$\nu_i = \mathbf{a}'\boldsymbol{\mu}_i/(\mathbf{a}'\boldsymbol{\Sigma}\mathbf{a})^{1/2}.$$

For a linear combination $\mathbf{a}'\mathbf{x}$, the measure of spread (3.9.13) reduces to a monotonic increasing function of

$$\mathbf{a}'\mathbf{B}_o\mathbf{a}/\mathbf{a}'\mathbf{\Sigma}\mathbf{a},$$

which can be expressed as

$$\begin{aligned}\mathbf{a}'\mathbf{B}_o\mathbf{a}/\mathbf{a}'\mathbf{\Sigma}\mathbf{a} &= \frac{1}{g-1}\sum_{i=1}^{g}(\nu_i - \bar{\nu})^2 \\ &= \frac{1}{g-1}\sum_{i=1}^{g}(\nu_{(i)} - \bar{\nu})^2. \qquad (3.9.15)\end{aligned}$$

Hence, maximizing the spread is equivalent to maximizing (3.9.15). This contrasts with the function (3.9.14) to be minimized in order to minimize the overall error rate. There are thus two different functions of the ordered values of $\nu_1,\ldots,\nu_g$ to be optimized. In the special case of $g = 2$, the two goals of optimization coincide because

$$\mathbf{a}'\mathbf{B}_o\mathbf{a}/\mathbf{a}'\mathbf{\Sigma}\mathbf{a} = \tfrac{1}{2}\{\Delta(\mathbf{a})\}^2$$

and

$$eo(\Psi_E(\mathbf{a})) = \Phi\{-\tfrac{1}{2}\Delta(\mathbf{a})\},$$

where

$$\Delta(\mathbf{a}) = |\mathbf{a}'(\boldsymbol{\mu}_1 - \boldsymbol{\mu}_2)|/(\mathbf{a}'\mathbf{\Sigma}\mathbf{a})^{1/2}.$$

It follows from Section 3.3.3, that $\Delta(\mathbf{a})$ is maximized by taking

$$\mathbf{a} \propto \mathbf{\Sigma}^{-1}(\boldsymbol{\mu}_1 - \boldsymbol{\mu}_2).$$

3.9.7 Allocatory Versus Separatory Solution

For $g > 2$ groups, the problems of optimizing the overall error rate of the Bayes rule based on $\mathbf{a}'\mathbf{x}$ and a measure of spread have different solutions in general. We have seen that $\mathbf{a} = \boldsymbol{\gamma}_1$ maximizes the measure of spread for the class of measures (3.9.13). It can be seen from (3.9.15) that this choice of $\mathbf{a}$ maximizes the sum of squared distances between the projections $\nu_i = \mathbf{a}'\boldsymbol{\mu}_i$ of the group means. Hence, it tends to make the large distances as large as possible at the risk of making the small distances zero. However, with respect to the choice of $\mathbf{a}$ for the minimization of the error rate, it can be seen from (3.9.14) that the optimal value of $\mathbf{a}$ attempts to make the small distances between the means as large as possible. To illustrate this point, we use the example of Habbema and Hermans (1977), which they presented in the context of the F-statistic versus the error rate as a criterion for variable selection in discriminant analysis. Let $\nu_{(1)} = \omega_1$, $\nu_{(2)} = \omega_2$, and $\nu_{(3)} = -\omega_1$, where $0 < \omega_2 < \omega_1$.

Then from (3.9.15),

$$\mathbf{a}'\mathbf{B}_o\mathbf{a}/\mathbf{a}'\mathbf{\Sigma}\mathbf{a} = \tfrac{1}{3}(\omega_1^2 + \omega_2^2/3),$$

and from (3.9.14),

$$eo(\Psi_E(\mathbf{a})) = (2/3)\left[\Phi\{-\tfrac{1}{2}(\omega_1 + \omega_2)\} + \Phi\{\tfrac{1}{2}(\omega_2 - \omega_1)\}\right].$$

For fixed ω_1, the minimum value of $eo(\Psi_E(\mathbf{a}))$ with respect to ω_2 occurs at $\omega_2 = 0$, at which the measure of spread (3.9.13) is minimized rather than maximized.

More recently, Schervish (1984) has presented an example in which the overall error rate $eo(\Psi E(\mathbf{a}))$ is not minimized by taking $\mathbf{a} = \gamma_1$. In his example, he considered $g = 3$ groups with equal prior probabilities, having bivariate normal distributions with common covariance matrix $\mathbf{\Sigma} = \mathbf{I}_2$ and means

$$\mu_1 = (-1,-1)', \qquad \mu_2 = (1,3)', \qquad \text{and} \qquad \mu_3 = (3,1)'.$$

For these values of the parameters, the Bayes rule $r_o(\mathbf{x};\Psi_E)$ using both feature variables in $\mathbf{x}$ has an overall error rate of $eo(\Psi_E) = 0.065$. However, insight may be required into which single linear combination $\mathbf{a}'\mathbf{x}$ of the features best separates the three groups so far as the associated error rate $eo(\Psi_E(\mathbf{a}))$ is concerned. Schervish (1984) showed that the latter has a minimum value of 0.20 at

$$\mathbf{a}_o = (0.989, 0.149)'.$$

His procedure for minimizing $eo(\Psi_E(\mathbf{a}))$ in the case of $g = 3$ is to be described shortly. The error rate of the Bayes rule based on $\mathbf{a}_o'\mathbf{x}$ is much larger than the corresponding rule based on the bivariate feature vector. However, it is appreciably less than the error rate of 0.344 for the univariate Bayes rule using the first canonical variate $\gamma_1'\mathbf{x}$. Here $\gamma_1 = (1/\sqrt{2}, 1/\sqrt{2})'$ is the eigenvector of $\mathbf{B}_o$, corresponding to its largest eigenvalue. In this example,

$$\mathbf{B}_o = \begin{pmatrix} 4 & 2 \\ 2 & 4 \end{pmatrix}.$$

With the first canonical variate $\gamma_1'\mathbf{x}$, the projected means $\gamma_1'\mu_2$ and $\gamma_1'\mu_3$ are equal, demonstrating the tendency of the first canonical variate to use the projection to increase the distances between the well-separated means at the risk of decreasing the distances between the badly separated means to zero. On the other hand, the projection $\mathbf{a}_o$ that minimizes the error rate attempts to increase the small distances between the badly separated means in the original feature space.

As remarked by Schervish (1984), the difference between the error rates of the Bayes rule based on the first canonical variate $\gamma_1'\mathbf{x}$ and the optimal linear combination $\mathbf{a}_o'\mathbf{x}$ is more extreme in his example than typically found in practice. Previously, Geisser (1977) had provided a less dramatic example of how the linear rule based on $\gamma_1'\mathbf{x}$ is suboptimal in terms of the error rate criterion.

3.9.8 Computation of Best Linear Combination

We now describe the procedure developed by Schervish (1984) for the minimization of $eo(\Psi_E(\mathbf{a}))$ with respect to $\mathbf{a}$ for $g = 3$ groups. In this case, it can be assumed without loss of generality that for the linear combination $\mathbf{a}'\mathbf{x}$,

$$\mathbf{a}'\boldsymbol{\mu}_1 \leq \mathbf{a}'\boldsymbol{\mu}_2 \leq \mathbf{a}'\boldsymbol{\mu}_3, \tag{3.9.16}$$

and hence that the Bayes rule has the form

$$r_o(\mathbf{a}'\mathbf{x}; \Psi_E(\mathbf{a})) = i \qquad \text{if} \quad \mathbf{x} \in R_i \qquad (i = 1,\ldots,3),$$

where R_1, R_2, and R_3 are the intervals

$$\begin{aligned} R_1 &= (-\infty, \tfrac{1}{2}\mathbf{a}'(\boldsymbol{\mu}_1 + \boldsymbol{\mu}_2)], \\ R_2 &= (\tfrac{1}{2}\mathbf{a}'(\boldsymbol{\mu}_1 + \boldsymbol{\mu}_2), \tfrac{1}{2}\mathbf{a}'(\boldsymbol{\mu}_2 + \boldsymbol{\mu}_3)], \\ R_3 &= (\tfrac{1}{2}\mathbf{a}'(\boldsymbol{\mu}_2 + \boldsymbol{\mu}_3), \infty). \end{aligned}$$

Let $u_1 = \|\boldsymbol{\mu}_1 - \boldsymbol{\mu}_2\|$, $u_2 = \|\boldsymbol{\mu}_3 - \boldsymbol{\mu}_2\|$, and

$$\begin{aligned} u_3 &= \cos^{-1}\{(\boldsymbol{\mu}_3 - \boldsymbol{\mu}_2)'(\boldsymbol{\mu}_1 - \boldsymbol{\mu}_2)/(u_1 u_2)\}, \\ u_4 &= \cos^{-1}\{\mathbf{a}'(\boldsymbol{\mu}_3 - \boldsymbol{\mu}_2)/u_2\}. \end{aligned}$$

Then Schervish (1984) showed that the minimization of $eo(\Psi_E(\mathbf{a}))$ reduces to minimizing the function

$$k(u_4) = \tfrac{2}{3}[1 + \Phi\{\tfrac{1}{2}u_1\cos(u_3 + u_4)\} - \Phi(\tfrac{1}{2}u_2\cos u_4)]$$

with respect to u_4. This is a straightforward numerical problem because $k(u_4)$ is convex over the interval

$$\max\{0, \tfrac{1}{2}(\pi - u_3)\} \leq u_4 \leq \min\{\tfrac{1}{2}\pi, \pi - u_3\},$$

implied by the condition (3.9.16).

In a similar manner, Schervish (1984) also showed for $g = 3$ groups how to choose $\mathbf{a}$ so as to minimize the maximum of the group-specific error rates of the linear rule based on $\mathbf{a}'\mathbf{x}$. However, difficulties remain in attempting to extend his results to $g > 3$ groups.

More recently, McCulloch (1986) has proposed some methods for approximating the $\mathbf{a}_o$ that minimizes the overall error rate $eo(\Psi_E(\mathbf{a}))$ for an arbitrary number g of groups. Roughly speaking, the idea is to work with normalized linear combinations of the first few canonical variates, retaining only those that correspond to the dominant eigenvalues of $\boldsymbol{\Sigma}^{-1}\mathbf{B}_o$. For instance, if $\lambda_{o,1}$ and $\lambda_{o,2}$ are large relative to the remaining $d-2$ nonzero eigenvalues of $\boldsymbol{\Sigma}^{-1}\mathbf{B}_o$, then the problem is to minimize the error rate of the Bayes rule using $a_1v_1 + a_2v_2$, where $a_1^2 + a_2^2 = 1$. McCulloch (1986) notes that this is a simple numerical problem. On putting $a_1 = \cos\omega$ and $a_2 = \sin\omega$, it requires the minimization of the error rate (3.9.14) with respect to the one variable ω $(0 \le \omega \le 2\pi)$. For linear combinations of more than the first two canonical variates, McCulloch (1986) describes a method for obtaining an approximate solution. He showed that for the aforementioned example of Schervish (1984), it yields a linear combination for which the Bayes rule has an error rate of 0.21, which is quite close to the rate of 0.20 for the optimal univariate rule.

3.10 LINEAR PROJECTIONS OF HETEROSCEDASTIC FEATURE DATA

3.10.1 Best Linear Rule for $g = 2$ Groups

In the previous section, we have considered the choice of the linear projection $\mathbf{C}_q$ of given rank q so as to minimize the overall error rate of the linear rule $r_o(\mathbf{C}_q\mathbf{x}; \Psi_E(\mathbf{C}_q))$. The latter is the Bayes rule based on $\mathbf{C}_q\mathbf{x}$ under the homoscedastic normal model (3.3.1). In this section, we consider the choice of $\mathbf{C}_q$ to produce the best linear rule, in the sense of having the smallest overall error rate, under the heteroscedastic normal model (3.2.1). Because the group-covariance matrices are now no longer assumed to be equal, $r_o(\mathbf{C}_q\mathbf{x}; \Psi_E(\mathbf{C}_q))$ is not the best linear rule based on $\mathbf{C}_q\mathbf{x}$ for the given projection $\mathbf{C}_q$.

Historically, this type of problem was first considered by Aoyama (1950) in the univariate case, for which $\mathbf{a}$ is a scalar and so can be taken to be one. This reduces the scope of the problem to the choice of the cutoff point, which was considered also by Stoller (1954), but from a nonparametric perspective.

In the multivariate case, the development of the best linear rule based on the linear combination $\mathbf{a}'\mathbf{x}$ for $g = 2$ heteroscedastic normal groups was considered independently by Riffenburgh and Clunies-Ross (1960), Clunies-Ross and Riffenburgh (1960), Anderson and Bahadur (1962), and Jennrich (1962). It follows from their work that optimal linear rules can be formed by consideration of the following class of admissible linear procedures. For given ω_1 and ω_2, where $\omega_1\mathbf{\Sigma}_1 + \omega_2\mathbf{\Sigma}_2$ is positive definite, consider the linear rule that assigns an entity with feature vector $\mathbf{x}$ to G_1 or G_2, according as to whether $\mathbf{a}'\mathbf{x}$ is greater or less than the cutoff point c, where $\mathbf{a}$ and c satisfy

$$\mathbf{a} = (\omega_1\mathbf{\Sigma}_1 + \omega_2\mathbf{\Sigma}_2)^{-1}(\boldsymbol{\mu}_1 - \boldsymbol{\mu}_2)$$

and

$$c = \mathbf{a}'\boldsymbol{\mu}_1 - \omega_1\mathbf{a}'\mathbf{\Sigma}_1\mathbf{a} = \mathbf{a}'\boldsymbol{\mu}_2 + \omega_2\mathbf{\Sigma}_2\mathbf{a}.$$

By appropriate choices of ω_1 and ω_2, the linear rule that (a) minimizes the overall error rate or (b) minimizes one error with the level of the other error specified or (c) minimizes the maximum of the two errors can be obtained; see T. W. Anderson (1984, Section 6.10) for details. Concerning (c), the minimax solution is given by $\omega_1 = \omega_2 = \omega_o$, where ω_o is the value of ω that maximizes

$$\Delta(\omega) = \mathbf{a}'(\boldsymbol{\mu}_1 - \boldsymbol{\mu}_2)/[\tfrac{1}{2}\{(\mathbf{a}'\boldsymbol{\Sigma}_1\mathbf{a})^{1/2} + (\mathbf{a}'\boldsymbol{\Sigma}_2\mathbf{a})^{1/2}\}]$$

with respect to ω $(0 < \omega < 1)$. Banjeree and Marcus (1965) have considered the computation of bounds for ω_o. The common value of the two error rates is $\Phi\{-\frac{1}{2}\Delta(\omega_o)\}$.

For the homoscedastic case of $\boldsymbol{\Sigma}_1 = \boldsymbol{\Sigma}_2 = \boldsymbol{\Sigma}$, $\Delta(\omega)$ is identically equal to

$$\Delta = \{(\boldsymbol{\mu}_1 - \boldsymbol{\mu}_2)'\boldsymbol{\Sigma}^{-1}(\boldsymbol{\mu}_1 - \boldsymbol{\mu}_2)\}^{1/2},$$

the Mahalanobis distance between two groups with a common covariance matrix. As noted by Anderson and Bahadur (1962), it suggests the use of $\Delta(\omega_o)$ as a measure of the distance between two groups with unequal covariance matrices. It has subsequently been considered in this way by Chaddha and Marcus (1968) and by Marks and Dunn (1974) and McLachlan (1975c) for group-covariance matrices that are proportional. Also, Chernoff (1952, 1972, 1973) has considered measures, including $\Delta(\omega_o)$, of the distance between two multivariate normal distributions with unequal covariance matrices.

A sample version of the optimal linear projections, and hence sample linear rules, as defined in the previous work can be obtained by replacing the unknown group means and covariance matrices with their unbiased estimates. Other sample linear rules have been proposed in the literature. For example, in the spirit of the approach of Glick (1969), Greer (1979, 1984) has given an algorithm for constructing the linear rule $\mathbf{a}'\mathbf{x}$, where $\mathbf{a}$ and its cutoff point c are chosen to maximize its apparent error rate when applied to the training data. A linear programming approach was adopted also by Freed and Glover (1981), where $\mathbf{a}$ and c are chosen from the training data by minimization of

$$\sum_{i=1}^{2}\sum_{j=1}^{n}(-1)^i z_{ij}(\mathbf{a}'\mathbf{x}_j - c)$$

with respect to $\mathbf{a}$ and c. Additional algorithms for linear discrimination can be found in Castagliola and Dubuisson (1989).

Linear procedures besides the sample NLDR $r_o(\mathbf{x}; \hat{\boldsymbol{\Psi}}_E)$ are not widely used in practice. As discussed by Marks and Dunn (1974), the so-called best linear rule offers little improvement over $r_o(\mathbf{x}; \hat{\boldsymbol{\Psi}}_E)$ in those situations where the latter is preferable to the quadratic rule $r_o(\mathbf{x}; \hat{\boldsymbol{\Psi}}_U)$, and it is still not as good as $r_o(\mathbf{x}; \hat{\boldsymbol{\Psi}}_U)$, where the latter is superior to $r_o(\mathbf{x}; \hat{\boldsymbol{\Psi}}_E)$.

3.10.2 Best Quadratic Rule

It was seen in the previous section that without the assumption of equal group-covariance matrices, it is not a trivial task to find the best linear rule in the case of $g = 2$ groups, even though a univariate rule suffices. For $g > 2$ groups, there are immense difficulties in attempting to choose the linear projection $\mathbf{C}_q$

of rank q of the feature vector so as to give the best linear rule based on $\mathbf{C}_q\mathbf{x}$. The problem becomes more tractable if the choice of $\mathbf{C}_q$ is made not for linear rules based on $\mathbf{C}_q\mathbf{x}$, but for the quadratic Bayes rule $r_o(\mathbf{C}_q\mathbf{x};\mathbf{\Psi}_U(\mathbf{C}_q))$. This requires the minimization of the overall error rate of the latter rule, $eo(\mathbf{\Psi}_U(\mathbf{C}_q))$, with respect to $\mathbf{C}_q$. As the Bayes rule, and hence its error rate, is invariant under a nonsingular transformation of the feature vector, it follows from this and the fact that $\mathbf{C}_q\mathbf{C}_q'$ is positive definite that we can assume $\mathbf{C}_q\mathbf{C}_q' = \mathbf{I}_q$. Hence, the problem is reduced to minimizing $eo(\mathbf{\Psi}_U(\mathbf{C}_q))$ over all $q \times p$ matrices $\mathbf{C}_q$ of rank q satisfying $\mathbf{C}_q\mathbf{C}_q' = \mathbf{I}_q$, which is compact. Since $eo(\mathbf{\Psi}_U(\mathbf{C}_q))$ is a continuous function of $\mathbf{C}_q$, a solution exists.

Guseman et al. (1975) have considered this problem. In the case of $q = 1$, they gave a method of obtaining a local minimum of $eo(\mathbf{\Psi}_U(\mathbf{C}_q))$. Motivated by the work of Odell (1979), Decell, Odell, and Coberly (1981), Tubbs, Coberly, and Young (1982), and Young and Odell (1984), Young, Marco, and Odell (1987) have considered this problem. They provided an approximate solution in the following way.

Let $\mathbf{M}$ be the $p \times s$ matrix defined by

$$\mathbf{M} = [\boldsymbol{\mu}_2 - \boldsymbol{\mu}_1 \mid \boldsymbol{\mu}_3 - \boldsymbol{\mu}_1 \mid \cdots \mid \boldsymbol{\mu}_g - \boldsymbol{\mu}_1 \mid \boldsymbol{\Sigma}_2 - \boldsymbol{\Sigma}_1 \mid \boldsymbol{\Sigma}_3 - \boldsymbol{\Sigma}_2 \mid \cdots \mid \boldsymbol{\Sigma}_g - \boldsymbol{\Sigma}_1],$$

where $s = (g-1)(p+1)$, and where it is assumed that $\boldsymbol{\Sigma}_i \neq \boldsymbol{\Sigma}_1$ for at least one value of i $(i = 2,\ldots,g)$. Further, let

$$\mathbf{M} = \mathbf{H}_m\mathbf{K}_m$$

be a full-rank decomposition of $\mathbf{M}$, where $\mathbf{H}_m$ is a $p \times m$ matrix, $\mathbf{K}_m$ is a $m \times s$ matrix, and m is the rank of $\mathbf{M}$. The pseudoinverse of $\mathbf{H}_m$ is denoted by $\mathbf{H}_m^-$, satisfying

$$\mathbf{H}_m\mathbf{H}_m^-\mathbf{H}_m = \mathbf{H}_m.$$

Then Young, Marco, and Odell (1987) showed that the Bayes rule based on $\mathbf{H}_m^-\mathbf{x}$ has the same error rate as the Bayes rule based on $\mathbf{x}$. Moreover, m is the smallest value of q for which a linear projection from $\mathbb{R}^p$ down to $\mathbb{R}^q$ $(q \leq p)$ does not increase the error rate of the Bayes rule. They provided an approximation to the solution to the problem of minimizing $eu(\mathbf{\Psi}_U(\mathbf{C}_q))$ over all projections $\mathbf{C}_q$ of rank $q < m$ by first finding the $p \times s$ matrix of rank q, denoted by $\mathbf{M}_q$, that best approximates $\mathbf{M}$, using the singular-value decomposition of $\mathbf{M}$. Then factoring $\mathbf{M}_q$ into a full-rank decomposition

$$\mathbf{M}_q = \mathbf{H}_q\mathbf{K}_q, \tag{3.10.1}$$

their approximate solution to the optimal linear projection of rank q is given by $\mathbf{H}_q^-$.

Concerning the actual computation of the approximation $\mathbf{M}_q$ to $\mathbf{M}$, let $\lambda_1,\ldots,\lambda_s$ be the eigenvalues of $\mathbf{M}'\mathbf{M}$ in order of decreasing size, where $\lambda_k = 0\,(k = m+1,\ldots,s)$. Then the singular-value decomposition of $\mathbf{M}$ is given by

$$\mathbf{M} = \mathbf{U}\mathbf{\Lambda}\mathbf{V}', \tag{3.10.2}$$

where $\mathbf{U}$ is an orthogonal $p \times p$ matrix consisting of the orthonormal eigenvectors corresponding to the eigenvalues $\lambda_1, \ldots, \lambda_p$ of $\mathbf{MM}'$, $\mathbf{V}$ is an $s \times p$ matrix with columns consisting of the orthonormal eigenvectors corresponding to the eigenvalues $\lambda_1, \ldots, \lambda_p$ of $\mathbf{M}'\mathbf{M}$, and

$$\mathbf{\Lambda} = \operatorname{diag}(\lambda_1, \ldots, \lambda_p).$$

On letting

$$\mathbf{\Lambda}_k = \operatorname{diag}(\lambda_1, \ldots, \lambda_k) \qquad (k = 1, \ldots, m),$$

the right-hand side of (3.10.2) can be partitioned accordingly as

$$\mathbf{M} = (\mathbf{U}_m, \mathbf{U}_{p-m}) \begin{pmatrix} \mathbf{\Lambda}_m & \mathbf{O} \\ \mathbf{O} & \mathbf{O} \end{pmatrix} \begin{pmatrix} V_m' \\ V_{p-m}' \end{pmatrix}.$$

The $p \times s$ matrix $\mathbf{M}_q$ that minimizes $\|\mathbf{M} - \mathbf{N}_q\|$ over all $p \times s$ matrices $\mathbf{N}_q$ of rank q is given by

$$\mathbf{M}_q = \mathbf{U}_q \mathbf{\Lambda}_q \mathbf{V}_q'. \tag{3.10.3}$$

Here $\|\mathbf{M} - \mathbf{N}_q\|$ is the usual Euclidean or Frobenius norm of the matrix $\mathbf{M} - \mathbf{N}_q$, given by

$$\|\mathbf{M} - \mathbf{N}_q\| = \left\{ \sum_{i=1}^{p} \sum_{j=1}^{s} (\mathbf{M} - \mathbf{N}_q)_{ij}^2 \right\}^{1/2},$$

which is equal to

$$\left(\sum_{k=q+1}^{m} \lambda_k^2 \right)^{1/2}$$

at $\mathbf{N}_q = \mathbf{M}_q$. By equating (3.10.1) with (3.10.3), it follows that the approximate solution $\mathbf{H}_q^-$ proposed by Young, Marco, and Odell (1987) is equal to $\mathbf{U}_q'$. As noted above, the Bayes rule is invariant under any nonsingular transformation of the feature vector. Hence, $\mathbf{J}_q \mathbf{U}_q'$ is also an approximate solution, where $\mathbf{J}_q$ is an arbitrary $q \times q$ nonsingular matrix.

In the case of equal group-covariance matrices, Young, Marco, and Odell (1987) noted that their method can still be used but with $\mathbf{M}$ defined now as the $p \times (g-1)$ matrix

$$\mathbf{M} = [\boldsymbol{\mu}_2 - \boldsymbol{\mu}_1 \mid \boldsymbol{\mu}_3 - \boldsymbol{\mu}_1 \mid \cdots \mid \boldsymbol{\mu}_g - \boldsymbol{\mu}_1].$$

This definition of $\mathbf{M}$ presupposes that the data have been transformed in the first instance so that the common group-covariance matrix is equal to the identity matrix.

With this method in practical situations where the $\boldsymbol{\mu}_i$ and $\boldsymbol{\Sigma}_i$ are unknown, the latter can be replaced by their sample analogues $\bar{\mathbf{x}}_i$ and $\mathbf{S}_i$ $(i = 1, \ldots, g)$; see Tubbs, Coberly, and Young (1982). Morgera and Datta (1984) have considered the particular case where the group-conditional means are specified to be the same. The predictive version of the problem has been considered by Young,

Marco, and Odell (1986). In an earlier paper, Young, Odell, and Marco (1985) extended the normal model to a general class of density functions known as θ-generalized normal densities.

This problem of reducing the dimension of the feature vector through the choice of a suitable linear projection $\mathbf{C}_q$ has been considered on the basis of other criteria besides the Bayes error rate. Decell and Marani (1976) used Bhattacharyya's distance measure, Decell and Mayekar (1977) used the Kullback–Leibler information, and in a more recent paper, Bidasaria (1987) used Jeffreys' measure of information. Peters (1979) invoked the mean-squared error criterion to select a linear projection of the feature variables. This problem has also been considered nonparametrically (Bryant and Guseman, 1979).

CHAPTER 4

Distributional Results for Discrimination via Normal Models

4.1 INTRODUCTION

In this chapter, we are to consider available results for the distribution of discriminant functions and associated rules of allocation in the case of multivariate normal group-conditional distributions for which the parameters may or may not be known. We will consider also for these models distributional results for the conditional error rates of sample-based rules, in particular of the plug-in sample version of the Bayes rule. It will be seen that analytical results are available only in special cases such as for $g = 2$ homoscedastic groups. Even then, the distributional problems are so complex that most have been tackled using asymptotic rather than exact methods. As is discussed in Chapter 10, asymptotic expansions of the means and variances of the conditional error rates can be of use in their parametric estimation.

4.2 DISTRIBUTION OF SAMPLE NLDF (W-STATISTIC)

4.2.1 Historical Review

It was shown in Section 3.3 that under the homoscedastic normal model (3.3.1) for $g = 2$ groups, the plug-in sample version $r_o(\mathbf{x}; \hat{\boldsymbol{\Psi}}_E)$ of the Bayes rule is based on the single sample NLDF (normal-based linear discriminant function),

$$\xi(\mathbf{x}; \hat{\boldsymbol{\theta}}_E) = \{\mathbf{x} - \tfrac{1}{2}(\bar{\mathbf{x}}_1 + \bar{\mathbf{x}}_2)\}' \mathbf{S}^{-1} (\bar{\mathbf{x}}_1 - \bar{\mathbf{x}}_2). \qquad (4.2.1)$$

It is obvious from (4.2.1) that $\xi(\mathbf{x};\hat{\boldsymbol{\theta}}_E)$ is invariant under a nonsingular linear transformation. To show that the group-conditional distributions of $\xi(\mathbf{x};\hat{\boldsymbol{\theta}}_E)$ depend only on the Mahalanobis distance Δ in addition to the sample sizes, we transform the feature vector $\mathbf{x}$ to

$$\mathbf{C}_2\mathbf{C}_1\{\mathbf{x}-\tfrac{1}{2}(\boldsymbol{\mu}_1+\boldsymbol{\mu}_2)\}, \tag{4.2.2}$$

where $\mathbf{C}_1$ is such that $\mathbf{C}_1\boldsymbol{\Sigma}\mathbf{C}_1' = \mathbf{I}_p$ (the $p \times p$ identity matrix), and $\mathbf{C}_2$ is an orthogonal matrix whose first row is

$$\{\mathbf{C}_1(\boldsymbol{\mu}_1-\boldsymbol{\mu}_2)\}'/\|\mathbf{C}_1(\boldsymbol{\mu}_1-\boldsymbol{\mu}_2)\|.$$

Here again $\|\mathbf{a}\|$ is used to denote the norm $(\mathbf{a}'\mathbf{a})^{1/2}$ of a vector $\mathbf{a}$. After transformation according to (4.2.2), it can be easily seen that the feature vector is distributed $N(\boldsymbol{\mu}_{io},\mathbf{I}_p)$ in group G_i, where

$$\boldsymbol{\mu}_{io} = ((-1)^{i+1}\tfrac{1}{2}\Delta,0,\ldots,0)' \qquad (i=1,2).$$

Hence, it can be assumed without loss of generality that

$$\boldsymbol{\mu}_1 = -\boldsymbol{\mu}_2 = (\tfrac{1}{2}\Delta,0,\ldots,0)' \qquad \text{and} \qquad \boldsymbol{\Sigma} = \mathbf{I}_p. \tag{4.2.3}$$

The distribution of the sample NLDF $\xi(\mathbf{x};\hat{\boldsymbol{\theta}}_E)$ is extremely complicated and has been the focus of many investigations over the years. Surveys of this work can be found in Raudys and Pikelis (1980) and Siotani (1982). The former paper is also a valuable source of references in the considerable Russian literature on this problem. The latter paper by Siotani (1982) provides a review of allocation statistics including $\xi(\mathbf{x};\hat{\boldsymbol{\theta}}_E)$. In particular, it lists all the available higher-order terms in the asymptotic expansions of the distributions of these statistics. The few exact results that have obtained apply mainly to certain univariate cases; see Schaafsma and van Vark (1977) and the references therein.

The limiting ith group-conditional distribution of $\xi(\mathbf{x};\hat{\boldsymbol{\theta}}_E)$, as $n_1,n_2 \to \infty$, is the same as that of $\xi(\mathbf{x};\boldsymbol{\theta}_E)$, which is

$$\xi(\mathbf{x};\boldsymbol{\theta}_E) \sim N((-\tfrac{1}{2})^{i+1}\Delta^2,\Delta^2) \quad \text{in} \quad G_i \qquad (i=1,2).$$

This limiting result, which was first given by Wald (1944), follows in a straightforward manner on noting that $\hat{\boldsymbol{\theta}}_E$ converges in probability to $\boldsymbol{\theta}_E$ as $n_1,n_2 \to \infty$; see, for example, T. W. Anderson (1984, Section 6.5).

Following Wald (1944), T. W. Anderson (1951), Harter (1951), and Sitgreaves (1952) studied the distribution of $\xi(\mathbf{X};\hat{\boldsymbol{\theta}}_E)$ and associated problems. For the case of known $\boldsymbol{\Sigma}$, John (1959, 1960b) derived the exact distribution of $\xi(\mathbf{X};\hat{\boldsymbol{\theta}}_E)$ and, in a later paper (John, 1961), expressed its expectation in terms of the cumulative distribution of the ratio of two independent but noncentral chi-squared variates. An explicit justification of this last result was given later by Moran (1975).

Bowker (1961) showed that $\xi(\mathbf{x};\hat{\boldsymbol{\theta}}_E)$ can be represented as a function of two independent Wishart matrices, one of which is noncentral. This representation was used by Sitgreaves (1961) to obtain an exact formula for the distribution of $\xi(\mathbf{X};\hat{\boldsymbol{\theta}}_E)$. It was used also by Bowker and Sitgreaves (1961) to obtain an asymptotic expansion of the distribution of $\xi(\mathbf{X};\hat{\boldsymbol{\theta}}_E)$ for $n_1 = n_2$. Elfving (1961) obtained a large-sample-size approximation to the distribution function of $\xi(\mathbf{X};\hat{\boldsymbol{\theta}}_E)$ for univariate feature data, and Teichroew and Sitgreaves (1961) considered an empirical approximation. These last five papers appeared simultaneously, each being a separate chapter in a volume edited by H. Solomon.

4.2.2 Asymptotic Expansions

As indicated by Sitgreaves (1961) at the time, her exact expression for the distribution of the sample NLDF $\xi(\mathbf{X};\hat{\boldsymbol{\theta}}_E)$ was too complicated to be used numerically. Although Estes (1965) was able to reduce this formula to a form more suitable for numerical computation, asymptotic expansions have been the main theoretical tool for providing insight into the distribution of $\xi(\mathbf{X};\hat{\boldsymbol{\theta}}_E)$.

Conditional on n_1 and n_2, Okamoto (1963, 1968) derived the asymptotic expansions of the group-conditional distributions of $\xi(\mathbf{X};\hat{\boldsymbol{\theta}}_E)$, appropriately normalized for each group, up to terms of order $O(N^{-2})$, where $N = n_1 + n_2 - 2$. More precisely, the error of these expansions is of order $O(n_o^{-3})$, where $n_o = \min(n_1, n_2, N)$. These expansions have since been used extensively in studies on the unconditional error rates of the rule based on $\xi(\mathbf{x};\hat{\boldsymbol{\theta}}_E)$.

Let $F_{W1}(w;\Delta,n_1,n_2)$ denote the probability

$$\mathrm{pr}\{(W - \tfrac{1}{2}\Delta^2)/\Delta < w \mid \mathbf{X} \in G_1, n_1, n_2\},$$

where $W = \xi(\mathbf{X};\hat{\boldsymbol{\theta}}_E)$. Then as $N \to \infty$ with $n_1/n_2 \to$ a finite positive limit, we have from Okamoto (1963) that

$$F_{W1}(w;\Delta,n_1,n_2) = \Phi(w) + \phi(w)(h_1/n_1 + h_2/n_2 + h_3/N) + O(N^{-2}), \tag{4.2.4}$$

where

$$h_1 = -\tfrac{1}{2}\Delta^{-2}\{w^3 + (p-3)w - p\Delta\},$$

$$h_2 = -\tfrac{1}{2}\Delta^{-2}\{w^3 + 2\Delta w^2 + (p - 3 + \Delta^2)w + (p-2)\Delta\},$$

and

$$h_3 = -\tfrac{1}{4}\{4w^3 + 4\Delta w^2 + (6p - 6 + \Delta^2)w + 2(p-1)\Delta\}.$$

As this expansion contains powers of Δ^{-2}, it may not provide a reliable approximation for values of Δ less than one if n_1 and n_2 are not large relative to p. Okamoto (1963, 1968) also gave the second-order terms in this expansion, and Siotani and Wang (1975, 1977) added the third-order terms. A single reference for these higher-order terms is Siotani (1982).

The asymptotic expansion of $F_{W2}(w;\Delta,n_1,n_2)$ can be obtained from (4.2.4) on using the result that

$$F_{W2}(w;\Delta,n_1,n_2) = 1 - F_{W1}(-w;\Delta,n_2,n_1). \tag{4.2.5}$$

To see this last result, we let $\hat{\boldsymbol{\theta}}_E^\dagger$ denote $\hat{\boldsymbol{\theta}}_E$ with the group labels interchanged. The distribution of $\xi(\mathbf{X};\hat{\boldsymbol{\theta}}_E^\dagger)$ in G_2 is clearly the same as the distribution of $\xi(\mathbf{X};\hat{\boldsymbol{\theta}}_E)$ in G_1, but with the group labels interchanged; that is, with only n_1 and n_2 interchanged, as Δ is invariant under an interchange of group labels. Hence, we can write

$$\begin{aligned} F_{W2}(w;\Delta,n_1,n_2) &= \mathrm{pr}\{(\xi(\mathbf{X};\boldsymbol{\theta}_E)+\tfrac{1}{2}\Delta^2)/\Delta < w \mid \mathbf{X}\in G_2,n_1,n_2\} \\ &= \mathrm{pr}\{-(\xi(\mathbf{X};\boldsymbol{\theta}_E^\dagger)-\tfrac{1}{2}\Delta^2)/\Delta < w \mid \mathbf{X}\in G_2,n_1,n_2\} \\ &= \mathrm{pr}\{(\xi(\mathbf{X};\boldsymbol{\theta}_E)-\tfrac{1}{2}\Delta^2)/\Delta > -w \mid \mathbf{X}\in G_1,n_2,n_1\} \\ &= 1-F_{W1}(-w;\Delta,n_2,n_1). \end{aligned}$$

These results, which are conditional on n_1 and n_2, apply directly to the case of separate sampling of the training data **t**. They can be easily modified to apply to mixture sampling where n_i has a binomial distribution with parameters n and π_i $(i=1,2)$; see, for example, Leung (1988a).

Kocherlakota, Kocherlakota, and Balakrishnan (1987) have provided a unified development of the asymptotic expansions available for the distributions of the error rates of the sample rule based on $\xi(\mathbf{x};\hat{\boldsymbol{\theta}}_E)$ in univariate nonnormal situations. The work in the papers referenced therein are to be reported in Section 5.6 in the context of the robustness of the sample NLDF $\xi(\mathbf{x};\hat{\boldsymbol{\theta}}_E)$.

4.2.3 Derivation of Asymptotic Expansions

In this section, we present an outline of Okamoto's (1963) derivation of the second-order expansion of the distribution function $F_{W1}(w;\Delta,n_1,n_2)$ of the normalized sample NLDF

$$\{\xi(\mathbf{X};\hat{\boldsymbol{\theta}}_E)-\tfrac{1}{2}\Delta^2\}/\Delta, \tag{4.2.6}$$

where $\mathbf{X}$ comes from G_1. It is an important theoretical result in discriminant analysis. Moreover, the differential operator used in its derivation has since been applied extensively in large-sample studies of the error rates of the normal-based linear discriminant rule (NLDR) based on $\xi(\mathbf{x};\hat{\boldsymbol{\theta}}_E)$.

Corresponding to the form (3.3.4) for $\xi(\mathbf{x};\boldsymbol{\theta}_E)$, we express $\xi(\mathbf{x};\hat{\boldsymbol{\theta}}_E)$ as

$$\xi(\mathbf{x};\hat{\boldsymbol{\theta}}_E)=\hat{\beta}_{0E}^o+\hat{\boldsymbol{\beta}}_E'\mathbf{x}, \tag{4.2.7}$$

where

$$\hat{\beta}_{0E}^o=-\tfrac{1}{2}(\bar{\mathbf{x}}_1+\bar{\mathbf{x}}_2)'\mathbf{S}^{-1}(\bar{\mathbf{x}}_1-\bar{\mathbf{x}}_2), \tag{4.2.8}$$

and

$$\hat{\boldsymbol{\beta}}_E=\mathbf{S}^{-1}(\bar{\mathbf{x}}_1-\bar{\mathbf{x}}_2). \tag{4.2.9}$$

We also need the following notation for the differential operator concerning the elements of $\bar{\mathbf{x}}_1$, $\bar{\mathbf{x}}_2$, and $\mathbf{S}$. Let

$$\partial_{i,j}=\partial/\partial(\bar{\mathbf{x}}_i)_j \qquad (i=1,2;\ j=1,\ldots,p)$$

and

$$\partial_{ij} = \partial_{ji}$$
$$= \tfrac{1}{2}(1+\delta_{ij})\partial/\partial(\mathbf{S})_{ij} \qquad (i \leq j = 1,\ldots,p),$$

where δ_{ij} is the Kronecker delta. We let ∂_i be the vector

$$\partial_i = (\partial_{i,1},\ldots,\partial_{i,p})'$$

and ∂ the matrix with typical element ∂_{ij}.

As demonstrated in Section 4.2.2, it can be assumed without loss of generality that

$$\boldsymbol{\mu}_1 = \boldsymbol{\mu}_{1o}, \qquad \boldsymbol{\mu}_2 = \boldsymbol{\mu}_{2o}, \qquad \boldsymbol{\Sigma} = \mathbf{I}_p, \tag{4.2.10}$$

where

$$\boldsymbol{\mu}_{1o} = -\boldsymbol{\mu}_{2o} = (\tfrac{1}{2}\Delta,0,\ldots,0)'.$$

Actually, Okamoto (1963) adopted a slightly different canonical form with $\boldsymbol{\mu}_{1o} = \mathbf{0}$ and

$$\boldsymbol{\mu}_{2o} = (\Delta,0,\ldots,0)'.$$

Okamoto (1963) expanded $F_{w1}(w;\Delta,n_1,n_2)$ by first expanding the characteristic function of

$$\{\xi(\mathbf{X};\hat{\boldsymbol{\theta}}_E) - \tfrac{1}{2}\Delta^2\}/\Delta, \tag{4.2.11}$$

where $\mathbf{X}$ belongs to G_1, and then inverting the resulting expansion. This derivation was effected by a Taylor series expansion about the point

$$(\bar{\mathbf{x}}_1 = \boldsymbol{\mu}_{1o},\ \bar{\mathbf{x}}_2 = \boldsymbol{\mu}_{2o},\ \mathbf{S} = \mathbf{I}_p). \tag{4.2.12}$$

On writing the sample NLDF $\xi(\mathbf{x};\hat{\boldsymbol{\theta}}_E)$ as W, the characteristic function $\psi_{W1}(t)$ of its normalized form (4.2.11) is defined by

$$\psi_{W1}(t) = E\{it\Delta^{-1}(W - \tfrac{1}{2}\Delta^2) \mid \mathbf{X} \in G_1\}, \tag{4.2.13}$$

where i is the complex operator. The expectation in (4.2.13) over the joint distribution of $\mathbf{X}$, $\overline{\mathbf{X}}_1$, $\overline{\mathbf{X}}_2$, and $\mathbf{S}$ can be effected by first conditioning on the latter three sample statistics to give

$$\psi_{W1}(t) = E\{\psi_{W1}(t;\overline{\mathbf{X}}_1,\overline{\mathbf{X}}_2,\mathbf{S})\}, \tag{4.2.14}$$

where

$$\psi_{W1}(t;\bar{\mathbf{x}}_1,\bar{\mathbf{x}}_2,\mathbf{S}) = E\{it\Delta^{-1}(W - \tfrac{1}{2}\Delta^2) \mid \mathbf{X} \in G_1,\bar{\mathbf{x}}_1,\bar{\mathbf{x}}_2,\mathbf{S}\}. \tag{4.2.15}$$

It can be seen from (4.2.7) that, conditional on $\bar{\mathbf{x}}_1$, $\bar{\mathbf{x}}_2$, and $\mathbf{S}$, the statistic $(W - \frac{1}{2}\Delta^2)/\Delta$ within group G_1 has a univariate normal distribution with mean

$$\Delta^{-1}(\hat{\beta}^o_{0E} + \hat{\boldsymbol{\beta}}'_E\boldsymbol{\mu}_{1o} - \tfrac{1}{2}\Delta^2)$$

and variance

$$\Delta^{-2}\hat{\boldsymbol{\beta}}'_E\hat{\boldsymbol{\beta}}_E$$

under the canonical form (4.2.10). Hence, (4.2.15) is equal to

$$\psi_{W1}(t;\bar{\mathbf{x}}_1,\bar{\mathbf{x}}_2,\mathbf{S}) = \exp\{it\Delta^{-1}(\hat{\beta}_{0E}^o + \hat{\boldsymbol{\beta}}_E'\boldsymbol{\mu}_{1o} - \tfrac{1}{2}\Delta^2) - \tfrac{1}{2}t^2\Delta^{-2}\hat{\boldsymbol{\beta}}_E'\hat{\boldsymbol{\beta}}_E\}.$$

On using the fact that the function $\psi_{W1}(t;\bar{\mathbf{x}}_1,\bar{\mathbf{x}}_2,\mathbf{S})$ is analytic about the point (4.2.12), Okamoto (1963) expressed it in the form

$$\psi_{W1}(t;\bar{\mathbf{x}}_1,\bar{\mathbf{x}}_2,\mathbf{S}) = \{\Theta(\bar{\mathbf{x}}_1,\bar{\mathbf{x}}_2,\mathbf{S})\psi_{W1}(t;\bar{\mathbf{x}}_1,\bar{\mathbf{x}}_2,\mathbf{S})\}|_o, \tag{4.2.16}$$

where $\Theta(\bar{\mathbf{x}}_1,\bar{\mathbf{x}}_2,\mathbf{S})$ is the differential operator defined formally by

$$\Theta(\bar{\mathbf{x}}_1,\bar{\mathbf{x}}_2,\mathbf{S}) = \exp[(\bar{\mathbf{x}}_1 - \boldsymbol{\mu}_{1o})'\partial_1 + (\bar{\mathbf{x}}_2 - \boldsymbol{\mu}_{2o})'\partial_2 + \mathrm{tr}\{(\mathbf{S} - \mathbf{I}_p)\partial\}].$$

In (4.2.16), the symbol $|_o$ implies evaluation of the differential operations at the point (4.2.12). After noting that the operations of expectation and differentiation can be interchanged, Okamoto (1963) proceeded to show that the expectation of (4.2.14) can be expressed as

$$\begin{aligned}\psi_{W1}(t) &= E[\{\Theta(\bar{\mathbf{X}}_1,\bar{\mathbf{X}}_2,\mathbf{S})\psi_{W1}(t;\bar{\mathbf{x}}_1,\bar{\mathbf{x}}_2,\mathbf{S})\}|_o] \\ &= \{\Theta\psi_{W1}(t;\bar{\mathbf{x}}_1,\bar{\mathbf{x}}_2,\mathbf{S})\}|_o,\end{aligned} \tag{4.2.17}$$

where

$$\Theta = E\{\Theta(\bar{\mathbf{X}}_1,\bar{\mathbf{X}}_2,\mathbf{S})\}.$$

As $\bar{\mathbf{X}}_1$, $\bar{\mathbf{X}}_2$, and $\mathbf{S}$ are independently distributed, we can write Θ as

$$\Theta = m_1(\partial_1)m_2(\partial_2)m_3(\partial), \tag{4.2.18}$$

where

$$m_i(\partial_i) = E[\exp\{(\bar{\mathbf{X}}_i - \boldsymbol{\mu}_{io})'\partial_i\}]$$

is the moment-generating function of $\bar{\mathbf{X}}_i - \boldsymbol{\mu}_{io}$ with ∂_i as the vector of dummy variables ($i = 1,2$), and

$$m_3(\partial) = E[\exp\{\mathrm{tr}(\mathbf{S} - \mathbf{I}_p)\partial\}]$$

is the moment-generating function of $\mathbf{S} - \mathbf{I}_p$ with ∂ as the matrix of dummy variables. These moment-generating functions are well-known, because

$$\bar{\mathbf{X}}_i - \boldsymbol{\mu}_{io} \sim N(\mathbf{O}, n_i^{-1}\mathbf{I}_p) \qquad (i = 1,2)$$

and

$$N\mathbf{S} \sim W(N,\mathbf{I}_p),$$

where $W(N,\mathbf{I}_p)$ denotes a Wishart distribution with N degrees of freedom and expectation matrix $N\mathbf{I}_p$. They are given by

$$m_i(\partial_i) = \exp(\tfrac{1}{2}n_i^{-1}\partial_i'\partial_i) \qquad (i = 1,2)$$

and

$$m_3(\partial) = \exp\{-\mathrm{tr}(\partial) - \tfrac{1}{2}N\log|\mathbf{I}_p - 2N^{-1}\partial|\}.$$

Substitution of these expressions into (4.2.18) yields

$$\Theta = \exp\{\tfrac{1}{2}n_1^{-1}\partial_1'\partial_1 + \tfrac{1}{2}n_2^{-1}\partial_2'\partial_2 - \operatorname{tr}(\partial) - \tfrac{1}{2}N\log|\mathbf{I}_p - 2N^{-1}\partial|\}. \tag{4.2.19}$$

On using the expansion

$$-\log|\mathbf{I} - \mathbf{C}| = \operatorname{tr}(\mathbf{C}) + \tfrac{1}{2}\operatorname{tr}(\mathbf{C}^2) + \tfrac{1}{3}\operatorname{tr}(\mathbf{C}^3) + \cdots$$

with $\mathbf{C} = (2/N)\partial$ in (4.2.19), Okamoto (1963) showed that Θ can be expanded as

$$\begin{aligned}\Theta &= \exp\left\{\frac{1}{2}n_1^{-1}\partial_1'\partial_1 + \frac{1}{2}n_2^{-1}\partial_2'\partial_2 + N^{-1}\operatorname{tr}(\partial^2) + (4/3)N^{-2}\operatorname{tr}(\partial^3) + \cdots\right\}\\
&= 1 + \sum\left[\frac{1}{2}n_1^{-1}\partial_{1,j}^2 + \frac{1}{2}n_2^{-1}\partial_{2,j}^2 + N^{-1}\partial_{ij}^2 + \frac{1}{8}n_1^{-2}\partial_{1,i}^2\partial_{1,j}^2 + \frac{1}{8}n_2^{-2}\partial_{2,i}^2\partial_{2,j}^2\right.\\
&\quad + \frac{1}{4}n_1^{-1}n_2^{-1}\partial_{1,i}^2\partial_{2,j}^2 + \frac{1}{2}n_1^{-1}N^{-1}\partial_{1,i}^2\partial_{jk}^2 + \frac{1}{2}n_2^{-1}N^{-1}\partial_{2,i}^2\partial_{jk}^2\\
&\quad \left. + N^{-2}\left\{\frac{1}{2}\partial_{ij}^2\partial_{kl}^2 + \frac{4}{3}\partial_{ij}\partial_{jk}\partial_{ki}\right\}\right] + O(N^{-3}).\end{aligned} \tag{4.2.20}$$

The second-order expansion of $\psi_{W1}(t)$, and hence of $F_{W1}(w;\Delta,n_1,n_2)$ after inversion, is obtained on evaluating (4.2.17) up to terms of the second order with Θ given by (4.2.20).

4.3 MOMENTS OF CONDITIONAL ERROR RATES OF SAMPLE NLDR

4.3.1 Unconditional Error Rates

Under separate sampling of the training data, we let $eu_i(\Delta,n_1,n_2)$ denote the unconditional error rate specific to group G_i $(i = 1,2)$ for the normal-based linear discriminant rule (NLDR) using the sample NLDF $\xi(\mathbf{x};\hat{\boldsymbol{\theta}}_E)$ with cut-off point k. For $k = \log(\pi_2/\pi_1)$, this rule corresponds to the plug-in sample version $r_o(\mathbf{x};\hat{\boldsymbol{\Psi}}_E)$ of the Bayes rule. Now

$$\begin{aligned}eu_1(\Delta,n_1,n_2) &= \operatorname{pr}\{\xi(\mathbf{X};\hat{\boldsymbol{\theta}}_E) < k \mid \mathbf{X}\in G_1,n_1,n_2\}\\
&= F_{W1}(w_o;\Delta,n_1,n_2),\end{aligned} \tag{4.3.1}$$

where

$$w_o = (k - \tfrac{1}{2}\Delta^2)/\Delta.$$

For a zero cutoff point k corresponding to equal group-prior probabilities, it follows from (4.2.4) on substituting $w_o = -\frac{1}{2}\Delta$ in (4.3.1) that

$$eu_1(\Delta, n_1, n_2) = \Phi(-\tfrac{1}{2}\Delta) + \{\phi(\tfrac{1}{2}\Delta)/16\}[\{\Delta + 12(p-1)\Delta^{-1}\}/n_1 + \{\Delta - 4(p-1)\Delta^{-1}\}/n_2 + 4(p-1)\Delta/N] + O(N^{-2}). \tag{4.3.2}$$

The corresponding error rate with respect to the second group $eu_2(\Delta, n_1, n_2)$ is given by interchanging n_1 and n_2 in (4.3.2). The leading term of order $O(1)$ in both these expansions is $\Phi(-\frac{1}{2}\Delta)$, which is the common value of the optimal error rate for each group in the case where the prior probabilities are specified to be equal. Concerning exact expressions for the unconditional error rates of the sample NLDR in special situations, in addition to the previously referenced work of John (1961) and Moran (1975) in the case of known Σ, there is the more recent paper of Streit (1979), who considered the case of unknown Σ but known group means μ_1 and μ_2. Srivastava (1973b) has evaluated the unconditional error rates under various states of knowledge of μ_1, μ_2, and Σ. Sayre (1980) has derived exactly the density of the overall conditional error rate in the case of univariate feature data. Fatti (1983) has considered the distribution theory for the random effects model where the group-conditional means are taken to be the realizations of a multivariate normal random vector.

In a recent study, Wyman, Young, and Turner (1990) drew attention to the existence in the Russian literature of three simply expressed asymptotic expansions for $eu_i(\Delta, n_1, n_2)$, by Deev (1972), Raudys (1972), and Kharin (1984). Wyman et al. (1990) compared the accuracy of these three approximations with that of widely available expansions, including Okamoto's (1963). It was found for the combinations of the parameters considered in the study that the approximation of Raudys (1972) has overall the best accuracy. The overall accuracy of the approximation of Deev (1972) was found to be very similar to that of Raudys (1972).

For a zero cutoff point, the asymptotic expansion of $eu_i(\Delta, n_1, n_2)$ by Raudys (1972), as reported in Wyman et al. (1990), is $\Phi(-\frac{1}{2}\Delta^2/c)$, where

$$c = \{(n\Delta^2 + 4p)/(n-p)\}^{1/2}.$$

4.3.2 Variances of Conditional Error Rates

As can be seen from their definition (1.10.2) in a general context, the conditional error rates of an allocation rule are themselves random variables, being functions of the realized training data $\mathbf{t}$. They also depend on the unknown group-conditional distributions and so must be estimated in practice. Their estimation is considered in Chapter 10.

In this section, we are concerned with the sampling distribution of the conditional error rates taken over the training data from which the rule has been formed. It has been seen in the previous work, that even for $g = 2$ groups under the homoscedastic normal model (3.3.1), the distribution of the sample NLDF $\xi(\mathbf{x};\hat{\boldsymbol{\theta}}_E)$ is quite complicated. The distributions of the conditional error rates of the rule using the sample NLDF $\xi(\mathbf{x};\hat{\boldsymbol{\theta}}_E)$ are even more difficult to consider analytically.

Under the homoscedastic normal model (3.3.1) for $g = 2$ groups, we let $ec_i(k,\Delta;\hat{\boldsymbol{\theta}}_E)$ denote the probability conditional on $\mathbf{t}$, and hence on $\hat{\boldsymbol{\theta}}_E$, that a randomly chosen entity from group G_i $(i = 1,2)$ is misallocated by the NLDR using the sample NLDF $\xi(\mathbf{x};\hat{\boldsymbol{\theta}}_E)$ with a cutoff point k. That is,

$$ec_{12}(k,\Delta;\hat{\boldsymbol{\theta}}_E) = \mathrm{pr}\{\xi(\mathbf{X};\hat{\boldsymbol{\theta}}_E) < k \mid \mathbf{X} \in G_1, \hat{\boldsymbol{\theta}}_E\} \tag{4.3.3}$$

and

$$ec_{21}(k,\Delta;\hat{\boldsymbol{\theta}}_E) = \mathrm{pr}\{\xi(\mathbf{X};\hat{\boldsymbol{\theta}}_E) > k \mid \mathbf{X} \in G_2, \hat{\boldsymbol{\theta}}_E\}. \tag{4.3.4}$$

Closed expressions are available for the conditional errors (4.3.3) and (4.3.4). It can be seen from (4.2.7) that the distribution of $\xi(\mathbf{x};\hat{\boldsymbol{\theta}}_E)$, conditional on $\hat{\boldsymbol{\theta}}_E$, is univariate normal within group G_i, with mean

$$\hat{\beta}_{0E}^{o} + \hat{\boldsymbol{\beta}}_E'\boldsymbol{\mu}_i \qquad (i = 1,2),$$

and common variance $\hat{\boldsymbol{\beta}}_E'\boldsymbol{\Sigma}\hat{\boldsymbol{\beta}}_E$. Thus,

$$ec_i(k,\Delta;\hat{\boldsymbol{\theta}}_E) = \Phi\{(-1)^{i+1}(k - \hat{\beta}_{0E}^{o} - \hat{\boldsymbol{\beta}}_E'\boldsymbol{\mu}_i)/(\hat{\boldsymbol{\beta}}_E'\boldsymbol{\Sigma}\hat{\boldsymbol{\beta}}_E)^{1/2}\} \qquad (i = 1,2). \tag{4.3.5}$$

It can be seen from (4.3.5) that these conditional errors are functions of the group-sample means $\bar{\mathbf{x}}_1$ and $\bar{\mathbf{x}}_2$ and the (unbiased) pooled sample covariance matrix $\mathbf{S}$. In order to study the distributions of the conditional error rates, McLachlan (1973a, 1974a) considered for the canonical form (4.2.12) Taylor series expansions of relevant functions of them about the point

$$(\bar{\mathbf{x}}_1 = \boldsymbol{\mu}_{1o},\ \bar{\mathbf{x}}_2 = \boldsymbol{\mu}_{2o},\ \mathbf{S} = \mathbf{I}_p), \tag{4.3.6}$$

before averaging the retained terms over the joint sampling distribution of $\bar{\mathbf{X}}_1$, $\bar{\mathbf{X}}_2$, and $\mathbf{S}$. The large-sample approximations so obtained are based on the following result, which was established by proceeding along the lines as in Cramér (1946, Chapter 27) on the expansion of the expectation of a function of sample moments.

Let $H(\bar{\mathbf{x}}_1,\bar{\mathbf{x}}_2,\mathbf{S})$ be a bounded function of $\bar{\mathbf{x}}_1$, $\bar{\mathbf{x}}_2$, and $\mathbf{S}$ that satisfies certain regularity conditions in a neighborhood of the point (4.3.6). For the present distributional problems, McLachlan (1974a) provided an adequate set of conditions, but it can be considerably generalized if so desired. Then if n_2/n_1 tends to a positive limit as $N = n_1 + n_2 - 2$ tends to infinity, the expectation of $H(\bar{\mathbf{x}}_1,\bar{\mathbf{x}}_2,\mathbf{S})$ can be expanded as

$$E\{H(\bar{\mathbf{x}}_1,\bar{\mathbf{x}}_2,\mathbf{S})\} = \{\Theta H(\bar{\mathbf{x}}_1,\bar{\mathbf{x}}_2,\mathbf{S})\}|_o + O(N^{-3}), \tag{4.3.7}$$

where Θ is the differential operator (4.2.20) obtained originally by Okamoto (1963) in his expansion of the distribution of the sample NLDF $\xi(\mathbf{x};\hat{\boldsymbol{\theta}}_E)$.

As the results that follow are for a zero cutoff point ($k = 0$), we abbreviate $ec_i(\Delta, 0; \hat{\boldsymbol{\theta}}_E)$ to $ec_i(\Delta; \hat{\boldsymbol{\theta}}_E)$ for brevity of notation. We let $eu_i(\Delta; n_1, n_2)$ denote the unconditional or expected error rates,

$$eu_i(\Delta; n_1, n_2) = E\{ec_i(\Delta; \hat{\boldsymbol{\theta}}_E) \mid n_1, n_2\} \qquad (i = 1, 2).$$

Concerning expansions of $eu_i(\Delta; n_1, n_2)$, we let

$$eu_i(\Delta; n_1, n_2) = e_i^{(h)} + O(N^{-(1+h)}) \qquad (i = 1, 2),$$

where $e_i^{(h)}$ denotes the expansion of $eu_i(\Delta; n_1, n_2)$ up to terms of order $O(N^{-h})$ ($h = 0, 1, 2$), as introduced in the previous section. For the overall error rate,

$$eu(\Delta; n_1, n_2) = \frac{1}{2}\sum_{i=1}^{2} eu_i(\Delta; n_1, n_2),$$

we let

$$eu(\Delta; n_1, n_2) = e^{(h)} + O(N^{-(1+h)}),$$

where

$$e^{(h)} = \tfrac{1}{2}(e_1^{(h)} + e_2^{(h)})$$

for $h = 0, 1, 2$.

As

$$eu_i(\Delta; n_1, n_2) = E\{ec_i(\Delta; \hat{\boldsymbol{\theta}}_E) \mid n_1, n_2\},$$

we can obtain $e_i^{(2)}$ by taking $H(\bar{\mathbf{x}}_1, \bar{\mathbf{x}}_2, \mathbf{S})$ to be $ec_i(\Delta; \hat{\boldsymbol{\theta}}_E)$ in (4.3.7). As the expansions of $e_1^{(2)}$ and $e_2^{(2)}$ had been derived already by Okamoto (1963), McLachlan (1974a) was concerned with the application of (4.3.7) to obtain approximations to the distributions of the conditional error rates. Previously, McLachlan (1972b) had shown that the variance of $ec_1(\Delta; \hat{\boldsymbol{\theta}}_E)$ can be expanded as

$$\operatorname{var}\{ec_1(\Delta; \hat{\boldsymbol{\theta}}_E)\} = v_1^{(2)} + O(N^{-3}),$$

where

$$\begin{aligned} v_1^{(2)} = \{\tfrac{1}{2}\phi(\tfrac{1}{2}\Delta)\}^2 \big[& 1/n_1 + 1/n_2 + \{3\Delta^2 + 8(3p-4) - (p-1)(16/\Delta^2)\}/(32n_1^2) \\ & + \{3\Delta^2 - 8p - (p-1)(16/\Delta^2)\}/(32n_2^2) \\ & + \{3\Delta^2 + 8(p-2) + (p-1)(48/\Delta^2)\}/(16n_1n_2) \\ & + (p-1)(\Delta^2 + 12)/(4n_1N) \\ & + (p-1)(\Delta^2 - 4)/(4n_2N) + (p-1)\Delta^2/(2N^2)\big]. \end{aligned}$$

This result can be obtained by evaluating

$$(\Theta H^2)|_o - \{(\Theta H)|_o\}^2 \tag{4.3.8}$$

up to terms of order $O(N^{-2})$ for $H(\bar{\mathbf{x}}_1,\bar{\mathbf{x}}_2,\mathbf{S}) = ec_1(\Delta;\hat{\boldsymbol{\theta}}_E)$. Similarly,

$$\text{var}\{ec_2(\Delta;\hat{\boldsymbol{\theta}}_E)\} = v_2^{(2)} + O(N^{-3}),$$

where $v_2^{(2)}$ is obtained by interchanging n_1 and n_2 in $v_1^{(2)}$. In the evaluation of (4.3.8) for $H(\bar{\mathbf{x}}_1,\bar{\mathbf{x}}_2,\mathbf{S}) = ec(\Delta;\hat{\boldsymbol{\theta}}_E)$ to obtain the second-order expansion of the variance of the overall conditional error rate, all the first-order partial derivatives are zero. Hence, the first-order expansion $v^{(1)}$ is zero. The second-order expansion is given by

$$\text{var}\{ec(\Delta;\hat{\boldsymbol{\theta}}_E)\} = v^{(2)} + O(N^{-3}),$$

where

$$v^{(2)} = [\{\phi(\tfrac{1}{2}\Delta)\}^2/128][\{\Delta^2 + (p-1)(16/\Delta^2)\}(1/n_1 + 1/n_2)^2$$
$$+ (p-1)(16/N)\{1/n_1 + 1/n_2 + \Delta^2/N\}].$$

The corresponding expansions of the variances of the overall and group-specific conditional error rates for $\xi(\mathbf{x};\hat{\boldsymbol{\theta}}_E)$ applied with an arbitrary cutoff point have been derived by Sayre (1980); see also Siotani et al. (1985, page 411).

4.3.3 Means and Variances of Conditional Error Rates: $g > 2$ Groups

Up to this point, the results presented in this chapter have been limited to the case of $g = 2$ groups, primarily because of the formidable nature of the distributional problems involved with discrimination. The case of $g > 2$ groups has been addressed by Schervish (1981a), who considered the conditional error rates of the plug-in sample version $r_o(\mathbf{x};\boldsymbol{\pi},\hat{\boldsymbol{\theta}}_E)$ of the Bayes rule for known but arbitrary prior probabilities under the homoscedastic normal model (3.3.1) for the group-conditional distributions. In this situation, it can be assumed without loss of generality that the common group-covariance matrix $\boldsymbol{\Sigma}$ is the identity matrix $\mathbf{I}_p$. Further, as the groups can be relabeled, there is no loss of generality in taking $j = g$ in the consideration of the conditional error rate $ec_{ij}(\boldsymbol{\Psi}_E;\hat{\boldsymbol{\theta}}_E)$ that a randomly chosen entity from G_i is assigned to G_j. Let $\mathbf{b}_{ig}$ be the $(g-1)$-dimensional vector with jth element

$$(\mathbf{b}_{ig})_j = \{\boldsymbol{\mu}_i - \tfrac{1}{2}(\bar{\mathbf{x}}_j + \bar{\mathbf{x}}_g)\}'\mathbf{S}^{-1}(\bar{\mathbf{x}}_j - \bar{\mathbf{x}}_g) + k_{ijg}$$

for $j = 1,\ldots,g-1$, where

$$k_{ijg} = \tfrac{1}{2}(\boldsymbol{\mu}_j - \boldsymbol{\mu}_g)'(\boldsymbol{\mu}_j - \boldsymbol{\mu}_g) - (\boldsymbol{\mu}_i - \boldsymbol{\mu}_g)'(\boldsymbol{\mu}_j - \boldsymbol{\mu}_g).$$

Also, let $\mathbf{C}_g$ be the $(g-1)\times(g-1)$ matrix whose (j,k)th element is

$$(\mathbf{C}_g)_{jk} = (\bar{\mathbf{x}}_j - \bar{\mathbf{x}}_g)'\mathbf{S}^{-1}\mathbf{S}^{-1}(\bar{\mathbf{x}}_k - \bar{\mathbf{x}}_g)$$

for $j,k = 1,\ldots,g-1$. It is assumed that the group means $\boldsymbol{\mu}_1,\ldots,\boldsymbol{\mu}_g$ lie in no space of dimension $g-2$ or less so that $\mathbf{C}_g$ is nonsingular with probability one.

Under the assumptions above, Schervish (1981a) noted that $ec_{ig}(\Psi_E;\hat{\boldsymbol{\theta}}_E)$ can be expressed as

$$ec_{ig}(\Psi_E;\hat{\boldsymbol{\theta}}_E) = (2\pi)^{-(g-1)/2}|\mathbf{C}_g|^{-1/2}\int_{R_{ig}} \exp\left\{-\frac{1}{2}(\mathbf{v}-\mathbf{b}_{ig})'\mathbf{C}_g^{-1}(\mathbf{v}-\mathbf{b}_{ig})\right\}d\mathbf{v}, \tag{4.3.9}$$

where

$$R_{ig} = \{\mathbf{v} \in \mathrm{R}^{g-1} : (\mathbf{v})_j \leq k_{ijg} - \log(\pi_j/\pi_g),\ j = 1,\ldots,g-1\}$$

for $i = 1,\ldots,g-1$. As demonstrated by Schervish (1981a), this representation (4.3.9) of $ec_{ig}(\Psi_E;\hat{\boldsymbol{\theta}}_E)$ is a convenient starting point for the expansions of its mean and variance. Using (4.3.9), he proceeded to give the first-order expansions of the mean and variance of $ec_{ig}(\Psi_E;\hat{\boldsymbol{\theta}}_E)$ for separate sampling of the training data from an arbitrary number g of groups. In the special case of $g = 2$ groups, Schervish (1981a) noted that his expansion of the mean of $ec_{12}(\Psi_E;\hat{\boldsymbol{\theta}}_E)$ reduces to the expansion available from the results of Okamoto (1963). Likewise, his first-order expansion of the variance of $ec_{12}(\Psi_E;\hat{\boldsymbol{\theta}}_E)$ for $g = 2$ groups with equal prior probabilities reduces to that provided originally by McLachlan (1972b).

4.4 DISTRIBUTIONS OF CONDITIONAL ERROR RATES OF SAMPLE NLDR

4.4.1 Large-Sample Approximations

We consider here some large-sample approximations to the distributions of the conditional error rates $ec_i(\Delta;\hat{\boldsymbol{\theta}}_E)$ of the NLDR using the sample NLDF $\xi(\mathbf{x};\hat{\boldsymbol{\theta}}_E)$ with a zero cutoff point. Initial work on this problem has included the investigations of John (1961, 1964, 1973).

Let $\psi_j(t)$ denote the characteristic function of the conditional error rate $ec_j(\Delta;\hat{\boldsymbol{\theta}}_E)$ specific to the jth group $G_j\,(j = 1,2)$, and $\psi(t)$ the characteristic function of the overall conditional error $ec(\Delta;\hat{\boldsymbol{\theta}}_E)$. Then on taking

$$H(\bar{\mathbf{x}}_1,\bar{\mathbf{x}}_2,\mathbf{S}) = \exp\{itec(\Delta;\hat{\boldsymbol{\theta}}_E)\}$$

in (4.3.7), where i is the complex operator, McLachlan (1974a) showed that for separate sampling of the training data,

$$\psi(t) = \exp\{ite^{(2)} - \tfrac{1}{2}t^2v^{(2)}\} + O(N^{-3}). \tag{4.4.1}$$

Similarly, he showed for the group-specific conditional error rates that

$$\begin{aligned}\psi_j(t) &= \exp\{ite_j^{(1)} - \tfrac{1}{2}t^2v_j^{(1)}\} + O(N^{-2})\\ &= \exp\{ite_j^{(2)} - \tfrac{1}{2}t^2v_j^{(2)} + (it)^3m\} + O(N^{-3}),\end{aligned} \tag{4.4.2}$$

for $j = 1, 2$, where

$$m = \{\phi(\tfrac{1}{2}\Delta)\}^3(\Delta/64)(n_1^{-1} + n_2^{-1})^2.$$

The second-order expansions $e^{(2)}$ and $v^{(2)}$ for the mean and variance, respectively, of $ec(\Delta, \hat{\boldsymbol{\theta}}_E)$ and those for the group-specific conditional error rates $ec_j(\Delta; \hat{\boldsymbol{\theta}}_E)$ have been defined in the previous two sections. The expansion corresponding to (4.4.2) for a nonzero cutoff point k has been given by Sayre (1980); see also Siotani et al. (1985, page 411).

The expansion (4.4.1) implies that up to terms of order $O(N^{-2})$, the overall conditional error rate $ec(\Delta; \hat{\boldsymbol{\theta}}_E)$ has a normal distribution with mean $e^{(2)}$ and variance $v^{(2)}$. However, it can be seen from (4.4.2) that the group-specific conditional error rates have a normal distribution only up to terms of order $O(N^{-1})$, and not $O(N^{-2})$, since the third moment of each about its mean is $6m + O(N^{-3})$.

4.4.2 Asymptotic Distributions of Group-Specific Conditional Error Rates

Concerning the asymptotic distributions of the group-specific conditional error rates $ec_i(\Delta; \hat{\boldsymbol{\theta}}_E)$ of the sample NLDR, it can be shown that if $n_1/n_2 \to$ a positive limit as $n = n_1 + n_2 \to \infty$, then

$$\{ec_i(\Delta; \hat{\boldsymbol{\theta}}_E) - e_i^{(0)}\}/\sqrt{v_i^{(1)}} \tag{4.4.3}$$

converges in distribution to that of a standard normal. We write

$$\{ec_i(\Delta; \hat{\boldsymbol{\theta}}_E) - e_i^{(0)}\}/\sqrt{v_i^{(1)}} \xrightarrow{\mathcal{L}} N(0,1) \qquad (i = 1, 2). \tag{4.4.4}$$

The leading term $e_i^{(0)}$ of order $O(1)$ in the expansion of $eu_i(\Delta; n_1, n_2)$ is the optimal error rate $eo_i(\Delta)$, which has a common value of $\Phi(-\frac{1}{2}\Delta)$ in the case here of a zero cutoff point. The expansion of the characteristic function $\psi_i(t)$ of $ec_i(\Delta; \hat{\boldsymbol{\theta}}_E)$ was described in the last section. In a similar manner, the limiting result (4.4.4) can be obtained by expanding the characteristic function of the normalized version (4.4.3) of $ec_i(\Delta; \hat{\boldsymbol{\theta}}_E)$. The corresponding results for an arbitrary cutoff point have been given by Sayre (1980).

The normalized version

$$\{ec(\Delta; \hat{\boldsymbol{\theta}}_E) - e^{(0)}\}/\sqrt{v^{(2)}} \tag{4.4.5}$$

of the overall conditional error rate does have a nondegenerate limiting distribution as $n \to \infty$, although it is not a standard normal. This is because all first-order partial derivatives in the Taylor series expansion of $ec(\Delta; \hat{\boldsymbol{\theta}}_E)$ are zero, and so the quadratic terms are dominant.

The asymptotic distribution of the overall conditional error rate of the sample NLDR under a mixture sampling scheme has been derived by Efron (1975) and O'Neill (1976, 1978) in the course of their work in comparing the relative efficiencies of sample-based rules in their estimation of the Bayes rule $r_o(\mathbf{x}; \boldsymbol{\Psi}_E)$. As shown by Sayre (1980), their results can be used to obtain the asymptotic distribution of the overall error rate for separate sampling of the training data. O'Neill (1976, 1980) has also derived the asymptotic distribution of the overall conditional error rate of a general allocation rule for $g = 2$ groups. We will first present this general result and then show how it can be specialized to give the asymptotic distributional result for the sample NLDR.

4.4.3 Asymptotic Distribution of Overall Conditional Error Rate of a Plug-In Sample Rule

For $g = 2$ groups, we consider here the asymptotic distribution of the plug-in sample version $r_o(\mathbf{x};\hat{\mathbf{\Psi}})$ of a general Bayes rule $r_o(\mathbf{x};\mathbf{\Psi})$ of the form (1.4.3). For $g = 2$, the Bayes rule $r_o(\mathbf{x};\mathbf{\Psi})$ is based on the posterior log odds

$$\begin{aligned}\eta(\mathbf{x};\mathbf{\Psi}) &= \log\{\tau_1(\mathbf{x};\mathbf{\Psi})/\tau_2(\mathbf{x};\mathbf{\Psi})\} \\ &= \log(\pi_1/\pi_2) + \log\{f_1(\mathbf{x};\boldsymbol{\theta}_1)/f_2(\mathbf{x};\boldsymbol{\theta}_2)\}. \qquad (4.4.6)\end{aligned}$$

In some situations, $\eta(\mathbf{x};\mathbf{\Psi})$ may be expressed as a function of $\mathbf{x}$ and $\boldsymbol{\alpha}$, where $\boldsymbol{\alpha}$ is a q-dimensional vector containing fewer parameters than $\mathbf{\Psi}$. That is, $\boldsymbol{\alpha}$ contains the so-called discriminant function coefficients. For economy of notation, we will still use η to denote this function of $\mathbf{x}$ and $\boldsymbol{\alpha}$. For example, if $f_1(\mathbf{x};\boldsymbol{\theta}_1)/f_2(\mathbf{x};\boldsymbol{\theta}_2)$ is linear in $\mathbf{x}$, then $\boldsymbol{\alpha}$ will contain $q = p+1$ parameters.

We suppose that $\hat{\boldsymbol{\alpha}}$ is a consistent estimator of $\boldsymbol{\alpha}$ formed from the training data $\mathbf{t}$, such that as $n \to \infty$,

$$\sqrt{n}(\hat{\boldsymbol{\alpha}} - \boldsymbol{\alpha}) \xrightarrow{\mathcal{L}} N(\mathbf{O}, A), \qquad (4.4.7)$$

where $\mathbf{A}$ is a $q \times q$ positive definite symmetric matrix. In order to derive the asymptotic distribution of the overall error rate of the plug-in sample rule $r_o(\mathbf{x};\hat{\mathbf{\Psi}})$, it is convenient to work with its overall conditional error rate $ec(\mathbf{\Psi};\hat{\boldsymbol{\alpha}})$, expressed as a function of the training data $\mathbf{t}$ through $\hat{\boldsymbol{\alpha}}$ rather than $\hat{\mathbf{\Psi}}$.

O'Neill (1976, 1980) derived the asymptotic distribution of

$$n\{ec(\mathbf{\Psi};\hat{\boldsymbol{\alpha}}) - eo(\mathbf{\Psi})\}$$

using the well-known asymptotic result, which can be stated as follows. Suppose $H(\boldsymbol{\alpha})$ is a twice continuously differentiable function of $\boldsymbol{\alpha}$ that satisfies

$$\nabla_\alpha H(\boldsymbol{\alpha}) = \mathbf{0}, \qquad (4.4.8)$$

where ∇_α is the gradient vector

$$\nabla_\alpha = (\partial/\partial\alpha_1, \ldots, \partial/\partial\alpha_q)'$$

and $\alpha_j = (\boldsymbol{\alpha})_j$ for $j = 1,\ldots,q$. Then for a sequence of estimators $\hat{\boldsymbol{\alpha}}$ satisfying (4.4.7), the asymptotic distribution of $n\{H(\hat{\boldsymbol{\alpha}}) - H(\boldsymbol{\alpha})\}$ is given by

$$n\{H(\hat{\boldsymbol{\alpha}}) - H(\boldsymbol{\alpha})\} \xrightarrow{\mathcal{L}} \mathbf{U}'\mathbf{J}\mathbf{U}, \qquad (4.4.9)$$

as $n \to \infty$, where $\mathbf{U} \sim N(\mathbf{O},\mathbf{A})$, and

$$\mathbf{J} = \tfrac{1}{2}\nabla_\alpha \nabla'_\alpha H(\boldsymbol{\alpha}).$$

The result (4.4.9) is obtained in a standard manner by a Taylor series expansion of $H(\hat{\boldsymbol{\alpha}})$ about the point $\boldsymbol{\alpha}$; see, for example, Lehmann (1983, Section 5.1).

O'Neill (1980) used (4.4.9) to give the asymptotic distribution of

$$n\{ec(\boldsymbol{\Psi};\hat{\boldsymbol{\alpha}}) - eo(\boldsymbol{\Psi})\}$$

under mixture sampling of the training data. In forming $\eta(\mathbf{x};\hat{\boldsymbol{\alpha}})$, the term $\log(\pi_1/\pi_2)$ in (4.4.6) was estimated as $\log(n_1/n_2)$; that is, the cutoff point was taken to be $-\log(n_1/n_2)$. As noted previously, under mixture sampling, n_i/n provides an estimate of π_i $(i = 1,2)$. Provided regularity conditions hold, O'Neill (1980) has shown that for a sequence of estimators $\hat{\boldsymbol{\alpha}}$ satisfying (4.4.7), the asymptotic distribution of $n\{ec(\boldsymbol{\Psi};\hat{\boldsymbol{\alpha}}) - eo(\boldsymbol{\Psi})\}$ is given as

$$n\{ec(\boldsymbol{\Psi};\hat{\boldsymbol{\alpha}}) - eo(\boldsymbol{\Psi})\} \xrightarrow{\mathcal{L}} \mathbf{U}'\mathbf{J}\mathbf{U}, \tag{4.4.10}$$

where $\mathbf{U} \sim N(\mathbf{O}, A)$, and where

$$\begin{aligned}\mathbf{J} &= \frac{1}{2}\nabla_\alpha \nabla'_\alpha ec(\boldsymbol{\Psi};\boldsymbol{\alpha}) \\ &= \frac{1}{4}\|\nabla_\alpha \eta(\mathbf{x};\boldsymbol{\alpha})\|^{-1} \int_R \{\nabla_\alpha \eta(\mathbf{x};\boldsymbol{\alpha})\}\{\nabla'_\alpha \eta(\mathbf{x};\boldsymbol{\alpha})\} f_X(\mathbf{x};\boldsymbol{\Psi})\,d\nu.\end{aligned} \tag{4.4.11}$$

In (4.4.11),

$$R = \{\mathbf{x} : \eta(\mathbf{x};\boldsymbol{\alpha}) = 0\}$$

and

$$f_X(\mathbf{x};\boldsymbol{\Psi}) = \sum_{i=1}^{2} \pi_i f_i(\mathbf{x};\boldsymbol{\theta}_i)$$

is the mixture density of $\mathbf{X}$. An adequate set of regularity conditions has been given by O'Neill (1980), who notes that (4.4.10) will hold under a variety of such conditions.

From (4.4.10), the mean of the asymptotic distribution of $n\{ec(\boldsymbol{\Psi};\hat{\boldsymbol{\alpha}}) - eo(\boldsymbol{\Psi})\}$ is equal to $E(\mathbf{U}'\mathbf{J}\mathbf{U})$, which can be written as

$$\begin{aligned}E(\mathbf{U}'\mathbf{J}\mathbf{U}) &= \mathrm{tr}\{E(\mathbf{U}'\mathbf{J}\mathbf{U})\} \\ &= \mathrm{tr}\{\mathbf{J}E(\mathbf{U}\mathbf{U}')\} \\ &= \mathrm{tr}(\mathbf{J}\mathbf{A}).\end{aligned} \tag{4.4.12}$$

Efron (1975) and O'Neill (1976, 1978, 1980) have used (4.4.12) to define the asymptotic relative efficiency of two rules based on different estimates $\hat{\boldsymbol{\alpha}}_1$ and $\hat{\boldsymbol{\alpha}}_2$ of $\boldsymbol{\alpha}$. With respect to the estimation of the Bayes rule $r_o(\mathbf{x};\boldsymbol{\Psi})$, they defined the asymptotic efficiency of a rule based on $\eta(\mathbf{x};\hat{\boldsymbol{\alpha}}_2)$ relative to a rule based on $\eta(\mathbf{x};\hat{\boldsymbol{\alpha}}_1)$ by the ratio

$$\text{tr}(\mathbf{J}\mathbf{A}_1)/\text{tr}(\mathbf{J}\mathbf{A}_2), \tag{4.4.13}$$

where $\mathbf{A}_i$ is the asymptotic covariance matrix of $\sqrt{n}(\hat{\boldsymbol{\alpha}}_i - \boldsymbol{\alpha})$ for $i = 1,2$. Efron (1975) evaluated the measure (4.4.13) in comparing the logistic regression rule with the sample NLDR under the homoscedastic normal model (3.3.1). O'Neill (1976, 1980) evaluated it for a comparison in nonnormal situations of the logistic regression rule with the sample-based rule using a fully efficient estimator of $\boldsymbol{\alpha}$. Their results are discussed in Chapter 8, where the logistic regression approach to discriminant analysis is presented.

A first-order expansion of the overall unconditional error rate $E\{ec(\boldsymbol{\Psi};\hat{\boldsymbol{\alpha}})\}$ can be obtained from (4.4.10) provided its asymptotic mean is equal to

$$\text{Lt}_{n\to\infty}E[n\{ec(\boldsymbol{\Psi};\hat{\boldsymbol{\alpha}}) - eo(\boldsymbol{\Psi})\}].$$

For finite n, we have that

$$n[E\{ec(\boldsymbol{\Psi};\hat{\boldsymbol{\alpha}}) - eo(\boldsymbol{\Psi})\}] = n[E\{ec(\boldsymbol{\Psi};\hat{\boldsymbol{\alpha}})\} - eo(\boldsymbol{\Psi})]. \tag{4.4.14}$$

Assuming that the limit of (4.4.14) is equal to the asymptotic mean (4.4.12), we can write

$$\text{Lt}_{n\to\infty}n[E\{ec(\boldsymbol{\Psi};\hat{\boldsymbol{\alpha}}\} - eo(\boldsymbol{\Psi})] = \text{tr}(\mathbf{J}\mathbf{A}),$$

and so

$$E\{ec(\boldsymbol{\Psi};\hat{\boldsymbol{\alpha}})\} = eo(\boldsymbol{\Psi}) + n^{-1}\text{tr}(\mathbf{J}\mathbf{A}) + o(n^{-1}). \tag{4.4.15}$$

It can be confirmed on evaluation of $\mathbf{A}$ and $\mathbf{J}$ for the homoscedastic normal model that (4.4.15) agrees with the first-order expansion given originally by Okamoto (1963) for the overall unconditional error rate of the sample NLDR. This is after allowance is made for the fact that (4.4.15) has been derived under mixture sampling of the training data $\mathbf{t}$, including the estimation of the cutoff point from $\mathbf{t}$ as $-\log(n_1/n_2)$.

More recently, O'Neill (1984a, 1984b) has considered expansions of the type (4.4.15), where the plug-in sample rule does not provide a consistent estimate of the Bayes rule. These expansions were used, for example, to investigate the large-sample performance of the sample NLDR under a normal but heteroscedastic model. O'Neill (1986) has also considered expansions of the variance of the overall conditional error rate in such situations.

4.4.4 Asymptotic Distribution of Overall Conditional Error Rate of Sample NLDR

We now specialize the asymptotic distributional result (4.4.10) of the previous section for a general sample rule to provide the asymptotic distribution of the overall conditional error rate of the sample NLDR for $g = 2$ homoscedastic normal groups. In the latter case, the sample NLDR given by the plug-in sample version of the Bayes rule is based on

$$\begin{aligned}\eta(\mathbf{x};\hat{\boldsymbol{\alpha}}_E) &= \hat{\boldsymbol{\alpha}}_E'(1,\mathbf{x}')' \\ &= \hat{\boldsymbol{\alpha}}_E'\mathbf{x}^+,\end{aligned}$$

where $\hat{\boldsymbol{\alpha}}_E = (\hat{\beta}_{0E},\hat{\boldsymbol{\beta}}_E')'$, and

$$\hat{\beta}_{0E} = \log(n_1/n_2) + \hat{\beta}_{0E}^o,$$

and where $\hat{\beta}_{0E}^o$ and $\hat{\boldsymbol{\beta}}_E$ are defined by (4.2.8) and (4.2.9), respectively.

Let $ec(\Psi_E;\hat{\boldsymbol{\alpha}}_E)$ denote the overall conditional error rate of the sample NLDR $r_o(\mathbf{x};\hat{\boldsymbol{\alpha}}_E)$. From (4.3.5), it follows that

$$ec(\Psi_E;\hat{\Psi}_E) = \sum_{i=1}^{2}\pi_i\Phi\{(-1)^i(\hat{\beta}_{0E}+\hat{\boldsymbol{\beta}}_E'\boldsymbol{\mu}_i)/(\hat{\boldsymbol{\beta}}_E'\boldsymbol{\Sigma}\hat{\boldsymbol{\beta}}_E)^{1/2}\}.$$

Although $ec(\Psi_E;\hat{\boldsymbol{\alpha}})$ depends on Ψ_E only through Δ and the vector $\boldsymbol{\pi} = (\pi_1, \pi_2)'$ of the group-prior probabilities, for notational convenience, we will continue to write it as a function of Ψ_E rather than of Δ and $\boldsymbol{\pi}$.

In order to specify the asymptotic distribution of

$$n\{ec(\Psi_E;\boldsymbol{\alpha}_E) - eo(\Psi_E)\}$$

under a mixture sampling scheme, it remains now to give the expressions $\mathbf{J}_E$ and $\mathbf{A}_E$ for $\mathbf{J}$ and $\mathbf{A}$, respectively, in (4.4.7) and (4.4.10) under the normal model (3.3.1) with equal group-covariance matrices. O'Neill (1976) has shown that the necessary regularity conditions are satisfied for (4.4.10) to be valid for the latter model.

From (4.4.11), $\mathbf{J}_E$ can be expressed as

$$\mathbf{J}_E = \frac{1}{4}\|\boldsymbol{\alpha}_E\|^{-1}\int_R \mathbf{x}^+\mathbf{x}^{+\prime}f_X(\mathbf{x};\Psi_E)\,d\nu, \tag{4.4.16}$$

where

$$f_X(\mathbf{x};\Psi_E) = \sum_{i=1}^{2}\pi_i\phi(\mathbf{x};\boldsymbol{\mu}_i,\boldsymbol{\Sigma})$$

is the mixture density of $\mathbf{X}$ under the homoscedastic normal model (3.3.1). O'Neill (1976) showed that (4.4.16) can be easily evaluated under the canonical form (4.2.10) to give

$$\mathbf{J}_E = \pi_1\phi(\tfrac{1}{2}\Delta + k_o\Delta^{-1})\begin{pmatrix} 1 & k_o\Delta^{-1} & \\ & & \mathbf{O} \\ k_o\Delta^{-1} & k_o^2\Delta^{-2} & \\ & \mathbf{O}' & \mathbf{I}_{p-1}\end{pmatrix},$$

where $k_o = \log(\pi_2/\pi_1)$, and $\mathbf{O}$ is a $2\times(p-1)$ matrix of zeros. Concerning the asymptotic covariance matrix $\mathbf{A}_E$ of $\sqrt{n}(\hat{\boldsymbol{\alpha}}_E - \boldsymbol{\alpha}_E)$, Efron (1975) has shown that it is given under the canonical form (4.2.10) by

$$\mathbf{A}_E = (\pi_1\pi_2)^{-1}\begin{pmatrix} 1+\frac{1}{4}\Delta^2 & \frac{1}{2}\Delta(\pi_1-\pi_2) & \mathbf{O} \\ \frac{1}{2}\Delta(\pi_1-\pi_2) & 1+2\pi_1\pi_2\Delta^2 & \\ & \mathbf{O}' & c\mathbf{I}_{p-1} \end{pmatrix},$$

where $c = 1+\pi_1\pi_2\Delta^2$, and $\mathbf{O}$ is a $2\times(p-1)$ matrix of zeros.

With separate sampling of the training data, interest usually lies in the distribution of the overall conditional error rate of the sample NLDR using $\xi(\mathbf{x};\hat{\boldsymbol{\theta}}_E)$ with a fixed cutoff point k. We can replace k by $k_o = \log(\pi_2/\pi_1)$, since it is equal to $\log(\pi_2/\pi_1)$ for some value of π_1 in the unit interval. Under separate sampling (Sayre, 1980), the asymptotic distribution of

$$n\{ec(\boldsymbol{\Psi}_E;\hat{\boldsymbol{\theta}}_E) - eo(\boldsymbol{\Psi}_E)\}$$

is as for mixture sampling, but with $\mathbf{A}_E$ slightly modified to allow for the fact that the cutoff point is no longer estimated from the data. The modified version of $\mathbf{A}_E$ is obtained by replacing its first diagonal element, namely,

$$(1+\tfrac{1}{4}\Delta^2)/\pi_1\pi_2,$$

by

$$\tfrac{1}{4}\Delta^2/\{(n_1/n)(n_2/n)\},$$

and by replacing π_i with n_i/n $(i = 1,2)$ in its other elements. In the special case of $k_o = 0$ and equal group-sample sizes $(n_1 = n_2 = \frac{1}{2}n)$, it can be so confirmed as $n \to \infty$ that for $p = 1$,

$$n\{ec(\boldsymbol{\Psi}_E;\hat{\boldsymbol{\alpha}}_E) - eo(\boldsymbol{\Psi}_E)\} \xrightarrow{\mathcal{L}} (\Delta/8)\phi(\tfrac{1}{2}\Delta)\chi_1^2,$$

and for multivariate feature data,

$$n\{ec(\boldsymbol{\Psi}_E;\hat{\boldsymbol{\alpha}}_E) - eo(\boldsymbol{\Psi}_E)\} \xrightarrow{\mathcal{L}} (\Delta/8)\phi(\tfrac{1}{2}\Delta)\{\chi_1^2 + (p-1)(\Delta+4\Delta^{-2})\chi_{p-1}^2\}.$$

4.5 CONSTRAINED ALLOCATION WITH THE SAMPLE NLDR

4.5.1 Constraint on One Unconditional Error

We consider here the problem of choosing the cutoff point k so as to control the error rates of the NLDR using the sample NLDF $\xi(\mathbf{x};\hat{\boldsymbol{\theta}}_E)$ under the homoscedastic normal model (3.3.1) for $g = 2$ groups. This problem in a general context was discussed in Section 1.10.

We first report the work of T. W. Anderson (1973a, 1973b), who showed how the cutoff point k can be chosen so that the unconditional error rate for a designated group is asymptotically equal to a prescribed level, say, ϵ. With Δ unknown, it is convenient to work with the Studentized version of $\xi(\mathbf{x};\hat{\boldsymbol{\theta}}_E)$,

$$W_{S1} = \{\xi(\mathbf{x};\hat{\boldsymbol{\theta}}_E) - \tfrac{1}{2}D^2\}/D,$$

where

$$D = \{(\bar{\mathbf{x}}_1 - \bar{\mathbf{x}}_2)'\mathbf{S}^{-1}(\bar{\mathbf{x}}_1 - \bar{\mathbf{x}}_2)\}^{1/2}$$

is the sample Mahalanobis distance.

T. W. Anderson (1973a) has shown that if $n_2/n_1 \to$ a positive limit as $N \to \infty$, then

$$\text{pr}\{W_{S1} < w \mid \mathbf{X} \in G_1, n_1, n_2\} = \Phi(w) - \phi(w)b_1(\Delta, w, n_1, n_2) + O(N^{-2}),$$

where

$$b_1(\Delta, w, n_1, n_2) = \{\tfrac{1}{2}w - (p-1)\Delta^{-1}\}/n_1 + \{\tfrac{1}{4}w^3 + (p - \tfrac{3}{4})w\}/N. \tag{4.5.1}$$

Using this result, T. W. Anderson (1973a, 1973b) showed that

$$\text{pr}\{W_{S1} < w_\epsilon \mid \mathbf{X} \in G_1, n_1, n_2\} = \epsilon + O(N^{-2}), \tag{4.5.2}$$

if

$$w_\epsilon(D, n_1, n_2) = q_\epsilon + b_1(D, q_\epsilon, n_1, n_2),$$

where

$$q_\epsilon = \Phi^{-1}(\epsilon).$$

It follows from (4.5.2) that if an entity is assigned to G_1 or G_2 according as the Studentized version W_{S1} of the sample NLDF is greater or less than $w_\epsilon(D, n_1, n_2)$, then under separate sampling, its unconditional error rate with respect to G_1 satisfies

$$eu_{S1}(\Delta, n_1, n_2) = \epsilon + O(N^{-2}).$$

Concerning the specification of the cutoff point so that the corresponding rate for G_2 rather than G_1 asymptotically achieves a prescribed level ϵ, we work with the sample NLDF studentized as

$$W_{S2} = \{\xi(\mathbf{x};\hat{\boldsymbol{\theta}}_E) + \tfrac{1}{2}D^2\}/D. \tag{4.5.3}$$

On noting that the negative of W_{S2} has the same distribution in G_2 as that of W_{S1} in G_1 but with n_1 and n_2 interchanged, we have that

$$\begin{aligned}\text{pr}\{W_{S2} > w \mid \mathbf{X} \in G_2, n_1, n_2\} &= \text{pr}\{W_{S1} < -w \mid \mathbf{X} \in G_1, n_2, n_1\}\\ &= \epsilon + O(N^{-2}),\end{aligned}$$

if

$$w = -w_\epsilon(D, n_2, n_1).$$

Hence, if an entity is assigned to G_1 or G_2 according as to whether W_{S2} is greater or less than $-w_\epsilon(D, n_2, n_1)$, then under separate sampling, the unconditional error rate with respect to G_2 satisfies

$$eu_{S2}(\Delta, n_1, n_2) = \epsilon + O(N^{-2}).$$

4.5.2 Confidence Bound on One Conditional Error

McLachlan (1977b) extended these results above to give a method that ensures that the conditional error rate of the sample NLDF for a designated group is less than a prescribed bound, say, M, with a prescribed confidence, say, α. Let

$$ec_{S1}(w, \Delta; \hat{\boldsymbol{\theta}}_E) = \mathrm{pr}\{W_{S1} < w \mid \mathbf{X} \in G_1, \mathbf{t}\}$$

denote the conditional error rate of the allocation rule based on the Studentized version W_{S1} of the sample NLDF applied with the cutoff point w. Under separate sampling of the training data $\mathbf{t}$, McLachlan (1977b) showed that if $n_2/n_1 \to$ a positive limit as $N \to \infty$, then

$$\mathrm{pr}\{ec_{S1}(w, \Delta; \hat{\boldsymbol{\theta}}_E) < M\} = \Phi\{(M - \nu_1)/\nu_2\} + O(N^{-2}), \tag{4.5.4}$$

where

$$\nu_1 = \Phi(w) - \phi(w) b_1(\Delta, w, n_1, n_2),$$

and

$$\nu_2^2 = \{\phi(w)\}^2 b_2(w, n_1, n_2),$$

and where

$$b_2(w, n_1, n_2) = \frac{1}{n_1} + \frac{1}{2} w^2 \frac{1}{N},$$

and $b_1(\Delta, w, n_1, n_2)$ is defined by (4.5.1). From (4.5.4), McLachlan (1977b) established that

$$\mathrm{pr}\{ec_{S1}(w, \Delta; \hat{\boldsymbol{\theta}}_E) < M\} = \alpha + O(N^{-2}) \tag{4.5.5}$$

for

$$\begin{aligned} w &= w_\alpha(\Delta, n_1, n_2) \\ &= q_M - k_1/N^{1/2} - k_2/N - k_3/N^{3/2}, \end{aligned} \tag{4.5.6}$$

where $q_M = \Phi^{-1}(M)$, and where, on writing $b_1(\Delta, q_M, n_1, n_2)$ as b_{1M} and $b_2(q_M, n_1, n_2)$ as b_{2M},

$$\begin{aligned} k_1 &= N^{1/2}(q_\alpha b_{2M}^{1/2}), \\ k_2 &= N\{-b_{1M} + \tfrac{1}{2} q_\alpha^2 q_M (b_{2M} - N^{-1})\}, \\ k_3 &= \{N^{1/2} q_\alpha b_{2M}^{1/2}\}[-N q_M b_{1M} + (q_M/8N^{1/2} b_{2M}^{1/2})(q_\alpha^2 q_M + 4N b_{1M}) \\ &\quad + (N q_\alpha^2 b_{2M}/3)(q_M^2 - 1) + (3q_M^2/4)(1 - q_\alpha^2) + \tfrac{1}{4} q_\alpha^2 + p - \tfrac{3}{4} + \tfrac{1}{2}(N/n_1)], \end{aligned}$$

and $q_\alpha = \Phi^{-1}(\alpha)$. The cutoff point $w_\alpha(\Delta, n_1, n_2)$ as it stands cannot be used in practice because it depends on the unknown Δ through the term $b_1(\Delta, q_M, n_1, n_2)$. However, as the latter appears only in the terms of order $O(N^{-1})$ and $O(N^{-3/2})$ in the expression for w_α, it can be shown that the result still holds to order $O(N^{-2})$ with w now taken to be the random variable $w_\alpha(D, n_1, n_2)$.

If we wish to bound the conditional error rate with respect to G_2 by M with an approximate prescribed confidence α, then we use the allocation rule based on the Studentized version (4.5.3) of the sample NLDF. An entity with feature vector $\mathbf{x}$ is assigned to G_1 or G_2 according as to whether W_{S2} is greater or less than $-w_\alpha(D, n_2, n_1)$.

4.5.3 Confidence Bounds on Both Conditional Errors

McLachlan (1977b) also considered the situation where it is desired to bound both conditional error rates simultaneously with an application of the sample NLDF. Let α_i be the desired level of confidence that the conditional error with respect to G_i does not exceed some prescribed bound M_i $(i = 1, 2)$. By combining the rules above for bounding separately each of the conditional error rates, McLachlan (1977b) proposed the allocation rule $r(\mathbf{x}; \hat{w}_1, \hat{w}_2, \hat{\boldsymbol{\theta}}_E)$ that assigns an entity with feature vector $\mathbf{x}$ to G_1 if

$$W_{S1} > \hat{w}_1 \quad \text{and} \quad W_{S2} > \hat{w}_2 \tag{4.5.7}$$

and to G_2 if

$$W_{S1} < \hat{w}_1 \quad \text{and} \quad W_{S2} < \hat{w}_2, \tag{4.5.8}$$

where

$$\hat{w}_1 = w_{\alpha_1}(D, n_1, n_2) \quad \text{and} \quad \hat{w}_2 = -w_{\alpha_2}(D, n_2, n_1).$$

The region (4.5.7) can be expressed as

$$\xi(\mathbf{x}; \hat{\boldsymbol{\theta}}_E) > \max(K_1, K_2)$$

and the region (4.5.8) as

$$\xi(\mathbf{x}; \hat{\boldsymbol{\theta}}_E) < \min(K_1, K_2),$$

where

$$K_i = D\{\hat{w}_i + (-1)^{i+1}\tfrac{1}{2}D\} \qquad (i = 1, 2).$$

Hence, in order to bound both errors simultaneously, the rule $r(\mathbf{x}; \hat{w}_1, \hat{w}_2, \hat{\boldsymbol{\theta}}_E)$ makes no allocation for an entity with feature vector $\mathbf{x}$ in the region

$$\{\mathbf{x} : \min(K_1, K_2) \leq \xi(\mathbf{x}; \hat{\boldsymbol{\theta}}_E) \leq \max(K_1, K_2)\}. \tag{4.5.9}$$

Let $ec_{Si}(\Delta; \hat{w}_1, \hat{w}_2, \hat{\boldsymbol{\theta}}_E)$ denote the conditional error rate of this rule specific to group G_i $(i = 1, 2)$. By the introduction of the doubtful region (4.5.9) of group membership for which a decision is deferred, it follows that

$$\text{pr}\{ec_{Si}(\Delta;\hat{w}_1,\hat{w}_2,\hat{\boldsymbol{\theta}}_E) < M_i\} \geq \alpha_i + O(N^{-2})$$

for $i = 1,2$ under separate sampling of the training data $\mathbf{t}$.

4.6 DISTRIBUTIONAL RESULTS FOR QUADRATIC DISCRIMINATION

4.6.1 Distribution of Sample NQDF

It has been seen for $g = 2$ groups under the heteroscedastic normal model (3.2.1) that the Bayes rule $r_o(\mathbf{x};\boldsymbol{\Psi}_U)$ is based on the NQDF (normal-based quadratic discriminant function)

$$\xi(\mathbf{x};\boldsymbol{\theta}_U) = -\tfrac{1}{2}(\mathbf{x}-\boldsymbol{\mu}_1)'\boldsymbol{\Sigma}_1^{-1}(\mathbf{x}-\boldsymbol{\mu}_1) + \tfrac{1}{2}(\mathbf{x}-\boldsymbol{\mu}_2)'\boldsymbol{\Sigma}_2^{-1}(\mathbf{x}-\boldsymbol{\mu}_2)$$

$$-\tfrac{1}{2}\log\{|\boldsymbol{\Sigma}_1|/\boldsymbol{\Sigma}_2|\}. \quad (4.6.1)$$

As $\xi(\mathbf{x};\boldsymbol{\theta}_U)$ is a quadratic function of the feature vector $\mathbf{x}$, it is not a straightforward exercise to give its group-conditional distribution as it is for the NLDF $\xi(\mathbf{x};\boldsymbol{\theta}_E)$ with equal group-covariance matrices. For bivariate feature data, Bayne and Tan (1981) have derived approximations to the error rates for $\xi(\mathbf{x};\boldsymbol{\theta}_U)$ in order to study the effects of correlation and unequal variances on them. More recently, Young, Turner, and Marco (1987) have provided conditions under which there exists simple forms for the error rates for $\xi(\mathbf{x};\boldsymbol{\theta}_U)$. Also, a simple bound on the overall error rate was derived. For the bivariate case, Bayne, Beauchamp, and Kane (1984) have provided an algorithm for the computation of the conditional error rates of a rule based on the sample NQDF $\xi(\mathbf{x};\hat{\boldsymbol{\theta}}_U)$ or any other specified quadratic discriminant function.

The distribution of $\xi(\mathbf{x};\hat{\boldsymbol{\theta}}_U)$ is very complicated and manageable analytical expressions were obtained initially only in special cases; see Hildebrandt, Michaelis, and Koller (1973). Okamoto (1961) considered the distribution of the plug-in sample version $\xi(\mathbf{x};\hat{\boldsymbol{\theta}}_U)$ in the case of $\boldsymbol{\mu}_1 = \boldsymbol{\mu}_2$. He gave an expansion of the group-conditional distribution of

$$(\mathbf{X}-\boldsymbol{\mu})'(\mathbf{S}_1^{-1} - \mathbf{S}_2^{-1})(\mathbf{X}-\boldsymbol{\mu}),$$

where $\boldsymbol{\mu}$ is the known common value of $\boldsymbol{\mu}_1$ and $\boldsymbol{\mu}_2$; see Siotani (1982) for further details.

Gilbert (1969), Han (1969, 1974), Hawkins and Raath (1982), and McLachlan (1975c) have studied the distribution of $\xi(\mathbf{x};\boldsymbol{\theta}_U)$ and its sample versions under the simplifying assumption that the group-covariance matrices are proportional, where

$$\boldsymbol{\Sigma}_2 = \kappa\boldsymbol{\Sigma}_1.$$

We can take $\kappa > 1$ without loss of generality.

In this special case, the complexity of the problem is reduced substantially, as $\xi(\mathbf{x};\boldsymbol{\theta}_U)$ can be expressed as

$$\xi(\mathbf{x};\boldsymbol{\theta}_U) = -\tfrac{1}{2}(1+\omega)^{-1}\{Q - \omega(1+\omega)\Delta_1^2\} + \tfrac{1}{2}p\log\kappa, \tag{4.6.2}$$

where $\omega = 1/(\kappa - 1)$, and

$$Q = \{\mathbf{x} - \boldsymbol{\mu}_1 - \omega(\boldsymbol{\mu}_1 - \boldsymbol{\mu}_2)\}'\boldsymbol{\Sigma}_1^{-1}\{\mathbf{x} - \boldsymbol{\mu}_1 - \omega(\boldsymbol{\mu}_1 - \boldsymbol{\mu}_2)\}$$

and

$$\Delta_1^2 = \{(\boldsymbol{\mu}_1 - \boldsymbol{\mu}_2)'\boldsymbol{\Sigma}_1^{-1}(\boldsymbol{\mu}_1 - \boldsymbol{\mu}_2)\}.$$

As noted by Han (1969), if $\mathbf{X}$ belongs to G_1, then Q is distributed as $\chi_p'^2(\omega^2\Delta_1^2)$, a noncentral chi-squared with p degrees of freedom and noncentrality parameter $\omega^2\Delta_1^2$. If $\mathbf{X}$ comes from G_2, then Q is distributed as $\kappa\chi_p'^2(\kappa\omega^2\Delta_1^2)$. Han (1969) proceeded to give the expansion of the group-conditional distribution of the sample NQDF $\xi(\mathbf{x};\hat{\boldsymbol{\theta}}_U)$ up to terms of the second order in the case of unknown $\boldsymbol{\mu}_1$ and $\boldsymbol{\mu}_2$, but known $\boldsymbol{\Sigma}$ and κ. In a later paper, Han (1974) gave the first-order expansion in the case where $\boldsymbol{\Sigma}$ is also unknown, but κ is still known. In the following year, McLachlan (1975c) derived the unconditional error rates of the rule based on a plug-in sample version of (4.6.2) in the general case where all the parameters are unknown. For analytical convenience, the plug-in estimates of $\boldsymbol{\Sigma}_1$ and κ used in this study differed from the maximum likelihood ones in that $\hat{\kappa}$ was taken to be

$$\{|\mathbf{S}_2|/|\mathbf{S}_1|\}^{1/p}.$$

The computation of the maximum likelihood estimates of $\boldsymbol{\Sigma}_1$ and κ is described in Section 5.4.2, where the use of proportional covariance matrices is discussed as a method of regularization with applications of $\xi(\mathbf{x};\hat{\boldsymbol{\theta}}_U)$ in situations where p is large relative to the group-sample sizes.

Recently, Fukunaga and Hayes (1989a) have derived manageable analytical expressions for the mean and variance for a certain class of functions of the training data that includes the conditional error rates for the sample NQDF $\xi(\mathbf{x};\hat{\boldsymbol{\theta}}_U)$ as well as for the sample NLDF $\xi(\mathbf{x};\hat{\boldsymbol{\theta}}_E)$. There is also the theoretical work of O'Neill (1984b) and, more recently, Wakaki (1990), who have derived asymptotic expansions of the unconditional error rates for $\xi(\mathbf{x};\hat{\boldsymbol{\theta}}_U)$ as part of their large-sample comparisons of the sample NQDR and NLDR under the heteroscedastic normal model (3.2.1). Their comparisons are discussed in Section 5.3. Marco, Young, and Turner (1987b) have derived asymptotic expansions of the unconditional error rates for $\xi(\mathbf{x};\hat{\boldsymbol{\theta}}_U)$ in the case of equal group-means and the uniform covariance structure (3.2.11).

4.6.2 Distribution of Z-Statistic

As shown in Section 3.4.2, the likelihood ratio criterion under the homoscedastic normal model (3.3.1) yields a rule based on the so-called Z-statistic,

$$Z = \frac{n_1}{n_1+1}\hat{\delta}_{1,E}(\mathbf{x}) - \frac{n_2}{n_2+1}\hat{\delta}_{2,E}(\mathbf{x}),$$

where

$$\hat{\delta}_{i,E}(\mathbf{x}) = (\mathbf{x} - \overline{\mathbf{x}}_i)'\mathbf{S}^{-1}(\mathbf{x} - \overline{\mathbf{x}}_i) \qquad (i = 1,2).$$

We let $r_Z(\mathbf{x};\hat{\boldsymbol{\theta}}_E)$ be the rule based on Z, which equals one or two according as to whether Z is less than or greater than zero. This scalar random variable should not be confused with the random vector $\mathbf{Z}$ consisting of the g zero-one indicator variables used to define the group membership of an entity with feature vector $\mathbf{x}$.

The limiting group-conditional of Z as $n_1, n_2 \to \infty$ is

$$Z \xrightarrow{\mathcal{L}} N((-1)^i\Delta^2, 4\Delta^2) \qquad \text{if} \quad \mathbf{X} \in G_i \qquad (i = 1,2).$$

Let

$$F_{Z1}(z;\Delta,n_1,n_2) = \text{pr}\{(2\Delta)^{-1}(Z + \Delta^2) < z \mid \mathbf{X} \in G_1, n_1, n_2\}.$$

For equal group-sample sizes,

$$Z = -2\frac{n_i}{n_i + 1}W,$$

where W denotes the sample NLDF defined by (4.2.1). Hence, in this case, an expansion of $F_{Z1}(z;\Delta,n_1,n_2)$ is easily obtained from the expansion (4.2.4) of $F_{W1}(w;\Delta,n_1,n_2)$ as derived by Okamoto (1963). The expansion of $F_{Z1}(z;\Delta, n_1,n_2)$ for unequal group-sample sizes ($n_1 \neq n_2$) can be obtained in a similar manner to that of $F_{W1}(w;\Delta,n_1,n_2)$. But it is more complicated, because the Z-statistic is quadratic in $\mathbf{x}$ and it also involves n_1 and n_2 explicitly. Memon and Okamoto (1971) showed for separate sampling under the assumed normal model that if $n_1/n_2 \to$ a positive limit as $n_1, n_2 \to \infty$, then

$$F_{Z1}(z;\Delta,n_1,n_2) = \Phi(z) + \phi(z)\{h_{Z1}/n_1 + h_{Z2}/n_2 + h_{Z3}/N\} + O(N^{-2}), \tag{4.6.3}$$

where

$$h_{Z1} = -\tfrac{1}{2}\Delta^{-2}\{z^3 - \Delta z^2 + (p-3)z + \Delta\},$$

$$h_{Z2} = -\tfrac{1}{2}\Delta^{-2}\{z^3 - \Delta z^2 + (p - 3 - \Delta^2)z + \Delta(\Delta^2 + 1)\},$$

and

$$h_{Z3} = -\tfrac{1}{4}\{4z^3 - 4\Delta z^2 + \Delta^2 z + 6(p-1)z - 2(p-1)\Delta\}.$$

Memon and Okamoto (1971) also gave the second-order terms in this expansion, and Siotani and Wang (1975, 1977) have since added the third-order terms; see Siotani (1982).

The expansion of

$$F_{Z2}(z;\Delta,n_1,n_2) = \text{pr}\{(2\Delta)^{-1}(Z - \Delta^2) < z \mid \mathbf{X} \in G_2, n_1, n_2\}$$

can be obtained from that of $F_{Z1}(z;\Delta,n_1,n_2)$ on using the result that

$$F_{Z2}(z;\Delta,n_1,n_2) = 1 - F_{Z1}(-z;\Delta,n_2,n_1).$$

This latter result can be established in a similar manner to the corresponding result (4.2.5) for the W-statistic.

If $eu_{Zi}(\Delta;n_1,n_2)$ denotes the unconditional error rate specific to group G_i $(i = 1,2)$ for the rule $r_Z(\mathbf{x};\hat{\boldsymbol{\theta}}_E)$ based on the Z-statistic with a zero cutoff point, then

$$\begin{aligned} eu_{Z1}(\Delta;n_1,n_2) &= \text{pr}\{Z > 0 \mid \mathbf{X} \in G_1\} \\ &= 1 - F_{Z1}(\tfrac{1}{2}\Delta;\Delta,n_1,n_2) \end{aligned} \tag{4.6.4}$$

and

$$\begin{aligned} eu_{Z2}(\Delta;n_1,n_2) &= \text{pr}\{Z < 0 \mid \mathbf{X} \in G_2\} \\ &= F_{Z2}(-\tfrac{1}{2}\Delta;\Delta,n_1,n_2) \\ &= 1 - F_{Z1}(\tfrac{1}{2}\Delta;\Delta,n_2,n_1). \end{aligned} \tag{4.6.5}$$

Hence, expansions of the unconditional error rates of $r_Z(\mathbf{x};\hat{\boldsymbol{\theta}}_E)$ are available from the expansion (4.6.3) of $F_{Z1}(z;\Delta,n_1,n_2)$.

Siotani (1980) has derived large-sample approximations to the distributions of the conditional error rates $ec_{Zi}(k,\Delta;\hat{\boldsymbol{\theta}}_E)$ of the rule based on the Z-statistic with an arbitrary cutoff point k. As with the conditional error rates of the sample NLDR rule, which uses the W-statistic, they have a univariate normal distribution if terms of order $O(N^{-2})$ are ignored. The first-order expansion of the mean of the conditional rate $ec_{Z1}(k,\Delta;\hat{\boldsymbol{\theta}}_E)$ specific to group G_1 is already available by using the expansion (4.6.3) of $F_{Z1}(z;\Delta,n_1,n_2)$ in (4.6.4) and (4.6.5), but where now $F_{Z1}(z;\Delta,n_1,n_2)$ is evaluated at $z = \frac{1}{2}(\Delta + \Delta^{-1}k)$ because the cutoff point k is not specified to be zero.

Concerning the variance of $ec_{Z1}(k,\Delta;\hat{\boldsymbol{\theta}}_E)$, Siotani (1980) showed that

$$\begin{aligned} \text{var}\{ec_{Z1}(k,\Delta;\hat{\boldsymbol{\theta}}_E)\} =& [\phi\{\tfrac{1}{2}(\Delta + \Delta^{-1}k)\}]^2(h_{Z4}/n_1 + h_{Z5}/n_2 + h_{Z6}/N) \\ &+ O(N^{-2}), \end{aligned} \tag{4.6.6}$$

where

$$h_{Z4} = \tfrac{1}{4}\Delta^{-4}(\Delta^2 + k)^2,$$

$$h_{Z5} = \tfrac{1}{4}\Delta^{-4}(\Delta^2 - k)^2,$$

and

$$h_{Z6} = \tfrac{1}{2}\Delta^{-2}k^2.$$

The first-order expansion of the variance of $ec_{Z2}(k,\Delta;\hat{\boldsymbol{\theta}}_E)$ is obtained by replacing k with $-k$ and interchanging n_1 and n_2 in (4.6.6).

4.6.3 Comparison of Rules Based on Z- and W-Statistics

We let

$$\begin{aligned} eu_{Z1}(\Delta; n_1, n_2) = \Phi(-\tfrac{1}{2}\Delta) &+ a_{Z1}/n_1 + a_{Z2}/n_2 + a_{Z3}/N \\ &+ a_{Z1}/n_1^2 + a_{Z2}/n_2^2 + a_{Z12}/(n_1 n_2) \\ &+ a_{Z13}/(n_1 N) + a_{Z23}/(n_2 N) + a_{Z33}/N^2 + O(N^{-3}) \end{aligned}$$

denote the second-order expansion, which can be computed from the second-order version of (4.6.3). Then it follows that the second-order expansion of the overall unconditional error rate,

$$eu_Z(\Delta; n_1, n_2) = \tfrac{1}{2}\{eu_{Z1}(\Delta; n_1, n_2) + eu_{Z2}(\Delta; n_1, n_2)\},$$

can be expressed as

$$\begin{aligned} eu_Z(\Delta; n_1, n_2) = \Phi(-\tfrac{1}{2}\Delta) &+ \tfrac{1}{2}(a_{Z1} + a_{Z2})(n_1^{-1} + n_2^{-1}) \\ &+ a_{Z3}N^{-1} + \tfrac{1}{2}(a_{Z11} + a_{Z22} + a_{Z12})(n_1^{-2} + n_2^{-2}) \\ &+ \tfrac{1}{2}(a_{Z13} + a_{Z23})(n_1^{-1} + n_2^{-1})N^{-1} \\ &+ a_{Z33}N^{-2} - \tfrac{1}{2}a_{Z12}(n_1^{-1} - n_2^{-1})^2 + O(N^{-3}). \end{aligned} \tag{4.6.7}$$

We let the corresponding expansion of the overall unconditional error rate $eu(\Delta; n_1, n_2)$ of the sample NLDR using the W-statistic with a zero cutoff point be given by (4.6.7), but with a_{Zi} and a_{Zij} replaced by a_i and a_{ij}, respectively ($i \leq j = 1, 2, 3$). As noted by Memon and Okamoto (1971),

$$a_3 = a_{Z3} \qquad \text{and} \qquad a_{33} = a_{Z33}. \tag{4.6.8}$$

Further, since Z is just a constant multiple of W for equal group-sample sizes, it follows on noting (4.6.8) that

$$(a_1 + a_2) = (a_{Z1} + a_{Z2}),$$

$$(a_{11} + a_{22} + a_{12}) = (a_{Z11} + a_{Z22} + a_{Z12}),$$

and

$$(a_{13} + a_{23}) = (a_{Z13} + a_{Z23}).$$

An obvious consequence of these relationships between the coefficients in the expansion (4.6.7) of $eu_Z(\Delta; n_1, n_2)$ and those in the expansion of $eu(\Delta; n_1, n_2)$ is that

$$\begin{aligned} eu_Z(\Delta; n_1, n_2) &= eu(\Delta; n_1, n_2) + O(N^{-2}) \\ &= eu(\Delta; n_1, n_2) - \tfrac{1}{2}(a_{Z12} - a_{12})(n_1^{-1} - n_2^{-1})^2 + O(N^{-3}). \end{aligned}$$

Memon and Okamoto (1971) used this result to show that there is no reduction in the overall error rate up to the first order as a consequence of using the sample rule based on Z and not W, but that there is a second-order reduction, as $a_{Z12} > a_{12}$.

Das Gupta (1965) proved that $r_Z(\mathbf{x};\hat{\boldsymbol{\theta}}_E)$ is admissible and minimax in the case of known $\boldsymbol{\Sigma}$ and also that it is admissible and minimax in the class of invariant rules in the case of unknown $\boldsymbol{\Sigma}$. Memon and Okamoto (1971) noted that this minimax property of $r_Z(\mathbf{x};\hat{\boldsymbol{\theta}}_E)$ manifests itself in that the values of the coefficients a_{Z1} and a_{Z2} are in general closer to each other than a_1 and a_2. They gave $p > \frac{1}{2}\Delta^2 + 1$ as the necessary and sufficient condition for this to be the case. Similarly, the coefficients a_{Z11} and a_{Z22} and the pair a_{Z13} and a_{Z23} tend to be closer to each other than the coefficients for the corresponding pairs in the expansion of the overall error rate $eu(\Delta;n_1,n_2)$ of the sample NLDR, which uses the W-statistic.

As explained by Siotani and Wang (1977), the group-sample sizes n_1 and n_2 have to be very large for the second-order expansions of $eu(\Delta;n_1,n_2)$ and $eu_Z(\Delta;n_1,n_2)$ to provide reliable approximations to the true error rates. For such large-sample sizes, the Z-and W-statistics are practically equivalent. Therefore, in order to provide a comparison of these two statistics for moderately sized samples, Siotani and Wang (1977) compared the third-order expansions of $eu(\Delta;n_1,n_2)$ and $eu_Z(\Delta;n_1,n_2)$. With the inclusion of the third-order terms, the Z-based rule $r_Z(\mathbf{x};\hat{\boldsymbol{\theta}}_E)$ is not uniformly superior to the sample NLDR, although it is generally superior. Siotani and Wang (1977) have provided tables that summarize those situations in which $eu_Z(\Delta;n_1,n_2)$ is smaller than $eu(\Delta;n_1,n_2)$.

4.6.4 Distribution of the Studentized Z-Statistic

Fujikoshi and Kanazawa (1976) considered the group-conditional distribution of the Studentized version of Z given by

$$Z_{Si} = (Z + (-1)^{i+1}D^2)/(2D)$$

for $\mathbf{X}$ belonging to group G_i $(i = 1,2)$. They showed that if $n_1/n_2 \to$ a positive limit as $n_1,n_2 \to \infty$, then

$$\text{pr}\{Z_{S1} < z \mid \mathbf{X} \in G_1\} = \Phi(z) - \phi(z)b_{Z1}(\Delta,z,n_1n_2) + O(N^{-2}), \qquad (4.6.9)$$

where

$$b_{Z1}(\Delta,z,n_1,n_2) = h_{S1}/n_1 + h_{S2}/n_2 + h_{S3}/N,$$

and where

$$h_{S1} = \tfrac{1}{2}\Delta^{-1}(\Delta z - z^2 + p - 1),$$
$$h_{S2} = \tfrac{1}{2}\Delta^{-1}\{(z-\Delta)^2 + p - 1\},$$

and

$$h_{S3} = \tfrac{1}{4}z(z^2 + 4p - 3).$$

The corresponding expansion for Z_{S2}, conditional on $\mathbf{X}$ belonging to G_2, can be obtained from (4.6.9) by using a relation similar to (4.6.5). Using the result (4.6.9), Kanazawa (1979) has shown that

$$\text{pr}\{Z_{S1} < z_\epsilon \mid \mathbf{X} \in G_1, n_1, n_2\} = \epsilon + O(N^{-2}),$$

if

$$z_\epsilon = q_\epsilon + b_{Z1}(D, q_\epsilon, n_1, n_2),$$

where $q_\epsilon = \Phi^{-1}(\epsilon)$.

CHAPTER 5

Some Practical Aspects and Variants of Normal Theory-Based Discriminant Rules

5.1 INTRODUCTION

In this chapter, we focus on some problems that arise with the estimation of normal (theory)-based discriminant rules in practical situations. One problem to be addressed is that of discriminant rules formed from estimates with too much variability, which arises in fitting models with too many parameters relative to the size n of the available training sample. Other problems to be considered in this chapter concern the performance of the sample normal-based linear and quadratic rules under departures from normality, and robust estimation of discriminant rules.

Regarding the first problem, by allowing the covariance matrices to be arbitrary in the specification of the multivariate normal densities for the group-conditional distributions, the consequent quadratic discriminant analysis requires a large number of parameters to be estimated if the dimension p of the feature vector or the number of groups g is not small. In such situations, a linear discriminant analysis is often carried out with the principle of parsimony as the main underlying thought. The reader is referred to Dempster (1972) for an excellent account of this principle, which suggests that parameters should be introduced sparingly and only when the data indicate they are required. This account was given in the introduction of his paper on covariance selection, in which the principle of parsimony is applied to the estimation of the covariance matrix of a multivariate normal distribution by setting selected elements of its inverse equal to zero. In this chapter, we consider estimation of the group-covariance matrices Σ_i for intermediate models between the overly diffuse heteroscedastic model and the overly rigid homoscedastic model.

A related approach to the estimation of the Σ_i in situations where p is large relative to n is the use of regularization methods, which are introduced in the next section.

5.2 REGULARIZATION IN QUADRATIC DISCRIMINATION

When the group-sample sizes n_i are small relative to p, the sample group-covariance matrices $\hat{\Sigma}_i$ and their bias-corrected versions $\mathbf{S}_i$ become highly variable. Moreover, when $n_i < p$, not all of the parameters are identifiable. The effect this has on the plug-in sample version $r_o(\mathbf{x}; \hat{\Psi}_U)$ of the normal-based quadratic discriminant rule (NQDR) can be seen by representing the $\mathbf{S}_i$ by their spectral decompositions

$$\mathbf{S}_i = \sum_{k=1}^{p} \lambda_{ik} \psi_{ik} \psi_{ik}', \qquad (i = 1, \ldots, g), \tag{5.2.1}$$

where λ_{ik} is the kth eigenvalue of $\mathbf{S}_i$ (ordered in decreasing size), and ψ_{ik} is the corresponding eigenvector of unit length ($k = 1, \ldots, p$). The inverse of $\mathbf{S}_i$ in this representation is

$$\mathbf{S}_i^{-1} = \sum_{k=1}^{p} \psi_{ik} \psi_{ik}' / \lambda_{ik} \qquad (i = 1, \ldots, g).$$

Hence, the log of the plug-in estimate of the ith group-conditional density under the heteroscedastic normal model (3.2.1) can be expressed as

$$\begin{aligned} \log f_i(\mathbf{x}; \hat{\boldsymbol{\theta}}_i) &= \log \phi(\mathbf{x}; \bar{\mathbf{x}}_i, \mathbf{S}_i) \\ &= -\frac{1}{2} \sum_{k=1}^{p} \{(\mathbf{x} - \bar{\mathbf{x}}_i)' \psi_{ik}\}^2 / \lambda_{ik} - \frac{1}{2} \sum_{k=1}^{p} \log \lambda_{ik} - \frac{1}{2} p \log(2\pi) \end{aligned}$$

for $i = 1, \ldots, g$.

It is well known that the estimates λ_{ik} of the eigenvalues of Σ_i are biased. The largest ones are biased toward values that are too high, and the smallest ones are biased too low. This bias is most pronounced when the eigenvalues of Σ_i tend toward equality, being less severe when they are highly disparate. In all situations, this phenomenon becomes more pronounced as n_i decreases. If $n_i \leq p$, then $\mathbf{S}_i$ is singular with rank $\leq n_i$. Its smallest $p - n_i + 1$ eigenvalues are estimated then to be zero, with their corresponding eigenvectors arbitrary subject to orthogonality constraints.

As explained by Friedman (1989), the net effect of this biasing phenomenon on the sample NQDR $r_o(\mathbf{x}; \hat{\Psi}_U)$ is to (sometimes dramatically) exaggerate the importance of the feature subspace spanned by the eigenvectors corresponding to the smallest eigenvalues. It is this exaggeration that accounts for much of the variance in the sampling distribution of $r_o(\mathbf{x}; \hat{\Psi}_U)$.

One way of attempting to provide more reliable estimates of the Σ_i is to correct for the eigenvalue distortion in the $\mathbf{S}_i$. James and Stein (1961), Stein, Efron, and Morris (1972), Stein (1973), Efron and Morris (1976), Olkin and Sellian (1977), Haff (1980, 1986), Lin and Perlman (1984), and Dey and Srinivasan (1985), among others, have adopted this approach to the estimation of a covariance matrix by seeking estimates that minimize given loss criteria (often some form of squared-error loss) on the eigenvalue estimates. However, as pointed out by Friedman (1989), none of these loss criteria that have been considered is related to the error rate of the discriminant rule subsequently formed from the estimated covariance matrices. Also, they nearly all require that the $\mathbf{S}_i$ be nonsingular.

Another approach is to use a regularization method. Regularization techniques have been applied with much success in the solution of ill- and poorly posed inverse problems; see Titterington (1985) and O'Sullivan (1986) for reviews. In the present context, quadratic discriminant analysis is ill-posed if the number n_i of classified entities from G_i is not greater than p for any i $(i = 1,\dots,g)$, and poorly posed if n_i is not appreciably larger than p. Regularization attempts to reduce the variances of highly unstable estimates by biasing them toward values that are deemed to be more physically plausible. The extent of the potential increase in bias depends on the aptness of the "plausible" values of the parameters. The trade-off between variance and bias is generally regulated by one or more parameters that control the strength of the biasing toward the plausible set of parameter values. This line of approach is pursued further in Section 5.5 with the description of regularized discriminant analysis as proposed by Friedman (1989).

It will be seen in subsequent sections that some of the discriminant rules used in practice can be viewed as regularized versions of the sample NQDR $r_o(\mathbf{x}; \hat{\Psi}_U)$. As an obvious example, the use of the parsimonious linear rule $r_o(\mathbf{x}; \hat{\Psi}_E)$ in the presence of heteroscedascticity can be viewed also as applying a high degree of regularization by attempting to improve the estimates $\mathbf{S}_i$ of the Σ_i by replacing each with the pooled estimate $\mathbf{S}$. If the Σ_i are disparate, then this method of regularization would introduce severe bias. A comparison of the sample linear rule $r_o(\mathbf{x}; \hat{\Psi}_E)$ with its quadratic counterpart $r_o(\mathbf{x}; \hat{\Psi}_U)$ is therefore of much practical relevance and is undertaken in the next section. As discussed there, it is desirable if the choice between $r_o(\mathbf{x}; \hat{\Psi}_E)$ and $r_o(\mathbf{x}; \hat{\Psi}_U)$ is made on the basis of the available training data. This is the case with regularized discriminant analysis as proposed by Friedman (1989). It avoids an outright choice between the alternatives of linear and quadratic sample rules, which is fairly restrictive, by providing a sophisticated compromise between them.

5.3 LINEAR VERSUS QUADRATIC NORMAL-BASED DISCRIMINANT ANALYSIS

5.3.1 Introduction

In this section, we consider the choice between the plug-in sample versions of the normal-based linear and quadratic discriminant rules, $r_o(\mathbf{x};\hat{\Psi}_E)$ and $r_o(\mathbf{x};\hat{\Psi}_U)$, respectively, under the assumption of multivariate normality for the group-conditional distributions. The sample NLDR $r_o(\mathbf{x};\hat{\Psi}_E)$ is asymptotically optimal if the covariance matrix $\mathbf{\Sigma}_i$ within group G_i is the same for all i $(i = 1,\ldots,g)$. However, in practice, it is perhaps unlikely that homoscedasticity will hold exactly. Further, even if a preliminary test does not reject the null hypothesis of homoscedasticity, the null hypothesis is really a proxy for a small neighborhood of the null parameter values. Therefore, it is of interest to assess the sample NLDR $r_o(\mathbf{x};\hat{\Psi}_E)$ under departures from homoscedasticity, in particular, its performance relative to the sample NQDR $r_o(\mathbf{x};\hat{\Psi}_U)$, which is asymptotically optimal in the case of heteroscedasticity.

This latter comparison is particularly relevant to the use of $r_o(\mathbf{x};\hat{\Psi}_E)$ in place of $r_o(\mathbf{x};\hat{\Psi}_U)$ as a method of regularization. If the group-covariance matrices are markedly different, then their estimation by the pooled estimate $\mathbf{S}$ will be a source of bias. However, the consequent decrease in variance may lead to $r_o(\mathbf{x};\hat{\Psi}_E)$ being overall superior to $r_o(\mathbf{x};\hat{\Psi}_U)$, in particular in small-sized samples. This, coupled with its good performance for discrete or mixed data in many situations, explains the versatility and consequent popularity of linear discriminant analysis based on $r_o(\mathbf{x};\hat{\Psi}_E)$.

Another reason for the wide use of normal-based linear discriminant analysis is ease of interpretation with the estimated posterior probabilities of group membership and the implied regions of allocation, arising from the simplicity of linearity. In the case of $g = 2$ groups, the boundary of the allocation regions is simply a straight line or hyperplane. With a quadratic rule, more complicated allocation regions can be obtained. For example, with bivariate feature data, the region of allocation into one group may be the interior of an ellipse or the region between two hyperbolas. In general with $r_o(\mathbf{x};\hat{\Psi}_U)$, the regions are defined by means of a quadratic function of the feature vector $\mathbf{x}$, which is not necessarily a positive definite quadratic form.

5.3.2 Comparison of Plug-In Sample Versions of NLDR and NQDR

There have been many investigations carried out on the relative performance of $r_o(\mathbf{x};\hat{\Psi}_E)$ and $r_o(\mathbf{x};\hat{\Psi}_U)$ under the heteroscedastic normal model (3.2.1). Many of these have been performed as part of wider studies on the behavior of these rules relative to their nonparametric or semiparametric competitors, where also multivariate normality may not apply. The results of those comparisons not relevant here are reported later where appropriate, for example, in Section 5.6 on the robustness of $r_o(\mathbf{x};\hat{\Psi}_E)$ and $r_o(\mathbf{x};\hat{\Psi}_U)$, and in Chapter 9 on nonparametric discrimination.

The comparisons of $r_o(\mathbf{x};\hat{\Psi}_E)$ and $r_o(\mathbf{x};\hat{\Psi}_U)$ in the literature have concentrated on the relative behavior of the plug-in sample versions of the normal-based linear discriminant function (NLDF), $\xi(\mathbf{x};\hat{\boldsymbol{\theta}}_E)$, and the normal-based quadratic discriminant function (NQDF), $\xi(\mathbf{x};\hat{\boldsymbol{\theta}}_U)$, applied with respect to $g = 2$ groups. Initial studies included those by Gilbert (1969), Marks and Dunn (1974), Van Ness and Simpson (1976), Aitchison et al. (1977), Wahl and Kronmal (1977), and Van Ness (1979). It can be seen from their work and more recent studies, such as Bayne et al. (1983), that the decision concerning whether to use the sample NLDR $r_o(\mathbf{x};\hat{\Psi}_E)$ or its quadratic counterpart $r_o(\mathbf{x};\hat{\Psi}_U)$ should be based on consideration of the sample sizes n_i relative to p, the degree of heteroscedasticity, and the separation between the groups. The size of the training data is an important initial consideration. Firstly, if there are adequate training data, then a preliminary assessment can be made of the question of homoscedasticity and also normality. The sample sizes are of interest in themselves as they often indicate a clear choice between the linear and quadratic rules. For instance, if the n_i are large relative to p and the assumption of homoscedasticity is not tenable, then the quadratic rule should be chosen. On the other hand, if the n_i are small relative to p, perhaps too small to allow a proper assessment of the presence of homoscedasticity in the training data, then the linear rule is preferable to the quadratic. For moderately sized n_i relative to p, the relative superiority of the linear and quadratic rules depends on the degree of heteroscedasticity and the amount of separation between the groups. For given sample sizes, the performance of the normal-based quadratic rule relative to the linear improves as the covariance matrices become more disparate and the separation between the groups becomes smaller. But how severe the heteroscedasticity must be in conjunction with how close the groups should be before the quadratic rule is preferable to the linear is difficult to resolve. It is a question of balance between the heteroscedastic normal model with its unbiased estimates and the homoscedastic model with fewer parameters having biased but less variable estimates.

It is therefore desirable if it can be left to the training data to decide between the linear or quadratic rules on the basis of some appropriate criterion such as the estimated overall error rate. For instance, Devroye (1988) has considered model selection in terms of the overall apparent error rate of a discriminant rule. His approach is discussed in Section 10.8.1.

Even if the appropriate choice between linear and quadratic rules is made in a given situation, the chosen rule may still have an error rate too large for it to be of practical use if the sample sizes are very small relative to p and the groups are not widely separated. However, as discussed in the subsequent sections, there are ways of obtaining a rule with an improved error rate in such situations. In particular, it will be seen that regularized discriminant analysis as proposed by Friedman (1989) provides a fairly rich class of regularized alternatives to the choice between homoscedastic and heteroscedastic models. For a Bayesian approach where an outright choice between linear and quadratic rules is avoided, the reader is referred to Smith and Spiegelhalter (1982).

5.3.3 Theoretical Results

The aforementioned results on the relative superiority of the plug-in sample versions of the NLDR and NQDR are of an empirical nature, because of the complexities involved with an analytical comparison. However, O'Neill (1984b, 1986) has presented some theoretical findings based on his asymptotic expansions of the expectation and variance of $ec(\Psi_U;\hat{\Psi}_E)$ and of $ec(\Psi_U;\hat{\Psi}_U)$, the overall conditional error rates associated with the linear and quadratic sample rules $r_o(\mathbf{x};\hat{\Psi}_E)$ and $r_o(\mathbf{x};\hat{\Psi}_U)$, respectively, under the heteroscedastic normal model (3.2.1). The derivation of these expansions was discussed in Section 3.2. O'Neill (1984b) has evaluated them in the case of $g = 2$ groups with

$$\boldsymbol{\mu}_i = (-1)^{i+1}\mu\mathbf{1}_p$$

and

$$\boldsymbol{\Sigma}_i = (1+\gamma)^{2-i}\mathbf{I}_p$$

for $i = 1,2$, where $\mathbf{1}_p$ is the $p\times 1$ vector of ones. He found that the unconditional error rate of $r_o(\mathbf{x};\hat{\Psi}_E)$ is asymptotically less than that of $r_o(\mathbf{x};\hat{\Psi}_U)$ even for quite large γ in moderately sized training samples. O'Neill (1986) subsequently compared the asymptotic variances of the overall conditional error rates. The leading term in the expansion of the variance of $ec(\Psi_U;\hat{\Psi}_E)$ is of order $O(N^{-1})$, and the corresponding term in the variance of $ec(\Psi_U;\hat{\Psi}_U)$ is of order $O(N^{-2})$. However, O'Neill (1986) found that this latter second-order term sufficiently dominates the first-order term in the expansion of the variance of $ec(\Psi_U;\hat{\Psi}_E)$ for $r_o(\mathbf{x};\hat{\Psi}_E)$ to be less variable than that of $r_o(\mathbf{x};\hat{\Psi}_U)$ for most values of the parameters in small- and moderate-size training samples. It was found in some situations that $ec(\Psi_U;\hat{\Psi}_E)$ has a smaller mean than that of $ec(\Psi_U;\hat{\Psi}_U)$, but a larger variance. But O'Neill (1984b) demonstrated that the extra variability of $ec(\Psi_U;\hat{\Psi}_E)$ in these situations is not large enough to warrant the use of $r_o(\mathbf{x};\hat{\Psi}_U)$ instead of $r_o(\mathbf{x};\hat{\Psi}_E)$ with its lower error rate.

Critchley et al. (1987) have considered the relative performance of normal-based linear and quadratic discriminant analyses in terms of the estimates they give in the case of $g = 2$ groups for the NQDF $\xi(\mathbf{x};\boldsymbol{\theta}_U)$, or, equivalently, the posterior log odds for known prior probabilities. They derived the approximate bias, conditional on the feature vector $\mathbf{x}$, of the sample NLDF $\xi(\mathbf{x};\hat{\boldsymbol{\theta}}_E)$ under the heteroscedastic normal model (3.2.1). As $n\to\infty$ with $n_1/n\to k$, the asymptotic conditional bias of $\xi(\mathbf{x};\hat{\boldsymbol{\theta}}_E)$ is given by

$$\begin{aligned}\text{bias}\{\xi(\mathbf{x};\hat{\boldsymbol{\theta}}_E)\} &= E\{\xi(\mathbf{x};\hat{\boldsymbol{\theta}}_E)\mid\mathbf{x}\} - \xi(\mathbf{x};\boldsymbol{\theta}_U)\\ &\approx \frac{1}{2}\sum_{i=1}^{2}(-1)^i\{\delta(\mathbf{x},\boldsymbol{\mu}_i;\boldsymbol{\Sigma}^{(k)}) - \log|\boldsymbol{\Sigma}_i|\} - \xi(\mathbf{x};\boldsymbol{\theta}_U),\end{aligned}$$

where

$$\boldsymbol{\Sigma}^{(k)} = k\boldsymbol{\Sigma}_1 + (1-k)\boldsymbol{\Sigma}_2.$$

As the bias of $\xi(\mathbf{x};\hat{\boldsymbol{\theta}}_U)$ is of order $O(1)$ in its estimation of $\xi(\mathbf{x};\boldsymbol{\theta}_U)$, the bias component in its mean-squared error remains essentially unchanged as

$n \to \infty$. Hence, its performance relative to the asymptotically unbiased estimator $\xi(\mathbf{x};\hat{\boldsymbol{\theta}}_U)$ rapidly deteriorates as $n \to \infty$. Critchley et al. (1987) also investigated the effect of misspecification of the homoscedastic model on confidence intervals for the posterior log odds provided by $\xi(\mathbf{x};\hat{\boldsymbol{\theta}}_E)$. The problem of assessing the reliability of the estimated log odds is addressed in Chapter 11.

Recently, Wakaki (1990) has compared the large-sample relative performance of the sample NLDF and NQDF applied with a zero cutoff point with respect to $g = 2$ groups in equal proportions under the heteroscedastic normal model (3.2.1). We let $ec(\boldsymbol{\theta}_U;\hat{\boldsymbol{\theta}}_E)$ and $ec(\boldsymbol{\theta}_U;\hat{\boldsymbol{\theta}}_U)$ denote the overall conditional error rates associated with this application of the sample NLDR and NQDR, respectively. Wakaki (1990) derived the first-order expansions of the overall unconditional error rates as given by the expectations of $ec(\boldsymbol{\theta}_U;\hat{\boldsymbol{\theta}}_E)$ and $ec(\boldsymbol{\theta}_U;\hat{\boldsymbol{\theta}}_U)$. As the sample NLDR is not Bayes risk consistent under heteroscedasticity, it follows that its overall unconditional error rate is greater than that of the sample NQDR for sufficiently large n.

Accordingly, in the special case of proportional group-covariance matrices $\Sigma_2 = \kappa\Sigma_1$ and equal group-sample sizes, Wakaki (1990) calculated the common value n_o of n_1 and n_2 at which the first-order expansions of the expectations of $ec(\boldsymbol{\theta}_U;\hat{\boldsymbol{\theta}}_U)$ and $ec(\boldsymbol{\theta}_U;\hat{\boldsymbol{\theta}}_E)$ are equal. That is, if the common value of the group-sample sizes is not less than n_o, then $ec(\boldsymbol{\theta}_U;\hat{\boldsymbol{\theta}}_U)$ is less than or equal to $ec(\boldsymbol{\theta}_U;\hat{\boldsymbol{\theta}}_E)$ on average, ignoring terms of the second order. Wakaki (1990) tabulated n_o for various combinations of p, κ, and μ, where, without loss of generality, it was assumed that $\boldsymbol{\Sigma}_1 = \mathbf{I}_p$, $\boldsymbol{\mu}_1 = 0$, and $\boldsymbol{\mu}_2 = (\mu, 0, \ldots, 0)'$. He also tabulated, as percentages, the corresponding values of the leading terms of order $O(1)$, $e_E^{(o)}$ and $e_U^{(o)}$, and of the first-order coefficients e_{1E} and e_{1U}, in the first-order asymptotic expansions of the expectations of $ec(\boldsymbol{\theta}_U;\hat{\boldsymbol{\theta}}_E)$ and $ec(\boldsymbol{\theta}_U;\hat{\boldsymbol{\theta}}_U)$, respectively. These results are reported in Table 5.1, where it can be seen that n_o increases, as the number p of feature variables increases or as $\kappa - 1$ decreases (that is, as the group-covariance matrices become more similar).

5.3.4 Loss of Efficiency in Using the Sample NQDF Under Homoscedasticity

Up to now, we have focused exclusively on sample normal-based linear discriminant analysis under heteroscedasticity, as there is a potential gain to be had over a quadratic analysis in certain situations. In the reverse case with a quadratic discriminant analysis in the case of homoscedasticity, there is of course nothing to be gained by having a model more general than is needed. Indeed, there is a loss in efficiency in having to estimate the superfluous parameters of the heteroscedastic model. To demonstrate this loss, Critchley et al. (1987) have considered the efficiency of $\xi(\mathbf{x};\hat{\boldsymbol{\theta}}_U)$ relative to $\xi(\mathbf{x};\hat{\boldsymbol{\theta}}_E)$ in terms of the ratio of their variances, conditional on $\mathbf{x}$. More precisely, they worked with the unbiased versions $\hat{\xi}_U(\mathbf{x})$ and $\hat{\xi}_E(\mathbf{x})$ as given by (3.2.10) and (3.3.16),

TABLE 5.1 Variation in Threshold n_o for Common Group-Sample Sizes with Respect to p, μ, and κ

p	μ	$\kappa - 1$	$e_E^{(o)}$	e_{1E}	$e_U^{(o)}$	e_{1U}	n_o
2	1.0	0.2	31.63	27.17	31.43	51.37	121.7
		0.6	32.74	28.60	31.35	57.52	20.8
	2.0	0.2	16.97	18.78	16.91	25.41	120.6
		0.6	18.66	20.10	18.25	27.83	18.6
	3.0	0.2	7.61	12.78	7.58	16.63	119.4
		0.6	9.23	14.40	9.01	18.22	17.3
4	1.0	0.2	31.63	72.92	31.13	187.51	229.1
		0.6	32.74	77.10	29.46	184.29	32.6
	2.0	0.2	16.97	44.18	16.80	83.52	240.1
		0.6	18.66	47.74	17.47	90.33	35.6
	3.0	0.2	7.61	28.11	7.54	44.24	219.8
		0.6	9.23	32.14	8.69	50.01	32.7
6	1.0	0.2	31.63	118.67	30.83	390.21	341.4
		0.6	32.74	125.61	27.79	343.40	44.0
	2.0	0.2	16.97	69.57	16.69	167.00	358.4
		0.6	18.66	75.38	16.73	174.52	51.3
	3.0	0.2	7.61	43.43	7.50	81.50	332.9
		0.6	9.23	49.88	8.37	91.49	48.4

Source: From Wakaki (1990).

respectively. We have from their results that as $n \to \infty$ with $n_1/n \to 1/2$,

$$\begin{aligned}\epsilon &= \operatorname{var}\{\hat{\xi}_E(\mathbf{x})\}/\operatorname{var}\{\hat{\xi}_U(\mathbf{x})\} \\ &\approx \nu_1/\nu_2,\end{aligned}$$

where

$$\begin{aligned}\nu_1 &= \{\xi(\mathbf{x};\boldsymbol{\theta}_E)\}^2 + (2 + \tfrac{1}{2}\Delta^2)\{\delta_{1,E}(\mathbf{x}) + \delta_{2,E}(\mathbf{x})\} - \tfrac{1}{4}\Delta^4, \\ \nu_2 &= 2\{\xi(\mathbf{x};\boldsymbol{\theta}_E)\}^2 + \tfrac{1}{2}\{\delta_{1,E}(\mathbf{x}) + \delta_{2,E}(\mathbf{x})\}^2 + 2p,\end{aligned}$$

and

$$\delta_{i,E}(\mathbf{x}) = (\mathbf{x} - \boldsymbol{\mu}_i)'\boldsymbol{\Sigma}^{-1}(\mathbf{x} - \boldsymbol{\mu}_i) \qquad (i = 1,2).$$

For example, in the case where $\boldsymbol{\mu}_1 = (2,0,\ldots,0)'$, $\boldsymbol{\mu}_2 = \mathbf{0}$, and $\boldsymbol{\Sigma} = \mathbf{I}_p$, the asymptotic value of ϵ at $\mathbf{x} = \mathbf{0}$ is equal to 0.8 and 0.67 for $p = 2$ and 4, respectively. The quantity $\epsilon^{-1} - 1$ represents the extra proportion of training observations needed in order to achieve the same precision with $\xi(\mathbf{x};\hat{\boldsymbol{\theta}}_U)$ as with $\xi(\mathbf{x};\hat{\boldsymbol{\theta}}_E)$ under homoscedasticity. Note that the results reported above on the relative performance of $r_o(\mathbf{x};\hat{\boldsymbol{\Psi}}_E)$ and $r_o(\mathbf{x};\hat{\boldsymbol{\Psi}}_U)$ are qualified by the assumption of multivariate normality for the group-conditional distributions. The effect of departures from normality on these rules is surveyed in Section 5.6.

5.4 SOME MODELS FOR VARIANTS OF THE SAMPLE NQDR

5.4.1 Equal Spherical Group-Covariance Matrices (Minimum-Euclidean Distance Rule)

As can be seen from (3.3.13), the use of the sample NLDR $r_o(\mathbf{x};\hat{\Psi}_E)$ with equal group-prior probabilities π_i is equivalent to the minimum-distance rule,

$$\min_i\{(\mathbf{x}-\bar{\mathbf{x}}_i)'\mathbf{S}^{-1}(\mathbf{x}-\bar{\mathbf{x}}_i)\}^{1/2}, \tag{5.4.1}$$

where the Mahalanobis distance is used as the metric. More generally, the use of (5.4.1) is equivalent to the sample plug-in version of the Bayes rule with equal π_i in the case of group-conditional distributions belonging to the same family of elliptic distributions having different locations but a common shape. This follows directly from the form (1.12.5) of the elliptic density.

In situations where the group-conditional distributions appear to have unequal covariance matrices, the minimum-distance rule is often modified to

$$\min_i\{(\mathbf{x}-\bar{\mathbf{x}}_i)'\mathbf{S}_i^{-1}(\mathbf{x}-\bar{\mathbf{x}}_i)\}^{1/2}. \tag{5.4.2}$$

It can be seen from (5.4.2) that this modified distance rule is not quite the same as the sample NQDR $r_o(\mathbf{x};\hat{\Psi}_U)$. It ignores the normalizing term $|\mathbf{S}_i|^{-1/2}$ in the plug-in estimate $\phi(\mathbf{x};\bar{\mathbf{x}}_i,\mathbf{S}_i)$ of the multivariate normal density adopted for the distribution of $\mathbf{X}$ in group G_i $(i = 1,\ldots,g)$. This term can make an important contribution in the case of disparate group-covariance matrices.

Another modification of the minimum-distance rule (5.4.2) is to completely ignore the covariance structure of $\mathbf{X}$ within a group and to base the allocation of an entity with feature $\mathbf{x}$ on

$$\min_i\{(\mathbf{x}-\bar{\mathbf{x}}_i)'(\mathbf{x}-\bar{\mathbf{x}}_i)\}^{1/2}. \tag{5.4.3}$$

That is, Euclidean distance is used as the metric. This rule is commonly applied in pattern recognition; see, for example, Raudys and Pikelis (1980). It is very easy to implement given that the estimates $\mathbf{S}_i$ of the group-covariance matrices do not have to be computed from the training data.

The minimum-Euclidean distance rule (5.4.3) is equivalent to the sample NLDR $r_o(\mathbf{x};\hat{\Psi}_E)$ formed under the assumption that the group-covariance matrices have a common spherical form,

$$\mathbf{\Sigma}_i = \sigma^2\mathbf{I}_p \qquad (i = 1,\ldots,g).$$

Thus, it can be viewed as a substantially regularized version of $r_o(\mathbf{x};\hat{\Psi}_E)$. It was in this spirit that Marco, Young, and Turner (1987a) have compared its performance relative to the Mahalanobis distance version, or, equivalently, the sample NLDR formed with a zero cutoff point (equal group-priors) in the case of $g = 2$ groups in which the feature vector has a multivariate normal distribution with a common covariance matrix. They concluded from their simulation experiments that the sample Euclidean distance rule is superior to the sample NLDR, not only for group-conditional distributions that are spherical normal, but also for some nonspherical parameter configurations where p is very large relative to n. In the latter case, they found that the relative superiority of these two rules is highly dependent on the ratio of the Mahalanobis distance to the Euclidean distance between the groups. Whenever this ratio was small in their experiments, the sample Euclidean distance rule tended to outperform the sample NLDR, whereas the reverse appeared to be true whenever the ratio was large. Note, however, that there is no need to make an outright choice between the sample Euclidean distance rule over the sample NLDR with applications in those aforementioned situations where the former has the potential to offer some improvement in error rate. We can use regularized discriminant analysis as proposed by Friedman (1989), whereby as part of the estimation of each group-covariance matrix, shrinkage toward a common multiple of the identity matrix is allowed. The amount of shrinkage is inferred from the training data. Before we describe this new method of regularization in Section 5.5, we proceed to consider some less sophisticated methods of regularization that have been employed in the past in discriminant analysis.

5.4.2 Proportional Group-Covariance Matrices

The choice between homoscedasticity and the general heteroscedastic model is a fairly restrictive one. Also, as argued by Hawkins and Raath (1982) and others, the general specification of the group-covariance matrices Σ_i under the heteroscedastic model is contrary to experience in real-life applications, where it is usual to find a degree of similarity of pattern between the Σ_i. They therefore proposed specifying the Σ_i to be proportional as an intermediate model between the overly diffuse heteroscedastic model and the overly rigid homoscedastic model. It results in $\frac{1}{2}(g-1)p(p-1)$ fewer parameters having to be estimated than with arbitrary group-covariance matrices.

Practical applications of this proportional model in discriminant analysis have been considered also by Switzer (1980), Dargahi-Noubary (1981), and Owen (1984), among others. From a theoretical point of view, we have already seen in Section 4.6.1 that the distribution of the sample NQDF is simplified considerably for proportional group-covariance matrices. On the use of proportional covariance matrices in general and not necessarily in a discriminant analysis context, Flury (1988, Chapter 5) has given an account, including some historical remarks.

The assumption of proportional group-covariance matrices can be represented as

$$\Sigma_i = \kappa_i^2 \Sigma_P \qquad (i = 1, \ldots, g), \tag{5.4.4}$$

where $\kappa_1^2 = 1$. We set $\hat{\kappa}_1^2 = 1$ and let $\hat{\boldsymbol{\Sigma}}_i$ be the sample covariance matrix of the training feature data from group G_i $(i = 1,\ldots,g)$. Then under the assumption of multivariate normality for the group-conditional distributions, the maximum likelihood estimates $\hat{\boldsymbol{\Sigma}}_P$ and $\hat{\kappa}_i$ of $\boldsymbol{\Sigma}_P$ and κ_i, respectively, satisfy

$$\hat{\boldsymbol{\Sigma}}_P = \sum_{i=1}^{g} n_i \hat{\boldsymbol{\Sigma}}_i/(n\hat{\kappa}_i^2) \tag{5.4.5}$$

and

$$\hat{\kappa}_i = \{\mathrm{tr}(\hat{\boldsymbol{\Sigma}}_P^{-1}\hat{\boldsymbol{\Sigma}}_i)/p\}^{1/2} \qquad (i = 2,\ldots,g). \tag{5.4.6}$$

These equations can be solved iteratively. Starting with $\hat{\kappa}_i = 1$ for all i, $\hat{\boldsymbol{\Sigma}}_P$ can be obtained from (5.4.5) and then used in (5.4.6) to produce new values for $\hat{\kappa}_2,\ldots,\hat{\kappa}_g$. This iterative method of solution is the same, or essentially the same, as proposed independently by Owen (1984), Eriksen (1987), and Manly and Rayner (1987). Eriksen (1987) also established the convergence of the process and the uniqueness of the maximum likelihood estimates. Jensen and Johansen (1987) proved existence and uniqueness of the maximum likelihood estimates using results on the convexity of the likelihood function. Flury (1986) used a different parameterization of the model (5.4.4) in terms of the eigenvectors and eigenvalues of the $\boldsymbol{\Sigma}_i$ instead of the $\boldsymbol{\Sigma}_i$ themselves.

The likelihood ratio test for proportional group-covariance matrices is described in Section 6.2.6, along with a test of the more general model that the group-conditional correlations between the variates of the feature vector are the same within each group. We now consider estimation of the group-covariance matrices under this last model.

5.4.3 Equal Group-Correlation Matrices

The model of equal correlation matrices within the groups can be represented in terms of the group-covariance matrices $\boldsymbol{\Sigma}_i$ as

$$\boldsymbol{\Sigma}_i = \mathbf{K}_i \boldsymbol{\Sigma}_C \mathbf{K}_i, \tag{5.4.7}$$

where $\mathbf{K}_1$ is the identity matrix, and

$$\mathbf{K}_i = \mathrm{diag}(\kappa_{i1},\ldots,\kappa_{ip}) \qquad (i = 2,\ldots,g).$$

Note that (5.4.4) is a special case of (5.4.7) with $\kappa_{iv} = \kappa_i$ $(v = 1,\ldots,p)$. For multivariate normal group-conditional distributions, the maximum likelihood estimates $\hat{\boldsymbol{\Sigma}}_C$ and $\hat{\mathbf{K}}_i$ of $\boldsymbol{\Sigma}_C$ and $\mathbf{K}_i$, respectively, satisfy

$$\hat{\boldsymbol{\Sigma}}_C = \sum_{i=1}^{g} (n_i/n)\hat{\mathbf{K}}_i^{-1}\hat{\boldsymbol{\Sigma}}_i\hat{\mathbf{K}}_i^{-1} \tag{5.4.8}$$

and

$$\hat{\kappa}_{iv} = \sum_{j=1}^{p} (\hat{\boldsymbol{\Sigma}}_C^{-1})_{jv}(\hat{\boldsymbol{\Sigma}}_i)_{jv}/\hat{\kappa}_{ij} \qquad (i = 2,\ldots,g;\ v = 1,\ldots,p).$$

These equations can be solved by iteration, starting with $\hat{\kappa}_{ij} = 1$ for $j = 1,\ldots,p$. Manly and Rayner (1987) note that this procedure has always converged in test data, although the number of iterations required has been quite large in some cases.

The models of proportional group-covariance matrices and equal group-correlation matrices as given above provide ways of exploiting similarities in the group-covariance matrices $\mathbf{\Sigma}_i$ and thereby reducing the number of parameters to be estimated. These two models complete a hierarchy of models for the $\mathbf{\Sigma}_i$ with their being nested between the lower level of homoscedasticity and the upper level of heteroscedasticity. Another hierarchy of models for the $\mathbf{\Sigma}_i$ is considered next.

5.4.4 Common Principal-Component Model

Flury (1984) proposed the common principal-component (CPC) model for incorporating similarities in the group-covariance matrices. This model has been studied in some depth in the monograph of Flury (1988, Chapter 7), who has outlined a hierarchical set of models for a collection of covariance matrices. Under the CPC model, the $\mathbf{\Sigma}_i$ are taken to have the same principal axes, but these axes can be of different sizes and rankings in the different groups. It is thus equivalent to the assumption that the $\mathbf{\Sigma}_i$ are all diagonalizable by the same orthogonal matrix; that is,

$$\mathbf{A}\mathbf{\Sigma}_i\mathbf{A}' = \mathbf{\Lambda}_i \qquad (i = 1,\ldots,g), \tag{5.4.9}$$

where $\mathbf{A}$ is an orthogonal $p \times p$ matrix, and the $\mathbf{\Lambda}_i$ are all diagonal matrices,

$$\mathbf{\Lambda}_i = \operatorname{diag}(\lambda_{i1},\ldots,\lambda_{ip}) \qquad (i = 1,\ldots,g).$$

The model of proportional group-covariance matrices introduced in the previous section can be viewed as an offspring of the CPC model obtained by imposing the constraints

$$\lambda_{ik} = \kappa_i^2\lambda_{1k} \qquad (i = 2,\ldots,g;\ k = 1,\ldots,p).$$

Flury (1988, Chapter 4) has given the maximum likelihood estimates of $\mathbf{A}$ and the $\mathbf{\Lambda}_i$ for the CPC model under the assumption of multivariate normal group-conditional distributions. By writing $\mathbf{A} = (\boldsymbol{\psi}_1,\ldots,\boldsymbol{\psi}_p)$, these maximum likelihood estimates of $\mathbf{A}$ and $\mathbf{\Lambda}_i$ satisfy

$$\boldsymbol{\psi}_k'\tilde{\mathbf{S}}_{km}\boldsymbol{\psi}_m = \mathbf{0} \qquad (k,m = 1,\ldots,p;\ k \neq m), \tag{5.4.10}$$

where

$$\tilde{\mathbf{S}}_{km} = \sum_{i=1}^{g}(n_i - 1)\frac{\lambda_{ik} - \lambda_{im}}{\lambda_{ik}\lambda_{im}}\mathbf{S}_i.$$

The equation system (5.4.10) has to be solved under the orthogonality constraints

$$\boldsymbol{\psi}_k'\boldsymbol{\psi}_m = \delta_{km},$$

where δ_{km} is the Kronecker delta. An algorithm for solving (5.4.10) has been proposed by Flury and Gautschi (1986).

If no distributional assumptions are made about the group-conditional distributions, then the least-squares estimates of $\mathbf{A}$ and the $\mathbf{\Lambda}_i$ can be computed using the routine provided by Clarkson (1988).

Schmid (1987) has investigated the performances of the usual plug-in sample versions $r_o(\mathbf{x}; \hat{\Psi}_E)$ and $r_o(\mathbf{x}; \hat{\Psi}_U)$ of the normal-based linear and quadratic rules relative to the quadratic versions formed with the group-covariance matrices $\mathbf{\Sigma}_i$ estimated under the models of proportional $\mathbf{\Sigma}_i$ and of common principal components. Some of his results and conclusions have been summarized by Flury (1988, Section 8.4). They indicate that the use of the CPC model in discriminant analysis may be worthwhile in the case of several groups and relatively high dimension.

Flury (1987a) has proposed a generalization of the CPC model, namely, the partial CPC model, in which only q out of the p eigenvectors specific to a group-covariance matrix are common to all g groups. He also proposed another generalization in the form of the common space model, in which the first q (or the last $p-q$) eigenvectors of $\mathbf{S}_i$ span the same subspace for each i $(i = 1,\ldots,g)$; see also Schott (1988). The common space model is an alternative to the suggestions of Krzanowski (1979b, 1982b, 1984b) in comparing the principal components of several groups.

As cautioned by Flury (1988, Section 7.3), principal-component analysis is scale-dependent, and so the hierarchy of principal-component models outlined above may not be meaningful if the feature variables are measured on disparate scales. As the models of proportional group-covariance matrices and equal group-correlation matrices considered in the previous section are both scale-dependent, they would appear to provide a preferable way of completing a hierarchy of models ranging from homoscedasticity to heteroscedasticity.

5.4.5 SIMCA and DASCO Methods

A method that has been especially developed for situations in which the number of features p is large relative to the training sample size n is SIMCA (soft independent modeling of class analogy). An initial description of this method was given by Wold (1976). Recent accounts can be found in Dröge and van't Klooster (1987), Frank and Friedman (1989), and Frank and Lanteri (1989). Many successful applications of SIMCA to chemical problems have been reported; see the references in Kowalski and Wold (1982), who have reviewed the use of pattern-recognition techniques, including SIMCA, in chemistry. Hence, SIMCA is widely used as a discriminant technique in chemometrics, where typically p is large relative to n; see, for example, Dröge et al. (1987).

With the SIMCA method, the feature vector $\mathbf{x}$ is represented in each group by a principal-component model. Generally, the number of principal components retained will be different for each group. An unclassified entity is allocated on the basis of the relative distance of its feature vector from these group models. More specifically, let $\psi_{i1},\ldots,\psi_{ip}$ be the eigenvectors of unit length corresponding to the eigenvalues $\lambda_{i1} \geq \cdots \geq \lambda_{ip}$ of $\mathbf{S}_i$ $(i = 1,\ldots,g)$. Suppose that p_i denotes the number of principal components retained in the model for the ith group G_i $(i = 1,\ldots,g)$. Then the SIMCA allocation rule assigns an entity with feature $\mathbf{x}$ on the basis of the minimum ratio of entity to average group-residual sum of squares,

$$\min_i \frac{d_i^2(\mathbf{x})/(p-p_i)}{\sum_{j=1}^n z_{ij} d_i^2(\mathbf{x}_j)/\{(p-p_i)(n_i-p_i-1)\}}, \tag{5.4.11}$$

where

$$d_i^2(\mathbf{x}) = \sum_{k=p_i+1}^{p} \{(\mathbf{x}-\bar{\mathbf{x}}_i)'\psi_{ik}\}^2 \qquad (i = 1,\ldots,g).$$

The term $d_i^2(\mathbf{x})$ is the sum of squares of the values of the omitted principal components of $\mathbf{x}-\bar{\mathbf{x}}_i$ in group G_i.

Frank and Friedman (1989) have shown that the SIMCA rule can be viewed as a minimum-distance rule of the form (5.4.1), based on the estimated Mahalanobis distance from each group-sample mean. To see this note that

$$\begin{aligned}\sum_{j=1}^n z_{ij} d_i^2(\mathbf{x}_j) &= \sum_{j=1}^n \sum_{k=p_i+1}^{p} z_{ij}\psi_{ik}'(\mathbf{x}_j-\bar{\mathbf{x}}_i)(\mathbf{x}_j-\bar{\mathbf{x}}_i)'\psi_{ik} \\ &= (n_i-1)\sum_{k=p_i+1}^{p} \psi_{ik}'\mathbf{S}_i\psi_{ik} \\ &= (n_i-1)\sum_{k=p_i+1}^{p} \lambda_{ik}.\end{aligned}$$

Thus (5.4.11) can be expressed as

$$\min_i (\mathbf{x}-\bar{\mathbf{x}}_i)'\tilde{\mathbf{\Sigma}}_i^{-1}(\mathbf{x}-\bar{\mathbf{x}}_i),$$

where

$$\tilde{\mathbf{\Sigma}}_i^{-1} = \sum_{k=p_i+1}^{p} \psi_{ik}\psi_{ik}'/c_i \tag{5.4.12}$$

and

$$c_i = \{(n_i-1)/(n_i-p_i-1)\} \sum_{k=p_i+1}^{p} \lambda_{ik}. \tag{5.4.13}$$

From (5.2.1),

$$\mathbf{S}_i^{-1} = \sum_{k=1}^{p} \psi_{ik}\psi_{ik}'/\lambda_{ik}.$$

On contrasting this with (5.4.12), it can be seen that SIMCA is a method of regularization whereby all the eigenvalues associated with the ith group primary subspace are estimated to be infinitely large, that is, $1/\lambda_{ik} = 0$ for $k = 1,\ldots,p_i$. The remaining eigenvalues λ_{ik} $(k = p_i + 1,\ldots,p)$ are all estimated by c_i.

As pointed out by Frank and Friedman (1989), the SIMCA method has two shortcomings. Firstly, by taking $1/\lambda_{ik} = 0$ for $k = 1,\ldots,p_i$ in the estimate of $\mathbf{\Sigma}_i^{-1}$, it ignores all information on group differences in the primary subspaces. Secondly, by considering only the Mahalanobis distance of $\mathbf{x}$ from each group-sample mean, it effectively ignores the normalizing term $|\mathbf{\Sigma}_i|^{-1/2}$ in the multivariate normal density for the distribution of $\mathbf{X}$ in G_i. As remarked earlier, this term can make an important contribution in the case of disparate group-covariance matrices. Frank and Friedman (1989) introduced a modified method called DASCO (discriminant analysis with shrunken covariances), which overcomes these two weaknesses of the SIMCA method. As with SIMCA, DASCO estimates the inverse of each $\mathbf{\Sigma}_i$ by partitioning the p-dimensional feature space in G_i into two subspaces: a primary subspace of dimension p_i and its complement (secondary subspace) of dimension $p - p_i$. In forming the estimate of $\mathbf{\Sigma}_i$, the $p - p_i$ eigenvalues associated with the latter subspace are all estimated by

$$\bar{\lambda}_i = \sum_{k=p_i+1}^{p} \lambda_{ik}/(p - p_i),$$

the average of the last $p - p_i$ eigenvalues of $\mathbf{S}_i$. It can be seen from (5.4.12) and (5.4.13) that this is almost the same as with SIMCA, because $\bar{\lambda}_i$ equals c_i apart from a multiplicative constant. However, in contrast to SIMCA, with DASCO, the eigenvalues associated with the primary subspace are not taken to be infinitely large. Rather they are taken to be the same as the p_i largest ones of $\mathbf{S}_i$. The estimate of $\mathbf{\Sigma}_i^{-1}$ so obtained is given by

$$\hat{\hat{\mathbf{\Sigma}}}_i^{-1} = \sum_{k=1}^{p_i} \psi_{ik}\psi_{ik}'/\lambda_{ik} + \sum_{k=p_i+1}^{p} \psi_{ik}\psi_{ik}'/\bar{\lambda}_i.$$

The DASCO allocation rule is then taken to be the NQDR $r_o(\mathbf{x};\mathbf{\Psi}_U)$ with $\boldsymbol{\mu}_i$ and $\mathbf{\Sigma}_i$ replaced by $\bar{\mathbf{x}}_i$ and $\hat{\hat{\mathbf{\Sigma}}}_i$ respectively $(i = 1,\ldots,g)$, and with the group-prior probabilities π_i assumed to be equal. It reduces under the latter assumption to the sample NQDR $r_o(\mathbf{x};\hat{\mathbf{\Psi}}_U)$ in the special case $p_i = p$ $(i = 1,\ldots,g)$.

The SIMCA and DASCO discriminant rules depend on the parameters $p_1,\ldots,p_g$, the primary subspace dimensionality for each of the groups. It is not computationally feasible to choose them jointly so as to minimize an estimate of the overall error rate. With the SIMCA method, each p_i is assessed separately by the value of p_i that minimizes a cross-validated estimate of

$$\left\| \mathbf{S}_i - \sum_{k=1}^{p_i} \lambda_{ik}\psi_{ik}\psi_{ik}' \right\|; \tag{5.4.14}$$

see Wold (1976, 1978) and Frank and Friedman (1989). Also, Eastment and Krzanowski (1972) have considered a similar cross-validation approach to the choice of the number of retained principal components. As explained by Frank and Friedman (1989), although (5.4.14) is not an unreasonable criterion, it is not directly related to the overall error rate of the consequent allocation rule. There is a wide variety of situations in which minimizing (5.4.14) gives very different results from minimizing the cross-validated error rate. An example occurs when the eigenvectors of the primary subspace are the same or similar for each group and the group differences occur along these directions. This latter information is ignored with the SIMCA rule, because it is based on only the secondary subspaces.

The cross-validation of (5.4.14) may be too expensive to undertake, at least with one observation omitted at a time, if n is large. In order to achieve computational feasibility, the DASCO method takes a different approach to the assessment of each p_i in order to achieve computational feasibility. Its assessment is the value of p_i that maximizes a cross-validated estimate of

$$\sum_{k=1}^{p_i} \lambda_{ik}/\text{tr}(\mathbf{S}_i), \tag{5.4.15}$$

where

$$\text{tr}(\mathbf{S}_i) = \sum_{k=1}^{p} \lambda_{ik}.$$

The quantity (5.4.15) is the fraction of the total variance associated with the primary subspace in G_i $(i = 1,\ldots,g)$.

5.5 REGULARIZED DISCRIMINANT ANALYSIS (RDA)

5.5.1 Formulation

Friedman (1989) has proposed regularized discriminant analysis (RDA) as a compromise between normal-based linear and quadratic discriminant analyses. With this approach, a two-parameter family of estimates of the $\mathbf{\Sigma}_i$ is considered, where one parameter controls shrinkage of the heteroscedastic estimates toward a common estimate. The other parameter controls shrinkage toward a multiple of a specified covariance matrix such as the identity matrix. Through these two parameters, a fairly rich class of regularized alternatives is provided. Further, with these two parameters assessed from the training set by minimization of the cross-validated estimate of the overall error rate, a compromise between sample normal-based linear and quadratic analyses is determined automatically from the available data.

More specifically, let

$$\hat{\mathbf{\Sigma}}_i(\lambda) = \{(1-\lambda)(n_i-1)\mathbf{S}_i + \lambda(n-g)\mathbf{S}\}/\{(1-\lambda)(n_i-1) + \lambda(n-g)\}, \tag{5.5.1}$$

where $\lambda(0 \leq \lambda \leq 1)$ is a regularization parameter controlling the degree of shrinkage toward the pooled estimate $\mathbf{S}$. Previously, Randles et al. (1978a) had proposed a weighted average of the sample NLDF and NQDF by using a weighted estimate of the form (5.5.1), except that it was expressed in terms of the inverses of the covariance matrices. The weight λ was chosen adaptively by basing it on Wilks' likelihood ratio statistic.

The regularization provided by (5.5.1) is still fairly limited. Firstly, it might not provide for enough regularization. If n is less than or comparable to p, then even linear discriminant analysis is ill- or poorly posed. Secondly, biasing the $\mathbf{S}_i$ toward their pooled value may not be the most effective way to shrink them. Friedman (1989) therefore proposed that the estimate of $\mathbf{\Sigma}_i$ be regularized further as

$$\hat{\mathbf{\Sigma}}_i(\lambda,\gamma) = (1-\gamma)\hat{\mathbf{\Sigma}}_i(\lambda) + \gamma c_i \mathbf{I}_p, \tag{5.5.2}$$

where $\mathbf{I}_p$ is the $p \times p$ identity matrix, and

$$c_i = \{\mathrm{tr}\hat{\mathbf{\Sigma}}_i(\lambda)\}/p.$$

For a given value of λ, the additional regularization parameter γ $(0 \leq \gamma \leq 1)$ controls shrinkage toward a multiple of the identity matrix. The multiplier c_i is just the average value of the eigenvalues of $\hat{\mathbf{\Sigma}}_i(\lambda)$. This shrinkage has the effect of decreasing the larger eigenvalues and increasing the smaller ones of $\hat{\mathbf{\Sigma}}_i(\lambda)$, thereby counteracting the bias inherent in the estimates provided by these eigenvalues. It is the same type of shrinkage as with ridge regression estimates of a covariance matrix that have been used in the context of discriminant analysis by Di Pillo (1976, 1977, 1979), Campbell (1980b), Peck and Van Ness (1982), Kimura et al. (1987), and Rodriguez (1988). A comparison of some of these biased methods for improving the error rate of the consequent sample quadratic discriminant rule has been given recently by Peck, Jennings, and Young (1988). Biased estimators of the group-covariance matrices such as these before are formally derivable by Bayes and empirical Bayes arguments. An example of this is the empirical Bayes formulation adopted by Greene and Rayens (1989) in providing a compromise rule between the sample NLDR and the NQDR.

Let $\hat{\mathbf{\Psi}}_U(\lambda,\gamma)$ be the estimate of $\mathbf{\Psi}_U$ obtained by replacing $\mathbf{S}_i$ with $\hat{\mathbf{\Sigma}}(\lambda,\gamma)$ for $i = 1,\ldots,g$. Then $r_o(\mathbf{x};\hat{\mathbf{\Psi}}_U(\lambda,\gamma))$ is the normal-based regularized discriminant rule (NRDR) as proposed by Friedman (1989). This rule provides a fairly rich class of regularization alternatives. The four corners defining the extremes of the λ,γ plane represent fairly well-known discriminant rules. The lower left-hand corner ($\lambda = 0,\ \gamma = 0$) gives the usual quadratic rule $r_o(\mathbf{x};\hat{\mathbf{\Psi}}_U)$, and the lower right-hand corner ($\lambda = 1,\ \gamma = 0$) gives the usual linear rule $r_o(\mathbf{x};\hat{\mathbf{\Psi}}_E)$. The upper right-hand corner ($\lambda = 1,\ \gamma = 1$) corresponds to the minimum-Euclidean distance rule

$$\min_i \{(\mathbf{x}-\bar{\mathbf{x}}_i)'(\mathbf{x}-\bar{\mathbf{x}}_i)\}^{1/2},$$

because

$$\hat{\Sigma}_i(1,1) = \{\text{tr}(\mathbf{S})/p\}\mathbf{I}_p.$$

Finally, the upper left-hand corner ($\lambda = 0,\ \gamma = 1$) corresponds to the minimum-distance rule

$$\min_i \{c_i(\mathbf{x}-\bar{\mathbf{x}}_i)'(\mathbf{x}-\bar{\mathbf{x}}_i)\}^{1/2}.$$

Holding γ fixed at zero and varying λ provides rules between $r_o(\mathbf{x};\hat{\mathbf{\Psi}}_E)$ and $r_o(\mathbf{x};\hat{\mathbf{\Psi}}_U)$. A ridge regression analogue of $r_o(\mathbf{x};\mathbf{\Psi}_E)$ is obtained by holding λ fixed at one and increasing γ.

The linear and quadratic rules $r_o(\mathbf{x};\hat{\mathbf{\Psi}}_E)$ and $r_o(\mathbf{x};\hat{\mathbf{\Psi}}_U)$ are scale-invariant. However, the regularized rule $r_o(\mathbf{x};\hat{\mathbf{\Psi}}_U(\hat{\lambda},\hat{\gamma}))$ is generally not. This lack of scale invariance results from the use of the shrinkage parameter γ. In the formulation (5.5.2) of $\hat{\Sigma}_i(\lambda,\gamma)$, shrinkage is toward a multiple of the identity matrix $\mathbf{I}_p$. As there is nothing special about the choice of $\mathbf{I}_p$, Friedman (1989) noted that one could consider more general regularizations of the form

$$\hat{\Sigma}_i(\lambda,\gamma) = (1-\gamma)\hat{\Sigma}_i(\lambda) + \gamma c_i \mathbf{M}, \tag{5.5.3}$$

where $\mathbf{M}$ is a prespecified positive definite matrix, and

$$c_i = \text{tr}\{\hat{\Sigma}_i(\lambda)\}/\text{tr}(\mathbf{M}).$$

Let $\mathbf{M} = \mathbf{M}_1\mathbf{M}_1'$ be the Cholesky factorization of $\mathbf{M}$, where $\mathbf{M}_1$ is a lower triangular matrix. Then this generalized version of RDA can be implemented by first transforming the feature data $\mathbf{x}_j$ to $\mathbf{M}_1^{-1}\mathbf{x}_j$ ($j = 1,\ldots,n$), so that $\mathbf{M}$ reduces to $\mathbf{I}_p$ in (5.5.3).

A common procedure is to standardize the feature data so that all p feature variables have the same (sample) variances. This can be achieved by specifying $\mathbf{M}$ as

$$\mathbf{M} = \text{diag}(m_1,\ldots,m_p),$$

where

$$m_k = \sum_{j=1}^{n}(\mathbf{x}_j - \bar{\mathbf{x}})_k^2/(n-1)$$

or $(\mathbf{S})_{kk}$, depending on whether the variance is computed from the training data $\mathbf{t}$ as a whole or is pooled within each group. The reader is referred to Friedman (1989) for further discussion on the choice of $\mathbf{M}$, including the case where the feature vector $\mathbf{x}$ corresponds to a signal or image.

5.5.2 Assessment of Regularization Parameters for RDA

Without complete knowledge of the group-conditional distributions, the optimal values of λ (the covariance matrix mixing parameter) and γ (the eigenvalue-shrinkage parameter) are unknown. Friedman (1989) recommends that in a given situation without this knowledge, the optimal values of λ and γ be assessed by $\hat{\lambda}$ and $\hat{\gamma}$, defined to be the values of λ and γ that minimize the cross-validated estimate $A^{(CV)}(\lambda,\gamma)$ of the overall error rate associated with $r_o(\mathbf{x};\hat{\mathbf{\Psi}}_U(\lambda,\gamma))$. On the basis of the available training data $\mathbf{t}$ as defined by (1.6.1),

$$A^{(CV)}(\lambda,\gamma) = \frac{1}{n}\sum_{i=1}^{g}\sum_{j=1}^{n} z_{ij}Q[z_{ij}, r_o(\mathbf{x};\hat{\mathbf{\Psi}}_{U_{(j)}}(\lambda,\gamma))]$$

where, for any u and v, $Q[u,v] = 0$ for $u = v$ and 1 for $u \neq v$, and where $\hat{\mathbf{\Psi}}_{U_{(j)}}(\lambda,\gamma)$ denotes the estimate $\hat{\mathbf{\Psi}}_U(\lambda,\gamma)$ formed from the training data $\mathbf{t}$ with $\mathbf{y}_j = (\mathbf{x}_j', \mathbf{z}_j')'$ omitted $(j = 1,\ldots,n)$. Cross-validation of the error rates of a sample discriminant rule is to be considered in some depth in Chapter 10.

The computation of the assessed regularization parameters $\hat{\lambda}$ and $\hat{\gamma}$ thus gives rise to a two-parameter numerical minimization problem. The strategy recommended by Friedman (1989) is to choose $\hat{\lambda}$ and $\hat{\gamma}$ on the basis of the smallest value of $A^{(CV)}(\lambda,\gamma)$ evaluated at each prescribed point on a grid of points on the λ,γ plane $(0 \leq \lambda \leq 1,\ 0 \leq \gamma \leq 1)$. Typically, the optimization grid is taken to be from 25 to 50 points. With this strategy, each grid point requires the calculation of the n estimates $\hat{\mathbf{\Psi}}_{U_{(1)}},\ldots,\hat{\mathbf{\Psi}}_{U_{(n)}}$. In order to reduce this computational burden to an acceptable level, Friedman (1989) developed updating formulas for the computation of $\hat{\mathbf{\Sigma}}_{i(j)}(\lambda,\gamma)$, the estimate $\hat{\mathbf{\Sigma}}_i(\lambda,\gamma)$ based on $\mathbf{t}$ with $\mathbf{y}_j$ omitted $(j = 1,\ldots,n;\ i = 1,\ldots,g)$. In the presentation of his RDA approach, Friedman (1989) allowed for robust versions of $\mathbf{S}_i$ and $\mathbf{S}$ to be used in the form (5.5.1) for $\hat{\mathbf{\Sigma}}_i(\lambda)$. Instead of the latter estimate, he used

$$\tilde{\mathbf{\Sigma}}_i(\lambda) = \{(1-\lambda)w_i\tilde{\mathbf{\Sigma}}_i + \lambda w\tilde{\mathbf{\Sigma}}\}/w_i(\lambda) \qquad (i = 1,\ldots,g),$$

where

$$\tilde{\mathbf{\Sigma}}_i = \sum_{j=1}^{n} u_j z_{ij}(\mathbf{x} - \tilde{\boldsymbol{\mu}}_i)(\mathbf{x}_j - \tilde{\boldsymbol{\mu}}_i)'/w_i,$$

$$\tilde{\mathbf{\Sigma}} = \sum_{i=1}^{g} w_i\tilde{\mathbf{\Sigma}}_i/w,$$

$$\tilde{\boldsymbol{\mu}}_i = \sum_{j=1}^{n} u_j z_{ij}\mathbf{x}_j/w_i,$$

$$w_i = \sum_{j=1}^{n} u_j z_{ij},$$

$$w = \sum_{i=1}^{g} w_i,$$

$$w_i(\lambda) = (1-\lambda)w_i + \lambda w,$$

and where u_j is the weight $(0 \leq u_j \leq 1)$ assigned to $\mathbf{x}_j (j = 1,\ldots,n)$. If $u_j = 1$ for $j = 1,\ldots,n$, then $\tilde{\mathbf{\Sigma}}_i(\lambda)$ reduces to $\hat{\mathbf{\Sigma}}_i(\lambda)$ for $i = 1,\ldots,g$.

The formulas given by Friedman (1989) apply to the use of the robust estimate $\tilde{\mathbf{\Sigma}}_i(\lambda,\gamma)$ defined by using $\tilde{\mathbf{\Sigma}}_i(\lambda)$ in place of $\hat{\mathbf{\Sigma}}_i(\lambda)$ in (5.5.2). Let $w_{i(j)}(\lambda)$ and $\tilde{\mathbf{\Sigma}}_{i(j)}(\lambda,\gamma)$ denote $w_i(\lambda)$ and $\tilde{\mathbf{\Sigma}}_i(\lambda,\gamma)$ with $\mathbf{y}_j$ omitted from the training data $\mathbf{t}$. Friedman (1989) showed that

$$w_{i(j)}(\lambda)\tilde{\mathbf{\Sigma}}_{i(j)}(\lambda,\gamma) = \mathbf{A}_{ij} - \mathbf{a}_{ij}\mathbf{a}'_{ij}, \tag{5.5.4}$$

where

$$\begin{aligned}
\mathbf{A}_{ij} &= w_i(\lambda)\tilde{\mathbf{\Sigma}}_i(\lambda,\gamma) - k_{ij}\mathbf{I}_p,\\
\mathbf{a}_{ij} &= \sqrt{(1-\gamma)c_{ij}}\sum_{h=1}^{g} z_{hj}(\mathbf{x}_j - \tilde{\boldsymbol{\mu}}_h),\\
k_{ij} &= \{\gamma(\mathbf{a}'_{ij}\mathbf{a}_{ij})\}/\{p(1-\gamma)\},\\
w_{i(j)}(\lambda) &= w_i(\lambda) - u_j\lambda^{1-z_{ij}},
\end{aligned}$$

and

$$c_{ij} = u_j\lambda^{1-z_{ij}}\sum_{h=1}^{g} z_{hj}w_h \Big/ \left(\sum_{h=1}^{g} z_{hj}w_h - u_j\right).$$

It can be seen from (5.5.4) that removing an observation from $\mathbf{t}$ is equivalent to downdating $\tilde{\mathbf{\Sigma}}_i(\lambda,\gamma)$ by a rank-one matrix plus a multiple of the identity matrix. From (5.5.4), it follows using a result of Bartlett (1951a) that the inverse of $\tilde{\mathbf{\Sigma}}_{i(j)}(\lambda,\gamma)$ can be computed as

$$w_{i(j)}(\lambda)[\mathbf{A}_{ij}^{-1} + \{(\mathbf{A}_{ij}^{-1}\mathbf{a}_{ij}\mathbf{a}'_{ij}\mathbf{A}_{ij}^{-1})/(1 - \mathbf{a}'_{ij}\mathbf{A}_{ij}^{-1}\mathbf{a}_{ij})\}].$$

The matrix $\mathbf{A}_{ij}$ is then inverted through its spectral decomposition to give

$$\mathbf{A}_{ij}^{-1} = \sum_{v=1}^{p} \boldsymbol{\psi}_{iv}\boldsymbol{\psi}'_{iv}/(\lambda_{iv} - k_{ij}),$$

where λ_{iv} is the vth eigenvalue of $w_i(\lambda)\tilde{\mathbf{\Sigma}}_i(\lambda,\gamma)$, and $\boldsymbol{\psi}_{iv}$ is its corresponding eigenvector ($v = 1,\ldots,p$; $i = 1,\ldots,g$). By using those results, the determinant of $\tilde{\mathbf{\Sigma}}_{i(j)}(\lambda,\gamma)$ can be computed as

$$\log|\tilde{\mathbf{\Sigma}}_{i(j)}(\lambda,\gamma)| = \sum_{v=1}^{p}\log(\lambda_{iv} - k_{ij}) + \log\left(1 - \sum_{v=1}^{p}\frac{a_{ijv}^2}{\lambda_{iv} - k_{ij}}\right) - p\log w_{i(j)}(\lambda),$$

where $a_{ijv} = (\mathbf{a}_{ij})_v$.

Finally, the estimate $\tilde{\boldsymbol{\mu}}_{i(j)}$ is given by $\tilde{\boldsymbol{\mu}}_i$ for $z_{ij} = 0$ and by

$$\tilde{\boldsymbol{\mu}}_{i(j)} = (w_i\tilde{\boldsymbol{\mu}}_i - u_j\mathbf{x}_j)/(w_i - u_j)$$

for $z_{ij} = 1$.

As noted recently by Rayens and Greene (1991) in a critical comparison of RDA and the empirical Bayes approach of Greene and Rayens (1989), their simulated values of the cross-validated error rate $A^{(CV)}(\lambda,\gamma)$ of $r_o(\mathbf{x};\hat{\Psi}_U(\lambda,\gamma))$ were often constant for a wide range of values of λ and γ. This implies that the optimal choice may not be uniquely determined. Thus, how ties are broken with RDA is obviously an important issue, as illustrated by Rayens and Greene (1991). As reported in their study, with RDA, ties are resolved by selecting among the set of points (λ,γ) with the smallest cross-validated error rate, the grid point with the largest value of γ from among the grid points with the largest value of λ. Breaking ties in this way amounts to maximizing the shrinkage of the resulting rule with λ given priority over γ.

5.5.3 Effectiveness of RDA

For the NRDR (normal-based regularized discriminant rule) $r_o(\mathbf{x};\hat{\Psi}_U(\hat{\lambda},\hat{\gamma}))$ to be of value in practice, the assessed values $\hat{\lambda}$ and $\hat{\gamma}$ of the regularization parameters need to lead to a high degree of regularization that substantially reduces the variability with the use of $\hat{\Psi}_U$.

In order to investigate the effectiveness of the RDA approach, Friedman (1989) compared the NRDR $r_o(\mathbf{x};\hat{\Psi}_U(\hat{\lambda},\hat{\gamma}))$ with the sample NLDR $r_o(\mathbf{x};\hat{\Psi}_E)$ and the sample NQDR $r_o(\mathbf{x};\hat{\Psi}_U)$ in terms of their simulated overall error rates. The simulated examples were designed to provide a fairly wide spectrum of situations in terms of the structure of the group means and covariance matrices. Some were chosen because they were highly unfavorable to regularization, and others were included because they were representative of situations where appreciable reduction in the error rate is possible through appropriate regularization. In each example, there were $g = 3$ groups and the training data consisted of $n = 40$ observations generated from a p-dimensional normal mixture of these groups in equal proportions ($p = 2$, 6, 10, 20, and 40). The optimization grid of (λ,γ) values was defined by the outer product of $\lambda = (0, 0.125, 0.354, 0.650, 1.0)$ and $\gamma = (0, 0.25, 0.5, 0.75, 1.0)$. When an $\mathbf{S}_i$ with the quadratic rule or the pooled estimate $\mathbf{S}$ with the linear rule happened to be singular, the zero eigenvalues were replaced by a small number just large enough to permit numerically stable inversion. This had the effect of producing an Euclidean-distance rule in the zero-variance subspace (the subspace spanned by the eigenvectors corresponding to the essentially zero eigenvalues). On each simulation trial, the overall conditional error rates of $r_o(\mathbf{x};\hat{\Psi}_U(\hat{\lambda},\hat{\gamma}))$, $r_o(\mathbf{x};\hat{\Psi}_E)$, and $r_o(\mathbf{x};\hat{\Psi}_U)$ were computed. In the notation of Section 1.10, these error rates are given by $ec(\Psi_U;\hat{\Psi}_U(\hat{\lambda},\hat{\gamma}))$, $ec(\Psi_U;\hat{\Psi}_E)$, and $ec(\Psi_U;\hat{\Psi}_U)$, respectively. The sample means of these conditional error rates over 100 simulation trials provided the simulated values of the corresponding unconditional rates. Friedman (1989) concluded from his simulations that the assessment of the regularization parameters λ and γ on the basis of cross-validation seems to perform surprisingly well. In each of the situations simulated, the optimal joint values of λ and γ were roughly known. The simulated distributions of the assessed values $\hat{\lambda}$ and $\hat{\gamma}$ were concentrated near these optimal values in each

TABLE 5.2 Equal, Highly Ellipsoidal Group-Covariance Matrices with Mean Differences in Low-Variance Subspace

	$p = 6$	$p = 10$	$p = 20$	$p = 40$
Error rate				
NRDR	0.07(0.04)	0.07(0.04)	0.27(0.07)	0.39(0.06)
NLDR	0.06(0.03)	0.06(0.03)	0.24(0.06)	0.59(0.07)
NQDR	0.17(0.08)	0.14(0.12)	0.60(0.07)	0.60(0.06)
$A^{(CV)}(\hat{\lambda},\hat{\gamma})$	0.05(0.04)	0.06(0.04)	0.21(0.07)	0.34(0.08)
CORR	0.19	0.0	0.0	0.16
$\hat{\lambda}$	0.77(0.32)	0.83(0.27)	0.75(0.30)	0.72(0.32)
$\hat{\gamma}$	0.02(0.08)	0.07(0.16)	0.19(0.27)	0.45(0.25)

Source: Adapted from Friedman (1989).

case. This explains why $r_o(\mathbf{x}; \hat{\Psi}_U(\hat{\lambda},\hat{\gamma}))$ seems to gain so much over $r_o(\mathbf{x}; \hat{\Psi}_E)$ and $r_o(\mathbf{x}; \hat{\Psi}_U)$ in favorable situations, yet to lose little to them in unfavorable ones. The reduction in error rate in situations where n is smaller relative to p is most encouraging. Indeed, as remarked by Friedman (1989), it is surprising how small the ratio n/p can be and still obtain fairly accurate allocation with $r_o(\mathbf{x}; \hat{\Psi}_U(\hat{\lambda},\hat{\gamma}))$.

As the assessed values $\hat{\lambda}$ and $\hat{\gamma}$ are taken to minimize the cross-validated estimate $A^{(CV)}(\lambda,\gamma)$ of the regularized rule for the given training data, $A^{(CV)}(\hat{\lambda},\hat{\gamma})$ will provide an optimistic assessment of the overall error rate $ec(\Psi_U; \hat{\Psi}_U(\hat{\lambda},\hat{\gamma}))$ of $r_o(\mathbf{x}; \hat{\Psi}_U(\hat{\lambda},\hat{\gamma}))$. In the simulated examples, $A^{(CV)}(\hat{\lambda},\hat{\gamma})$ underestimated $ec(\Psi_U; \hat{\Psi}_U(\hat{\lambda},\hat{\gamma}))$ by around 20% on average. A surprising result was the low correlation between $A^{(CV)}(\hat{\lambda},\hat{\gamma})$ and $ec(\Psi_U; \hat{\Psi}_U(\hat{\lambda},\hat{\gamma}))$. This implies that $A^{(CV)}(\hat{\lambda},\hat{\gamma})$ is providing an assessment of the unconditional error rate of $r_o(\mathbf{x}; \hat{\Psi}_U(\hat{\lambda},\hat{\gamma}))$ rather than its conditional error rate for the realized training data $\mathbf{t}$.

To illustrate the simulation results of Friedman (1989), we tabulate here the results for two of his simulation examples. The first here was the most difficult of the simulation examples from the point of view of RDA. The group-covariance matrices were the same and highly ellipsoidal, and the differences between the group means were concentrated in the low-variance subspace. The simulated overall unconditional error rates of the NRDR $r_o(\mathbf{x}; \hat{\Psi}_U(\hat{\lambda},\hat{\gamma}))$, of the sample NLDR $r_o(\mathbf{x}; \hat{\Psi}_E)$, and of the sample NQDR $r_o(\mathbf{x}; \hat{\Psi}_U)$ are reported in Table 5.2. Also listed in Table 5.2 are the averages of the assessed regularization parameters $\hat{\lambda}$ and $\hat{\gamma}$ and the minimum value $A^{(CV)}(\hat{\lambda},\hat{\gamma})$ of the cross-validated estimate of the overall error rate of $r_o(\mathbf{x}; \hat{\Psi}_U(\lambda,\gamma))$. The entry for CORR refers to the simulated correlation between $A^{(CV)}(\hat{\lambda},\hat{\gamma})$ and the apparent error rate of $r_o(\mathbf{x}; \hat{\Psi}_U(\hat{\lambda},\hat{\gamma}))$ in its application to $m = 100$ test observations generated subsequent to the training data $\mathbf{t}$. Standard errors are listed in parentheses in Table 5.2.

TABLE 5.3 Unequal, Highly Ellipsoidal Group-Covariance Matrices with Zero Mean Differences

	$p = 6$	$p = 10$	$p = 20$	$p = 40$
Error rate				
NRDR	0.21(0.06)	0.15(0.06)	0.12(0.05)	0.12(0.06)
NLDR	0.61(0.06)	0.58(0.06)	0.58(0.06)	0.63(0.06)
NQDR	0.19(0.06)	0.35(0.13)	0.44(0.10)	0.43(0.07)
$A^{(CV)}(\hat{\lambda},\hat{\gamma})$	0.17(0.06)	0.13(0.05)	0.11(0.05)	0.12(0.06)
CORR	0.03	−0.03	0.09	0.25
$\hat{\lambda}$	0.03(0.05)	0.04(0.06)	0.06(0.07)	0.05(0.07)
$\hat{\gamma}$	0.17(0.16)	0.27(0.18)	0.46(0.17)	0.60(0.15)

Source: Adapted from Friedman (1989).

It can be seen from Table 5.2 that the sample NLDR $r_o(\mathbf{x};\hat{\Psi}_E)$ performs slightly better in all but the highest dimension, where none of the three rules does particularly well. This situation, as constructed, is ideal for $r_o(\mathbf{x};\hat{\Psi}_E)$ because any shrinkage away from the point $(\lambda = 1, \gamma = 0)$ is strongly counterproductive. As the assessed values of $\hat{\lambda}$ and $\hat{\gamma}$ are concentrated in this corner of the λ,γ plane, the incurred increases in the average error rate from using the NRDR instead of the sample NLDR is only slight in this most unfavorable situation. In Table 5.2, the average value of $\hat{\gamma}$ increases with p, so that at the highest level of $p = 40$, considerable shrinkage is needed to damp the variance even though this introduces substantial bias.

For the second simulation example reported here in Table 5.3, the group-covariance matrices were highly ellipsoidal and very unequal. As the group means were taken to be all the same, it represents a situation where the NLDR does very poorly. It can be seen from Table 5.3 that for the lowest dimension of $p = 6$, the NRDR is slightly worse than the NQDR, but for the other values of p, the NRDR is substantially better. As in the first example, the assessed values $\hat{\lambda}$ and $\hat{\gamma}$ of the regularization parameters appear to be appropriate. The average value of $\hat{\lambda}$ is near to zero for each p, indicating that very little covariance matrix mixing takes place at any dimension.

Further simulation results on the RDA approach can be found in the recent study by Rayens and Greene (1991). They also describe the development of another discriminant rule, which combines ideas from RDA and the approach of Greene and Rayens (1989).

5.5.4 Examples

Friedman (1989) also analysed a real data set to demonstrate the effectiveness of regularization through the use of $r_o(\mathbf{x};\hat{\Psi}_U(\hat{\lambda},\hat{\gamma}))$. This data set consisted of $p = 14$ sensory characteristics measured on $n = 38$ wines originating from $g = 3$ different geographical regions: $n_1 = 9$ from California, $n_2 = 17$ from the

Pacific Northwest, and $n_3 = 12$ from France. Friedman (1989) performed two analyses of this data set in which the group-prior probabilities were always taken to be equal to 1/3. In the first, the regularized, linear, and quadratic rules, $r_o(\mathbf{x};\hat{\Psi}(\hat{\lambda},\hat{\gamma}))$, $r_o(\mathbf{x};\hat{\Psi}_E)$, and $r_o(\mathbf{x};\hat{\Psi}_U)$, respectively, were formed from the entire set. Their overall error rates as estimated by the 0.632 estimator of Efron (1983), which is defined in Section 10.4.3, were 0.18, 0.26, and 0.36, respectively. The cross-validated estimate $A^{(CV)}(\lambda,\gamma)$ of $r_o(\mathbf{x};\hat{\Psi}_U(\lambda,\gamma))$ had a minimum value of 0.14 at $(\hat{\lambda},\hat{\gamma}) = (0.35, 0.04)$.

In the second analysis, the data set was split randomly into half samples of size $n = 19$. The regularized, linear, and quadratic rules were formed from one sample and were applied to the other sample with apparent error rates of 0.21, 0.50, and 0.59, respectively. As noted by Friedman (1989), this data set does not appear to be favorable to regularization through the use of the linear rule $r_o(\mathbf{x};\hat{\Psi}_E)$. Even though the latter is fairly well-posed in the first analysis with $p = 14$ and $n = 38$, the regularized rule $r_o(\mathbf{x};\hat{\Psi}_U(\hat{\lambda},\hat{\gamma}))$ is substantially superior. In the second analysis, where the size n of the training data is halved to 19, $r_o(\mathbf{x};\hat{\Psi}_E)$ appears to collapse completely, whereas the estimated error rate of $r_o(\mathbf{x};\hat{\Psi}_U(\hat{\lambda},\hat{\gamma}))$ is increased surprisingly little.

Another way of proceeding in situations where the number of feature variables p is large relative to the training sample size n is to use only a subset of the p observed feature variables. There is a variety of variable-selection techniques available for this purpose and they are reviewed in Chapter 12, which is devoted to the topic of feature selection. The variable-subset-selection approach, which is really another method of regularization, can be effective if the main differences between the group means and covariance matrices happen to occur for a very small number of the original feature variables. However, as argued by Friedman (1989), the influential subset of features has to be surprisingly small for subset-selection techniques to be competitive with other regularization methods, or even no regularization at all; see Copas (1983).

5.6 ROBUSTNESS OF SAMPLE NLDR AND NQDR

5.6.1 Continuous Feature Data

Under the assumption of multivariate normality, the sample NLDR $r_o(\mathbf{x};\hat{\Psi}_E)$ and the sample NQDR $r_o(\mathbf{x};\hat{\Psi}_U)$ are asymptotically optimal in the presence of homoscedasticity and heteroscedasticity, respectively. Given that in real life there is no such thing as a "true model," it is of interest to consider the behavior of these two rules under departures of varying degrees from normality. Moreover, these rules are widely applied in practice in situations even where the group-conditional distributions of the feature data are obviously nonnormal.

There have been many studies on the robustness of $r_o(\mathbf{x};\hat{\Psi}_E)$ and $r_o(\mathbf{x};\hat{\Psi}_U)$, mainly in the case of $g = 2$ groups, where these two rules are based on the plug-in sample versions of the NLDF $\xi(\mathbf{x};\hat{\boldsymbol{\theta}}_E)$ and of the NQDF $\xi(\mathbf{x};\hat{\boldsymbol{\theta}}_U)$, respectively.

Concerning the robustness of the sample NLDF $\xi(\mathbf{x};\hat{\boldsymbol{\theta}}_E)$, it was shown in Section 3.3.3 that it can be obtained also, at least up to the coefficients of the feature variables in $\mathbf{x}$, by choosing the linear combination of the features (appropriately normalized) that maximizes the sample index (3.3.8). This is regardless of the distribution of $\mathbf{X}$. Hence, when the true log likelihood ratio $\xi(\mathbf{x})$ is linear in $\mathbf{x}$,

$$\begin{aligned}\xi(\mathbf{x}) &= \log\{f_1(\mathbf{x})/f_2(\mathbf{x})\}\\ &= \beta_0^o + \boldsymbol{\beta}'\mathbf{x},\end{aligned} \tag{5.6.1}$$

the sample NLDF $\xi(\mathbf{x};\hat{\boldsymbol{\theta}}_E)$ should provide a reasonable allocation rule. In some situations, the aim of a discriminant analysis is not the outright allocation of an unclassified entity or entities. Rather it is the estimation of the discrimination function coefficients in order either to assess the relative importance of the feature variables or to form reliable estimates of the posterior probabilities of group membership. In these situations, the use of the normal-based estimates $\hat{\beta}_{0E}^o$ and $\hat{\boldsymbol{\beta}}_E$ of the discriminant function coefficients can lead to quite misleading inferences being made. As discussed in Chapter 8, one way to proceed under the assumption of linearity of the log ratio of nonnormal group-conditional densities is to use logistic discrimination. This approach allows consistent estimation of the discriminant function coefficients, and, hence, of the posterior probabilities of group membership.

For the allocation problem, however, the sample NLDF $\xi(\mathbf{x};\hat{\boldsymbol{\theta}}_E)$ is fairly robust if (5.6.1) holds approximately. It is not robust against interactions between the feature variables, unless the interaction structure is essentially the same for each group.

For the remainder of this section, we are to focus on studies that have been carried out on the robustness of the allocation rules based on the sample NLDF and NQDF for continuous feature data. The case where some or all of the feature variables are discrete is considered in the next two sections.

Lachenbruch, Sneeringer, and Revo (1973) used Johnson's system of distributions to study the robustness of the sample NLDR and NQDR against lognormal, logitnormal, and inverse hyperbolic sine normal group-conditional distributions. They found that both these rules can be severely affected by such departures from normality. In the light of subsequent results in the literature that the sample NLDR is not badly affected by at least mild skewness and kurtosis in the group-conditional distributions, Lachenbruch and Goldstein (1979) commented that its poor performance in the aforementioned study may be a consequence of the fact that the lognormal distribution used for each feature vector had extremely large skewness and kurtosis. Later Chinganda and Subrahmaniam (1979) investigated the robustness of the sample NLDR against the same class of distributions. They concluded that in practice one should first attempt where possible to transform the feature data to normality before constructing the sample NLDF. In a parallel study, Subrahmaniam and Chinganda (1978) considered the robustness of the sample NLDR against several skewed distributions of the Edgeworth series type, representing a situation where the feature data are not amenable to being transformed to normality. The sample NLDR was found not to be affected to any great extent by this skewness.

Using the techniques developed by Subrahmaniam and Chinganda (1978), Balakrishnan and Kocherlakota (1985) investigated the robustness of the sample NLDR by modeling the group-conditional distributions as univariate normal mixtures. The normal mixture model had been used previously by Ashikaga and Chang (1981) to study the asymptotic error rate of this rule under departures from normality. These studies suggested that a moderate amount of skewness does not unduly affect the performance of the NLDR, especially if the group-conditional distributions are similar in shape. In more recent work on the robustness of the sample NLDR, Amoh and Kocherlakota (1986) have derived its error rates for inverse normal group-conditional distributions, and Kocherlakota, Balakrishnan, and Kocherlakota (1987) have investigated the effect of truncation on the error rates.

Ahmed and Lachenbruch (1975, 1977) have studied the performance of the sample NLDR when scale contamination is present in the training data. The effect of contaminated training data on the sample NLDR was pursued further by Broffitt, Clarke, and Lachenbruch (1981) under a random-shift-location contamination model, where each variate was allowed to be contaminated independently of the other variates in the feature vector. Only minimal effects on the error rates resulted.

On the robustness of the sample NQDR, Clarke, Lachenbruch, and Broffitt (1979) concluded from their simulation experiments that it was robust to nonnormality provided the group-conditional distributions were not highly skewed. Heavy kurtosis caused no problems. In all cases studied, the overall error rate was relatively stable, whereas the group-specific error rates exhibited considerable variability. Broffitt, Clarke, and Lachenbruch (1980) subsequently assessed the improvement in performance of the sample NQDR by Huberizing and trimming the estimated group means and covariance matrices before forming the sample NLDF. The computation of Huber M-estimates of the means and covariance matrices of the group-conditional distributions is considered in Section 5.7.1. The use of these robust estimates provides protection against any harmful effects of contamination of the training data, and also of misclassification as discussed in Section 2.5.

As part of a wider study on the performances of the sample NLDR and NQDR relative to other discriminant rules under three types of nonnormality for bivariate group-conditional distributions, Bayne et al. (1983) reported that the robustness of the sample NQDR depends upon the type of nonnormality present. Also, they found its performance to be sensitive to the sample size. This last factor was seen in Section 5.3.3 to be crucial in determining whether the sample NQDF was superior to the sample NLDF under normality. More recent investigations, where the sample NLDR and NQDR are evaluated as part of wider studies on various sample discriminant rules in nonnormal situations, include Rawlings and Faden (1986) and Joachimsthaler and Stam (1988).

On the basis of the published results and their own work on the robustness of the sample NLDR and NQDR, Fatti, Hawkins, and Raath (1982) made the following generalizations. If the group-conditional distributions have lighter tails than the normal, then the sample NLDR and NQDR should still perform adequately. However, if the distributions are heavy-tailed and skewed, then the sample NLDR and NQDR will perform very poorly. But if the distributions are heavy-tailed but essentially symmetric, then the sample NQDR may perform reasonably well in terms of its overall error rate. This is provided the sample size is sufficiently large to avoid excessive sampling errors in the estimated group means and covariance matrices.

On this last point of the robustness the sample NQDR to heavy-tailed but symmetric group-conditional distributions, it is the ellipsoidal symmetry associated with the multivariate normal assumption for the group-conditional distributions that appears to be the most important aspect rather than its detailed shape. It underscores the importance of normalizing transformations before constructing the sample NQDF. Even if a transformation may not achieve normality, it can still play a valuable role through the removal of skewness from the group-conditional distributions.

5.6.2 Discrete Feature Data

In this section, we consider the performance of the sample NLDR $\xi(\mathbf{x};\hat{\boldsymbol{\theta}}_E)$ in situations where some or all of the feature variables are discrete. As noted in the previous section, for allocation purposes, the sample NLDF $\xi(\mathbf{x};\hat{\boldsymbol{\theta}}_E)$ is quite robust to departures from normality if the true log ratio $\xi(\mathbf{x})$ of the group-conditional densities is (approximately) linear in $\mathbf{x}$, that is, if any interactions present between the feature variables are essentially the same within each group.

The performance of the sample NLDF in its application to discrete and mixed feature data has been surveyed in some detail by Krzanowski (1977). One of the initial investigations was undertaken by Moore (1973), who demonstrated how nonlinearity in the true log likelihood ratio $\xi(\mathbf{x})$ with discrete feature data limits the usefulness of the sample NLDF in providing an allocation rule. He considered the case where all the features are binary variables taking on the values of zero or one. He adopted a second-order approximation to the Lazarsfeld–Bahadur reparameterization of the multinomial distribution. For given i $(i = 1,\ldots,g)$, let

$$f_{ij} = \mathrm{pr}\{X_j = 1 \mid \mathbf{X} \in G_i\}$$

and

$$X_{j,i} = (X_j - f_{ij})/\{f_{ij}(1 - f_{ij})\}^{1/2}$$

for $j = 1,\ldots,p$, and

$$\rho_{i,jk\ldots m} = E(X_{j,i}X_{k,i}\ldots X_{m,i} \mid \mathbf{X} \in G_i)$$

for $j,k,\ldots,m = 1,\ldots,p$. Then Lazarsfeld (1956, 1961) and Bahadur (1961) independently showed that $f_i(\mathbf{x})$ can be reparameterized as

$$f_i(\mathbf{x}) = \prod_{j=1}^{p} f_{ij}^{x_j}(1-f_{ij})^{1-x_j}\left[1+\sum_{j<k}\rho_{i,jk}x_{j,i}x_{k,i}\right.$$
$$\left. +\cdots+\sum_{j<k<\cdots<m}\rho_{i,jk\ldots m}x_{j,i}x_{k,i}\ldots x_{m,i}\right]. \qquad (5.6.2)$$

Moore (1973) generated the feature vector $\mathbf{x}$ of binary variables, using the second-order approximation to (5.6.2),

$$f_i(\mathbf{x}) = \prod_{j=1}^{p} f_{ij}^{x_j}(1-f_{ij})^{1-x_j}\left[1+\sum_{j<k}\rho_{i,jk}x_{j,i}x_{k,i}\right]. \qquad (5.6.3)$$

He considered the case of $g = 2$ groups, where, for a given i, the f_{ij} were the same for all j, and the $\rho_{i,jk}$ were the same for all j and k. As a consequence, the true log likelihood ratio

$$\xi(\mathbf{x}) = \log\{f_1(\mathbf{x})/f_2(\mathbf{x})\}$$

had a simple dependence on the sum of the feature variables

$$\sum_{j=1}^{p} x_j;$$

that is, on the number of positive feature variables.

If $\xi(\mathbf{x})$ does not increase monotonically with the number of positive feature variables, then it is said to undergo a reversal. In such situations, the sample NLDF $\xi(\mathbf{x};\hat{\boldsymbol{\theta}}_E)$ or indeed any linear function of the feature variables is unable to follow the reversal, and so will perform poorly relative to the true discriminant function $\xi(\mathbf{x})$. In situations without reversals in $\xi(\mathbf{x})$, the sample NLDF $\xi(\mathbf{x};\hat{\boldsymbol{\theta}}_E)$ was found to give good results.

This explains why the sample NLDF was found to perform well in the earlier study by Gilbert (1968). There were no reversals in $\xi(\mathbf{x})$ for the binary feature variables generated according to her first-order interaction model for which $\xi(\mathbf{x})$ was a linear function of $\mathbf{x}$. That is, in each group, the feature variables had common marginal distributions, and their first-order interactions were the same for both groups.

To illustrate a reversal in $\xi(\mathbf{x})$, we consider the example presented in Moore (1973) on the study of Yerushalmy et al. (1965) of infant maturity, using the indices of birth weight and gestation. The latter comprised the $p = 2$ binary feature variables, with $x_1 = 0$ or 1 corresponding to a low or high birth weight, and $x_2 = 0$ or 1 corresponding to a short or long gestation period. The two groups G_1 and G_2 represent "normal" or abnormal babies. The combination of a low birth weight and short gestation period ($x_1 = 0,\ x_2 = 0$) or a high birth weight and a long gestation ($x_1 = 1,\ x_2 = 1$) is suggestive of G_1. Other combinations of x_1 and x_2 are more consistent with G_2 than G_1. It is thus evident that $\xi(\mathbf{x})$ would not be an increasing function of $x_1 + x_2$. Further, it can be seen that a linear discriminant function cannot allow for this reversal in $\xi(\mathbf{x})$. For let

$$\xi(x_1, x_2; \boldsymbol{\alpha}) = \beta_0^o + \boldsymbol{\beta}'\mathbf{x}$$

be any linear discriminant function, where

$$\boldsymbol{\alpha} = (\beta_0^o, \boldsymbol{\beta}')',$$

and where an entity is assigned to G_1 or G_2, according as to whether $\xi(x_1, x_2; \boldsymbol{\alpha})$ is greater or less than some cutoff point k. Now

$$\xi(1,1;\boldsymbol{\alpha}) = \xi(0,1;\boldsymbol{\alpha}) + \xi(1,0;\boldsymbol{\alpha}) - \xi(0,0;\boldsymbol{\alpha}). \tag{5.6.4}$$

Hence, if $\mathbf{x} = (1,1)'$ is assigned to G_1, that is,

$$\xi(1,1;\boldsymbol{\alpha}) > k,$$

and both $\mathbf{x} = (0,1)'$ and $\mathbf{x} = (1,0)'$ are assigned to G_2, that is,

$$\xi(0,1;\boldsymbol{\alpha}) < k, \qquad \xi(1,0;\boldsymbol{\alpha}) < k,$$

then it follows from (5.6.4) that

$$\xi(0,0;\boldsymbol{\alpha}) < k.$$

Thus, an entity with $\mathbf{x} = (0,0)'$ is assigned to G_2 and not to G_1 as desired in this example.

In this example, X_1 and X_2 are positively correlated in G_1, but negatively correlated in G_2, and so the assumption of equal group-covariance matrices under which the sample NLDF $\xi(\mathbf{x}; \hat{\boldsymbol{\theta}}_E)$ is formed does not hold. However, Dillon and Goldstein (1978) have demonstrated that reversals can occur even in situations where the group-covariance matrices are equal. They generated binary feature data using the second-order form (5.6.3) of the Lazarsfeld–Bahadur representation. The group-correlation matrices were made equal by specifying $\rho_{1,jk} = \rho_{2,jk}$ for all j,k. Equality of the group-covariance matrices was achieved then by taking $f_{2,j} = 1 - f_{1,j}$ for each j. Although the group-covariance matrices are the same with this specification, it can be seen from (5.6.3) that interactions between these binary feature variables are not equal within each group, and so $\xi(\mathbf{x})$ is not linear in $\mathbf{x}$.

There have been a number of empirical studies carried out in recent times to investigate the sample NLDR and NQDR, along with other discriminant rules, in situations where some or all of the feature variables are discrete. One such study was that by Titterington et al. (1981), who considered the application of several discriminant rules to a prognosis problem involving a series of

1000 patients with severe head injury. The results of these broad investigations are reported in Chapter 9 after the various semiparametric and nonparametric competitors of the sample NLDR and NQDR have been defined.

Confining remarks now to the latter two normal-based rules, Titterington et al. (1981) found on the basis of the data sets analyzed that the sample NLDF is quite robust in its provision of an allocation rule for discrete feature data. As they also found the independence-based rule to be robust, it suggests that any interactions between the discrete data analyzed are similar for each of the two groups. As remarked earlier, the sample NLDF is not robust against dissimilar interaction structures of the feature variables within each group. This has been confirmed in the simulations and case studies by Krzanowski (1977), Knoke (1982), Vlachonikolis and Marriott (1982), and Schmitz, Habbema, and Hermans (1985), among several others. Although the sample NQDF will generally be preferable to the sample NLDF in the presence of first-order interactions, it may not provide a satisfactory allocation rule because of the presence of higher-order interactions between the discrete feature. Discriminant rules that have been specifically designed for discrete and mixed features are presented in Chapter 7.

5.6.3 Mixed Feature Data

As in the case of all continuous or all discrete feature variables, the allocatory performance of the sample NLDF for mixed feature variables depends on the similarity of the interaction structure of the features within each group. From their simulations with mixed binary and continuous feature variables, Schmitz et al. (1985) concluded that the interaction structure of the continuous features has a greater impact on the performance of the sample NLDF than that of the binary features. A further consideration with mixed feature data is the similarity of the interaction structure between the discrete and continuous feature variables within each group. High within-group correlations between the discrete and continuous feature variables, unless similar for each group, are suggestive of a situation in which the sample NLDF may perform poorly. Linear transformations of the continuous feature variables that attempt to achieve zero correlations between the discrete and continuous features within each group are described in Section 7.6.1. With unequal interaction structures, the sample version of the NQDF instead of the NLDF might be considered. However, in situations in which some of the feature variables are discrete, interactions of order higher than the first may have to be incorporated into the discriminant function if a satisfactory rule is to be formed. In Section 7.6.2, we discuss ways in which the performance of the sample NLDF can be improved in the presence of interactions by augmenting the feature vector to include appropriate products of the feature variables. This approach appears to be preferable to the use of the sample NQDF (Knoke, 1982; Vlachonikolis and Marriott, 1982).

As cautioned in the previous section, although the sample NLDF may provide a satisfactory allocation rule under departures from normality that preserve approximately the linearity of the log likelihood ratio, its use can lead to misleading inferences for at least some of the discriminant function coefficients and hence for the posterior probabilities of group membership. To illustrate this in the present situation of mixed feature data, we now consider the case in which the feature vector $\mathbf{X}$ consists of a single binary variable X_1 and $(p-1)$ continuous variables $\mathbf{X}^{(2)} = (X_2, \ldots, X_{p-1})'$. We let

$$q_{i1} = \text{pr}\{X_1 = 1 \mid \mathbf{X} \in G_i\} \qquad (i = 1,2).$$

The distribution of $\mathbf{X}^{(2)}$ given $X_1 = x_1$ is assumed to be

$$\mathbf{X}^{(2)} \mid X_1 = x_1 \sim N(\boldsymbol{\mu}_i^{(2)} + x_1\boldsymbol{\omega}, \boldsymbol{\Sigma}^{(2)}) \quad \text{in} \quad G_i \qquad (i = 1,2). \tag{5.6.5}$$

This is a special version of the location model defined in Chapter 7, where parametric discriminant rules for the specific case of mixed features are to be considered.

For the model (5.6.5), the covariance between X_1 and $\mathbf{X}^{(2)}$ in G_i is equal to

$$\text{cov}(X_1, \mathbf{X}^{(2)}) = q_{i1}(1-q_{i1})\boldsymbol{\omega}' \qquad (i = 1,2).$$

But as the interactions between X_1 and the elements of $\mathbf{X}^{(2)}$ are the same in both G_1 and G_2, the log likelihood ratio $\xi(\mathbf{x})$ is linear in $\mathbf{x}$. It is given by

$$\xi(\mathbf{x}) = \beta_0^o + \beta_1 x_1 + \boldsymbol{\beta}^{(2)'}\mathbf{x}^{(2)},$$

where

$$\beta_0^o = \log\{(1-q_{11})/(1-q_{21})\} - \tfrac{1}{2}(\boldsymbol{\mu}_1^{(2)} + \boldsymbol{\mu}_2^{(2)})'\boldsymbol{\Sigma}^{(2)^{-1}}(\boldsymbol{\mu}_1^{(2)} - \boldsymbol{\mu}_2^{(2)}),$$

$$\beta_1 = \log[\{q_{11}/(1-q_{11})\}/\{q_{21}/(1-q_{21})\}] - \boldsymbol{\omega}'\boldsymbol{\Sigma}^{(2)^{-1}}(\boldsymbol{\mu}_1^{(2)} - \boldsymbol{\mu}_2^{(2)}),$$

and

$$\boldsymbol{\beta}^{(2)} = \boldsymbol{\Sigma}^{(2)^{-1}}(\boldsymbol{\mu}_1^{(2)} - \boldsymbol{\mu}_2^{(2)}).$$

As $\xi(\mathbf{x})$ is linear in $\mathbf{x}$, the sample NLDF $\xi(\mathbf{x}; \hat{\boldsymbol{\theta}}_E)$ should suffice here for allocation purposes. To examine this further, we consider the asymptotic form of $\xi(\mathbf{x}; \hat{\boldsymbol{\theta}}_E)$ under mixture sampling of the training data in proportions π_1 and π_2 from G_1 and G_2. Corresponding to the partition $(x_1, \mathbf{x}^{(2)'})'$ of $\mathbf{x}$, we express $\xi(\mathbf{x}; \hat{\boldsymbol{\theta}}_E)$ as

$$\xi(\mathbf{x}; \hat{\boldsymbol{\theta}}_E) = \hat{\beta}_{0E}^o + (\hat{\beta}_{1E}, \hat{\boldsymbol{\beta}}_E^{(2)'})'\mathbf{x}, \tag{5.6.6}$$

where from (4.2.8) and (4.2.9),

$$\hat{\beta}_{0E}^o = -\tfrac{1}{2}(\bar{\mathbf{x}}_1 + \bar{\mathbf{x}}_2)'\mathbf{S}^{-1}(\bar{\mathbf{x}}_1 - \bar{\mathbf{x}}_2)$$

and

$$(\hat{\beta}_{1E}, \hat{\boldsymbol{\beta}}_E^{(2)'})' = \mathbf{S}^{-1}(\bar{\mathbf{x}}_1 - \bar{\mathbf{x}}_2).$$

The expectation and covariance matrix of $\mathbf{X}$ in group G_i are equal to

$$\boldsymbol{\mu}_i = (q_{i1}, (\boldsymbol{\mu}_i^{(2)} + q_{i1}\boldsymbol{\omega})')'$$

and

$$\boldsymbol{\Sigma}_i = \begin{pmatrix} q_{i1}(1-q_{i1}) & q_{i1}(1-q_{i1})\boldsymbol{\omega}' \\ q_{i1}(1-q_{i1})\boldsymbol{\omega} & \boldsymbol{\Sigma}^{(2)} + q_{i1}(1-q_{i1})\boldsymbol{\omega}\boldsymbol{\omega}' \end{pmatrix}$$

respectively, for $i = 1,2$. It follows that as $n \to \infty$, $\overline{\mathbf{x}}_i$ and $\mathbf{S}$ converge in probability to $\boldsymbol{\mu}_i$ and $\boldsymbol{\Sigma}$, respectively, where

$$\boldsymbol{\Sigma} = \pi_1\boldsymbol{\Sigma}_1 + \pi_2\boldsymbol{\Sigma}_2.$$

On substitution of these limiting values in (5.6.6), the asymptotic form of $\xi(\mathbf{x};\hat{\boldsymbol{\theta}}_E)$ can be expressed as

$$\xi(\mathbf{x};\hat{\boldsymbol{\theta}}_E) \approx (\beta_0^o + b_0) + (\beta_1 + b_1)x_1 + \boldsymbol{\beta}^{(2)\prime}\mathbf{x}^{(2)}, \tag{5.6.7}$$

where

$$\begin{aligned} b_0 = &-\tfrac{1}{2}\{(q_{11} + q_{21})(q_{11} - q_{21})\}/\{\pi_1 q_{11}(1-q_{11}) + \pi_2 q_{21}(1-q_{21})\} \\ &- \log\{(1-q_{11})/(1-q_{21})\} \end{aligned}$$

and

$$\begin{aligned} b_1 = &(q_{11} - q_{21})/\{\pi_1 q_{11}(1-q_{11}) + \pi_2 q_{21}(1-q_{21})\} \\ &- \log[\{q_{11}/(1-q_{11})\}/\{q_{21}/(1-q_{21})\}]. \end{aligned}$$

This result is available from Hosmer, Hosmer, and Fisher (1983a), who derived it in their study on the sample NLDF for mixed binary and continuous feature variables. They also gave the generalization of the result in the case of an arbitrary number of binary variables in the feature vector. In other closely related work, O'Hara et al. (1982) and Hosmer, Hosmer, and Fisher (1983b) have used simulation experiments to study the problem for a training sample of finite size.

It can be seen from (5.6.7) that the use of the normal-based estimates $\hat{\beta}_{0E}^o$ and $\hat{\boldsymbol{\beta}}_E$ provides consistent estimation of $\boldsymbol{\beta}^{(2)}$ containing the discriminant function coefficients of the continuous feature variables. However, the constant term β_0^o and the coefficient β_1 of the binary feature variable are estimated inconsistently, their asymptotic biases being given by b_0 and b_1, respectively. This inconsistent estimation of β_1 would be of concern if an aim of the analysis was to assess the discriminatory importance of the binary feature variable x_1. For example, the two groups might correspond to the presence or absence of some disease in an individual and x_1 might be used to denote whether in the past the individual was exposed or not to some carcinogenic agent.

But for the purposes of constructing an allocation rule, the inconsistent estimation of β_0^o and β_1 would not be of practical concern. This is because β_0^o and β_1 serve only to locate the hyperplane in the continuous feature variables, that is, the cutoff point for the linear discriminant function $\boldsymbol{\beta}^{(2)\prime}\mathbf{x}^{(2)}$ formed from the continuous variables in the feature vector. Although this cutoff point has a bearing on the balance between the two errors of allocation, the overall error rate is not greatly affected. Hence, as $\boldsymbol{\beta}^{(2)}$ is estimated consistently by the normal-based estimate $\hat{\boldsymbol{\beta}}_E^{(2)}$, the performance of the sample NLDF $\xi(\mathbf{x};\hat{\boldsymbol{\theta}}_E)$ should not be too far below a consistent sample version of the optimal discriminant function $\xi(\mathbf{x})$.

5.7 ROBUST ESTIMATION OF GROUP PARAMETERS

5.7.1 *M*-Estimates

We consider here the computation of robust M-estimates of the location vector and scatter matrix of the feature vector $\mathbf{X}$ in a given group. These estimates give full weight to observations from the main body of the data, but automatically give reduced weight to any observation assessed as being atypical of its group of origin. Huber (1964) developed a theory of robust estimation of a location parameter using M-estimates. It was later extended to the multivariate case by taking an elliptically symmetric density and then associating it with a contaminated normal density; see Maronna (1976), Huber (1977, 1981, Chapter 8), and Collins (1982). Further references on this approach and subsequent modifications can be found in Collins and Wiens (1985), Hampel et al. (1986, Chapter 5), and Wiens and Zheng (1986).

We let $\mathbf{x}_{ij}$ $(j = 1,\ldots,n_i)$ denote the n_i feature vectors in $\mathbf{t}$ that belong to G_i $(i = 1,\ldots,g)$. For the ith group-conditional distribution of $\mathbf{X}$, the M-estimates of $\boldsymbol{\mu}_i$ and $\boldsymbol{\Sigma}_i$ as proposed by Maronna (1976) are defined by the equations

$$\sum_{j=1}^{n_i} u_1(\hat{d}_{ij})(\mathbf{x}_{ij} - \hat{\boldsymbol{\mu}}_i) = 0 \tag{5.7.1}$$

and

$$\sum_{j=1}^{n_i} u_2(\hat{d}_{ij}^2)(\mathbf{x}_{ij} - \hat{\boldsymbol{\mu}}_i)(\mathbf{x}_{ij} - \hat{\boldsymbol{\mu}}_i)'/n_i = \hat{\boldsymbol{\Sigma}}_i, \tag{5.7.2}$$

where, for convenience, $\delta(\mathbf{x}_{ij},\hat{\boldsymbol{\mu}}_i;\hat{\boldsymbol{\Sigma}}_i)$ is written as $\hat{d}_{ij}^2$, and where $u_1(s)$ and $u_2(s)$ are nonnegative weight functions. Under fairly general conditions on $u_1(s)$ and $u_2(s)$, Maronna (1976) established the existence and uniqueness of the solution of (5.7.1) and (5.7.2). He also showed that it is consistent and asymptotically normal under the assumption that $f_i(\mathbf{x};\boldsymbol{\theta}_i)$ is a member of the family of p-dimensional elliptically symmetric densities

$$|\boldsymbol{\Sigma}_i|^{-1/2} f_S[\{\delta(\mathbf{x},\boldsymbol{\mu}_i;\boldsymbol{\Sigma}_i)\}^{1/2}], \tag{5.7.3}$$

where $f_S(\|\mathbf{x}\|)$ is any spherically symmetric density function. Under (5.7.3), $\boldsymbol{\Sigma}_i$ is a scalar multiple of the covariance matrix of $\mathbf{X}$. One of the conditions on the weight functions, that $su_1(s)$ and $su_2(s)$ be bounded, ensures that these estimates of $\boldsymbol{\mu}_i$ and $\boldsymbol{\Sigma}_i$ will be robust. If

$$u_1(s) = -s^{-1}\partial \log f_S(s)/\partial s$$

and $u_2(s^2) = u_1(s)$ for $s > 0$, then (5.7.1) and (5.7.2) give the maximum likelihood estimates of $\boldsymbol{\mu}_i$ and $\boldsymbol{\Sigma}_i$ $(i = 1,\ldots,g)$.

More general to (5.7.2), Huber (1977, 1981, page 213) observed that an affinely invariant estimate of $\boldsymbol{\Sigma}_i$ can be defined by

$$\sum_{j=1}^{n_i} u(\hat{d}_{ij})(\mathbf{x}_{ij} - \hat{\boldsymbol{\mu}}_i)(\mathbf{x}_{ij} - \hat{\boldsymbol{\mu}}_i)' \Big/ \sum_{j=1}^{n_i} v(\hat{d}_{ij}) \tag{5.7.4}$$

for arbitrary functions u and v, and noted it was "particularly attractive" to take $u \equiv v$. The M-estimate of a covariance matrix in high dimensions has a low breakdown point α_0 (which, roughly speaking, is the limiting proportion of bad outliers that can be tolerated by the estimate). For example, Maronna (1976) established that $\alpha_0 \leq 1/(p+1)$ for (5.7.2), where u_2 is monotone increasing and $u_2(0) = 0$, and Huber (1977) showed that this could be improved only to $\alpha_0 \leq 1/p$ in the more general framework (5.7.4). Recently, Huber (1985) has reported projection pursuit (PP) estimators of $\boldsymbol{\mu}_i$ and $\boldsymbol{\Sigma}_i$ that are both affine equivariant and whose breakdown point approaches 1/2 in large samples. These PP estimators are defined by replacing $\hat{d}_{ij}$ in (5.7.1) and (5.7.2) by

$$\tilde{d}_{ij} = \sup_{\mathbf{a}}(\mathbf{a}'x_{ij} - m_i)/M_i,$$

where m_i is the median, and M_i is the mean absolute deviation of $\mathbf{a}'\mathbf{x}_{ij}$ $(j = 1,\ldots,n_i)$ for $i = 1,\ldots,g$.

One proposal by Maronna (1976) for the weight function $u_1(s)$ in (5.7.1) is

$$u_1(s) = \psi(s)/s,$$

where $\psi(s)$ is Huber's (1964) ψ-function. It is defined by $\psi(s) = -\psi(-s)$, where, for $s > 0$,

$$\psi(s) = \begin{cases} s, & 0 \leq s \leq k_1(p), \\ k_1(p), & s > k_1(p), \end{cases} \tag{5.7.5}$$

for an appropriate choice of the "tuning" constant $k_1(p)$, which is written here as a function of p to emphasize its dependence on the dimension of $\mathbf{x}$.

An associated choice for a weight function in the estimation of $\boldsymbol{\Sigma}_i$ is to take

$$u(s) = v(s) = \{u_1(s)\}^2$$

in (5.7.4) to give

$$\hat{\boldsymbol{\Sigma}}_i = \sum_{j=1}^{n_i}\{u_1(\hat{d}_{ij})\}^2(\mathbf{x}_{ij} - \hat{\boldsymbol{\mu}}_i)(\mathbf{x}_{ij} - \hat{\boldsymbol{\mu}}_i)' \Big/ \sum_{j=1}^{n_i}\{u_1(\hat{d}_{ij})\}^2. \tag{5.7.6}$$

Note that if $u_1(\hat{d}_{ij})$ and not its square were used as a weight in (5.7.6), then the influence of grossly atypical observations would not be bounded.

The value of the tuning constant $k_1(p)$ in Huber's ψ-function (5.7.5) depends on the amount of contamination, and so $k_1(p)$ is chosen to give estimators with reasonable performances over a range of situations. In the univariate case, $k_1(1)$ is generally taken to be between 1 and 2. For $p > 1$, Devlin,

Gnanadesikan, and Kettenring (1981) took $k_1(p)$ to be the square root of the 90th percentile of the χ_p^2 distribution, $\sqrt{\chi^2_{p;0.9}}$, because $\hat{d}_{ij}^2$ is asymptotically χ_p^2 under normality. Campbell (1984c, 1985) recommends $k_1(1) = 2$ and that in the multivariate case, $k_1(p)$ be computed corresponding to $k_1(1) = 2$ by taking $\hat{d}_{ij}^2$ to be χ_p^2 and then using the approximation of Wilson and Hilferty (1931),

$$\chi^2_{p;\alpha} \approx p\left\{1 - \frac{2}{9p} + \left(\frac{2}{9p}\right)^{1/2} q_\alpha\right\}^3 \tag{5.7.7}$$

for the αth quantile of the χ_p^2 distribution. In (5.7.7), q_α denotes the αth quantile of the $N(0,1)$ distribution. Because $k_1(1) = 2$ is approximately equal to the square root of $\chi^2_{1;0.95}$, this leads to

$$k_1(p) = \left[p\left\{1 - \frac{2}{9p} + \left(\frac{2}{9p}\right)^{1/2} 1.645\right\}^3\right]^{1/2}, \tag{5.7.8}$$

on setting $k_1(p) = \sqrt{\chi^2_{p;0.95}}$ and then approximating the latter according to (5.7.7). Hawkins and Wicksley (1986) have discussed power transformations of chi-squared to normality other than (5.7.7), suggesting the fourth root when p is small. Broffitt et al. (1980) observed from their simulations on the performance of the sample QDF using M-estimates of $\boldsymbol{\mu}_i$ and $\boldsymbol{\Sigma}_i$ that the choice of $k_1(p)$ is not critical.

If it is desired that observations extremely atypical of G_i should have zero weight for values of $\hat{d}_{ij}$ above a certain level (rejection point), then a redescending ψ-function can be used, for example, Hampel's (1973) piecewise linear function. It is defined by $\psi(s) = -\psi(-s)$, where, for $s > 0$,

$$\psi(s) = \begin{cases} s, & s \leq k_1(p), \\ k_1(p), & k_1(p) < s \leq k_2(p), \\ k_1(p)\dfrac{\{k_3(p) - s\}}{\{k_3(p) - k_2(p)\}}, & k_2(p) < s \leq k_3(p), \\ 0, & s > k_3(p), \end{cases} \tag{5.7.9}$$

where $k_1(p)$ is chosen in the same way as $k_1(p)$ in Huber's ψ-function (5.7.5). Care must be taken in choosing the remaining tuning constants $k_2(p)$ and $k_3(p)$ to ensure that ψ does not descend too steeply, as cautioned by Huber (1981, Chapter 4), who also warns that redescending estimators are susceptible to underestimation of scale and that there may be multiple solutions. It can be seen that using Huber's nondescending ψ-function with manual rejection of grossly atypical observations beforehand is almost equivalent to the use of a redescending function, since either procedure removes very extreme observations. The difference lies in the treatment of the observations over which there is some doubt as to their retention in the data. With a redescending

ψ-function, this step is carried out automatically by having ψ redescend to zero from $k_2(p)$. For further discussion on the comparative properties of non-descending and redescending ψ-functions, the reader is referred to Goodall (1983), who also considers the various smooth versions subsequently proposed for Hampel's original ψ-function (5.7.9).

For convenience of use, Campbell (1984c, 1985) advocates (5.7.9) with tuning constants $k_1(1) = 2$, $k_2(1) = 3$, and $k_3(1) = 5$ for $p = 1$. In the multivariate case, he recommends defining $k_1(p)$ by (5.7.8) and $k_2(p)$ and $k_3(p)$ by replacing 1.645 with 2.8 and 5, respectively, in the right-hand side of (5.7.8). With this choice, $k_1(p)$ and $k_2(p)$ are approximately equal to the square root of $\chi^2_{p;0.95}$ and $\chi^2_{p;0.9974}$, respectively.

5.7.2 Use of a Rank-Cutoff Point

Randles et al. (1978b) have considered robust versions of the normal-based linear and quadratic discriminant rules in the case of $g = 2$ groups. One proposal was to form the linear discriminant function $\mathbf{a}'\mathbf{x}$, where $\mathbf{a}$ is chosen to maximize an appropriate function of the separation index (3.3.6) introduced by Fisher (1936). This index can, of course, be regarded as a projection index in the spirit of projection pursuit (Huber, 1985). Another robust proposal of Randles et al. (1978b) is to use the linear or quadratic discriminant rules formed with Huber-type M-estimates in conjunction with a rank-cutoff point. The use of the latter provides an extra degree of robustness as it provides some control over the relative size of the two unconditional error rates. For nonnormal feature data, severe imbalances can occur between the error rates of the sample NLDR or NQDR when applied with a fixed cutoff point.

We now describe the allocation procedure using a rank-cutoff point with the sample NLDF $\xi(\mathbf{x};\hat{\boldsymbol{\theta}}_E)$. We let $\mathbf{y} = (\mathbf{x}',\mathbf{z}')'$, where $\mathbf{z}$ is the unknown vector of zero-one indicators denoting the group membership of $\mathbf{x}$. Corresponding to $\mathbf{x}$ belonging to G_1 or G_2, we let

$$\mathbf{y}^{(1)} = (\mathbf{x}',1,0)'$$

and

$$\mathbf{y}^{(2)} = (\mathbf{x},0,1)'.$$

The estimate of $\boldsymbol{\theta}_E$ based on $\mathbf{t}$ and $\mathbf{y}^{(i)}$ is denoted by $\hat{\boldsymbol{\theta}}_E(\mathbf{t},\mathbf{y}^{(i)})$ for $i = 1,2$. The rank of $\xi(\mathbf{x};\hat{\boldsymbol{\theta}}_E(\mathbf{t},\mathbf{y}^{(1)}))$ when included among the $\xi(\mathbf{x}_{1j};\hat{\boldsymbol{\theta}}_E(\mathbf{t},\mathbf{y}^{(1)}))$ for $j = 1,\ldots,n_1$ is denoted by $R_1(\mathbf{x};\mathbf{t})$, and $R_2(\mathbf{x};\mathbf{t})$ denotes the rank of $-\xi(\mathbf{x};\hat{\boldsymbol{\theta}}_E(\mathbf{t},\mathbf{y}^{(2)}))$ when included among the $-\xi(\mathbf{x}_{2j};\hat{\boldsymbol{\theta}}_E(\mathbf{t},\mathbf{y}^{(2)}))$ for $j = 1,\ldots,n_2$. As before, $\mathbf{x}_{ij}$ $(j = 1,\ldots,n_i)$ denote the n_i feature vectors in $\mathbf{t}$ that belong to G_i $(i = 1,2)$. With the rank-cutoff point as proposed initially by Broffit et al. (1976) and used subsequently by Randles et al. (1978a, 1978b), an entity with feature vector $\mathbf{x}$ is assigned to G_1 if

$$R_1(\mathbf{x};\mathbf{t})/(n_1+1) > k\{R_2(\mathbf{x};\mathbf{t})/(n_2+1)\},$$

and to G_2 if

$$R_1(\mathbf{x};\mathbf{t})/(n_1+1) < k\{R_2(\mathbf{x};\mathbf{t})/(n_2+1)\}, \tag{5.7.10}$$

where k is a constant that reflects the desired balance between the error rates. For example, k is set equal to one if the intent is to have comparable error rates. The rationale behind this procedure is that the greater the value of $R_1(\mathbf{x};\mathbf{t})$, the more likely it is to be from G_1 rather than G_2. Similarly, the greater the value of $R_2(\mathbf{x};\mathbf{t})$, the more likely it is to be from G_2 rather than G_1. Moreover, Broffit et al. (1976) showed that if the unclassified entity with feature vector $\mathbf{x}$ comes from G_i, then $R_i(\mathbf{X};\mathbf{T})$ has a uniform distribution over $1,\ldots,n_i+1$ with mass $1/(n_i+1)$ at each of these values $(i=1,2)$. This implies that allocation according to (5.7.10) is based on the relative size of the P-values for the affinity of $\mathbf{x}$ with G_1 and G_2, respectively.

5.7.3 MML Estimates

We consider here the approach of Tiku (1983), Tiku and Balakrishnan (1984), Balakrishnan, Tiku, and El Shaarawi (1985), and Balakrishnan and Tiku (1988) to the construction of a robust discriminant rule in the case of $g=2$ groups. Their approach is to use the NLDF

$$\xi(\mathbf{x};\boldsymbol{\theta}_E) = \{\mathbf{x} - \tfrac{1}{2}(\boldsymbol{\mu}_1+\boldsymbol{\mu}_2)\}'\boldsymbol{\Sigma}^{-1}(\boldsymbol{\mu}_1-\boldsymbol{\mu}_2),$$

where $\boldsymbol{\mu}_1$, $\boldsymbol{\mu}_2$, and $\boldsymbol{\Sigma}$ are replaced by the modified maximum likelihood (MML) estimates of Tiku (1967, 1980).

To define the MML estimate, we consider first the univariate case $p=1$. Let $x_{i(1)},\ldots,x_{i(n_i)}$ denote the n_i order statistics for the training data x_{ij} $(j=1,\ldots,n_i)$ from group G_i $(i=1,2)$. Then the MML estimate of $\hat{\mu}_i$ is defined by

$$\hat{\mu}_i = \frac{1}{m_i}\sum_{j=h_i+1}^{n_i-h_i} x_{i(j)} + h_i\beta_i\{x_{i(h_i+1)} + x_{i(n_i-h_i)}\} \qquad (i=1,2), \tag{5.7.11}$$

where

$$m_i = n_i - 2h_i(1-\beta_i),$$

$$\beta_i = -\{\phi(q_{2i})/q_{1i}\}\{q_{2i} - \phi(q_{2i})/q_{1i}\},$$

and where $q_{1i} = h_i/n_i$, and $q_{2i} = \Phi(-q_{1i})$ for $i=1,2$. The MML estimate of σ^2 is given by

$$\hat{\sigma}^2 = (M_{11}+M_{12}-2)^{-1}\sum_{i=1}^{2}(M_{1i}-1)\hat{\sigma}_i^2, \tag{5.7.12}$$

where

$$\hat{\sigma}_i^2 = \tfrac{1}{2}\{M_{2i} + (M_{2i}^2 + 4M_{1i}M_{3i})^{1/2}\}/\{A_i(A_i-1)\}^{1/2},$$

and where

$$M_{1i} = n_i - 2h_i,$$

$$M_{2i} = h_i\alpha_i\{x_{i(n_i-h_i)} - x_{i(h_i+1)}\},$$

$$M_{3i} = \sum_{j=h_i+1}^{n_i-h_i} [x_{i(j)}^2 + h_i\beta_i\{x_{i(h_i+1)}^2 + x_{i(n_i-h_i)}^2\}] - m_i\hat{\mu}_i^2,$$

and

$$\alpha_i = \{\phi(q_{2i})/q_{1i}\} - \beta_i q_{2i}$$

for $i = 1,2$.

Note that $0 < \alpha_i < 1$ and $0 < \beta_i < 1$ $(i = 1,2)$. For example, for $q_i = 0.1$, $\alpha_i = 0.690$ and $\beta_i = 0.831$. For $h_i = 0$ $(i = 1,2)$, $\hat{\mu}_1$, $\hat{\mu}_2$, and $\hat{\sigma}^2$ reduce to the sample means $\bar{\mathbf{x}}_1$, $\bar{\mathbf{x}}_2$, and the pooled sample variance s^2 corrected for bias. The choice of h_i is based on robustness considerations. For distributions most prevalent in practice, h_i is usually chosen to be the greatest integer less than or equal to $0.5 + (n_i/10)$; see Tiku, Tan, and Balakrishnan (1986) for details regarding various properties and efficiency of MML estimators.

As defined above, the MML estimates of μ_1, μ_2, and σ^2 can be viewed as approximate maximum likelihood estimates for symmetric Type II censoring, where the h_i smallest and the h_i largest observations in the sample on the entities from G_i are censored $(i = 1,2)$. For extremely skewed distributions, a more appropriate procedure is obtained by censoring on only one side, in the direction of the long tail (Tiku et al., 1986, page 113).

To extend the definition of the MML estimates to the bivariate case of $p = 2$, Tiku and Balakrishnan (1984) noted that instead of allocating an entity on the basis of its feature vector in its original form, $\mathbf{x} = (x_1, x_2)'$, one can equivalently use

$$\mathbf{x}^* = (x_1, x_{2.1})',$$

where

$$x_{2.1} = x_2 - bx_1,$$

and

$$b = (\mathbf{S})_{12}/(\mathbf{S})_{11}.$$

The sample discriminant function is taken then to be

$$\hat{\xi}_E^{(R)}(\mathbf{x}^*) = \{\mathbf{x}^* - \tfrac{1}{2}(\hat{\boldsymbol{\mu}}_1^* + \hat{\boldsymbol{\mu}}_2^*\}'\mathbf{S}^{*^{-1}}(\hat{\boldsymbol{\mu}}_1^* - \hat{\boldsymbol{\mu}}_2^*), \tag{5.7.13}$$

where $\hat{\boldsymbol{\mu}}_i^*$ and $\mathbf{S}^*$ denote the MML estimates of the mean and covariance matrix, respectively, of $\mathbf{X}^*$ within G_i $(i = 1,2)$. They are defined as follows.

Let $\mathbf{x}_{ij}^*$ $(j = 1,\ldots,n_i)$ denote the n_i replications of $\mathbf{X}^*$ corresponding to the original observations $\mathbf{x}_{ij}$ $(j = 1,\ldots,n_i)$ from G_i $(i = 1,2)$. Also, let $x_{iv(1)}^*,\ldots, x_{iv(n_i)}^*$ denote the n_i-order statistics of $(\mathbf{x}_{i1}^*)_v,\ldots,(\mathbf{x}_{in_i}^*)_v$ for $i = 1,2$ and $v = 1,2$. Then as $\mathbf{S}^*$ is diagonal, $\hat{\boldsymbol{\mu}}_i^*$ and $\mathbf{S}^*$ can be calculated by considering the ordered observations $x_{iv(1)}^*,\ldots,x_{iv(n_i)}^*$ separately for each variate v $(v = 1,2)$.

Thus, for each v ($v = 1,2$), $(\hat{\boldsymbol{\mu}}_i^*)_v$ and $(\mathbf{S}^*)_{vv}$ are given by (5.7.11) and (5.7.12) with the $x_{i(j)}$ replaced by the $x_{iv(j)}^*$ ($i = 1,2$; $j = 1,\ldots,n_i$).

We let $ec_i^{(R)}(F_i;\mathbf{t})$ for $i = 1,2$ denote the group-specific error rates of the rule based on the robust version $\hat{\xi}_E^{(R)}(\mathbf{x}^*)$ of the NLDF given by (5.7.13). Tiku and Balakrishnan (1984) and Balakrishnan et al. (1985) have investigated the performance of $\hat{\xi}_E^{(R)}(\mathbf{x}^*)$ in applications where the cutoff point is chosen so that $ec_1^{(R)}(F_1;\mathbf{t})$ achieves, at least asymptotically, a specified level. They concluded that $\hat{\xi}_E^{(R)}(\mathbf{x}^*)$ is quite robust in the sense that the values of $ec_1^{(R)}(F_1;\mathbf{t})$ are more stable from distribution to distribution than those of the sample NLDF $\xi(\mathbf{x};\hat{\boldsymbol{\theta}}_E)$. Also, $\hat{\xi}_E^{(R)}(\mathbf{x}^*)$ appeared to be more efficient than $\xi(\mathbf{x};\hat{\boldsymbol{\theta}}_E)$ and the nonparametric discriminant functions studied in that it produced lower values for the error rate with respect to the second group. Balakrishnan and Tiku (1988) have pointed out that the MML estimates can be easily generalized to the $p > 2$ case by proceeding along the same lines as in Tiku and Singh (1982). More recently, Tiku and Balakrishnan (1989) have considered the use of MML estimates in providing a robust allocation rule with respect to two univariate distributions with unequal variances.

CHAPTER 6

Data Analytic Considerations with Normal Theory-Based Discriminant Analysis

6.1 INTRODUCTION

With a parametric formulation of problems in discriminant analysis, there is a number of items in need of attention before the final version of the sample discriminant rule is formed and applied to the problem at hand. These preliminary items are associated with the actual fitting of the postulated models for the group-conditional distributions of the feature vector **X**. There is the detection of apparent outliers among the training data and their possible effect on the estimation process. One option for handling extremely atypical observations is to use a manual rejection procedure in conjunction with a robust estimation procedure whereby an observation assessed as atypical of its group of origin is given reduced weight in the formation of the estimates of the group parameters.

On the assessment of model fit, there are several tests of univariate and multivariate normality that can be applied to assess the validity of a normal model for the group-conditional distributions. The question of model fit is particularly relevant in situations in which reliable estimates of the posterior probabilities of group membership are required. In some cases where a normal model does not appear to provide an adequate fit, the training data may be able to be transformed to achieve approximate normality.

After the question of model fit has been addressed, there is still another item to be considered before completing the probabilistic or outright allocation of an unclassified entity. There is the question of whether the assumption that the entity belongs to one of the g specified groups is apparently valid. In some situations, the question does not arise, as the specified groups by their very definition are exhaustive for the origin of the unclassified entity. However, in other situations, this may not be known with certainty. For example, in the case where the g specified groups represent the known causes of a disease, they may not be completely exhaustive as there may well be additional causes as yet unrecognized and classified. In situations like this, there is particular interest in the monitoring of the typicality with respect to the specified groups of the feature vector $\mathbf{x}$ on the unclassified entity.

These aforementioned aspects of detection of apparent outliers among the training data and robust estimation, assessment of model fit before and after the application of data-based transformations, and the monitoring of typicality of feature data on unclassified entities are discussed and illustrated in the following sections in the context of parametric discrimination via normal theory-based models.

In the second half of the chapter, attention is devoted to the graphical representation of the feature data on the classified entities in the training set. As the human capacity for pattern recognition is limited to low dimension, highly revealing low-dimensional representations of the feature data are required if much light is to be shed on such matters as the underlying group structure.

For homoscedastic feature data, a well-used method of portraying the group structure is to use the sample analogues of the population canonical variates introduced in Section 3.9.2. In association with this problem, we are to consider the construction of confidence regions for the group means that take into consideration the sampling variability in the sample versions of the canonical variates. Also, we are to consider a test for the number of canonical variates, that is, the number of discriminant functions needed, with feature data from multiple groups. Another problem to be discussed is the generalization of canonical variate analysis to the case of heteroscedastic feature data. The role of principal components in discriminant analysis is to be examined, too.

At the end of the chapter, we present some examples of discriminant analyses in less than straightforward situations, including a case in which there is not an unequivocal classification of the training data with respect to the underlying groups and a case in which there is some doubt as to the actual group structure. Problems of this type, which are on the interface of discriminant and cluster analyses, occur frequently in practice and can be difficult to handle.

6.2 ASSESSMENT OF NORMALITY AND HOMOSCEDASTICITY

6.2.1 Introduction

With the parametric approach to discriminant analysis, an important consideration in practice is the applicability of the adopted forms for the group-conditional densities of the feature vector $\mathbf{X}$. This can be carried out by assessing the fitted group-conditional densities for each group in turn. On the question of fit for a normal model for the distribution of $\mathbf{X}$ in a given group, there is now an extensive literature on both formal and informal methods for assessing univariate and multivariate normality. They are surveyed in the comprehensive reviews by Gnanadesikan (1977, Section 5.2), Cox and Small (1978), and Mardia (1980). More recent work on this topic is described in the compendium by Koziol (1986) and the reviews by Looney (1986) and Baringhaus, Danschke, and Henze (1989); see also Csörgö (1986) and the references therein.

6.2.2 Hawkins' Test for Normality and Homoscedasticity

In discriminant analysis where more than one group-conditional distribution is under assessment, there is also the question of whether the group-conditional distributions have a common covariance matrix. A convenient test in this framework is the procedure proposed by Hawkins (1981), which can be used to test simultaneously for multivariate normality and homoscedasticity. It can be easily incorporated as a routine diagnostic test in a computer program. Also, it has very good power compared with standard inferential procedures such as the likelihood ratio test statistic for homoscedasticity used in conjunction with the multivariate coefficients of skewness and kurtosis (Mardia, 1970, 1974) for normality. We now give a brief description of Hawkins' (1981) test, as the quantities used in its construction are to be used in Section 6.4 as measures of typicality of the training feature vectors. For a more detailed account, the reader is referred to Fatti, Hawkins, and Raath (1982) and McLachlan and Basford (1988, Section 2.5).

Corresponding to the homoscedastic normal model (3.3.1) for the group-conditional distributions, the problem is to test the hypothesis

$$H_0 : \mathbf{X} \sim N(\boldsymbol{\mu}_i, \boldsymbol{\Sigma}) \quad \text{in} \quad G_i \qquad (i = 1, \ldots, g)$$

on the basis of the training data $\mathbf{t}$. It is more convenient here if we relabel the n feature vectors $\mathbf{x}_1, \ldots, \mathbf{x}_n$ in $\mathbf{t}$ to explicitly denote their group of origin. We let $\mathbf{x}_{ij}$ ($j = 1, \ldots, n_i$) denote the n_i training feature vectors from group G_i, that is, those with $z_{ij} = 1$ ($i = 1, \ldots, g$).

For a given group G_i, Hawkins' (1981) test considers the squared Mahalanobis distance between each $\mathbf{x}_{ij}$ ($j = 1, \ldots, n_i$) and the mean of the sample from G_i, with respect to the (bias-corrected) pooled sample covariance matrix $\mathbf{S}$. In forming this distance, or equivalently Hotelling's T^2, each $\mathbf{x}_{ij}$ is deleted first from the sample in case it severely contaminates the estimates of the mean and covariance matrix of G_i ($i = 1, \ldots, g$). Accordingly, the squared Mahalanobis distance

$$\delta(\mathbf{x}_{ij}, \bar{\mathbf{x}}_{i(ij)}; \mathbf{S}_{(ij)}) \tag{6.2.1}$$

is computed, where $\bar{\mathbf{x}}_{i(ij)}$ and $\mathbf{S}_{(ij)}$ denote the resulting values of $\bar{\mathbf{x}}_i$ and $\mathbf{S}$, respectively, after the deletion of $\mathbf{x}_{ij}$ from the data. Under H_0, it follows that

$$c(n_i, \nu)\delta(\mathbf{x}_{ij}, \bar{\mathbf{x}}_{i(ij)}; \mathbf{S}_{(ij)}) \tag{6.2.2}$$

is distributed according to an F-distribution with p and $\nu = n - g - p$ degrees of freedom, where

$$c(n_i, \nu) = (n_i - 1)\nu / (n_i p)(\nu + p - 1). \tag{6.2.3}$$

Note that for economy of notation, we are using ν elsewhere to denote the underlying measure on $\mathbb{R}^p$ appropriate for the mixture density of the feature vector $\mathbf{X}$.

The quantity (6.2.2) can be computed using the result that

$$c(n_i, \nu)\delta(\mathbf{x}_{ij}, \overline{\mathbf{x}}_{i(ij)}; \mathbf{S}_{(ij)}) = \frac{(\nu n_i / p)\delta(\mathbf{x}_{ij}, \overline{\mathbf{x}}_i; \mathbf{S})}{(\nu + p)(n_i - 1) - n_i \delta(\mathbf{x}_{ij}, \overline{\mathbf{x}}_i; \mathbf{S})}, \tag{6.2.4}$$

which avoids the recomputation of $\overline{\mathbf{x}}_i$ and $\mathbf{S}$ after the deletion of each $\mathbf{x}_{ij}$ from the data. If a_{ij} denotes the area to the right of the observed value of (6.2.4) under the $F_{p,\nu}$ distribution, then under H_0 we have that

$$H_{0i} : a_{i1}, \ldots, a_{in_i} \overset{iid}{\sim} U(0,1) \qquad (i = 1, \ldots, g) \tag{6.2.5}$$

holds approximately, where $U(0,1)$ denotes the uniform distribution on the unit interval. The result (6.2.5) is only approximate as for a given i, the a_{ij} are only independent exactly as $n_i \to \infty$, due to the presence of estimates of $\boldsymbol{\mu}_i$ and $\boldsymbol{\Sigma}$ in the formation of (6.2.1). Hawkins (1981) has reported empirical evidence that suggests that subsequent steps in his test that treat (6.2.5) as if it were an exact result should be approximately valid. He also noted that (6.2.5) can be valid for nonnormal group-conditional distributions, but it is likely that such cases would be very rare.

A close inspection of the tail areas a_{ij} including Q-Q plots can be undertaken to detect departures from the g hypotheses H_{0i}, and hence from the original hypothesis H_0. In conjunction with this detailed analysis, Hawkins (1981) advocated the use of the Anderson–Darling statistic for assessing (6.2.5), as this statistic is particularly sensitive to the fit in the tails of the distribution. It can be computed for the sample of n_i values a_{ij} $(j = 1, \ldots, n_i)$ by

$$W_i = -n_i - \frac{1}{n_i} \sum_{j=1}^{n_i} (2j - 1)\{\log a_{i(j)} + \log(1 - a_{i(n_i - j + 1)})\} \qquad (i = 1, \ldots, g),$$

where for each i, $a_{i(1)} \leq a_{i(2)} \leq \cdots < a_{i(n_i)}$ denote the n_i order statistics of the a_{ij}. In the asymptotic resolution of each W_i into standard normal variates W_{ik} according to

$$W_i = \sum_{k=1}^{\infty} W_{ik}^2 / k(k+1) \qquad (i = 1, \ldots, g),$$

attention is focused on the first two components

$$W_{i1} = -(3/n_i)^{1/2} \sum_{j=1}^{n_i} (2a_{i(j)} - 1)$$

and

$$W_{i2} = -(5/n_i)^{1/2} \sum_{j=1}^{n_i} \frac{1}{2}\{3(2a_{i(j)} - 1)^2 - 1\}.$$

Similarly, the Anderson–Darling statistic W_T and its first two components W_{T1} and W_{T2} can be computed for the single sample where all the a_{ij} are combined.

Caution should be exercised in any formal test based on the Anderson–Darling statistics W_i and W_T and their components in view of the reliance on asymptotic theory in establishing their validity. However, Hawkins (1981) has provided simulations to demonstrate that the asymptotic theory will not lead one seriously astray and errs on the side of conservatism. Note that further conservatism is introduced into the problem if the data have been transformed first according to a data-based transformation; see Linnett (1988).

The primary role of the statistics W_i and W_T and their components as proposed by Hawkins (1981) is to provide quick summary statistics for interpreting qualitatively departures from the null hypothesis H_0 of multivariate normality and homoscedasticity. Nonnormality in the form of heavy tails in the data leads to a U-shaped distribution with an excess of the a_{ij} near zero or one for a given i. The second component, W_{i2}, which is essentially the sample variance of the a_{ij} $(j = 1,\ldots,n_i)$, is useful in detecting this (and other departures from normality). The presence of heteroscedasticity in normally distributed data generally will result in large positive values for the first component of one of the W_i, say, W_{h1}, and large negative values for the first component of another W_i, say, W_{u1}. For nonnormal data, W_{h1} and W_{u1} may not necessarily have different signs, but there should be large differences between them to indicate heteroscedasticity. If a normal model provides an adequate fit, then the imbalances in the tail areas due to heteroscedasticity tend to cancel each other out in the formation of W_T and its components from the totality of the n areas a_{ij} $(i = 1,\ldots,g;\ j = 1,\ldots,n_i)$. Thus, significant values of W_T and its components can be taken as a fair indication of nonnormality.

6.2.3 Some Other Tests of Normality

The question of multivariate normality for some data may be considered in terms of radii (distances of the data points from the center) and angles (directions of the data points from the center). Andrews, Gnanadesikan, and Warner (1973) provided an informal graphical procedure for assessing multivariate normality in terms of the radii and angles. Hjort (1986a, Section 11.2) has since given a rigorous test based on them. Fatti et al. (1982) are of the view that tests of model fit in discriminant analysis should be directed toward the detection of heavy-tailed and skewed departures from normality, as the performance of discriminant rules based on normal models can be severely affected by such departures. In this pursuit, they suggested that the multivariate sample measures of skewness and kurtosis as proposed by Mardia (1970, 1974) should be useful. For the training data $\mathbf{x}_{ij}$ $(j = 1,\ldots,n_i)$ from group G_i $(i = 1,\ldots,g)$, the sample measures of skewness and kurtosis are given by

$$b_{1i,p} = \frac{1}{n_i^2}\sum_{j=1}^{n_i}\sum_{k=1}^{n_i}\{(\mathbf{x}_{ij}-\bar{\mathbf{x}}_i)'\mathbf{S}_i^{-1}(\mathbf{x}_{ik}-\bar{\mathbf{x}}_i)\}^3$$

and

$$b_{2i,p} = \frac{1}{n_i}\sum_{j=1}^{n_i}\{(\mathbf{x}_{ij}-\bar{\mathbf{x}}_i)'\mathbf{S}_i^{-1}(\mathbf{x}_{ij}-\bar{\mathbf{x}}_i)\}^2,$$

respectively. Note that in contrast with the scalar coefficient of skewness, $b_{1i,p}$ can be used only to assess whether skewness is present, not whether it is in any particular direction.

Under the heteroscedastic normal model,

$$\mathbf{X} \sim N(\boldsymbol{\mu}_i, \boldsymbol{\Sigma}_i) \quad \text{in} \quad G_i \qquad (i = 1,\ldots,g), \tag{6.2.6}$$

we have from Mardia (1970) that asymptotically,

$$(n_i/6)b_{1i,p} \sim \chi_d^2, \tag{6.2.7}$$

where $d = (p/6)(p+1)(p+2)$, and

$$\{b_{2i,p} - p(p+2)\}/\{(8/n_i)p(p+2)\}^{1/2} \sim N(0,1). \tag{6.2.8}$$

On a cautionary note, it would appear that the exact significance level of the test based on the large-sample result (6.2.8) may be quite different from the assumed nominal level (Gnanadesikan, 1977, Section 5.4).

Fatti et al. (1982) have pointed out how these statistics can be easily combined to give composite test statistics for skewness and kurtosis. One way is to use the asymptotic results that under (6.2.6),

$$\sum_{i=1}^{g}(n_i/6)b_{1i,p} \sim \chi_{gd}^2$$

and

$$\left\{\sum_{i=1}^{g}(n_i/n)b_{2i,p} - p(p+2)\right\} \Big/ \{(8/n)p(p+2)\}^{1/2} \sim N(0,1).$$

Another way is to use Fisher's method for combining tests, which they note is more powerful but not as easy to apply. For example, for this method applied to the testing of skewness in the group-conditional distributions, we use the result that under (6.2.6),

$$-2\sum_{i=1}^{g}\log P_{1,i} \sim \chi_{2g}^2, \tag{6.2.9}$$

where, for each i, $P_{1,i}$ denotes the P-value for the test based on $b_{1i,p}$, that is, the area to the right of the realized value of $(n_i/6)b_{1i,p}$ under the χ_d^2 distribution. Note that (6.2.9) does not hold exactly, as (6.2.7) is only a large-sample result. If there were prior grounds for assuming homoscedasticity, then consideration might be given to using the pooled estimate **S** of the common group-covariance matrix in the formation of the $b_{1i,p}$ and the $b_{2i,p}$. However, as it is not clear what effect this would have on the distribution of the consequent test statistics, the question of model fit is usually considered separately for each group-conditional distribution.

One explanation for skewness or heavy-tailed departures from normality for a particular group-conditional distribution is that it may be a finite mixture of normal or some other distributions, as discussed in Section 2.11. In such situations, the assessment of multivariate normality for, say, the ith group-conditional distribution, may lead to the consideration of a test for $g_i = 1$ versus $g_i = 2$, or possibly a greater value, where g_i denotes the number of components in a g_i-component normal mixture. This problem is far from straightforward, as explained in McLachlan (1987a). However, it can be tackled by bootstrapping the likelihood ratio test statistic as outlined in Section 1.10. Note that the presence of a normal mixture is not always the only explanation for departures from behavior for a single normal distribution. For example, a single log normal density can be well approximated by a mixture of two univariate densities with a common variance; see Titterington et al. (1985, page 30). However, where the sole aim is to provide an adequate model for a group-conditional distribution, the question of whether a group is a genuine mixture of two or more subgroups, or has an inherently skewed distribution, is not of primary interest; see Schork and Schork (1988) for a discussion of this point.

6.2.4 Absence of Classified Training Data

It has been assumed in the preceding work that each n_i is sufficiently large to provide reasonable estimates of the unknown parameters in the group-conditional densities $f_i(\mathbf{x};\boldsymbol{\theta}_i)$, $i = 1,\ldots,g$. If some or all of the n_i are very small, some unclassified data may be used, too, in forming these estimates. Then the assessment procedure is essentially the same as for an unclassified sample. The problem of testing for multivariate normality and homoscedasticity in this situation is covered in McLachlan and Basford (1988, Chapter 2). They follow the approach of Hawkins, Muller, and ten Krooden (1982) in suggesting that the unclassified data be clustered first by fitting a mixture of normal but heteroscedastic component distributions. Then Hawkins' (1981) test be applied to the resulting clusters as if they represent a true classification of the data. As cautioned by McLachlan and Basford (1988), it is self-serving to cluster the data under the assumption of normality when one is subsequently to test for it, but it is brought about by the limitations on such matters as hypothesis testing in the absence of classified data.

6.2.5 Likelihood Ratio Test for Homoscedasticity

The standard method for testing homoscedasticity under the assumption of multivariate normality uses the likelihood ratio test, although it is sensitive to nonnormality in the data (Layard, 1974). Therefore, it is wise to first check that a normal model is a viable assumption before applying this test. One attractive feature of this test is that the test statistic can be decomposed to provide some explanation as to the nature of any heteroscedasticity that may be detected. Information obtained in this way can be exploited in the formulation of the final model to be adopted for the formation of the sample discriminant rule; see Section 5.4.

For the normal model (6.2.6), the likelihood ratio test for homoscedasticity

$$H_0 : \mathbf{\Sigma}_i = \mathbf{\Sigma}(i = 1,\ldots,g), \tag{6.2.10}$$

is based on the statistic

$$Q = \sum_{i=1}^{g} n_i \log\{|\hat{\mathbf{\Sigma}}|/|\hat{\mathbf{\Sigma}}_i|\}, \tag{6.2.11}$$

where $\hat{\mathbf{\Sigma}}_i$ is the sample covariance matrix of the training feature data from group G_i $(i = 1,\ldots,g)$, and

$$\hat{\mathbf{\Sigma}} = \sum_{i=1}^{g} (n_i/n)\hat{\mathbf{\Sigma}}_i$$

is the pooled within-group sample covariance matrix; see, for example, T. W. Anderson (1984, Section 10.3) and Seber (1984, Section 9.2). Note that each n_i must exceed p in order for Q to be formed. Under H_0, Q is distributed asymptotically as chi-squared with $\frac{1}{2}(g-1)p(p+1)$ degrees of freedom.

Fatti et al. (1982) have commented on the convenience of being able to decompose the likelihood ratio test statistic to explore the nature of any heteroscedasticity present. There is the obvious decomposition from the definition of Q, where the ith term in its definition,

$$-n_i \log\{|\hat{\mathbf{\Sigma}}_i|/|\hat{\mathbf{\Sigma}}|\},$$

can be interpreted directly as a standardized measure of how the generalized sample variance within group G_i compares with the generalized pooled within-group sample variance. They exhibited another decomposition of Q based on the result that the determinant of a covariance matrix can be decomposed into a product of partial variances. The resultant decomposition of Q focuses on the individual variates in the feature vector. The vth component in this decomposition of Q provides a measure of the additional heteroscedasticity arising from the vth feature variate conditional on the first $v-1$ features; see Fatti et al. (1982, page 60).

6.2.6 Hierarchical Partitions of Likelihood Ratio Test Statistic for Homoscedasticity

Recently, Manly and Rayner (1987) have shown how the likelihood ratio test can be made more informative by hierarchically partitioning the test statistic

Q into three components, Q_0, Q_1, and Q_2, for testing the nested hypotheses H_i $(i = 0, 1, 2, 3)$. The hypothesis H_0 of homoscedasticity is defined by (6.2.10), and H_3 denotes the hypothesis of heteroscedasticity. Using the same notation as in Section 5.4, the hypotheses H_1 and H_2 are defined by

$$H_1 : \mathbf{\Sigma}_i = \kappa_i^2 \mathbf{\Sigma}_P \qquad (i = 1, \ldots, g),$$

where $\kappa_1 = 1$, and by

$$H_2 : \mathbf{\Sigma}_i = \mathbf{K}_i \mathbf{\Sigma}_C \mathbf{K}_i \qquad (i = 1, \ldots, g),$$

where $\mathbf{K}_i = \operatorname{diag}(\kappa_{i1}, \ldots, \kappa_{ip})$ for $i = 2, \ldots, g$, and $\mathbf{K}_1$ is the identity matrix. The component statistic Q_i is minus twice the log likelihood ratio for the test of H_i versus H_{i+1}, and so is evaluated in terms of the maximum likelihood estimates of the parameters under H_i and H_{i+1} $(i = 0, 1, 2)$. The latter estimates were given in Section 5.4. It follows that

$$Q_0 = n \log\{|\hat{\mathbf{\Sigma}}|/|\hat{\mathbf{\Sigma}}_P|\} - 2p \sum_{i=2}^{g} n_i \log \hat{\kappa}_i,$$

$$Q_1 = n \log\{|\hat{\mathbf{\Sigma}}_P|/|\hat{\mathbf{\Sigma}}_C|\} - 2 \sum_{i=2}^{g} n_i \left\{ \sum_{v=1}^{p} \log \hat{\kappa}_{iv} - p \log \hat{\kappa}_i \right\},$$

and

$$Q_2 = \sum_{i=1}^{g} n_i \log\{|\hat{\mathbf{\Sigma}}_C|/|\hat{\mathbf{\Sigma}}_i|\} + 2 \sum_{i=2}^{g} n_i \sum_{v=1}^{p} \log \hat{\kappa}_{iv}.$$

The hypothesis H_i may be tested against H_{i+1}, using the result that Q_i is asymptotically distributed under H_i, as chi-squared with d_i degrees of freedom $(i = 0, 1, 2)$, where $d_0 = g - 1$, $d_1 = (p-1)(g-1)$, and $d_2 = \frac{1}{2}(g-1)p(p-1)$. The hierarchical testing proceeds as follows. First, Q_2 is assessed to see whether it is significantly large. If so, then the within-group correlations are concluded to vary between the groups and testing stops. If Q_2 is not significant, then Q_1 is assessed for significance. If a significant result is obtained, then the within-group variances are concluded to be different but the correlations the same, and testing ceases. If Q_1 is not significant, then Q_0 is assessed for significance. If a significant result is obtained, then the group-covariance matrices are assumed to be proportional; otherwise they are assumed to be equal.

The likelihood ratio test statistic is usually based on the modified statistic

$$Q^* = -\sum_{i=1}^{g} (n_i - 1) \log\{|\mathbf{S}_i|/|\mathbf{S}|\},$$

where

$$\mathbf{S}_i = n_i \hat{\mathbf{\Sigma}}_i/(n_i - 1)$$

is the bias-corrected estimate of $\mathbf{\Sigma}_i$ $(i = 1, \ldots, g)$, and

$$\mathbf{S} = n\hat{\mathbf{\Sigma}}/(n - g)$$

is the bias-corrected estimate of the common group-covariance matrix Σ under homoscedasticity. The replacement in Q of the group-sample size n_i by the degrees of freedom $n_i - 1$ $(i = 1,\ldots,g)$ and the total sample size n by the total degrees of freedom $n-g$ is the multivariate analogue of Bartlett's (1938) suggested modification to Q in the univariate case. When the n_i are unequal, it is well known that the test based on Q is biased. However, Perlman (1980) has shown that the test based on Q^* is unbiased. As noted by Manly and Rayner (1987), the decomposition of Q^* can be undertaken in the same manner as for Q, simply by replacing n_i by $n_i - 1$ $(i = 1,\ldots,g)$ and n by $n-g$ in the process. Greenstreet and Connor (1974) showed how the statistic Q or Q^* can be modified by a multiplicative constant to give a test with essentially the same power as before, but with the actual size closer to the nominal level. In the case of Q^*, the multiplicative constant is defined to be

$$C = 1 - \left[\left\{\sum_{i=1}^{g}\left(\frac{1}{n_i - 1} - \frac{1}{n-g}\right)\right\}(2p^3 + 3p - 1)/6(g-1)(p+1)\right].$$

On the use of the likelihood ratio statistic for testing homoscedasticity under a normal model, McKay (1989) has employed it as the basis of a simultaneous procedure for isolating all subsets of the variates in the feature vector that have the same covariance matrix within each of two groups. If the group-covariance matrices are taken to be the same, McKay (1989) showed how his approach can be applied to isolating subsets of the feature variates for which the corresponding group-covariance matrix is diagonal. The case of a diagonal covariance matrix within each group corresponds to the conditional independence factor model (Aitkin, Anderson, and Hinde, 1981). That is, the observed correlations between the features result from the grouped nature of the data, and that within the underlying groups, the features are independently distributed.

Flury (1987b, 1988) has shown how the likelihood test statistic can be partitioned hierarchically for testing his proposed hierarchy of models containing the principal-component models for similarities in the group-covariance matrices Σ_i. It was shown in Section 5.4 that this hierarchy differs from the four-level hierarchy considered by Manly and Rayner (1987) in that the model of equal group-correlation matrices at the third level between the models of proportionality and heteroscedasticity is replaced by the common principal component (CPC), partial CPC, and common space models. Further details can be found in Flury (1988, Chapter 7). We give in what follows the components of the likelihood ratio statistic Q for the test of the hypothesis H_0 of homoscedasticity versus H_{CPC} and the test of H_{CPC} versus the hypothesis H_3 of heteroscedasticity, where H_{CPC} denotes the hypothesis of common principal components as stated by (5.4.9).

A test of H_{CPC} versus H_3 can be undertaken on the basis that

$$Q_4 = \sum_{i=1}^{p} n_i \log\{|\hat{\Lambda}_i|/|\hat{A}|^2|\hat{\Sigma}_i|\}$$

is asymptotically distributed under H_{CPC} as chi-squared with degrees of freedom equal to $\frac{1}{2}(g-1)p(p-1)$. The computation of the estimates $\hat{\mathbf{\Lambda}}$ and $\hat{\mathbf{\Lambda}}_i$ was described in Section 5.4. If required, a test of H_0 versus H_{CPC} can be undertaken using

$$Q_5 = n\log|\hat{\mathbf{\Sigma}}| - \sum_{i=1}^{g} n_i \log\{|\hat{\mathbf{\Lambda}}_i|/|\hat{\mathbf{\Lambda}}|^2\},$$

which has an asymptotic null distribution of chi-squared with $p(g-1)$ degrees of freedom.

6.3 DATA-BASED TRANSFORMATIONS OF FEATURE DATA

6.3.1 Introduction

Although a normal model can be assessed as providing an inadequate fit to the training data **t**, one may well decide in the end to still use discriminant procedures based on normality. As seen in Section 5.6, moderate departures from normality are not necessarily serious in their consequences for normal theory-based discriminant rules. This point has been emphasized also by Hjort (1986a, Section 11.2). Indeed, only a rough description of the group-conditional densities is needed to produce an effective allocation rule for widely separated groups. On the other hand, the question of model fit for the group-conditional densities is important in the case where reliable estimates of the posterior probabilities of group membership are sought. One option, if it is concluded that the normal model does not provide an adequate fit, is to seek a transformation of the feature data in an attempt to achieve normality. This course of action would be particularly relevant in the last case.

6.3.2 Box–Cox Transformation

An account of transformations can be found in Gnanadesikan (1977, Section 5.3). One approach discussed there is the multivariate generalization by Andrews, Gnanadesikan, and Warner (1971) of the univariate power transformations of Box and Cox (1964). In the present discriminant analysis context with the training data $\mathbf{x}_{ij}$ $(i=1,\ldots,g;\ j=1,\ldots,n_i)$, the vth variate of each $\mathbf{x}_{ij}$, x_{ijv} $(v=1,\ldots,p)$ is transformed into

$$x_{ijv}^{(\zeta_v)} = \begin{cases} (x_{ijv}^{\zeta_v}-1)/\zeta_v, & \zeta_v \neq 0, \\ \log x_{ijv}, & \zeta_v = 0. \end{cases} \tag{6.3.1}$$

Let $\boldsymbol{\zeta} = (\zeta_1,\ldots,\zeta_p)'$ be the vector of powers. Consistent with our previous notation, we let $\boldsymbol{\theta}_E$ contain the elements of $\boldsymbol{\mu}_1,\ldots,\boldsymbol{\mu}_g$ and the distinct elements of $\Sigma_1,\ldots,\Sigma_g$ under the constraint $\Sigma_1 = \cdots = \Sigma_g = \Sigma$, where now $\boldsymbol{\mu}_i$ and Σ_i denote the mean and covariance matrix, respectively, in the multivariate normal distribution assumed for the ith group-conditional distribution of the transformed feature vector. We can estimate ζ by maximum likelihood. Let $L(\zeta, \boldsymbol{\theta}_E; \mathbf{t})$ denote the likelihood function for ζ and $\boldsymbol{\theta}_E$ formed from $\mathbf{t}$ under the assumption of multivariate normality and homoscedasticity for the feature data when transformed according to (6.3.1). Then ζ can be estimated by $\hat{\zeta}$, obtained by maximization with respect to ζ of

$$\log L(\zeta;\mathbf{t}) = -\frac{1}{2}n\log|\hat{\boldsymbol{\Sigma}}| + \sum_{v=1}^{p}(\zeta_v - 1)\sum_{i=1}^{g}\sum_{j=1}^{n_i}\log x_{ijv}, \tag{6.3.2}$$

where $\hat{\boldsymbol{\Sigma}}$ is the pooled within-group sample covariance matrix of the transformed feature data. Here $L(\zeta;\mathbf{t})$ denotes (up to an additive constant) the maximum value of $L(\zeta, \boldsymbol{\theta}_E; \mathbf{t})$ with respect to $\boldsymbol{\theta}_E$ for specified ζ. This initial maximization, therefore, replaces $\boldsymbol{\theta}_E$ in $L(\zeta, \boldsymbol{\theta}_E; \mathbf{t})$ by its maximum likelihood estimate for specified ζ, which amounts to replacing the group means and common group-covariance matrix of the transformed feature data by their sample analogues.

Asymptotic theory yields an approximate $100(1-\alpha)\%$ confidence region for ζ given by

$$2\{\log L(\hat{\zeta};\mathbf{t}) - \log L(\zeta;\mathbf{t})\} \le \chi^2_{p;1-\alpha},$$

where $\chi^2_{p;1-\alpha}$ denotes the quantile of order $1-\alpha$ of the χ^2_p distribution. This approach also yields another assessment of multivariate normality as the likelihood ratio statistic

$$2\{\log L(\hat{\zeta};\mathbf{t}) - \log L(\mathbf{1}_p;\mathbf{t})\} \tag{6.3.3}$$

can be used to test the null hypothesis that $\zeta = \mathbf{1}_p = (1,\ldots,1)'$. The asymptotic P-value is given by the area to the right of (6.3.3) under the χ^2_p density.

If heteroscedasticity is apparent in the transformed data and the adoption of homoscedasticity is not crucial to the discriminant procedure to be used, then before making a final choice of ζ, one might consider its maximum likelihood estimate under the heteroscedastic normal model (3.2.1). For the latter model, the term

$$-\tfrac{1}{2}n\log|\hat{\boldsymbol{\Sigma}}|$$

in (6.3.2) is replaced by

$$-\frac{1}{2}\sum_{i=1}^{g} n_i \log|\hat{\boldsymbol{\Sigma}}_i|,$$

where $\hat{\boldsymbol{\Sigma}}_i$ denotes the sample covariance matrix of the transformed training data from group G_i $(i = 1,\ldots,g)$.

The maximization of (6.3.2) with respect to ζ involves much computation. A simpler method is to focus only on the marginal distributions of the variates in the feature vector. Then $\hat{\zeta}_v$ is estimated by maximizing

$$-\frac{1}{2}n\log(\hat{\boldsymbol{\Sigma}})_{vv} + (\zeta_v - 1)\sum_{i=1}^{g}\sum_{j=1}^{n_i}\log x_{ijv}$$

separately for each ζ_v ($v = 1,\ldots,p$). Although normal marginals certainly do not ensure joint normality, empirical evidence suggests it may suffice in practical applications; see Andrews et al. (1971), Johnson and Wichern (1982, Section 4.7), and Bozdogan and Ramirez (1986). This last approach, which chooses $\hat{\zeta}$ on the basis of enhancing marginal normality of each of the p variates in the feature vector, is thus concerned with only p directions, namely, those in the directions of the original coordinate axes. In contrast the first approach chooses $\hat{\zeta}$ to improve the joint normality of the p variates of the feature vector and so is concerned with all possible directions. Gnanadesikan (1977, Section 5.3) has described a method of transformation that identifies directions (not necessarily confined to prespecified directions such as the coordinate axes) of possible nonnormality and then selects a power transformation so as to improve normality of the projections of the original observations onto these directions.

A requirement for the use of the transformation (6.3.1) is that the original feature variates x_{ijv} be nonnegative. A simple way of ensuring this is to shift all the x_{ijv} by an arbitrary amount to make them all nonnegative. Another way is to use the more general shifted-power class of transformations, where the shift for each of the p variates is treated as an additional parameter to be estimated (Gnanadesikan, 1977, Section 5.3).

6.3.3 Some Results on the Box–Cox Transformation in Discriminant Analysis

Dunn and Tubbs (1980) have investigated the use of the transformation (6.3.1) as a means of enhancing homoscedasticity in the classical *Iris* data set (Fisher, 1936). Although heteroscedasticity was reduced as a consequence of the transformation, it was still assessed as highly significant. This is to be expected in general as it is an ambitious requirement for a transformation to achieve both approximate normality and homoscedasticity for a number of groups if the original data are heteroscedastic. Hermans et al. (1981) have drawn attention to this point. However, even if a transformation does not induce homoscedasticity, nor even near normality, it will have played a useful role if it has been able to produce good symmetry in the data.

The transformation (6.3.1), along with more robust methods, has been considered by Lesaffre (1983), who surveyed normality tests and transformations for use, in particular, in discriminant analysis. In the context of $g = 2$ groups, he used a modified version of the transformation (6.3.1), where ζ was allowed to be different for each group G_i. The ith group-specific value of ζ, $\hat{\zeta}_i$, was obtained by maximization of the log likelihood formed separately from the transformed data from G_i ($i = 1,2$). It was proposed that the group-conditional distributions be simultaneously normalized by taking $\hat{\zeta}$ to be a value of ζ in the overlap of the 95% confidence regions for ζ_1 and ζ_2. This approach, where ζ is taken to be different for each group, is perhaps most useful in situations where there would appear to be no ζ for achieving at least approximate normality in the transformed data. By subsequently allowing ζ to be different for each group, the group-specific estimates of ζ can serve as useful diagnostic tools in indicating particular departures from normality. Knowledge of these group-specific departures can be used in postulating more adequate models for the group-conditional distributions.

Beauchamp and Robson (1986) have investigated the use of the transformation (6.3.1) in discriminant analysis in terms of the asymptotic error rates of the sample normal-based linear and quadratic discriminant rules. For $g = 2$ groups with univariate or bivariate feature data, they compared the asymptotic error rates of these discriminant rules formed before and after transformation of the data according to (6.3.1) in which the appropriate value of ζ was used. Their results, therefore, serve as a guide to the maximum improvement that can be expected in the error rate for data-based versions of the transformation (6.3.1). Previously, Beauchamp, Folkert, and Robson (1980) had considered this problem in the special case of the logarithmic transformation applied to a univariate feature vector.

In the context of a single group, Bozdogan and Ramirez (1986) have presented an algorithm for transforming data according to transformation (6.3.1) if multivariate normality is assessed as untenable. The latter assessment is undertaken using the multivariate coefficients of skewness and kurtosis of Mardia (1970, 1974). Their algorithm, which can be routinely incorporated into any preprocessing analysis, also has the provision for assessing the transformed data for normal model assumptions.

6.4 TYPICALITY OF A FEATURE VECTOR

6.4.1 Assessment for Classified Feature Data

The detection of outliers is a challenging problem, particularly in multivariate data. Generally, an identification of observations that appear to stand apart from the bulk of the data is undertaken with a view to taking some action to reduce their effect in subsequent processes such as model fitting, parameter estimation, and hypothesis testing. In some instances, however, the atypical observations can be of much interest in their own right. Extensive surveys on the subject can be found in the books of Barnett and Lewis (1978) and Hawkins (1980), and in the review paper of Beckman and Cook (1983). More recent references include Cléroux, Helbling, and Ranger (1986), Bacon-Shone and Fung (1987), and Marco, Young, and Turner (1987c).

We consider now the detection of apparent outliers among the training data $\mathbf{t}$. It is still convenient to denote the n feature vectors in $\mathbf{t}$ as $\mathbf{x}_{ij}$ ($i = 1,\ldots,g$; $j = 1,\ldots,n_i$). For each feature vector $\mathbf{x}_{ij}$, we have, from the assessment of multivariate normality and homoscedasticity described in the previous section, the associated tail area a_{ij} computed to the right of the (normalized) squared Mahalanobis distance

$$c(n_i,\nu)\delta(\mathbf{x}_{ij},\bar{\mathbf{x}}_{i(ij)};\mathbf{S}_{(ij)}) \qquad (6.4.1)$$

under the $F_{p,\nu}$ distribution. If a_{ij} is close to zero, then $\mathbf{x}_{ij}$ is regarded as atypical of group G_i. Under the restriction to tests invariant under arbitrary full-rank linear transformations (Hawkins, 1980, page 107), (6.4.1) is equivalent to the optimal test statistic for a single outlier for both the alternative hypotheses

$$H_{1i} : \mathbf{X}_{ij} \sim N(\boldsymbol{\mu}_i^{\dagger}, \boldsymbol{\Sigma})$$

and

$$H_{2i} : \mathbf{X}_{ij} \sim N(\boldsymbol{\mu}_i, \kappa_i \boldsymbol{\Sigma}),$$

where $\boldsymbol{\mu}_i^{\dagger}$ is some vector not equal to $\boldsymbol{\mu}_i$, and κ_i is some positive constant different from unity. A more relevant question than whether an observation is atypical is whether it is statistically unreasonable when assessed as an extreme (a discordant outlier), for example, with univariate data, whether, say, the largest of the n_i observations from G_i is unreasonably large as an observation on the n_ith order statistic for a random sample of size n_i. However, generalizations to multivariate data are limited as the notion of order then is ambiguous and ill-defined; see Barnett (1983).

If it were concluded that it is reasonable to take the group-distributions to be multivariate normal but heteroscedastic, that is,

$$\mathbf{X} \sim N(\boldsymbol{\mu}_i, \boldsymbol{\Sigma}_i) \quad \text{in} \quad G_i \ (i = 1, \ldots, g), \tag{6.4.2}$$

then the a_{ij} should be recomputed under (6.4.2) before being used to assess whether there are any atypical observations among the training data. Under (6.4.2), we modify the squared Mahalanobis distance (6.4.1) by replacing $\mathbf{S}_{(ij)}$ with the ith group-specific estimate $\mathbf{S}_{i(ij)}$. Consequently, a_{ij} would be taken to be the area to the right of

$$c(n_i, \nu_i)\delta(\mathbf{x}_{ij}, \overline{\mathbf{x}}_{i(ij)}; \mathbf{S}_{i(ij)}) \tag{6.4.3}$$

under the F_{p,ν_i} distribution, where $\nu_i = n_i - p - 1$. The value of (6.4.3) can be computed from (6.2.4), where ν is replaced by ν_i and $\mathbf{S}$ by $\mathbf{S}_i$.

The fitting of the group-conditional distributions depends on how the presence of apparently atypical observations among the training data is handled. If the aim is to undertake estimation of the group-conditional distributions with all atypical feature observations deleted from the training data, then $\mathbf{x}_{ij}$ would be a candidate for deletion if $a_{ij} < \alpha$, where α is set at a conventionally small level, say, $\alpha = 0.05$. A second approach is to eliminate only those feature observations assessed as being extremely atypical, say, $\alpha = 0.01$ or 0.005. Protection against the possible presence of less extreme outliers can be provided by the use of robust M-estimators, as discussed in Section 5.7. A third course of action might be to use redescending M-estimators to accommodate all discordant outliers, no matter how atypical. Another approach might be to proceed as in Aitkin and Tunnicliffe Wilson (1980) and allow for another group(s) through the fitting of a mixture model. More recently, Butler (1986) has examined the role of predictive likelihood inference in outlier theory.

On other approaches to the detection of atypical observations among the training data, Campbell (1978) and Radhakrishnan (1985) have considered the use of the influence function as an aid in this pursuit. A Bayesian approach to the problem of assessing influence can be found in Geisser (1987).

6.4.2 Assessment for Unclassified Feature Data

A preliminary matter for consideration before an outright or probabilistic allocation is made for an unclassified entity with feature vector $\mathbf{x}$ is whether $\mathbf{x}$ is consistent with the assumption that the entity belongs to one of the target groups $G_1, \ldots, G_g$. One way of approaching the assessment of whether the feature $\mathbf{x}$ on an unclassfied entity is typical of a mixture of $G_1, \ldots, G_g$ is to assess how typical $\mathbf{x}$ is of each G_i in turn. For a given $\mathbf{x}$, the squared Mahalanobis distance

$$\delta(\mathbf{x}, \overline{\mathbf{x}}_i; \mathbf{S}_i) \tag{6.4.4}$$

is computed for $i = 1, \ldots, g$. Then under the heteroscedastic normal model (6.4.2),

$$c(n_i + 1, \nu_i + 1)\delta(\mathbf{x}, \overline{\mathbf{x}}_i; \mathbf{S}_i) \tag{6.4.5}$$

has the F_{p,ν_i+1} distribution, where, as before, $\nu_i = n_i - p - 1$, and the function $c(\cdot,\cdot)$ is defined by (6.2.3). The situation here is more straightforward than with the assessment of typicality for a classified entity, because $\mathbf{x}$ is the realization of a random vector distributed independently of $\overline{\mathbf{x}}_i$ and $\mathbf{S}_i$ $(i = 1, \ldots, g)$. An assessment of how typical $\mathbf{x}$ is of G_i is given by $a_i(\mathbf{x}; \mathbf{t})$, the tail area to the right of (6.4.5) under the F_{p,ν_i+1} distribution. For the homoscedastic normal model, $a_i(\mathbf{x}; \mathbf{t})$ is defined similarly, but where $\mathbf{S}_i$ is replaced by the pooled within-group estimate $\mathbf{S}$ in (6.4.4) and (6.4.5), and, as a consequence, ν_i is replaced by $\nu = n - g - p$ in (6.4.5). A feature vector can be assessed as being atypical of the mixture G if

$$a(\mathbf{x}; t) = \max_i a_i(\mathbf{x}; \mathbf{t}) \leq \alpha, \tag{6.4.6}$$

where α is some specified threshold. Although this measure of typicality $a(\mathbf{x}; \mathbf{t})$ is obtained under the assumption of normality, it should still be useful provided the group-conditional densities are at least elliptically symmetric.

The work of Cacoullos (1965a, 1965b) and Srivastava (1967b) is related to this problem.

6.4.3 Typicality Index Viewed as a *P*-Value

Concerning the use of $a(\mathbf{x}; \mathbf{t})$ as a measure of typicality of $\mathbf{x}$ with respect to the groups $G_1, \ldots, G_g$, it can be interpreted also as a P-value for a test of the compatibility of $\mathbf{x}$ with the g groups, as specified by

$$H_0^{(C)} : \mathbf{X} \sim N(\boldsymbol{\mu}_i, \boldsymbol{\Sigma}_i) \qquad \text{for} \quad i = 1, 2, \ldots, \text{ or } g.$$

For under the null hypothesis $H_0^{(C)}$ that $\mathbf{x}$ is a realization of the feature vector $\mathbf{X}$ on an entity from one of the g groups $G_1,\ldots,G_g$, the distribution of $a_i(\mathbf{X};\mathbf{T})$ should be approximately uniform on the unit interval. Indeed, the distribution of $a(\mathbf{X};\mathbf{T})$ would be precisely uniform on (0,1) if the allocation rule $r_a(\mathbf{x};\mathbf{t})$ based on the $a_i(\mathbf{x};\mathbf{t})$ were infallible, where

$$r_a(\mathbf{x};\mathbf{t}) = i \qquad \text{if} \quad a_i(\mathbf{x};\mathbf{t}) \geq a_h(\mathbf{x};\mathbf{t}) \qquad (h = 1,\ldots,g;\ h \neq i). \tag{6.4.7}$$

Of course, this rule is fallible, but its overall error rate will be small if the groups $G_1,\ldots,G_g$ are well-separated. For example, it can be seen from (3.4.7) that under the homoscedastic normal model (3.3.1), (6.4.7) defines the allocation rule based on the likelihood ratio criterion. Hence, depending on the extent to which the G_i are separated, we have approximately that

$$a_i(\mathbf{X};\mathbf{T}) \sim U(0,1).$$

Suppose that $H_0^{(C)}$ holds and that without loss of generality the entity comes from the hth group, that is, the associated feature vector $\mathbf{X}$ is distributed $N(\boldsymbol{\mu}_h, \boldsymbol{\Sigma}_h)$. Then as

$$\text{pr}\{a(\mathbf{X};\mathbf{T}) \leq \alpha\} \leq \text{pr}\{a_i(\mathbf{X};\mathbf{T}) \leq \alpha\} \qquad (i = 1,\ldots,g),$$

we have

$$\begin{aligned} \text{pr}\{a(\mathbf{X};\mathbf{T}) \leq \alpha \mid H_0^{(C)}\} &\leq \text{pr}\{a_h(\mathbf{X};\mathbf{T}) \leq \alpha \mid H_0^{(C)}\} \\ &= \alpha. \end{aligned}$$

Thus, the probability of a Type I error of the test of compatibility with a rejection region defined by (6.4.6) does not exceed α. For $g = 2$ groups with homoscedastic normal distributions, McDonald et al. (1976) showed that this test is not too conservative, as the probability of a Type I error does not fall below $\frac{1}{2}\alpha$. More recently, Hjort (1986a, Chapter 6) has considered the use of (6.4.6) and slightly modified versions as measures of typicality.

Another measure of typicality of $\mathbf{x}$ with respect to the g groups $G_1,\ldots,G_g$ is

$$\sum_{i=1}^{g} \hat{z}_i a_i(\mathbf{x};\mathbf{t}), \tag{6.4.8}$$

where $\hat{\mathbf{z}} = (\hat{z}_1,\ldots,\hat{z}_g)'$ denotes the allocation of the entity with feature $\mathbf{x}$ to one of these groups on the basis of the plug-in sample version of the normal-based Bayes rule, that is, on the basis of the sample NLDR or NQDR. The extent of the agreement between $a(\mathbf{x};t)$ and (6.4.8) depends on the separation between the groups, the disparity of the group-sample sizes and group-priors, and the severity of the heteroscedasticity. For equal group-sample sizes, the homoscedastic versions of these two measures coincide provided the sample NLDR is formed with equal group-prior probabilities.

6.4.4 Predictive Assessment of Typicality

Another way of monitoring the typicality of the feature $\mathbf{x}$ with respect to a given group G_i $(i = 1,\ldots,g)$ is to adopt the index of Aitchison and Dunsmore (1975, Chapter 11). Their index of typicality for G_i of $\mathbf{x}$ is defined by

$$a_i^{(P)}(\mathbf{x}) = 1 - \int_{H_i(\mathbf{x})} f_i(\mathbf{w};\boldsymbol{\theta}_i)d\mathbf{w}, \tag{6.4.9}$$

where

$$H_i(\mathbf{x}) = \{\mathbf{w} : f_i(\mathbf{w};\boldsymbol{\theta}_i) > f_i(\mathbf{x};\boldsymbol{\theta}_i)\} \tag{6.4.10}$$

is the set of all observations more typical of G_i than $\mathbf{x}$. In order to provide an assessment of $a_i^{(P)}(\mathbf{x})$, Aitchison and Dunsmore (1975) replaced the ith group-conditional density $f_i(\mathbf{x};\boldsymbol{\theta}_i)$ by its predictive estimate $\hat{f}_i^{(P)}(\mathbf{x})$ in (6.4.9) and (6.4.10). If $\hat{a}_i^{(P)}(\mathbf{x})$ denotes the estimate of $a_i^{(P)}(\mathbf{x})$ so obtained, then it can be shown under the heteroscedastic normal model that

$$\hat{a}_i^{(P)}(\mathbf{x}) = a_i(\mathbf{x};\mathbf{t}) \qquad (i = 1,\ldots,g). \tag{6.4.11}$$

That is, $\hat{a}_i^{(P)}(\mathbf{x})$ can be interpreted in a frequentest sense as the level of significance associated with the test of the compatibility of $\mathbf{x}$ with G_i on the basis of the squared Mahalanobis distance $\delta_i(\mathbf{x},\bar{\mathbf{x}}_i;\mathbf{S}_i)$, or, equivalently, Hotelling's T^2; see Moran and Murphy (1979). The result (6.4.11) also holds under homoscedasticity.

Note that if $f_i(\mathbf{x};\boldsymbol{\theta}_i)$ were replaced in (6.4.9) and (6.4.10) by its estimative version $\phi(\mathbf{x};\bar{\mathbf{x}}_i,\mathbf{S}_i)$, a multivariate normal density with mean $\bar{\mathbf{x}}_i$ and covariance matrix $\mathbf{S}_i$, then this assessment of $a_i^{(P)}(\mathbf{x})$ is equal to the area to the right of $\delta_i(\mathbf{x},\bar{\mathbf{x}}_i;\mathbf{S}_i)$ under the χ_p^2 distribution. This is equivalent to taking $\bar{\mathbf{x}}_i = \boldsymbol{\mu}_i$ and $\mathbf{S}_i = \boldsymbol{\Sigma}_i$ in considering the distribution of $\delta_i(\mathbf{X},\bar{\mathbf{X}}_i;\mathbf{S}_i)$, and so gives a cruder assessment than the predictive approach. Of course, although the estimative approach can be used to assess the group-conditional densities for the provision of an allocation rule, it is another matter to proceed further with this approximation in the calculation of the typicality index (6.4.9), thereby completely ignoring the variation in the sample estimates $\bar{\mathbf{x}}_i$ and $\mathbf{S}_i$.

6.5 SAMPLE CANONICAL VARIATES

6.5.1 Introduction

As discussed in Section 3.9.1, exploratory data analysis of multivariate feature observations is often facilitated by representing them in a lower-dimensional space. In Section 3.9.2, a canonical variate analysis was introduced for providing the optimal linear projection of the feature vector $\mathbf{x}$ from $\mathbb{R}^p$ to $\mathbb{R}^q$ $(q < p)$ for a certain class of measures of spread of g homoscedastic groups, based on the nonzero eigenvalues of $\boldsymbol{\Sigma}^{-1}\mathbf{B}_o$, where

$$\mathbf{B}_o = \frac{1}{g-1}\sum_{i=1}^{g}(\boldsymbol{\mu}_i - \overline{\boldsymbol{\mu}})(\boldsymbol{\mu}_i - \overline{\boldsymbol{\mu}})',$$

and

$$\overline{\boldsymbol{\mu}} = \sum_{i=1}^{g}\boldsymbol{\mu}_i/g.$$

In the present situation where $\mathbf{B}_o$ and $\boldsymbol{\Sigma}$ are unknown, this class of measures of spread can be based on the nonzero eigenvalues of $\mathbf{S}^{-1}\mathbf{B}$, where $\mathbf{S}$ is the pooled within-group sample covariance matrix corrected for bias and where

$$\mathbf{B} = \frac{1}{g-1}\sum_{i=1}^{g}n_i(\overline{\mathbf{x}}_i - \overline{\mathbf{x}})(\overline{\mathbf{x}}_i - \overline{\mathbf{x}})'.$$

In a multivariate analysis of variance, $\mathbf{B}$ is the between-group sums of squares and products matrix on its degrees of freedom. As $n \to \infty$ under mixture sampling of the training data $\mathbf{T}$, $\mathbf{B}/n$ converges in probability to $\mathbf{B}_o/g$ in the special case of equal group-prior probabilities. In some situations where the group-sample sizes n_i are extremely disparate, one may wish to use

$$\frac{1}{g-1}\sum_{i=1}^{g}(\overline{\mathbf{x}}_i - \overline{\mathbf{x}})(\overline{\mathbf{x}}_i - \overline{\mathbf{x}})'$$

in place of $\mathbf{B}$.

We let

$$\begin{aligned}\mathbf{v}_j &= \boldsymbol{\Gamma}_d\mathbf{x}_j \\ &= (\boldsymbol{\gamma}_1'\mathbf{x}_j, \ldots, \boldsymbol{\gamma}_d'\mathbf{x}_j)'\end{aligned}$$

be the vector of canonical variates for $\mathbf{x}_j$ $(j, \ldots, n)$, where $\boldsymbol{\Gamma}_d$ is defined as previously by (3.9.2), but now with $\mathbf{B}_0$ and $\boldsymbol{\Sigma}$, replaced by $\mathbf{B}$ and $\mathbf{S}$, respectively. That is, $\boldsymbol{\gamma}_1, \ldots, \boldsymbol{\gamma}_d$ satisfy

$$\boldsymbol{\gamma}_h'\mathbf{S}\boldsymbol{\gamma}_k = \delta_{hk} \qquad (h,k = 1, \ldots, d), \tag{6.5.1}$$

and are the eigenvectors of $\mathbf{S}^{-1}\mathbf{B}$ corresponding to $\lambda_{B,1}, \ldots, \lambda_{B,d}$, the d nonzero eigenvalues of $\mathbf{S}^{-1}\mathbf{B}$, ordered in decreasing size, where $d = \min(p, b)$, and b is the rank of $\mathbf{B}$. Generally, in practice, $b = g - 1$. It is assumed here that $n - g \geq p$ so that $\mathbf{S}$ is nonsingular. Note that we could equally use $(g-1)\mathbf{B}$ and $(n-g)\boldsymbol{\Sigma}$ instead of $\mathbf{B}$ and $\boldsymbol{\Sigma}$, as suggested sometimes in the literature. The eigenvectors $\boldsymbol{\gamma}_1, \ldots, \boldsymbol{\gamma}_d$ are unchanged, whereas the eigenvalues are now multiplied by $(g-1)/(n-g)$.

By direct analogy with the results in Section 3.9.2 in the case of known parameters, the sample canonical variate analysis finds those linear combinations $\boldsymbol{\gamma}'\mathbf{x}$ of the original p feature variables in $\mathbf{x}$ that successively maximize the ratio

$$\gamma'\mathbf{B}\gamma/\gamma'\mathbf{S}\gamma. \tag{6.5.2}$$

Applied to the training data $\mathbf{t}$, the n original p-dimensional feature vectors $\mathbf{x}_j$ $(j = 1,\ldots,n)$ are transformed to the scores on the canonical variates $\mathbf{\Gamma}_d\mathbf{x}_j$ $(j = 1,\ldots,n)$, which can be represented in d-dimensional Euclidean space (the canonical variate space). The first q dimensions of this space provide the optimal (for the aforementioned class of measures of spread) q-dimensional configuration of the training data in which to highlight the differences between the groups. In particular, plotting the first two canonical variates of the group-sample means against each other, that is, $\gamma_2'\bar{\mathbf{x}}_i$ versus $\gamma_1'\bar{\mathbf{x}}_i$ $(i = 1,\ldots,g)$, gives the best two-dimensional representation of the differences that exist between the group-sample means. In terms of the canonical variates $\mathbf{\Gamma}_q\mathbf{x}$ $(q \leq d)$, the sample analogue of the result (3.9.12) shows that the sample NLDR (formed with equal group-prior probabilities) assigns an unclassified entity to the group whose sample mean is closest in the canonical variate space to $\mathbf{\Gamma}_q\mathbf{x}$.

As noted by Fatti et al. (1982), another use of canonical variates when $p \geq g$ is in testing whether an unclassified entity with feature vector $\mathbf{x}$ comes from one of the g groups or a mixture of them. For if it does, it follows from (3.9.8) and (3.9.9) that, asymptotically as $n \to \infty$,

$$\gamma_k'(\mathbf{X} - \overline{\mathbf{X}}) \sim N(0,1)$$

for $k = d+1,\ldots,p$, and so

$$\sum_{k=d+1}^{p} \{\gamma_k'(\mathbf{X} - \overline{\mathbf{X}})\}^2 \tag{6.5.3}$$

is the realization of a random variable with an asymptotic chi-squared distribution having $p-d$ degrees of freedom. Hence, an approximate test can be constructed that rejects this hypothesis if (6.5.3) is sufficiently large. The result (6.5.3) requires the assumption of multivariate normality in addition to homoscedasticity for the group-conditional distributions.

The computation of the population, and hence by analogy the sample, canonical variates, was discussed in Section 3.9.2. Additional references on this topic include Campbell (1980b, 1982). In the first of these papers, he proposed a ridge-type adjustment to $\mathbf{S}$, and in the second, he developed robust M-estimates using a functional relationship model. The model-based approach is pursued in Campbell (1984a).

6.5.2 Tests for Number of Canonical Variates

We saw in the previous section that under homoscedasticity, the number of sample canonical variates required, that is, the number of sample linear discriminant functions required, is given by $d = \min(p,b)$, where b is the rank of $\mathbf{B}$. Let b_o denote the rank of the population analogue of $\mathbf{B}$. Then the number of population discriminant functions is $\min(p,b_o)$. In practice, we can assess

b_o by b and so use $d = \min(p,b)$ discriminant functions. Rather than basing d as such on a point estimate of b_o, we may wish to infer b_o by carrying out a test of significance.

If the population group means are collinear, then $b_o = 1$ and so a single discriminant function will be adequate. A test that b_o equals some specified integer not less than one has been constructed by analogy to the test of $b_o = 0$. The latter value is equivalent to the usual null hypothesis of equality of group means

$$H_0 : \boldsymbol{\mu}_1 = \boldsymbol{\mu}_2 = \cdots = \boldsymbol{\mu}_g$$

in a one-way multivariate analysis of variance (MANOVA). The hypothesis H_0 can be tested using Wilks' likelihood ratio statistic

$$\Lambda = |\mathbf{W}|/|\tilde{\mathbf{B}} + \mathbf{W}|,$$

where

$$\tilde{\mathbf{B}} = (g-1)\mathbf{B} \tag{6.5.4}$$

is the between-group matrix of sums of squares and products, and

$$\mathbf{W} = (n-g)\mathbf{S} \tag{6.5.5}$$

is the within-group matrix. As assumed previously, $n - g \geq p$ so that $\mathbf{W}$ is nonsingular. It follows that Λ can be expressed in terms of the nonzero eigenvalues $\lambda_{\tilde{B},1},\ldots,\lambda_{\tilde{B},d}$ of $\mathbf{W}^{-1}\tilde{\mathbf{B}}$ as

$$\Lambda = -\log \prod_{j=1}^{d} (1 + \lambda_{\tilde{B},j}).$$

The statistic Λ has Wilks' $\Lambda(n-1,p,g-1)$ distribution under H_0; see, for example, T. W. Anderson (1984, Chapter 8) and Kshirsagar (1972, pages 202–204). The assumption of multivariate normality is implicitly assumed in all the distributional results stated here.

Concerning the exact null distribution of Λ in special cases, for $d_o = \min(p, g-1) = 1$,

$$\frac{1-\Lambda}{\Lambda} \cdot \frac{c_2+1}{c_1+1} \sim F_{2c_1+2,2c_2+2},$$

and for $d_o = 2$,

$$\frac{1-\Lambda^{1/2}}{\Lambda^{1/2}} \cdot \frac{2c_2+2}{2c_1+3} \sim F_{4c_1+6,4c_2+4},$$

where

$$c_1 = \tfrac{1}{2}(|g-1-p|-1),$$

and

$$c_2 = \tfrac{1}{2}(n-g-p-1).$$

In general, Bartlett (1938) has shown that

$$-\{n-1-\tfrac{1}{2}(p+g)\}\log\Lambda \sim \chi^2_{p(g-1)} \tag{6.5.6}$$

holds approximately under H_0; see Seber (1984, Section 2.5.4) for other approximations to the null distribution of Λ.

We return now to the problem of testing that b_o is equal to a specified integer not less than one. For the hypothesis $H_0^{(b_o)} : b_o = s \ (s \geq 1)$, where $p > s$, Bartlett (1939, 1947) proposed by analogy with (6.5.6) the use of the statistic

$$-\left\{n-1-\frac{1}{2}(p+g)\right\}\log \prod_{j=s+1}^{d} (1+\lambda_{\tilde{B},j}),$$

which under $H_0^{(b_o)}$ is distributed approximately as chi-squared with $(p-s)(g-1)$ degrees of freedom.

Bartlett (1951b) and Williams (1952, 1955, 1967) have derived an exact goodness-of-fit test for the adequacy of an arbitrarily specified discriminant function, say, $\mathbf{a}'\mathbf{x}$, to discriminate between the g groups. In order for this single discriminant function $\mathbf{a}'\mathbf{x}$ to be adequate, the g means must be collinear (that is, $b_o = 1$ in the terminology above) and $\mathbf{a}$ must be an eigenvector of $\mathbf{W}^{-1}\tilde{\mathbf{B}}$, corresponding to the only nonzero eigenvalue of $\mathbf{W}^{-1}\tilde{\mathbf{B}}$. Thus, there are two aspects of the hypothesis that $\mathbf{a}'\mathbf{x}$ is an adequate discriminant function: (a) collinearity of the group means and (b) the direction $\mathbf{a}$ of the specified discriminant function.

Let Λ_a denotes Wilks' statistic based on the data transformed as $\mathbf{a}'\mathbf{x}$. Then

$$\Lambda_a = |\mathbf{a}'\mathbf{W}\mathbf{a}|/|\mathbf{a}'\tilde{\mathbf{B}}\mathbf{a} + \mathbf{a}'\mathbf{W}\mathbf{a}|.$$

Bartlett (1951b) proposed that the residual form Λ_R of Wilks' Λ statistic after allowance for the relationship between the groups and $\mathbf{a}'\mathbf{x}$,

$$\Lambda_R = \Lambda/\Lambda_a,$$

be used to test the null hypothesis that $\mathbf{a}'\mathbf{x}$ is an adequate discriminant function. The null distribution of Λ_R is Wilks' $\Lambda(n-2, p-1, g-1)$.

In the case of a significant result, it is useful to know the source of the significance. Bartlett (1951b) showed that this can be done by factorizing Λ_R into two factors,

$$\Lambda_R = \Lambda_D \Lambda_{(C|D)},$$

where

$$\Lambda_D = \frac{\{1-\mathbf{a}'\tilde{\mathbf{B}}(\tilde{\mathbf{B}}+\mathbf{W})^{-1}\tilde{\mathbf{B}}\mathbf{a}\}/\mathbf{a}'\tilde{\mathbf{B}}\mathbf{a}}{\mathbf{a}'\mathbf{W}\mathbf{a}/\mathbf{a}'(\tilde{\mathbf{B}}+\mathbf{W})\mathbf{a}}$$

is the "direction" factor, and $\Lambda_{(C|D)}$ is the "partial collinearity" factor. The statistic Λ_D can be used to test the direction aspect of $\mathbf{a}'\mathbf{x}$, and $\Lambda_{(C|D)}$ can be used to test the collinearity aspect, after elimination of the direction aspect. Under the null hypothesis, Λ_D and $\Lambda_{(C|D)}$ are distributed independently according to $\Lambda(n-1, p-1, 1)$ and $\Lambda(n-2, p-1, g-2)$, respectively.

Bartlett (1951b) also gave an alternative factorization of Λ_R,

$$\Lambda_R = \Lambda_C \Lambda_{(D|C)},$$

where

$$\Lambda_C = \frac{|\mathbf{W}|}{|\tilde{\mathbf{B}} + \mathbf{W}|}\left(1 + \frac{\mathbf{a}'\tilde{\mathbf{B}}\mathbf{W}^{-1}\tilde{\mathbf{B}}\mathbf{a}}{\mathbf{a}'\tilde{\mathbf{B}}\mathbf{a}}\right)$$

is the collinearity factor, and

$$\Lambda_{(D|C)} = \Lambda_R/\Lambda_C$$

is the partial direction factor. The null distribution of Λ_C is $\Lambda(n-1, p-1, g-2)$, independent of $\Lambda_{(D|C)}$, which has a null distribution of $\Lambda(n-g+1, p-1, 1)$.

These factors were derived by Bartlett (1951b) by a geometrical method. An alternative analytical method of derivation has been given since by Kshirsagar (1964, 1970). Williams (1961, 1967) has extended the factorization of Wilks' Λ to the goodness-of-fit test of $b_o > 1$ specified discriminant functions; see also Radcliffe (1966, 1968) and Kshirsagar (1969). Williams (1967) and Kshirsagar (1971) have considered the problem where the discriminant functions are specified not as functions of **x**, but in terms of dummy variates with respect to the g groups. Additional discussion of the results presented in this section can be found in Kshirsagar (1972, Chapter 9), who has provided an excellent account of this topic.

6.5.3 Confidence Regions in Canonical Variate Analysis

It is usually desirable with plots of the group-sample means if some indication of their sampling variability can be given. To this end, let $\overline{v}_{ik}(\mathbf{t})$ denote the sample mean of the scores on the kth canonical variate computed for the training data from group G_i, that is,

$$\overline{v}_{ik}(\mathbf{t}) = \gamma_k'\overline{\mathbf{x}}_i \qquad (i = 1,\ldots,g;\ k = 1,\ldots,d).$$

The expectation of $\overline{v}_{ik}(\mathbf{T})$ is given by

$$\begin{aligned}\zeta_{ik} &= E\{\overline{v}_{ik}(\mathbf{T})\} \\ &= E\{\overline{\mathbf{X}}_i'\gamma_k(\mathbf{T})\},\end{aligned}$$

where we have written γ_k as $\gamma_k(\mathbf{T})$ to denote explicitly that it is a random vector, being a function of the training data $\mathbf{T}$.

Put

$$\overline{\mathbf{v}}_i(\mathbf{t}) = (\overline{v}_{i1}(\mathbf{t}),\ldots,\overline{v}_{id}(\mathbf{t}))'$$

and

$$\zeta_i = (\zeta_{i1},\ldots,\zeta_{id})'.$$

The traditional confidence region for ζ_i is constructed using the result that

$$n_i\{\overline{\mathbf{v}}_i(\mathbf{T}) - \zeta_i\}'\{\overline{\mathbf{v}}_i(\mathbf{T}) - \zeta_i\} \sim \chi_d^2, \qquad (6.5.7)$$

which ignores the fact that the canonical variates were computed using the sample matrices **B** and **S** rather than their unknown population analogues. For two-dimensional plots of the canonical group means, the approximation (6.5.7) yields confidence circles centered at the respective group-sample means, with squared radii equal to the appropriate upper percentage point of the χ_2^2 distribution, divided by the relevant group-sample sizes; see, for example, Seber (1984, page 273). Note that instead of confidence regions for a mean, tolerance regions for a group can be constructed. These are regions within which a given percentage of the group is expected to lie. They are constructed in the same way as confidence regions, but since the variability being assessed is that due to individuals rather than group means, the relevant dispersion is increased by the factor n_i. That is, the tolerance region for the ith group G_i is given by (6.5.7), but with n_i replaced by unity $(i = 1,\ldots,g)$. One of the earliest examples of tolerance circles in canonical variate diagrams was given by Ashton, Healy, and Lipton (1957).

Recently, Krzanowski (1989), through a series of approximations to the exact sampling distribution of the canonical variates, has derived confidence ellipses for the canonical group means, which appear from empirical assessment to have much more accurate probability content than the traditional circles. As explain by Krzanowski (1989), the asymptotic distributions of the $\bar{v}_{ik}(\mathbf{T})$ would be available from T. W. Anderson (1984, page 545) if $(g-1)\mathbf{B}$ followed a Wishart distribution. However, it follows a noncentral Wishart distribution. This led Krzanowski (1989) to approximate $(g-1)\mathbf{B}$ by a random matrix having a central Wishart distribution with the same degrees of freedom and expectation as the noncentral Wishart distribution of $(g-1)\mathbf{B}$ and with a covariance matrix that differs only slightly. It is analogous to the approximation of a noncentral chi-squared distribution by an appropriate multiple of a central chi-squared distribution.

Under separate sampling of $\mathbf{T}$, Krzanowski (1989) obtained an approximation $\mathbf{C}_i$ to the covariance matrix of $\bar{\mathbf{v}}_i(\mathbf{T}) - \boldsymbol{\zeta}_i$ in the situation where $n-g$ and $g-1$ each tend to infinity in such a way that $(g-1)/(n-g) \to \kappa > 0$. We let $\hat{\mathbf{C}}_i$ denote the estimate of $\mathbf{C}_i$ obtained by using $\bar{v}_{ik}$ for ζ_{ik} and $\lambda_{B,k}$ for its population analogue in the expression for $\mathbf{C}_i$ $(i = 1,\ldots,g;\ k = 1,\ldots,d)$. The estimate $\hat{\mathbf{C}}_i$ so obtained is given by

$$(\hat{\mathbf{C}}_i)_{kk} = \frac{1}{n_i} + \left\{ \frac{1}{2}\bar{v}_{ik}^2 + \sum_{m \neq k = 1}^{d} \frac{\omega_k(\omega_m + \kappa\omega_k)}{\kappa(\omega_k - \omega_m)^2}\bar{v}_{im}^2 \right\} \Big/ (n-g) \qquad (k = 1,\ldots,d),$$

and

$$(\hat{\mathbf{C}}_i)_{km} = -\frac{\omega_k\omega_m(1+\kappa)}{\kappa(n-g)(\omega_k - \omega_m)^2}\bar{v}_{ik}\bar{v}_{im} \qquad (k \neq m = 1,\ldots,d),$$

where

$$\omega_k = 1 + \lambda_{B,k}/(g-1).$$

An approximate confidence region for the ith group canonical mean $\boldsymbol{\zeta}_i$ can be constructed using the large sample result

$$\{\bar{\mathbf{v}}_i(\mathbf{T}) - \boldsymbol{\zeta}_i\}'\hat{\mathbf{C}}_i^{-1}\{\bar{\mathbf{v}}_i(\mathbf{T}) - \boldsymbol{\zeta}_i\} \sim \chi_d^2.$$

In particular, it allows confidence ellipses to be imposed on the plot of the group-sample means of the first two canonical variates. Denoting by v_1 and v_2 the coordinate values along the first two canonical axes, respectively, the equation of this confidence ellipse for the ith group-canonical mean can be written as

$$c_{i22}(v_1 - \overline{v}_{i1})^2 - 2c_{i12}(v_1 - \overline{v}_{i1})(v_2 - \overline{v}_{i2}) + c_{i11}(v_2 - \overline{v}_{i2})^2$$
$$= (c_{i11}c_{i22} - c_{i12}^2)\chi^2_{2;1-\alpha},$$

where $c_{ikm} = (\hat{\mathbf{C}}_i^{-1})_{km}$ for $k,m = 1,2$ $(i = 1,\ldots,g)$.

Krzanowski and Radley (1989) have described an alternative approach to that above for adjusting confidence region shapes in a practical situation. Their approach allows for sampling variability in the canonical variates through a nonparametric resampling approach with either the jackknife or bootstrap schemes. These two resampling schemes are considered in some detail in Chapter 10, where they are employed in the context of error-rate estimation. We now briefly describe the bootstrap scheme in its application to the present problem; the jackknife scheme is implemented in a similar manner.

The bootstrap resampling approach yields an ellipsoidal confidence region of the form

$$(\overline{\mathbf{v}}_i^* - \zeta_i)'\mathbf{S}_i^{*-1}(\overline{\mathbf{v}}_i^* - \zeta_i) \leq c_B, \tag{6.5.8}$$

where $\overline{\mathbf{v}}_i^*$ and $\mathbf{S}_i^*$ are the sample mean and covariance matrix of the bootstrap replications (say, K in number) of $\overline{\mathbf{v}}_i(\mathbf{T})$ for $i = 1,\ldots,g$. As proposed by Krzanowski and Radley (1989), a bootstrap replicate of $\overline{\mathbf{v}}_i(\mathbf{T})$ is obtained by sampling n_i feature vectors randomly with replacement from the training data from G_i and then performing a sample canonical variate analysis on them in conjunction with the original training data from the other $g-1$ groups. In order to allow for varying frames of reference between the replicates, each replicate configuration is rotated to position of best fit with respect to the original configuration by means of a Procrustes analysis. For a $100(1-\alpha)\%$ confidence region, a suggested value for the constant c_B in (6.5.8) is

$$c_B = 2(K-1)/K(K-2)F_{2,K-2;1-\alpha},$$

where $F_{2,K-2;1-\alpha}$ denotes the $(1-\alpha)$th quantile of the $F_{2,K-2}$ distribution. This choice of c_B corresponds to the use of the normal approximation to the distribution of the bootstrap replicates of $\overline{\mathbf{v}}_i(\mathbf{T})$.

To illustrate their resampling approach, Krzanowski and Radley (1990) applied it to the well-known set of *Iris* data as originally collected by E. Anderson (1935) and first analyzed by Fisher (1936). It consists of measurements of the length and width of both sepals and petals of 50 plants for each of three types of *Iris* species, *virginica*, *versicolor*, and *setosa*. Figure 6.1 shows the two-dimensional canonical variate diagram, in which points A, B, and C, representing the sample means of the *virginica, versicolor*, and *setosa* species, respectively, are surrounded by tolerance regions computed using the classical approach based on (6.5.7) and the bootstrap and jackknife resampling ap-

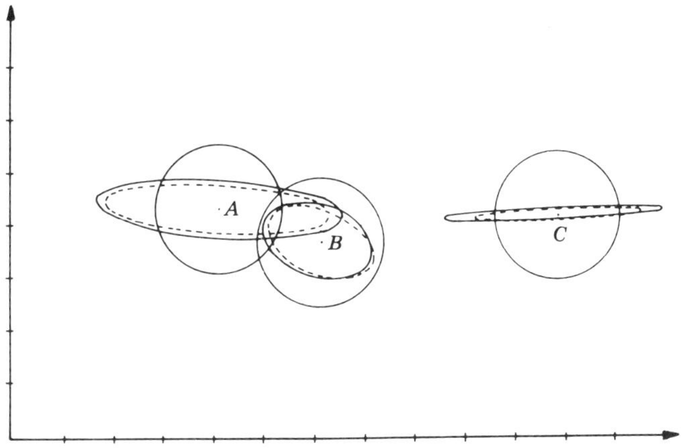

FIGURE 6.1. Tolerance regions for Fisher's *Iris* data. Circles represent the classical approach, solid-line ellipses represent the jackknife regions, and dash-line ellipses represent the bootstrap regions. From Krzanowski and Radley (1989), with permission from the Biometric Society.

proaches. The classical regions are circles, the jackknife regions are the solid-line ellipses, and the bootstrap regions are the dash-line ellipses.

It can be seen from Figure 6.1 that the bootstrap and jackknife regions are very similar for each group, but the former are consistently slightly smaller than the latter. For the *setosa* species, both the bootstrap and jackknife regions are very long and very thin, differing greatly from the classical circle. The nonparametric regions for the other extreme species *(virginica)* in the diagram are also both longer along the first axis than the second, whereas the corresponding regions for the middle species *(versicolor)* are most similar to the classical circle.

6.5.4 Alternative Formulation of Canonical Variates

As an alternative to the derivation of the canonical variates by direct computation of the canonical variates of $\mathbf{S}^{-1}\mathbf{B}$, we consider now a formulation that we will see in the next section is useful in the development of canonical variate analysis in the case of unequal group-covariance matrices.

An obvious way of carrying out the maximization of (6.5.2) subject to (6.5.1) is to first transform the $\mathbf{x}_j$ as

$$\mathbf{u}_j = \mathbf{\Lambda}_S^{-1/2}\mathbf{A}_S\mathbf{x}_j \qquad (j = 1,\ldots,n), \tag{6.5.9}$$

where

$$\mathbf{A}_S = (\mathbf{a}_{S,1},\ldots,\mathbf{a}_{S,p})',$$

and $\mathbf{a}_{S,1},\ldots,\mathbf{a}_{S,p}$ are the eigenvectors of $\mathbf{S}$ corresponding to the eigenvalues $\lambda_{S,1},\ldots,\lambda_{S,p}$ of $\mathbf{S}$ in order of decreasing size. Also,

$$\mathbf{\Lambda}_S = \mathrm{diag}(\lambda_{S,1},\ldots,\lambda_{S,p}).$$

Hence, the original feature data $\mathbf{x}_j$ are first rotated so that the axes coincide with the principal ones of $\mathbf{S}$. Then they are scaled so that the (bias-corrected) pooled within-group sample covariance matrix for the transformed data is the $p \times p$ identity matrix. The probem of maximizing (6.5.2) for the transformed data reduces to maximizing the transformed between-group matrix

$$(\mathbf{\Lambda}_S^{-1/2}\mathbf{A}_S)\mathbf{B}(\mathbf{\Lambda}_S^{-1/2}\mathbf{A}_S)',$$

which can be expressed as

$$\frac{1}{g-1}\sum_{i=1}^{g} n_i(\overline{\mathbf{u}}_i - \overline{\mathbf{u}})(\overline{\mathbf{u}}_i - \overline{\mathbf{u}})', \tag{6.5.10}$$

where

$$\overline{\mathbf{u}}_i = \sum_{j=1}^{n} z_{ij}\mathbf{u}_j/n_i \qquad (i = 1,\ldots,g),$$

and

$$\overline{\mathbf{u}} = \sum_{i=1}^{g} n_i\overline{\mathbf{u}}_i/n.$$

To this end, let $\mathbf{a}_{B,1},\ldots,\mathbf{a}_{B,d}$ denote the eigenvectors corresponding to the d nonzero eigenvalues of (6.5.10), in order of decreasing size. Then the d principal components of $\mathbf{u}_j$,

$$(\mathbf{a}_{B,1}'\mathbf{u}_j,\ldots,\mathbf{a}_{B,d}'\mathbf{u}_j)' \qquad (j = 1,\ldots,n),$$

give the scores $\mathbf{v}_j$ on the d-dimensional canonical variates of the n original feature observations $\mathbf{x}_j$ $(j = 1,\ldots,n)$; see, for example, Campbell and Atchley (1981). It follows that

$$\boldsymbol{\gamma}_k = \mathbf{A}_S'\mathbf{\Lambda}_S^{-1/2}\mathbf{a}_{B,k} \qquad (k = 1,\ldots,d).$$

Another way in which the canonical variates of the group means can be computed is by performing a principal-coordinate analysis (metric scaling) of the $g \times g$ dissimilarity matrix whose elements give the squared Mahalanobis distances

$$D_{ij}^2 = (\overline{\mathbf{x}}_i - \overline{\mathbf{x}}_j)'\mathbf{S}^{-1}(\overline{\mathbf{x}}_i - \overline{\mathbf{x}}_j)$$

between the group-sample means $(i,j = 1,\ldots,g)$. This follows from Gower (1966); see also Jolliffe (1986; Section 5.2).

6.5.5 Heteroscedastic Case

The generalization of canonical variate analysis to the case of unequal group-covariance matrices $\mathbf{\Sigma}_i$ is considered now. One approach is to ignore the differences between the $\mathbf{\Sigma}_i$ and compute the canonical variates as in the homoscedastic case. However, it can lead to unrealiable results if the $\mathbf{\Sigma}_i$ are sufficiently disparate. There appears to be few results in the literature on techniques that specifcally allow for the heteroscedasticity. One generalization of canonical variate analysis to the heteroscedastic case has been suggested by Campbell (1984b), who makes no assumptions about the similarities of the $\mathbf{\Sigma}_i$.

Recently, Krzanowski (1990) has considered generalizations of canonical variate analysis in the heteroscedastic case by working with the common principal-component (CPC) model to exploit any similarities in the $\mathbf{\Sigma}_i$. This model was introduced in Section 5.4 and the likelihood ratio test for its validity was presented earlier in this chapter. Under the common principal-component model, we have that

$$\mathbf{A}\mathbf{\Sigma}_i\mathbf{A}' = \mathbf{\Lambda}_i \qquad (i = 1,\ldots,g),$$

where $\mathbf{A}$ is an orthogonal $p \times p$ matrix, and the $\mathbf{\Lambda}_i$ are all diagonal matrices,

$$\mathbf{\Lambda}_i = \text{diag}(\lambda_{i1},\ldots,\lambda_{ig}) \qquad (i = 1,\ldots,g).$$

We let $\hat{\mathbf{A}}$ and $\hat{\mathbf{\Lambda}}$ be estimates of $\mathbf{A}$ and $\mathbf{\Lambda}_i$ computed as described in Section 5.4.4. It was suggested there that maximum likelihood or least squares can be used according as multivariate normality does or does not hold for the group-conditional distributions.

Krzanowzki (1990) proposed one generalization of canonical variates for the CPC model by basing it on the alternative formulation that obtains the solution through a principal-component analysis of the matrix of total sums of squares and products for the feature data after appropriate transformation. Proceeding analogously to this formulation, the original feature axes are first rotated so that they align with the common principal axes of the $\mathbf{\Sigma}_i$. The feature data so transformed are then scaled, where, given the heterogenetiy of the $\mathbf{\Sigma}_i$, the scaling is now specific to their group of origin. Accordingly,

$$\mathbf{u}_j = \sum_{i=1}^{g} z_{ij}\hat{\mathbf{\Lambda}}_i^{-1/2}\hat{\mathbf{A}}\mathbf{x}_j \qquad (j = 1,\ldots,n). \tag{6.5.11}$$

The generalized canonical variates for $\mathbf{x}_j$ are defined to be

$$(\mathbf{a}_{B,1}'\mathbf{u}_j,\ldots,\mathbf{a}_{B,d}'\mathbf{u}_j)' \qquad (j = 1,\ldots,n),$$

where $\mathbf{a}_{B,1},\ldots,\mathbf{a}_{B,d}$ are the eigenvectors corresponding to the ordered (nonzero) eigenvalues of the between-group matrix for the transformed feature data, given by (6.5.10), but where now $\mathbf{u}_j$ is defined by (6.5.11).

The other generalization of canonical variates proposed by Krzanowski (1990) for the CPC model uses the principal-coordinate formulation. As noted in the previous section, with this approach in the homoscedastic case, the canonical variates for the group-sample means can be obtained through a principal-coordinate analysis of the dissimilarity matrix whose elements contain the squared Mahalanobis distances between the group means. As discussed in Section 1.12, the Mahalanobis distance is an appropriate measure between not only multivariate normal distributions with a common covariance matrix, but also between distributions belonging to the same family of elliptically symmetric distributions with common shape. However, in the present situation, the group-conditional distributions have unequal covariance matrices.

For the CPC model under the additional assumption of multivariate normality for the group-conditional distributions, Krzanowski (1990) suggested that the intergroup distances be given by

$$\{2(1-\hat{\rho}_{ij})\}^{1/2}. \tag{6.5.12}$$

In (6.5.12), $\hat{\rho}_{ij}$ denotes the plug-in estimate of ρ_{ij}, which is taken to be Matusita's (1956) measure of affinity between the $N(\boldsymbol{\mu}_i, \boldsymbol{\Sigma}_i)$ and $N(\boldsymbol{\mu}_i, \boldsymbol{\Sigma}_j)$ distributions under the CPC model. Krzanowski (1990) showed that ρ_{ij}, which is defined in general by (1.12.1), reduces in this particular case to

$$\rho_{ij} = c_{ij}\exp\{-\tfrac{1}{4}(\boldsymbol{\mu}_i - \boldsymbol{\mu}_j)'\mathbf{A}'(\boldsymbol{\Lambda}_i + \boldsymbol{\Lambda}_j)^{-1}\mathbf{A}(\boldsymbol{\mu}_i - \boldsymbol{\mu}_j)\},$$

where

$$c_{ij} = 2^{p/2}\{|\boldsymbol{\Lambda}_i\boldsymbol{\Lambda}_j|^{1/4}/|\boldsymbol{\Lambda}_i + \boldsymbol{\Lambda}_j|^{1/2}\}.$$

Note that another way to proceed is to use the sample analogue of the feature-reduction method of Young et al. (1987), as described in Section 3.10.2.

6.6 SOME OTHER METHODS OF DIMENSION REDUCTION TO REVEAL GROUP STRUCTURE

6.6.1 Mean-Variance Plot

For $g = 2$ groups, W. C. Chang (1987) has proposed a two-dimensional display of the training data, which is known as a mean-variance plot or graph. This is because the first coordinate reflects mainly differences between the group-sample means, and the second coordinate reflects the separation between the groups solely due to their covariance matrices. The mean-variance plot is intended as an improvement of Sammon's graph. Sammon (1970) and Foley and Sammon (1975) considered a plot in which the first coordinate is used to display the separation of the two groups based on the sample NLDF, or, alternatively, the best linear discriminant rule in the case of unequal covariance matrices, as defined in Section 3.10.1. The second coordinate is chosen by finding a vector orthogonal to the first one that maximizes the difference between the group means. However, as pointed out by W. C. Chang (1987), with this method of defining the second coordinate, any differences between the group-covariance matrices are not captured. He therefore proposed that the second coordinate be chosen for the latter purpose; see also Fukunaga and Ando (1977). W. C. Chang (1987) proposed that the first coordinate v_1 be defined as

$$\begin{aligned} v_1 = & \{\mathbf{a}_1'(\mathbf{x} - \overline{\mathbf{x}}_2) - \mathbf{a}_1'(\overline{\mathbf{x}}_1 - \overline{\mathbf{x}}_2)\}^2/\mathbf{a}_1'\mathbf{S}_1\mathbf{a}_1 \\ & - \{\mathbf{a}_1'(\mathbf{x} - \overline{\mathbf{x}}_2)\}^2/\mathbf{a}_1'\mathbf{S}_2\mathbf{a}, \end{aligned}$$

where

$$\mathbf{a}_1 = \mathbf{S}_\omega^{-1}(\overline{\mathbf{x}}_1 - \overline{\mathbf{x}}_2),$$

and

$$\mathbf{S}_\omega = \omega\mathbf{S}_1 + (1-\omega)S_2$$

for $0 < \omega < 1$. For $\omega = (n_1 - 1)/(n-2)$, $\mathbf{S}_\omega = \mathbf{S}$, and $\mathbf{a}_1$ corresponds to Fisher's (1936) choice of a linear discriminant function. Another value of ω that might be considered is one that defines the best linear rule, as considered in Section 3.10.1.

Concerning the second coordinate, W. C. Chang (1987) defined it as

$$v_2 = c_1 - c_2,$$

where

$$c_i = \{\mathbf{H}(\mathbf{x} - \overline{\mathbf{x}}_2)\}'(\mathbf{H}\mathbf{S}_i\mathbf{H}')^{-1}\{\mathbf{H}(\mathbf{x} - \overline{\mathbf{x}}_2)\} \qquad (i = 1, 2),$$

and where

$$\mathbf{H} = (\mathbf{a}_2, \ldots, \mathbf{a}_p)'.$$

Here $\mathbf{a}_2, \ldots, \mathbf{a}_p$ are orthogonal to $\overline{\mathbf{x}}_1 - \overline{\mathbf{x}}_2$, so that

$$\mathbf{a}_i'(\overline{\mathbf{x}}_1 - \overline{\mathbf{x}}_2) = 0 \qquad (i = 2, \ldots, p).$$

Thus, the sample mean of v_2 is zero both in G_1 and G_2. This coordinate therefore provides information on the covariance structure of the feature data in each group.

A simple way to compute $\mathbf{H}$ is to take $\mathbf{a}_2, \ldots, \mathbf{a}_p$ to be the eigenvectors of $(\overline{\mathbf{x}}_1 - \overline{\mathbf{x}}_2)(\overline{\mathbf{x}}_1 - \overline{\mathbf{x}}_2)'$ corresponding to its $p-1$ zero eigenvalues. This matrix has only one nonzero eigenvalue, being of rank one.

6.6.2 Principal Components: Classified Feature Data

As shown in Section 5.4.5, principal components play a fundamental role in the SIMCA method of discriminant analysis, where essentially a separate principal-component analysis is undertaken for each group. Also, we saw in Section 5.4 how a hierarchy of principal-component models can be formulated for incorporating similarities in the group-covariance matrices. However, a single principal-component analysis of the pooled within-group sample covariance matrix of the training feature data $\mathbf{x}_j$ is generally of limited value in discriminant analysis. Indeed, the use of a principal-component analysis as a procedure for dimensionality reduction is not based on any discrimination considerations. To examine this in more depth, we consider the partition

$$(n-1)\mathbf{V} = \tilde{\mathbf{B}} + \mathbf{W}$$

of the matrix $(n-1)\mathbf{V}$ of the sums of squares and products for all the training feature data $\mathbf{x}_j$ $(j = 1,\ldots,n)$ into the between-group matrix $\tilde{\mathbf{B}}$ and the within-group matrix $\mathbf{W}$, as defined by (6.5.4) and (6.5.5), respectively.

A principal-component analysis of $\mathbf{W}$, or, equivalently, $\mathbf{S} = \mathbf{W}/(n-g)$, is not going to assist in finding those directions in which the differences between the group-sample means are greatest, unless the first few principal axes of the analysis based on $\mathbf{W}$ fortuitously coincide with the dominant axes of $\tilde{\mathbf{B}}$. This can be seen from the principal-component formulation of a canonical variate analysis, as described in Section 6.5.4; see also Jolliffe (1986, Section 9.1).

A principal-component analysis performed on $\mathbf{V}$ can actually be more useful. For if there are only a few groups, and they are well-separated, and the between-group variation dominates the within-group variation, then projections of the training feature data $\mathbf{x}_j$ onto the first few principal axes should portray the group structure. However, a principal-component analysis of $\mathbf{V}$ may not always be useful. To illustrate this point, W. C. Chang (1983) calculated the squared Mahalanobis distance between $g = 2$ groups G_1 and G_2 after the feature data have been transformed according to $\mathbf{A}_V\mathbf{x}_j$ $(j = 1,\ldots,n)$, where

$$\mathbf{A}_V = (\mathbf{a}_{V,1},\ldots,\mathbf{a}_{V,p})' \tag{6.6.1}$$

and $\mathbf{a}_{V,1},\ldots,\mathbf{a}_{V,p}$ are the eigenvectors of unit length corresponding to the eigenvalues $\lambda_1,\ldots,\lambda_p$ of $\mathbf{V}$, in order of decreasing size. W. C. Chang (1983) effectively considered the case of an infinitely large training sample of size n from a mixture in proportions π_1 and π_2 of two distributions with a common covariance matrix $\boldsymbol{\Sigma}$. As $n \to \infty$, $\mathbf{V}$ converges in probability to the covariance matrix of the mixture distribution,

$$\boldsymbol{\Sigma} + \pi_1\pi_2(\boldsymbol{\mu}_1 - \boldsymbol{\mu}_2)(\boldsymbol{\mu}_1 - \boldsymbol{\mu}_2)'.$$

Let Δ_k^2 denote the squared Mahalanobis distance between G_1 and G_2 using the kth principal component of $\mathbf{x}, \mathbf{a}_{V,k}'\mathbf{x}$, that is,

$$\Delta_k^2 = \{\mathbf{a}_{V,k}'(\boldsymbol{\mu}_1 - \boldsymbol{\mu}_2)\}^2/\mathbf{a}_{V,k}'\boldsymbol{\Sigma}\mathbf{a}_{V,k}.$$

W. C. Chang (1983) showed that

$$\Delta_k^2 = \omega_k^2/(1 - \pi_1\pi_2\omega_k^2), \tag{6.6.2}$$

where

$$\omega_k^2 = \{\mathbf{a}_{V,k}'(\boldsymbol{\mu}_1 - \boldsymbol{\mu}_2)\}^2/\lambda_k.$$

It can be seen from (6.6.2) that the principal component of the feature vector that provides the best separation between the two groups in terms of Mahalanobis distance is not necessarily the first component $\mathbf{a}_{V,1}'\mathbf{x}$, corresponding to the largest eigenvalue λ_1 of $\mathbf{V}$. Rather it is the component that maximizes ω_k^2 over $k = 1,\ldots,p$.

W. C. Chang (1983) also derived an expression for $\Delta_{i_1,\ldots,i_q}^2$, the squared Mahalanobis distance between G_1 and G_2, using q (not necessarily the first) principal components of $\mathbf{x}$. Here the subscripts $i_1,\ldots,i_q$ are a subset of $\{1,2,\ldots,p\}$.

He showed that

$$\Delta^2_{i_1,\ldots,i_q} = \left(\sum_{k=1}^{q} \omega^2_{i_k}\right) \Big/ \left(1 - \pi_1\pi_2 \sum_{k=1}^{q} \omega^2_{i_k}\right).$$

The last expression implies that, in terms of the Mahalanobis distance, the best subset of size q of the principal components is obtained by choosing the principal components with the q largest values of ω_k^2.

Some recent results on the case of principal components in the presence of group structure can be found in Krzanowski (1987b), Malina (1987), Duchene and Leclercq (1988), and Biscay, Valdes, and Pascual (1990). In the role of reducing the dimension of the feature vector from p to $q(q < p)$, the modified selection method of Krzanowski (1987b) has the added advantage in that only q of the original p feature variables are needed to define the new q feature variables in the subsequent analysis. Note that in the pattern-recognition literature, dimension reduction by the use of principal components tends to be referred to as linear feature selection via the discrete Karhunen–Loève expansion; see Fukunaga (1972, Section 8.2).

The work of W. C. Chang (1983) underscores the ineffectiveness of a principal-component analysis in a typical discriminant analysis situation. However, it does suggest how it can be useful in an exploratory data analysis in situations where there are no classified training data available. This is pursued further in the next section.

6.6.3 Principal Components: Unclassified Feature Data

In this section, we briefly outline how a principal-component analysis using the sample covariance matrix of the unclassified feature data can be employed to elicit information on the group structure in the absence of classified data. In exploring high-dimensional data sets for group structure, it is typical to rely on "second-order" multivariate techniques, in particular, principal component analysis; see Huber (1985) and Jones and Sibson (1987) and the subsequent discussions for an excellent account of available exploratory multivariate techniques. It is common in practice to rely on two-dimensional plots to exhibit any group structure. No doubt three-dimensional plots can do the task more effectively, as demonstrated by the now well-known example in Reaven and Miller (1979) considered in Section 6.8. However, the technology for three-dimensional plots is not as yet widely available, although this situation is changing rapidly with the development of dynamic graphics programs such as MacSpin (Donoho, Donoho, and Gasko, 1985); see Becker, Cleveland and Wilks (1987), Friedman (1987), and Huber (1987).

We proceed here on the basis that any obvious nonnormal structure in the feature data such as substantial skewness in the marginal distributions as evidenced from preliminary scatter plots has been removed by appropriate transformations on the highly structured variates. Such transformations were discussed in Section 6.3. After suitable transformation of the data, clusters are often essentially elliptical, and so then they can be characterized by the location of their centers and the scatter matrix of the points lying within each. To this end, let $\mathbf{v}_j$ denote the principal components of the original feature vector $\mathbf{x}_j$, where

$$\mathbf{v}_j = \mathbf{A}_V \mathbf{x}_j \qquad (j = 1, \ldots, n),$$

and $\mathbf{A}_V$ is the matrix (6.6.1) containing the eigenvectors of the (bias-corrected) sample covariance matrix $\mathbf{V}$ of the $\mathbf{x}_j$, corresponding to the eigenvalues of $\mathbf{V}$ in order of decreasing size. Given the sensitivity of $\mathbf{V}$ to outliers, a robust version might be used, as considered in Section 5.7. Also, we may wish to perform a principal-component analysis on the correlation matrix, particularly in the case in which the variates are measured on disparate scales.

Suppose for the moment that there are $g = 2$ underlying groups in each of which the feature vector $\mathbf{X}$ has the same covariance matrix. Let $v_{(1)}, \ldots, v_{(p)}$ denote the principal components in $\mathbf{v}$ ranked in order of decreasing size of the Mahalanobis distance for each of these components. Then it follows from the results of W. C. Chang (1983) given in the last section that the most informative q-dimensional ($q < p$) scatter plot of the principal components for revealing the two-group structure is provided by the plot of the scores for $v_{(1)}, \ldots, v_{(q)}$. Of course, in practice, the Mahalanobis distance for each principal component is not known to effect the ranking $v_{(1)}, \ldots, v_{(p)}$. However, this ranking can be based on the estimated Mahalanobis distances. Equivalently, it can be taken to correspond to the ranking of the values $-2\log\lambda_k$ $(k = 1, \ldots, p)$ in order of decreasing size. Here λ_k is the likelihood ratio statistic for the test of a single normal distribution versus a mixture of two normals with a common variance, performed using the data on just the kth principal component $(k = 1, \ldots, p)$. Rather than actually fitting a univariate normal mixture to the data on each principal component in turn, a Q-Q plot, or a $\Phi - P$ versus Q plot as proposed by Fowlkes (1979), can often indicate which components show marked departure from a single normal distribution, and hence which are most useful in revealing a mixture of groups; see also McLachlan and Jones (1988).

In general situations where the presence of at most two groups and also homoscedasticity cannot be assumed as above, two- and three-dimensional plots involving all the principal components need to be considered. At least all $\binom{p}{2}$ scatter plots of the principal components should be inspected provided p is not too large. A useful graphical tool would be the scatter plot matrix; see Becker et al. (1987). Commonly in practice, p is between 4 and 10, so that the examination of all two-dimensional scatter plots is then a feasible exercise. Also, with most statistical packages, it is a simple matter to request all such plots.

6.6.4 Projection Pursuit: Unclassified Feature Data

We saw in the previous section how the principal axes are not necessarily the optimal directions in which to view the data. This has led to projection pursuit as considered by Wright and Switzer (1971), Friedman and Tukey (1974) and, more recently, by Huber (1985), Friedman (1987), and Jones and Sibson (1987), among others. In carrying out projection pursuit as recommended in the latter references, the data are first sphered to give

$$\mathbf{v}_j = \boldsymbol{\Lambda}_V^{-1/2}\mathbf{A}_V(\mathbf{x}_j - \bar{\mathbf{x}}) \qquad (j = 1,\ldots,n), \tag{6.6.3}$$

where $\mathbf{A}_V$ is the matrix (6.6.1), containing the eigenvectors of $\mathbf{V}$, and $\boldsymbol{\Lambda}_V$ is a diagonal matrix containing its eigenvalues.

One- and two-dimensional projections of the $\mathbf{v}_j$ are then considered that maximize some index that reflects interesting departures from multivariate normality; see the aforementioned references for details.

On contrasting (6.6.3) with the transformation (6.5.9) in the alternative formulation of a canonical variate analysis, it can be seen why sphering of the data is not really designed for the detection of groups. This is because the (bias-corrected) sample covariance matrix $\mathbf{V}$ of all the feature data is used instead of the pooled within-group estimate $\mathbf{S}$. The latter is, of course, not available in the present situation of only unclassified feature data.

6.7 EXAMPLE: DETECTION OF HEMOPHILIA A CARRIERS

6.7.1 Preliminary Analysis

In this example, we illustrate the updating of a sample discriminant rule using unclassified feature data in conjunction with the classified training observations. The data set under consideration is that analyzed originally by Hermans and Habbema (1975) in the context of genetic counseling. The problem concerns discriminating between normal women and hemophilia A carriers on the basis of two feature variables, $x_1 = \log_{10}$ (AHF activity) and $x_2 = \log_{10}$ (AHF-like antigen). We let G_1 be the group corresponding to the noncarriers, or normals, and G_2 the group corresponding to the carriers.

The classified training data consist of $n_1 = 30$ observations $\mathbf{x}_j$ $(j = 1,\ldots,n_1)$ on known noncarriers and $n_2 = 22$ observations $\mathbf{x}_j$ $(j = n_1 + 1,\ldots,n = n_1 + n_2)$ on known obligatory carriers. They are plotted in Figure 6.2, and the group-sample means, variances, and correlations are displayed in Table 6.1. There are also $m = 19$ unclassified women with feature vector $\mathbf{x}_j$ and with specified prior probability π_{1j} of noncarriership $(j = n + 1,\ldots,n + m)$.

As concluded by Hermans and Habbema (1975) from the scatter plot of the classified training data, it is reasonable to assume that the group-conditional distributions are bivariate normal. To confirm this, we applied Hawkins' (1981) test for normality and homoscedasticity, whereby for each classified feature vector $\mathbf{x}_j$ from G_i, the tail area a_{ij} to the right of the statistic corresponding to (6.4.3) was computed. The results are summarized in Table 6.2. As explained in the presentation of Hawkins' (1981) test in Section 6.4, the Anderson–Darling statistic and its first two asymptotic $N(0,1)$ components for the a_{ij} from G_i

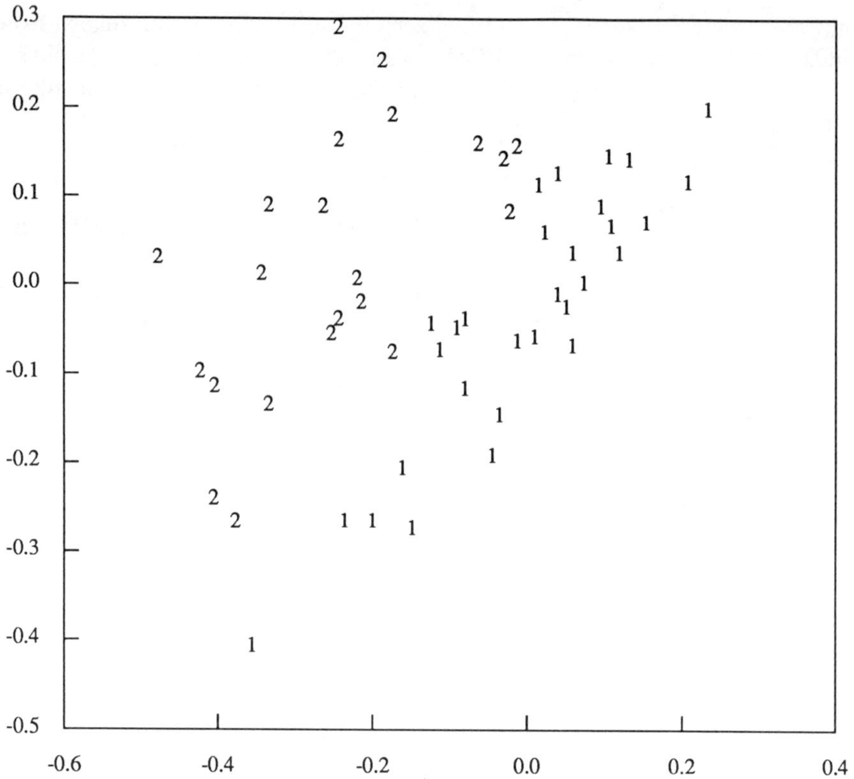

FIGURE 6.2. Hemophilia data with 1 and 2 representing noncarriers and carriers, respectively. (From Hermans and Habbema, 1975.)

TABLE 6.1 Sample Means, Standard Deviations, and Correlations for the Classified Feature Data by Groups with Updated Estimates in Parentheses

	Sample estimates				
Source	Means		Standard Deviations		Correlation
G_1	−0.006 (−0.016)	−0.039 (−0.048)	0.134 (0.121)	0.147 (0.136)	0.894 (0.878)
G_2	−0.248 (−0.226)	0.026 (0.040)	0.134 (0.140)	0.149 (0.141)	0.576 (0.607)

and for the totality of the a_{ij} are useful in interpreting qualitatively departures from the null hypothesis. The nonsignificance of the Anderson–Darling statistic and its first two components for the totality of the a_{ij} here gives a fair indication that bivariate normality is a tenable assumption for the group-conditional distributions of **X**. The difference in sign of the first components

TABLE 6.2 Results of Hawkins' (1981) Test for Normality and Homoscedasticity Applied to Classified Data ($n_1 = 30$ from G_1 and $n_2 = 22$ from G_2)

Source	Anderson–Darling Statistic	Components of Anderson–Darling Statistic	
		First	Second
G_1	1.18	−1.23	1.24
G_2	2.00	1.89	0.12
Totality	0.42	0.30	1.02

of the Anderson–Darling statistics for the individual groups and the significance of one of these components suggest the presence of some heteroscedasticity. It can be seen from Table 6.1 that any heteroscedasticity present must be attributable to the difference between the group-sample correlations, as the sample standard deviations are practically the same within each group.

6.7.2 Fit Based on Classified Feature Data

Consistent with our previous notation, we let $\boldsymbol{\theta}_U$ contain the elements of $\boldsymbol{\mu}_1$ and $\boldsymbol{\mu}_2$ and the distinct elements of Σ_1 and Σ_2 in the two bivariate normal group-conditional distributions to be fitted, and $\boldsymbol{\theta}_E$ denotes $\boldsymbol{\theta}_U$ under the constraint of equal group-covariance matrices $\Sigma_1 = \Sigma_2$. We let $\hat{\boldsymbol{\theta}}_E$ and $\hat{\boldsymbol{\theta}}_U$ be the fits obtained in fitting a bivariate model under homoscedasticity and heteroscedasticity, respectively, to the $n = 52$ classified feature observations. The estimated group means common to both fits and the estimated standard deviations and correlations for the heteroscedastic fit are given by the sample quantities in Table 6.1. In forming $\hat{\boldsymbol{\Psi}}_U = (\hat{\boldsymbol{\pi}}', \hat{\boldsymbol{\theta}}_U')'$ and $\hat{\boldsymbol{\Psi}}_E = (\hat{\boldsymbol{\pi}}', \hat{\boldsymbol{\theta}}_E')'$ for assessing the fit of the model to the classified feature data, the estimate of the vector $\boldsymbol{\pi} = (\pi_1, \pi_2)'$ containing the group-prior probabilities was taken to be $\hat{\boldsymbol{\pi}} = (30/52, 22/52)'$ in each instance, in the absence of any information on the prior probabilities for the classified women.

For a classified feature vector $\mathbf{x}_j$ from group G_i, we computed its measure a_{ij} of typicality of G_i, along with its estimated posterior probability of membership of G_i for both fits $\hat{\boldsymbol{\Psi}}_E$ and $\hat{\boldsymbol{\Psi}}_U$. This information is reported in Table 6.3 for those classified feature vectors $\mathbf{x}_j$ from G_i with either a measure of typicality of G_i less than 0.05 or an estimated posterior probability of membership of G_i less than 0.85 under either the homoscedastic or heteroscedastic versions of the normal model. They represent the cases of most interest in an examination of the fit of the postulated model for the group-conditional distributions. In Table 6.3, we have written a_{ij} as either $a_{ij,E}$ or $a_{ij,U}$ to explicitly denote whether it was computed from (6.4.1), as appropriate under the homoscedastic normal model, or from (6.4.3), as under the heteroscedastic version.

TABLE 6.3 Fitted Posterior Probabilities of Group Membership and Typicality Values for Some of the Classified Feature Data

j	$a_{1j}(\hat{\Psi}_E)$	$a_{1j}(\hat{\Psi}_U)$	$\tau_1(\mathbf{x}_j;\hat{\Psi}_E)$	$\tau_1(\mathbf{x}_j;\hat{\Psi}_U)$
4	0.63	0.43	0.94	0.86
13	0.42	0.16	0.82	0.59
15	0.68	0.46	0.94	0.87
18	0.70	0.48	0.95	0.88
23	0.02	0.02	0.96	0.91
29	0.40	0.17	0.93	0.85
30	0.37	0.13	0.89	0.74
	$a_{2j}(\hat{\Psi}_E)$	$a_{2j}(\hat{\Psi}_U)$	$\tau_2(\mathbf{x}_j;\hat{\Psi}_E)$	$\tau_2(\mathbf{x}_j;\hat{\Psi}_U)$
33	0.02	0.09	1.00	1.00
44	0.18	0.40	0.35	0.68
46	0.08	0.17	0.16	0.39
52	0.02	0.09	1.00	1.00

It can be seen from Table 6.3 that under the heteroscedastic normal model, there is only one feature vector ($\mathbf{x}_{23}$ from G_1) with a typicality value less than 0.05. Under the homoscedastic version of the model, there are also two other feature vectors with the same typicality value of 0.02 as $\mathbf{x}_{23}$, namely, $\mathbf{x}_{33}$ and $\mathbf{x}_{52}$ from G_2. In Figure 6.2, $\mathbf{x}_{23}$ is the observation from G_1, with the smallest values of both feature variables x_1 and x_2, and $\mathbf{x}_{52}$ has the smallest value of x_1 and $\mathbf{x}_{33}$ the largest value of $\mathbf{x}_2$, among the observations from G_2.

It can be seen from Figure 6.2 that there is little overlap between the classified feature observations from each of the two groups. Hence, the sample NQDR $r_o(\mathbf{x};\hat{\Psi}_U)$ and sample NLDR $r_o(\mathbf{x};\hat{\Psi}_E)$ are able to allocate these classified data to their groups of origin with at most two misallocations. The sample NQDR misallocates only one observation, $\mathbf{x}_{46}$, from G_2, and the sample NLDR also misallocates another observation, $\mathbf{x}_{44}$, from G_2.

Although the fitted estimates $\tau_i(\mathbf{x}_j;\hat{\Psi}_U)$ and $\tau_i(\mathbf{x}_j;\hat{\Psi}_E)$ imply the same outright allocation of the feature vectors $\mathbf{x}_j$ except for $j = 44$, it can be seen from Table 6.3 that there are practical differences between these estimates for some of the other feature vectors. For example, although there is little doubt over the group of origin of observations $\mathbf{x}_{13}$, $\mathbf{x}_{30}$, and $\mathbf{x}_{46}$ according to their fitted posterior probabilities of group membership under homoscedasticity, there is quite some doubt under their heteroscedastic estimates.

6.7.3 Updating Assessed Posterior Probabilities of Noncarriership

We proceed now to the assessment of the posterior probabilities of group membership for the $m = 19$ unclassified feature observations $\mathbf{x}_j$ with specified vector $\boldsymbol{\pi}_j = (\pi_{1j}, \pi_{2j})'$ of prior probabilities of group membership ($j = 53,\ldots,71$). These $\mathbf{x}_j$ and π_{1j} are listed in Table 6.4, along with their assessed typicality $a(\mathbf{x}_j;\hat{\boldsymbol{\theta}}_U)$ and their assessed posterior probability $\tau_1(\mathbf{x}_j;\boldsymbol{\pi}_j,\hat{\boldsymbol{\theta}}_U)$

TABLE 6.4 Assessed Posterior Probability of Noncarriership for an Unclassified Woman with Feature Vector $\mathbf{x}_j$ and Prior Probability π_{1j} of Noncarriership

j	x_{1j}	x_{2j}	π_{1j}	$a(\mathbf{x}_j;\hat{\boldsymbol{\theta}}_U)$	$\tau_1(\mathbf{x}_j;\boldsymbol{\pi}_j,\hat{\boldsymbol{\theta}}_U)$	$\tau_1(\mathbf{x}_j;\boldsymbol{\pi}_j,\hat{\hat{\boldsymbol{\theta}}}_U)$
53	−0.210	−0.044	0.75	0.72	0.03	0.02
54	−0.250	0.095	0.75	0.85	0.00	0.00
55	−0.043	−0.052	.67	0.91	0.98	0.98
56	−0.210	−0.090	0.67	0.50	0.16	0.14
57	0.064	0.012	0.67	0.84	0.99	0.99
58	−0.059	−0.068	0.50	0.88	0.95	0.95
59	−0.050	−0.098	0.50	0.92	0.98	0.98
60	−0.094	−0.113	0.50	0.80	0.95	0.96
61	−0.112	−0.279	0.50	0.08	0.98	0.99
62	−0.287	0.137	0.50	0.52	0.00	0.00
63	−0.002	0.206	0.50	0.18	0.01	0.01
64	−0.371	−0.133	0.50	0.53	0.00	0.00
65	−0.009	0.037	0.50	0.51	0.90	0.87
66	0.030	0.224	0.50	0.11	0.04	0.02
67	−0.357	0.031	0.33	0.60	0.00	0.00
68	−0.123	−0.143	0.33	0.68	0.89	0.91
69	−0.162	0.162	0.33	0.66	0.00	0.00
70	0.069	0.192	0.33	0.06	0.44	0.30
71	0.002	−0.075	0.33	0.80	0.98	0.98

of membership of G_1 (noncarriership) for $j = 53,\ldots,71$. The typicality $a(\mathbf{x}_j;\hat{\boldsymbol{\theta}}_U)$ of each unclassified $\mathbf{x}_j$ of the mixture of G_1 and G_2 was assessed from (6.4.8).

Also listed in Table 6.4 is the updated assessment of noncarriership $\tau_1(\mathbf{x}_j;\boldsymbol{\pi}_j,\hat{\hat{\boldsymbol{\theta}}}_U)$, where $\hat{\hat{\boldsymbol{\theta}}}_U$ denotes the estimate of $\boldsymbol{\theta}_U$ calculated on the basis of the classified data in conjunction with the $m = 19$ unclassified feature observations. The updated estimates of the group means, standard deviations, and correlations are listed in parentheses in Table 6.1. The updated estimate $\hat{\hat{\boldsymbol{\theta}}}_U$ is the maximum likelihood estimate of $\boldsymbol{\theta}_U$ obtained by fitting a mixture of two bivariate normal heteroscedastic components to the combined data set of $n + m = 71$ feature vectors. This can be carried out, using the EM algorithm as described in Section 3.8, for a partially unclassified sample. The only modification needed here is to allow for the fact that the group-prior probabilities are specified for each of the unclassified feature observations.

By comparing the estimated posterior probability of noncarriership with the updated assessment for each of the unclassified observations in Table 6.4, it can be seen that where there is a difference, the effect of the updating is to give a more extreme estimate, that is, a less doubtful assessment. This difference is greatest for observation $\mathbf{x}_{70}$, for which the updated assessment of the posterior probability of noncarriership is 0.30, which, although still in the doubtful category, is not a borderline case as implied by the assessment of 0.44 based on just the classified feature data. It can be seen from Table 6.4 that $\mathbf{x}_{70}$ has the lowest typicality value of the 19 unclassified feature vectors, with $a(\mathbf{x}_{70};\hat{\theta}_U) = 0.06$.

6.8 EXAMPLE: STATISTICAL DIAGNOSIS OF DIABETES

6.8.1 Graphical Representation of Data

In Section 2.10, the situation was discussed in which an unequivocal classification of the training feature data may not be available. An example cited there concerned medical applications in which initially only a provisional grouping is specified, based on a clinical diagnosis using perhaps just one of the feature variables measured on the patients. We now give such an example, using the data set analyzed by Reaven and Miller (1979) in their investigation of the relationship between chemical diabetes and overt diabetes in 145 nonobese adult subjects. Attention in their study was focused on three variables: (1) the area under the plasma glucose curve for the three-hour oral glucose-tolerance test (OGTT), (2) the area under the plasma insulin curve for the OGTT, and

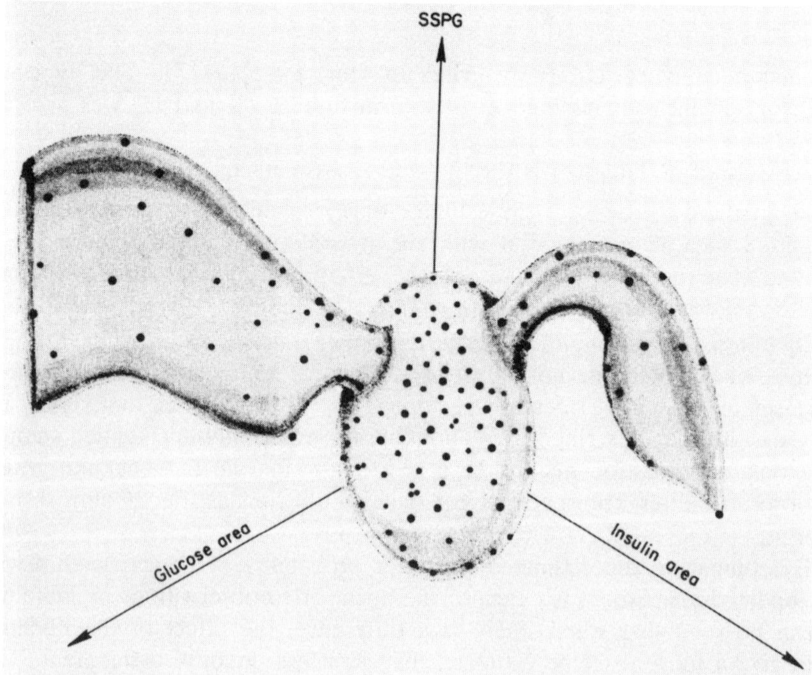

FIGURE 6.3. Artist's rendition of data as seen in three dimensions. View is approximately along 45° line as seen through the PRIM-9 computer program. (From Reaven and Miller, 1979.)

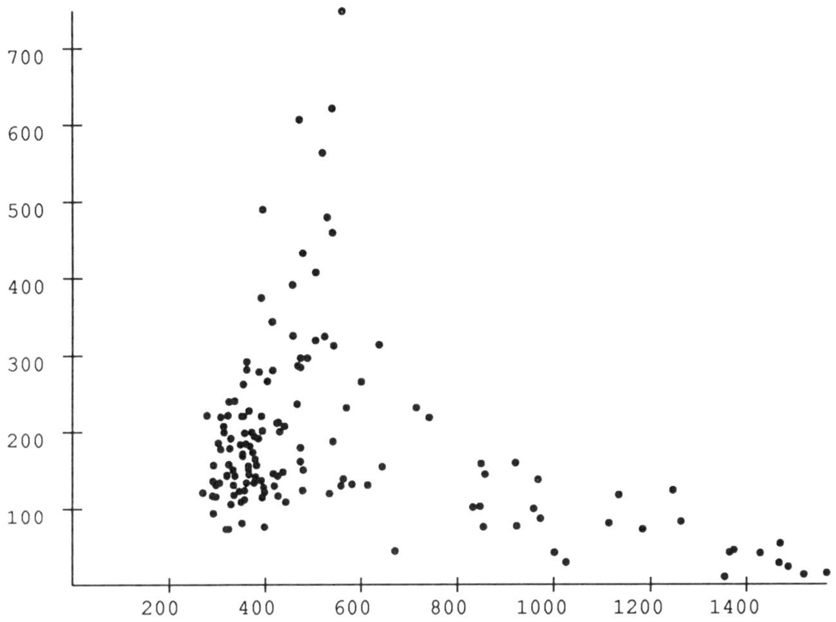

FIGURE 6.4. Scatter plot of the first two principal components (second versus first) of metabolic data.

(3) the steady-state plasma glucose response (SSPG). These data were examined with the now-defunct PRIM-9 program at the Stanford Linear Accelerator Computation Center. This program permits any three variables to be selected and then displayed as a two-dimensional image of the projection of the points along any direction. By continuously moving the direction, the three-dimensional configuration of the points is revealed. From a study of the various combinations of three variables, the configuration displayed in Figure 6.3 (an artist's rendition of the data cloud) emerged. This figure was considered to be very interesting to medical investigators. The plump middle of the points roughly corresponds to normal patients, the right arm to chemical diabetes, and the left arm to overt diabetes. It was noted by Reaven and Miller (1979) that this apparent separation between subjects with chemical and overt diabetes may explain why patients with chemical diabetes rarely develop overt diabetes.

As commented by Huber (1985), this work of Reaven and Miller (1979) represents one of the first successful applications of three-dimensional displays to statistical data. The "rabbit head" structure of the diabetes data is strikingly visible in three dimensions and it then assists with the understanding of two-dimensional projections. To illustrate this point, we have plotted in Figure 6.4 the two-dimensional projection of the data as given by the first two principal components, computed using the sample covariance matrix of the data. This

TABLE 6.5 Clinical Classification of 145 Subjects into Three Groups on the Basis of the Oral Glucose Tolerance Test with Results from Fitting a Normal Mixture Model Given in Parentheses

		Metabolic Characteristics (mean ± SD)		
Group	No.	Glucose Area (mg/100 mL-hr)	Insulin Area (μU/mL-hr)	SSPG (mg/100 mL)
G_1(normal)	76 (76)	350 ± 37 (356 ± 44)	173 ± 69 (165 ± 52)	114 ± 58 (105 ± 43)
G_2(chemical diabetes)	36 (32)	494 ± 55 (477 ± 73)	288 ± 158 (344 ± 151)	209 ± 60 (244 ± 37)
G_3(overt diabetes)	33 (37)	1044 ± 309 (939 ± 358)	106 ± 93 (104 ± 60)	319 ± 88 (285 ± 106)

two-dimensional plot appears to represent two wings emanating from a central core, and the resemblance to the artist's three-dimensional depiction of the data set is apparent.

Using conventional clinical criteria, Reaven and Miller (1979) initially classified the 145 subjects on the basis of their plasma glucose levels into three groups, G_1 (normal subjects), G_2 (chemical diabetes), and G_3 (overt diabetes); see Table 6.5. They subsequently clustered the 145 subjects into three clusters corresponding to these three groups in order to develop a computer classification that took into account all three metabolic variables and was independent of *a priori* clinical judgments. Their clustering criterion utilized the trace of the pooled within-group sample covariance matrix, supplemented by typical central values for groups from an earlier set of 125 patients that were similar to the groups represented in Figure 6.3. Additional clustering criteria have since been applied to this data set by Symons (1981).

6.8.2 Fitting of Normal Mixture Model

To provide another objective division of these 145 subjects into three clusters, we fitted a mixture of three trivariate normal components. Their covariance matrices were allowed to be different in view of the heteroscedasticity present in the clusters in Figure 6.3. This model was fitted iteratively using the EM algorithm as described in Section 3.8, with the initial estimates of the group parameters based on their clinical classification. We let $\hat{\Psi}_U$ now denote the estimate of the vector of unknown parameters obtained by fitting a mixture model with heteroscedastic normal components. The estimates of the group means and standard deviations obtained are listed in parentheses in Table 6.5.

6.8.3 Statistical Diagnosis

Let CC denote the clinical classification in Reaven and Miller (1979) of the jth subject with feature vector $\mathbf{x}_j$ ($j = 1,\ldots,145$). The corresponding diagnosis here is given by the value of the sample NQDR $r_o(\mathbf{x}_j;\hat{\Psi}_U)$ defined in terms of the relative size of the fitted posterior probabilities of group membership $\tau_i(\mathbf{x}_j;\hat{\Psi}_U)$. In an outright assignment of the 145 subjects to the three groups according to this statistical diagnosis, 76 are assigned to group G_1, 32 to G_2, and 37 to G_3. More precisely, if $\mathbf{N}_i = (N_{i1}, N_{i2}, N_{i3})'$, where N_{ih} is the number of clinically classified subjects from G_i assigned to G_h by $r_o(\mathbf{x};\hat{\Psi}_U)$, then

$$\mathbf{N}_1 = (68, 8, 0)',$$

$$\mathbf{N}_2 = (8, 22, 6)',$$

$$\mathbf{N}_3 = (0, 2, 31)'.$$

It was found that the clinical classification and outright statistical diagnoses are different for 24 subjects. For the latter, we have listed in Table 6.6 their feature vectors $\mathbf{x}_j$, their clinical classification, and statistical diagnosis in terms of their outright allocation given by $r_o(\mathbf{x}_j;\hat{\Psi}_U)$ and their probabilistic grouping given by the fitted posterior probabilities of group membership $\tau_i(\mathbf{x}_j;\hat{\Psi}_U)$ for $i = 1$, 2, and 3.

A comparison of the feature vectors $\mathbf{x}_j$ in Table 6.6 with the estimates of the group means in Table 6.5 sheds light on why the statistical diagnosis that takes into account all three metabolic variables differs from the clinical classification. For example, $r_o(\mathbf{x}_j;\hat{\Psi}_U)$ assigns subjects $j = 131$ and 136 to G_2 (chemical diabetes) rather than to G_3 (overt diabetes) as their second variable (insulin area) is atypically high for membership of G_3.

6.8.4 Validity of Normal Model for Group-Conditional Distributions

The nonellipsoidal nature of the two noncentral clusters in Figure 6.3 suggests that this is an example where the normal-based mixture model may be put to the practical test. To examine the assumption of multivariate normality for the group-conditional distributions, we applied Hawkins' (1981) test for multivariate normality (but not also homoscedasticity) to the clusters implied by the fitted normal mixture model, proceeding as if they represented a correct partition of the data with respect to the three possible groups. No significant departures from normality were obtained. Of course, as cautioned in Section 6.2.4, it is self-serving to first cluster the data under the assumption of the model whose validity is to be tested subsequently.

As discussed in Section 6.4, Hawkins' (1981) test also provides a measure of typicality of a feature observation with respect to the group to which it is specified to belong. This measure, which is given by (6.4.8) for unclassified

TABLE 6.6 Statistical Diagnosis for 24 Subjects with Different Clinical Classification (CC)

Subject No.	Metabolic Characteristics				Statistical Diagnosis			
j	x_{1j}	x_{2j}	x_{3j}	CC	$r_o(\mathbf{x}_j; \hat{\boldsymbol{\Psi}}_U)$	$\tau_1(\mathbf{x}_j; \hat{\boldsymbol{\Psi}}_U)$	$\tau_2(\mathbf{x}_j; \hat{\boldsymbol{\Psi}}_U)$	$\tau_3(\mathbf{x}_j; \hat{\boldsymbol{\Psi}}_U)$
26	365	228	235	1	2	0.20	0.68	0.12
51	313	200	233	1	2	0.13	0.50	0.37
62	439	208	244	1	2	0.04	0.84	0.12
65	472	162	257	1	2	0.00	0.78	0.22
68	391	137	248	1	2	0.02	0.89	0.09
69	390	375	273	1	2	0.00	1.00	0.00
71	413	344	270	1	2	0.00	1.00	0.00
75	403	267	254	1	2	0.04	0.95	0.01
77	426	213	177	2	1	0.87	0.07	0.06
78	364	156	159	2	1	0.96	0.01	0.03
79	391	221	103	2	1	0.99	0.00	0.01
80	356	199	59	2	1	0.99	0.00	0.01
81	398	76	108	2	1	0.99	0.00	0.01
85	465	237	111	2	1	0.95	0.00	0.05
88	540	188	211	2	3	0.00	0.41	0.59
96	477	124	60	2	3	0.41	0.00	0.59
105	442	109	157	2	1	0.79	0.03	0.18
107	580	132	155	2	3	0.00	0.01	0.99
109	562	139	198	2	3	0.00	0.09	0.91
110	423	212	156	2	1	0.95	0.01	0.04
111	643	155	100	2	3	0.00	0.00	1.00
112	533	120	135	2	3	0.01	0.01	0.98
131	538	460	320	3	2	0.00	1.00	0.00
136	636	314	220	3	2	0.00	1.00	0.00

data, revealed an extremely atypical observation, namely,

$$\mathbf{x}_j = (558, 748, 122)'$$

on subject $j = 86$, who is both clinically and statistically diagnosed as coming from group G_2 (chemical diabetes). This subject has relatively a very large reading for its second feature variable (insulin area); indeed, so large, that the measure (6.4.8) of typicality is zero. Concerning other subjects with a typicality measure less than 0.05, there were four in the first cluster (subjects $j = 25$, 39, 57, and 74) and one in the third cluster (subject $j = 145$), with the typicality measure equal to 0.04, 0.02, 0.04, 0.04, and 0.02, respectively.

A robust version of a normal mixture model can be implemented without difficulty by incorporating the robust estimation of the group-conditional distributions as considered in Section 5.7; see also McLachlan and Basford (1988, Section 2.8). When fitted here, it was found to give similar results as before with the ordinary version. Not surprisingly, because of the aforementioned presence of one observation in the second group with an extremely atypical value for the second feature variable, the estimated standard deviation for the latter variable was reduced; it fell from 151 to 139. So far as the clustering produced, there were different outright allocations for two subjects ($j = 40, 59$) from group G_1 (normal) according to their clinical classification. On the basis of their posterior probabilities fitted robustly, they are marginally assigned to G_3 (overt diabetes), whereas previously they were marginally put in the same group G_1 as their clinical classification.

6.9 EXAMPLE: TESTING FOR EXISTENCE OF SUBSPECIES IN FISHER'S *IRIS* DATA

6.9.1 Testing for Subspecies in *Iris Virginica* Data

We consider an example in which we wish to test for the existence of further grouping to that supplied for the training data. We consider Fisher's (1936) *Iris* data set, which was described in Section 6.5.3. As pointed out by Wilson (1982), the *Iris* data were collected originally by E. Anderson (1935) with the view to seeing whether there was "evidence of continuing evolution in any group of plants." In keeping with this spirit, Wilson (1982) applied her aural approach to data analysis to explore each of the three species separately for signs of heterogeneity. She concluded that the *setosa* sample was homogeneous apart from an outlier, which she noted could be explained by a possible misprint/misrecording in one of the sepal width values. However, her analysis suggested that both *versicolor* and *virginica* species should be split into two subspecies.

To examine this further, we fitted a four-dimensional mixture of two heteroscedastic normal components, firstly to the 50 *virginica* observations. To explore these 50 four-dimensional observations for the presence of clusters, we applied the projection pursuit algorithm of Friedman (1987). The two-dimensional projection of the data as given by the first solution of this algorithm is displayed in Figure 6.5, where the observations are labeled 1 to 50 in order of their listing in Table 1.1 in Andrews and Herzberg (1985). This plot may be of use in searching for group structure as a basis for the specification of the initial values of the posterior probabilities of component membership of the mixture to be fitted.

The fitting of a two-component normal mixture produces a cluster containing the five observations numbered 6, 18, 19, 23, and 31, with the remaining 45 observations in a second cluster. This two-component clustering is portrayed in Figure 6.6, where the cluster labels are attached to the two-dimensional projection pursuit representation of these 50 *virginica* observations. It differs from that obtained by Wilson (1982), who also put the four observations labeled 8,

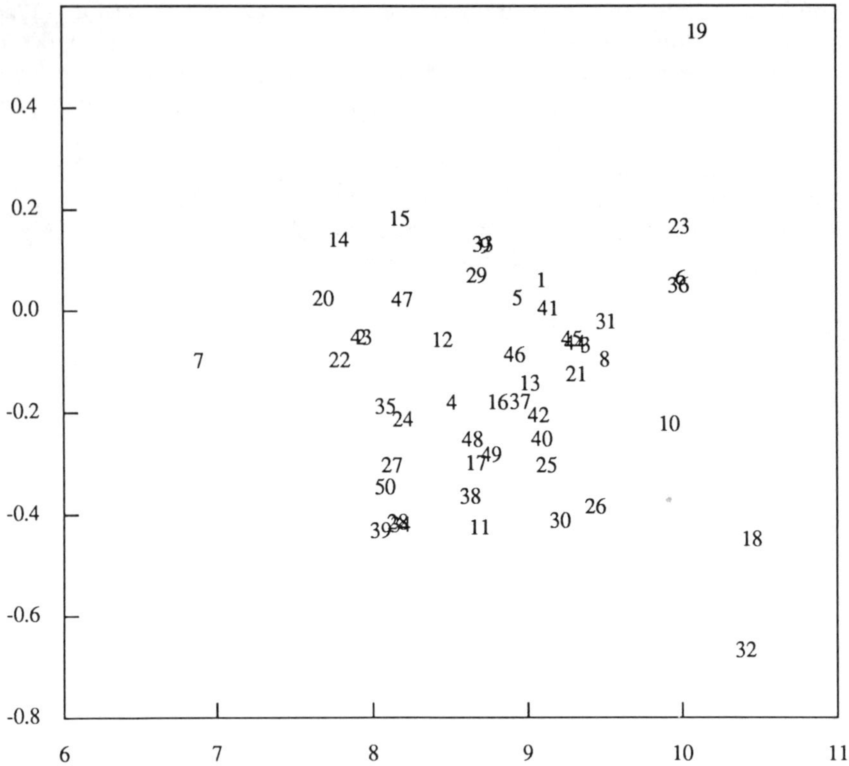

FIGURE 6.5. Two-dimensional projection of data on *Iris virginica* as given by first solution of Friedman's (1987) projection pursuit algorithm.

26, 30, and 32 with those five in the first cluster as above. This latter clustering is obtained by fitting the two-component normal mixture model, using the solution of the likelihood equation that corresponds to the second largest of the local maxima that we located.

The likelihood ratio statistic λ was calculated for the test of a single component (one *virginica* species) versus two components (two *virginica* subspecies). An assessment of the P-value for this test can be obtained by resampling of $-2\log\lambda$, as described in Section 2.10. Ninety-nine replications of $-2\log\lambda$ were generated, and the original value of 46.37 for $-2\log\lambda$ fell between the 85th and 86th smallest replicated values. This implies a P-value of approximately 0.15. Thus, at any conventional level of significance, the null hypothesis of a single *virginica* species would not be rejected in favor of two subspecies.

A common way of proceeding with the assessment of P-values for tests performed on the basis of $-2\log\lambda$ is to use the approximation of Wolfe (1971), whereby the null distribution of $-2\log\lambda$ is taken to be chi-squared, but with

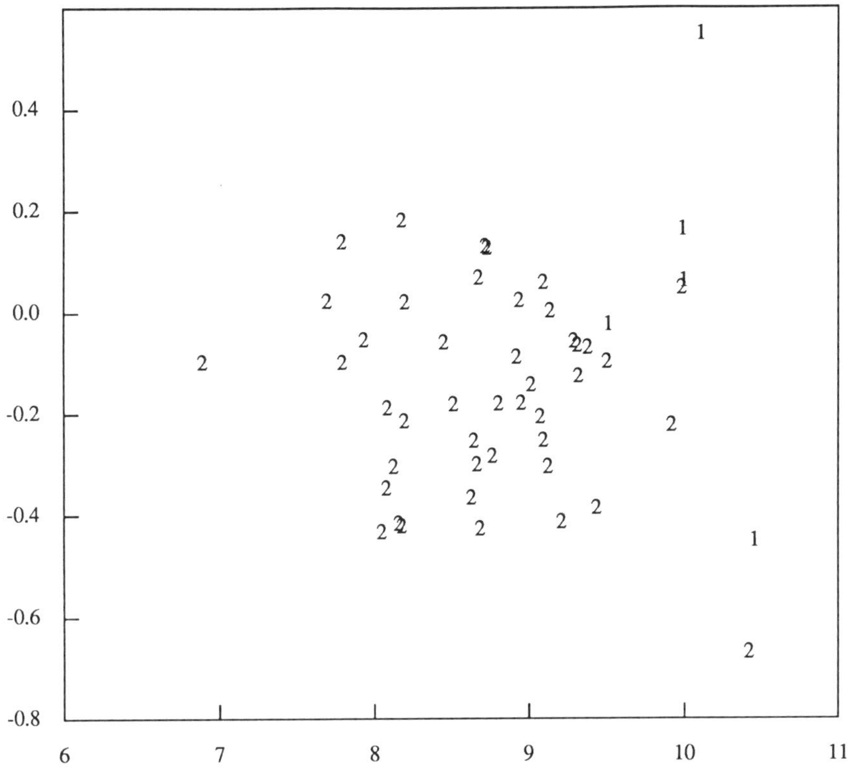

FIGURE 6.6. Two-group clustering imposed on two-dimensional projection of data on *Iris virginica*, as given by the first solution of Friedman's (1987) projection pursuit algorithm.

the degrees of freedom modified from the number that would apply if regularity conditions held. The degrees of freedom are taken to be twice the difference between the number of parameters under the null and alternative hypotheses, ignoring the mixing proportions. As cautioned by McLachlan (1987a) and McLachlan and Basford (1988), this approximation is applicable essentially only for univariate normal components with a common variance, and then only if the sample size is large. Otherwise, this approximation underestimates the P-value and so leads to too many groups being inferred from the data. This has been demonstrated further by McLachlan, Basford, and Green (1990). With the present application of testing for the presence of two subspecies within the *Iris virginica* data, Wolfe's approximation takes the null distribution of $-2\log\lambda$ to be χ^2_{28}. It implies a P-value of 0.016 for the realized value of 46.37 for $-2\log\lambda$, as obtained above in testing for a single versus a two-component normal mixture. Hence, at the 2% level, it leads to the rejection of the hypothesis that *Iris virginica* is a homogeneous species.

6.9.2 Testing for Subspecies in *Iris Versicolor* Data

We now consider the existence of subspecies in the 50 *Iris versicolor* observations. Their two-dimensional projection as given by the first solution of Friedman's projection pursuit algorithm is displayed in Figure 6.7. The observations are labeled 1 to 50 in order of their listing in Table 1.1 of Andrews and Herzberg (1985). The fitting of a four-dimensional mixture of two heteroscedastic normal components to the 50 *versicolor* observations produced one cluster containing 13 observations labeled 1, 3, 5, 9, 16, 19, 23, 25, 26, 27, 28, 37, and 38, and with the other cluster containing the remaining 37 *versicolor* observations. This clustering, which is portrayed in Figure 6.8, is different from that of Wilson (1982), who split the 50 versicolor observations into a cluster containing the twelve observations labeled 8, 10, 12, 15, 17, 21, 35, 36, 39, 46, 47, and 49, and with the remaining 38 observations in another cluster. Concerning the significance of the clustering effected by fitting a two-component normal mixture model, the P-value of the likelihood ratio test of one versus two components was assessed by resampling on the basis of 99

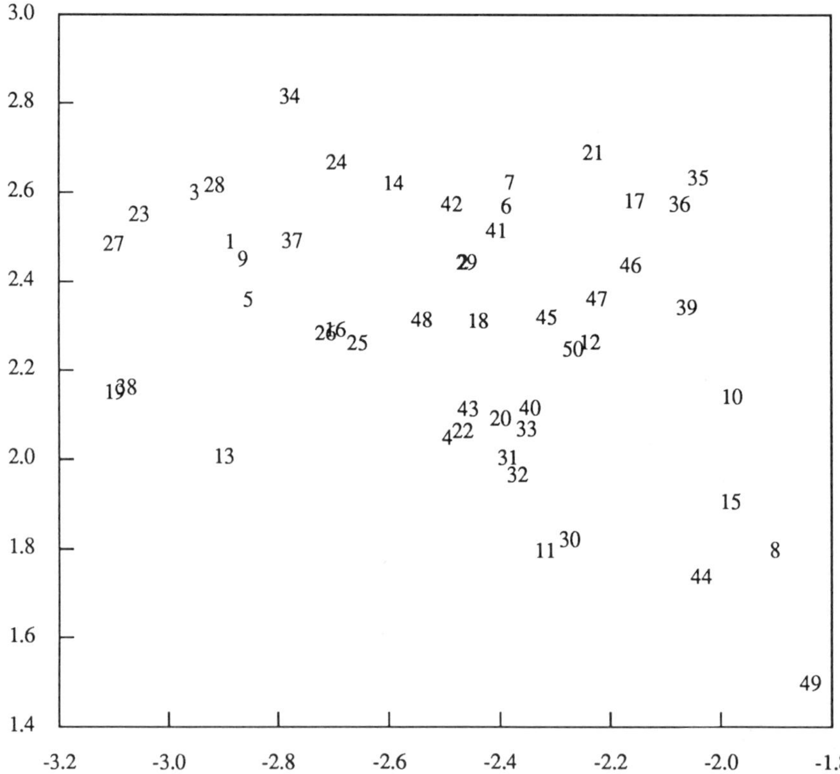

FIGURE 6.7. Two-dimensional projection of data on *Iris versicolor*, as given by the first solution of Friedman's (1987) projection pursuit algorithm.

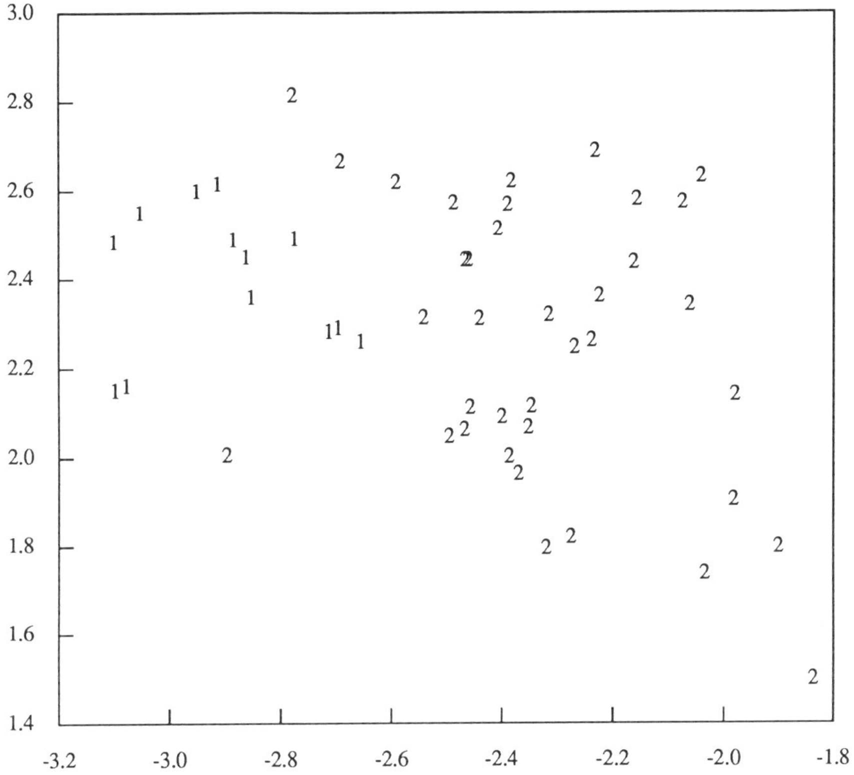

FIGURE 6.8. Two-group clustering imposed on two-dimensional projection of data on *Iris veriscolor*, as given by the first solution of Friedman's (1987) projection pursuit algorithm.

replications of $-2\log\lambda$ to be approximately 0.4. It suggests at any conventional level of significance that the *versicolor* species can be regarded still as being homogeneous.

CHAPTER 7

Parametric Discrimination via Nonnormal Models

7.1 INTRODUCTION

In this chapter, we consider the use of nonnormal models to fit the group-conditional distributions of the feature vector in the parametric construction of discriminant rules. The case in which all the feature variables are discrete is to be considered first. The parametric models to be considered in the discrete case can be viewed as attempts to smooth the nonparametric estimates obtained with the multinomial model.

In the case of mixed feature variables in which some are discrete and some are continuous, we will highlight the role of the location model. It assumes that within each group, the continuous feature variables have a multivariate normal distribution with a common covariance matrix, conditional on each distinct realization of the discrete feature variables. Given its computational simplicity, we will also consider ways in which the performance of the sample NLDF $\xi(\mathbf{x};\hat{\boldsymbol{\theta}}_E)$ can be improved for applications to mixed feature data. Finally in this chapter, a brief account is given of some nonnormal models that have been adopted for discrimination in the case of all continuous feature variables.

7.2 DISCRETE FEATURE DATA

7.2.1 Multinomial and Independence Models

In this and the next few sections, it is supposed that the feature vector $\mathbf{X}$ consists of variables that are all discrete. Suppose that the kth feature variable in $\mathbf{X}$ takes on, after appropriate truncation if necessary, d_k distinct values ($k = 1,\ldots,p$). Then there are

$$m = \prod_{k=1}^{p} d_k$$

distinct realizations of the random feature vector $\mathbf{X}$. Although we are concerned in this chapter with parametric estimation of the group-conditional distributions, we will first briefly introduce the multinomial model. This is because the parametric models to be considered are equivalent in their fully saturated form to the multinomial.

The m distinct realizations of $\mathbf{X}$ define m multinomial cells with $f_i(\mathbf{x})$ denoting the probability of the realization $\mathbf{x}$ in G_i ($i = 1,\ldots,g$). The maximum likelihood estimate of $f_i(\mathbf{x})$ under the nonparametric multinomial model is the naive or counting estimate,

$$\hat{f}_i(\mathbf{x}) = n_i(\mathbf{x})/n_i,$$

where $n_i(\mathbf{x})$ is the number of training feature vectors from G_i equal to $\mathbf{x}$ ($i = 1,\ldots,g$). Unless n is large relative to m, several of the frequencies $n_i(\mathbf{x})$ will be zero. In which case, other estimates of $f_i(\mathbf{x})$ have to be considered, using some form of smoothing. Cox (1972b) has provided a concise summary of the analysis of multivariate discrete data.

The independence model is perhaps the simplest way to smooth the multinomial estimates. Under this model, the p feature variables are taken to be independent, so that the ith group-conditional density of $\mathbf{X}$ is given by

$$f_i(\mathbf{x}) = \prod_{k=1}^{p} f_{ik}(x_k),$$

where

$$f_{ik}(x_k) = \mathrm{pr}\{X_k = x_k \mid \mathbf{X} \in G_i\}$$

for $i = 1,\ldots,g$ and $k = 1,\ldots,p$. This model has been applied by Warner et al. (1961), among others. The maximum likelihood estimate of $f_{ik}(x_k)$ is

$$\hat{f}_{ik}(x_k) = n_i(x_k)/n_i \qquad (i = 1,\ldots,g;\ k = 1,\ldots,p), \tag{7.2.1}$$

where $n_i(x_k)$ is the number of training feature vectors from G_i with the kth element equal to x_k. A modified version of (7.2.1) is

$$\tilde{f}_{ik}(x_k) = \prod_{k=1}^{p} \{n_i(x_k) + d_k^{-1}\}/(n_i + 1), \tag{7.2.2}$$

which employs a zero-avoiding device; see Hilden and Bjerregaard (1976), Titterington et al. (1981), and Schmitz et al. (1985). In the first two of these references, a small amount of smoothing is imposed on $\tilde{f}_{ik}$ by taking its square root or using some other fractional power of it. As noted by Titterington et al. (1981), an attractive feature of the independence model in addition to its simplicity is the ease with which it can deal with incomplete data.

The independence model, however, can impose too extreme a degree of smoothing. In some situations, models intermediate between it and the multinomial may be more appropriate. One way of fitting an intermediate model is to postulate a parametric model for $f_i(\mathbf{x})$ in each group G_i $(i = 1,\ldots,g)$. If the postulated models contain as many parameters as the number of cells m, then it is fully saturated and so will yield the multinomial estimates. But by fitting a model with a reduced number of parameters, a degree of smoothing is applied to the multinomial estimates.

7.3 PARAMETRIC FORMULATION FOR DISCRETE FEATURE DATA

7.3.1 Log-Linear Models

One parametric approach to the estimation of the multinomial probabilities $f_i(\mathbf{x})$ for discrete data $\mathbf{x}$ is to adopt a log-linear representation. In a given group G_i $(i = 1,\ldots,g)$, $\log f_i(\mathbf{x})$ is modeled as

$$\log f_i(\mathbf{x}) = \zeta_i'\mathbf{u}(\mathbf{x}), \tag{7.3.1}$$

where $\mathbf{u}(\mathbf{x})$ is a vector of ones and minus ones, and ζ_i is a vector of parameters. In the fully saturated form of this model, ζ_i consists of m parameters and so maximum likelihood estimation of ζ_i leads to the multinomial estimate for $f_i(\mathbf{x})$. A reduced version of the model can be formulated by eliminating those parameters in ζ_i corresponding to interactions higher than a certain order, for instance, the first. Then the retained parameters in ζ_i represent the main effects of the feature variables and their first-order interactions. If also the first-order interactions are specified to be zero, then the log-linear model is equivalent to the independence model. If we specify the first- and higher-order interaction terms to be the same in each group, then we obtain the logistic model to be considered in the next chapter.

To examine more closely the log-linear representation (7.3.1), suppose that $p = 3$ and that the realization $\mathbf{x}$ of $\mathbf{X} = (X_1, X_2, X_3)'$ corresponds to, say, the rth level of X_1, the sth level of X_2, and the tth level of X_3. On writing $f_i(\mathbf{x})$ as f_{irst} now to denote this, the representation (7.3.1) implies that

$$\log f_{irst} = \zeta_i + \zeta_{i1,r} + \zeta_{i2,s} + \zeta_{i3,t} + \zeta_{i12,rs} + \zeta_{i13,rt} + \zeta_{i23,st} + \zeta_{i123,rst}, \tag{7.3.2}$$

where the parameters on the right-hand side of (7.3.2) satisfy the constraints

$$\zeta_{ij(+)} = 0 \qquad (j = 1,2,3),$$

$$\zeta_{ijk(+)} = 0 \qquad (j,k = 1,2,3;\ j < k),$$

and

$$\zeta_{i123(+)} = 0.$$

In these constraints,

$$\zeta_{ij(+)} = \sum_{v=1}^{d_j} \zeta_{ij,v} \qquad (j = 1,2,3),$$

and $\zeta_{ijk(+)}$ and $\zeta_{i123(+)}$ are defined similarly. This is the fully saturated version of the log-linear model for $p = 3$. A reduced version with $(d_1 - 1)(d_2 - 1)(d_3 - 1)$ less parameters can be fitted by setting

$$\zeta_{i123(rst)} = 0$$

for $r = 1,\ldots,d_1$; $s = 1,\ldots,d_2$; $t = 1,\ldots,d_3$. That is, second-order interactions between the feature variables are taken to be zero.

One way to compute the maximum likelihood estimates of the parameters of the log-linear model in a reduced form is to use iterative proportional fitting, as described in Bishop, Fienberg, and Holland (1975, Section 3.5) and Fienberg (1980, Chapter 3). An alternative approach based on the Newton–Raphson method is described by Haberman (1974, 1978, 1979). The widely available package GLIM (1986) uses Newton-Raphson; see Krzanowski (1988, Section 10.2). The adequacy of the fitted model can be tested using the goodness-of-fit statistics, as described in the aforementioned references.

Brunk and Pierce (1974) have provided a Bayesian analysis of an alternative to the log-linear model with interactions expressed as multiplicative adjustments to the independent binary model. Recently, Wernecke et al. (1989) have developed an algorithm for model selection for discrimination with categorical feature variables. It is based on the hierarchical application of four models (multinominal, log-linear with retention of first-order interactions, logistic, and independence).

7.3.2 Lancaster Models

In the last section, the multinomial probabilities were represented by log-linear models, which used the factorial-type structure of the features when recorded in the form of discrete variables. Another way with this structure of formulating parametrically a full range of models ranging from the basic independence model to the full multinomial is by the use of the models of Lancaster (1969). In the case in which all the feature variables are binary, the Lancaster models are equivalent to the Lazarsfeld–Bahadur representation, as given by (5.6.2).

Lancaster's (1969) definition of interaction can be used to give an expression for $f_i(\mathbf{x})$ explicitly in terms of the marginal distributions of the feature variables, under the presupposition that all interactions higher than a certain order vanish. Assuming that the hth and higher-order interactions vanish, Zentgraf (1975) has given an expression for $f_i(\mathbf{x})$ using only the hth and lower-order marginal distributions of the feature variables. To demonstrate this, we consider now the case of $p = 3$ discrete variables. We let f_{irst} be defined as in the previous section when p was set equal to 3. Then under the assumption that all second-order interactions vanish within group G_i ($i = 1,\ldots,g$), the f_{irst} must satisfy for all r, s, and t ($r = 1,\ldots,d_1$; $s = 1,\ldots,d_2$; $t = 1,\ldots,d_3$),

$$f_{irst} = f_{irso}f_{ioot} + f_{iost}f_{iroo} + f_{irot}f_{ioso} - 2f_{iroo}f_{ioso}f_{ioot} \qquad (i = 1,\ldots,g),$$

where, for instance, f_{irso} and f_{iroo} denote the two- and one-dimensional marginal probabilities, that is, in group G_i, f_{irso} is the joint probability that X_1 takes on its rth value and X_2 its sth value, and f_{iroo} is the marginal probability that X_1 takes on its rth value. As f_{irst} is specified by the two- and one-dimensional marginal probabilities, it suffices to estimate just the latter in order to estimate it. These marginal probabilities can be estimated simply by the relative frequencies, although this does not in general provide the maximum likelihood estimates of the f_{irst}. Also, the estimates so obtained can be negative; see Moore (1973) and Trampisch (1976).

The treatment of missing values with Lancaster models is straightforward since, as with independence models, the estimates of the probabilities are based on the nonmissing feature vectors. Additional references to those above on the use of Lancaster models in discriminant analysis include Victor, Trampisch, and Zentgraf (1974), Titterington et al. (1981), and Trampisch (1983).

7.3.3 Latent Class Models

With latent class models (Lazarsfeld and Henry, 1968), the ith group-conditional density $f_i(\mathbf{x})$ is modeled as a mixture of the same C densities,

$$f_i(\mathbf{x}) = \sum_{h=1}^{C} \omega_{ih} f_{oh}(\mathbf{x}),$$

where C denotes the number of classes, and $f_{oh}(\mathbf{x})$ denotes the density of $\mathbf{X}$ in the hth class within any of the groups. In latent class analysis, conditional independence of the feature variables with each class in a group is widely assumed, so that

$$f_{oh}(\mathbf{x}) = \prod_{k=1}^{p} f_{kh}(x_k),$$

where $f_{kh}(x_k)$ is the density of the kth feature variable in the hth class of any of the groups. The f_{kh} and the ω_{ih} can be estimated using the EM algorithm of Dempster, Laird, and Rubin (1977); see, for example, Skene (1978), Everitt (1984, Section 4.3), and McLachlan and Basford (1988, Section 3.5), and the references therein.

7.4 LOCATION MODEL FOR MIXED FEATURES

7.4.1 Introduction

In this section, attention is focused on the case of mixed feature variables, where some are discrete and some are continuous. Suppose that the p-dimensional feature vector $\mathbf{x}$ consists of p_1 discrete variates and $p_2 = p - p_1$ continuous variates. We assume here that the variates are labeled so that $\mathbf{x} = (\mathbf{x}^{(1)\prime}, \mathbf{x}^{(2)\prime})'$, where the subvector $\mathbf{x}^{(1)}$ contains the p_1 discrete features and $\mathbf{x}^{(2)}$ the p_2 continuous features in $\mathbf{x}$. We further assume without loss of generality that the p_1 discrete variates are all binary, each taking on zero or one values. For example, a discrete random variable taking on d distinct values with nonzero probabilities can be replaced equivalently by $d-1$ dummy binary random variables, where some combinations of them are constrained not to be observable; see Krzanowski (1982a). If the values taken on by the discrete random variable are ordered, then information will be lost. If it is essential to retain this information, then one might consider treating such a variable as continuous. Krzanowski (1982a) considers this to be the best course of action if there is a moderate number of distinct values and if there is good reason to assume that the log likelihood ratio is linear with this respect to this variable. On the other hand, it is certainly advisable to replace this variable by binary ones if nonlinear behavior is suspected or apparent. An example of nonlinearity occurs when the discrete variable takes on three distinct values, corresponding to the categories of absent, mild, or severe of a medical condition, and where "absent" and "severe" are prevalent in group G_1, but "mild" is prevalent in group G_2.

Each particular realization of the p_1 binary feature variables $X_1, \ldots, X_{p_1}$ in the subvector $\mathbf{X}_1^{(1)}$ gives rise to $m = 2^{p_1}$ different multinomial cells. As noted by Krzanowski (1976), we can uniquely order the cells by letting the cell number $c(\mathbf{x}^{(1)})$ corresponding to the realization $\mathbf{x}^{(1)}$ be given by

$$c(\mathbf{x}^{(1)}) = 1 + \sum_{v=1}^{p_1} x_v^{(1)} 2^{(v-1)}, \tag{7.4.1}$$

where $x_v^{(1)} = (\mathbf{x}^{(1)})_v$ for $v = 1, \ldots, p_1$.

The location model as introduced by Olkin and Tate (1961), and as used in discriminant analysis initially by Chang and Afifi (1974) and Krzanowski (1975), assumes that in group G_i, the conditional distribution of $\mathbf{X}^{(2)}$ given $\mathbf{X}^{(1)} = \mathbf{x}^{(1)}$ is

$$\mathbf{X}^{(2)} \mid \mathbf{X}^{(1)} = \mathbf{x}^{(1)} \sim N(\boldsymbol{\mu}_{ic}^{(2)}, \boldsymbol{\Sigma}^{(2)}) \quad \text{in} \quad G_i \; (i = 1, \ldots, g), \tag{7.4.2}$$

where $c = c(\mathbf{x}^{(1)})$ is defined by (7.4.1). With this model (7.4.2), the conditional mean of the subvector $\mathbf{X}^{(2)}$ of continuous feature variables is modeled within each group as an arbitrary p_2-dimensional vector for each of the $m = 2^{p_1}$ cells $c(\mathbf{x}^{(1)})$ corresponding to the different realizations $\mathbf{x}^{(1)}$ of the subvector $\mathbf{X}^{(1)}$ of binary features. The within-group covariance matrix of $\mathbf{X}^{(2)}$, however, is specified to be the same for all cells and for all groups.

The location model (7.4.2) with $g = 2$ groups was applied by Chang and Afifi (1974) in the special case of $p_1 = 1$, and Krzanowski (1975) extended its use for an arbitrary number p_1 of binary features. Actually, in the formulation of the problem by Chang and Afifi (1974), the within-cell covariance matrix of $\mathbf{X}^{(2)}$ was allowed to be different for each of the two levels of the single binary feature variable. The location model in a discriminant analysis context has since been studied in a series of papers by Krzanowski (1976, 1977, 1980, 1982a, 1983b, 1986), including for $g > 2$ groups.

Under (7.4.2), the ith group-conditional density of $\mathbf{X} = (\mathbf{X}^{(1)\prime}, \mathbf{X}^{(2)\prime})'$ can therefore be expressed as

$$f_i(\mathbf{x}; q_{ic}, \boldsymbol{\theta}_{E,c}^{(2)}) = q_{ic}\phi(\mathbf{x}^{(2)}; \boldsymbol{\mu}_{ic}^{(2)}, \boldsymbol{\Sigma}^{(2)}), \tag{7.4.3}$$

where $c = c(\mathbf{x}^{(1)})$ is defined by (7.4.1), and where

$$q_{ic} = \operatorname{pr}\{\mathbf{X}^{(1)} = \mathbf{x}^{(1)} \mid \mathbf{X} \in G_i\} \qquad (i = 1, \ldots, g).$$

In the left-hand side of (7.4.3), the vector of parameters $\boldsymbol{\theta}_{E,c}^{(2)}$ is defined as $\boldsymbol{\theta}_E$ was for the homoscedastic normal model (3.3.1), that is, $\boldsymbol{\theta}_{E,c}^{(2)}$ denotes $\boldsymbol{\theta}_E$ when $\boldsymbol{\mu}_i$ is replaced by $\boldsymbol{\mu}_{ic}^{(2)}$ for $i = 1, \ldots, g$ and $\boldsymbol{\Sigma}$ by $\boldsymbol{\Sigma}^{(2)}$. Similarly, corresponding to

$$\boldsymbol{\Psi}_E = (\boldsymbol{\pi}', \boldsymbol{\theta}_E')'$$

in Section 3.3.1, we let

$$\boldsymbol{\Psi}_{E,c}^{(2)} = (\boldsymbol{\pi}', \boldsymbol{\theta}_{E,c}^{(2)\prime})'.$$

In the case of $g = 2$ groups, Afifi and Elashoff (1969) have considered the problem of testing whether

$$(\mathbf{q}_1', \boldsymbol{\mu}_{11}^{(2)\prime}, \ldots, \boldsymbol{\mu}_{1m}^{(2)\prime})' = (\mathbf{q}_2', \boldsymbol{\mu}_{21}^{(2)\prime}, \ldots, \boldsymbol{\mu}_{2m}^{(2)\prime})',$$

where

$$\mathbf{q}_i = (q_{i1}, \ldots, q_{im})'.$$

7.4.2 Optimal Rule

The Bayes rule of allocation for the location model (7.4.2) is denoted here by $r_o(\mathbf{x}; \boldsymbol{\pi}, \mathbf{q}, \boldsymbol{\theta}_{LO}^{(2)})$, where

$$\mathbf{q} = (\mathbf{q}_1', \ldots, \mathbf{q}_g')',$$

and

$$\boldsymbol{\theta}_{LO}^{(2)} = (\boldsymbol{\theta}_{E,1}^{(2)\prime}, \ldots, \boldsymbol{\theta}_{E,m}^{(2)\prime})'.$$

This location-model-based discriminant rule (LODR), $r_o(\mathbf{x}; \boldsymbol{\pi}, \mathbf{q}, \boldsymbol{\theta}_{LO}^{(2)})$, is specified by (3.3.2), where (7.4.3) is used for the ith group-conditional density of $\mathbf{X}$. It therefore assigns an entity with feature vector $\mathbf{x} = (\mathbf{x}^{(1)\prime}, \mathbf{x}^{(2)\prime})'$ to G_g if

$$\log(q_{ic}/q_{gc}) + \eta_{ig}(\mathbf{x}^{(2)}; \boldsymbol{\Psi}_{E,c}^{(2)}) \le 0 \qquad (i = 1, \ldots, g-1) \tag{7.4.4}$$

is satisfied, where the label c of the location cell is determined from $\mathbf{x}^{(1)}$ by (7.4.1), and where

$$\eta_{ig}(\mathbf{x}^{(2)}; \boldsymbol{\Psi}_{E,c}^{(2)}) = \log(\pi_i/\pi_g) + \xi_{ig}(\mathbf{x}^{(2)}; \boldsymbol{\theta}_{E,c}^{(2)}), \tag{7.4.5}$$

and $\xi_{ig}(\mathbf{x}^{(2)};\boldsymbol{\theta}_{E,c}^{(2)})$ is defined according to (3.3.3). If (7.4.4) does not hold, then the entity is assigned to G_h if

$$\log(q_{ic}/q_{gc}) + \eta_{ig}(\mathbf{x}^{(2)};\boldsymbol{\Psi}_{E,c}^{(2)}) \leq \log(q_{hc}/q_{gc}) + \eta_{hg}(\mathbf{x}^{(2)};\boldsymbol{\Psi}_{E,c}^{(2)})$$

for $i = 1,\ldots,g-1$; $i \neq h$.

7.4.3 Maximum Likelihood Estimation: Fully Saturated Model

The location model (7.4.2) has

$$(m-1) + mgp_2 + \tfrac{1}{2}p_2(p_2+1)$$

parameters to be estimated from the training data $\mathbf{t}$. This is in addition to the prior probabilities of the groups.

Let $\mathbf{x}_j = (\mathbf{x}_j^{(1)\prime}, \mathbf{x}_j^{(2)\prime})'$ for $j = 1,\ldots,n$ denote the classified data in $\mathbf{t}$. It is convenient here if the $\mathbf{x}_j$ are relabeled as $\mathbf{x}_{ij,c}(i = 1,\ldots,g;\ j = 1,\ldots,n_{ic};\ c = 1,\ldots,m)$, where the $\mathbf{x}_{ij,c}(j = 1,\ldots,n_{ic})$ denote those $\mathbf{x}_j$ from G_i, n_{ic} in number, which give rise to the cell $c = c(\mathbf{x}_{ij}^{(1)})$. If n is very large, then the naive estimates of the parameters (that is, the maximum likelihood estimates) will suffice. They are defined by

$$\hat{q}_{ic} = n_{ic}/n_i \tag{7.4.6}$$

and

$$\begin{aligned}\hat{\mu}_{ic}^{(2)} &= \bar{\mathbf{x}}_{ic}^{(2)} \\ &= \sum_{j=1}^{n_{ic}} \mathbf{x}_{ij,c}^{(2)}/n_{ic}\end{aligned} \tag{7.4.7}$$

for $c = 1,\ldots,m$ and $i = 1,\ldots,g$, and

$$\hat{\boldsymbol{\Sigma}}^{(2)} = \sum_{i=1}^{g}\sum_{c=1}^{m}\sum_{j=1}^{n_{ic}}(\mathbf{x}_{ij,c}^{(2)} - \bar{\mathbf{x}}_{ic}^{(2)})(\mathbf{x}_{ij,c}^{(2)} - \bar{\mathbf{x}}_{ic}^{(2)})'/n. \tag{7.4.8}$$

The bias-corrected version of $\hat{\boldsymbol{\Sigma}}^{(2)}$ is given by

$$\mathbf{S}^{(2)} = \{n/(n-gm)\}\hat{\boldsymbol{\Sigma}}^{(2)}. \tag{7.4.9}$$

Recently, in the special case of $g = 2$, $p_1 = 1$, and $p_2 = 1$ or 2, Balakrishnan and Tiku (1988) have shown how a robust sample version of the LODR can be formed, using the modified maximum likelihood (MML) estimates as defined in Section 5.7.3.

The location model (7.4.2) can be generalized by allowing the within-cell covariance matrix to be different within each group. A further generalization would be to allow the within-cell covariance matrix to be different not only within each group, but also for each cell. However, these generalizations exacerbate the problem of precision estimation of a large number of parameters from training sets of limited size. The first generalization requires an additional $\frac{1}{2}p_2(p_2+1)(g-1)$ parameters to be estimated and, with the second, the number of parameters increases dramatically by $\frac{1}{2}p_2(p_2+1)(mg-1)$.

7.4.4 Maximum Likelihood Estimation: Reduced Model

In practice, estimation under the location model is generally only feasible if the size n of the training set $\mathbf{t}$ is very large. If n is not very large, then for a given group G_i, n_{ic} may be zero for several of the cells. For parsimonious estimation of the $\boldsymbol{\mu}_{ic}^{(2)}$ and the q_{ic}, Krzanowski (1975) has proposed fitting, say, a second-order log-linear model to the q_{ic} and a second-order model, in the binary variables, to the conditional means $\boldsymbol{\mu}_{ic}^{(2)}$ of the continuous variables. The linear model for the estimation of the $\boldsymbol{\mu}_{ic}^{(2)}$ has the form

$$\boldsymbol{\mu}_{ic}^{(2)} = \mathbf{K}_i\mathbf{u}(\mathbf{x}^{(1)}), \tag{7.4.10}$$

where $\mathbf{K}_i$ is a $p \times m$ matrix of parameters, and $\mathbf{u}(\mathbf{x}^{(1)})$ is a $m \times 1$ vector of ones or zeros defined as

$$\mathbf{u}(\mathbf{x}^{(1)}) = (1, x_1^{(1)}, \ldots, x_{p_1}^{(1)}, x_1^{(1)}x_2^{(1)}, \ldots, x_{p_1-1}^{(1)}x_{p_1}^{(1)}, x_1^{(1)}x_2^{(1)}x_3^{(1)}, \ldots, x_1^{(1)}x_2^{(1)}\cdots x_{p_1}^{(1)})'. \tag{7.4.11}$$

The m cell probabilities q_{ic} can be modeled using the log-linear representation (7.3.1) discussed in Section 7.3.1.

These models for the q_{ic} and $\boldsymbol{\mu}_{ic}^{(2)}$ are fully saturated, and so as they stand, they provide no reduction in the number of parameters to be estimated. They provide the naive estimates of the q_{ic}, $\boldsymbol{\mu}_{ic}^{(2)}$, and $\boldsymbol{\Sigma}^{(2)}$, as given by (7.4.6) to (7.4.8). However, a reduced form of (7.4.2) can be formulated by eliminating all products in $\mathbf{u}(\mathbf{x}^{(1)})$ of, say, third or higher order. This corresponds to taking all second- or higher-order interactions to be zero. Let $\mathbf{u}_{ij} = \mathbf{u}(\mathbf{x}_{ij}^{(1)})$ for $i = 1, \ldots, g$ and $j = 1, \ldots, n_i$, when third- and higher-order terms in $\mathbf{u}(\mathbf{x}_{ij}^{(1)})$ are deleted. That is, $\mathbf{u}_{ij}$ is of dimension

$$h = 1 + p_1 + \tfrac{1}{2}p_1(p_1 - 1).$$

Then the maximum likelihood estimate of $\mathbf{K}_i$ is given as

$$\hat{\mathbf{K}}_i = \mathbf{C}_{1i}\mathbf{C}_{2i}^{-1},$$

where

$$\mathbf{C}_{1i} = \sum_{j=1}^{n_i} \mathbf{x}_{ij}^{(2)}\mathbf{u}_{ij}',$$

and

$$\mathbf{C}_{2i} = \sum_{j=1}^{n_i} \mathbf{u}_{ij}\mathbf{u}_{ij}'$$

for $i = 1,\ldots,g$. For this reduced model, an unbiased estimate of $\mathbf{\Sigma}^{(2)}$ is given by

$$\hat{\mathbf{S}}^{(2)} = \sum_{i=1}^{g}\sum_{j=1}^{n_i}(\mathbf{x}_{ij}^{(2)} - \hat{\mathbf{K}}_i\mathbf{u}_{ij})(\mathbf{x}_{ij}^{(2)} - \hat{\mathbf{K}}_i\mathbf{u}_{ij})'/(n-gh). \tag{7.4.12}$$

A log-linear model in which second- or higher-order interactions are set equal to zero can be fitted to the cell probabilities q_{ic}, as described in the previous section for discrete feature variables in general, and not necessarily binary as in the present context.

Krzanowski (1982a) has given the sample version of the reduced LODR in the case where the likelihood ratio criterion, as defined in Section 3.4.1, is used to estimate the optimal form of the rule. He also gave the associated computations required to implement economically the cross-validation method for the estimation of the error rates. Little and Schluchter (1985) have considered the handling of missing values for the location model.

7.4.5 Minimum-Distance Rules for Location Model

Krzanowski (1983a, 1984a, 1987a) has considered the distance between two groups under a variety of assumptions about the group-conditional distributions. In this work attention was focused on the use of Matusita's (1956) distance measure, as defined in Section 1.12. It was seen there that in the case of $g = 2$ groups in which the feature vector $\mathbf{X}$ has group-conditional distribution functions $F_1(\mathbf{x})$ and $F_2(\mathbf{x})$, the Matusita distance is given by

$$\delta_M(F_1, F_2) = \{2(1-\rho)\}^{1/2},$$

where

$$\rho = \int \{f_1(\mathbf{x})f_2(\mathbf{x})\}^{1/2}\, d\nu$$

is the affinity between F_1 and F_2. For the location model (7.4.2), Krzanowski (1983a) has shown that the affinity ρ_{ij} between the distributions of $\mathbf{X}$ in G_i and G_j, respectively, is given by

$$\rho_{ij} = \sum_{c=1}^{m}(q_{ic}q_{jc})^{1/2}\rho_{ijc},$$

where ρ_{ijc} is the affinity between the distributions of $\mathbf{X}^{(2)}$ in G_i and G_j, conditional on $\mathbf{X}^{(1)} = \mathbf{x}^{(1)}$, and where the cell label c is determined from $\mathbf{x}^{(1)}$ by (7.4.1). That is, ρ_{ijc} is the affinity between the distributions $N(\boldsymbol{\mu}_{ic}^{(2)}, \mathbf{\Sigma}^{(2)})$ and $N(\boldsymbol{\mu}_{jc}^{(2)}, \mathbf{\Sigma}^{(2)})$ for $i \neq j = 1,\ldots,g$. The squared Mahalanobis distance between these two multivariate normal distributions is

$$\Delta_{ijc}^2 = (\boldsymbol{\mu}_{ic}^{(2)} - \boldsymbol{\mu}_{jc}^{(2)})'\mathbf{\Sigma}^{(2)-1}(\boldsymbol{\mu}_{ic}^{(2)} - \boldsymbol{\mu}_{jc}^{(2)}).$$

From (1.12.4), the affinity ρ_{ijc} is given by

$$\rho_{ijc} = \exp(-\Delta_{ijc}^2/8).$$

Krzanowski (1986) proposed that as an alternative allocation procedure to the plug-in sample version $r_o(\mathbf{x};\hat{\boldsymbol{\pi}},\hat{\mathbf{q}},\hat{\boldsymbol{\theta}}_{LO}^{(2)})$ of the Bayes rule, an entity be assigned on the basis of the minimum distance

$$\min_i \delta_M(F_{\mathbf{x}}, F_i), \tag{7.4.13}$$

where $F_{\mathbf{x}}$ is the degenerate distribution that places mass one at the point $\mathbf{x}$. He showed that

$$\delta_M(F_{\mathbf{x}}, F_i) = [2\{1 - \rho(F_{\mathbf{x}}, F_i)\}]^{1/2},$$

where

$$\rho(F_{\mathbf{x}}, F_i) = \{q_{ic}\phi(\mathbf{x}^{(2)}; \boldsymbol{\mu}_{ic}^{(2)}, \boldsymbol{\Sigma}^{(2)})\}^{1/2}, \tag{7.4.14}$$

and $c = c(\mathbf{x}^{(1)})$ is defined by (7.4.1). It can be seen from (7.4.14) that the minimum-distance rule defined by (7.4.13) is equivalent to the plug-in sample version $r_o(\mathbf{x};\hat{\boldsymbol{\pi}},\hat{\mathbf{q}},\hat{\boldsymbol{\theta}}_{LO}^{(2)})$ of the Bayes rule with $\hat{\boldsymbol{\pi}}$ specifying equal prior probabilities for the groups.

A sample version of (7.4.13),

$$\min_i \delta_M(F_{\mathbf{x}}, \hat{F}_i), \tag{7.4.15}$$

can be obtained by plugging in estimates of the unknown parameters. Dillon and Goldstein (1978) highlighted some unsatisfactory behavior with this sample plug-in version of the minimum Matusita distance rule in the case of multinomial group-conditional distributions. They therefore proposed an alternative principle for distance-based allocation. Let $\hat{F}_i(\mathbf{x})$ denote the usual plug-in estimate $F_i(\mathbf{x};\hat{\boldsymbol{\theta}}_i)$ of the distribution function $F_i(\mathbf{x};\boldsymbol{\theta}_i)$ of $\mathbf{X}$ in group G_i, where $\hat{\boldsymbol{\theta}}_i$ is based on the classified training data $\mathbf{t}$ $(i = 1,\ldots,g)$. Further, let $\hat{F}_{i,\mathbf{x}}(\mathbf{x})$ denote the plug-in estimate of $F_i(\mathbf{x};\boldsymbol{\theta}_i)$, where $\boldsymbol{\theta}_i$ is estimated from $\mathbf{t}$ and also $\mathbf{x}$ with the latter specified as belonging to $G_i (i = 1,\ldots,g)$. The entity with feature $\mathbf{x}$ is assigned then on the basis of

$$\min_i \delta_M(F_{\mathbf{x}}, \hat{F}_{i,\mathbf{x}}). \tag{7.4.16}$$

For $g = 2$ groups, Krzanowski (1987a) showed that the rules (7.4.15) and (7.4.16) have the same asymptotic form and, in the case of the homoscedastic normal model (3.3.1), are asymptotically equivalent. This last result follows immediately on noting that the minimum-distance rule (7.4.16), as modified by Dillon and Goldstein (1978), is the same as the rule (3.4.6) obtained with the likelihood ratio criterion under the homoscedastic normal model (3.3.1). Krzanowski (1987a) also established the asymptotic equivalence of the minimum-distance rules (7.4.15) and (7.4.16) in the case of multinomial group-conditional distributions. Because of the asymptotic equivalence of these two distance rules in the situations in which the feature variables are either all discrete or have a multivariate normal distribution with the same covariance matrix in each group, Krzanowski (1987a) speculated that their asymptotic equivalence should carry over under the location model to mixed feature variables. As cautioned by Krzanowski (1987a), these results have been obtained asymptotically, and so may not always be applicable in small to medium-size training samples. For example, as discussed in Section 9.2.3, Dillon and Goldstein (1978) have provided evidence that their modification (7.4.16) of (7.4.15) yields a superior rule in certain circumstances with discrete feature data.

Another distance-based rule for the location model can be constructed by modifying the definition of the Mahalanobis distance to allow for discrete variates in the feature vector; see Krusińska (1987).

7.4.6 Predictive Approach

In the previous sections, we have described the so-called estimative approach to the fitting of the location model for mixed binary and continuous feature vectors. Recently, Vlachonikolis (1990) has presented the predictive approach to the fitting of this model in the case of $g = 2$ groups. As with the predictive approach for the homoscedastic normal model (3.3.1), Vlachonikolis (1990) adopted the vague prior density for the $\boldsymbol{\mu}_{ic}$ and $\boldsymbol{\Sigma}$,

$$p(\{\boldsymbol{\mu}_{ic}^{(2)}\}, \boldsymbol{\Sigma}^{(2)}) \propto |\boldsymbol{\Sigma}^{(2)}|^{-(1/2)(p_2+1)}. \tag{7.4.17}$$

For the vector of cell probabilities, $\mathbf{q}_i = (q_{i1}, \ldots, q_{im})'$, he adopted a prior density of the Dirichlet form,

$$p(\mathbf{q}_i) \propto \prod_{i=1}^{m} q_{ic}^{a_{ic}-1} \qquad (i = 1,2), \tag{7.4.18}$$

where the a_{ic} are positive constants. As suggested by Vlachonikolis (1990), they may be chosen so that they reflect previous cell frequencies or intuitive impressions about the frequencies of the cells. When no prior information exists about the location cells, we can adopt a vague prior with $a_{ic} = a_i$ ($i = 1,2$) for all $c = 1, \ldots, m$. The choice $a_i = 1/2$ leads to the standard Jeffreys' prior density proportional to

$$\left(\prod_{c=1}^{m} q_{ic} \right)^{-1/2}.$$

This and other approaches are discussed in Vlachonikolis (1990).

With the prior densities (7.4.17) and (7.4.18), the ith group-conditional density of $\mathbf{X} = (\mathbf{X}^{(1)\prime}, \mathbf{X}^{(2)\prime})'$ can be expressed as

$$\hat{f}_i^{(P)}(\mathbf{x}) = \tilde{q}_{ic} f_{St}(\mathbf{x}^{(2)}; n - 2m - p_2 + 1, p_2, \bar{\mathbf{x}}_{ic}^{(2)}, (1 + n_{ic}^{-1}) M \mathbf{S}^{(2)}), \tag{7.4.19}$$

where $M = (n-2m)/(n-2m-p_2+1)$, $c = c(\mathbf{x}^{(1)})$ is defined by (7.4.1), $\mathbf{S}^{(2)}$ is defined by (7.4.9), and where

$$\tilde{q}_{ic} = \left\{ (n_{ic} + a_{ic}) \Big/ \left(n_i + \sum_{k=1}^{m} a_{ik} \right) \right\} \{n_{ic}/(n_{ic}+1)\}^{p_2/2}. \qquad (7.4.20)$$

The last term on the right-hand side of (7.4.19) is the multivariate t-density defined by (3.5.4).

As explained by Vlachonikolis (1990), the predictive estimates of the $\boldsymbol{\mu}_{ic}$ will, like their estimative counterparts, be poor in situations with sparse data. The predictive estimates of the cell probabilities q_{ic} will be less affected by sparse data than the estimative ones, since they are smoothed by the presence of the a_{ic} terms introduced through the Dirichlet prior density. Vlachonikolis (1990) consequently derived the predictive estimates of the $\boldsymbol{\mu}_{ic}^{(2)}$ and $\boldsymbol{\Sigma}^{(2)}$, where the number of parameters under estimation is reduced by the use of the linear model (7.4.10). Again, we let h be the dimension of $\mathbf{u}(\mathbf{x}^{(1)})$ after elimination of terms, say, those of third- or higher-order, in (7.4.11).

For the case of $g = 2$ groups, Vlachonikolis (1990) adopted, following Tiao and Zellner (1964), the vague prior density

$$p(\mathbf{K}_1, \mathbf{K}_2, \boldsymbol{\Sigma}^{(2)}) \propto |\boldsymbol{\Sigma}^{(2)}|^{-(1/2)(p_2+1)}$$

for $\mathbf{K}_1$, $\mathbf{K}_2$, and $\boldsymbol{\Sigma}^{(2)}$. He proceeded to show that the predictive estimate of the ith group-conditional density of $\mathbf{X}$ is

$$\hat{f}_i^{(P)}(\mathbf{x}) = \tilde{q}_{ic} b_{1ic} f_{St}(\mathbf{x}^{(2)}; n-2h-p_2+1, p_2, \hat{\mathbf{K}}_i \mathbf{u}(\mathbf{x}^{(1)}), b_{2ic} M \mathbf{S}^{(2)}), \qquad (7.4.21)$$

where

$$b_{1ic} = |\mathbf{C}_{2i}\mathbf{C}_{3i}|^{-(1/2)p_2} |b_{2ic}(n-2h)\mathbf{S}^{(2)}|^{1/2},$$

$$b_{2ic} = 1 + \mathbf{u}(\mathbf{x}^{(1)})' \mathbf{C}_{2i}^{-1} \mathbf{u}(\mathbf{x}^{(1)}),$$

and

$$\mathbf{C}_{3i} = \mathbf{C}_{2i} + u(\mathbf{x}^{(1)}) u(\mathbf{x}^{(1)})'$$

for $i = 1,2$, and where now $M = (n-2h)/(n-2h-p_2+1)$, and $\mathbf{S}^{(2)}$ is defined by (7.4.12). It can be seen that for $h = m$, (7.4.21) is equal to the predictive density (7.4.19) obtained previously under the full model.

As mentioned above, the estimate (7.4.20) of q_{ic} as obtained under the full model may be satisfactory. Suppose, however, that a reduced model for the q_{ic} is to be fitted. One straightforward approach as proposed by Vlachonikolis (1990) is to adopt a vague prior with

$$a_{ic} = a_i + 1 \qquad (c = 1, \ldots, m)$$

for $i = 1,2$, where the a_i are positive constants. This leads to a posterior density for q_{ic}, proportional to

$$\prod_{c=1}^{m} q_{ic}^{n_{ic}+a_i} \qquad (i = 1,2).$$

A second-order log-linear model can be fitted to the cell frequencies taken as $n_{ic} + a_i$.

Vlachonikolis (1990) performed some simulation results to compare the predictive and estimative approaches to allocation under the location model for mixed binary and continuous feature vectors. He concluded that, except perhaps in those cases in which the binary variables provided no discrimination between the two groups, the predictive rule did not provide lower error rates.

7.5 ERROR RATES OF LOCATION MODEL-BASED RULES

7.5.1 Expressions for Optimal and Conditional Errors ($g = 2$ Groups)

Since conditional on $\mathbf{x}^{(1)}$, the location-model-based discriminant rule (LODR) given by $r_o(\mathbf{x};\boldsymbol{\pi},\mathbf{q},\boldsymbol{\theta}_{LO}^{(2)})$ is linear in the subvector $\mathbf{x}^{(2)}$ containing the continuous feature variables, it is straightforward to give closed expressions for the optimal error rates and for the conditional error rates of its plug-in sample version $r_o(\mathbf{x};\hat{\boldsymbol{\pi}},\hat{\mathbf{q}},\hat{\boldsymbol{\theta}}_{LO}^{(2)})$ in the case of $g = 2$ groups. Corresponding to the definition (3.3.4) of the NLDF, we write $\xi_{ig}(\mathbf{x}^{(2)};\boldsymbol{\theta}_{E,c}^{(2)})$ in the case of $g = 2$ as

$$\begin{aligned}\xi_{ig}(\mathbf{x}^{(2)};\boldsymbol{\theta}_{E,c}^{(2)}) &= \xi(\mathbf{x}^{(2)};\boldsymbol{\theta}_{E,c}^{(2)}) \\ &= \beta_{0E,c}^{o} + \boldsymbol{\beta}_{E,c}'\mathbf{x}^{(2)}, \end{aligned} \qquad (7.5.1)$$

where

$$\beta_{0E,c}^{o} = -\tfrac{1}{2}(\boldsymbol{\mu}_{1c}^{(2)} + \boldsymbol{\mu}_{2c}^{(2)})'\boldsymbol{\Sigma}^{(2)^{-1}}(\boldsymbol{\mu}_{1c}^{(2)} - \boldsymbol{\mu}_{2c}^{(2)}),$$

and

$$\boldsymbol{\beta}_{E,c} = \boldsymbol{\Sigma}^{(2)^{-1}}(\boldsymbol{\mu}_{1c}^{(2)} - \boldsymbol{\mu}_{2c}^{(2)})$$

for $c = 1,\ldots,m$. We let

$$\Delta_c = \{(\boldsymbol{\mu}_{1c}^{(2)} - \boldsymbol{\mu}_{2c}^{(2)})'\boldsymbol{\Sigma}^{(2)^{-1}}(\boldsymbol{\mu}_{1c}^{(2)} - \boldsymbol{\mu}_{2c}^{(2)})\}^{1/2}$$

denote the Mahalanobis distance between G_1 and G_2 in the cth location cell $(c = 1,\ldots,m)$.

We consider the LODR $r_o(\mathbf{x};k,\mathbf{q},\boldsymbol{\theta}_{LO}^{(2)})$ with a general cutoff point k, that is, $\log(\pi_1/\pi_2)$ is replaced by $-k$ in (7.4.5). Then by proceeding conditionally on $\mathbf{x}^{(1)}$, or, equivalently, the location cell c, the error rate of the LODR with respect to G_i is given by

$$eo_{iLO} = \sum_{c=1}^{m} q_{ic} eo_i(k_c, \Delta_c) \qquad (i = 1,2),$$

where $eo_i(k_c, \Delta_c)$ is the optimal error rate of the NLDF $\xi(\mathbf{x}^{(2)}; \boldsymbol{\theta}_{E,c}^{(2)})$ applied with cutoff point

$$k_c = k - \log(q_{1c}/q_{2c}) \tag{7.5.2}$$

with respect to G_1 and G_2, separated by a Mahalanobis distance $\Delta_c (c = 1,\ldots, m)$. That is,

$$eo_i(k_c, \Delta_c) = \Phi[\Delta_c^{-1}\{(-1)^{i+1}k_c - \tfrac{1}{2}\Delta_c^2\}] \qquad (c = 1,\ldots,m).$$

In a similar manner, we can obtain expressions for the conditional error rates of the plug-in sample version of the LODR, using the naive estimates (7.4.6) to (7.4.8) obtained with the fully saturated version of the location model. Corresponding to equations (4.2.7) to (4.2.9) defining the sample NLDF $\xi(\mathbf{x}; \hat{\boldsymbol{\theta}}_E)$, we let

$$\xi(\mathbf{x}^{(2)}; \hat{\boldsymbol{\theta}}_{E,c}) = \hat{\beta}_{0E,c}^{o} + \hat{\boldsymbol{\beta}}_{E,c}' \mathbf{x}^{(2)},$$

where

$$\hat{\beta}_{0E,c}^{o} = -\tfrac{1}{2}(\bar{\mathbf{x}}_1^{(2)} + \bar{\mathbf{x}}_2^{(2)})' \mathbf{S}^{(2)^{-1}} (\bar{\mathbf{x}}_1^{(2)} - \bar{\mathbf{x}}_2^{(2)}),$$

and

$$\hat{\boldsymbol{\beta}}_{E,c} = \mathbf{S}^{(2)^{-1}} (\bar{\mathbf{x}}_1^{(2)} - \bar{\mathbf{x}}_2^{(2)})$$

for $c = 1,\ldots,m$. Then the conditional error rate with respect to G_i of the naive sample version $r_o(\mathbf{x}; k, \hat{\mathbf{q}}, \hat{\boldsymbol{\theta}}_{LO}^{(2)})$ of the LODR with cutoff point k is given by

$$ec_{iLO} = \sum_{c=1}^{m} q_{ic} ec_i(\hat{k}_c, \boldsymbol{\theta}_{E,c}^{(2)}; \hat{\boldsymbol{\theta}}_{E,c}^{(2)}),$$

where $ec_i(\hat{k}_c, \boldsymbol{\theta}_{E,c}^{(2)}; \hat{\boldsymbol{\theta}}_{E,c}^{(2)})$ denotes the ith group-specific conditional error rate of $\xi(\mathbf{x}^{(2)}; \hat{\boldsymbol{\theta}}_{E,c}^{(2)})$ applied with cutoff point $\hat{k}_c$ with respect to G_1 and G_2, and where $\hat{k}_c$ denotes k_c with q_{ic} replaced by $\hat{q}_{ic}$ $(i = 1,2)$ in (7.5.2). A closed expression for $ec_i(\hat{k}_c, \boldsymbol{\theta}_{E,c}^{(2)}; \hat{\boldsymbol{\theta}}_{E,c}^{(2)})$ is available from (4.3.5).

7.5.2 Asymptotic and Empirical Results for Unconditional Errors

In the case of separate sampling, the unconditional error rates of the sample LODR $r_o(\mathbf{x}; \hat{\mathbf{q}}, \hat{\boldsymbol{\theta}}_{LO}^{(2)})$ are given by

$$eu_{iLO} = E\{ec_{iLO} \mid n_1, n_2\} \qquad (i = 1,2).$$

By adopting a similar approach to that used by Okamoto (1963) and described in Section 4.2, Vlachonikolis (1985) obtained under the location model (7.4.2) for $g = 2$ groups the second-order expansion $eu_{iLO}^{(2)}$ of eu_{iLO} for $i = 1, 2$. The principal term of order $O(1)$ in the expansion $eu_{iLO}^{(2)}$ is eo_{iLO}, because $eu_{iLO} \rightarrow eo_{iLO}$, as $n_1, n_2 \rightarrow \infty$ $(i = 1, 2)$. In this and a later study (Vlachonikolis, 1986), he performed some simulation experiments to assess the reliability of these asymptotic expansions in situations where n is not very large.

In both of these studies, the asymptotic and simulated values of the unconditional error rates of the sample LODR were examined for combinations of the parameters similar to those chosen by Krzanowski (1977) in his earlier investigation of the optimal error rates. In the Monte Carlo study of Vlachonikolis (1986), the simulations were conducted for a total of 324 combinations with $p_1 = 2, 3$; $p_2 = 1, 3, 5$; $n_1 = n_2 = 50, 100, 200$; and with $q_{ic} = q_i$ $(c = 1, \ldots, m)$, where q_i was chosen in the range $0.3 \leq q_i \leq 0.7$ $(i = 1, 2)$. The p_1 binary feature variables were taken to be independent. Three different settings with respect to the cell Mahalanobis distances Δ_c were considered: (a) $\Delta_c^2 = 1$; (b) Δ_c^2 equal to the mean of the cth-order statistic of a sample of size c from the standard half-normal distribution; (c) $\Delta_c^2 = 1$ $(c = 1, \ldots, \frac{1}{2}m)$ and 2 $(c = \frac{1}{2}m + 1, \ldots, m)$. For those combinations of the parameters with $p_1 = 3$ binary variables, a reduced version of the LODR was formed by fitting a second-order log-linear model to the cell probabilities and a second-order regression model, in the original binary variables, to the means of the continuous feature variables. As the group-prior probabilities were taken to be equal, the unconditional overall error rate is given by

$$eu_{LO} = \frac{1}{2} \sum_{i=1}^{2} eu_{iLO},$$

and its second-order asymptotic approximation by

$$eu_{LO}^{(2)} = \frac{1}{2} \sum_{i=1}^{2} eu_{iLO}^{(2)}.$$

A comparison of the asymptotic and empirical values of the unconditional overall error rate revealed that $eu_{LO}^{(2)}$ is greater than the corresponding simulated value in each instance. Hence, it is apparent that $eu_{LO}^{(2)}$ provides a conservative assessment of the exact unconditional error rate eu_{LO}. Overall, it was found that at least for moderately sized training samples, $eu_{LO}^{(2)}$ provides a good approximation to the unconditional overall error rate of not only of the fully saturated version of the sample LODR, but also of its reduced version.

Concerning the behavior of the true unconditional overall error rate eu_{LO}, it was noted that both asymptotic and empirical assessments exhibited the following patterns across the combinations of the parameters specified.

1. They increase as p_2 increases, although some exceptions were noted for combinations with small n;
2. they both decrease as the Δ_c increase, for example, situation (c) against (a), as above;
3. for large n, they both decrease as p_1 increases, except for combinations with $q_1 = q_2$ and certain combinations with larger values of p_2.

It was noticed that these patterns are similar to those for the optimal overall error rate, apart from (1), of course, because the latter does not depend on p_2.

Balakrishnan, Kocherlakota, and Kocherlakota (1986) have derived exact expressions for the unconditional error rates of the LODR in the special case of $g = 2$, $p_1 = p_2 = 1$, and $q_{11} = q_{21}$. The difference between the group-conditional means of the continuous variable X_2 was taken to be the same within each of the two location cells (that is, $\mu_{21}^{(2)} - \mu_{11}^{(2)} = \mu_{22}^{(2)} - \mu_{12}^{(2)}$), but the group-conditional variance of X_2 was allowed to be different within each of the two location cells, although still common to each group. For the same case, Balakrishnan, Kocherlakota, and Kocherlakota (1988) have obtained asymptotic expressions for the unconditional error rates in situations in which the group-conditional distributions of $X^{(2)}$ are not univariate normal. They considered nonnormal distributions represented by the truncated normal and normal mixtures (contamination model) in the course of their investigation of the robustness of the LODR. Also for the same case, Tiku, Balakrishnan, and Amagaspitiya (1989) have used asymptotic and simulative methods to investigate the error rates of the robust sample version of the LODR by Balakrishnan and Tiku (1988), using modified maximum likelihood (MML) estimates. Various nonnormal group-conditional distributions were adopted. The extension to $p_2 = 2$ continuous variables was considered.

Previously, Tu and Han (1982) considered the error rates of a sample version of the LODR in the special case of $g = 2$ groups and a single ($p_1 = 1$) binary feature variable. Plug-in estimates were obtained by a double-inverse sampling scheme, whereby feature observations are recorded sequentially at random and sampling is continued until $n_{ic} \geq n_{ico}$, where n_{ico} is a constant integer greater than p_2. This is to ensure the sample covariance matrix for each of the two location cells is nonsingular. For this sample version of the LODR, Tu and Han (1982) obtained asymptotic expansions of its error rates in the case in which the binary variable is independent of the continuous variables. The error rates in the dependent case were studied using simulations.

7.6 ADJUSTMENTS TO SAMPLE NLDR FOR MIXED FEATURE DATA

7.6.1 Linear Transformations

Given its computational simplicity, it is not surprising that the sample NLDR $r_o(\mathbf{x}; \hat{\boldsymbol{\Psi}}_E)$ is still widely used in situations in which the full feature vector obviously does not have a multivariate normal group-conditional distribution, such as with mixed feature variables. The robustness of the sample NLDR to departures from normality has been reported in Section 5.6, where it was seen that it is not robust to dissimilar interaction structures for the feature variables within each group. In this section, we consider the use of linear transformations of the continuous feature variables to improve the performance of the sample NLDR with mixed feature data under the location model.

In the previous chapter, we considered linear transformations of the feature vector $\mathbf{x}$ in the case in which all its components are continuous. In the case in which all the feature variables are binary, there is a transformation analogous to a principal component analysis, as discussed by Bloomfield (1974). The transformed binary variables, which are linear functions of the original ones, are such that their frequencies of occurrence can be described by a log-linear model containing as few interaction terms as possible.

In order to apply the aforementioned techniques in the present situation of mixed feature variables, we either have to treat the binary features as continuous or dichotomize the continuous variables. The former step can result in transformed variables that are not readily interpretable, and the latter may result in loss of information. This led Krzanowski (1979a) to consider for $g = 2$ groups linear transformations that are directly applicable to mixed feature variables and so which pay due regard to the inherent structure of the feature data.

As explained in Krzanowski (1979a), in order for the sample NLDR $r_o(\mathbf{x}; \hat{\boldsymbol{\Psi}}_E)$ to perform well relative to the sample LODR $r_o(\mathbf{x}; \hat{\boldsymbol{\pi}}, \hat{\mathbf{q}}, \hat{\boldsymbol{\theta}}_{LO}^{(2)})$ under the location model, the means of the continuous feature variables conditional on the binary features should be as homogeneous as possible within groups. The homogeneity of those conditional means is equivalent to zero correlations between binary and continuous feature variables. This can be seen from the result that under the location model (7.4.2), the covariance between the uth binary variable in $\mathbf{X}^{(1)}$ and the vth continuous variable in $\mathbf{X}^{(2)}$ in group $G_i (i = 1,2)$ is given by

$$\operatorname{cov}\{(\mathbf{X}^{(1)})_u, (\mathbf{X}^{(2)})_v\} = \sum q_{ic}(\mu_{ivc}^{(2)} - \overline{\mu}_{iv}^{(2)}) \tag{7.6.1}$$

for $u = 1, \ldots, p_1$ and $v = 1, \ldots, p_2$, where $c = c(\mathbf{x}^{(1)})$ is as defined by (7.4.1), the summation in (7.6.1) is over all values of $\mathbf{x}^{(1)}$ with $(\mathbf{x}^{(1)})_u = 1$,

$$\mu_{ivc}^{(2)} = E\{(\mathbf{X}^{(2)})_v \mid \mathbf{x}^{(1)}, \mathbf{X} \in G_i\},$$

and where

$$\overline{\mu}_{iv}^{(2)} = \sum_{c=1}^{m} q_{ic}\mu_{ivc}^{(2)}.$$

Krzanowski (1979a) suggested transforming the continuous feature variables as $\mathbf{a}'\mathbf{x}^{(2)}$, where $\mathbf{a}$ is chosen to minimize

$$\mathbf{a}'\mathbf{H}_1\mathbf{a}/\mathbf{a}'\boldsymbol{\Sigma}^{(2)}\mathbf{a},$$

where

$$\mathbf{H}_1 = \sum_{i=1}^{2} \sum_{c=1}^{m} (\boldsymbol{\mu}_{ic}^{(2)} - \overline{\boldsymbol{\mu}}_i^{(2)})(\boldsymbol{\mu}_{ic}^{(2)} - \overline{\boldsymbol{\mu}}_i^{(2)})',$$

and where

$$\boldsymbol{\mu}_{ic}^{(2)} = (\mu_{i1c}^{(2)}, \ldots, \mu_{ip_2c}^{(2)})',$$

and

$$\overline{\boldsymbol{\mu}}_i^{(2)} = (\overline{\mu}_{i1}^{(2)}, \ldots, \overline{\mu}_{ip_2}^{(2)})'$$

for $i = 1, 2$. By analogy with canonical variate analysis, the solution is given by the eigenvector vector corresponding to the smallest eigenvalue of $\boldsymbol{\Sigma}^{(2)^{-1}}\mathbf{H}_1$.

Another suggestion of Krzanowski (1979a) for the choice of $\mathbf{a}$ is to take $\mathbf{a}$ so as to minimize

$$\mathbf{a}'\mathbf{H}_1\mathbf{a}/\mathbf{a}'\mathbf{H}_2\mathbf{a},$$

where

$$\mathbf{H}_2 = \sum_{c=1}^{m} (\boldsymbol{\mu}_{1c}^{(2)} - \boldsymbol{\mu}_{2c}^{(2)})(\boldsymbol{\mu}_{1c}^{(2)} - \boldsymbol{\mu}_{2c}^{(2)})'.$$

Krzanowski (1979a) noted that a slight adjustment to the above analyses will ensure that the variables in $\mathbf{a}'\mathbf{X}^{(2)}$ are uncorrelated unconditionally, although it may disturb their correlation structure with the binary variables. The adjustment is to use the weighted between-cell matrix,

$$\mathbf{H}_3 = \sum_{i=1}^{2} \sum_{c=1}^{m} q_{ic} (\boldsymbol{\mu}_{ic}^{(2)} - \overline{\boldsymbol{\mu}}_i^{(2)})(\boldsymbol{\mu}_{ic}^{(2)} - \overline{\boldsymbol{\mu}}_i^{(2)})',$$

in place of $\mathbf{H}_1$. In practice, the matrices $\mathbf{H}_1$, $\mathbf{H}_2$, $\mathbf{H}_3$, and $\boldsymbol{\Sigma}$ are unknown and must be replaced in the analyses by their sample analogues computed from the training data $\mathbf{t}$.

Krzanowski (1979a) has given the likelihood ratio test for the effectiveness of the discriminant rule based on the location model over the sample NLDR. As pointed out by Krzanowski (1979a), this test requires estimation to be performed under the location model, and so may not be appropriate in situations in which the aim is to reduce the computational effort through the possible use of the sample NLDR. However, the test would be worthwhile if the discriminant rule were being designed for extensive future use.

7.6.2 Augmenting the Sample NLDF

Another approach to the problem of improving the performance of the sample NLDF in its application to mixed feature data is to augment the original feature vector with appropriate products of the binary and continuous variables. Before proceeding to show how this can be effected, we consider briefly the applicability of the original form $\xi(\mathbf{x};\hat{\boldsymbol{\theta}}_E)$ of the sample NLDF under the location model. Vlachonikolis and Marriott (1982) have given a lucid account of this topic.

It can be seen from (7.5.1) that the sample LODR $r_o(\mathbf{x};k,\hat{\mathbf{q}},\hat{\boldsymbol{\theta}}_{LO}^{(2)})$ effects an allocation of an entity with mixed feature vector by fitting a separate hyperplane

$$\hat{\boldsymbol{\beta}}_{E,c}'\mathbf{x}^{(2)} \tag{7.6.2}$$

in the region of the continuous feature variables. These hyperplanes (7.6.2) are in general nonparallel, their direction being different for each location cell c, that is, for each distinct realization $\mathbf{x}^{(1)}$. Also, for the cutoff point k, they have different positions, as determined by

$$\hat{k}_c - \hat{\beta}_{0E,c}^o$$

for the cth location cell, where k_c is given by (7.5.2). On the other hand with the use of the sample NLDF $\xi(\mathbf{x};\hat{\boldsymbol{\theta}}_E)$, allocation is effected by using the same hyperplane

$$\sum_{v=p_1+1}^{p} \hat{\beta}_{vE}x_v^{(2)}$$

in all location cells, where

$$\hat{\beta}_{vE} = (\hat{\boldsymbol{\beta}}_E)_v \qquad \text{for} \quad v = 1,\ldots,p.$$

Its position is specified by

$$k - \hat{\beta}_0^o - \sum_{v=1}^{p_1} \hat{\beta}_{vE}x_v^{(1)},$$

that is, by a weighted sum of the binary feature variables. When this additivity in the binary feature variables and the implied assumption of parallelism of the hyperplanes in the continuous feature variables do not apply, that is, when the interactions between the binary features and between the binary and continuous features are different for each group, the sample NLDF will not perform well.

Knoke (1982), following the approach of Truett, Cornfield, and Kannel (1967), and Vlachonikolis and Marriott (1982) have proposed various modifications of the sample NLDF $\xi(\mathbf{x};\hat{\boldsymbol{\theta}}_E)$ in its application to mixed feature data from $g = 2$ groups. Their approach is to augment the feature vector $\mathbf{x}$ with appropriate products of the feature variables to allow for interactions between the binary features and between the binary and continuous features that are not common to both groups. Suppose the p_1 binary variables in $\mathbf{x}^{(1)}$ are replaced by $m = 2^{p_1}$ variables $w_1,\ldots,w_m$, where $w_u = \delta_{uc}$, and c is determined from $\mathbf{x}^{(1)}$ by (7.4.1). To avoid linear dependence between these variables w_u, only $m-1$ of them, say, $w_1,\ldots,w_{m-1}$, should be used. In order to allow for interactions between the binary feature variables, the sample NLDF is formed with the feature vector augmented by the inclusion of the $m-1$ variables in

$$\mathbf{w} = (w_1, \ldots, w_{m-1})'.$$

Let

$$\mathbf{x}^+ = (\mathbf{w}', \mathbf{x}^{(2)\prime})'$$

be the feature vector so augmented, and let $\mathbf{x}_j^+$ be the jth observation on $\mathbf{x}^+$ corresponding to $\mathbf{x}_j$ $(j = 1, \ldots, n)$. The sample NLDF formed from $\mathbf{x}_1^+, \ldots, \mathbf{x}_n^+$ is given by

$$\xi(\mathbf{x}^+; \hat{\boldsymbol{\theta}}_E^+) = \{\mathbf{x}^+ - \tfrac{1}{2}(\overline{\mathbf{x}}_1^+ + \overline{\mathbf{x}}_2^+)\}'(\mathbf{S}^+)^{-1}(\overline{\mathbf{x}}_1^+ - \overline{\mathbf{x}}_2^+),$$

where

$$\overline{\mathbf{x}}_i^+ = \sum_{j=1}^{n} z_{ij}\mathbf{x}_j^+ / n_i \qquad (i = 1, 2),$$

and

$$\mathbf{S}^+ = (n_1 + n_2 - 2)^{-1} \sum_{i=1}^{2} \sum_{j=1}^{n_i} z_{ij}(\mathbf{x}_j^+ - \overline{\mathbf{x}}_i^+)(\mathbf{x}_j^+ - \overline{\mathbf{x}}_i^+)'.$$

Here $\hat{\boldsymbol{\theta}}_E^+$ denotes $\hat{\boldsymbol{\theta}}_E$ with $\overline{\mathbf{x}}_1$, $\overline{\mathbf{x}}_2$, and $\mathbf{S}$ replaced by $\overline{\mathbf{x}}_1^+$, $\overline{\mathbf{x}}_2^+$, and $\mathbf{S}^+$, respectively.

With the use of this augmented sample NLDF $\xi(\mathbf{x}^+; \hat{\boldsymbol{\theta}}_E^+)$, the same hyperplane is still used in each location cell, but now its position is determined separately for each cell. The sample NLDF $\xi(\mathbf{x}^+; \hat{\boldsymbol{\theta}}_E^+)$ can be augmented further by letting $\mathbf{x}^+$ consist of the elements of $\mathbf{w}$, $\mathbf{x}^{(2)}$, and all products of the form $w_u x_v^{(2)}$ $(u = 1, \ldots, m-1;\ v = 1, \ldots, p_2)$. The latter products are to allow for interactions between the binary and continuous feature variables. As $\mathbf{x}^+$ is now of dimension $m(p_2 + 1) - 1$, the augmented sample NLDF $\xi(\mathbf{x}^+; \hat{\boldsymbol{\theta}}_E^+)$ so obtained is now analogous to the sample location model-based discriminant function in that it fits a separate hyperplane at each location cell. It is more general than the location model in that it does not assume the covariances of the continuous variables to be the same within each cell, although this homoscedasticity can be easily incorporated into the estimation process. Under the location model, it is assumed that the common covariance matrix of the continuous feature variables within each location cell is also common to each group. If this assumption is relaxed, it can be handled by forming the sample NLDF with the feature vector augmented with also quadratic terms in the continuous feature variables. As remarked in Section 5.6.3, Schmitz et al. (1985) have concluded that interactions between the continuous feature variables have a greater impact on the performance of the sample NLDF than the interactions involving the binary feature variables.

As with the location model, this approach of augmenting the sample NLDF involves too many parameters unless p_1 is small. Vlachonikolis and Marriott (1982) proposed that this problem be tackled in a similar manner to that with the location model. They noted that $(w_1,\ldots,w_{m-1})'$ has an equivalent representation as

$$(x_1^{(1)},\ldots,x_{p_1}^{(1)}, x_1^{(1)}x_2^{(1)},\ldots,x_{p_1-1}^{(1)}x_{p_1}^{(1)}, x_1^{(1)}x_2^{(1)}x_3^{(1)},\ldots,x_1^{(1)}x_2^{(1)}\ldots x_{p_1}^{(1)})'.$$

This suggests the formation of a reduced sample version of the NLDF, where only terms corresponding to main effects and interactions up to a certain order are retained. Thus, to allow for first-order interactions between the binary variables and also between the continuous variables and the binary features taken both separately and in pairs, the augmented sample NLDF can be formed by taking the feature vector $\mathbf{x}^+$ to consist of $\mathbf{x}$ and all products of the form $x_r^{(1)}x_s^{(1)}$, $x_r^{(1)}x_v^{(2)}$, and $x_r^{(1)}x_s^{(1)}x_v^{(2)}$ $(r \neq s = 1,\ldots,p_1;\ v = 1,\ldots,p_2)$. This reduction in the number of parameters to be fitted may not be sufficient or it may not be applicable. There may be some higher-order interactions that have a significant effect on allocation. In such circumstances, the feature vector can be augmented by stepwise selection. There are major statistical packages (Section 3.3.3) for the fitting of the sample NLDF, in a stepwise manner if so required. Indeed, this is the main advantage of this approach of augmenting the sample NLDF. It is easily implemented in practice and so provides a quick way of identifying those interaction terms that are needed in the provision of a satisfactory allocation rule. Hence, it is quite useful, whether in the role of improving the allocatory performance of the sample NLDF or in the preliminary screening of the interaction terms to be retained in the fitting of a reduced version of the location model.

In Section 5.6.3, it was demonstrated in a simple case of the location model in which a single binary feature variable has a common interaction structure with the continuous variables in each group that the use of the sample NLDF does not yield consistent estimates of all the discriminant function coefficients. Similarly, in the presence of unequal interaction structures within the groups, the fully augmented sample NLDF will not, in general, provide consistent estimates of the discriminant function coefficients, and, hence, of the posterior probabilities of group membership under the location model. Thus, it is stressed that the analogy of the fully augmented sample NLDF with the location-model-based approach is limited only to the allocation problem. This is not surprising as it has been seen that the use of the augmented sample NLDF has been motivated by ad hoc considerations. For instance, it is formed as if the augmented feature vector $\mathbf{x}^+$ has a multivariate normal distribution with a common covariance matrix in each group, although these assumptions are obviously not met. If consistent estimation is required, then the location-model-based approach provides a parametric way of proceeding. A semiparametric approach using logistic discrimination and nonparametric approaches to this problem are described in the next two chapters. Irrespective of what approach is adopted to the estimation problem, the augmented sample NLDF can still be of value initially in providing a quick and cheap way computationwise of identifying those terms to be included either in the reduced form of the location model or in the logistic formulation.

7.7 SOME NONNORMAL MODELS FOR CONTINUOUS FEATURE DATA

7.7.1 Introduction

Cooper (1962a, 1962b, 1963, 1965) and Day (1969) have investigated the applicability of linear discriminant rules for nonnormal group-conditional distributions. Day (1969) showed that the optimal discriminant rule is linear in the feature vector $\mathbf{x}$ for group-conditional distributions in any linear exponential family. This result can be generalized.

Suppose that the ith group-conditional density is a member of the family of p-dimensional elliptically symmetric densities

$$|\boldsymbol{\Sigma}_i|^{-1/2} f_S[\{\delta(\mathbf{x}, \boldsymbol{\mu}_i; \boldsymbol{\Sigma}_i)\}^{1/2}], \tag{7.7.1}$$

where, as previously, $f_S(\|\mathbf{x}\|)$ is any spherically symmetric density function. Also, as before,

$$\delta(\mathbf{x}, \boldsymbol{\mu}_i; \boldsymbol{\Sigma}_i) = (\mathbf{x} - \boldsymbol{\mu}_i)' \boldsymbol{\Sigma}_i^{-1} (\mathbf{x} - \boldsymbol{\mu}_i).$$

Then for group-conditional distributions of the form (7.7.1) with equal prior probabilities, the optimal rule is linear if the $\boldsymbol{\Sigma}_i$ are all equal and f_S is strictly monotonically decreasing (Glick, 1976).

For unequal $\boldsymbol{\Sigma}_i$, Cooper (1963) considered some distributions of the type (7.7.1), for which the optimal discriminant rule is of the same form as the NQDR. These particular distributions are multivariate extensions of the Pearson Types II and VII distributions. Cooper (1965) studied the likelihood ratio statistics for these group-conditional distributions.

As discussed in Section 2.11, one parametric approach to discriminant analysis when normality does not appear to be a tenable assumption for a group-conditional distribution is to model it as a finite mixture of, for example, multivariate normal components. An example of this is presented shortly in Section 7.8. But, first, we briefly report some other nonnormal models that have been suggested for parametric discrimination.

7.7.2 θ-Generalized Normal Model

Chhikara and Odell (1973) proposed a parametric family of multivariate density functions known as r-normed exponential densities for the group-conditional distributions in discriminant analysis. Goodman and Kotz (1973) investigated the same family of densities and labeled it the family of θ-generalized normal densities.

The family of univariate θ-generalized normal densities can be represented as

$$f(x; \theta, \mu, a) = \{2a\Gamma(1 + \theta^{-1})\}^{-1} \exp\{-|(x - \mu)/a|^{\theta}\}, \tag{7.7.2}$$

for all real x, where a, θ ($a > 0, \theta > 0$), and μ are fixed real numbers. This family becomes the Laplace for $\theta = 1$, the normal for $\theta = 2$, and approaches the uniform on $(\mu - a, \mu + a)$ as θ tends to infinity. As θ tends to zero, (7.7.2) defines an improper uniform distribution over the whole real line. It can be verified for any $\theta > 0$ that if U is distributed according to a gamma distribution with parameter $1/\theta$, then $U^{1/\theta}$ has the density given by (7.7.2), appropriately modified by being restricted to the positive half of the real line.

The multivariate θ-generalized normal density is given by

$$f(\mathbf{x};\theta,\boldsymbol{\mu},\mathbf{A}) = K(p,\theta,\mathbf{A})\exp\left\{-\sum_{k=1}^{p}|\mathbf{B}_k(\mathbf{x}-\boldsymbol{\mu})|^{\theta}\right\}, \tag{7.7.3}$$

where $\mathbf{A}$ is a $p \times p$ nonsingular matrix, $\boldsymbol{\mu} = (\mu_1,\ldots,\mu_p)'$, $\theta > 0$, and

$$K(p,\theta,\mathbf{A}) = [\{2\Gamma(1+\theta^{-1})\}^p|\mathbf{A}|]^{-1},$$

and where $\mathbf{B}_k$ is the kth row of $\mathbf{B} = \mathbf{A}^{-1}$. The covariance matrix $\boldsymbol{\Sigma}$ of this p-dimensional random vector is given by

$$\boldsymbol{\Sigma} = c(\theta)\mathbf{A}\mathbf{A}',$$

where

$$c(\theta) = \Gamma(3/\theta)/\Gamma(1/\theta).$$

For $\theta = 2$, (7.7.3) reduces to the p-variate normal distribution with mean $\boldsymbol{\mu}$ and covariance matrix $\boldsymbol{\Sigma}$.

7.7.3 Exponential Model

The exponential distribution appears as a natural model in problems of life testing and reliability analysis. For this model, the ith group-conditional density of the feature variable X is given by

$$f_i(x;\mu_i) = \mu_i^{-1}\exp(-x/\mu_i)I_{(0,\infty)}(x) \qquad (i = 1,\ldots,g),$$

where the indicator function $I_{(0,\infty)}(x) = 1$ for $x > 0$ and zero elsewhere. The implied optimal discriminant rule is linear in x, because

$$\log\{f_i(x;\mu_i)/f_g(x;\mu_g)\} = (\mu_g^{-1} - \mu_i^{-1})x - \log(\mu_i/\mu_g) \qquad (i = 1,\ldots,g-1). \tag{7.7.4}$$

In the case of the group means μ_i being unknown, a sample discriminant rule can be formed by plugging into (7.7.4) the maximum likelihood estimate of μ_i, given by the sample mean $\bar{\mathbf{x}}_i$ of the training feature data from group G_i $(i = 1,\ldots,g)$. The use of the optimal discriminant rule and its sample plug-in version for exponential group-conditional distributions has been considered by Bhattacharya and Das Gupta (1964) and Basu and Gupta (1974), among others.

7.7.4 Inverse Normal Model

The inverse normal distribution is frequently used in applications involving, for example, Brownian motion, sequential analysis, reliability theory, and electronics. It is a good representative of a positively skewed long-tailed distribution and is often recommended as an alternative to the log normal distribution; see Folks and Chhikara (1978) and Chhikara and Folks (1989). As reported in Section 5.6.1, Amoh and Kocherlakota (1986) modeled the group-conditional distributions of the feature variable as inverse normal in their study of the robustness of the sample NLDR under departures from normality.

Let μ_i and σ_i^2 denote the mean and variance, respectively, of the univariate feature variable X in group G_i $(i = 1,\ldots,g)$. Also, put

$$\lambda_i = \mu_i^3/\sigma_i^2,$$

where it is supposed that $\mu_i > 0$ $(i = 1,\ldots,g)$. Then with the group-conditional distributions of X modeled by the family of inverse normal distributions, the ith group-conditional density of X is given by

$$f_i(x;\boldsymbol{\theta}_i) = \left(\frac{\lambda_i}{2\pi x^3}\right)^{1/2} \exp\left\{-\frac{\lambda_i(x-\mu_i)^2}{2\mu_i^2 x}\right\} I_{(0,\infty)}(x) \qquad (i = 1,\ldots,g), \tag{7.7.5}$$

where $\boldsymbol{\theta}_i = (\mu_i, \lambda_i)'$. For small values of λ_i this distribution is highly positively skewed. As λ_i tends to infinity, the distribution approaches the normal.

Amoh and Kocherlakota (1986) have derived the n^{-1}-order asymptotic expansions of the unconditional error rates of the sample plug-in version of the optimal discriminant rule in the special case of $\pi_1 = \pi_2 = 0.5$ and $\lambda_1 = \lambda_2 = \lambda$, where λ is known. More recently, El Khattabi and Streit (1989) have derived the asymptotic unconditional error rates of the sample plug-in rule in the special case of $\pi_1 = \pi_2 = 0.5$ and $\mu_1 = \mu_2 = \mu$, where μ is known.

We let $\eta(x;\boldsymbol{\Psi})$ denote the posterior log odds for $g = 2$ groups, so that

$$\eta(x;\boldsymbol{\Psi}) = \log\{\tau_1(x;\boldsymbol{\Psi})/\tau_2(x;\boldsymbol{\Psi})\}, \tag{7.7.6}$$

where

$$\boldsymbol{\Psi} = (\pi_1, \pi_2, \mu_1, \lambda_1, \mu_2, \lambda_2)'.$$

Then, on substituting the inverse normal form (7.7.5) for $f_i(x;\theta_i)$ in (7.7.6), we have that

$$\eta(x;\boldsymbol{\Psi}) = \log(\pi_1/\pi_2) + x^{-1}(\alpha_2 x^2 + \alpha_1 x + \alpha_0), \tag{7.7.7}$$

where

$$\alpha_2 = \tfrac{1}{2}(\lambda_2\mu_2^{-2} - \lambda_1\mu_1^{-2}),$$

$$\alpha_1 = (\lambda_1\mu_1^{-1} - \lambda_2\mu_2^{-1}) + \tfrac{1}{2}\log(\lambda_1/\lambda_2),$$

and

$$\alpha_0 = \tfrac{1}{2}(\lambda_2 - \lambda_1).$$

It can be seen from (7.7.7) that the optimal discriminant rule is based on the quadratic discriminant function

$$\alpha_2 x^2 + \{\alpha_1 + \log(\pi_1/\pi_2)\}x + \alpha_0.$$

This reduces to a linear discriminant function if $\alpha_2 = 0$, that is, if

$$\lambda_1/\lambda_2 = \mu_1^2/\mu_2^2.$$

In the case of unknown μ_i and λ_i, a sample discriminant rule can be formed by plugging in their maximum likelihood estimates from the training data,

$$\hat{\mu}_i = \bar{\mathbf{x}}_i$$

and

$$\hat{\lambda}_i = \left\{\sum_{j=1}^{n} z_{ij}(x_j^{-1} - \bar{\mathbf{x}}_i^{-1})/n_i\right\}^{-1}$$

for $i = 1,2$.

7.7.5 Multivariate *t*-Model

Recently, Sutradhar (1990) used the classical multivariate t-density to model the group-conditional distributions in situations in which they may have tails longer than those of the multivariate normal. If the ith group-conditional density $f_i(\mathbf{x};\boldsymbol{\theta}_i)$ is modeled as the p-dimensional t-density $f_{St}(\mathbf{x};m,p,\boldsymbol{\mu}_i,\boldsymbol{\Sigma})$ with m degrees of freedom, as given by (3.5.4), then it has mean $\boldsymbol{\mu}_i$ and covariance matrix $\{m/(m-2)\}\boldsymbol{\Sigma}$ $(i = 1,\ldots,g)$. As the degrees of freedom m tends to infinity, this t-density tends to the multivariate normal density with mean $\boldsymbol{\mu}_i$ and covariance matrix $\boldsymbol{\Sigma}$.

Sutradhar (1990) actually modeled $f_i(\mathbf{x};\boldsymbol{\theta}_i)$ as

$$f_i(\mathbf{x};\boldsymbol{\theta}_i) = (m-2)^{(1/2)m}c(m,p)|\boldsymbol{\Sigma}|^{-1/2}\{(m-2) + \delta(\mathbf{x},\boldsymbol{\mu}_i;\boldsymbol{\Sigma})\}^{-(1/2)(m+p)}$$
$$(i = 1,\ldots,g), \qquad (7.7.8)$$

where the constant term $c(m,p)$ is as defined in Section 3.5.1. This was achieved after a reparameterization so that the common group-covariance matrix does not depend on the degrees of freedom m. Here $\boldsymbol{\theta}_i$ consists of the elements of $\boldsymbol{\mu}_i$, the distinct elements of $\boldsymbol{\Sigma}$, and the degrees of freedom m $(i = 1,\ldots,g)$. Also, Sutradhar (1990) gave a model for the joint density of the n_i training observations from G_i in the situation in which they are not independent, but are pairwise uncorrelated with marginal density (7.7.8).

TABLE 7.1 Common Value of the Group-Specific Error Rates of the Optimal Rule Based on Group-Conditional Multivariate t-Densities in the Case of Equal Group-Prior Probabilities

	m						
Δ	3	4	5	6	8	10	∞
0	0.5000	0.5000	0.5000	0.5000	0.5000	0.5000	0.5000
1	0.2251	0.2593	0.2375	0.2814	0.2897	0.2942	0.3085
2	0.0908	0.1151	0.1266	0.1333	0.1408	0.1449	0.1587
3	0.0403	0.0506	0.0553	0.0579	0.0607	0.0622	0.0668
4	0.0114	0.0121	0.0117	0.0111	0.0101	0.0095	0.0062
7	0.0045	0.0039	0.0032	0.0026	0.0019	0.0015	0.0002

Source: From Sutradhar (1990), with permission from the Biometric Society.

For $g = 2$ groups, the error rates of the optimal discriminant rule $r_o(\mathbf{x};\mathbf{\Psi})$ are easily computed, as demonstrated by Sutradhar (1990). In the case of equal group-prior probabilities, both group-specific error rates are given by

$$eo_i(\mathbf{\Psi}) = (m-2)^{(1/2)m} c(m,1) \int_{-\infty}^{-(1/2)\Delta} \{(m-2)+u^2\}^{-(1/2)(m+1)} du \qquad (i = 1,2).$$

Their common values are displayed in Table 7.1 for selected combinations of the Mahalanobis distance Δ and m, along with the values of $\Phi(-\frac{1}{2}\Delta)$, the error rate of the NLDR, which corresponds to $m = \infty$. It can be seen that for small Δ, the t-based discriminant rule has a smaller error rate than that of the NLDR. This arises directly from the heaviness of the tails of the t-density.

Sutradhar (1990) illustrated the use of the multivariate t-density for modeling the group-conditional distributions by fitting this model to some bivariate data on two species of flea beetles, as given in Lubischew (1962). The mean $\boldsymbol{\mu}_i$ and the common group-covariance matrix $\mathbf{\Sigma}$ were estimated by the sample mean $\bar{\mathbf{x}}_i$ and the (bias-corrected) pooled sample covariance matrix $\mathbf{S}$, respectively. The method of moments was used to estimate the degrees of freedom m. This estimate of m based on the fourth moment was given by

$$\hat{m} = 2(3c_1 - 2c_2)/(3 - c_2),$$

where

$$c_1 = \sum_{k=1}^{p} (\mathbf{S})_{kk}^2,$$

and

$$c_2 = \sum_{i=1}^{2} \sum_{j=1}^{n_i} \sum_{k=1}^{p} (\mathbf{x}_{ij} - \bar{\mathbf{x}}_i)_k^4 / n.$$

7.8 CASE STUDY OF RENAL VENOUS RENIN IN HYPERTENSION

7.8.1 Description of Data

The case study undertaken by McLachlan and Gordon (1989) is reconsidered here using a subset of their data. The purpose of this analysis is twofold. One is to give an illustration of a case in which the group-conditional distributions are modeled as a finite mixture (here a normal mixture). The other is to give a practical example of a partially classified sample, where the classified feature vectors do not represent an observed random sample from the mixture distribution of the feature vector.

The data under analysis were collected as part of a study carried out at the Endocrine-Hypertension Research Unit, Greenslopes Hospital, Brisbane, between 1973 and 1985 on patients suffering from hypertension. Patients were referred to the study from all over Queensland and Northern New South Wales with either (a) suggestive features of renal artery stenosis; or (b) severe hypertension, either not sufficiently well controlled with high-dose medications or sufficiently well controlled but with unacceptable side effects; or (c) an arteriogram suggestive of renal artery stenosis. Catheters were introduced via the femoral veins into the right and left renal veins, and samples of renal venous blood were obtained with the patient recumbent (X_{L1}, X_{R1}), then tilted 45 degrees (X_{L2}, X_{R2}), and then after pharmacological stimulation by intravenous bolus injection of Diazoxide either 150 mg or 300 mg (X_{L3}, X_{R3}). Peripheral blood samples (X_{A1}, X_{A2}, X_{A3}) were also taken in the three situations, and are considered equivalent to artery renin levels. Hence, the full feature vector for a patient consists of the nine measurements:

$$(X_{L1}, X_{R1}, X_{L2}, X_{R2}, X_{L3}, X_{R3}, X_{A1}, X_{A2}, X_{A3})'.$$

A renal venous renin ratio (RVRR) greater than 1.5 (that is, X_{Lv}/X_{Rv} or X_{Rv}/X_{Lv} greater than 1.5 for any v from 1 to 3) was considered suggestive of renovascular hypertension, which is potentially curable by surgery. Patients with an RVRR greater than 1.5 were subjected to arteriography and, on the basis of the X-rays so obtained, were categorized as having renal artery stenosis (RAS) or not. A ratio of 1.5 is often used as a cutoff point for further investigations; see also Sealey et al. (1973) and the references therein. Hence, there are two groups, with G_2 denoting the group of patients with RAS and G_1 the group of those without RAS, as diagnosed by an arteriogram. It should be noted that an arteriogram does not reveal with absolute certainty the absence or presence of RAS in a patient.

Of the 163 patients considered in this analysis, 60 had an RVRR greater than 1.5 and had subsequently undergone arteriography; 50 were diagnosed as having RAS, and the results for the other 10 are not available. For the remaining 103 patients with an RVRR less than 1.5, there is no arteriogram-based diagnosis, although for 70 patients, there is supplementary evidence suggestive of their not having RAS. They had a normal creatinine level, which is consis-

TABLE 7.2 Summary of Information on $M = 163$ Patients with Hypertension

	Presence of RAS			
RVRR	Unknown	No	Yes	Total
< 1.5	103	0	0	103
> 1.5	10	0	50	60
Total	$m = 113$	$n_1 = 0$	$n_2 = 50$	$M = 163$

tent with normal kidney function. For the remaining 33 patients, the creatinine level was either abnormal or not available.

The partial classification of the patients is summarized in Table 7.2. On the basis of this information, we wish to construct a discriminant rule for use as a guide to the diagnosis of future patients suffering from hypertension due to RAS. As there is an element of risk associated with arteriography, there is some concern not to use the procedure needlessly. On the other hand, if a patient has RAS, then it is potentially curable by surgery. With respect to the patients in the study, of particular interest are the diagnoses for those 10 patients with a high RVRR (greater than 1.5) for which the results of the performed arteriography are not available.

7.8.2 Modeling the Distribution of Renal Venous Renin Ratio (RVRR)

A preliminary inspection of the data suggested taking logs of the measurements for the RVRR to induce normality. The feature vector $\mathbf{X}$ is taken to be $(X_1, X_2, X_3)'$, where

$$X_v = \log(X_{Lv}/X_{Rv}) \qquad (v = 1,2,3).$$

In group G_1 corresponding to the absence of RAS, each X_i should be distributed symmetrically about zero. As an initial choice for the density of $\mathbf{X}$ in G_1, we took

$$f_1(\mathbf{x};\boldsymbol{\theta}_1) = \phi(\mathbf{x};\mathbf{0},\boldsymbol{\Sigma}_1),$$

the (trivariate) normal density with zero mean and arbitrary covariance matrix $\boldsymbol{\Sigma}_1$. Correspondingly, the density of $\mathbf{X}$ in G_2 was taken as

$$f_2(\mathbf{x};\boldsymbol{\theta}_2) = \tfrac{1}{2}\{\phi(\mathbf{x};\boldsymbol{\mu}_2,\boldsymbol{\Sigma}_2) + \phi(\mathbf{x};-\boldsymbol{\mu}_2,\boldsymbol{\Sigma}_2)\}, \tag{7.8.1}$$

a mixture in equal proportions of two normal component densities with a common covariance matrix and with the mean of one component equal to the negative of the other. This reflects the left and right kidneys having the same chance of developing RAS.

From preliminary analyses along the lines described in the previous study of McLachlan and Gordon (1989), it was concluded that a single normal distribution is not adequate for modeling the distribution of the feature vector $\mathbf{X}$ in

G_1. This is because of the presence of some values with relatively large dispersion about the origin in the data on the 70 patients for whom, as mentioned before, it is reasonable to assume do not have RAS, at least for the purposes of this exercise. Accordingly, a two-component normal mixture model is adopted for the density $f_1(\mathbf{x};\boldsymbol{\theta}_1)$ of $\mathbf{X}$ in G_1, namely,

$$f_1(\mathbf{x};\boldsymbol{\theta}_1) = \sum_{h=1}^{2} \alpha_{1h} f_{1h}(\mathbf{x};\boldsymbol{\theta}_1),$$

where

$$f_{1h}(\mathbf{x};\boldsymbol{\theta}_1) = \phi(\mathbf{x};\mathbf{0},\boldsymbol{\Sigma}_{1h}) \qquad (h = 1,2),$$

and $\alpha_{11} + \alpha_{12} = 1$. That is, the group G_1 is decomposed into two subgroups, G_{11} and G_{12}, where, with probability α_{1h}, $\mathbf{X}$ has density $f_{1h}(\mathbf{x};\boldsymbol{\theta}_1)$ in G_{1h} $(h = 1,2)$.

Likewise, plots of the data from the 50 patients known to have RAS clearly suggest that G_2 consists of two subgroups, in one of which the observations are widely dispersed in a direction away from the origin. The density of $\mathbf{X}$ in G_2 was taken, therefore, to have the form

$$f_2(\mathbf{x};\boldsymbol{\theta}_2) = \sum_{h=1}^{2} \alpha_{2h} f_{2h}(\mathbf{x};\boldsymbol{\theta}_2),$$

where

$$f_{2h}(\mathbf{x};\boldsymbol{\theta}_2) = \tfrac{1}{2}\{\phi(\mathbf{x};\boldsymbol{\mu}_{2h},\boldsymbol{\Sigma}_{2h}) + \phi(\mathbf{x};-\boldsymbol{\mu}_{2h},\boldsymbol{\Sigma}_{2h})\},$$

and $\alpha_{21} + \alpha_{22} = 1$. That is, G_2 is decomposed into two subgroups, G_{21} and G_{22}, where, with probability α_{2h}, $\mathbf{X}$ has density $f_{2h}(\mathbf{x};\boldsymbol{\theta}_2)$ in G_{2h} $(h = 1,2)$.

With a two-component normal mixture model for the feature vector $\mathbf{X}$ in each separate group G_i $(i = 1,2)$, the density of an observation in the mixture G of the two groups G_1 and G_2 is therefore modeled as a four-component normal mixture,

$$f_X(\mathbf{x};\boldsymbol{\Psi}) = \sum_{h=1}^{2}\{\pi_{1h} f_{1h}(\mathbf{x};\boldsymbol{\theta}_1) + \pi_{2h} f_{2h}(\mathbf{x};\boldsymbol{\theta}_2)\}, \tag{7.8.2}$$

where

$$\sum_{h=1}^{2}(\pi_{1h} + \pi_{2h}) = 1.$$

The parameters π_{11}, π_{12}, π_{21}, and π_{22} are the proportions in which the subgroups G_{11}, G_{12}, G_{21}, and G_{22} occur in the mixture G of G_1 and G_2, and it follows that

$$\alpha_{ih} = \pi_{ih}/(\pi_{i1} + \pi_{i2}) \qquad (i,h = 1,2).$$

In this example, the classified data are only on patients with an RVRR greater than 1.5. However, as shown in Section 2.8, we can effectively ignore this conditioning in forming the likelihood function on the basis of the classified and unclassified feature data and the observed sample size as given in Table 7.2. That is, we can treat the observations on the 50 classified patients as if they were an observed random sample from the mixture distribution of $\mathbf{X}$. We proceed here on the basis that the patients have been relabeled so that $\mathbf{x}_j$ $(j = 1,\ldots,n)$ denote the n classified feature vectors and $\mathbf{x}_j$ $(j = n+1,\ldots,M = n+m)$ the m unclassified ones. We assume for the moment that there are no classified data. Then by a straightforward extension of the application of the EM algorithm as described in Section 3.8, it follows that

$$\hat{\pi}_{ih} = \sum_{j=1}^{M} \hat{\tau}_{ih}(\mathbf{x}_j)/M \qquad (i = 1,2), \tag{7.8.3}$$

$$\hat{\boldsymbol{\Sigma}}_{1h} = \sum_{j=1}^{M} \hat{\tau}_{1h}(\mathbf{x}_j)\mathbf{x}_j\mathbf{x}_j'/(M\hat{\pi}_{1h}), \tag{7.8.4}$$

$$\hat{\boldsymbol{\mu}}_{2h} = \sum_{j=1}^{M} \hat{\tau}_{2h}(\mathbf{x}_j)[\hat{\omega}_h(\mathbf{x}_j)\mathbf{x}_j - \{1-\hat{\omega}_h(\mathbf{x}_j)\}\mathbf{x}_j]/(M\hat{\pi}_{2h}), \tag{7.8.5}$$

and

$$\begin{aligned}\hat{\boldsymbol{\Sigma}}_{2h} = \sum_{j=1}^{M} \hat{\tau}_{2h}(\mathbf{x}_j)[&\hat{\omega}_h(\mathbf{x}_j)(\mathbf{x}_j - \hat{\boldsymbol{\mu}}_{2h})(\mathbf{x}_j - \hat{\boldsymbol{\mu}}_{2h})' \\ &+ \{1-\hat{\omega}_h(\mathbf{x}_j)\}(\mathbf{x}_j + \hat{\boldsymbol{\mu}}_{2h})(\mathbf{x}_j + \hat{\boldsymbol{\mu}}_{2h})']/(M\hat{\pi}_{2h})\end{aligned} \tag{7.8.6}$$

for $h = 1,2$. In these equations, $\hat{\omega}_h(\mathbf{x}_j) = \omega_h(\mathbf{x}_j;\hat{\boldsymbol{\Psi}})$, where

$$\omega_h(\mathbf{x};\boldsymbol{\Psi}) = \phi(\mathbf{x};\boldsymbol{\mu}_{2h},\boldsymbol{\Sigma}_{2h})/\{\phi(\mathbf{x};\boldsymbol{\mu}_{2h},\boldsymbol{\Sigma}_{2h}) + \phi(\mathbf{x};-\boldsymbol{\mu}_{2h},\boldsymbol{\Sigma}_{2h})\}.$$

Also, $\hat{\tau}_{ih}(\mathbf{x}_j) = \tau_{ih}(\mathbf{x}_j;\hat{\boldsymbol{\Psi}})$, where

$$\tau_{ih}(\mathbf{x};\boldsymbol{\Psi}) = \pi_{ih}f_{ih}(\mathbf{x};\boldsymbol{\theta}_i)/f_X(\mathbf{x};\boldsymbol{\Psi})$$

is the posterior probability that the patient belongs to subgroup G_{ih} $(i = 1,2;$ $h = 1,2)$. If $\hat{\omega}_h(\mathbf{x}_j)$ is set equal to 1 in (7.8.3) to (7.8.6), then they reduce to those for a four-component normal mixture. The presence of $\hat{\omega}_h(\mathbf{x}_j)$ in (7.8.5) and (7.8.6) is a consequence of the assumption that in G_{2h} $(h = 1,2)$, $\mathbf{X}$ is a mixture in equal proportions of two normal densities with a common covariance matrix and with the mean of one component equal to the negative of the other.

Equations (7.8.3) to (7.8.6) have to be modified to reflect the fact that the group of origin of 50 obsevations $\mathbf{x}_j$ $(j = 1,\ldots,50)$ is known to be G_1. The modification is effected by replacing $\hat{\tau}_{1h}(\mathbf{x}_j)$ with

$$\hat{\tau}_{1h}(\mathbf{x}_j)/\{\hat{\tau}_{11}(\mathbf{x}_j) + \hat{\tau}_{12}(\mathbf{x}_j)\},$$

and $\hat{\tau}_{2h}(\mathbf{x}_j)$ by zero $(h = 1,2)$ for $j = 1,\ldots,50$. Equations (7.8.3) to (7.8.5) so modified can be solved iteratively by substituting some initial values for the

TABLE 7.3 Estimates of Parameters in Four-Component Mixture Model Fitted to the RVRR for Two Groups (RAS, no RAS) of Patients with Hypertension

Group	Parameters	Estimates		
G_1 (no RAS)	π_{11}, π_{12}	0.16	0.48	
	$\sigma_{11,11}$	0.03		
	$\sigma_{11,21}, \sigma_{11,22}$	−0.00	0.02	
	$\sigma_{11,31}, \sigma_{11,32}, \sigma_{11,33}$	−0.02	0.01	0.02
	$\sigma_{12,11}$	0.02		
	$\sigma_{12,12}, \sigma_{12,22}$	0.01	0.03	
	$\sigma_{12,31}, \sigma_{12,32}, \sigma_{12,33}$	0.01	0.01	0.03
G_2 (RAS)	π_{21}, π_{22}	0.18	0.18	
	$(\mu_{21})_1, (\mu_{21})_2, (\mu_{21})_3$	0.45	0.69	0.61
	$(\mu_{22})_1, (\mu_{22})_2, (\mu_{22})_3$	1.01	1.16	1.33
	$\sigma_{21,11}$	0.05		
	$\sigma_{21,21}, \sigma_{21,22}$	0.04	0.13	
	$\sigma_{21,31}, \sigma_{21,32}, \sigma_{21,33}$	0.01	0.04	0.04
	$\sigma_{22,11}$	0.34		
	$\sigma_{22,21}, \sigma_{22,22}$	0.28	0.34	
	$\sigma_{22,31}, \sigma_{22,32}, \sigma_{22,33}$	0.16	0.25	0.37

estimates into the right-hand side to produce new estimates on the left-hand side, and so on; see McLachlan and Gordon (1989) for further details, including choice of initial values.

To provide some guide as to the validity of the adopted model (7.8.2), we first clustered the data with respect to the four subgroups. Hawkins' (1981) test for multivariate normality was then applied to the resulting clusters as if they represented a correct partition of the data with respect to the subgroups. For the two subgroups of the RAS group G_2, we worked with the absolute values of the data for if the density $f_{2h}(\mathbf{x})$ of $\mathbf{X}$ in subgroup G_{2h} has the specialized normal mixture form (7.8.1), then $(|X_1|, |X_2|, |X_3|)'$ should be essentially trivariate normal with mean μ_{2h} and covariance matrix Σ_{2h} in G_{2h} $(h = 1, 2)$. This is because in group G_2, X_1, X_2, and X_3 tend to have the same sign. No significant departures from normality were obtained for the subgroup-conditional distributions.

7.8.3 Fitted Mixture Distribution of RVRR

We report in Table 7.3 the estimates obtained for the unknown parameters in the mixture density (7.8.2) for the feature vector $\mathbf{X}$ containing the RVRR as measured on a patient while recumbent, tilted at 45 degrees, and after pharmacological stimulation. In this table, $\sigma_{ih,rs}$ denotes the (r,s)th element of Σ_{ih}.

We have plotted in Figure 7.1 the fit obtained for the marginal density of X_3 in G_1 and in G_2.

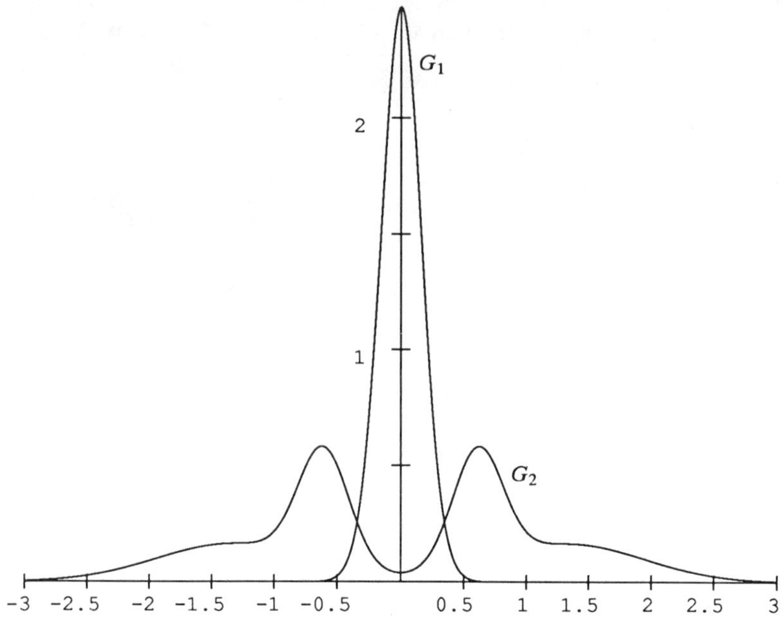

FIGURE 7.1. Marginal density of log($RVRR$) after pharmacological stimulation (X_3) in the no RAS group (G_1) and in the RAS group (G_2).

7.8.4 Diagnosis of RAS on Basis of Estimated Posterior Probabilities

Once the mixture density (7.8.2) has been fitted, a patient with feature vector $\mathbf{x}$ can be assigned to G_2 (diagnosed as having RAS) if

$$\hat{\tau}_2(\mathbf{x}) = \sum_{h=1}^{2} \hat{\pi}_{2h}\hat{\tau}_{2h}(\mathbf{x}) > \hat{\tau}_1(\mathbf{x}) = \sum_{h=1}^{2} \hat{\pi}_{1h}\hat{\tau}_{1h}(\mathbf{x}).$$

As explained previously, there was no arteriogram-based diagnosis for RAS in 113 of the 163 patients in this study. Of particular interest are the estimated posterior probabilities for the presence of RAS in the 10 patients with an RVRR greater than 1.5 (more appropriately here, log RVRR > 0.405 in magnitude), but for whom the results of the subsequent arteriography are not available. In Table 7.4, we list for those 10 patients their log(RVRR) as measured while recumbent (x_{1j}), tilted 45 degrees (x_{2j}), and after pharmacological stimulation (x_{3j}). The patients have been relabeled here for convenience so that $j = 1,\ldots,10$.

It can be seen from Table 7.4 that all but one of these 10 patients would be put in the RAS group G_2 in an outright assignment on the basis of the relative sizes of $\hat{\tau}_1(\mathbf{x}_j)$ and $\hat{\tau}_2(\mathbf{x}_j)$. For this patient ($j = 6$) and patient numbered $j = 5$, the diagnosis is not clear-cut. It is not suggested that a diagnosis should be made solely on the base of these estimated posterior probabilities. However,

TABLE 7.4 Estimated Posterior Probabilities for Absence/Presence of RAS in 10 Patients with Absolute Value of log(RVRR) > 0.405

				No RAS			RAS		
				Subgroups		Group	Subgroups		Group
				G_{11}	G_{12}	G_1	G_{21}	G_{22}	G_2
Patient No.	log(RVRR) x_{1j}	x_{2j}	x_{3j}	$\hat{\tau}_{11}(\mathbf{x}_j)$	$\hat{\tau}_{12}(\mathbf{x}_j)$	$\hat{\tau}_1(\mathbf{x}_j)$	$\hat{\tau}_{21}(\mathbf{x}_j)$	$\hat{\tau}_{22}(\mathbf{x}_j)$	$\hat{\tau}_2(\mathbf{x}_j)$
1	1.29	1.62	1.93	0.00	0.00	0.00	0.00	1.00	1.00
2	0.34	1.14	1.65	0.00	0.00	0.00	0.00	1.00	1.00
3	0.56	1.09	0.27	0.00	0.00	0.00	0.92	0.08	1.00
4	0.67	0.40	−0.02	0.00	0.00	0.00	0.08	0.92	1.00
5	0.42	0.32	0.40	0.00	0.28	0.28	0.66	0.06	0.72
6	−0.44	−0.21	−0.17	0.00	0.60	0.60	0.30	0.10	0.40
7	−0.48	−0.47	−0.82	0.00	0.00	0.00	0.70	0.30	1.00
8	0.07	−0.32	−0.52	0.00	0.02	0.02	0.73	0.25	0.98
9	0.88	0.62	0.61	0.00	0.00	0.00	0.52	0.48	1.00
10	1.07	1.02	1.00	0.00	0.00	0.00	0.02	0.98	1.00

they do provide the clinician with a quantitative guide as to the presence of RAS.

Of the 103 patients with log(RVRR) less than 0.405 in magnitude, all but four were clearly diagnosed as not having RAS on the basis of the relative size of the estimated posterior probabilities $\hat{\tau}_1(\mathbf{x}_j)$ and $\hat{\tau}_2(\mathbf{x}_j)$. We also computed the estimated posterior probabilities for the 50 patients diagnosed by arteriography to have RAS. All 50 were predicted to have RAS, although two were borderline cases.

7.9 EXAMPLE: DISCRIMINATION BETWEEN DEPOSITIONAL ENVIRONMENTS

7.9.1 Mass-Size Particle Data

As considered by Novotny and McDonald (1986), model selection can be undertaken using discriminant analysis. It is in this spirit that we now describe an example in which the feature vector used in the formation of the discriminant rule contains the fitted parameters of a postulated parametric family of distributions. It concerns the use of the mass-size particle distribution of a soil sample in an attempt to determine its depositional environment. Suppose that

$$\mathcal{X}_k = (a_k, b_k)$$

for $k = 1, \ldots, v$ denote the v intervals into which the diameter width of a particle is divided. Let

$$w = w_1 + \cdots + w_v$$

be the total weight of a soil sample, where w_k denotes the weight of particles in it with diameters in $\mathcal{X}_k$ $(k = 1,\ldots,v)$. This situation in which the percentage of particle sizes by weight is observed, and not the explicit number of particles, occurs frequently in studies of size phenomena. It is because of this that by far the most common method for measuring particle sizes in geological applications is to pass a sample of soil through a sequence of sieves with progressively finer meshes. The particles trapped on a particular sieve have size in a known interval and can be weighed.

One approach to the problem of characterizing and then discriminating between specified depositional environments is to take the feature vector equal to $(w_1,\ldots,w_v)'$. However, this approach ignores the implicit ordering of the intervals and is not applicable if different sets of intervals are used for different samples. An approach that avoids these two shortcomings is to model the mass-size distribution. Let $m(u;\zeta)$, where ζ is a vector of parameters, denote the probability density function chosen to model the theoretical mass-size distribution. Following Barndorff–Nielsen (1977), ζ can be estimated by $\hat{\zeta}$, obtained by maximizing the expression

$$\sum_{k=1}^{v} u_k \log P_k(\zeta), \tag{7.9.1}$$

where

$$P_k(\zeta) = \int_{\mathcal{X}_k} m(u;\zeta)\,du,$$

and $u_k = w_k/w$ $(k = 1,\ldots,v)$. The maximization of (7.9.1) is equivalent to the minimization of

$$\delta_{KL}\{\mathbf{u},\mathbf{P}(\zeta)\} = \sum_{k=1}^{v} u_k \log\{u_k/P_k(\zeta)\},$$

which is the Kullback–Leibler distance function between the vector of the empirical mass-size relative frequencies, $\mathbf{u} = (u_1,\ldots,u_v)'$, and its theoretical counterpart,

$$\mathbf{P}(\zeta) = (P_1(\zeta),\ldots,P_v(\zeta))'.$$

Barndorff–Nielsen (1977) termed this estimation procedure maximum likeness as it coincides with maximum likelihood in the case where the random vector $\mathbf{U}$, corresponding to the realization $\mathbf{u}$, is distributed for some positive integer M as

$$M\mathbf{U} \sim \text{Mult}(M,\mathbf{P}(\zeta)),$$

a multinomial distribution consisting of M independent draws on v categories with probabilities specified by $\mathbf{P}(\zeta)$. A multinomial distribution would apply if the explicit number of particles were observed rather than their mass-size relative frequencies in the present situation. With only the latter being observed, there are difficulties in providing standard errors and carrying out tests of statistical hypotheses; see Jones and McLachlan (1989) for a discussion of this.

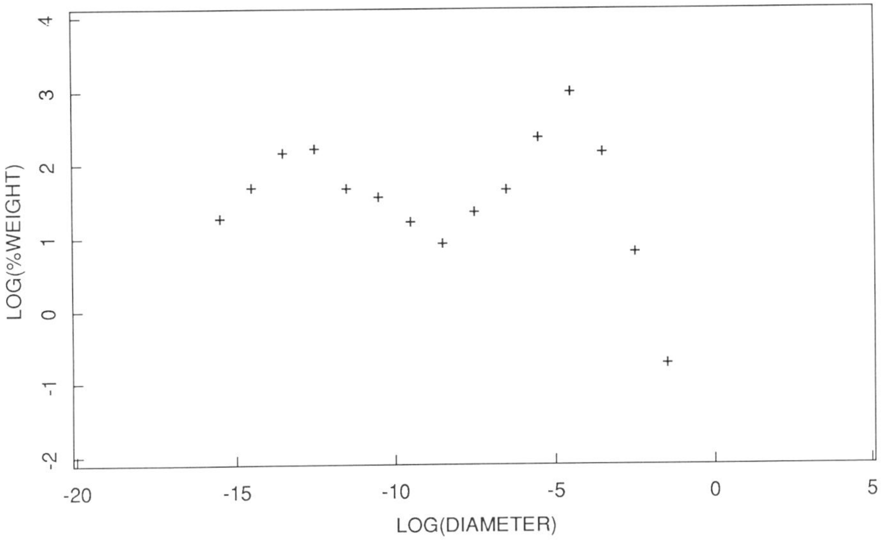

FIGURE 7.2. Empirical mass-size distribution of a soil sample.

Before proceeding with the characterization and allocation of depositional environments, using the fit $\hat{\zeta}$ as the feature vector, we briefly consider some methods of modeling mass-size distributions.

7.9.2 Modeling Mass-Size Particle Distributions

To illustrate some of the models postulated for mass-size distributions, we consider the data set analyzed by Jones and McLachlan (1989). It was taken from Walker and Chittleborough (1986) and consisted of the percentages by weight of particles falling in diameter ranges of 1 unit on a log scale for three horizons labelled Bt, BC, and C for each of two soil profiles labeled C and Q. An example of these data is given in Figure 7.2 where, for horizon BC of profile C, the natural log of mass-size relative frequencies expressed as percentages is plotted against the log to the base 2 of the diameter (in mm) of particles. This type of plot was used by Bagnold (1941) and subsequently by Barndorff–Nielsen (1977) in his introduction of the log hyperbolic distribution to model mass-size distributions of sand deposits. The hyperbolic density is given by

$$m_H(u;\zeta) = c\exp[-\tfrac{1}{2}(\psi+\gamma)\surd\{\alpha^2+(u-\mu)^2\} + \tfrac{1}{2}(\psi-\gamma)(u-\mu)], \tag{7.9.2}$$

where

$$c = \omega/\alpha\kappa K_1(\alpha\kappa),$$

$$\omega = 1/(\psi^{-1}+\gamma^{-1}),$$

$$\kappa = \sqrt{\psi\gamma},$$

and $K_1(\cdot)$ is the modified Bessel function of the third kind and with index $\nu = 1$. Bagnold and Barndorff–Nielsen (1980) have given a geological interpretation of these parameters for wind and water deposited sources. The normal distribution with mean μ and variance σ^2 is a limiting case of (7.9.2), when $\psi = \gamma$ and $\alpha \to \infty$ while $\alpha\sqrt{\psi\gamma} \to \sigma^2$.

It would appear from the bimodality exhibited in Figure 7.2 that a mixture model is appropriate for this profile. Jones and McLachlan (1989) showed how a mixture of two log hyperbolic components can be fitted satisfactorily, at least to the midpoints of the intervals (with the two extreme intervals discarded), using a generalized EM algorithm. However, they found for the data sets they analyzed that the α parameter in each log hyperbolic component is essentially zero or the likelihood is rather flat in it. This latter behavior of the likelihood was observed, too, by Fieller, Gilbertson, and Olbright (1984) in their modeling of sand particle sizes by the log hyperbolic family.

In these situations, there is, therefore, little to be gained, for the considerable additional computation, in adopting the hyperbolic distribution over its limiting form as α tends to zero, which is

$$\begin{aligned} m(u;\zeta) &= \omega \exp\{-\gamma|u-\mu|\} \qquad (u > \mu) \\ &= \omega \exp\{-\psi|u-\mu|\} \qquad (u < \mu). \end{aligned}$$

This limiting form is the skew Laplace or asymmetric double exponential distribution. It can be fitted to mass-size data in their original categorized form.

Jones and McLachlan (1989) concluded that mixtures of two log skew Laplace components gave an excellent fit to their data. This is evident from Figure 7.3, in which the relative frequencies and the fitted mixture of two log skew Laplace components are displayed for six profile/horizon combinations. The estimates of the parameters for these fits are listed in Table 7.5, and $\hat{\pi}_1$ and $\hat{\pi}_2 = 1 - \hat{\pi}_1$ denote the estimates of the proportions in which the two log skew Laplace components are mixed. Note that elsewhere in this book the use of $\hat{\pi}_i$ and $\hat{\mu}_i$ is reserved solely for the estimates of the prior probability and mean, respectively, of the ith group G_i $(i = 1,\ldots,g)$.

7.9.3 Discrimination on Basis of Mass-Size Fit

It was shown in the present example how a horizon/profile combination can be characterized by its mass-size distribution, as modeled by a mixture of two log skew Laplace components. Therefore, for the purposes of assigning a soil sample, say, of unidentified horizon within a known profile, we can take the feature vector $\mathbf{x}$ to be the fit

$$\hat{\zeta} = (\hat{\pi}_1, \hat{\mu}_1, \hat{\psi}_1, \hat{\gamma}_1, \hat{\mu}_2, \hat{\psi}_2, \hat{\gamma}_2)',$$

obtained by fitting a mixture of two log skew hyperbolic components in proportions π_1 and π_2. A sample discriminant rule with $\hat{\zeta}$ as feature vector can be formed provided there are replicate samples from each of the specified horizons within a given profile to enable classified training replications of the fit $\hat{\zeta}$ to be calculated. Because $\hat{\zeta}$ is a minimum-distance estimator, it will have asymptotically under the usual regularity conditions a multivariate normal dis-

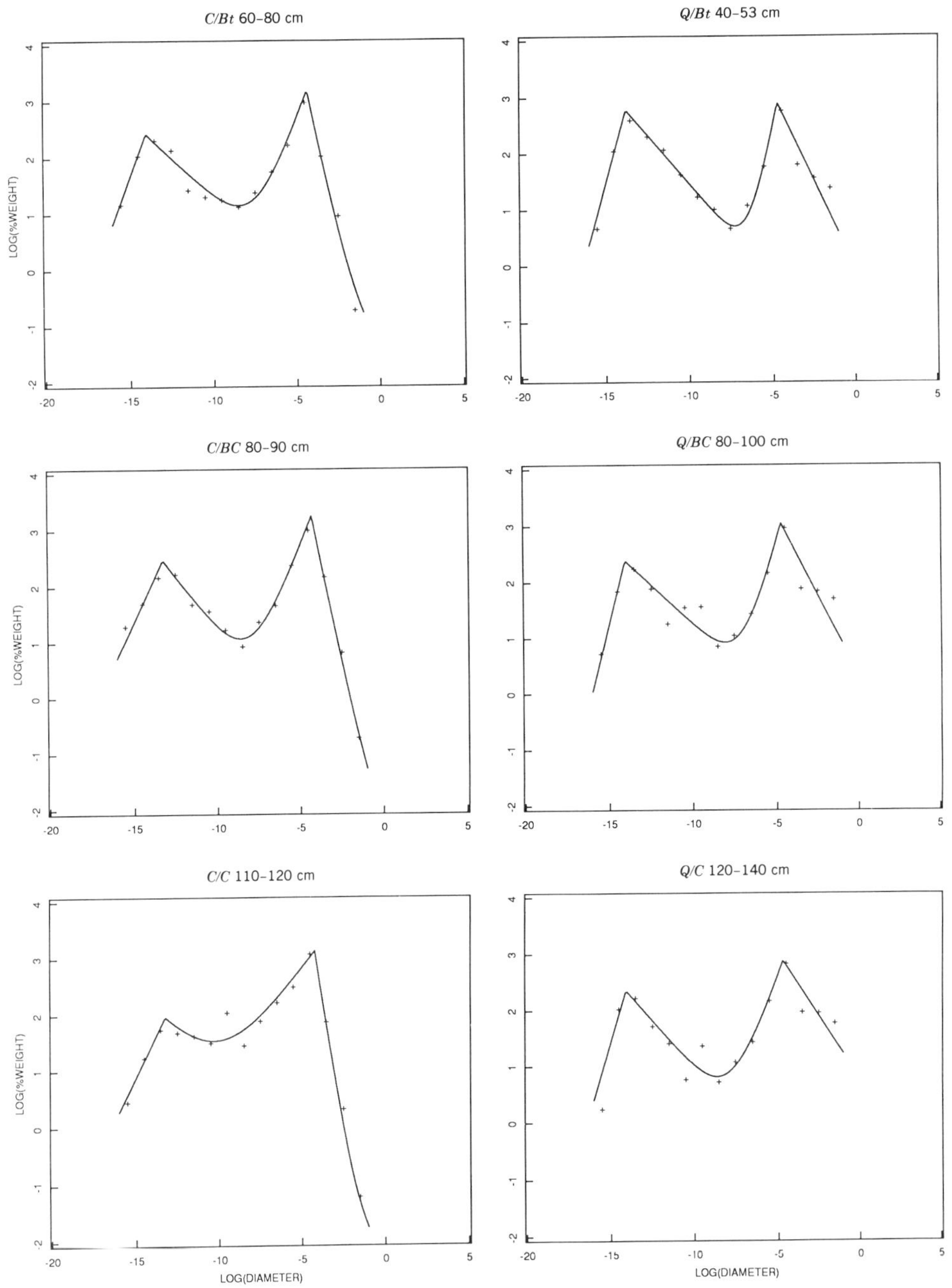

FIGURE 7.3. Mass-size histograms for profiles *C* and *Q* of Walker and Chittleborough (1986) along with fitted log skew Laplace mixtures. (From Jones and McLachlan (1989), by courtesy of Marcel Dekker, Inc.)

TABLE 7.5 Parameter Estimates for Log Skew Laplace Mixture Model Fitted to Data from Horizons of Two Soil Profiles

Horizon	$\hat{\pi}_1$	$\hat{\mu}_1$	$\hat{\psi}_1$	$\hat{\gamma}_1$	$\hat{\mu}_2$	$\hat{\psi}_2$	$\hat{\gamma}_2$
			Profile C				
Bt	0.54	−14.02	0.81	0.29	−4.35	0.80	1.44
BC	0.47	−13.25	0.64	0.39	−4.28	0.75	1.50
C	0.30	−13.23	0.65	0.32	−4.19	0.38	1.96
			Profile Q				
Bt	0.61	−13.77	1.08	0.35	−4.73	1.52	0.63
BC	0.48	−14.04	1.19	0.28	−4.72	1.10	.059
C	0.41	−14.07	1.01	0.34	−4.69	0.84	0.45

Source: Adapted from Jones and McLachlan (1989).

tribution for a given horizon/profile combination. Thus, the use of the sample NQDR with $\hat{\zeta}$ as feature vector can be justified in the present context, at least asymptotically.

In other applications of this type, Wyrwoll and Smyth (1985) used the sample NLDR to discriminate between three Australian dune settings: dune sides, dune crests, and interdunal corridor. The feature vector was taken to be

$$\mathbf{x} = (\hat{\mu}, \log\hat{\sigma})',$$

where $\hat{\zeta} = (\hat{\mu}, \hat{\sigma}^2)'$ is the fit obtained from modeling the mass-size distribution of a sand sample by a single log normal component with mean μ and variance σ^2. They noted that the logarithmic transformation of $\hat{\sigma}$ in the feature vector was to make more tenable the assumption of multivariate normality with a common covariance matrix for the group-conditional distribution of $\mathbf{X}$, which is implicit in the use of the sample NLDR. They compared the apparent error rate of this rule with that of the same rule, but with the feature vector $\mathbf{x}$ based on the fit of a single log hyperbolic component,

$$\mathbf{x} = (\hat{\mu}, \log\hat{\alpha}, \log\hat{\psi}, \log\hat{\gamma})'.$$

The three group-specific apparent error rates were actually increased rather than decreased with the fitting of this more complicated model. Fieller, Flenley, and Olbright (1990) noted a similar result in their comparison of the apparent error rates of the sample NLDR with feature vector based on the log hyperbolic model with those of the sample rule based on the simpler log normal model for discriminating between beach and dune samples collected near the eastern shore of the Hebridean island of Oronsay. On the other hand in their comparative study, the use of the log skew Laplace model in forming the feature vector resulted in smaller apparent error rates.

For other approaches to problems of the type above, the reader is referred to Aitchison (1986), who has given a systematic account of statistical methods designed to handle the special nature of compositional data.

CHAPTER 8

Logistic Discrimination

8.1 INTRODUCTION

8.1.1 Formulation of Logistic Model

In Chapters 2 to 7, we have considered a fully parametric approach to discriminant analysis, in which the group-conditional densities $f_i(\mathbf{x})$ are assumed to have specified functional forms except for a finite number of parameters to be estimated. Logistic discrimination can be viewed as a partially parametric approach, as it is only the ratios of the densities $\{f_i(\mathbf{x})/f_j(\mathbf{x}),\ i \neq j\}$ that are being modeled. For simplicity of discussion, we consider first the case of $g = 2$ groups.

The fundamental assumption of the logistic approach to discrimination is that the log of the group-conditional densities is linear, that is,

$$\log\{f_1(\mathbf{x})/f_2(\mathbf{x})\} = \beta_0^o + \boldsymbol{\beta}'\mathbf{x}, \tag{8.1.1}$$

where β_0^o and $\boldsymbol{\beta} = (\beta_1,\ldots,\beta_p)'$ constitute $p+1$ parameters to be estimated. Let

$$\beta_0 = \beta_0^o + \log(\pi_1/\pi_2).$$

The assumption (8.1.1) is equivalent to taking the log (posterior) odds to be linear, as under (8.1.1), we have that the posterior probability of an entity with $\mathbf{X} = \mathbf{x}$ belonging to group G_1 is

$$\tau_1(\mathbf{x}) = \exp(\beta_0 + \boldsymbol{\beta}'\mathbf{x})/\{1 + \exp(\beta_0 + \boldsymbol{\beta}'\mathbf{x})\}, \tag{8.1.2}$$

and so

$$\begin{aligned}\text{logit}\{(\tau_1(\mathbf{x})\} &= \log\{\tau_1(\mathbf{x})/\tau_2(\mathbf{x})\} \\ &= \beta_0 + \boldsymbol{\beta}'\mathbf{x}.\end{aligned} \tag{8.1.3}$$

Conversely, (8.1.1) is implied by the linearity (8.1.3) of the log odds. The linearity here is not necessarily in the basic variables; transforms of these may be taken.

The earliest applications of the logistic model were in prospective studies of coronary heart disease by Cornfield (1962) and Truett, Cornfield, and Kannel (1967), although there the parameters were estimated subsequently under the assumption of normality. The estimation problem in a general context was considered by Cox (1966, 1970), Day and Kerridge (1967), and Walker and Duncan (1967). For reviews of the logistic model in discriminant analysis, the reader is referred to J. A. Anderson (1982) and Albert and Lesaffre (1986). More generally, van Houwelingen and le Cessie (1988) have given the development of logistic regression as a multipurpose statistical tool. There is also the recent monograph of Hosmer and Lemeshow (1989) and the expanded and substantially revised second edition (Cox and Snell, 1989) of Cox(1970).

To define the logistic model in the case of $g > 2$ groups, let

$$\boldsymbol{\beta}_i = (\beta_{1i}, \ldots, \beta_{pi})' \qquad (i = 1, \ldots, g-1)$$

be a vector of p parameters. Corresponding to the conventional but arbitrary choice of G_g as the base group, the logistic model assumes that

$$\log\{f_i(\mathbf{x})/f_g(\mathbf{x})\} = \beta_{0i}^o + \boldsymbol{\beta}_i'\mathbf{x} \qquad (i = 1, \ldots, g-1), \tag{8.1.4}$$

or, equivalently, that

$$\tau_i(\mathbf{x}) = \exp(\beta_{0i} + \boldsymbol{\beta}_i'\mathbf{x}) \Big/ \sum_{h=1}^{g-1} \{1 + \exp(\beta_{0h} + \boldsymbol{\beta}_h'\mathbf{x})\} \qquad (i = 1, \ldots, g-1), \tag{8.1.5}$$

where

$$\beta_{0i} = \beta_{0i}^o + \log(\pi_i/\pi_g) \qquad (i = 1, \ldots, g-1).$$

As emphasized by J. A. Anderson (1982), logistic discrimination is broadly applicable, as a wide variety of families of distributions and perturbations of them satisfy (8.1.4), including (i) multivariate normal distributions with equal covariance matrices; (ii) multivariate discrete distributions following the log-linear model with equal interaction terms; (iii) joint distributions of continuous and discrete random variables following (i) and (ii), but not necessarily independent; and (iv) truncated versions of the foregoing.

8.1.2 Applicability of Logistic Model

J. A. Anderson (1972) did note for discrete feature variables that the logistic assumption (8.1.1) of linearity in $\mathbf{x}$ is quite severe for all but binary variables. Consider, for example, the trichotomous variable x, which takes on the values of 0, 1, and 2. In some cases, it will be reasonable to assume that the log likelihood ratio is linear in x. In other cases, this will not be so and, in order to apply the logistic model, x must be transformed into two binary variables,

x_1 and x_2, where $x = 0$ if and only if $x_1 = 0$ and $x_2 = 0$; $x = 1$ if and only if $x_1 = 1$ and $x_2 = 0$; and $x = 2$ if and only if $x_1 = 0$ and $x_2 = 1$. Analogously, each variable with more than $r \geq 2$ levels can be replaced by $r - 1$ binary variables.

The logistic model has been introduced before from a discriminant analysis point of view. However, the logistic form (8.1.2) for the posterior probabilities of group membership provides a way of modeling the relationship between a g-valued categorical response variable and a vector $\mathbf{x}$ of explanatory variables or covariates. In this regression context, the groups $G_1, \ldots, G_g$ can be viewed as corresponding to the different categories of the response variable. Multiple logistic regression as such has been used extensively in epidemiology. The covariate vector represents levels of risk factors thought to be related to a disease under study and the categorical response variable denotes the disease status. For example, in a study of the effect of exposure to smoking, the disease categories might be the absence of lung cancer and two or three histologies of lung cancer. For a binary response indicating the presence of a disease (G_1) or the absence (G_2), the logistic formulation (8.1.2) implies a log-linear model for the posterior odds for an individual with covariate $\mathbf{x}$ compared with a baseline individual with covariate $\mathbf{x}_0$. Under (8.1.2),

$$\log\left\{\frac{\tau_1(\mathbf{x})}{\tau_2(\mathbf{x})}\Big/\frac{\tau_1(\mathbf{x}_0)}{\tau_2(\mathbf{x}_0)}\right\} = \boldsymbol{\beta}'(\mathbf{x} - \mathbf{x}_0).$$

Often the disease under study is fairly rare, so that $\tau_2(\mathbf{x})$ and $\tau_2(\mathbf{x}_0)$ are close to one, and hence the odds ratio is approximately

$$\tau_1(\mathbf{x})/\tau_1(\mathbf{x}_0),$$

which is the relative risk of contracting the disease. This makes the odds ratio an even more natural quantity to model in epidemiological studies. For an account of the logistic model and its further developments in the latter context, the reader is referred to Breslow and Day (1980) and Kleinbaum, Kupper, and Chambless (1982). Some of the developments, for example, have been concerned with the use of the logistic model where there is a pairing or matching. Indeed, many medical case-control studies involve the close matching of controls (individuals free of the disease) to cases (diseased individuals), for instance, on age and sex, rather than using regression modeling to allow for such effects.

Other choices besides the logistic transform are possible in (8.1.3) for relating $\mathbf{x}$ to the posterior probability $\tau_1(\mathbf{x})$. For example, Albert and Anderson (1981) considered the concept of probit discrimination by replacing logit$\{\tau_1(\mathbf{x})\}$ with

$$\text{probit}\{\tau_1(\mathbf{x})\} = \Phi^{-1}\{\tau_1(\mathbf{x})\},$$

in (8.1.3), where $\Phi(\cdot)$ denotes the standard normal distribution function. However, the logistic and probit models are virtually indistinguishable for practical purposes, since

$$\text{logit}\{\tau_1(\mathbf{x})\} \simeq c\Phi^{-1}\{\tau_1(\mathbf{x})\}, \tag{8.1.6}$$

where $c = \sqrt{8/\pi} = 1.6$. The approximation (8.1.6) has a maximum error of 0.01767 (Page, 1977). The logistic model is usually preferred for its computational convenience. Albert, Chapelle, and Smeets (1981) have described one of the few applications of the probit approach.

A more general approach to logistic or probit discrimination is to use the class of generalized additive models proposed by Hastie and Tibshirani (1986, 1990). The linear form (8.1.3) is generalized to a sum of smooth functions $\sum_{j=1}^{p} s_j((\mathbf{x})_j)$, where the $s_j(\cdot)$'s are unspecified functions that are estimated using a scatter plot smoother, in an iterative procedure called the local scoring algorithm.

It was assumed above that the groups are qualitatively distinct or, in the regression context, that the categories of the response variable are unordered. However, as explained by J. A. Anderson (1982), the groups or categories can be quantitatively defined. For example, they may correspond to different levels (high or low) of some test scores. In which case, the groups are not qualitatively distinct as the difference between them is one of degree. For $g = 2$ groups or categories, Albert and Anderson (1981) showed that the probit or logistic models apply equally well to ordered groups or categories. Anderson and Philips (1981) provided an extension to the case of $g > 2$ groups or categories, where the latter may be viewed as corresponding to the grouping of an unobservable continuous variable. This extended model does not belong to the exponential family. J. A. Anderson (1984) subsequently showed how the logistic model may be applied to his proposed stereotype model for assessed ordered categorical variables. The latter are viewed as being generated by an assessor who provides a judgment of the grade of the ordered variable. This ordinal model was independently proposed by Greenland (1985). Campbell and Donner (1989) have investigated the impact of incorporating an ordinality assumption into the logistic model in the case where ordinality is indeed a correct assumption. They derived the efficiency of the ordinal logistic approach of J. A. Anderson (1984) relative to the ordinary logistic approach in terms of the asymptotic error rates of the implied allocation rules.

8.1.3 Quadratic Logistic Discrimination

J. A. Anderson (1982) in reviewing his series of papers on the logistic model has stressed that the fundamental assumption (8.1.4) of the logistic model is not restricted to linearity of the log ratios of the group-conditional densities in the basic variables. On the contrary, any specified functions of these can be included as **x**-variates. Perhaps most common are the log, square, and square root transformations, which arise for the same reasons as in linear regression.

There is a type of quadratic transformation introduced for a rather different reason. Suppose that

$$\mathbf{X} \sim N(\boldsymbol{\mu}_i, \boldsymbol{\Sigma}_i) \quad \text{in} \quad G_i \qquad (i = 1, \ldots, g).$$

Then

$$\log\{f_i(\mathbf{x})/f_g(\mathbf{x})\} = \beta_{0i}^o + \boldsymbol{\beta}_i'\mathbf{x} + \mathbf{x}'\boldsymbol{\Lambda}_i\mathbf{x}, \tag{8.1.7}$$

where $\mathbf{\Lambda}_i$ is a $p \times p$ symmetric matrix. Now (8.1.7) is linear in the coefficients given by β_{0i}^o, the elements of $\boldsymbol{\beta}_i$, and the distinct elements of $\mathbf{\Lambda}_i$, so that it can be written in the form of (8.1.4) with $p + 1 + \frac{1}{2}p(p+1)$ parameters.

J. A. Anderson (1975) has discussed the logistic model in some detail for (8.1.7). He suggested some approximations for the quadratic form $\mathbf{x}'\mathbf{\Lambda}_i\mathbf{x}$, which enables estimation to proceed when the number of basic variables p is not small ($p > 4$ or 5, say). The simplest of his approximations is

$$\mathbf{\Lambda}_i \simeq \lambda_{1i}\boldsymbol{\psi}_{1i}\boldsymbol{\psi}_{1i}',$$

where λ_{1i} is the largest eigenvector of $\mathbf{\Lambda}_i$, and $\boldsymbol{\psi}_{1i}$ is the corresponding eigenvector ($i = 1,\ldots,g-1$). As explained by J. A. Anderson (1975), the need for a quadratic discriminant rule is by no means restricted to situations where all the variables in $\mathbf{x}$ are continuous. For example, if the variables are binary and the first-order interactions on the log-linear scale are not equal in each group, then (8.1.7) holds provided the higher-order interactions are the same for all groups.

8.2 MAXIMUM LIKELIHOOD ESTIMATION OF LOGISTIC REGRESSION COEFFICIENTS

8.2.1 x-Conditional Sampling

We consider now maximum likelihood estimation of the parameters in the logistic model on the basis of the observed training data $\mathbf{t}$, where, as before,

$$\mathbf{t}' = (\mathbf{y}_1,\ldots,\mathbf{y}_n),$$

and $\mathbf{y}_j = (\mathbf{x}_j',\mathbf{z}_j')'$ for $j = 1,\ldots,n$. We let

$$\mathbf{t}_x' = (\mathbf{x}_1,\ldots,\mathbf{x}_n)$$

be the matrix containing only the feature observations and

$$\mathbf{t}_z' = (\mathbf{z}_1,\ldots,\mathbf{z}_n)$$

the matrix containing only the associated group-indicator vectors.

In principle, we have to distinguish between the sampling design under which $\mathbf{t}$ was realized. In practice, however, it turns out that there is a minor consideration with the same function being maximized for three important study designs: $\mathbf{x}$-conditional sampling, joint or mixture sampling, and $\mathbf{z}$-conditional or separate sampling. They correspond respectively to sampling from the distribution of $\mathbf{Z}$ conditional on $\mathbf{x}$, sampling from the joint distribution of $\mathbf{Y} = (\mathbf{X}',\mathbf{Z}')'$, and to sampling from the distribution of $\mathbf{X}$ conditional on $\mathbf{z}$.

The first design applies to a multiple logistic regression where it is assumed that conditional on $\mathbf{x}_1,\ldots,\mathbf{x}_n$, $\mathbf{Z}_1,\ldots,\mathbf{Z}_n$ are distributed independently with

$$\mathbf{Z}_j \mid \mathbf{x}_j \sim \text{Mult}_g(1,\boldsymbol{\tau}_j) \qquad (j = 1,\ldots,n), \tag{8.2.1}$$

where $\boldsymbol{\tau}_j = (\tau_{1j}, \ldots, \tau_{gj})'$ is the vector of posterior probabilities for $\mathbf{x}_j$, and

$$\tau_{ij} = \tau_i(\mathbf{x}_j; \boldsymbol{\alpha}) \qquad (i = 1, \ldots, g).$$

Here

$$\boldsymbol{\alpha} = (\boldsymbol{\alpha}_1', \ldots, \boldsymbol{\alpha}_{g-1}')' \tag{8.2.2}$$

denotes the vector of the logistic regression coefficients, where

$$\boldsymbol{\alpha}_i = (\beta_{0i}, \boldsymbol{\beta}_i')' \qquad (i = 1, \ldots, g-1).$$

The dimension of the vector $\boldsymbol{\alpha}$ is thus $(g-1)(p+1)$. The use of this design is exemplified by a dose-response investigation in bioassay in which the response is noted at each of a series of dose levels x.

From (8.2.1), we have that the log likelihood function for $\boldsymbol{\alpha}$, formed from $\mathbf{z}_1, \ldots, \mathbf{z}_n$ conditional on $\mathbf{x}_1, \ldots, \mathbf{x}_n$, is

$$\log L(\boldsymbol{\alpha}; \mathbf{t}_z \mid \mathbf{t}_x) = \sum_{j=1}^{n} \log L(\boldsymbol{\alpha}; \mathbf{z}_j \mid \mathbf{x}_j), \tag{8.2.3}$$

where

$$\log L(\boldsymbol{\alpha}; \mathbf{z}_j \mid \mathbf{x}_j) = \sum_{i=1}^{g} z_{ij} \log \tau_i(\mathbf{x}_j; \boldsymbol{\alpha}) \tag{8.2.4}$$

is the contribution to the log likelihood from the single response vector $\mathbf{z}_j$ conditional on $\mathbf{x}_j$ $(j = 1, \ldots, n)$. It can be written in the exponential family form

$$\log L(\boldsymbol{\alpha}; \mathbf{t}_z \mid \mathbf{t}_x) = \sum_{i=1}^{g-1} \boldsymbol{\alpha}_i' \mathbf{u}_i - k(\boldsymbol{\alpha}), \tag{8.2.5}$$

where, on putting $\mathbf{x}_j^+ = (1, \mathbf{x}_j')'$,

$$\mathbf{u}_i = \sum_{j=1}^{n} z_{ij} \mathbf{x}_j^+ \qquad (i = 1, \ldots, g-1),$$

and

$$k(\boldsymbol{\alpha}) = \sum_{j=1}^{n} \log \left(1 + \sum_{h=1}^{g-1} \boldsymbol{\alpha}_h' \mathbf{x}_j^+ \right).$$

With multiple logistic regression, it is this likelihood that is maximized. However, we will see that it is also appropriate for both mixture and separate sampling schemes, which are more relevant in discriminant analysis.

8.2.2 Mixture Sampling

We now consider maximum likelihood for the logistic model under a mixture sampling scheme, which is common in prospective studies and diagnostic situations. In a prospective study design, a sample of individuals is followed and their responses recorded. Under a mixture sampling scheme, $\mathbf{x}_1,\ldots,\mathbf{x}_n$ are the observed values of a random sample drawn from a mixture of the g groups $G_1,\ldots,G_g$. Hence, the associated group-indicator vectors are distributed unconditionally as

$$\mathbf{Z}_1,\ldots,\mathbf{Z}_n \overset{iid}{\sim} \text{Mult}_g(1,\boldsymbol{\pi}), \tag{8.2.6}$$

where $\boldsymbol{\pi} = (\pi_1,\ldots,\pi_g)'$.

The log likelihood function for $\boldsymbol{\alpha}$ formed from $\mathbf{t}$ under mixture sampling is

$$\log L(\boldsymbol{\alpha};\mathbf{t}) = \log L(\boldsymbol{\alpha};\mathbf{t}_z \mid \mathbf{t}_x) + \sum_{j=1}^{n} \log f_X(\mathbf{x}_j), \tag{8.2.7}$$

where

$$f_X(\mathbf{x}) = \sum_{i=1}^{g} \pi_i f_i(\mathbf{x})$$

is the mixture density of $\mathbf{X}$. As no assumption is made about the functional form for $f_X(\mathbf{x})$ under the logistic model, the maximum likelihood estimate of $\boldsymbol{\alpha}$ obtained by maximizing (8.2.7), is the same value of $\boldsymbol{\alpha}$ that maximizes $L(\boldsymbol{\alpha};\mathbf{t}_z \mid \mathbf{t}_x)$ in the regression situation.

From (8.2.6), an estimate of π_i is provided by

$$\hat{\pi}_i = n_i/n \qquad (i = 1,\ldots,g),$$

where

$$n_i = \sum_{j=1}^{n} z_{ij} \qquad (i = 1,\ldots,g).$$

As

$$\beta_{0i}^{o} = \beta_{0i} - \log(\pi_i/\pi_g), \tag{8.2.8}$$

it can be estimated as

$$\hat{\beta}_{0i}^{o} = \hat{\beta}_{0i} - \log(n_i/n_g). \tag{8.2.9}$$

8.2.3 Separate Sampling

Maximum likelihood estimation for the logistic model under a separate sampling scheme is not as straightforward at first sight. Under a separate sampling scheme, the training data $\mathbf{t}$ have been obtained by sampling n_i observations separately from each group G_i $(i = 1,\ldots,g)$. Hence, it is appropriate to retrospective studies, which are common in epidemiological investigations. For example, with the simplest retrospective case-control study of a disease, one sample is taken from the cases that occur during the study period and the other sample is taken from the group of individuals who remained free of the disease. As many disease conditions are rare and even a large prospective study may produce few diseased individuals, retrospective sampling can result in important economies in cost and study duration.

As the β_{0i} are not estimable from $\mathbf{t}$ under separate sampling, the vector of parameters to be estimated from $\mathbf{t}$ is

$$\boldsymbol{\alpha}^o = (\boldsymbol{\alpha}_1^{o\prime},\ldots,\boldsymbol{\alpha}_{g-1}^{o\prime})',$$

where $\boldsymbol{\alpha}_i^o = (\beta_{0i}^o,\boldsymbol{\beta}_i')'$ for $i = 1,\ldots,g-1$. Under separate sampling, the log likelihood formed from $\mathbf{t}$ by proceeding conditionally on $\mathbf{t}_z$ is

$$\begin{aligned}\log L(\boldsymbol{\alpha}^o;\mathbf{t}_x \mid \mathbf{t}_z) &= \sum_{i=1}^{g}\sum_{j=1}^{n} z_{ij}\log f_i(\mathbf{x}_j)\\ &= \sum_{j=1}^{n}\left\{\log f_g(\mathbf{x}_j) + \sum_{i=1}^{g-1} z_{ij}(\beta_{0i}^o + \boldsymbol{\beta}_i'\mathbf{x}_j)\right\} \qquad (8.2.10)\end{aligned}$$

under the assumption (8.1.4). It can be seen from (8.2.10) that the difficulty here is that a full specification of the likelihood $L(\boldsymbol{\alpha}^o;\mathbf{t}_x \mid \mathbf{t}_z)$ depends also on the unknown density function $f_g(\mathbf{x})$ in addition to $\boldsymbol{\alpha}^o$.

Suppose that $\mathbf{X}$ is a discrete random variable, taking on a finite number, say d, of distinct values with nonzero probabilities $f_{g1},\ldots,f_{gd}$ in G_g, where the only assumption on these probabilities is that they sum to one. Let $\hat{\beta}_{0i}$ and $\hat{\boldsymbol{\beta}}_i$ denote the estimates of β_{0i} and $\boldsymbol{\beta}_i$ obtained by maximization of the log likelihood function $L(\boldsymbol{\alpha};\mathbf{t} \mid \mathbf{t}_x)$, and let $\hat{\beta}_{0i}^o$ be the corresponding estimate of β_{0i}^o defined by (8.2.8), $i = 1,\ldots,g-1$. Then J. A. Anderson (1972, 1974, 1982) and Anderson and Blair (1982) showed that the estimates of β_{0i}^o and $\boldsymbol{\beta}_i$ obtained by maximization of $L(\boldsymbol{\alpha};\mathbf{t}_x \mid \mathbf{t}_z)$ over $\boldsymbol{\alpha}^o$ and the $f_{gh}(h = 1,\ldots,d)$ are given by $\hat{\beta}_{0i}^o$ and $\hat{\boldsymbol{\beta}}_i(i = 1,\ldots,g-1)$. Thus, the estimate of $\boldsymbol{\alpha}^o$ is precisely the same as obtained under a mixture sampling scheme. Extrapolation to the case where all or some of the observations $\mathbf{x}_1,\ldots,\mathbf{x}_n$ are continuous was given using a discretization argument. However, Prentice and Pyke (1979) showed there was no need to make the sample space finite in order to establish that the maximum likelihood estimates $\hat{\boldsymbol{\beta}}_i$ of the $\boldsymbol{\beta}_i(i = 1,\ldots,g-1)$ under mixture sampling are also the maximum likelihood estimates under separate sampling, with the $\hat{\boldsymbol{\beta}}_i$ having similar asymptotic properties under either scheme. These results also follow from Cosslett (1981a, 1981b); see Scott and Wild (1986). Anderson and Blair (1982) showed that the mixture sampling solution can be obtained under separate sampling by using a penalized maximum likelihood procedure.

Using conditional likelihood derivations similar to those given by Cox (1972a) and Prentice and Breslow (1978), Farewell (1979) eliminated the terms $f_g(\mathbf{x}_j)$ in the log likelihood $L(\boldsymbol{\alpha}^o;\mathbf{t}_x \mid \mathbf{t}_z)$ formed under separate sampling by considering the conditional probability that $\mathbf{T} = \mathbf{t}$ given $n_1,\ldots,n_g$ and the set $\mathcal{C}$ of the values obtained for $\mathbf{x}_1,\ldots,\mathbf{x}_n$. The log of the conditional likelihood so formed can be expressed as

$$\sum_{i=1}^{g-1}\sum_{j=1}^{n} z_{ij}\boldsymbol{\beta}_i'\mathbf{x}_j - \log\sum_{\tilde{z}_{ij}}\prod_{i=1}^{g-1}\prod_{j=1}^{n}\exp(\tilde{z}_{ij}\boldsymbol{\beta}_i'\mathbf{x}_j), \qquad (8.2.11)$$

where the zero-one variables $\tilde{z}_{ij}$ satisfy

$$\sum_{i=1}^{g} \tilde{z}_{ij} = 1 \; (j = 1,\ldots,n) \qquad \text{and} \qquad \sum_{j=1}^{n} \tilde{z}_{ij} = n_i \; (i = 1,\ldots,g). \tag{8.2.12}$$

The summation with respect to the $\tilde{z}_{ij}$ in (8.2.11) is over the $n!/(n_1!\cdots n_g!)$ possible assignments of zero-one values to the $\tilde{z}_{ij}$ so that (8.2.12) is satisfied always. Estimation of $\boldsymbol{\beta}_1,\ldots,\boldsymbol{\beta}_{g-1}$ from (8.2.11) is computationally feasible in circumstances in which at most one of $n_1,\ldots,n_g$ is large; see Efron (1977) and Prentice and Pyke (1979).

8.2.4 Computation of Maximum Likelihood Estimate

Day and Kerridge (1967) and J. A. Anderson (1972) originally suggested using the Newton–Raphson procedure to maximize $L(\boldsymbol{\alpha};\mathbf{t}_z \mid \mathbf{t}_x)$ in obtaining the maximum likelihood estimator $\hat{\boldsymbol{\alpha}}$ of $\boldsymbol{\alpha}$. J. A. Anderson (1982) subsequently recommended quasi-Newton methods, as they combine the Newton property of speed of convergence near the optimum with the advantage possessed by the steepest descent method with poor starting values (Gill and Murray, 1972). A further advantage of the quasi-Newton methods is that they require only the first-order derivatives at each iteration while giving an estimate of the matrix of second-order derivatives at the maximizer. On the choice of starting value for $\hat{\boldsymbol{\alpha}}$, both J. A. Anderson (1972) and Albert and Lesaffre (1986) have reported very satisfactory results with the Newton–Raphson procedure started from $\hat{\boldsymbol{\alpha}} = \mathbf{0}$. The latter authors found that convergence usually takes no more than a dozen iterations, except in those circumstances to be described shortly, where the likelihood $L(\boldsymbol{\alpha};\mathbf{t}_z \mid \mathbf{t}_x)$ does not have a maximum for finite $\boldsymbol{\alpha}$. The major statistical packages listed in Section 3.3.3 all offer some form of logistic regression analysis; see Panel on Discriminant Analysis, Classification, and Clustering (1989, Section 4.2).

The asymptotic covariance matrix of $\hat{\boldsymbol{\alpha}}$ under $\mathbf{x}$-conditional or mixture sampling can be estimated as usual by the inverse of the observed information matrix,

$$\operatorname{cov}(\boldsymbol{\alpha}) \simeq \hat{\mathbf{I}}^{-1}, \tag{8.2.13}$$

where $\hat{\mathbf{I}} = \mathbf{I}(\hat{\boldsymbol{\alpha}})$ is the observed information matrix and

$$\mathbf{I}(\boldsymbol{\alpha}) = -\partial^2 \log L(\boldsymbol{\alpha};\mathbf{t}_z \mid \mathbf{t}_x)/\partial\boldsymbol{\alpha}\,\partial\boldsymbol{\alpha}'.$$

Prentice and Pyke (1979) showed that the appropriate rows and columns of $\hat{\mathbf{I}}^{-1}$ also provide a consistent assessment of the covariance matrix of $(\hat{\boldsymbol{\beta}}_1',\ldots,\hat{\boldsymbol{\beta}}_{g-1}')'$ under separate sampling. Concerning an assessment of standard errors and covariances involving the $\hat{\beta}_{0i}^{o}$, J. A. Anderson (1972) showed for discrete $\mathbf{x}$ under separate sampling that

$$\operatorname{cov}(\boldsymbol{\alpha}^{o}) = \hat{\mathbf{I}}^{-1} - n_g^{-1}\mathbf{C}, \tag{8.2.14}$$

where $\mathbf{C}$ is a $(g-1)(p+1)$ square matrix with all elements zero, except for some elements in the $1+(r-1)(p+1)$ row $(r=1,\dots,g-2)$. In this row, the element in the $1+(s-1)(p+1)$ column is $1+(n_g/n_r)$ for $s=r$ and is 1 for $s \neq r$ $(s=1,\dots,g-2)$; all other elements in the row are zero. In particular,

$$\text{var}(\hat{\beta}_{0i}^{o}) = \text{var}(\hat{\beta}_{0i}) - n_i^{-1} - n_g^{-1} \qquad (i=1,\dots,g-1).$$

The effect of collinearity of the data on maximum likelihood estimation for the logistic model has been studied by Schaefer (1986). He presented some simple transformations of the maximum likelihood estimator $\hat{\alpha}$ that reduce the effect of the collinearity. More recently, Duffy and Santner (1989) have described the computation of norm-restricted maximum likelihood estimators of the logistic regression coefficients.

8.2.5 Existence of Maximum Likelihood Solution

We consider now the pattern of data points for which the likelihood appropriate under the logistic model, $L(\boldsymbol{\alpha};\mathbf{t}_z \mid \mathbf{t}_x)$, does not have a maximum for finite $\boldsymbol{\alpha}$. It is convenient here to write those observations from $\mathbf{x}_1,\dots,\mathbf{x}_n$ belonging to G_i (that is, those with $z_{ij}=1$) as $\mathbf{x}_{ij}$ $(i=1,\dots,g;\ j=1,\dots,n_i)$.

Although the likelihood $L(\boldsymbol{\alpha};\mathbf{t}_z \mid \mathbf{t}_x)$ is bounded above by one for all $\boldsymbol{\alpha}$, some configurations of the data can result in a nonunique maximum on the boundary of the parameter space at infinity. It is necessary, therefore, to recognize those data configurations that lead to infinite estimates, in order to avoid unnecessary iterations in the optimization process. Albert and Lesaffre (1986) have explored this problem and, by considering the possible patterns of the data $\mathbf{x}_1,\dots,\mathbf{x}_n$, have proved existence theorems for the maximum likelihood estimates. The patterns fall essentially into three distinct categories: complete separation, quasi-complete separation, and overlap. Albert and Lesaffre (1986) have noted since that this categorization does not cover all possibilities, and that other data configurations can cause problems.

The first category of complete separation was recognized by Day and Kerridge (1967). By definition, there is complete separation in the sample points if there exists $\boldsymbol{\alpha}$ belonging to $\mathbb{R}^d$ such that for a given i $(i=1,\dots,g)$,

$$\tau_i(\mathbf{x}_{ij};\boldsymbol{\alpha}) > \tau_h(\mathbf{x}_{ij};\boldsymbol{\alpha}) \qquad (j=1,\dots,n_i;\ h=1,\dots,g;\ h\neq i). \quad (8.2.15)$$

Here $d=(g-1)(p+1)$ denotes the dimension of $\boldsymbol{\alpha}$. To examine this more closely for $g=2$ groups, suppose that $\tilde{\beta}_{01}+\tilde{\boldsymbol{\beta}}_1'\mathbf{x}$ is a separating hyperplane, whose existence is implied by (8.2.15). Suppose that $\tilde{\beta}_{01}+\tilde{\boldsymbol{\beta}}_1'\mathbf{x}>0$ for $\mathbf{x}$ lying on the same side of the hyperplane as the $\mathbf{x}_{1j}$ from G_1, so that $\tilde{\beta}_{01}+\tilde{\boldsymbol{\beta}}_1'\mathbf{x}<0$ for the $\mathbf{x}_{2j}$ from G_2. Let $\boldsymbol{\alpha}=(c\tilde{\beta}_{01},c\tilde{\boldsymbol{\beta}}_1')'$. Then, as $c\to\infty$,

$$L(\boldsymbol{\alpha};\mathbf{t}_z \mid \mathbf{t}_x) \to 1.$$

Hence, there is a maximum of $L(\boldsymbol{\alpha};\mathbf{t}_z \mid \mathbf{t}_x)$ at infinity. Although this implies that $\boldsymbol{\alpha}$ cannot be estimated with much precision, the discriminant function corresponding to any separating hyperplane will have zero overall misallocation rate in its application to the training data. Although the apparent error rate of a discriminant rule can provide an optimistic assessment of the rule in its application to data outside the training data, it is desirable that a discriminant rule with a zero apparent error rate be found if it exists; see Greer (1979). In fitting a logistic model, it is quick and simple to check whether there is complete separation at any each stage of the iterative process; see Day and Kerridge (1967) and J. A. Anderson (1972) for further details.

The second category of quasi-complete separation occurs if there exists a nonzero $\boldsymbol{\alpha}$ belonging to $\mathbb{R}^d$ such that for a given i ($i = 1,\ldots,g$),

$$\tau_i(\mathbf{x}_{ij};\boldsymbol{\alpha}) \geq \tau_h(\mathbf{x}_{ij};\boldsymbol{\alpha}) \qquad (j = 1,\ldots,n_i;\ h = 1,\ldots,g;\ h \neq i),$$

with equality for at least one (h,i,j)-triplet. A special case of this general concept as defined above by Albert and Anderson (1984) was noted by J. A. Anderson (1974) in the case of $g = 2$ groups, where one of the variables, say, the first element of $\mathbf{x}$, is binary and where $(\mathbf{x}_{1j})_1 = 0$ $(j = 1,\ldots,n_1)$ and $(\mathbf{x}_{2j})_1 = 1$ for at least one value of j $(j = 1,\ldots,n_2)$. Then the upper bound of one is attained by $L(\boldsymbol{\alpha};\mathbf{t}_z \mid \mathbf{t}_x)$ as $(\boldsymbol{\beta}_1)_1 \to -\infty$. Further discussion of this category is found in Albert and Lesaffre (1986), who also discuss some intermediate situations, neither complete nor quasicomplete separation, where the maximum likelihood estimate of $\boldsymbol{\alpha}$ does not exist. These are referred to as partial separation. A simple example of such a data pattern for $g = 3$ groups occurs where two of the groups overlap, but the other is completely separated from these two. This concept can be generalized to several groups, as considered by Lesaffre and Albert (1989a). Geometrically, partial separation is defined as complete separation of clusters of groups.

The third category is overlap and it occurs if neither complete, quasi-complete, nor (quasi)-partial separation exists in the sample points. The maximum likelihood estimate of $\boldsymbol{\alpha}$ exists and is unique if and only if there is overlap of $\mathbf{x}_1,\ldots,\mathbf{x}_n$. The work of Albert and Anderson (1984) on the existence and uniqueness of the maximum likelihood estimate of $\boldsymbol{\alpha}$ in the logistic regression model has been expanded recently by Santner and Duffy (1986). They have given a linear program that determines whether the configuration of the data belongs to category one, two, or three as defined above. Lesaffre and Albert (1989a) have developed a general algorithm to help to distinguish separation from multicollinearity given divergence in the maximum likelihood estimation procedure.

8.2.6 Predictive Logistic Discrimination

In the predictive framework of Aitchison and Dunsmore (1975), the estimated posterior probability of membership of the first of two groups,

$$\tau_1(\mathbf{x};\boldsymbol{\alpha}) = \exp(\boldsymbol{\alpha}'\mathbf{x}^+)/\{1+\exp(\boldsymbol{\alpha}'\mathbf{x}^+)\},$$

is replaced by

$$\hat{\tau}_1^{(P)}(\mathbf{x}) = \int \tau_1(\mathbf{x};\boldsymbol{\alpha})\phi(\boldsymbol{\alpha};\hat{\boldsymbol{\alpha}},\hat{\mathbf{V}})\,d\boldsymbol{\alpha}, \tag{8.2.16}$$

where, corresponding to the Bayesian version of maximum likelihood theory, the posterior density of $\boldsymbol{\alpha}$ is taken to be multivariate normal with mean $\hat{\boldsymbol{\alpha}}$ and with covariance matrix equal to an estimate $\hat{\mathbf{V}}$ of the covariance matrix of $\hat{\boldsymbol{\alpha}}$. Using the approximation (8.1.6), Aitken (1978) showed that (8.2.16) reduces to

$$\hat{\tau}_1^{(P)}(\mathbf{x}) \simeq \Phi\{(c+\mathbf{x}^{+\prime}\hat{\mathbf{V}}\mathbf{x}^{+})^{-1/2}\hat{\boldsymbol{\alpha}}'\mathbf{x}^{+}\}. \tag{8.2.17}$$

Other values of c besides $c = 1.6$ can be used in the approximation (8.1.6) leading to (8.2.17); see Aitchison and Begg (1976) and Lauder (1978). Aitken (1978) has compared this predictive version of the logistic discriminant rule with several other rules for the case of multivariate binary data.

8.3 BIAS CORRECTION OF MLE FOR $g = 2$ GROUPS

8.3.1 Introduction

It is well known that the bias of maximum likelihood estimators can be of some concern in practice in small samples. As in the previous sections, $\hat{\boldsymbol{\alpha}}$ refers to the estimate of the vector $\boldsymbol{\alpha}$ of parameters in the logistic model, obtained by maximization of $L(\boldsymbol{\alpha};\mathbf{t}_z \mid \mathbf{t}_x)$. One way of correcting $\hat{\boldsymbol{\alpha}}$ for bias is by an application of the bootstrap as introduced by Efron (1979). The use of the bootstrap to assess the sampling distribution of the posterior probabilities of group membership is described in Chapter 11 in a general context, where the logistic model need not necessarily be assumed to hold.

We therefore focus here on results specific to the logistic model. Most of the results in the literature are for $g = 2$ groups and so we will confine our attention to this case. As a consequence, we will write β_{01} and $\boldsymbol{\beta}_1$ as β_0 and $\boldsymbol{\beta}$, suppressing the subscript one which is now superfluous. Hence, now

$$\begin{aligned}\boldsymbol{\alpha} &= (\alpha_1,\ldots,\alpha_{p+1})' \\ &= (\beta_0,\boldsymbol{\beta}')'.\end{aligned}$$

In most situations, it is the bias of $\tau_1(\mathbf{x};\hat{\boldsymbol{\alpha}})$, conditional on $\mathbf{x}$, that is of interest. It is analytically more convenient to work with the log odds $\eta(\mathbf{x};\hat{\boldsymbol{\alpha}})$ and to consider bias correction of $\hat{\boldsymbol{\alpha}} = (\hat{\beta}_0,\hat{\boldsymbol{\beta}}')'$ since, from (8.1.3),

$$\begin{aligned}\eta(\mathbf{x};\hat{\boldsymbol{\alpha}}) &= \log\{\tau_1(\mathbf{x};\hat{\boldsymbol{\alpha}})/\tau_2(\mathbf{x};\hat{\boldsymbol{\alpha}})\} \\ &= \hat{\beta}_0 + \hat{\boldsymbol{\beta}}'\mathbf{x}.\end{aligned}$$

Moreover, in an epidemiological context, it is the estimation of $\boldsymbol{\alpha}$ that is of central interest. However, even if the unbiased estimation of $\tau_1(\mathbf{x};\boldsymbol{\alpha})$ were of prime concern, bias correction of $\hat{\boldsymbol{\alpha}}$ should still be a useful exercise, in particular for the region

$$|\hat{\beta}_0 + \hat{\boldsymbol{\beta}}'\mathbf{x}| \leq 3,$$

because there $\tau_i(\mathbf{x};\hat{\boldsymbol{\alpha}})$ is approximately linear in $\eta(\mathbf{x};\hat{\boldsymbol{\alpha}})$; see Cox (1970).

We consider the bias correction of $\hat{\boldsymbol{\alpha}}$ firstly in the regression situation, where $L(\boldsymbol{\alpha}, \mathbf{t}_z \mid \mathbf{t}_x)$ is the relevant likelihood, as under the **x**-conditional or mixture sampling schemes. For brevity of expression, we will henceforth write this likelihood as $L(\boldsymbol{\alpha})$, suppressing the notation that denotes that it has been formed from $\mathbf{z}_1, \ldots, \mathbf{z}_n$ conditional on $\mathbf{x}_1, \ldots, \mathbf{x}_n$, respectively.

8.3.2 Asymptotic Bias for Both x-Conditional and Mixture Sampling Schemes

From the general expansion for the asymptotic bias of a maximum likelihood estimator under regularity conditions as given, for example, by Cox and Snell (1968), the bias of $\hat{\boldsymbol{\alpha}}$ can be expanded in the form

$$\text{bias}(\hat{\alpha}_s) = \tfrac{1}{2}\mathcal{I}^{rs}\mathcal{I}^{tu}(K_{rtu} + 2J_{t,ru}) + o(1/n) \qquad (s = 1, \ldots, p+1), \qquad (8.3.1)$$

where

$$K_{rtu} = E(\partial^3 \log L(\boldsymbol{\alpha})/\partial\alpha_r\partial\alpha_t\partial\alpha_u),$$

and

$$J_{r,tu} = E\{(\partial \log L(\boldsymbol{\alpha})/\partial\alpha_t)(\partial^2 \log L(\boldsymbol{\alpha})/\partial\alpha_r\partial\alpha_u)\},$$

and where the summation convention applies to multiple suffixes. The superscripts in $\mathcal{I}^{rs}$ denote matrix inversion of the information matrix $\mathcal{I}$, so that $\mathcal{I}^{rs} = (\mathcal{I}^{-1})_{rs}$, where

$$\begin{aligned}(\mathcal{I})_{rs} &= E\{(\partial \log L(\boldsymbol{\alpha})/\partial\alpha_r)(\partial \log L(\boldsymbol{\alpha})/\partial\alpha_s)\}\\ &= -E(\partial^2 \log L(\boldsymbol{\alpha})/\partial\alpha_r\partial\alpha_s) \qquad (r,s = 1,\ldots,p+1).\end{aligned}$$

There is no need for expectations to be taken in (8.3.1), as we may use the approximations

$$E(\partial^3 \log L(\boldsymbol{\alpha})/\partial\alpha_r\partial\alpha_t\partial\alpha_u) \approx \partial^3 \log L(\hat{\boldsymbol{\alpha}})/\partial\alpha_r\partial\alpha_t\partial\alpha_u$$

and

$$\begin{aligned}&E\{(\partial \log L(\boldsymbol{\alpha})/\partial\alpha_t)(\partial^2 \log L(\boldsymbol{\alpha})/\partial\alpha_r\partial\alpha_u)\}\\ &\quad\approx \sum_{j=1}^{n}(\partial \log L(\hat{\boldsymbol{\alpha}};\mathbf{z}_j \mid \mathbf{x}_j)/\partial\alpha_t)(\partial^2 \log L(\hat{\boldsymbol{\alpha}};\mathbf{z}_j \mid \mathbf{x}_j)/\partial\alpha_r\partial\alpha_u),\end{aligned}$$

where $L(\boldsymbol{\alpha};\mathbf{z}_j \mid \mathbf{x}_j)$ given by (8.2.4) is the likelihood formed from the single response $\mathbf{z}_j$ conditional on $\mathbf{x}_j$. This device has been used extensively in maximum likelihood estimation to provide standard errors.

However, in this particular instance where E in (8.3.1) refers to expectation under the **x**-conditional sampling scheme as specified by (8.2.1), the expansion (8.3.1) is simplified considerably by evaluating the expectations. To this end, let $\mathbf{t}_x^+$ be the $n \times (p+1)$ matrix defined by

$$\mathbf{t}_x^{+\prime} = (\mathbf{x}_1^+, \ldots, \mathbf{x}_n^+),$$

so that $\mathbf{t}_x^+ = (\mathbf{1}_n, \mathbf{t}_x)$, where $\mathbf{1}_n$ denotes the n-dimensional vector with each element one and, as before,

$$\mathbf{t}_x' = (\mathbf{x}_1, \ldots, \mathbf{x}_n).$$

From the exponential family form (8.2.5) for $L(\boldsymbol{\alpha})$, it can be seen for $g = 2$ that

$$\begin{aligned} \mathcal{I} &= \operatorname{cov}(\mathbf{U}_1) \\ &= \mathbf{t}_x^{+\prime}\mathbf{H}\mathbf{t}_x, \end{aligned} \tag{8.3.2}$$

where $\mathbf{H} = \operatorname{diag}(h_1, \ldots, h_n)$, and

$$h_j = \tau_1(\mathbf{x}_j; \boldsymbol{\alpha})\tau_2(\mathbf{x}_j; \boldsymbol{\alpha}) \qquad (j = 1, \ldots, n).$$

Also, $J_{t,ru}$ is zero and K_{rtu} reduces to

$$K_{rtu} = -\sum_{j=1}^{n} x_{jr}^+ x_{jt}^+ x_{ju}^+ \tau_1(\mathbf{x}_j; \boldsymbol{\alpha})\tau_2(\mathbf{x}_j; \boldsymbol{\alpha})\{1 - 2\tau_1(\mathbf{x}_j; \boldsymbol{\alpha})\}, \tag{8.3.3}$$

where $x_{jr}^+ = (\mathbf{x}_j^+)_r$ $(j = 1, \ldots, n;\ r = 1, \ldots, p+1)$. On substituting (8.3.2) and (8.3.3) into (8.3.1), it can be verified that the asymptotic bias of $\hat{\boldsymbol{\alpha}}$ can be expressed in the form

$$\operatorname{bias}(\hat{\boldsymbol{\alpha}}) \approx \mathbf{B}(\boldsymbol{\alpha}; \mathbf{t}_x), \tag{8.3.4}$$

where

$$\mathbf{B}(\boldsymbol{\alpha}; \mathbf{t}_x) = -\tfrac{1}{2}(\mathbf{t}_x^{+\prime}\mathbf{H}\mathbf{t}_x^+)^{-1}\mathbf{t}_x^{+\prime}\mathbf{H}\mathbf{a}, \tag{8.3.5}$$

and where

$$(\mathbf{a})_j = \{1 - 2\tau_1(\mathbf{x}_j; \boldsymbol{\alpha})\}\mathbf{x}_j^{+\prime}(\mathbf{t}_x^{+\prime}\mathbf{H}\mathbf{t}_x^+)^{-1}\mathbf{x}_j^+ \qquad (j = 1, \ldots, n). \tag{8.3.6}$$

As pointed out by McLachlan (1985) and Richardson (1985), the result (8.3.4) is available from the work of Byth and McLachlan (1978), Anderson and Richardson (1979), and McLachlan (1980a); see also Schaefer (1983). Walter (1985) showed that (8.3.4) reduces, in the special case of a single dichotomous covariate variable, to an earlier result of Haldane (1956), for the estimation of a single log-odds ratio.

From (8.3.4), $\boldsymbol{\alpha}$ may be corrected for bias in the **x**-conditional sampling design, as

$$\hat{\boldsymbol{\alpha}} - \mathbf{B}(\hat{\boldsymbol{\alpha}}; \mathbf{t}_x). \tag{8.3.7}$$

This bias-corrected version of $\hat{\boldsymbol{\alpha}}$ applies also under mixture sampling. We saw in Section 8.2 that $\hat{\boldsymbol{\alpha}}$ is also the maximum likelihood estimate of $\boldsymbol{\alpha}$ under mixture sampling, as the latter estimate is obtained by maximization of the same function $L(\boldsymbol{\alpha})$ as for $\mathbf{x}$-conditional sampling. It follows that (8.3.4) can be used to correct $\hat{\boldsymbol{\alpha}}$ for bias under mixture sampling.

8.3.3 Asymptotic Bias for Separate Sampling Scheme

Concerning the case of separate sampling, Anderson and Richardson (1979) performed some simulations to support their conjecture that (8.3.4) can be used for separate sampling. McLachlan (1980a) subsequently confirmed on a theoretical basis the applicability of (8.3.4), with minor adjustment, to separate sampling. With the latter scheme, the estimable parameters are β_0^o and $\boldsymbol{\beta}$. Of course, if the prior probabilities π_i are known, or estimates of them are available from another source, β_0 can be estimated from the relation

$$\beta_0^o = \beta_0 - \log(\pi_1/\pi_2).$$

We let $\hat{\boldsymbol{\alpha}}^o = (\hat{\beta}_0^o, \hat{\boldsymbol{\beta}})'$, where

$$\hat{\beta}_0^o = \hat{\beta}_0 - \log(n_1/n_2),$$

and where, as before, $\hat{\beta}_0$ and $\hat{\boldsymbol{\beta}}$ are the maximum likelihood estimates for both $\mathbf{x}$-conditional and mixture sampling schemes. It was seen in Section 8.2.3 that the use of $\hat{\boldsymbol{\alpha}}^o$ applies also under separate sampling. We let $b_i(\pi_1, \pi_2, \boldsymbol{\alpha}^o)$ denote the bias of $\hat{\beta}_i$ $(i = 0, 1, \ldots, p)$ under mixture sampling. Its asymptotic assessment is denoted by $B_i(\boldsymbol{\alpha}; \mathbf{t}_x)$, the $(i+1)$th element of $\mathbf{B}(\boldsymbol{\alpha}; \mathbf{t}_x)$ as defined by (8.3.5). The bias of $\hat{\beta}_0^o$ and of $\hat{\beta}_i$ under separate sampling is denoted by $b_{0S}^o(q_1, q_2, \boldsymbol{\alpha}^o)$ and by $b_{iS}(q_1, q_2; \boldsymbol{\alpha}^o)$, respectively $(i = 1, \ldots, p)$, where $q_1 = 1 - q_2 = n_1/n$. McLachlan (1980a) showed that

$$b_{0S}^o(q_1, q_2, \boldsymbol{\alpha}^o) = b_0(q_1, q_2, \boldsymbol{\alpha}^o) + (q_2 - q_1)/(2q_1q_2n) + o(1/n) \tag{8.3.8}$$

and

$$b_{iS}(q_1, q_2, \boldsymbol{\alpha}^o) = b_i(q_1, q_2, \boldsymbol{\alpha}^o) + o(1/n) \qquad (i = 1, \ldots, p). \tag{8.3.9}$$

It was noted above that $B_i(\hat{\boldsymbol{\alpha}}; \mathbf{t}_x)$ can be used under mixture sampling to assess the bias of $\hat{\beta}_i$, which we have expressed as $b_i(\pi_1, \pi_2, \boldsymbol{\alpha}^o)$, for $i = 0, 1, \ldots, p$. As $B_i(\hat{\boldsymbol{\alpha}}; \mathbf{t}_x)$ is formed from a sample of size nq_i from G_i $(i = 1, 2)$ and q_i converges in probability to π_i, as $n \to \infty$, for mixture sampling, it follows that $B_i(\hat{\boldsymbol{\alpha}}; \mathbf{t}_x)$ can be used to assess $b_i(q_1, q_2, \boldsymbol{\alpha}^o)$. Hence, from (8.3.8) and (8.3.9), $\hat{\beta}_0^o$ and $\hat{\beta}_i$ can be corrected for bias under separate sampling as

$$\hat{\beta}_0^o - B_0(\hat{\boldsymbol{\alpha}}; \mathbf{t}_x) - (q_2 - q_1)/(2q_1q_2n)$$

and

$$\hat{\beta}_i - B_i(\hat{\boldsymbol{\alpha}}; \mathbf{t}_x),$$

respectively ($i = 1,\ldots,p$). Thus, the bias correction of $\hat{\beta}_i$ is the same as for **x**-conditional or mixture sampling schemes.

In addition to the asymptotic theory to support the case of bias correction of $\hat{\boldsymbol{\alpha}}$, Anderson and Richardson (1979) provided empirical evidence from studies using simulated and real data. They found that a reduction in bias was achieved by and large without a corresponding increase in variability and thus with a reduction in the mean-squared error.

8.3.4 Effect of Bias Correction on Discrimination

Besides their work on the asymptotic bias of $\hat{\boldsymbol{\alpha}}$, Byth and McLachlan (1978) also considered the asymptotic biases of the posterior probabilities $\tau_i(\mathbf{x};\hat{\boldsymbol{\alpha}})$ in a discriminant analysis context. They showed that unconditionally over $\mathbf{X}$, where

$$\mathbf{X} \sim N(\boldsymbol{\mu}_i, \boldsymbol{\Sigma}) \quad \text{with prob. } \pi_i \text{ in } G_i \qquad (i = 1,2), \tag{8.3.10}$$

the bias of $\tau_1(\mathbf{X};\hat{\boldsymbol{\alpha}})$ is $o(1/n)$. By contrast, the unconditional bias of $\tau_1(\mathbf{X};\hat{\boldsymbol{\alpha}}_E)$ is $O(1/n)$, where $\hat{\boldsymbol{\alpha}}_E$ is the maximum likelihood estimator of $\boldsymbol{\alpha}$ obtained from the full likelihood function formed from $\mathbf{t}$ under the normal mixture model (8.3.10) with equal component-covariance matrices. The full maximum likelihood estimator $\hat{\boldsymbol{\alpha}}_E$ is as defined in Section 4.4.4.

More generally, Byth and McLachlan (1978) were able to show that the unconditional bias of $\tau_1(\mathbf{X};\hat{\boldsymbol{\alpha}})$ is $o(1/n)$ for mixture sampling from all continuous distributions that are valid under the logistic model (8.1.1) and that satisfy certain regularity conditions. These results, where the bias of $\tau_1(\mathbf{X};\hat{\boldsymbol{\alpha}})$ is taken over the mixture distribution of $\mathbf{X}$ in addition to over the sampling distribution of $\hat{\boldsymbol{\alpha}}$, are, of course, only relevant in global considerations of the estimated posterior probabilities of group membership.

Of associated interest is the combined effect of bias correction of $\hat{\boldsymbol{\alpha}}$ on the sampling properties of the misallocation rates of the logistic discriminant rule $r_o(\mathbf{x};\hat{\boldsymbol{\alpha}})$. This effect was investigated asymptotically by McLachlan (1980a) for mixture sampling of $\mathbf{T}$ under (8.3.10). In this particular case, he established that bias correction of $\hat{\boldsymbol{\alpha}}$ produces no first-order reduction in the overall unconditional error rate of $r_o(\mathbf{x};\hat{\boldsymbol{\alpha}})$. The individual unconditional error rates do undergo a first-order change under bias correction for unequal prior probabilities. It was concluded from this study that the effect of bias correction is to generally decrease the rate for the group whose prior probability is greater than 0.5, but at the same time to increase by more the rate for the other group.

8.4 ASSESSING THE FIT AND PERFORMANCE OF LOGISTIC MODEL

8.4.1 Assessment of Model Fit

Several chi-squared goodness-of-fit tests have been proposed for evaluating the fit of the logistic regression model. Among these are the methods considered by Hosmer and Lemeshow (1980), Tsiatis (1980), Lemeshow and Hosmer (1982), Hosmer and Lemeshow (1985), and Hosmer, Lemeshow, and Klar (1988). A review of some of these goodness-of-fit statistics has been given recently by Green (1988), who has proposed a simple measure of the discriminatory power of the fitted logistic regression model, based on maximization of Youden's J index.

Indeed, in recent times, much attention has been devoted to the assessment of model fit for binary logistic regression. For example, there are the regression diagnostics of Pregibon (1981), the resistant techniques considered in Pregibon (1982), Johnson (1985), and Stefanski, Carroll, and Rupert (1986), and the graphical methods of Landwehr, Pregibon, and Shoemaker (1984). Kay and Little (1986) have illustrated the developments in model choice and assessment in terms of a case study. Fowlkes (1987) has discussed some diagnostic tools that use smoothing techniques. More recently, Lesaffre and Albert (1989b) have extended the logistic regression diagnostics of Pregibon (1981) to $g > 2$ groups. They have developed diagnostics that measure the influence of each observation on the performance of the sample logistic rule $r_o(\mathbf{x};\hat{\boldsymbol{\alpha}})$. Copas (1988) has considered the nature of outliers in the context of binary regression models. He has proposed a simple model that allows for a small number of the binary responses being misrecorded.

For logistic regression models, Kay and Little (1987) have investigated how to improve the fit to the data by transformations of the explanatory variables. The generalized additive models of Hastie and Tibshirani (1986, 1990) can be used to suggest suitable transformations of $\mathbf{x}$.

There is also the approach of Copas (1983), based on smoothed estimates of the log odds, for assisting model choice in terms of functions of $\mathbf{x}$. For univariate data x_j with $(\mathbf{z}_j)_1 = z_{1j}$ $(j = 1,\ldots,n)$, $\tau_1(x;\boldsymbol{\alpha})$ is estimated by

$$\tilde{\tau}_1(x) = \frac{\sum_{j=1}^{n} z_{1j}\phi(h^{-1}(x - x_j))}{\sum_{j=1}^{n} \phi(h^{-1}(x - x_j))}, \tag{8.4.1}$$

where the standard normal density $\phi(x)$ is taken as the smooth kernel, and h is a suitably chosen smoothing constant. That is, $\tilde{\tau}_1(x)$ is the observed proportion of observations from G_1, smoothed over neighboring values of x. This smoothing is needed if x is discrete and n is not large or if x is continuous so that $x_1,\ldots,x_n$ are essentially distinct. Plotting $\text{logit}\{\tilde{\tau}_1(x)\}$ against x will provide information about how x should be included in the logistic model. A linear trend, for example, would confirm that $\beta_1 x$ is the appropriate form to use. As noted by Copas (1981), this plot can be drawn both retrospectively (using the original data) and prospectively (using new data). In the former plot, lack of linearity indicates inadequacies in the model; in the latter, a flattening of the line from 45 degrees indicates the loss of discriminatory power as one goes from original to new data. Some examples of these plots are given in Copas (1983). For multivariate $\mathbf{x}$, one can conduct this plot for a particular component, say, $(\mathbf{x})_r$, by splitting the data according to $(\mathbf{x})_s$ $(s = 1,\ldots,p;\ s \neq r)$ and assessing the form of $\tau_1((\mathbf{x})_r)$ separately within these groups; see Kay and Little (1986). The method can be extended also to model validation for more than $g = 2$ groups (Copas, 1981).

In a discriminant analysis context, the suitability of the logistic model may be approached through consideration of the group-conditional distributions of **X**. For example, if the latter appear to be multivariate normal with unequal covariance matrices, then quadratic terms need to be included in the model as described in Section 8.1.3; see also Kay and Little (1987). However, even in a regression context, Rubin (1984) favors this approach of checking the logistic model by consideration of its implied distribution for **X** given **Z**. His proposal is based on the result of Rosenbaum and Rubin (1983) that **X** and **Z** are conditionally independent given the value of τ_1 of $\tau_1(\mathbf{x};\boldsymbol{\alpha})$, that is,

$$f(\mathbf{x} \mid \tau_1, \mathbf{X} \in G_1) = f(\mathbf{x} \mid \tau_1, \mathbf{X} \in G_2).$$

If $\hat{\tau}_1 = \tau_1(\mathbf{x};\hat{\boldsymbol{\alpha}})$ is a good estimate of $\tau_1(\mathbf{x};\boldsymbol{\alpha})$, then the regression of $\mathbf{x}$ on $\hat{\tau}_1$ should be the same for the data in both groups G_1 and G_2. To check this, Rubin (1984) suggests

1. plot x versus $\hat{\tau}_1$ using different symbols for the points from G_1 than those from G_2;
2. regress x on $\hat{\tau}_1$ for the x_j from G_1 and then separately for the x_j from G_2, and compare fits and residuals;
3. categorize $\hat{\tau}_1$ and compare the distribution of X for $(\mathbf{z})_1 = 1$ and $(\mathbf{z})_1 = 0$ within categories of $\hat{\tau}_1$.

In this list of suggestions, x refers to a component of $\mathbf{x}$, a scalar function of $\mathbf{x}$, or a vector of components of $\mathbf{x}$, depending on the context and purpose. As explained by Rubin (1984), evidence of inequality in these regressions of x on $\hat{\tau}_1$ is evidence of inadequacy in the estimated logistic regression.

8.4.2 Apparent Error Rate of a Logistic Regression

After the logistic model has been fitted to the data, there is the question of its capacity to accurately predict future responses, that is, in a discriminant analysis context, to correctly allocate future entities. Under the logistic model (8.1.5), we write the Bayes rule of allocation as $r_o(\mathbf{x};\boldsymbol{\alpha})$ to denote explicitly its dependence on the vector $\boldsymbol{\alpha}$ containing the logistic regression coefficients. We will refer to $r_o(\mathbf{x};\hat{\boldsymbol{\alpha}})$ as the logistic discriminant rule (LGDR). An obvious guide to the performance of the LGDR $r_o(\mathbf{x};\hat{\boldsymbol{\alpha}})$ is its apparent error rate given by

$$A(\mathbf{t}) = \frac{1}{n}\sum_{i=1}^{g}\sum_{j=1}^{n} z_{ij} Q[i, r_o(\mathbf{x};\hat{\boldsymbol{\alpha}})],$$

where, for any u and v, $Q[u,v] = 0$ if $u = v$ and 1 for $u \neq v$. As the same training data $\mathbf{t}$ are being used both to fit the model and then to assess its accuracy, the apparent error rate tends to provide an optimistic assessment of the performance of $r_o(\mathbf{x};\hat{\boldsymbol{\alpha}})$.

The overall conditional error rate of $r_o(\mathbf{x};\hat{\boldsymbol{\alpha}})$ is given by

$$\begin{aligned} ec(F;t) &= \sum_{i=1}^{g} E\{Z_i\, Q[i, r_o(\mathbf{X};\hat{\boldsymbol{\alpha}})] \mid \mathbf{t}\} \\ &= \sum_{i=1}^{g} \pi_i E\{Q[i, r_o(\mathbf{X};\hat{\boldsymbol{\alpha}})] \mid \mathbf{t},\, \mathbf{X} \in G_i\}, \end{aligned}$$

where $\mathbf{Y} = (\mathbf{X}', \mathbf{Z}')'$ is distributed independently of $\mathbf{T}$, and where F denotes the distribution function of $\mathbf{Y}$. The bias of $A(\mathbf{t})$ in its estimation of $ec(F;\mathbf{t})$ is

$$b(F) = E\{A(\mathbf{T}) - ec(F;\mathbf{T})\}.$$

Error-rate estimation for an arbitrary sample-based discriminant rule is covered in Chapter 10. The general methodology described there, including the bootstrap and cross-validation, can be applied directly here to correct the apparent error rate of the LGDR $r_o(\mathbf{x};\hat{\boldsymbol{\alpha}})$ for bias. Hence, this topic is not pursued here. Rather we present some theoretical results for the bias of the apparent error rate, which are specific to the logistic model in a regression context.

By conditioning on $\mathbf{t}_x$, Efron (1986) has derived the asymptotic bias of the apparent error rate $A(\mathbf{t})$ in its estimation of

$$ec_x(\boldsymbol{\alpha};\hat{\boldsymbol{\alpha}},\mathbf{t}_x) = \frac{1}{n}\sum_{i=1}^{g}\sum_{j=1}^{n} E\{Z_{ij} Q[i, r_o(\mathbf{x}_j;\hat{\boldsymbol{\alpha}})] \mid \hat{\boldsymbol{\alpha}}, \mathbf{t}_x\}.$$

It can be seen that $ec_x(\boldsymbol{\alpha};\hat{\boldsymbol{\alpha}},\mathbf{t}_x)$ is the average of the conditional error of $r_o(\mathbf{x};\hat{\boldsymbol{\alpha}})$ in its application to n future entities with feature observations fixed at $\mathbf{x}_1,\ldots,\mathbf{x}_n$, respectively.

Let $b(\boldsymbol{\alpha};\mathbf{t}_x)$ denote the asymptotic bias of $A(\mathbf{t})$ in its estimation of $ec_x(\boldsymbol{\alpha};\hat{\boldsymbol{\alpha}}, \mathbf{t}_x)$, so that

$$b(\boldsymbol{\alpha};\mathbf{t}_x) \approx E[\{A(\mathbf{T}) - ec_x(\boldsymbol{\alpha};\hat{\boldsymbol{\alpha}},\mathbf{t}_x)\} \mid \mathbf{t}_x].$$

Efron (1986) showed in the case of $g = 2$ that

$$b(\boldsymbol{\alpha};\mathbf{t}_x) = -\frac{2}{n}\sum_{j=1}^{n} \tau_1(\mathbf{x}_j;\boldsymbol{\alpha})\tau_2(\mathbf{x}_j;\boldsymbol{\alpha})\phi(a_j^{-1/2}\boldsymbol{\alpha}'\mathbf{x}_j^+)a_j^{1/2}, \tag{8.4.2}$$

where a_j is the jth element of $\mathbf{a}$ defined by (8.3.6), divided by $\{1 - 2\tau_1(\mathbf{x}_j;\boldsymbol{\alpha})\}$.

As explained by Efron (1986), the most obvious use of (8.4.2) is to correct $A(\mathbf{t})$ for bias to give

$$A(\mathbf{t}) - b(\hat{\boldsymbol{\alpha}};\mathbf{t}_x).$$

The estimated bias $b(\hat{\boldsymbol{\alpha}};\mathbf{t}_x)$ is also an interesting measure of how vulnerable $r_o(\mathbf{x};\hat{\boldsymbol{\alpha}})$ is to overfitting. A large absolute value of $b(\hat{\boldsymbol{\alpha}};\mathbf{t}_x)$ or $b(\hat{\boldsymbol{\alpha}};\mathbf{t}_x)/A(\mathbf{t})$ suggests retreating to a more parsimonious rule.

TABLE 8.1 The First 10 Trials and Summary Statistics for 100 Trials of a Simulation Experiment for the Bias of the Overall Apparent Error Rate of a Logistic Regression and its Bootstrap Correction

Trial	$ec_x(\boldsymbol{\alpha};\hat{\boldsymbol{\alpha}},\mathbf{t}_x)$	$A(\mathbf{t})$	$A(\mathbf{t})-b(\hat{\boldsymbol{\alpha}};\mathbf{t}_x)$	$b(\hat{\boldsymbol{\alpha}};\mathbf{t}_x)$
1	0.364	0.300	0.409	−0.109
2	0.302	0.300	0.405	−0.105
3	0.378	0.250	0.324	−0.074
4	0.276	0.200	0.289	−0.089
5	0.320	0.250	0.335	−0.085
6	0.369	0.200	0.284	−0.084
7	0.296	0.200	0.278	−0.078
8	0.437	0.100	0.177	−0.077
9	0.336	0.350	0.450	−0.100
10	0.354	0.150	0.234	−0.084
100 trials				
Average	0.342	0.254	0.346	−0.093
(S.D.)	(0.055)	(0.094)	(0.105)	(0.015)

Source: Adapted from Efron (1986).

To illustrate the bias of the apparent error rate of the LGDR $r_o(\mathbf{x};\hat{\boldsymbol{\alpha}})$, we have listed in Table 8.1 the values of the overall error rate $ec_x(\boldsymbol{\alpha};\hat{\boldsymbol{\alpha}},\mathbf{t}_x)$, the apparent error rate $A(\mathbf{t})$, and its bias-corrected version on the first 10 trials of a simulation experiment as reported in Efron (1986) for $n = 20$ observations drawn from a mixture of $g = 2$ bivariate normal distributions under the canonical form (4.2.3) with $\Delta = 1$.

Efron (1986) also derived the asymptotic bias of $A(\mathbf{t})$ in its estimation of $ec_x(\boldsymbol{\alpha};\hat{\boldsymbol{\alpha}},\mathbf{t}_x)$ in the case where the LGDR $r_o(\mathbf{x};\hat{\boldsymbol{\alpha}})$ is based on a possibly inadequate parametric model where some of the variables in $\mathbf{x}$ are omitted in the fitting. Suppose that $\boldsymbol{\alpha}$ is partitioned into $(\boldsymbol{\alpha}_1',\boldsymbol{\alpha}_2')'$ and, correspondingly, $\mathbf{x}_j = (\mathbf{x}_{j1}',\mathbf{x}_{j2}')'$. Let $\mathbf{x}_{j1}^+ = (1,\mathbf{x}_{j1}')'$ for $j = 1,\ldots,n$. If $\hat{\boldsymbol{\alpha}}_1$ denotes the maximum likelihood estimate of $\boldsymbol{\alpha}_1$ under the reduced logistic model where $\boldsymbol{\alpha}_2$ is assumed to be zero, Efron (1986) established that the asymptotic bias of the apparent error rate of $r(\mathbf{x};\hat{\boldsymbol{\alpha}}_1)$ is

$$b(\boldsymbol{\alpha};\mathbf{t}_x) = -\frac{2}{n}\sum_{j=1}^{n}\tau_1(\mathbf{x}_j;\boldsymbol{\alpha})\tau_2(\mathbf{x}_j;\boldsymbol{\alpha})\phi(\tilde{a}_j^{-1/2}\boldsymbol{\alpha}'\mathbf{x}_j^+)(a_{j1}\tilde{a}_j^{-1/2}),$$

where

$$a_{j1} = \mathbf{x}_{j1}^{+\prime}\mathbf{K}_1^{-1}\mathbf{x}_{j1}^+$$

and

$$\tilde{a}_j = \mathbf{x}_{j1}^{+\prime}K_1^{-1}\tilde{K}K_1^{-1}\mathbf{x}_{j1}^+$$

for $j = 1,\ldots,n$, and where

$$\mathbf{K}_1 = \sum_{j=1}^{n} \tau_1(\mathbf{x}_{j1};\boldsymbol{\alpha}_1)\tau_2(\mathbf{x}_{j1};\boldsymbol{\alpha}_1)\mathbf{x}_{j1}^{+}\mathbf{x}_{j1}^{+\prime}$$

and

$$\tilde{\mathbf{K}} = \sum_{j=1}^{n} \tau_1(\mathbf{x}_j;\boldsymbol{\alpha})\tau_2(\mathbf{x}_j;\boldsymbol{\alpha})\mathbf{x}_{j1}^{+}\mathbf{x}_{j1}^{+\prime}.$$

Actually, Efron (1986) considered the bias of the apparent error rate in a wider context for general exponential family linear models and general measures of the apparent error besides the proportion of misallocated entities.

8.4.3 β-Confidence Logistic Discrimination

In these situations where attention is focused on the allocation of an entity with a particular observation $\mathbf{x}$, it is the sampling variation in the estimated posterior probabilities of group membership, conditional on $\mathbf{x}$, which is of central interest. An assessment of the reliability of the estimated posterior probability provided by the logistic model can be undertaken using the methods described in Chapter 11. There the problem of assessing the variability in the estimated posterior probabilities of group membership is addressed for a general discriminant rule. Rather than work directly with the posterior probabilities of group membership, for the logistic model, it is mathematically more convenient, as noted in Section 8.3.1, to work with the log odds, given its linearity in $\boldsymbol{\alpha}$.

In a study of patients with severe head injuries, Stablein et al. (1980) made use of this linearity of the log odds in the logistic model to construct a confidence interval for the discriminant score on each patient and to check whether the upper and lower limits of this interval were correctly allocated. In proceeding in this way, they were able to assess the sharpness of their prognosis for an individual patient. Lesaffre and Albert (1988) extended this work of Stablein et al. (1980) to multiple groups ($g > 2$).

Corresponding to the logistic model (8.1.4), let

$$\begin{aligned}\hat{w}_i(\mathbf{x}) &= \log\{\tau_i(\mathbf{x};\hat{\boldsymbol{\alpha}})/\tau_g(\mathbf{x};\hat{\boldsymbol{\alpha}})\}\\ &= \hat{\boldsymbol{\alpha}}_i'\mathbf{x}^{+}\end{aligned}$$

denote the discriminant score with respect to group G_i ($i = 1,\ldots,g-1$). For given $\mathbf{x}$, the vector of estimated discriminant scores is denoted by $\hat{\mathbf{w}} = (\hat{w}_1,\ldots,\hat{w}_{g-1})'$ and $\mathbf{w}$ is the true score vector corresponding to $\hat{\mathbf{w}}$ with $\hat{\boldsymbol{\alpha}}$ replaced by $\boldsymbol{\alpha}$. For known $\boldsymbol{\alpha}$, an entity with $\mathbf{X} = \mathbf{x}$ is allocated to group G_i if and only if

$$w_i \geq w_h \qquad (h = 1,\ldots,g;\ h \neq i), \tag{8.4.3}$$

where $w_g = 0$. The β-confidence logistic discriminant rule proposed by Lesaffre and Albert (1988) assigns an entity with $\mathbf{X} = \mathbf{x}$ to G_i if and only if (8.4.3) holds for all $\mathbf{w}$ belonging to the ellipsoidal region

$$R_\beta(\hat{\mathbf{w}}) = \{\mathbf{w} : (\mathbf{w} - \hat{\mathbf{w}})'\hat{\mathbf{V}}^{-1}(\mathbf{w} - \hat{\mathbf{w}}) \leq c(\beta)\}, \tag{8.4.4}$$

where $c(\beta)$ denotes the βth quantile of the chi-squared distribution with $g-1$ degrees of freedom, and where $\hat{\mathbf{V}}$ denotes an estimate of the covariance matrix of $\hat{\mathbf{W}}$ conditional on $\mathbf{x}$, given by

$$(\mathbf{V})_{ij} = \mathbf{x}^{+\prime}\text{cov}(\hat{\alpha}_i, \hat{\alpha}_j)\mathbf{x}^+ \qquad (i, j = 1, \ldots, g-1). \tag{8.4.5}$$

From (8.4.5), an estimate of $\mathbf{V}$ can be formed from an available stimate of the covariance matrix of $\hat{\boldsymbol{\alpha}}$, say, $\hat{\mathbf{I}}^{-1}$, the inverse of the observed information matrix for $\boldsymbol{\alpha}$ as considered in Section 8.2.4. This is provided there is no complete, quasi-complete, or partial separation of the data points where the latter estimate breaks down (Albert and Lesaffre, 1986). It follows from the asymptotic theory of maximum likelihood estimation that (8.4.4) defines asymptotically a β-level confidence region for the vector $\mathbf{w}$ of true scores at $\mathbf{x}$. They showed that (8.4.3) holds for $\mathbf{w}$ satisfying (8.4.4) if and only if

$$(\zeta_h'\hat{\mathbf{w}})^2 - c(\beta)(\zeta_h'\hat{\mathbf{V}}\zeta_h) > 0 \qquad (h = 1, \ldots, g;\ h \neq i),$$

where

$$\zeta_h = (\delta_{1i} - \delta_{1h}, \ldots, \delta_{g-1,i} - \delta_{g-1,h})',$$

and δ_{ij} is the usual Kronecker delta. They noted a more convenient approach is to proceed as follows with the allocation of an entity with given $\mathbf{x}$. If the entity is assigned to some group G_i in the ordinary sense (that is, with $\beta = 0$ in the β-confidence rule), determine the maximum value β_m of β for which the entity is still allocated to G_i. It follows that β_m is given by the area to the left of c_m under the χ^2_{g-1} density, where c_m is the maximum value of

$$(\zeta_h'\hat{\mathbf{w}})^2/(\zeta_h'\mathbf{V}\zeta_h)$$

over $h = 1, \ldots, g-1;\ h \neq i$. A value of β_m close to zero gives little weight to the assignment, whereas a high value implies a reliable decision about the group membership.

8.5 LOGISTIC VERSUS NORMAL-BASED LINEAR DISCRIMINANT ANALYSIS

8.5.1 Relative Efficiency of Logistic Approach Under Normality

The asymptotic efficiency of logistic discriminant analysis (LGDA) relative to sample normal-based linear discriminant analysis (NLDA) has been derived by Efron (1975) in the case of $g = 2$ groups under the normal mixture model (8.3.10) with mixture sampling of the training data. As previously, $\hat{\boldsymbol{\alpha}}_E$ denotes the maximum likelihood estimate of $\boldsymbol{\alpha}$ obtained by maximization of the full likelihood function formed from $\mathbf{t}$ under (8.3.10). Hence, when the latter assumption is valid, $\hat{\boldsymbol{\alpha}}_E$ is asymptotically efficient. By contrast then, the logistic estimate $\hat{\alpha}$ is not fully efficient, as it is computed from the conditional likelihood function for α, $L(\alpha;\mathbf{t}_z \mid \mathbf{t}_x)$.

Under (8.3.10), let $ec(\Psi_E;\hat{\alpha})$ denote the overall conditional error rate of the LGDR $r_o(\mathbf{x};\hat{\alpha})$. Then $ec(\Psi_E;\hat{\alpha}_E)$ is the corresponding rate for the sample NLDR $r_o(\mathbf{x};\hat{\alpha}_E)$. As in the previous work, Ψ_E denotes the vector of parameters associated with the normal mixture distribution of $\mathbf{X}$, as specified by (8.3.10). The efficiency of LGDA relative to sample NLDA can be defined by the ratio

$$\epsilon = [E\{ec(\Psi_E;\hat{\alpha}_E)\} - eo(\Psi_E)]/[E\{ec(\Psi_E;\hat{\alpha})\} - eo(\Psi_E)], \qquad (8.5.1)$$

where $eo(\Psi_E)$ is the common limiting value (the optimal error rate) of the overall unconditional error rates $E\{ec(\Psi_E;\hat{\alpha}_E\}$ and $E\{ec(\Psi_E;\hat{\alpha}\}$, as $n \to \infty$. As explained in Section 4.4.3, Efron (1975) derived the asymptotic relative efficiency (ARE) of LGDA in terms of the ratio of the means of the asymptotic distributions of $n\{ec(\Psi_E;\hat{\alpha}_E) - eo(\Psi_E)\}$ and $n\{ec(\Psi_E;\hat{\alpha}) - eo(\Psi_E)\}$. For this problem, it is equivalent to evaluating the numerator and denominator of (8.5.1) up to terms of order $O(1/n)$. The values of the ARE so obtained for LGDA are displayed below for some values of the Mahalanobis distance Δ with $\pi_1 = \pi_2 = 0.5$, the case most favorable to the logistic procedure.

Δ	0	0.5	1	1.5	2	2.5	3	3.5
ARE	1.000	1.000	0.995	0.968	0.899	0.786	0.641	0.486

It can be seen that the LGDR compares quite favorably with the sample NLDR for groups having a small Mahalanobis distance Δ between them ($\Delta \leq 1.5$, say), but, as the groups become more widely separated, the ARE falls off sharply.

To investigate further the difference between the LGDR and the sample NLDR under (8.3.10), McLachlan and Byth (1979) derived the asymptotic expansions of the individual unconditional error rates $E\{ec_i(\Psi_E;\hat{\alpha})\}$ and $E\{ec_i(\Psi_E;\hat{\alpha}_E)\}$ specific to group G_i for $i = 1,2$. On the basis of these expansions, it was found that the unconditional error rates of the LGDR are always close in size to the corresponding error rates of the sample NLDR, even for widely separated groups. Thus, although the terms of order $O(1/n)$ in the asymptotic unconditional error rates of the LGDR may be approximately two to three times as large as the corresponding first-order terms with the sample NLDR, when Δ is large, the actual differences between the error rates of the two procedures are quite small in absolute terms.

More recently, Bull and Donner (1987) have evaluated the ARE of LGDA in the case of more than $g = 2$ normal groups. The ARE was defined by the ratio of the asymptotic variances of the slope parameters. In order to study the ARE of LGDA in situations other than (8.3.10), O'Neill (1980) had previously derived the asymptotic distribution of the overall conditional error rate associated with the LGDR $r_o(\mathbf{x};\hat{\alpha})$ for a general exponential family with respect to Lebesgue measure. The ARE of LGDA was evaluated for $g = 2$ groups with univariate normal distributions having unequal variances and with bivariate exponential distributions. In both cases, the inefficiency of LGDA relative to the full maximum likelihood procedure is more marked than for groups with homoscedastic normal distributions.

8.5.2 LGDA Versus NLDA Under Nonnormality

In this section, logistic discriminant analysis (LGDA) is compared with sample normal-based linear discriminant analysis (NLDA) in situations where a homoscedastic normal model does not apply for the group-conditional distributions of the feature vector. In these situations, NLDA does not provide a consistent estimator of $\boldsymbol{\alpha}$ and so the more robust LGDA is preferable, at least asymptotically, if the logistic model still holds.

Halperin, Blackwelder, and Verter (1971) compared LGDA with NLDA for nonnormal data. In particular, they showed that the estimator $\hat{\boldsymbol{\alpha}}_E$ from NLDA can be quite biased for data consisting entirely of binary variables as well as of a mixture of binary and continuous variables. Further work by O'Hara et al. (1982) and Hosmer, et al. (1983a, 1983b) also showed that $\hat{\boldsymbol{\alpha}}_E$ can be severely biased for mixed continuous and discrete variables. This last point was discussed in Section 5.6.3, where an example of the bias of $\hat{\boldsymbol{\alpha}}_E$ for mixed feature data was given. Amemiya and Powell (1983), however, found in the models they considered that the degree of inconsistency in the error rate of the sample NLDR is surprisingly small. For $\mathbf{x}$ consisting of binary, mutually independent variables, they compared LGDA and NLDA asymptotically in terms of the mean-squared error of estimation and the unconditional error rates of the associated discriminant rules. Their main asymptotic conclusion is that NLDA does quite well in prediction (allocation) and reasonably well in estimation of $\boldsymbol{\alpha}$; see also Amemiya (1985, Chapter 9), Halperin et al. (1985), and Brooks et al. (1988).

Press and Wilson (1978) found that LGDA generally outperformed NLDA in their application to two data sets with both continuous and discrete variables, although the differences in their apparent error rates were not large. On the basis of a simulation study, Crawley (1979) concluded that LGDA is preferable to NLDA when the group-conditional distributions are clearly nonnormal or their dispersion matrices are clearly unequal.

For some univariate normal distributions that satisfy the logistic model, Byth and McLachlan (1980) compared the LGDR with the sample NLDR in terms of the ratio

$$\epsilon = \mathrm{MSE}\{\tau_1(\mathbf{X};\hat{\boldsymbol{\alpha}})\}/\mathrm{MSE}\{\tau_1(\mathbf{X};\hat{\boldsymbol{\alpha}}_E)\}, \qquad (8.5.2)$$

where the numerator and denominator of (8.5.2) are evaluated up to terms of order $O(n^{-1})$. In (8.5.2),

$$\mathrm{MSE}\{\tau_1(\mathbf{X};\hat{\boldsymbol{\alpha}})\} = E\{\tau_1(\mathbf{X};\hat{\boldsymbol{\alpha}}) - \tau_1(\mathbf{X};\boldsymbol{\alpha})\}^2,$$

where E refers to expectation over the distribution of $\hat{\boldsymbol{\alpha}}$ as well as over the mixture distribution of $\mathbf{X}$; similarly, for $\mathrm{MSE}\{\tau_1(\mathbf{X};\hat{\boldsymbol{\alpha}}_E)\}$. As noted by Day and

Kerridge (1967), the logistic model is valid if

$$f_i(\mathbf{x}) = c_i\zeta(\mathbf{x})\exp\{-\tfrac{1}{2}(\mathbf{x}-\boldsymbol{\mu}_i)'\boldsymbol{\Sigma}^{-1}(\mathbf{x}-\boldsymbol{\mu}_i)\} \qquad (i = 1,2), \qquad (8.5.3)$$

where c_i is a normalizing constant, $\boldsymbol{\mu}_i$ is a vector of parameters, $\boldsymbol{\Sigma}$ is a positive definite symmetric matrix of parameters, and $\zeta(\mathbf{x})$ is a nonnegative integrable, scalar function of $\mathbf{x}$. If $\zeta(\mathbf{x}) \equiv 1$, then (8.5.3) reduces to the multivariate normal density with mean $\boldsymbol{\mu}_i$ and covariance matrix $\boldsymbol{\Sigma}$ $(i = 1,2)$. For $\zeta(\mathbf{x}) \not\equiv 1$, $\epsilon \to 0$ as $n \to \infty$, because $\hat{\boldsymbol{\alpha}}_E$ is not consistent, but $\hat{\boldsymbol{\alpha}}$ is. For univariate continuous x, Byth and McLachlan (1980) evaluated ϵ for three specific forms of $\zeta(x)$, corresponding to various levels of skewness and truncation in the group-conditional densities. They concluded that if the distance between the groups is small, then the LGDR is preferable to the sample NLDR whenever the mixture distribution has a reasonably high proportion of members from the more heavily truncated or distorted group. This behavior appears to be true in truncated situations even when the groups are widely separated. However, for groups whose distributions have only been distorted and not so fundamentally altered as under truncation, the relative performance of the LGDR and the sample NLDR tends to resemble that under normality unless the distortion to both distributions is quite severe.

Many of the investigations referenced in Sections 5.6.2 and 5.6.3 on the robustness of the sample NLDR for discrete and mixed feature data have included the LGDR for comparative purposes. Also, the LGDR has been included mostly in the various comparative studies of nonparametric discriminant rules, which are referenced in Section 9.6. A common conclusion of these studies is that the allocatory performance of the LGDR is similar to that of the sample NLDR.

Concerning comparisons of the logistic estimator $\hat{\boldsymbol{\alpha}}$ with other estimators of $\boldsymbol{\alpha}$ besides $\hat{\boldsymbol{\alpha}}_E$ for the dichotomous $(g = 2)$ logistic model, Amemiya (1980) found for many examples, both real and artificial, that the minimum chi-squared estimator of $\boldsymbol{\alpha}$ has smaller n^{-2}-order mean-squared error than that of $\hat{\boldsymbol{\alpha}}$. However, L. Davis (1984) subsequently showed that the minimum chi-squared estimator of $\boldsymbol{\alpha}$ has smaller n^{-2}-order mean-squared error only for certain designs and parameter values.

8.6 EXAMPLE: DIFFERENTIAL DIAGNOSIS OF SOME LIVER DISEASES

We consider here the differential diagnosis of four liver diseases from a enzyme laboratory profile, as envisaged in Plomteux (1980) and analyzed further by Albert and Lesaffre (1986), Lesaffre and Albert (1988, 1989b). The three analytes, aspartate aminotransferase (AST), alanine aminotransferase (ALT), and glutamate dehydrogenase (GIDH), all expressed in International units per liter (IU/L), were taken to be the feature variables. The training data consisted of observations on $n_1 = 57$ patients from group G_1 (viral hepatitis),

TABLE 8.2 Maximum Likelihood Estimates of the Logistic Regression Coefficients with Their Standard Errors in Parentheses

	$\hat{\alpha}_1$	$\hat{\alpha}_2$	$\hat{\alpha}_3$
Constant	−1.930 (0.800)	1.140 (0.630)	−3.500 (0.500)
AST	−0.055 (0.010)	−0.057 (0.010)	−0.021 (0.006)
ALT	0.073 (0.010)	0.063 (0.010)	0.042 (0.009)
GIDH	−0.170 (0.079)	−0.240 (0.070)	0.061 (0.019)

Source: From Lesaffre and Albert (1988).

$n_2 = 44$ from G_2 (persistent chronic hepatitis), $n_3 = 40$ from G_3 (aggressive chronic hepatitis), and $n_4 = 77$ from G_4 (postnecrotic cirrhosis). These data are reproduced in Albert and Harris (1987). The empirical group-conditional distributions of the three feature variables exhibit marked positive skewness. Thus, in a fully parametric approach such as NLDA, the first action would be to transform the data in an attempt to remove the skewness. However, in fitting a semiparametric model such as the logistic, it is debatable whether to apply the transformation to the data; see Kay and Little (1987) and Lesaffre and Albert (1989b) on this point.

In Table 8.2, we display the maximum likelihood fit of the logistic model, using the original scale of the feature variables and with G_4 taken as the base group. As reported by Albert and Lesaffre (1986), the overall apparent error rate of this logistic rule when it is applied to the training data is 83%. It compares with 83% and 82% for the apparent and cross-validated rates, respectively, for the sample NLDR.

This fit of the logistic model to the four groups of hepatitis was used by Lesaffre and Albert (1988) to illustrate their β-confidence logistic discriminant rule. They applied it to a patient with feature vector $\mathbf{x} = (165, 330, 21)'$. The estimates of the group-posterior probabilities were equal to 0.64, 0.11, 0.25, and 0.0, respectively. The value of β_m for this patient is equal to 0.28, indicating a fair degree of uncertainity about the assignment. Lesaffre and Albert (1988) also considered the allocation of a second patient with feature vector $\mathbf{x}$ equal to $(75, 125, 50)'$, for which $\tau_i(\mathbf{x}; \hat{\alpha})$ is equal to 0.09, 0.15, 0.64, and 0.12 for $i = 1$, 2, 3, and 4, respectively. The value of β_m for this $\mathbf{x}$ is equal to 0.72. Thus, although in this instance the highest posterior probability of group membership also equals 0.64, the reliability of the assignment is much higher, implying a clearer decision. As remarked by Lesaffre and Albert (1988), this can be explained to some extent by the fact that the feature vector for the first patient lies somewhat extreme in the feature space, where prediction is generally less stable.

The previous data set was used also by Lesaffre and Albert (1989b) to illustrate the regression diagnostics they developed for the multiple-group logistic model. As part of their illustration, they also analyzed an augmented version of this set, where a control group (G_5) of $n_5 = 82$ presumably healthy individuals was added to the four existing hepatitis groups. The application of logistic

TABLE 8.3 Allocation Rates for the Logistic Model with log(AST), log(ALT), and log(GIDH) Fitted to the Five-Group Hepatitis Data Set

	True Group				
Allocated Group	G_1	G_2	G_3	G_4	G_5
G_1	53 (93%)	4 (9.1%)	1 (2.5%)	0 (0%)	0 (0%)
G_2	2 (3.5%)	38 (86.4%)	2 (5%)	1 (1%)	2 (2%)
G_3	2 (3.5%)	0 (0%)	23 (57.5%)	10 (13%)	0 (0%)
G_4	0 (0%)	1 (2.2%)	14 (3.5%)	66 (86%)	1 (1.2%)
G_5	0 (0%)	1 (2.2%)	0 (0%)	0 (0%)	79 (96.3%)
Total	57	44	40	77	82

Source: From Lesaffre and Albert (1988).

regression diagnostics to this five-group data set revealed a number of influential and outlying points. Following Kay and Little (1987), Lesaffre and Albert (1989b) logarithmically transformed all three feature variables. It resulted in an overall apparent rate of 86.3%, compared to 86.7% for the logistic model based on the untransformed data. The apparent allocation rates for the logistic model fitted to the transformed data are displayed in Table 8.3. In can be seen that there is some overlap between the diseases, but that most patients from G_1, G_2, and G_4 are correctly allocated. The controls are well separated from the diseased patients, with only three cases, 221, 245, and 268, misallocated.

The logarithmic transformation of the variables diminished the outlying character of most points, except for one on a control patient numbered 268. Lesaffre and Albert (1989b) subsequently found that the logistic model fitted to log(AST), log(ALT), and to GIDH rather than log(GIDH) no longer contained any influential observations. They expressed their surprise that the logarithmic transformation of variable GIDH, suggested by its group-conditional distribution (Kay and Little, 1987), resulted in such a clearly distinct influential observation. However, in order to quantify the impact of this case on the allocatory performance of the logistic discriminant rule, Lesaffre and Albert (1989b) deliberately chose to work with the model fitted to the logarithms of all three variables. In so doing, they found that case 268 influenced all the $\hat{\alpha}_i$ $(i = 1,2,3)$, but none of the $\hat{\alpha}_i - \hat{\alpha}_h$ $(i,h = 1,2,3,4;\ i \neq h)$. Thus, only allocation between the diseased cases and the controls is affected. However, because each $\hat{\alpha}_i$ occurs in each $\tau_i(\mathbf{x};\hat{\alpha})$, it is not obvious in what way case 268 affects the estimates of the posterior probabilities of group membership. In checking on this, Lesaffre and Albert (1989b) computed the change in the estimate of $\tau_i(\mathbf{x};\alpha)$ as a consequence of deleting case 268. That is, they com-

TABLE 8.4 Effect of Removing Case 268 on the Estimate of the Posterior Probability of Membership of the True Group for the Most Affected Cases Under the Logistic Model with log(AST), log(ALT), and log(GIDH) Fitted to the Five-Group Hepatitis Data Set

Case j	True Group Label i	$\tau_i(\mathbf{x}_j;\hat{\boldsymbol{\alpha}})$	$\tau_i(\mathbf{x}_j;\hat{\boldsymbol{\alpha}}_{(268)})$
60	2	0.51	0.14
75	2	0.72	0.96
97	2	0.46	0.84
147	4	0.73	0.84
221	5	0.12	0.38
237	5	0.43	0.71
245	5	0.46	0.77
268	5	0.42	0.00

Source: From Lesaffre and Albert (1989b).

puted the difference

$$\tau_i(\mathbf{x};\hat{\boldsymbol{\alpha}}) - \tau_i(\mathbf{x};\hat{\boldsymbol{\alpha}}_{(268)}),$$

where $\hat{\boldsymbol{\alpha}}_{(268)}$ denotes the estimate of $\boldsymbol{\alpha}$ when case 268 is deleted from the training data $\mathbf{t}$.

This difference is listed in Table 8.4 for the group of origin of the most affected cases. On the whole, it was found that the effect of observation 268 is minor except for cases lying on the edge between G_2 and G_5, a relatively sparse region. Concerning the cases in Table 8.4, it can be seen that only for cases 60 and 268 is the estimate of the posterior probability of membership of the true group not improved (that is, prediction deteriorates) as a consequence of the deletion of case 268 from the training data $\mathbf{t}$.

CHAPTER 9

Nonparametric Discrimination

9.1 INTRODUCTION

Up to now, we have been concerned with approaches to discriminant analysis that can be viewed as being parametric or semiparametric as with the logistic model in the last chapter. It has been seen that, although these approaches were developed by the explicit assumption of some model, some of them with little or no modification are fairly robust to certain departures from the assumed model. However, there is a need also for discriminant procedures that can be used regardless of the underlying group-conditional distributions.

In this chapter, we are to focus on discriminant procedures that have been developed without postulating models for the group-conditional distributions. In this sense they are referred to as being distribution-free. As T. W. Anderson (1966) has pointed out, an allocation procedure cannot be distribution-free in a literal sense. For if it were, then its error rates would not depend on the group-conditional distributions of the feature vector and would be constants even when the group-conditional distributions were identical (by a continuity argument). That is, the error rates would correspond to a pure randomization procedure.

Given the obvious application of the multinomial distribution in this role of providing a nonparametric approach to discrimination, we will first consider multinomial-based discriminant rules. As mentioned in Section 7.2.1, unless the number m of multinomial cells is very small or the sample size n is extremely large, some of the cells will contain no observations and so some form of smoothing has to be applied.

As remarked in Section 7.2.1, the nonparametric estimates of the multinomial cell probabilities under the assumption of independence of all the p

feature variables may impose too high a degree of smoothing on the multinomial estimates. Indeed, as discussed in Section 5.6.2, the independence-based discriminant rule is not robust against different interaction structures of the feature variables within the groups. Therefore, as with the parametric approaches to smoothing considered in Chapter 7, the aim with the nonparametric smoothing to be considered is to provide a degree of smoothing intermediate to that obtained under the extremes of the full multinomial and independence models.

With the kernel method dominating the nonparametric density estimation literature, the emphasis in this chapter is on so-called kernel discriminant analysis, whereby the group-conditional densities are replaced by their kernel estimates in the defining expressions for the posterior probabilities of group membership and consequent allocation rule. As to be discussed, the discrete kernel estimates can be viewed as smoothed estimates of the multinomial probabilities, obtained by a linear prescription. The recent advances to be described in kernel discriminant analysis essentially mirror the advances made with the kernel method of density estimation, in particular on the crucial problem of selecting the degree of smoothing to be incorporated into the kernel estimates.

9.2 MULTINOMIAL-BASED DISCRIMINATION

9.2.1 Multinomial Discriminant Rule (MDR)

The most universally applicable nonparametric discriminant rule is that based on the multinomial model for the group-conditional distributions of the feature vector **X**. This is because even if a feature variable is continuous, it can be discretized. Cochran and Hopkins (1961) examined the loss of discriminating power when continuous feature variables are discretized. They obtained asymptotic results for the best points of partition and the error rates when a large number of independent normal variates is partitioned to form discrete ones. More recently, Hung and Kshirsagar (1984) have considered the optimal partitioning of a continuous feature variable.

As in Section 7.2.1, where the multinomial model was first introduced, we suppose that there are m distinct realizations of the feature vector **X**, defining m multinomial cells. We let $q_{i1},\ldots,q_{im}$ denote the associated cell probabilities in group G_i $(i = 1,\ldots,g)$. That is, if the realization $\mathbf{x}$ corresponds to the cth cell, then $q_{ic} = f_i(\mathbf{x})$ for $i = 1,\ldots,g$. We let $r_o(\mathbf{x};\boldsymbol{\pi},\mathbf{q})$ denote the Bayes rule of allocation, where $\mathbf{q} = (\mathbf{q}_1',\ldots,\mathbf{q}_g')'$, $\mathbf{q}_i = (q_{i1},\ldots,q_{im})'$, and where as before, $\boldsymbol{\pi} = (\pi_1,\ldots,\pi_g)'$ is the vector of group-prior probabilities. Its sample plug-in version $r_o(\mathbf{x};\hat{\boldsymbol{\pi}},\hat{\mathbf{q}})$ is formed with q_{ic} replaced by

$$q_{ic} = n_{ic}/n_i, \tag{9.2.1}$$

where in the training data $\mathbf{t}$, n_{ic} is the number of feature vectors corresponding to the cth cell out of the n_i observations from G_i $(i = 1,\ldots,g)$. Under mixture sampling of the training data $\mathbf{t}$, n_i/n provides an estimate of π_i $(i = 1,\ldots,g)$.

9.2.2 Error Rates of Sample MDR

By using the theorems of Glick (1972), the plug-in sample version $r_o(\mathbf{x}; \hat{\boldsymbol{\pi}}, \hat{\mathbf{q}})$ of the multinomial-based allocation rule (MDR) can be established to be Bayes risk consistent. That is, the overall conditional error rate $ec(\boldsymbol{\pi}, \mathbf{q},; \hat{\boldsymbol{\pi}}, \hat{\mathbf{q}})$ converges in probability to the optimal or Bayes error $eo(\boldsymbol{\pi}, \mathbf{q})$, as $n \rightarrow \infty$. Glick (1973a) subsequently showed that this convergence rate is at least exponential for the sample plug-in MDR. Let m_o denote the number of points $\mathbf{x}$ with $f_1(\mathbf{x}) \neq f_2(\mathbf{x})$, and let d_o be the infimum of $|\sqrt{f_1(\mathbf{x})} - \sqrt{f_2(\mathbf{x})}|$ over these m_o points $\mathbf{x}$.

Glick (1973a) showed that

$$0 \leq 1 - \operatorname{pr}\{ec(\boldsymbol{\pi}, \mathbf{q},; \hat{\boldsymbol{\pi}}, \hat{\mathbf{q}}) = eo(\boldsymbol{\pi}, \mathbf{q})\} \leq \tfrac{1}{2} m_o (1 - d_o^2)^n$$

and

$$0 \leq eu(\boldsymbol{\pi}, \mathbf{q}) - eo(\boldsymbol{\pi}, \mathbf{q}) \leq \{\tfrac{1}{2} - eo(\boldsymbol{\pi}, \mathbf{q})\}(1 - d_o^2)^n,$$

where $eu(\boldsymbol{\pi}, \mathbf{q})$ denotes the overall unconditional error rate of the plug-in sample MDR.

Glick (1973a) also considered the bias of the apparent error rate $A = ec(\hat{\boldsymbol{\pi}}, \hat{\mathbf{q}}; \hat{\boldsymbol{\pi}}, \hat{\mathbf{q}})$ of the sample plug-in MDR. If the sample space contains no point $\mathbf{x}$ such that $\pi_1 f_1(\mathbf{x}) = \pi_2 f_2(\mathbf{x})$, then Glick (1973a) established that

$$0 \leq eo(\mathbf{q}, \boldsymbol{\pi}) - E(A) \leq \tfrac{1}{2} m_o n^{-1/2} (1 - d_o^2)^n. \qquad (9.2.2)$$

This exponential convergence contrasts with the convergence rate of n^{-1} for the apparent error rate of the normal-based sample linear discriminant rule under the homoscedastic normal model (3.3.1); see Section 10.2.1.

As conjectured by Hills (1966), intuitively it might be expected that the mean of the apparent error rate of a sample-based allocation rule is less than that of its unconditional error rate. Hills (1966) proceeded to show that his conjecture is indeed valid for the sample plug-in MDR. Also, for this rule, he showed in the special case of $m = 2$ multinomial cells that

$$E(A) < eo(\boldsymbol{\pi}, \mathbf{q}) < eu(\boldsymbol{\pi}, \mathbf{q}).$$

An algebraic expression for the exact bias of the apparent error rate of the sample MDR was obtained by Goldstein and Wolf (1977), who tabulated it under various combinations of n, m, and the cell probabilities. Their results demonstrated that the bound (9.2.2) is generally quite loose.

9.2.3 Minimum-Distance Rule for Multinomial Feature Data

Krzanowski (1983a, 1984a, 1987a) has considered the distance between two groups under the full multinomial model with cell probabilities q_{ic} ($c = 1, \ldots, m$) in G_i ($i = 1, 2$). The affinity ρ between the multinomial distributions of $\mathbf{X}$ in G_1 and G_2 is given by

$$\delta_M(F_1, F_2) = \{2(1 - \rho)\}^{1/2},$$

where

$$\rho = \sum_{c=1}^{m} (q_{1c} q_{2c})^{1/2}.$$

As an alternative allocation procedure to the Bayes rule $r_o(\mathbf{x}; \boldsymbol{\pi}, \mathbf{q})$ under the multinomial model, an entity with feature vector $\mathbf{x}$ can be assigned on the basis of the minimum distance

$$\min_i \delta_M(F_{\mathbf{x}}, F_i),$$

where $F_{\mathbf{x}}$ is the degenerate distribution that places mass one at the point $\mathbf{x}$. If cell c corresponds to the realization $\mathbf{x}$, then

$$\delta_M(F_{\mathbf{x}}, F_i) = \{2(1 - q_{ic}^{1/2})\}^{1/2}. \tag{9.2.3}$$

It can be seen from (9.2.3) that this minimum-distance rule is equivalent to the Bayes rule. However, it does provide a distinct alternative rule for the simultaneous allocation of $n_u > 1$ unclassified entities where all are known to be from the same group. Let n_{uc} denote the number of these entities with feature vector corresponding to the cth cell ($c = 1, \ldots, m$). Then if $\hat{F}_{n_u}$ denotes the empirical distribution function based on these n_u unclassified entities, the latter are assigned collectively on the basis of

$$\min_i \delta_M(\hat{F}_{n_u}, F_i), \tag{9.2.4}$$

where

$$\delta_M(\hat{F}_{n_u}, F_i) = \left[2\left\{1 - \sum_{c=1}^{m} q_{ic}^{1/2} (n_{uc}/n_u)^{1/2}\right\}\right]^{1/2}.$$

In the case of unknown group-conditional distributions, the use of the empirical distribution function $\hat{F}_i$ for F_i formed from the training data $\mathbf{t}$ leads to the replacement in (9.2.4) of q_{ic} by n_{ic}/n_i, as defined by (9.2.1). It can be seen that it does not matter whether the estimates of the cell probabilities are plugged in before or after the calculation of the Matusita distance. The sample version of the rule (9.2.4) was proposed and studied by Matusita (1956, 1964, 1967a, 1971), who also obtained lower bounds for its error rate and approximations to the latter when n is large.

As mentioned in Section 7.4.5, for the allocation of a single entity, Dillon and Goldstein (1978) highlighted some unsatisfactory behavior with the sample plug-in version of the Bayes rule $r_o(\mathbf{x}; \hat{\boldsymbol{\pi}}, \hat{\mathbf{q}})$, or, equivalently, of the minimum- distance rule (9.2.3). They therefore proposed an alternative rule obtained by modifying the minimum-distance rule (9.2.3). As before, we let $\hat{F}_i$ denote the empirical distribution function formed from the classified training data in $\mathbf{t}$ belonging to the ith group G_i. We now let $\hat{F}_{i,\mathbf{x}}$ denote the empirical distribution function based on the latter data and also $\mathbf{x}$, with the latter specified as belonging to G_i $(i = 1,\ldots,g)$. The entity with feature vector $\mathbf{x}$ is assigned to G_1 if

$$\delta_M(\hat{F}_{1,\mathbf{x}}, \hat{F}_2) > \delta_M(\hat{F}_1, \hat{F}_{2,\mathbf{x}}), \tag{9.2.5}$$

and to G_2, otherwise. That is, the entity is assigned to the group that maximizes the consequent estimate of the Matusita distance between the two groups.

With cell c corresponding to the realization $\mathbf{x}$, the inequality (9.2.5) can be written as

$$\left\{(n_{1c}+1)n_{2c} + \sum_{h\neq c} n_{1h}n_{2h}\right\}^{1/2} \Big/ \left\{n_{1c}(n_{2c}+1) + \sum_{h\neq c} n_{1h}n_{2h}\right\}^{1/2}$$

$$< [\{(n_1+1)n_2\}/\{n_1(n_2+1)\}]^{1/2}. \tag{9.2.6}$$

It can be seen that for $n_1 = n_2$, (9.2.6) holds if and only if

$$n_{2c} < n_{1c};$$

that is, if and only if

$$\hat{q}_{2c} < \hat{q}_{1c}.$$

Thus, (9.2.4) is equivalent to the Bayes rule under the multinomial model for equal group-sample sizes. However, it is distinct from the Bayes rule for unequal group-sample sizes. As explained by Goldstein and Dillon (1978, Section 2.4), the sample minimum-distance rule defined by (9.2.5) does alleviate one unattractive property of the plug-in sample version of the multinomial-based Bayes rule in the case of disparate group-sample sizes, which arises as a consequence of the sparseness of the training data. For example, suppose n_1 is much smaller than n_2, and that $n_{1c} = 0$. Then an entity with feature vector $\mathbf{x}$ will be assigned to G_2 if $n_{2c} > 0$, no matter how small n_{2c} is and how much greater n_2 is than n_1. However, with the sample minimum-distance rule (9.2.5), the entity in some situations can be assigned to G_1. It will be assigned to G_1 if

$$n_{2c}^{1/2} < \sum_{h\neq c}^{m} (n_{1h}n_{2h})^{1/2} \left[\left\{\frac{n_2(n_1+1)}{n_1(n_2+1)}\right\}^{1/2} - 1\right].$$

Recently, Krzanowski (1987a) has established that these rules are asymptotically equivalent, as n_1, n_2 tend to infinity.

9.2.4 Convex Smoothing of Multinomial Probabilities

In the context of discrimination between g groups, the maximum likelihood estimate $\hat{\mathbf{q}}_i = (\hat{q}_{i1},\ldots,\hat{q}_{im})'$ for the ith group-conditional multinomial distribution with cell probabilities given by $\mathbf{q}_i = (q_{i1},\ldots,q_{im})'$ can be smoothed as

$$\tilde{\mathbf{q}}_i = (1-\omega_i)\hat{\mathbf{q}}_i + \omega_i\boldsymbol{\zeta}_i, \tag{9.2.7}$$

where ζ_i is a vector of probabilities, and ω_i is a smoothing parameter ($0 \leq \omega_i \leq 1$) to be selected from the data. The smoothing parameter ω_i typically has the property that it tends to zero, as $n_i \to \infty$. In a Bayesian approach to the problem, ζ_i may be specified as the prior distribution of $\mathbf{q}_i$. A possible choice for ζ_i when there is no prior knowledge on the distribution of $\mathbf{q}_i$ is the uniform prior for which

$$\zeta_i = (1/m, \ldots, 1/m)'.$$

Good (1965); Fienberg and Holland (1973); Sutherland, Fienberg, and Holland (1974); M. Stone (1974); Leonard (1977); Van Ryzin and Wang (1978); Wang and Van Ryzin (1981); and Wang (1986a) have considered estimators of the form (9.2.7). Fienberg and Holland (1973) showed that $\tilde{\mathbf{q}}_i$ can often be better than the maximum likelihood estimate $\hat{\mathbf{q}}_i$ on the basis of risk.

Titterington (1980) has discussed various ways of choosing the smoothing parameter ω_i. For example, the value of ω_i that minimizes the mean-squared error is given by

$$\omega_i = C_i \Big/ \left\{ C_i + n_i \sum_{c=1}^{m} (q_{ic} - \zeta_{ic})^2 \right\},$$

where

$$C_i = 1 - \sum_{c=1}^{m} q_{ic}^2.$$

A data-based smoothing parameter $\hat{\omega}_i$ is then obtained by replacing q_{ic} by $\hat{q}_{ic}$.

Titterington (1980) and Titterington and Bowman (1985) have demonstrated how the discrete kernel estimator (9.3.12) of Aitchison and Aitken (1976) for a single unordered categorical variable with m categories can be expressed in the form (9.2.7), where

$$\omega_i = (1 - h_i)\{m/(m-1)\}$$

and

$$\zeta_i = (1/m, \ldots, 1m)'.$$

Discrete kernel estimators are to be considered, commencing in Section 9.3.5.

The link between (9.2.7) and other discrete kernels in the literature is pursued in Titterington and Bowman (1985). They also considered other forms of smoothing for ordered categorical data, including Bayes-based methods and penalized minimum-distance methods. Hall and Titterington (1987) have developed asymptotic theory for the problem of smoothing sparse multinomial data.

9.2.5 Smoothing of Multinomial Probabilities by Orthogonal Series Methods

Another way of smoothing multinomial probabilities is to use orthogonal series methods. Orthogonal series methods for the estimation of a density for continuous data were introduced by Whittle (1958), Čencov (1962), Schwartz (1967), and Kronmal and Tarter (1968). Ott and Kronmal (1976) suggested using Walsh series to estimate a density for multivariate binary data. More recently, Hall (1983b) has shown how orthogonal series methods can be applied to density estimation for discrete data not necessarily binary and for mixed data.

We now outline the approach of Ott and Kronmal (1976) to the estimation of the group-conditional densities in discriminant analysis with multivariate binary data. Corresponding to the feature vector $\mathbf{X}$ consisting of p zero-one random variables, we let $\mathbf{c} = (c_1, \ldots, c_p)'$, where each c_v is a zero-one variable $(v = 1, \ldots, p)$, be an index of the realizations of $\mathbf{X}$.

The ith group-conditional density $f_i(\mathbf{x})$ can be represented then in the form

$$f_i(\mathbf{x}) = 2^{-p} \sum_{\mathbf{c}} a_{i\mathbf{c}} \psi_{\mathbf{c}}(\mathbf{x}), \tag{9.2.8}$$

where the summation in (9.2.8) is over all values of $\mathbf{c}$, and where $\psi_{\mathbf{c}}(\mathbf{x})$ is the $\mathbf{c}$th Walsh function defined by

$$\psi_{\mathbf{c}}(\mathbf{x}) = (-1)^{\mathbf{x}'\mathbf{c}}.$$

The orthogonality of these functions $\psi_{\mathbf{c}}(\mathbf{x})$ is given by

$$\begin{aligned} \sum_{\mathbf{x}} \psi_{\mathbf{c}_1}(\mathbf{x}) \psi_{\mathbf{c}_2}(\mathbf{x}) &= \sum_{\mathbf{x}} (-1)^{2\mathbf{x}'\mathbf{c}_1} = 2^p \qquad \text{for} \quad \mathbf{c}_1 = \mathbf{c}_2 \\ &= \sum_{\mathbf{x}} (-1)^{\mathbf{x}'(\mathbf{c}_1 + \mathbf{c}_2)} = 0 \qquad \text{for} \quad \mathbf{c}_1 \neq \mathbf{c}_2, \end{aligned} \tag{9.2.9}$$

as demonstrated in Ott and Kronmal (1976).

It follows from (9.2.9) that the coefficients $a_{i\mathbf{c}}$ in (9.2.8) are given by

$$\begin{aligned} a_{i\mathbf{c}} &= \sum_{\mathbf{x}} \psi_{\mathbf{c}}(\mathbf{x}) f_i(\mathbf{x}) \\ &= E\{\psi_{\mathbf{c}}(\mathbf{x})\}. \end{aligned}$$

The maximum likelihood estimate of $a_{i\mathbf{c}}$ is

$$\hat{a}_{i\mathbf{c}} = 2^{-p} n_i^{-1} \sum_{j=1}^{n_i} \psi_{\mathbf{c}}(\mathbf{x}_{ij}) n_i(\mathbf{x}),$$

where $n_i(\mathbf{x})$ is the number of feature vectors equal to $\mathbf{x}$ out of those $\mathbf{x}_{i1}, \ldots, \mathbf{x}_{in_i}$ from G_i $(i = 1, \ldots, g)$. Let $\hat{f}_i(\mathbf{x})$ be the estimate of $f_i(\mathbf{x})$ obtained on replacing $a_{i\mathbf{c}}$ by $\hat{a}_{i\mathbf{c}}$ in the full-rank expansion (9.2.8); that is, $\hat{f}_i(\mathbf{x})$ is the unsmoothed multinomial estimate $n_i(\mathbf{x})/n_i$.

A degree of smoothing is imposed by taking some of the coefficients $a_{i\mathbf{c}}$ in (9.2.8) to be zero. A basic method for deciding which coefficients to be set equal to zero is to work in terms of the mean integrated squared error (MISE)

$$E\left[\sum_{\mathbf{x}}\{\hat{f}_i(\mathbf{x})-f_i(\mathbf{x})\}^2\right].$$

It can be shown that the MISE is reduced by putting $a_{i\mathbf{c}}$ equal to zero if

$$2^{-p}n_i^{-1}\{1-(n_i+1)a_{i\mathbf{c}}^2\} \tag{9.2.10}$$

is positive; see, for example, Goldstein and Dillon (1978, Section 2.3). An estimate of (9.2.10) is

$$2^{-p}(n_i-1)^{-1}\{2-(n_i+1)\hat{a}_{i\mathbf{c}}^2\}.$$

On the basis of this assessment, a reduction in the MISE is achieved by setting a_{ic} to zero if

$$\hat{a}_{i\mathbf{c}}^2 < 2/(n_i+1). \tag{9.2.11}$$

Of course, as $n_i \to \infty$, (9.2.11) will not be satisfied by any estimated coefficient so that we end up with the full-rank expansion (9.2.8), that is, the unsmoothed multinomial estimate.

The approach above was adopted by Goldstein (1977). In their original proposal, which was in the context of mixture sampling from $g = 2$ groups, Ott and Kronmal (1976) proceeded in a slightly different manner. They used the orthogonal series method to represent the density of $\mathbf{Y} = (\mathbf{X}', \mathbf{Z}')$ and also the difference between the densities of $(\mathbf{X}', 1, 0)'$ and $(\mathbf{X}', 0, 1)'$. Here, as before, $\mathbf{Z}$ is the vector of zero-one indicator variables defining the group of origin of the unclassified entity with feature vector $\mathbf{X}$.

Ott and Kronmal (1976) noted that another representation of $f_i(\mathbf{x})$ is obtained by working with $(2x_\nu - 1)$ and taking $\psi_{\mathbf{c}}(\mathbf{x})$ to be

$$\psi_{\mathbf{c}}(\mathbf{x}) = \prod_{\nu=1}^{p}(2x_\nu - 1)^{c_\nu}.$$

This orthogonal series expansion had been used previously by Martin and Bradley (1972) to represent $\{f_i(\mathbf{x}) - f(\mathbf{x})\}/f(\mathbf{x})$ for multivariate binary feature data for $g = 2$ groups. A third representation noted by Ott and Kronmal (1976) is to make use of Rademacher functions and Gray code. They showed that these representations are all equivalent to the discrete Fourier transform for binary variables.

A drawback of orthogonal series methods is that they have not yet been adapted to cope with missing values or high-dimensional data (Titterington et al., 1981). Also, they can produce smoothed estimates that are negative. Further discussion on this last point may be found in Hand (1981a, Section 5.3), who contrasts this approach with those based on log-linear models and the Lazarsfeld–Bahadur representation, as presented in Section 7.3.

As mentioned in the introduction to this section, Hall (1983b) has shown how orthogonal series methods can be used for density estimation with all types of data. An application of his work to a discrimination problem with mixed feature variables has been given by Wojciechowski (1988).

9.3 NONPARAMETRIC ESTIMATION OF GROUP-CONDITIONAL DENSITIES

9.3.1 Introduction

A common approach to nonparametric discriminant analysis with either continuous or discrete feature data is to substitute nonparametric estimates of the group-conditional densities, in particular kernel estimates, into the definition (1.4.4) of the Bayes rule $r_o(\mathbf{x};F)$. Reviews of nonparametric density estimation have been given by Wegman (1972a, 1972b) and Fryer (1977), and Wertz and Schneider (1979) have given a comprehensive bibliography. More recently, entire monographs have been devoted to the topic, including those by Tapia and Thompson (1978), Prakasa Rao (1983), Silverman (1986), Devroye and Györfi (1985), and Devroye (1987). There are also the monographs of Hand (1982) and Coomans and Broeckaert (1986), which focus exclusively on discriminant analysis via kernel estimates of the group-conditional densities.

Fix and Hodges (1951, 1952) introduced a number of ideas that have proved to be basic to the development of nonparametric density estimation and nonparametric discrimination. Recently, Silverman and Jones (1989) have provided a commentary of the first of these two papers, placing it in context and interpreting its ideas in the light of more modern developments. As noted in Silverman (1986) and Silverman and Jones (1989), it was for the express purpose of discriminant analysis that Fix and Hodges (1951) introduced their nonparametric procedures. They were responsible for introducing the widely used class of nearest neighbor allocation rules for nonparametric estimation. They also first introduced two popular methods for nonparametric density estimation: the kernel density and, rather briefly, the nearest neighbor density. We focus firstly on the kernel method, as it is the most widely used nonparametric method of density estimation.

9.3.2 Definition of Kernel Method of Density Estimation

To define the kernel method, it is more convenient if we first relabel the n training feature observations $\mathbf{x}_1,\ldots,\mathbf{x}_n$ to explicitly denote their group of origin. Accordingly, we let $\mathbf{x}_{ij}$ $(j = 1,\ldots,n_i)$ denote the n_i observations from G_i $(i = 1,\ldots,g)$. Without loss of generality, we focus for the present on the density estimation for one of the groups, namely, G_i. For a continuous p-dimensional feature matrix $\mathbf{X}$, a nonparametric estimate of the ith group-conditional density $f_i(\mathbf{x})$ provided by the kernel method is

$$\hat{f}_i^{(K)}(\mathbf{x}) = n_i^{-1}h_i^{-p}\sum_{j=1}^{n_i} K_p\left(\frac{\mathbf{x}-\mathbf{x}_{ij}}{h_i}\right), \tag{9.3.1}$$

where K_p is a kernel function that integrates to one, and h_i is a smoothing parameter. The analogue of (9.3.1) for a discrete feature vector is considered in Section 9.3.5. The smoothing parameter h_i is known also as the bandwidth or window width. Although the notation does not reflect it, h_i is a function of the ith group-sample size n_i. With most applications, the kernel K_p is fixed and the smoothing parameter h_i is specified as a function of the data. Usually, but not always, the kernel K_p is required to be nonnegative and symmetric, that is,

$$K_p(\mathbf{x}) \geq 0, \qquad \mathbf{x} \in \mathrm{R}^p, \tag{9.3.2}$$

and

$$K_p(\mathbf{x}) = K_p(-\mathbf{x}), \qquad \mathbf{x} \in \mathrm{R}^p. \tag{9.3.3}$$

If (9.3.2) holds, the kernel density estimate $\hat{f}_i^{(K)}(\mathbf{x})$ can be interpreted as a mixture of n component densities in equal proportions.

The use of the kernel method in forming estimates of the group-conditional densities is demonstrated in Figure 9.1, which gives the kernel density (9.1.1) formed from a simulated data set of size $n = 200$ for three values of the smoothing parameter with the univariate normal density as kernel. Figure 9.2 gives the true density, which is a mixture in equal proportions of two univariate normal densities with means -1 and 1 and common variance 1.

The kernel density estimator originally suggested by Fix and Hodges (1951) in the univariate case corresponds to the form (9.3.1), where $p = 1$, $h_i = 1$, and where K_1 is the uniform density over a prescribed neighborhood $\mathcal{N}_i$ of zero. This gives the univariate version of what is now often called the naive kernel density estimate. If, for example, the neighborhood $\mathcal{N}_i$ is taken to be the hypercube $[-k_i, k_i]^p$, then the naive estimate becomes (9.3.1) with smoothing parameter $h_i = k_i$ and with $K_p(\mathbf{x})$ the uniform density over $[-1,1]^p$. Since $\hat{f}_i^{(K)}$ inherits the smoothness properties of K_p, it is now more common for exploratory and presentation purposes to use a smooth kernel. When plotted out, a naive estimator has a locally irregular appearance; see Silverman (1986, Figure 2.3).

The potential for more general kernels was recognized by Rosenblatt (1956), which was the first published paper on kernel density estimation. There is now a vast literature concerning the theory and practical applications of the kernel density estimator. Silverman (1986) gives a good discussion of many important applied aspects of this estimator. A lead into the theoretical literature is provided by Prakasa Rao (1983) and Devroye and Györfi (1985).

By virtue of its definition, the kernel density approach to estimation is resistant to the effect of outliers. This is because $K_p\{(\mathbf{x} - \mathbf{x}_{ij})/h_i\}$ must become small if $\mathbf{x}_{ij}$ is far from $\mathbf{x}$. The reader is referred to Silverman (1986, Section 3.5) for computational aspects of kernel density estimation. As advised there, it is much faster to notice that the kernel estimate is a convolution of the data with the kernel and to use Fourier transforms to perform the convolution,

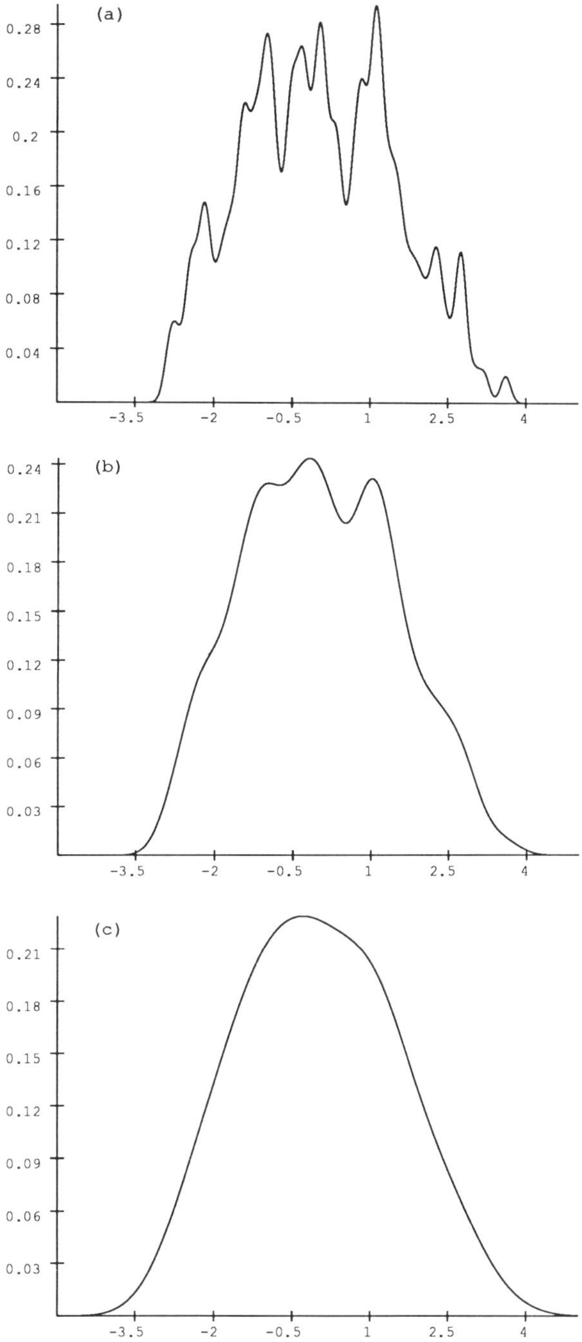

FIGURE 9.1. Kernel density estimate $\hat{f}_i^{(K)}(x)$ for 200 simulated data points for various values of the smoothing parameter h_i: (a) $h_i = 0.1$; (b) $h_i = 0.3$; and (c) $h_i = 0.6$.

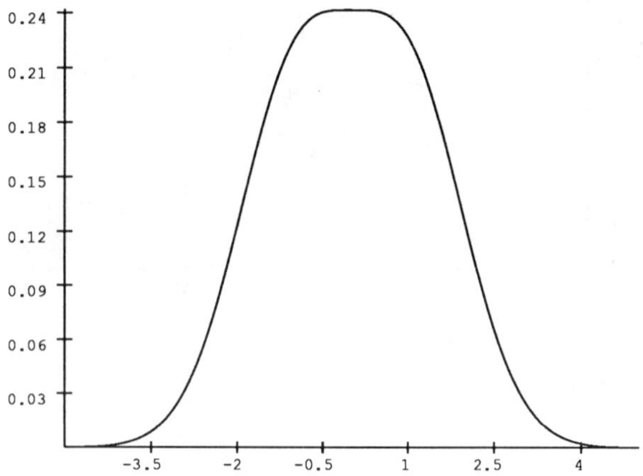

FIGURE 9.2. True density $f_i(x)$ underlying simulated data used to produce kernel density estimates in Figure 9.1.

rather than to directly compute it from its defining formula (9.3.1). The use of the fast Fourier transform makes it possible to find direct and inverse Fourier transforms very quickly.

In the widely available statistical computer packages, there are routines for nonparametric density estimation, at least in the univariate case; see, for example, IMSL (1984) subroutines NDKER (for kernel density estimation) and NDMDLE (for maximum penalized likelihood density estimation), the S package (Becker, Chambers, and Wilks, 1988) for kernel density estimation with a choice of cosine, normal, rectangular, and triangular kernels, and the SAS (1990) package for kernel and nearest neighbor methods of estimation. Concerning computer packages designed specifically for discriminant analysis, there is the algorithm ALLOC1 of Habbema, Hermans, and van den Broeck (1974a). Hermans, Habbema, and Schaefer (1982) have given an extension of this algorithm ALLOC80, which is in wide practical use. The procedures in this algorithm use the kernel method to nonparametrically estimate the group-conditional densities. In particular, this algorithm has the facility for the selection of feature variables, and it is discussed further in that role in Section 12.4.5.

9.3.3 Large-Sample Properties of Kernel Estimates

The large-sample properties of the kernel estimator were investigated initially by Rosenblatt (1956) and Parzen (1962) for the univariate case and by Cacoullos (1966) and Epanechnikov (1969) for the multivariate case. They showed that conditions can be imposed on K_p and h_i to ensure that $\hat{f}_i^{(K)}(\mathbf{x})$ is asymptotically unbiased and pointwise consistent in mean-squared error. For example, suppose that

$$\sup_{\mathbf{x}\in\mathbf{R}^p} |K_p(\mathbf{x})| < \infty, \tag{9.3.4}$$

$$\|\mathbf{x}\|^p K_p(\mathbf{x}) \to 0, \qquad \text{as} \quad \|\mathbf{x}\| \to \infty, \tag{9.3.5}$$

$$\int |K_p(\mathbf{x})| d\nu < \infty, \tag{9.3.6}$$

and, as stated above,

$$\int K_p(\mathbf{x}) d\nu = 1. \tag{9.3.7}$$

If conditions (9.3.4) to (9.3.7) hold and $h_i \to 0$ as $n_i \to \infty$, then as $n_i \to \infty$,

$$E\{\hat{f}_i^{(K)}(\mathbf{x})\} \to f_i(\mathbf{x}).$$

Suppose that in addition (9.3.2) and (9.3.3) hold, and that $h_i \to 0$ and $n_i h_i^p \to \infty$, as $n_i \to \infty$. Then as $n_i \to \infty$,

$$E\{\hat{f}_i^{(K)}(\mathbf{x}) - f_i(\mathbf{x})\}^2 \to 0,$$

wherever f_i is continuous in $\mathbf{x}$; in particular, $\hat{f}_i^{(K)}(\mathbf{x})$ is consistent for $f_i(\mathbf{x})$.

9.3.4 Product Kernel

We now consider the choice of kernel function in the definition (9.3.1) of the kernel density estimator. The more crucial problem of how to select the smoothing parameter for a given kernel is to be addressed in Section 9.4.

Epanechnikov (1969) and Deheuvels (1977) used an asymptotic argument to show that there is very little to choose between different kernel functions. Among the various kernels considered by Cacoullos (1966) was the so-called product kernel,

$$K_p(\mathbf{x}) = \prod_{v=1}^{p} K_1(x_v),$$

where K_1 is a univariate probability density function. This yields

$$\hat{f}_i^{(K)}(\mathbf{x}) = n_i^{-1} h_i^{-p} \sum_{j=1}^{n_i} \prod_{v=1}^{p} K_1\left(\frac{x_v - x_{ijv}}{h_i}\right), \tag{9.3.8}$$

where $x_v = (\mathbf{x})_v$, and $x_{ijv} = (\mathbf{x}_{ij})_v$ for $v = 1,\ldots,p$. A common choice for the univariate kernel $K_1(\mathbf{x})$ is the univariate standard normal density function. With this choice, $\hat{f}_i^{(K)}(\mathbf{x})$ is estimated via a spherical normal kernel,

$$\hat{f}_i^{(K)}(\mathbf{x}) = \sum_{j=1}^{n_i} n_i^{-1} \phi(\mathbf{x} - \mathbf{x}_{ij}; \mathbf{0}, \mathbf{M}_i), \tag{9.3.9}$$

where $\mathbf{M}_i = h_i^2\mathbf{I}_p$, and, as before, $\phi(\mathbf{x};\boldsymbol{\mu},\boldsymbol{\Sigma})$ denotes the multivariate normal density function with mean $\boldsymbol{\mu}$ and covariance matrix $\boldsymbol{\Sigma}$. The form (9.3.8), and hence (9.3.9), assumes that the feature data have been standardized. Otherwise,

$$\mathbf{M}_i = h_i^2\text{diag}(s_{i11},\ldots,s_{ipp}),$$

where $s_{ivv} = (\mathbf{S}_i)_{vv}$, and $\mathbf{S}_i$ denotes the (bias-corrected) sample covariance matrix of $\mathbf{x}_{i1},\ldots,\mathbf{x}_{in_i}$ $(i = 1,\ldots,g)$.

9.3.5 Kernels for Multivariate Binary Data

Suppose for the present that each variable in the feature vector is a zero-one variable. In this situation, Aitchison and Aitken (1976) proposed using a binomial kernel, whereby $\hat{f}_i(\mathbf{x})$ is estimated by

$$\hat{f}_i^{(K)}(\mathbf{x}) = n_i^{-1}\sum_{j=1}^{n_i} K_p(\mathbf{x};\mathbf{x}_{ij},h_i), \tag{9.3.10}$$

where

$$K_p(\mathbf{x};\mathbf{x}_{ij},h_i) = h_i^{p-d_{ij}^2}(1-h_i)^{d_{ij}^2}$$

with $\frac{1}{2} \leq h_i \leq 1$, and

$$\begin{aligned} d_{ij}^2 &= \|\mathbf{x}-\mathbf{x}_{ij}\|^2 \\ &= (\mathbf{x}-\mathbf{x}_{ij})'(\mathbf{x}-\mathbf{x}_{ij}). \end{aligned}$$

The squared Euclidean distance d_{ij}^2 between $\mathbf{x}$ and $\mathbf{x}_{ij}$ is simply the number of disagreements in their corresponding elements.

If we put $h_i = 1$, then $\hat{f}_i^{(K)}(\mathbf{x})$ reduces to the multinomial estimate $n_i(\mathbf{x})/n_i$, where $n_i(\mathbf{x})$ is the number of sample points with $\mathbf{x}_{ij} = \mathbf{x}$ $(j = 1,\ldots,n_i)$. As h_i decreases from one, the smoothing of the multinomial estimates increases, so that at $h_i = 1/2$, it puts equal mass $1/2^p$ at each possible realization of $\mathbf{X}$.

Further insight into this form of the kernel can be obtained if we rewrite $K_p(\mathbf{x};\mathbf{x}_{ij},h_i)$ in the form of a product of binomial kernels,

$$K_p(\mathbf{x};\mathbf{x}_{ij},h_i) = \prod_{v=1}^{p} h_i^{1-w_{ijv}}(1-h_i)^{w_{ijv}}, \tag{9.3.11}$$

where

$$w_{ijv} = |(\mathbf{x})_v - (\mathbf{x}_{ij})_v|.$$

It can be seen from (9.3.11) that $K_p(\mathbf{x};\mathbf{x}_{ij},h_i)$ gives weight h_i^p at $\mathbf{x}$ itself and $h_i^{p-1}(1-h_i)$ at those cells that differ from $\mathbf{x}$ in only one feature, $h_i^{p-2}(1-h_i)^2$ at those that differ from $\mathbf{x}$ in two features, and so on. Hence, it is required that $h_i \geq 1/2$.

9.3.6 Kernels for Unordered Categorical Data

Discrete feature variables with more than two categories can be converted to binary data, as explained in Section 7.4.1. Alternatively, Aitchison and Aitken (1976) have provided an extension of their kernel estimator for categorical feature data. Suppose that all the features are represented by unordered categorical variables, where c_v denotes the number of categories for the vth feature variable X_v $(v = 1,\ldots,p)$. Then Aitchison and Aitken (1976) proposed that (9.3.11) be modified to

$$K_p(\mathbf{x};\mathbf{x}_j,h_i) = \prod_{v=1}^{p} h_i^{1-w_{ijv}}\{(1-h_i)/(c_v-1)\}^{w_{ijv}}, \tag{9.3.12}$$

where

$$\max_v c_v^{-1} \leq h_i \leq 1.$$

This restriction ensures that the kernel has the desirable property of being a probability density function with its mode at $\mathbf{x} = \mathbf{x}_j$.

In the case where the smoothing parameter h_i is allowed to be different for each feature variable, the kernel (9.3.12) has the form

$$K_p(\mathbf{x};\mathbf{x}_j,\mathbf{h}_i) = \prod_{v=1}^{p} h_{iv}^{1-w_{ijv}}\{(1-h_{iv})/(c_v-1)\}^{w_{ijv}}, \tag{9.3.13}$$

where $\mathbf{h}_i = (h_{i1},\ldots,h_{iv})'$ is the vector containing the p smoothing parameters corresponding to each feature, and $c_v^{-1} \leq h_{iv} \leq 1$ $(v = 1,\ldots,p)$. This general approach is advocated by Titterington (1980), because of the possibly different variances of the feature variables. He notes that it is not practicable with categorical data to first standardize the variables, as is often done with multivariate continuous data when a common smoothing parameter h_i is specified. Also, this general approach allows the smoothing parameter h_{iv} to be conveniently chosen on the basis of the marginal feature data x_{ijv} $(j = 1,\ldots,n_i)$. Smoothing parameters chosen in this manner, however, do tend to provide less smoothing than those selected simultaneously on the basis of the joint distribution of all the feature data (Titterington, 1980). This was particularly noticeable in the study of Titterington et al. (1981), where marginally chosen smoothing parameters gave density estimates too sharp from a multivariate point of view.

9.3.7 Kernels for Ordered Categorical Data

The discrete kernel (9.3.13) can be modified to allow for an ordered categorical variable. This was demonstrated by Aitchison and Aitken (1976) for an ordered trinomial variable and has been considered further by Titterington and Bowman (1985), among others. Suppose that X_v is an ordered trinomial variable taking on the values of 0, 1, and 2. Then, under the suggestion of Aitchison and Aitken (1976), the vth term in the product on the right-hand

TABLE 9.1 Values of Contribution $K_1^{(v)}(x_v; x_{ijv}, h_{iv})$ to the Product Kernel for an Ordered Trinomial Variable X_v

x_{ijv}	$x_v = 0$	$x_v = 1$	$x_v = 2$
0	h_{iv}^2	$2h_{iv}(1-h_{iv})$	$(1-h_{iv})^2$
1	$\frac{1}{2}(1-h_{iv}^2)$	h_{iv}^2	$\frac{1}{2}(1-h_{iv}^2)$
2	$(1-h_{iv})^2$	$2h_{iv}(1-h_{iv})$	h_{iv}^2

side of (9.3.13) is replaced by $K_1^{(v)}(x_v; x_{ijv}, h_{iv})$, as defined in Table 9.1, where $h_{iv} \geq 2/3$. From this table, it can be seen how the contribution to the kernel falls away from h_{iv}^2 as $|x_v - x_{ijv}|$ increases from zero.

9.3.8 Kernels for Categorical Data with an Infinite Number of Cells

Aitken (1983) proposed a kernel function for univariate data in the form of counts. It is of the form

$$\hat{f}_i^{(K)}(x) = n_i^{-1} \sum_{j=1}^{n_i} K_1(x; x_{ij}, h_i),$$

where

$$\begin{aligned} K_1(x; x_{ij}, h_i) &= h_i, & x &= x_{ij}, \\ &= 2^{-(d_{ij}+1)}(1-h_i), & d_{ij} &\leq x, \\ &= 2^{-d_{ij}}(1-h_i), & d_{ij} &> x, \end{aligned} \qquad (9.3.14)$$

and where $d_{ij} = |x - x_{ij}|$, and $1/3 \leq h_i \leq 1$.

Habbema, Hermans, and Remme (1978a) discussed a kernel function of the form

$$K_1(x; x_{ij}, h_i) = h_i^{d_{ij}^2} / C(h_i) \qquad (9.3.15)$$

for each type of feature variable, continuous, binary, ordered and unordered categorical. The distance measure $d_{ij} = \|x - x_{ij}\|$ and the normalizing constant $C(h_i)$ depend on the type of variable; see Schmitz et al. (1983b, Appendix A) for further details. Aitken (1983) showed how (9.3.15) can be adapted to data in the form of counts.

9.3.9 Kernels for Mixed Feature Data

For mixed feature data as described in Section 7.4.1, Aitchison and Aitken (1976) suggested a sum of the product of normal and discrete kernels corresponding to the continuous and discrete subvectors, respectively. As stressed by them, this factorization does no way imply independence of the continuous and discrete feature variables. Suppose that the feature variables have been

TABLE 9.2 Values of Contribution $K_1^{(m)}(x_v; x_{ijv}, h_{iv})$ to the Product Kernel for Missing Data on an Unordered or Ordered Binomial Variable X_v

x_{ijv}	$x_v = 0$	$x_v = 1$	$x_v = m$
0	h_{iv}^2	$h_{iv}(1-h_{iv})$	$1-h_{iv}$
1	$h_{iv}(1-h_{iv})$	h_{iv}^2	$1-h_{iv}$
m	$\frac{1}{2}(1-h_{iv}^2)$	$\frac{1}{2}(1-h_{iv}^2)$	h_{iv}^2

labeled so that $\mathbf{x} = (\mathbf{x}^{(1)\prime}, \mathbf{x}^{(2)\prime})'$, where the subvector $\mathbf{x}^{(1)}$ contains the p_1 discrete features and $\mathbf{x}^{(2)}$ the p_2 continuous feature variables. Then a mixed kernel estimate of $f_i(\mathbf{x})$ is given by

$$\hat{f}_i^{(K)}(\mathbf{x}) = n_i^{-1} h_{2i}^{-p_2} \sum_{j=1}^{n_i} K_{p_1}^{(D)}(\mathbf{x}^{(1)}; \mathbf{x}_{ij}^{(1)}, h_{1i}) K_{p_2}^{(C)}\left(\frac{\mathbf{x}^{(2)} - \mathbf{x}_{ij}^{(2)}}{h_{2i}}\right), \tag{9.3.16}$$

where $K_{p_1}^{(D)}$ is a kernel and h_{1i} a smoothing parameter for the subvector $\mathbf{x}^{(1)}$ of discrete variables, and $K_{p_2}^{(C)}$ is a kernel and h_{2i} a smoothing parameter for the subvector $\mathbf{x}^{(2)}$ of continuous variables.

9.3.10 Incomplete Feature Data

In the case of a missing value for x_{ijv} in multivariate binary data, Titterington (1977) suggested that one way of dealing with this problem is to set h_{iv} equal to 1/2 in the vth term in the product on the right-hand side of (9.3.13). For $h_{iv} = 1/2$, each term in the product is equal to 1/2, regardless of the value of w_{ijv}. In the assessment of the smoothing parameters h_{iv}, Titterington (1977) pointed out that for a training observation $\mathbf{x}_{ij}$ with some missing elements, $\hat{f}_i^{(K)}(\mathbf{x}_{ij})$ can be evaluated as

$$\hat{f}_i^{(K)}(\mathbf{x}_{ij}) = \sum \hat{f}_i^{(K)}(\tilde{\mathbf{x}}_{ij}),$$

where the summation is over all possible completions $\tilde{\mathbf{x}}_{ij}$ of $\mathbf{x}_{ij}$.

Titterington (1976) and Murray and Titterington (1978) have suggested other ways of handling missing data for kernel estimation from discrete data. One way is to treat "missing" as an extra category which, for ordered categorical variables, destroys the ordered nature of the data. Another way is to adopt a more genuine missing-data kernel (Murray and Titterington, 1978). They proposed that for the binary variable X_v with $c_v = 2$, the three categories can be sensibly treated as unordered or ordered with "missing" in the middle, using the missing-data kernel $K_1^{(m)}(x_v; x_{ijv}; h_{iv})$, as given in Table 9.2, in place of the vth term in the product on the right-hand side of (9.3.13). In Table 9.2, m refers to the "missing" category.

The general form of this kernel for c_v categories is

$$\begin{aligned}
K^{(m)}(x_v; x_{ijv}, h_{iv}) &= h_{iv} K_{c_v}(x_v; x_{ijv}, h_{iv}) && \text{if} \quad x_v \neq m \quad \text{and} \quad x_{ijv} \neq m, \\
&= 1 - h_{iv} && \text{if} \quad x_v = m \quad \text{and} \quad x_{ijv} \neq m, \\
&= h_{iv}^{c_v+1} && \text{if} \quad x_v = m \quad \text{and} \quad x_{ijv} = m, \\
&= (1 - h_i^{c_v+1})/c_v && \text{if} \quad x_v \neq m \quad \text{and} \quad x_{ijv} = m,
\end{aligned}$$

where $K_{c_v}(x_v; x_{ijv}, h_{iv})$ is a c_v-category kernel. Murray and Titterington (1978) consider constraints on the smoothing parameters in order for the missing-data kernel as given above to have its mode at $x_v = x_{ijv}$. More recently, Titterington and Mill (1983) have considered the problem of forming kernel estimates from missing data in continuous variables.

9.4 SELECTION OF SMOOTHING PARAMETERS IN KERNEL ESTIMATES OF GROUP-CONDITIONAL DENSITIES

9.4.1 Separate Selection for Each Group

As the choice of kernel is not as important as it might first seem, the usual approach in forming $\hat{f}_i^{(K)}(\mathbf{x})$ is to fix the kernel K_p in (9.3.1) and then to assess the smoothing parameter h_i from the observed data. The effective performance of the kernel estimator is crucially dependent on the value of the smoothing parameter h_i. With the traditional approach to discriminant analysis via kernel density estimates, the smoothing parameters h_i $(i = 1, \ldots, g)$ are determined separately for each group-conditional density estimate $\hat{f}_i^{(K)}(\mathbf{x})$ rather than jointly. This approach is pursued first, and a joint approach is described in Sections 9.4.7 to 9.4.9.

The choice of a value for h_i is crucial in the estimation process. If h_i is too small, then $\hat{f}_i^{(K)}(\mathbf{x})$ will have sharp peaks at the sample points $\mathbf{x}_{ij}$ and be ragged elsewhere. If h_i is too large, excessive smoothing will occur resulting in bias. Thus, the aim is to choose h_i so as to give a suitable degree of smoothness relative to the amount of bias that can be tolerated. The latter is influenced by the purpose for which the density estimate is to be used. For use of $\hat{f}_i^{(K)}(\mathbf{x})$ in an exploratory data analysis role, it may be sufficient to choose h_i subjectively, as described in Silverman (1986, Section 3.4). But for routine problems in discriminant analysis, an automatic choice of the smoothing parameter h_i is appropriate.

The vital dependence of the performance of the kernel estimator on its smoothing parameter has led to many proposals for its selection from the observed data. We now consider some of these proposals.

9.4.2 Minimization of Mean Integrated Squared Error (MISE)

If $f_i(\mathbf{x})$ were known, then h_i could be chosen to minimize the integrated squared error (ISE),

$$\text{ISE}(h_i) = \int \{\hat{f}_i^{(K)}(\mathbf{x}) - f_i(\mathbf{x})\}^2 \, d\nu, \tag{9.4.1}$$

whose minimizer is denoted by $\hat{h}_i(\text{ISE})$. Alternatively, h_i could be chosen to minimize the mean integrated squared error (MISE),

$$\text{MISE}(h_i) = \text{E}\{\text{ISE}(h_i)\}, \tag{9.4.2}$$

whose minimizer is denoted by $\hat{h}_i(\text{MISE})$.

There is some controversy concerning which of $\hat{h}_i(\text{ISE})$ or $\hat{h}_i(\text{MISE})$ should be called the "correct" smoothing parameter. This is discussed by Marron (1989), who notes that there is, for all reasonable sample sizes, a very considerable difference between these goals (Hall and Marron, 1987a). In his recent investigation into the method of the choice of the smoothing parameter in the univariate case, Marron (1989) chose to work with MISE for a number of reasons, including its sample stability. MISE admits, under certain technical assumptions, which include that $K_p(\mathbf{x})$ is a radially symmetric density function, the asymptotic representation

$$\text{AMISE}(h_i) = n_i^{-1} h_i^{-p} \gamma_1 + \tfrac{1}{4} h_i^4 \gamma_2^2 \gamma_{3i}, \tag{9.4.3}$$

as $n_i \to \infty$, $h_i \to 0$, with $n_i h_i^p \to \infty$; see, for example, Silverman (1986, page 85). In (9.4.3),

$$\gamma_1 = \int \{K_p(\mathbf{x})\}^2 \, d\nu,$$

$$\gamma_2 = \int x_1^2 K_p(\mathbf{x}) \, d\nu,$$

and

$$\gamma_{3i} = \int \{\nabla^2 f_i(\mathbf{x})\}^2 \, d\nu,$$

where ∇ is the gradient vector with vth component $\partial/\partial x_v$ $(v = 1, \ldots, p)$.

The minimizer of $\text{AMISE}(h_i)$ is

$$\hat{h}_i(\text{AMISE}) = C(f_i, K_p) n_i^{-1/(p+4)}, \tag{9.4.4}$$

where

$$C(f_i, K_p) = \{p\gamma_1 / (\gamma_2^2 \gamma_{3i})\}^{1/(p+4)}. \tag{9.4.5}$$

Hall and Marron (1987a) have shown that $\hat{h}_i(\text{ISE})$, $\hat{h}_i(\text{MISE})$, and $\hat{h}_i(\text{AMISE})$ all come together in the limit.

The result (9.4.4) suggests one possible way of selecting h_i, by evaluating (9.4.5) for an estimate of $f_i(\mathbf{x})$ in γ_{3i}; see Devroye (1989).

9.4.3 Pseudo-Likelihood Cross-Validation Method

As $h_i \to 0$, $\hat{f}_i^{(K)}(\mathbf{x})$ approaches zero at all $\mathbf{x}$ except at $\mathbf{x} = \mathbf{x}_{ij}$, where it is $1/n_i$ times the Dirac delta function. This precludes choosing h_i by maximization of the pseudo-log likelihood

$$\sum_{j=1}^{n_i} \log \hat{f}_i^{(K)}(\mathbf{x}_{ij})$$

with respect to h_i. It led Habbema et al. (1974a) and Duin (1976) to choose h_i by maximizing the cross-validated pseudo-log likelihood,

$$\sum_{j=1}^{n} \log \hat{f}_{i(ij)}^{(K)}(\mathbf{x}_{ij})$$

with respect to h_i, where $\hat{f}_{i(ij)}^{(K)}(\mathbf{x})$ denotes the kernel density estimate $\hat{f}_i^{(K)}(\mathbf{x})$ formed from $\mathbf{x}_{ik}$ $(k = 1, \ldots, n_i;\ k \neq j)$. Let $\hat{h}_i(\text{PLCV})$ denote the value of h_i so obtained.

This procedure has been shown by Titterington (1980) to be cross-validatory in the sense of M. Stone (1974). It is not consistent for compactly supported kernels with infinitely supported densities (Schuster and Gregory, 1981), but it is consistent with compactly supported densities (Chow, Geman, and Wu, 1983).

Bowman, Hall, and Titterington (1984) showed that the use of $\hat{h}_i(\text{PLCV})$ leads to asymptotic minimization of the Kullback–Leibler loss in the case of univariate discrete data. The Kullback–Leibler loss in estimating $f_i(\mathbf{x})$ by $\hat{f}_i^{(K)}(\mathbf{x})$ is defined to be

$$\delta_{KL}(\hat{f}_i^{(K)}, f_i) = \int f_i(\mathbf{x}) \log\{f_i(\mathbf{x}) / \hat{f}_i^{(K)}(\mathbf{x})\}\, d\nu.$$

In the univariate continuous case, Hall (1987a) has demonstrated that the use of $\hat{h}_i(\text{PLCV})$ can lead to infinite Kullback loss. In a further study of this problem, Hall (1987b) provided an explicit account of how Kullback–Leibler loss and pseudo-likelihood cross-validation are influenced by interaction between tail properties of the kernel K_1 and of the unknown density $f_i(x)$. He showed that if the tails of the kernel K_1 are sufficiently thick for tail-effect terms not to play a role in determining minimum Kullback–Leibler loss, then the use of $\hat{h}_i(\text{PLCV})$ does lead to asymptotic minimization of Kullback–Leibler loss. Hall (1986c, 1987b) noted that the tails of the standard normal kernel are too thin for most purposes and even the double exponential is not always suitable. He suggested that a practical alternative is the kernel

$$K_1(\mathbf{x}) = 0.1438 \exp[-\tfrac{1}{2}\{\log(1 + |x|)\}^2], \qquad -\infty < x < \infty,$$

whose tails decrease more slowly than $\exp(-|x|^a)$ for any $a > 0$.

These theoretical results are supported by the empirical work of Remme, Habbema, and Hermans (1980), who investigated the use of the kernel method in estimating the group-conditional densities in discriminant analysis. They observed that while the use of the PLCV method of selection led to very good results for short-tailed distributions, it was an inappropriate method for long-tailed distributions.

9.4.4 Least-Squares Cross-Validation

With the least-squares cross-validation (LSCV) method, h_i is chosen to be $\hat{h}_i(\text{LSCV})$, the minimizer of

$$\int \{\hat{f}_i^{(K)}(\mathbf{x})\}^2 \, d\nu - 2n_i^{-1} \sum_{j=1}^{n_i} \hat{f}_{i(ij)}^{(K)}(\mathbf{x}_{ij}). \tag{9.4.6}$$

This method was suggested by Rudemo (1982) and Bowman (1984); see also Bowman et al. (1984) and Hall (1983a). It is motivated by the fact that the expectation of (9.4.6), minus the term

$$\int \{f_i(\mathbf{x})\}^2 \, d\nu,$$

which obviously does not depend on h_i, is an unbiased estimator of MISE(h_i), as defined by (9.4.2). The method is asymptotically optimal for all bounded f_i and all bounded compact support kernels K_1 (C. J. Stone, 1984). Also, it seems to be consistent whenever

$$\int \{f_i(\mathbf{x})\}^2 \, d\nu < \infty;$$

see Devroye (1989). He notes that $\hat{h}_i(\text{LSCV})$ is probably much too small whenever the data show clustering around one or more points.

Also, as demonstrated by Marron (1989) in the univariate case, the sample variability in $\hat{h}_i(\text{LSCV})$ is very high. Previously, Hall and Marron (1987a) had shown under certain technical conditions that

$$\hat{h}_i(\text{LSCV})/\hat{h}_i(\text{MISE}) - 1 = O_P(n_i^{-1/10}).$$

As commented by Marron (1989), this very slow rate of convergence goes a long way toward explaining why $\hat{h}_i(\text{LSCV})$ has not performed as well as anticipated in simulation studies, such as those by Kappenman (1987) discussed in the next section.

9.4.5 Moment Cross-Validation

The value of the smoothing parameter h_i proposed by Kappenman (1987) is the solution of the equation

$$\int \{\hat{f}_i^{(K)}(\mathbf{x})\}^2 \, d\nu = n_i^{-1} \sum_{j=1}^{n_i} \hat{f}_{i(ij)}^{(K)}(\mathbf{x}_{ij}). \tag{9.4.7}$$

We refer to this method as moment cross-validation (MCV), as it is analogous to the method of moments applied to the estimation of the mean of the density, after cross-validation.

As pointed out by Marron (1989), a potential flaw to this reasoning is that estimation of the mean of the density (the integrated squared density) is a different goal than estimation of the density itself (Hall and Marron, 1987b). Indeed, Marron (1989) showed that $\hat{h}_i(\text{MCV})$ provides a value of h_i that is too small. More precisely, he showed under certain technical conditions in the univariate case that, as $n_i \to \infty$, $\hat{h}_i(\text{MCV})$ essentially tends to zero at the rate $n_i^{-1/3}$ and, from (9.4.4), the appropriate rate is $n_i^{-1/5}$. However, in the simulation study conducted by Kappenman (1987), the performance of $\hat{h}_i(\text{MCV})$ was found to be superior relative to $\hat{h}_i(\text{LSCV})$, as obtained by least-squares cross-validation. This relative superiority of $\hat{h}_i(\text{MCV})$, in spite of its bias, can be explained by the large variance of $\hat{h}_i(\text{LSCV})$, which was referred to in the previous section. As illustrated by Marron (1989), $\hat{h}_i(\text{MCV})$ has a bias that becomes worse for larger sample sizes n_i. On the other hand, $\hat{h}_i(\text{LSCV})$ suffers from a high degree of variability, which, for large enough n_i, is eventually less debilitating than the bias of $\hat{h}_i(\text{MCV})$. The points where the trade-off occurs depend on the true density f_i, but except for the standard normal, it seems to happen typically for n_i around 100. Marron (1989) also considered a way of removing the bias inherent in $\hat{h}_i(\text{MCV})$. In the particular case of a standard normal kernel, he found that the bias-corrected estimate is the same as $\hat{h}_i(\text{LSCV})$.

9.4.6 Minimization of MISE for Multivariate Binary Features

A common way of choosing the smoothing parameter h_i in the discrete kernel density estimator (9.3.10) is to use pseudo-likelihood cross-validation, as defined in Section 9.4.3. As shown by Aitchison and Aitken (1976), if cross-validation is not employed here with this pseudo-likelihood approach, then the smoothing parameter h_i in (9.3.10) is estimated to be one. Hall (1981a) has shown that even with pseudo-likelihood cross-validation, if several cells are empty and all other cells contain two or more observations, then the value of h_i is likely to be estimated as one. To avoid this potential difficulty with multivariate binary data, Hall (1981a) has suggested that the smoothing parameter h_i be chosen by finding the value of h_i that minimizes a truncated expansion, in powers of $(1-h_i)$, of the discrete analogue of the mean integrated squared error of $\hat{f}_i^{(K)}(\mathbf{x})$, as given by (9.4.2). If terms of order $(1-h_i)^2$ are retained in this expansion and the unknown probabilities replaced by their maximum likelihood estimates, then the value of h_i is given by

$$\hat{h}_i(\text{IMSE}) = 1 - c_{i1}/c_{i2},$$

where

$$c_{i1} = n_i^2 p + \sum n_i(\mathbf{x})\{n_{i1}(\mathbf{x}) - p n_i(\mathbf{x})\}, \tag{9.4.8}$$

and

$$c_{i2} = n_i \sum n_i(\mathbf{x})\{2n_{i2}(\mathbf{x}) + p(p+1)n_i(\mathbf{x}) - 2pn_{i1}(\mathbf{x})\}, \qquad (9.4.9)$$

and where $n_{ik}(\mathbf{x})$ is the number of the n_i feature vectors $\mathbf{x}_{ij}$ $(j = 1,\ldots,n_i)$ from G_i for which

$$\|\mathbf{x} - \mathbf{x}_{ij}\|^2 = k \qquad (k = 1,2).$$

In (9.4.8) and (9.4.9), summation is over all possible 2^p realizations of $\mathbf{X}$.

To cut down further on the calculations, Hall (1981a) also gave a reduced form of the discrete kernel density estimator (9.3.10),

$$\hat{f}_i^{(K)}(\mathbf{x}) = n_i^{-1} n_i(\mathbf{x}) + n_i^{-1}(1-h_i)\{n_{i1}(\mathbf{x}) - n_i(\mathbf{x})\}.$$

It is obtained from (9.3.10) by eliminating terms of order $(1-h_i)^2$ and smaller. This reduced estimator ignores the effect of observations distant more than one unit from $\mathbf{x}$ on the full form (9.3.10) of the estimator. As noted by Hall (1981a), $\hat{f}_i^{(K)}(\mathbf{x})$ is thus an alternative to the nearest-neighbor estimator of order one suggested by Hills (1967). In a later paper, Hall (1981b) showed how to construct an adaptive weighted version of Hills' (1967) estimator using the MISE criterion.

9.4.7 Joint Selection of Smoothing Parameters

In the preceding sections, we have considered the selection of the smoothing parameter h_i separately for each group G_i $(i = 1,\ldots,g)$. In discriminant analysis, the essential aim of estimation of the individual group-conditional densities is to produce estimates of the posterior probabilities of group membership and/or a discriminant rule. This implies that the smoothing parameters should be selected jointly, rather than separately, using loss functions directly related to the discrimination problem at hand. This approach has been considered by Van Ness and Simpson (1976) and Van Ness (1979) for continuous feature data and by Tutz (1986, 1988, 1989) for discrete feature variables. Also, Hall and Wand (1988) have considered a joint selection approach that is applicable for either continuous, discrete, or mixed feature variables in the case of $g = 2$ groups.

The approach of Tutz (1986) applies to the use of the discrete kernel estimator of Aitchison and Aitken (1976). The smoothing parameters are selected simultaneously by minimization of the leave-one-out or cross-validated estimate of the overall error rate of the sample version of the Bayes rule formed by plugging in these kernel estimates of the group-conditional densities. The cross-validated estimate of the error rate of a sample discriminant rule is to be considered in Section 10.2.2. Tutz (1986) showed that his approach leads to a Bayes risk consistent rule. However, the cross-validated estimate of the error rate is a discontinuous function of the smoothing parameters, and itself requires smoothing for effective implementation; see Tutz (1988, 1989).

9.4.8 Joint Selection via Estimation of Differences Between Group-Conditional Densities

For the allocation problem in the case of $g = 2$ groups, Hall and Wand (1988) have proposed that the smoothing parameters be chosen by applying least-squares cross-validation with respect to the estimate of the difference between the two group-conditional densities. We consider first the case of discrete feature variables, taking on only a finite number of distinct values with nonzero probabilities. As noted in Section 7.4.1, it can be assumed without loss of generality that each X_v is a zero-one variable. That is, the sample feature space is the binary space $\{0,1\}^p$, the set of all p-tuples of zeros and ones. For this situation, Hall and Wand (1988) worked with the discrete kernel density of Aitchison and Aitken (1976),

$$\hat{f}_i^{(K)}(\mathbf{x}) = n_i^{-1}\sum_{j=1}^{n_i} h_i^{p-d_{ij}^2}(1-h_i)^{d_{ij}^2} \qquad (i = 1,2), \tag{9.4.10}$$

where $d_{ij}^2 = \|\mathbf{x}-\mathbf{x}_{ij}\|^2$ for $j = 1,\ldots,n_i$ and $i = 1,2$.

From (1.4.4), the optimal $r_o(\mathbf{x};F)$ assigns an entity with feature vector $\mathbf{x}$ to either G_1 or G_2, according as

$$\pi_1 f_1(\mathbf{x}) \geq \pi_2 f_2(\mathbf{x}) \tag{9.4.11}$$

holds or not. The inequality (9.4.11) is equivalent to

$$\pi_1 f_1(\mathbf{x}) - \pi_2 f_2(\mathbf{x}) \geq 0.$$

Thus, for the allocation problem, it led Hall and Wand (1988) to consider the estimation of the difference

$$\nu(\mathbf{x}) = \pi_1 f_1(\mathbf{x}) - \pi_2 f_2(\mathbf{x})$$

for fixed π_1 and π_2. Let

$$\hat{\nu}^{(K)}(\mathbf{x}) = \pi_1 \hat{f}_1^{(K)}(\mathbf{x}) - \pi_2 \hat{f}_2^{(K)}(\mathbf{x})$$

be an estimate of $\nu(\mathbf{x})$, formed using the discrete kernel estimate (9.4.10). Then with their approach, a variant of cross-validation is used as a tool to choose h_1 and h_2 jointly to minimize

$$\text{MISE}(h_1,h_2) = E\sum_{\mathbf{x}}\{\hat{\nu}^{(K)}(\mathbf{x}) - \nu(\mathbf{x})\}^2, \tag{9.4.12}$$

where the summation in (9.4.12) is over all values of $\mathbf{x}$.

Hall and Wand (1988) found the values of h_1 and h_2 that minimize (9.4.12), as n_1 and n_2 tend to infinity. They noted that one or other of these asymptotic optimal values of h_1 and h_2 may be negative. However, as they explained, this is not so absurd as might at first appear, for it is the difference between two densities that is being estimated and not an individual density. However, a negative smoothing parameter does lead to kernel weights that oscillate in sign as the distance from the cell at which the estimator is evaluated increases.

In practice, Hall and Wand (1988) suggested that h_1 and h_2 be given by their values that minimize

$$\sum_{\mathbf{x}}\{\hat{\upsilon}^{(K)}(\mathbf{x})\}^2 - 2\left[\sum_{i=1}^{2}\pi_i^2 n_i^{-1}\sum_{j=1}^{n_i}\hat{f}_{i(ij)}^{(K)}(\mathbf{x}_{ij}) - \pi_1\pi_2\sum_{i=1}^{2}n_i^{-1}\sum_{j=1}^{n_i}\hat{f}_i^{(K)}(\mathbf{x}_{ij})\right], \tag{9.4.13}$$

where $f_{i(ij)}^{(K)}(\mathbf{x})$ denotes the kernel density estimate $\hat{f}_i^{(K)}(\mathbf{x})$ formed from $\mathbf{x}_{ik}$ $(k = 1,\ldots,n_i;\ k \neq j)$. A simplification of this selection procedure is to carry out the minimization of (9.4.13) with the restriction $h_1 = h_2$ imposed from the outset. Apart from a term not involving h_1 and h_2, (9.4.13) is an unbiased estimator of MISE(h_1,h_2). Thus, their method has the same motivation of least-squares cross-validation in its application to the separate choice of the smoothing parameters in the individual estimation of the group-conditional densities.

For continuous feature data, Hall and Wand (1988) based their approach on the product kernel estimator (9.3.8), which supposes that the data have been standardized first. The values of the smoothing parameters h_1 and h_2 are again taken to be those that minimize (9.4.13), but where now the first-term sum in (9.4.13) is replaced by the integral

$$\int\{\hat{\upsilon}^{(K)}(\mathbf{x})\}^2\, d\nu \tag{9.4.14}$$

and where the product kernel estimator (9.3.8) is used throughout in place of the discrete kernel estimator (9.3.10). The integral in (9.4.14) can be calculated explicitly if the standard normal kernel is used.

As noted by Hall and Wand (1988), their procedure for selecting the smoothing parameters could be framed in terms of the ratio of the two group-conditional densities, rather than their difference. However, they warn that such a technique is hardly practicable, because of serious problems when the denominator of the ratio is close to zero. Some asymptotic results on the kernel approach applied to the problem of estimating the ratio of the group-conditional densities have been given recently by Ćwik and Mielniczuk (1989).

9.4.9 Example: Separate Versus Joint Selection of Smoothing Parameters

We present here one of the examples given by Hall and Wand (1988) to demonstrate their approach. It concerns the diagnosis of keratoconjunctivitis sicca (KCS) in people suffering from rheumatoid arthritis. The feature vector consists of $p = 10$ binary variables, each corresponding to the presence or absence of a particular symptom. The training data consist of $n_1 = 40$ observations on patients with KCS (group G_1) and $n_2 = 37$ observations on patients without KCS (group G_2). On the basis of these training data and with the group-prior probabilities each taken equal to half, Hall and Wand (1988) formed the plug-in sample version of the Bayes rule with the group-conditional

TABLE 9.3 Smoothing Parameters and Group-Specific Cross-Validated Error Rates for Each of Three Selection Methods

Selection Method	Smoothing Parameters		Cross-Validated Error Rates	
	h_1	h_2	G_1	G_2
LSCV (difference between densities)	0.216	0.012	4/40	2/37
PLCV	0.157	0.040	4/40	1/37
LSCV	0.195	0.008	4/40	3/37

densities replaced by their discrete kernel estimates (9.4.10). The smoothing parameters h_1 and h_2 were selected by three methods, including their joint selection approach. The other two methods selected h_1 and h_2, separately, using pseudo-likelihood cross-validation (PLCV) and least-squares cross-validation (LSCV). The values of the smoothing parameters obtained with each method are displayed in Table 9.3, along with the cross-validated error rates of the group-specific conditional error rates of the sample-based rule corresponding to each method. It can be seen that in terms of these estimated error rates, the method of Hall and Wand (1988) performs in between the methods of pseudo-likelihood cross-validation and least-squares cross-validation applied to the individual estimation of each group-conditional density.

9.5 ALTERNATIVES TO FIXED KERNEL DENSITY ESTIMATES

9.5.1 Introduction

As discussed by Silverman (1986, Chapter 5), a practical drawback of the kernel method of density estimation is its inability to deal satisfactorily with the tails of distributions without oversmoothing the main part of the density. Two possible adaptive approaches to this problem discussed in the next sections are the nearest neighbor method and the adaptive kernel method. Of course, there are various other approaches, including maximum penalized likelihood, as proposed initially by Good and Gaskins (1971). As noted in Section 9.4.3, the pseudo-likelihood can be made arbitrarily large by letting the smoothing parameter h_i in the kernel estimate tend to zero. With the maximum penalized approach, a penalty term is added to the log likelihood, which reflects the roughness, in some sense, of the likelihood under consideration; see Silverman (1986, Section 5.4) for further details. In a related approach, Silverman (1978) estimated the ratio of two densities by consideration of a penalized conditional log likelihood.

Other nonparametric methods of density estimation not pursued here include those based on projection pursuit and splines. Exploratory projection pursuit as described in Section 6.6.4 can be incorporated into a density estimation procedure (Friedman, Stuetzle, and Schroeder, 1984). This approach for the purposes of discrimination has been considered recently by Flick et al. (1990). On the use of splines, Villalobos and Wahba (1983) have proposed a nonparametric estimate of the posterior probabilities of group membership, using multivariate thin-plate splines. The degree of smoothing is determined from the data using generalized cross-validation.

9.5.2 Nearest Neighbor Method

As noted in Section 9.4, Fix and Hodges (1951) introduced the naive kernel density estimator formed by counting the number of sample points $M_i(\mathbf{x})$ lying in some prescribed neighborhood $\mathcal{N}_i$ of $\mathbf{x}$. They also briefly mentioned the complementary approach of nearest neighbor density estimation. Rather than fixing $\mathcal{N}_i$ and counting $M_i(\mathbf{x})$, they proposed that $M_i(\mathbf{x})$ be fixed at some value, say, k, and $\mathcal{N}_i$ be found just large enough to include the k nearest points to $\mathbf{x}$. The first mention of this method in the published literature is by Loftsgaarden and Quesenberry (1965). It subsequently formed the basis of the nonparametric discriminant procedure proposed by Pelto (1969).

Let $d_{i,k}(\mathbf{x})$ be the Euclidean distance from $\mathbf{x}$ to the kth nearest point among $\mathbf{x}_{ij}$ $(j = 1,\ldots,n_i)$. Then the nearest neighbor density estimate of order k is defined by

$$\hat{f}_i^{(NN)}(\mathbf{x}) = kn_i^{-1}/C_i(\mathbf{x}),$$

where

$$C_i(\mathbf{x}) = v_p\{d_{i,k}(\mathbf{x})\}^p,$$

and v_p is the volume of the unit sphere in p dimensions (so that $v_1 = 2$, $v_2 = \pi$, $v_3 = (4/3)\pi$, etc.). Recently, Chanda (1990) has derived asymptotic expansions of the error rates for the sample-based rule obtained by using $\hat{f}_i^{(NN)}(\mathbf{x})$ in place of $f_i(\mathbf{x})$. The new allocation method proposed by Patrick (1990), called The Outcome Adviser, evolved from this type of density estimate (Patrick and Fischer, 1970).

The generalized kth nearest neighbor estimate is defined by

$$\hat{f}_i^{(GNN)}(\mathbf{x}) = n_i^{-1}\{d_{i,k}(\mathbf{x})\}^{-p}\sum_{j=1}^{n_i} K_p[\{d_{i,k}(\mathbf{x})\}^{-1}(\mathbf{x} - \mathbf{x}_{ij})],$$

which reduces to the nearest neighbor estimate if

$$\begin{aligned} K_p(\mathbf{x}) &= v_p^{-1}, \qquad \|\mathbf{x}\| \leq 1, \\ &= 0, \qquad \text{otherwise.} \end{aligned}$$

For multivariate binary feature data, Aitchison and Aitken (1976) showed how a generalized kth nearest neighbor estimate of $f_i(\mathbf{x})$ can be defined in terms of their discrete kernel function. It is given by (9.3.10) with

$$K_p(\mathbf{x};\mathbf{x}_{ij},h_i) = \{h_i^{p-d_{ij}^2}(1-h_i)^{d_{ij}^2}\}/C(h_i,k), \qquad \text{if} \quad d_{ij}^2 \le k,$$
$$= 0, \qquad \text{otherwise}, \tag{9.5.1}$$

where $\frac{1}{2} \le h_i \le 1$, and where

$$C(h_i,k) = \sum_{\nu=0}^{k} \binom{p}{\nu} h_i^{p-\nu}(1-h_i)^{\nu}.$$

They noted that the nearest neighbor estimate of order k proposed by Hills (1967) is obtained from (9.5.1) on putting $h_i = 1/2$.

The estimates obtained by the nearest neighbor method are not very satisfactory. They are prone to local noise and also have very heavy tails and infinite integral; see Silverman (1986, Section 5.2) for further discussion.

9.5.3 Adaptive Kernel Method

An adaptive kernel estimator is given by

$$\hat{f}_i^{(K)}(\mathbf{x}) = n_i^{-1}\sum_{j=1}^{n_i} h_{ij}^{-p} K_p\left(\frac{\mathbf{x}-\mathbf{x}_{ij}}{h_{ij}}\right), \tag{9.5.2}$$

where the n_i smoothing parameters h_{ij} $(j = 1,\ldots,n_i)$ are based on some pilot estimate of the density $f_i(\mathbf{x})$. The adaptive kernel method has been shown to have good practical and theoretical properties and to be more accurate in the tails than either the fixed kernel or nearest neighbor methods; see Bowman (1985) and Worton (1989) for recent comparative studies of fixed versus adaptive kernel estimates. In a discriminant analysis context, Remme et al. (1980) reported in the case of log normal group-conditional distributions, that the poor performance of the sample discriminant rule based on fixed kernel density estimates was improved by using a variable kernel.

Silverman (1986, Section 5.3) has suggested a useful algorithm for obtaining adaptive kernel estimates. The smoothing parameters h_{ij} are specified as $h_i a_{ij}$, where h_i is a global smoothing parameter, and the a_{ij} are local smoothing parameters given by

$$a_{ij} = \{\tilde{f}_i(\mathbf{x}_{ij})/C_i\}^{-\alpha} \qquad (j = 1,\ldots,n_i),$$

where

$$\log C_i = n_i^{-1}\sum_{j=1}^{n_i} \log \tilde{f}_i(\mathbf{x}_{ij}),$$

and $\tilde{f}_i(\mathbf{x})$ is a pilot estimate of $f_i(\mathbf{x})$ that satisfies

$$\tilde{f}_i(\mathbf{x}_{ij}) > 0 \qquad (j = 1,\ldots,n_i),$$

and where α_i is the sensitivity parameter, satisfying $0 \leq \alpha_i \leq 1$. Silverman (1986) has reported that the adaptive kernel method with $\alpha_i = 1/2$ has good practical and theoretical properties. The choice of the pilot estimate $\tilde{f}_i(\mathbf{x})$ is not crucial, as the adaptive estimate appears to be insensitive to the final detail of the pilot estimate. With using this approach, there is only one smoothing parameter to select as with the fixed kernel method.

To illustrate the differences between fixed and adaptive kernel methods, Worton (1989) compared the estimates produced for data sets of size $n = 100$ simulated from four different bivariate densities, (i) the standard bivariate normal, (ii) a log normal-normal, (iii) a light-tailed density, and (iv) the standard bivariate Cauchy. The density in case (iii) was

$$\begin{aligned} f_i(\mathbf{x}) &= 3\pi^{-1}(1-\mathbf{x}'\mathbf{x})^2, \qquad &&\text{if} \quad \mathbf{x}'\mathbf{x} < 1, \\ &= 0, &&\text{otherwise.} \end{aligned}$$

The contour plots of these four densities are given in Figure 9.3, and the contour plots of their fixed and adaptive kernel estimates are given in Figures 9.4. and 9.5, respectively. In forming these kernel density estimates, Worton (1989) used the optimal choice of the smoothing parameter for the bivariate version of the kernel of Epanechnikov (1969), given by

$$\begin{aligned} K_2(\mathbf{x}) &= 2\pi^{-1}(1-\mathbf{x}'\mathbf{x}), \qquad &&\text{if} \quad \mathbf{x}'\mathbf{x} < 1, \\ &= 0, &&\text{otherwise.} \end{aligned}$$

The true density was used as the pilot estimate in forming each adaptive estimate.

A comparison of the kernel density estimates in Figures 9.4 and 9.5 with the true densities in Figure 9.3 demonstrate how the adaptive produces a clearer picture than does the fixed method, particularly in the tails of densities that are not light-tailed.

Breiman, Meisel, and Purcell (1977) considered a particular case of (9.5.2) by taking the pilot estimate of $f_i(\mathbf{x})$ as the kth nearest neighbor estimate (using a fairly large value of k) and setting the sensitivity parameter α equal to $1/p$. Their variable-kernel estimate is, therefore, of the form

$$\hat{f}_i^{(K)}(\mathbf{x}) = n_i^{-1}h_i^{-p}\sum_{j=1}^{n_i}\{d_{i,k}(\mathbf{x}_{ij})\}^{-p}K_p[(\mathbf{x}-\mathbf{x}_{ij})/\{h_i d_{i,k}(\mathbf{x}_{ij})\}]. \tag{9.5.3}$$

The order k of nearest neighbor has to be chosen, along with the smoothing parameter h_i.

Aitken (1983) showed how his discrete kenel for univariate data in the form of counts can be adapted to a variable form like (9.5.3). The variable-kernel analogue of (9.3.14) is obtained by replacing h_i with $h_i/\{d_{i,k}(\mathbf{x})\}^{\alpha}$. The factor α (>0) is introduced so that the kernel satisfies the requirements of monotonicity.

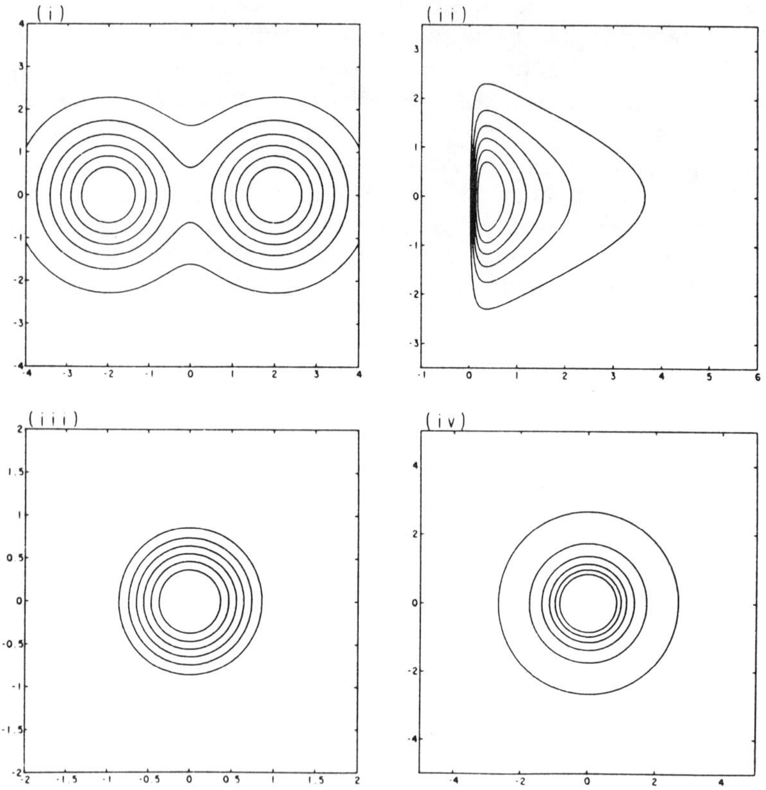

FIGURE 9.3. Contour plots of true bivariate densities (i) to (iv) underlying simulated data used to produce kernel density estimates in Figures 9.4 and 9.5. From Worton (1989).

9.6 COMPARATIVE PERFORMANCE OF KERNEL-BASED DISCRIMINANT RULES

9.6.1 Introduction

With the usual kernel approach to discriminant analysis, the unknown group-conditional densities in the expressions for the posterior probabilities of group membership are replaced by kernel density estimates formed from the training data **t**. The associated sample-based discriminant rule is therefore a plug-in version of the Bayes rule with kernel density estimates used in place of the group-conditional densities. It will be a Bayes risk consistent rule provided the kernel adopted satisfies certain regularity conditions as discussed in Section 9.3.3; see, for example, Marron (1983) and, more recently, Chanda and Ruymgaart (1989). For a more direct approach to the problem of estimating the posterior probabilities of group membership, the reader is referred to Lauder (1983), who considers direct kernel assessment of these probabilities.

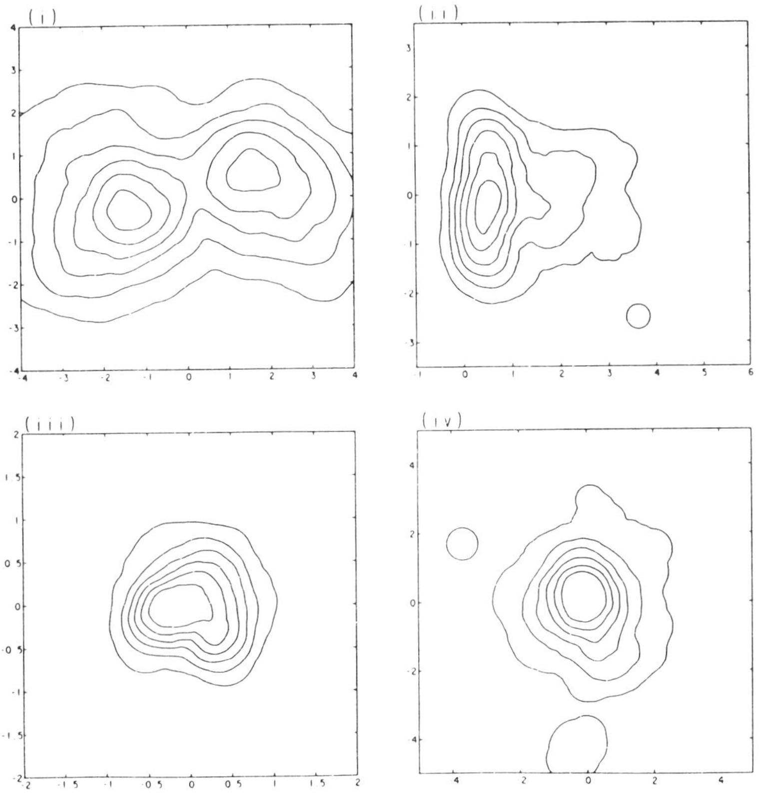

FIGURE 9.4. Contour plots of fixed kernel density estimates for 100 data points simulated from the densities (i) to (iv) with the optimal choice of smoothing parameter. From Worton (1989).

There has been a number of comparisons reported in the literature between kernel-based discriminant rules (KDRs) and other sample-based discriminant rules. Hand (1982) has provided a comprehensive summary of such comparisons published prior to his monograph. We first consider here those comparisons undertaken for continuous feature data.

9.6.2 Continuous Feature Data

Initial studies on the performance of KDRs for continuous feature variables include those in the case of $g = 2$ groups by Gessaman and Gessaman (1972), Goldstein (1975), Hermans and Habbema (1975), Van Ness and Simpson (1976), Koffler and Penfield (1979), Van Ness (1979), and Remme et al. (1980). The latter study was an extensive one and considered not only the allocatory performance of KDR rules, but also the estimates provided by the kernel approach for the posterior probabilities of group membership, or, equi-

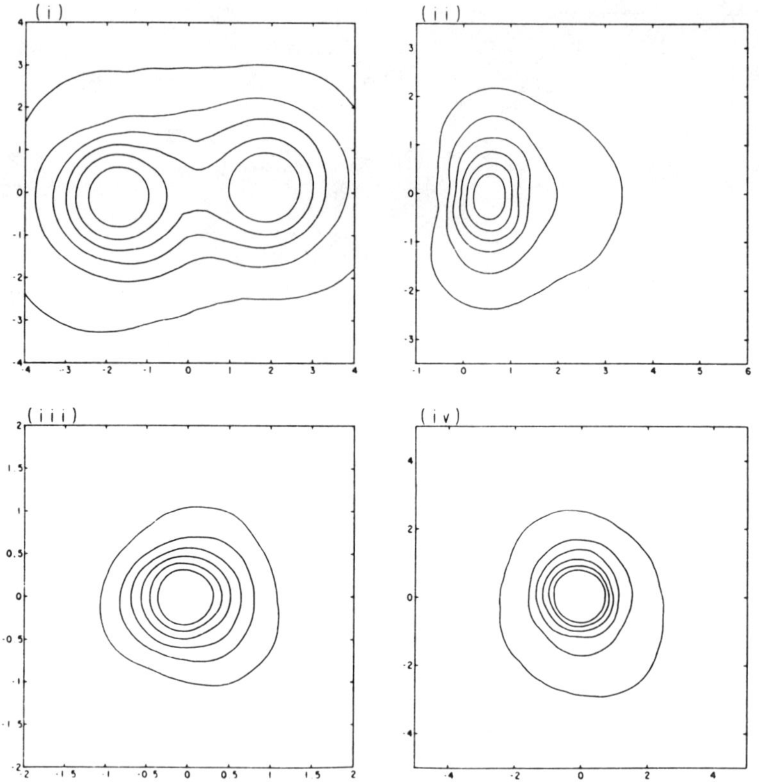

FIGURE 9.5. Contour plots of adaptive kernel density estimates for 100 data points simulated from the densities (i) to (iv) with the optimal choice of smoothing parameter. From Worton (1989).

valently, the posterior log odds. In some of the simulation experiments performed in these studies, the product normal KDR was quite superior to the sample NLDR. However, many of the experiments were carried out under conditions favorable to the product normal KDR. This was discussed by Hand (1982) and has been taken up more recently by Murphy and Moran (1986).

9.6.3 Loss of Generality of Product Normal Kernel Estimates Under Canonical Transformation

Murphy and Moran (1986) showed explicitly how the product normal KDR is not invariant under the linear transformation (4.2.2) that achieves the usual canonical form (4.2.3) without loss of generality for the sample NLDR applied under the homoscedastic normal model (3.3.1) for $g = 2$ groups.

More specifically, the product normal kernel estimate of the ith group-conditional density $f_i(\mathbf{x})$ is given by

$$\hat{f}_i^{(K)}(\mathbf{x}) = n_i^{-1} \sum_{j=1}^{n_i} \phi(\mathbf{x} - \mathbf{x}_{ij}; \mathbf{0}, h_i \mathbf{I}_p) \tag{9.6.1}$$

for feature data that have been standardized. For the linear transformation (4.2.2),

$$\tilde{\mathbf{x}} = \mathbf{C}_2\mathbf{C}_1\{\mathbf{x} - \tfrac{1}{2}(\boldsymbol{\mu}_1 + \boldsymbol{\mu}_2)\},$$

the estimate $\hat{f}_i^{(K)}(\mathbf{x})$ becomes

$$\hat{f}_i^{(K)}(\tilde{\mathbf{x}}) = n_i^{-1} \sum_{j=1}^{n_i} \phi(\tilde{\mathbf{x}} - \tilde{\mathbf{x}}_{ij}; \mathbf{0}, h_i \mathbf{M}), \tag{9.6.2}$$

where

$$\mathbf{M} = \mathbf{C}_2\mathbf{C}_1\mathbf{C}_1'\mathbf{C}_2',$$

and

$$\tilde{\mathbf{x}}_{ij} = \mathbf{C}_2\mathbf{C}_1\{\mathbf{x}_{ij} - \tfrac{1}{2}(\boldsymbol{\mu}_1 + \boldsymbol{\mu}_2)\} \qquad (j = 1, \ldots, n_i).$$

As $\mathbf{M}$ is not a diagonal matrix in general, (9.6.2) shows that $\hat{f}_i^{(K)}(\tilde{\mathbf{x}})$ is not a product normal kernel estimate of $f_i(\mathbf{x})$.

Hence, the canonical form (4.2.3) cannot be assumed without loss of generality in studies of the product normal KDR. It means that in those studies that have adopted the canonical form (4.2.3), the performances of the kernel density estimates are favored since they are formed using a product kernel that is appropriate for independent feature variables as specified by (4.2.3). Remme et al. (1980) did qualify their optimistic conclusions for KDRs using product kernels by noting they were limited to problems in which the feature variables were not highly correlated.

Another factor that would have contributed to the optimistic results of Van Ness and Simpson (1976) for KDRs with product normal and Cauchy kernels was that there were additional observations of known origin beyond the training set for use in choosing the smoothing parameters h_i in (9.6.1). These parameters were chosen so as to maximize the overall error rate of the product normal KDR in its application to the additional classified data. This resulted in much larger values of h_i than would have been obtained without these extra data. As shown by Specht (1967), the nonparametric density estimate (9.3.8) with either a product normal or Cauchy kernel approaches, for large values of the smoothing parameter h_i, the Euclidean distance rule. As the latter is the same as the sample normal-based rule formed with the knowledge that $\boldsymbol{\Sigma} = \mathbf{I}_p$ (or $\boldsymbol{\Sigma} = \sigma^2\mathbf{I}_p$), KDRs with product normal or Cauchy kernels have an advantage over rules using the general form of the sample group-covariance matrices.

Murphy and Moran (1986) thus performed a series of simulation experiments in which the product kernel estimates of the group-conditional densities did not have this advantage and in which the feature variables were not always specified to be independent. They concluded that for independent or moderately positively correlated feature variables with multivariate normal group-conditional distributions, sample discriminant rules based on product kernel density estimates are superior to sample normal-based rules with regard to both allocation and estimation of the posterior log odds for group membership. However, in the presence of other patterns of correlation, the kernel approach gave poor allocation or poor estimation of the posterior log odds, or both, depending on the underlying parametric configuration.

9.6.4 Discrete Feature Data

Comparative studies of KDRs in the case of discrete feature data have been undertaken by Aitken (1978), Titterington et al. (1981), and Hand (1983). These studies have been augmented by those for mixed feature data, where some variables are discrete and some are continuous. The latter studies are described in the next section. With the exception of the study by Titterington et al. (1981) in which there were $g = 3$ groups, the studies above were concerned with the case of $g = 2$ groups.

Aitken (1978) studied the KDR with the group-conditional densities estimated by using the Aitchison and Aitken (1976) type kernel, where for each group, the smoothing parameter was taken to be the same for each feature variable, corresponding to (9.3.10), and where it was allowed to be different for each variable, corresponding to (9.3.13) with $c_v = 2$. These two KDRs, along with the nearest neighbor procedure of Hills (1967), were compared with the predictive logistic discriminant rule (LGDR), as defined through (8.2.17), and the independence-based rule, as based on (7.2.2). These sample discriminant rules were applied to two real data sets consisting of multivariate binary feature variables. For one data set, the KDRs had the smallest cross-validated error rate, with the KDR based on (9.3.10) and the nearest neighbor rule of Hills (1967) being preferable to the KDR based on (9.3.13). However, the latter was the better of the two KDRs for the other data set, for which the predictive LGDR was the best.

The extensive comparative study by Titterington et al. (1981) was referred to in Section 5.6.2 on its reported findings in relation to the robustness of the sample NLDR and NQDR for discrete feature data. The feature variables in their study were all categorical and were either binary or ordered. Besides the sample NLDR and NQDR, the study considered several other sample discriminant rules, including those based on the independence, Lancaster, latent class, and logistic models, and the nonparametric kernel approach. For the latter, the basic kernel adopted was of the form (9.3.13) for an unordered categorical feature variable X_v and was modified as in Section 9.3.7 if X_v were an ordered categorical variable. Two techniques were used to handle missing data. One involved using the kernel specifically designed for missing data (for ordered or unordered variables), as defined in Table 9.2, whereas the other treated missing as an extra category. For a given group, the smoothing parameters were selected by various methods, which included marginal and multivariate choices and pseudo-Bayes choice, and which allowed both common and variable-specific smoothing.

In their comparisons between the KDRs and the other types of sample discriminant rules, Titterington et al. (1981) found that the performances of the KDRs were disappointing. The lack of success of the kernel methods, particularly with high-dimensional data, was noted. Among the KDRs, the one with a common smoothing parameter for the feature variables in the formation of each group-conditional density and with missing treated as an extra category was noticeably the best.

Hand (1983) reported the results of a comparison, along the lines of Aitken (1978) and Titterington et al. (1981). The kernel method was compared to the sample NLDR in their applications to two real data sets involving multivariate binary feature variables. The kernel (9.3.10) of Aitchison and Aitken (1976) was used and the smoothing parameter h_i common to each feature was selected according to two methods, pseudo-likelihood cross-validation and minimization of an expansion of the MISE, as proposed by Hall (1981a); see Section 9.4.6. Hand (1983) concluded that there was little to choose between the sample NLDR and either KDR, and between the two methods for selecting the smoothing parameters in the kernel density estimates.

9.6.5 Mixed Feature Data

Vlachonikolis and Marriott (1982); Schmitz, Habbema, and Hermans (1983a); and Schmitz et al. (1983b, 1985) have carried out comparative studies of KDRs in the case of mixed feature variables. In the first of these studies, the performance of the kernel and some other sample discriminant rules were compared on the basis of their performances when applied to two real data sets with binary and continuous feature variables. The KDR was formed by using a kernel of the form (9.3.16), where a product normal kernel was used for the continuous variables and, effectively, an Aitchison and Aitken (1976) type kernel for the binary variables. The other sample discriminant rules in the study were the sample NLDR, the sample NLDR with the sample NLDF augmented as described in Section 7.6.2, the logistic discriminant rule (LGDR), and the sample location model-based discriminant rule (LODR). They reported that the KDR was less effective than the other sample rules. This study demonstrated that provided the sample NLDF is augmented with the appropriate interaction terms, it is a good choice of a sample allocation rule. A similar conclusion was reached by Knoke (1982), although his study did not include KDRs. As stressed in Section 7.6.2, the usefulness of this approach of augmenting the sample NLDF is in its role of improving the allocatory performance of the sample NLDF and in the preliminary screening of interaction terms to be retained in the fitting of a reduced version of the location model or in the formulation of the logistic model. It is not applicable to the problem of providing estimates of the discriminant function coefficients, and hence of the posterior probabilities of group membership. Schmitz et al. (1983a) have pointed out that with a large number of feature variables, detection of relevant interaction terms for augmenting the sample NLDF may require much effort. In such instances, they suggest that one may prefer to use the kernel method, which is an option that automatically attempts to take into account the interaction structures.

In their study, Schmitz et al. (1983a) considered the sample NLDR and NQDR, LGDR, and KDR, where the latter used a mixed kernel based on the variable kernel (9.3.15) of Habbema et al. (1978). These sample discriminant rules were evaluated on the basis of their application to a myocardial infarction data set, consisting of three binary and nine continuous variables. The KDR, sample NLDR, and LGDR were found to perform nearly identically and better than the sample NQDR. Schmitz et al. (1983a) remarked on how the optimistic bias of the apparent error rate of the KDR was greater than that of the other sample rules. Nonparametric methods of reducing the bias of the apparent error rate of a sample rule are discussed in Section 10.2. This tendency of the kernel method to overfit the data has been noted in other studies, for example, Hand (1983). It is exacerbated by small group-sample sizes n_i relative to the number p of feature variables; see Schmitz et al. (1985) on the sensitivity of KDRs to the group-sample sizes n_i.

The sample discriminant rules in the previous study were compared further by Schmitz et al. (1983b), using simulated mixed data that were generated from a four-dimensional normal distribution for each group, before three of the variables were discretized. The relative performance of the KDR depended on the similarity of the two group-covariance matrices. In the unequal situation, the KDR performed as well as or better than the sample NQDR, which was superior to the sample NLDR. For equal group-covariance matrices, the sample NLDR was found to be best with the KDR again being better or as good as the sample NQDR. As in the other studies referred to here and in the previous section, the LGDR provided comparable allocation to that of the sample NLDR.

These four sample discriminant rules, along with the independence-based rule, were investigated further by Schmitz et al. (1985) in a simulation study with mixed feature variables. Whereas the earlier study of Schmitz et al. (1983b) indicated a preference for the KDR over the sample NQDR, the results of this later study reversed this preference. Schmitz et al. (1983b) did comment that the fact that the sample NQDR was never the best rule in their study may be explained by the absence of substantial nonlinearities in the likelihood ratios for the generated data. Indeed, in the later study of Schmitz et al. (1985), the sample NQDR generally outperformed the KDR in the presence of nonlinearities in the likelihood ratios.

Schmitz et al. (1985) observed that remarkably good results were obtained by the sample rule that corresponded to the better of the sample NLDR and NQDR in any given situation. This in itself is a strong recommendation for the regularized discriminant approach of Friedman (1989), as formulated in Section 5.5. For as seen there, this rule is essentially a compromise between the sample NLDR and NQDR, as decided automatically from the data.

9.7 NEAREST NEIGHBOR RULES

9.7.1 Introduction

As noted by Fix and Hodges (1951), nearest neighbor allocation is based on a variant of nearest neighbor density estimation of the group-conditional densities. Suppose that $f_1(\mathbf{x})$ is estimated from (9.3.1) by the naive kernel density estimator $f_1^{(K)}(\mathbf{x})$, where K_p is the uniform density over the neighborhood $\mathcal{N}_1$ taken large enough so as to contain a given number k of points in the combined sample $\mathbf{x}_{ij}$ $(i = 1,2;\ j = 1,\ldots,n_i)$. The same neighborhood $\mathcal{N}_1$ is then used to construct the naive kernel density $\hat{f}_2^{(K)}(\mathbf{x})$ of $f_2(\mathbf{x})$. Then it can be seen that the allocation rule based on the relative size of $\hat{f}_1^{(K)}(\mathbf{x})$ and $\hat{f}_2^{(K)}(\mathbf{x})$ is equivalent to that based on the relative size of k_1/n_1 and k_2/n_2, where k_i is the number out of the k points from the combined sample in the neighborhood $\mathcal{N}_1$ of $\mathbf{x}$, that come from G_i $(i = 1,2)$.

For $k = 1$, the rule based on the relative size of k_1/n_1 and k_2/n_2 is precisely the 1-NN discriminant rule (nearest neighbor rule of order 1), which assigns an entity with feature vector $\mathbf{x}$ to the group of its nearest neighbor in the training set. For $k > 1$, their approach incorporates a variety of so-called k-NN discriminant rules, including simple majority vote among the groups of a point's k nearest neighbors and modifications that take differing group-sample sizes into account; see Silverman and Jones (1989).

9.7.2 Definition of a k-NN Rule

We now give a formal definition of a kth nearest neighbor (k-NN) rule within the regression framework, as adopted by C. J. Stone (1977) in his important paper on consistent nonparametric regression. For an unclassified entity with feature vector $\mathbf{x}$, consider estimates of its posterior probabilities of group membership having the form

$$\hat{\tau}_i(\mathbf{x};\mathbf{t}) = \sum_{j=1}^{n} w_{nj}(\mathbf{x};\mathbf{t}_x)z_{ij} \qquad (i = 1,\ldots,g), \tag{9.7.1}$$

where $w_{nj}(\mathbf{x};\mathbf{t}_x)$ are nonnegative weights that sum to unity and that may depend on $\mathbf{x}$ and the training feature vectors $\mathbf{x}_1,\ldots,\mathbf{x}_n$, but not their associated group-indicator vectors $\mathbf{z}_1,\ldots,\mathbf{z}_n$. As previously, $\mathbf{t}_x' = (\mathbf{x}_1,\ldots,\mathbf{x}_n)$ and $\mathbf{t}' = (\mathbf{y}_1,\ldots,\mathbf{y}_n)$, where $\mathbf{y}_j = (\mathbf{x}_j',\mathbf{z}_j')'$ for $j = 1,\ldots,n$. Rank the n training observations $\mathbf{y}_1,\ldots,\mathbf{y}_n$ according to increasing values of $\|\mathbf{x}_j - \mathbf{x}\|$ to obtain the n indices $R_1,\ldots,R_n$. The training entities j with $R_j \leq k$ define the k nearest neighbors of the entity with feature vector $\mathbf{x}$. Ties among the $\mathbf{x}_j$ can be broken by comparing indices, that is, if $\|\mathbf{x}_{j_1} - \mathbf{x}\| = \|\mathbf{x}_{j_2} - \mathbf{x}\|$, then $R_{j_1} < R_{j_2}$ if $j_1 < j_2$; otherwise, $R_{j_1} > R_{j_2}$. In C. J. Stone's (1977) work, any weight attached to kth nearest neighbors is divided equally when there are ties.

If w_{nj} is a weight function such that $w_{nj}(\mathbf{x};\mathbf{t}_x) = 0$ for all $R_j > k$, it is called a k-NN weight function. The sample rule obtained by using the estimate $\hat{\tau}_i(\mathbf{x};\mathbf{t})$ with these weights in the definition (1.4.3) of the Bayes rule is a k-NN rule. For example, if $k = 1$, then

$$\begin{aligned} w_{nj}(\mathbf{x};\mathbf{t}) &= 1, \qquad \text{if} \quad R_j = 1, \\ &= 0, \qquad \text{if} \quad R_j > 1, \end{aligned}$$

so that

$$\hat{\tau}_i(\mathbf{x};\mathbf{t}) = z_{iR_j} \qquad (i = 1,\ldots,g).$$

These estimates of the posterior probabilities of group membership imply that in an outright allocation of the unclassified entity with feature vector $\mathbf{x}$, it is assigned to the group of its nearest neighbor in the training set.

In the case of $k > 1$ with uniform weights, where

$$\begin{aligned} w_{nj}(\mathbf{x};\mathbf{t}) &= 1/k, \qquad \text{if} \quad R_j \leq k, \\ &= 0, \qquad \text{if} \quad R_j > k, \end{aligned}$$

we have that

$$\hat{\tau}_i(\mathbf{x};\mathbf{t}) = \sum_{j=1}^{k} z_{iR_j}/k,$$

implying that the entity with feature vector $\mathbf{x}$ is allocated on the basis of simple majority voting among its k nearest neighbors.

The obvious metric on $\mathbb{R}^p$ to use is the Euclidean metric. This metric, however, is inappropriate if the feature variables are measured in dissimilar units. In which case, the feature variables should be scaled before applying the Euclidean metric. A recent study on the influence of data transformations and metrics on the k-NN rule is Todeschini (1989). Also, Myles and Hand (1990) have considered the choice of metric in NN rules for multiple groups.

Nearest neighbor rules with a reject option were introduced by Hellman (1970); see Loizou and Maybank (1987) for some recent results on the error rates of such rules. Luk and MacLeod (1986) have proposed an alternative nearest neighbor rule in which neighbors are examined sequentially in order of increasing distance from the unclassified feature data. In some other work on NN rules, Wojciechowski (1987) has described rules of this type for mixed feature data, and Davies (1988) has investigated whether NN rules take proper account of the group-prior probabilities.

9.7.3 Asymptotic Results for Error Rates of k-NN Rules

We let $ec(F,k;\mathbf{t})$ and $eu(F,k)$ denote the overall conditional and unconditional error rate, respectively, of a k-NN allocation rule formed from the training data **t**. Devroye and Wagner (1982) have provided a concise account of nearest neighbor methods in discrimination. As they point out, most of the results dealing with nearest neighbor rules are of the asymptotic variety, concerned with the limiting behavior of $ec(F,k;\mathbf{t})$, as n tends to infinity. The usefulness of these results lies in the fact that if $ec(F,k;\mathbf{t})$ converges to a value favourable relative to the error of the Bayes rule, then there is hope that the rule will perform well with large-sized training sets.

The first result of the type above, and certainly the best known, is that of Cover and Hart (1967), who showed that

$$eu(F,k) \to e_k, \tag{9.7.2}$$

as $n \to \infty$, if the posterior probability of ith group membership $\tau_i(\mathbf{x})$ has an almost everywhere continuous version for $i = 1,\ldots,g$. In (9.7.2), e_k is a constant satisfying

$$eo(F) \le e_k \le 2eo(F)\{1 - eo(F)\} \le 2eo(F) \tag{9.7.3}$$

for $k = 1$, where $eo(F)$ denotes the optimal or Bayes error rate. For arbitrary k, e_k satisfies

$$eo(F) \le e_k \le \alpha_k eo(F)\{1 - eo(F)\} \le 2eo(F), \tag{9.7.4}$$

where $\alpha_k \downarrow 1$ as $k \to \infty$. For these same assumptions, it is also known that

$$ec(F,k;\mathbf{T}) \to e_k \tag{9.7.5}$$

in probability, as $n \to \infty$ (Wagner, 1971), with convergence being with probability one for $k = 1$ (Fritz, 1975).

The result (9.7.3) shows that any other allocation rule based on a training sample of infinite size can cut the error rate of the 1-NN rule at most by half. In this sense, half of the available information in a training set of infinite size is contained in the nearest neighbor.

If k is allowed to vary with n, then C. J. Stone (1977) showed that for any distribution function F of $Y = (\mathbf{X}', \mathbf{Z}')'$,

$$ec(F,k;\mathbf{T}) \to eo(F) \text{ in probability, as } n \to \infty, \tag{9.7.6}$$

if $k = k_n \to \infty$ and $k_n/n \to 0$, as $n \to \infty$. This distribution-free result extends to a large class of nearest neighbor rules, as discussed by C. J. Stone (1977). As remarked by Devroye and Wagner (1982), because of its sheer technical achievement, the result (9.7.6) rivals the original accomplishment of Fix and Hodges (1951), who proved Bayes risk consistency of nearest neighbor rules (with uniform neighbor weights) under analytic assumptions on the distribution of Y.

Devroye (1981a) has since shown that if one also assumes that

$$k_n/(\log n) \to \infty,$$

as $n \to \infty$, then (9.7.6) holds with the convergence being with probability one. Following the distribution-free work of C. J. Stone (1977), Devroye (1981b) showed that the results (9.7.4) and (9.7.5) for fixed k can be established without any condition on the distribution of Y. For additional results on the large-sample behavior of error rates of k-NN rules, the reader is referred to Devroye and Wagner (1982) and Devijver and Kittler (1982, Chapter 3) and the references therein.

9.7.4 Finite Sample Size Behavior of Error Rates of k-NN Rules

For a training sample size of finite size, a k-NN rule can have an error rate that is much more than twice the optimal error $eo(F)$. Recently, Fukunaga and Hummels (1987a) have considered $eu(F,k) - eo(F)$ in training samples of finite size. They termed this difference the bias of $ec(F,k;\mathbf{T})$, which it is if the conditional error rate $ec(F,k;\mathbf{T})$ of the k-NN rule of interest is viewed as an estimator of the optimal error $eo(F)$. Up until the work of Fukunaga and Hummels (1987a), most of the work in reducing the size of the bias had been concentrated on the selection of an optimal metric (Short and Fukunaga, 1981; Fukunaga and Flick, 1984).

Fukunaga and Hummels (1987a) were able to give expressions that relate the bias of 1-NN and 2-NN errors to the sample size n, dimensionality, metric, and group-conditional distributions. These expressions isolated the effect of the sample size n from that of the distributions, giving an explicit relation showing how the bias changes as n is increased. It was shown that, in many cases, increasing n is not an effective way of estimating the asymptotic errors of NN rules. Fukunaga and Hummels (1987b) have considered ways of modifying k-NN and kernel-based rules to reduce this bias.

Previously, Cover (1968) had investigated the size of $eu(F,k) - e_k$ in the univariate case of $p = 1$. He showed that it is bounded by a function of order $O(n^{-2})$, assuming that $\tau_i(x)$ is continuous almost everywhere ($i = 1, \ldots, g$). Wagner (1971) tackled this problem by providing an exponential bound on

$$\text{pr}\{|ec(F,k;\mathbf{T}) - e_k| > c\},$$

where c is a constant. This bound was improved by Fritz (1975) under significantly reduced conditions on the underlying group-conditional distributions. As the actual error $ec(F,k;\mathbf{t})$ is unknown in practice, Rogers and Wagner (1978) and Devroye and Wagner (1979a, 1979b) considered bounds on the difference between the asymptotic error rate e_k of a k-NN rule and its apparent error rate before and after cross-validation. Devroye (1982) has given a generalization of the aforementioned work of Cover (1968).

Bailey and Jain (1978) proved that the asymptotic error rate of the unweighted k-NN rule (i.e., the k-NN rule with uniform weights) is smaller than that of any weighted k-NN rule (i.e., with nonuniform weights). On this problem of unweighted versus weighted k-NN rules for finite n, Fukunaga and Flick (1985) used both distance-weighted and unweighted distance measures in NN estimation of the Bayes error, and obtained lower error rates when using the weighted measures. More recently, MacLeod, Luk, and Titterington (1987) argued intuitively that for finite n, a weighted k-NN rule may in some cases have a lower error rate than the traditional unweighted rule. They confirmed this conclusion analytically in a particular example.

Goin (1984) has studied the group bias of k-NN rules, defined to be $\frac{1}{2} - eu(F,k)$, where the latter is evaluated in the particular case of no group differences.

9.7.5 Choice of k

It was seen that the result (9.7.6) holds only if k is chosen such that $k \to \infty$ and $k/n \to 0$, as $n \to \infty$. Hence, in practice, with small to moderate size training samples, the choice of k is important (Fukunaga and Hostetler, 1973). More recently, Enas and Choi (1986) looked at this problem in a simulation study performed to assess the sensitivity of the k-NN rule to the choice of k in the case of $g = 2$ groups. In their study, k was set equal to $[j/n]$ for $j = 0, 1, \ldots, 5$, when $[u]$ denotes the closest odd integer to u. They reported that their study suggests the following rough guidelines in selecting k for optimal performance of the k-NN rule with uniform weights. For comparable group-sample sizes, choose k to be approximately $n^{3/8}$ or $n^{2/8}$, depending on whether there are small or large differences between the group-covariance matrices. These two recommended choices of k, depending on the similarity of the underlying group-covariance structures, are reversed for disparate group-sample sizes.

On the algorithmic problem of searching for nearest neighbors, Fukunaga and Narendra (1975) used a branch and bound algorithm to increase the speed of computing the nearest neighbors. Other algorithms for finding nearest neighbors have been suggested by Friedman, Baskett, and Shustek (1975), among several others. More recently, Niemann and Goppert (1988) and Bryant (1989) have described efficient algorithms for the implementation of nearest neighbor rules.

9.8 TREE-STRUCTURED ALLOCATION RULES

9.8.1 Introduction

In practice, missing data and noisy feature variables can have a detrimental effect on the performance of nearest neighbor rules. Also, as remarked in Panel on Discriminant Analysis, Classification, and Clustering (1989, Section 3.2.4), another criticism that might be made of estimates of the form (9.7.1) is that the weights are insufficiently adaptive because they ignore the known group-indicator variables $\mathbf{z}_1, \ldots, \mathbf{z}_n$ of the training feature data $\mathbf{x}_1, \ldots, \mathbf{x}_n$. An extension to the case where the weights $w_{nj}(\mathbf{x}; \mathbf{t})$ depend on $\mathbf{z}_1, \ldots, \mathbf{z}_n$ as well as $\mathbf{x}_1, \ldots, \mathbf{x}_n$ was made by Gordon and Olshen (1980). They considered tree-structured recursive partitioning rules, which are now discussed.

A rather different approach to the allocation problem as considered up to now is to portray the rule in terms of a binary tree. The tree provides a hierarchical representation of the feature space. An allocation is effected by proceeding down the appropriate branches of the tree.

Tree-structured rules are the subject of Breiman et al. (1984). The earlier work is often referred to as AID (automatic interaction detection), and the contributions by Breiman et al. (1984) are known by the acronym CART (classification and regression trees). The software CART (1984) is available for implementing these latter contributions. Their line of development follows work started by Morgan and Sonquist (1963) and Morgan and Messenger (1973). Other relevant references include T. W. Anderson (1966), Friedman (1977), and Gordon and Olshen (1978, 1980). The evolution of tree-structured rules has taken place in the social sciences and the fields of electrical engineering, pattern recognition, and most recently, artificial intelligence. As discussed in Breiman et al. (1984), CART adds a flexible nonparametric tool to the data analyst's arsenal. Its potential advantages over traditional approaches are in the analysis of complex nonlinear data sets with many variables.

As tree-structure rules are treated in depth in Breiman et al. (1984), we will only give a brief sketch of them here. We will also outline a new method of tree-structured allocation FACT (fast algorithm for classification trees), as proposed by Loh and Vanichsetakul (1988).

9.8.2 Basic Terminology

Tree-structured rules or, more correctly, binary tree-structured rules are constructed by repeated splits of subsets of the feature space into two descendant subsets, commencing with the feature space itself. The terminal subsets form a partition of the feature space. Each terminal subset s is associated with a group assignment $r(s)$. If $r(s) = i$, then an entity with feature vector $\mathbf{x}$ leading to terminal subset s will be assigned to group G_i $(i = 1,\ldots,g)$. There can be two or more terminal subsets giving the same allocation.

In the terminology of tree theory, the subsets of the feature space are called nodes. Those subsets that are not split in the final version of the allocation tree are called terminal nodes. The root node s_1 is the entire feature space. The set of splits, together with the order in which they are used, determines what is called a binary tree S.

In order to describe the splitting procedure and the subsequent pruning of the tree, we need the following notation. Following Breiman et al. (1984), let

$$\hat{P}_{is} = \pi_i n_i(s)/n_i, \tag{9.8.1}$$

where

$$n_i(s) = \sum_{j=1}^{n} z_{ij} I_s(\mathbf{x}_j),$$

and $I_s(\mathbf{x}_j)$ equals one or zero, according as the feature vector $\mathbf{x}_j$ falls in node s or not. That is, $n_i(s)$ is the number of the entities in the training set that are from group G_i and that have feature observations in the node s. Hence, $\hat{P}_{is}$ can be viewed as the resubstitution estimate of the joint probability that an entity belongs to G_i and has a feature vector in node s.

The quantity

$$\hat{P}_s = \sum_{i=1}^{g} \hat{P}_{is}$$

is thus an estimate of the marginal probability that the feature vector will fall in node s. The corresponding estimate of the conditional probability that an entity belongs to G_i, given its feature vector is in node s, is, therefore,

$$\hat{\tau}_i(s) = \hat{P}_{is}/\hat{P}_s \qquad (i = 1,\ldots,g),$$

which reduces to

$$\hat{\tau}_i(s) = n_i(s)/n(s) \qquad (i = 1,\ldots,g), \tag{9.8.2}$$

where $n(s) = n_1(s) + \cdots + n_g(s)$, on estimating π_i by n_i/n in (9.8.1).

9.8.3 Outline of CART

Given the training data $\mathbf{t}$, CART constructs the binary decision tree by recursively partitioning the training data, and hence the feature space, in a forward/backward stepwise manner. The idea is to "grow" a large tree in the first instance, which ensures against stopping too early. The bottom nodes are then recombined or "pruned" upward to give the final tree. The degree of pruning is determined by cross-validation using a cost-complexity function that balances the apparent error rate with the tree size. Because CART recursively partitions the training data into subsets of ever decreasing size, it requires n to be very large in order to maintain reasonable sample sizes at each successive node. Breiman et al. (1984) proved that under mild regularity conditions, rules based on recursive partitioning are Bayes risk consistent.

With CART, an initial tree is produced as follows. Starting at the root node s_1, the best splitting feature variable, say, x_v, and its cutoff point k_v are chosen to maximize the group purity of the two daughter nodes defined by the split

$$x_v < k_v. \tag{9.8.3}$$

This same procedure is then recursively applied to the subset of the training observations associated with the left split (those that satisfy (9.8.3) at the root) and to the subset associated with the right split (those that do not satisfy (9.8.3) at the root), producing four new nodes. Each of these nodes is then split, and so on. The splitting can be continued until all nodes remaining to be split are pure (i.e., they contain observations all from the same group). The allocation rule $r(s)$ at each terminal node s is taken to be the Bayes rule formed with the posterior probabilities of group membership estimated by $\hat{\tau}_1(s),\ldots,\hat{\tau}_g(s)$, as defined by (9.8.2).

In problems where linear structure is suspected, (9.8.3) can be replaced by splits of the form

$$\mathbf{a}'\mathbf{x}_v < k$$

for some vector $\mathbf{a}$ and cutoff point k. Here $\mathbf{x}_v$ is a subvector containing a subset of the continuous variables in the full feature vector $\mathbf{x}$.

In other variations of (9.8.3), the allowable splits can be extended to include Boolean combinations where a Boolean structure is suspected. For example, in medical diagnosis, a decision is often made on the basis of the presence or absence of a large number of symptoms, which tend to occur in certain combinations. Another example is in the allocation of chemical compounds through the peaks in their mass spectra; see Breiman et al. (1984, Section 5.2.). Also, Marshall (1986) has tackled some of the difficulties in recursive partitioning in the presence of a Boolean combination structure.

A decision tree constructed in the manner above will produce an accurate allocation rule only if the underlying group-conditional distributions do not overlap. If this is not so, then the splits at the lower levels of the tree will be determined mainly by sampling fluctuations, with a consequent deterioration in allocatory capacity. To guard against this a backward recursive node recombination strategy is employed on the tree obtained in the first instance by the successive splits into pure terminal nodes.

It can be seen that the crux of the problem is how to determine the splits and the terminal nodes. In CART, splitting procedures are proposed on the basis of an impurity measure $m(s)$ associated with each node s. If $\tilde{S}$ denotes the current set of terminal nodes of the tree S, set

$$M(s) = m(s)\hat{P}_s.$$

Then the tree impurity M_S is defined by

$$M_S = \sum_{s\in\tilde{S}} M(s) = \sum_{s\in\tilde{S}} m(s)\hat{P}_s. \tag{9.8.4}$$

Breiman et al. (1984) showed that selecting the splits to minimize (9.8.4) is equivalent to selecting the splits to maximize

$$m(s) - \hat{P}_{s_R} m(s_R) - \hat{P}_{s_L} m(s_L), \tag{9.8.5}$$

where s_R and s_L denote the splits of each node s of the tree S. For the impurity measure $m(s)$, they used the so-called Gini index, which for zero-one costs of misallocation is

$$m(s) = \sum_{h\neq i} \hat{\tau}_h(s)\hat{\tau}_i(s).$$

Splitting according to this criterion is continued until all terminal nodes are either small or pure or contain only identical feature measurements. Let $S_{\max}$ denote a tree obtained in this manner. The next step is to prune this tree $S_{\max}$.

9.8.4 Minimal Cost-Complexity Pruning

The initial intent is to produce a sequence of subtrees of $S_{\max}$, which eventually collapse to the tree $\{s_1\}$ consisting of the root node. To describe the pruning of a tree as adopted by Breiman et al. (1984), we need to introduce the following terms.

A branch S_s of a tree S with root node $s \in S$ consists of the node s and all descendants of s in S. Pruning a branch S_s from a tree S consists of deleting from S all descendants of s, that is, cutting all of S_s except its root node. If S^* is obtained from S by successively pruning off branches, then S^* is called a pruned subtree of S and denoted by $S^* \prec S$. It has the same root node as S.

The complexity of any subtree S of $S_{\max}$ is defined to be $|\tilde{S}|$, the number of terminal nodes in S. The cost complexity of S is defined to be

$$C_\alpha(S) = A(S) + \alpha|\tilde{S}|,$$

where A is the apparent error rate of the S-based rule in its application to the training data $\mathbf{t}$, and α is a positive real number called the complexity parameter.

The smallest minimizing subtree $S(\alpha)$ for complexity parameter α is defined by the conditions

$$C_\alpha(S(\alpha)) = \min_{S \preceq S_{\max}} C_\alpha(S) \qquad (9.8.6)$$

and

$$\text{if} \quad C_\alpha(S) = C_\alpha(S(\alpha)), \qquad \text{then} \quad S(\alpha) \preceq S. \qquad (9.8.7)$$

This definition breaks ties in minimal cost complexity by selecting the smallest minimizer of C_α.

Breiman et al. (1984) showed that for every α, there exists a smallest minimizing subtree as defined by (9.8.6) and (9.8.7). Although α runs through a continuum of values, there are at most a finite number of subtrees of $S_{\max}$. They showed how a decreasing sequence of subtrees

$$S_1 \succ S_2 \succ S_3 \succ \cdots \succ \{s_1\}$$

can be obtained, where there is a sequence $\{\alpha_k\}$ with $\alpha_1 = 0$ and $\alpha_k < \alpha_{k+1}$ $(k \geq 1)$ for which

$$S(\alpha) = S(\alpha_k) = S_k \qquad \text{for} \quad \alpha_k \leq \alpha < \alpha_{k+1}.$$

The problem is now reduced to selecting one of these subtrees. This can be performed on the basis of their apparent error rates. Thus, if S_{k_o} is the optimum subtree in this sense, then

$$A(S_{k_o}) = \min_k A(S_k).$$

As the apparent error rate produces too optimistic an estimate of the actual error of a discriminant rule, the choice of subtree is made after cross-validation of the apparent error rates. For this task, it is common to use cross-validation with 10 or 25 observations removed at a time. Because of the dual role of cross-validation in error-rate estimation and tree construction, it has been noted that the assessed error rate of the rule based on the final pruned tree is not genuine (Loh and Vanichsetakul, 1988, page 176). However, Breiman et al. (1984, page 81) and Breiman and Friedman (1988) have reported that the relative effect is minor. As considered by Breiman and Friedman (1988), the optimal tree-pruning algorithm described above is probably the most important contribution of Breiman et al. (1984) to the evaluation of tree-structured methodology, for it intends to produce right-sized trees reliably.

In practice, there may be missing measurements on some of the feature variables of the classified entities in the training set or the unclassified entities to be allocated. CART handles both of these problems using surrogate splits to proceed down a tree. If the best split of node s is the split u on the feature variable x_v, find the split u^* on the feature variables other than x_v that is most similar to u. The split u^* is called the best surrogate for u. Similarly, define the second best surrogate, and so on. If an entity has a missing value for x_v, the best surrogate split is used. If it is missing an observation on the variable defining the best surrogate split, the second best surrogate split is used, and so on. The reader should consult Breiman et al. (1984, Chapter 5) for further information on surrogate splits and also other aspects of CART, such as the question of variable ranking.

9.8.5 Tree-Structured Rules via FACT Method

Loh and Vanichsetakul (1988) have proposed a new method of producing a tree-structured allocation rule, known as FACT. Its splitting rule uses normal-based linear discriminant analysis with F ratios to decide when to split and when to stop splitting. By sacrificing CART's thorough nonparametric approach with the use of this normal theory-based splitting rule, execution speed is greatly increased.

To avoid near singular group-covariance matrices, a principal-component analysis, described in Section 6.6.2, is performed at each node. Those principal components whose eigenvalues exceed c times the largest eigenvalue ($c = 0.05$ in the examples of Loh and Vanichsetakul, 1988), are retained. A split is selected at the root node using the sample NLDR, as defined by (3.3.2) and (3.3.6). The only difference here is that the full feature vector $\mathbf{x}$ and its sample group-means and common covariance matrix are replaced by the vector of retained principal components and corresponding sample moments. Similarly, a split is made at a subsequent node s, except that the ith group-prior probability π_i is now replaced by the estimate $\hat{\tau}_{is}$ defined by (9.8.2), which is the estimated posterior probability that an entity belongs to G_i, given its feature vector is in node s.

The F ratio for between- to within-group sums of squares (see Section 6.5.2) is used to select the variable for splitting. A split is made on the variable with the largest F ratio provided it exceeds some threshold F_o. Loh and Vanichsetakul (1988) commented that $F_o = 4$ usually, because it coincides with the F-to-enter value in the stepwise discriminant analysis program in BMDP. If the largest F ratio is not greater than F_o, suggesting that the group means are relatively close, then Loh and Vanichsetakul (1988) suggest looking for splits based on dispersion. To this end, each variable is replaced by its absolute distance from the sample mean of its group of origin. Polar coordinate splits are introduced to deal with possible radial symmetry.

A c-category categorical variable is converted into $c-1$ zero-one variables, and then replaced by the first canonical variate for this $c-1$ dimensional space. Missing data are handled by analogy to maximum likelihood under the assumption of normality.

9.8.6 Example on CART and FACT Methods

Loh and Vanichsetakul (1988) have compared their FACT method with the CART method of Breiman et al. (1984); see also the response of Breiman and Friedman (1988). As part of their comparison, Loh and Vanichsetakul (1988) contrasted CART and FACT in their application to the 1970 Boston housing data gathered by Harrison and Rubinfeld (1978). This example is reported here to illustrate the practical use of these two methods for producing tree-structured rules.

The Boston housing data became well known when they were extensively used in Belsley, Kuh, and Welsch (1980). Breiman et al. (1984) fitted a regression tree to these data using CART. This set consists of $n = 506$ cases (census tracts) and 14 variables, including the median value (MV) of homes in thousands of dollars. This variable was used to define $g = 3$ groups: low (G_1) if $\log(\text{MV}) \leq 9.84$, high ($G_2$) if $\log(\text{MV}) > 10.075$, and medium (G_2) otherwise. The $n = 506$ cases belong in roughly equal proportions to these three groups. The $p = 13$ feature variables are CRIM, crime rate; DIS, the weighted distance to employment centres; ZN, the percentage of land zoned for lots; CHAS = 1 if on Charles River and CHAS = 0, otherwise; AGE, the percentage built before 1940; $B = (Bk - 0.63)^2$, where Bk is the proportion of blocks in the population; INDUS, the percentage of nonretail business; RAD, accessibility to radial highways; RM, the average number of rooms; NOX, nitrogen oxide concentration; TAX, tax rate; LSTAT, the percentage of lower-status population; and P/T, the pupil/teacher ratio.

The tree rules produced by CART and FACT in their application to this data set are presented in Figures 9.6 and 9.7. In Figure 9.6, there is the binary tree rule produced by the CART with pruning by tenfold cross-validation and with equal costs of misallocation. In this figure, nonterminal nodes are indicated by circles and terminal nodes by squares. The corresponding tree

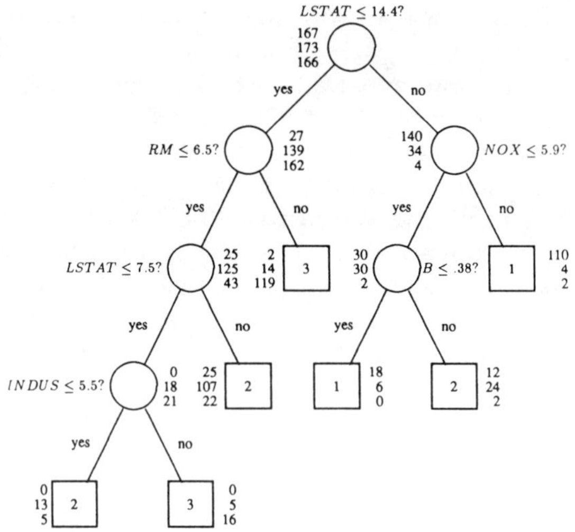

FIGURE 9.6. Tree from the CART method. Triples beside nodes are the node decompositions; for example, there are 167 G_1 cases, 173 G_2 cases, and 166 G_3 cases in the root node. (From Loh and Vanichsetakul, 1988.)

obtained with FACT is given in Figure 9.7, using $F_0 = 4$ and equal misallocation costs. The CART and FACT trees took 8.5 minutes and 30 seconds, respectively, to build.

Both trees, which were obtained using univariate splits, split on the variable LSTAT first. This variable effectively splits the feature space into two pieces in Figure 9.6 and into five pieces in Figure 9.7. The next variables split are RM and NOX with CART and RM and AGE with FACT. The CART tree splits NOX if LSTAT > 14.4, and FACT splits AGE if LSTAT > 15.7. This observation, along with the variable importance rankings given in Table 13 of Loh and Vanichsetakul (1988), suggests that NOX and AGE are proxies for each other if LSTAT > 15.

Figure 9.8 gives the FACT tree after transformation of five of the feature variables, according as log(LSTAT), RM^2, NOX^2, log(RAD), and log(DIS). The CART tree after transformation is indistinguishable from that in Figure 9.6, to the accuracy shown. This is because CART is essentially invariant of monotone transformations in the ordered variables when univariate splits are used. The FACT2 tree is shorter than the FACT1 tree and one of the cut points at the root node is the same as the cut point at the root node of the CART tree.

The apparent allocation rates of the CART and FACT rules in their application to the training data are displayed in Table 9.4. In this table, FACT1 refers to the tree rule in Figure 9.7 and FACT2 to the rule in Figure 9.8. It can be seen the apparent error rates are similar, with the CART tree having slightly the lowest overall rate. However, as pointed out by Breiman and

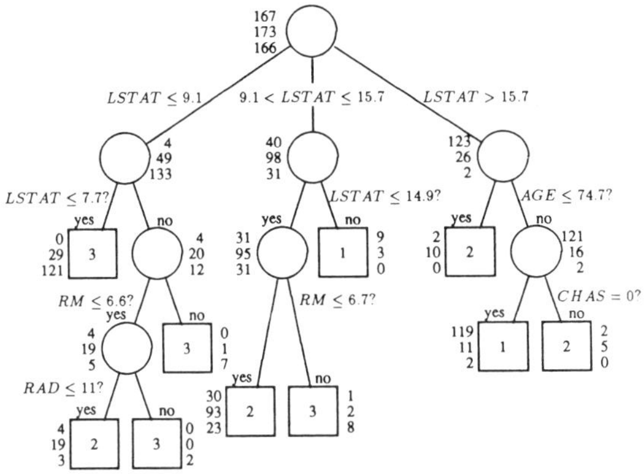

FIGURE 9.7. Tree from the FACT method. Triples beside nodes are the node decomposition. (From Loh and Vanichsetakul, 1988.)

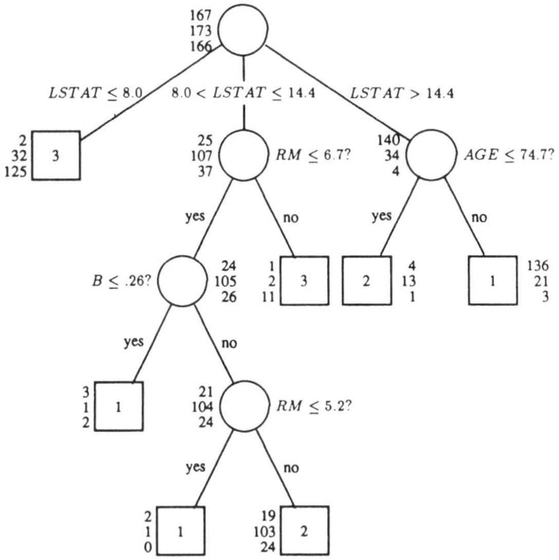

FIGURE 9.8. Tree from the FACT method, using transformed feature variables. (From Loh and Vanichsetakul, 1988.)

Friedman (1988) in their comment on the comparative study of CART and FACT made by Loh and Vanichsetakul (1988), the data sets used in this study should not be considered as serious test beds for accuracy comparisons, as they are ones on which almost any allocation rule would do fairly well. For example, the Boston housing data are essentially a three/four variable data set.

TABLE 9.4 Apparent Allocation Rates for the Three Tree Rules CART, FACT1, and FACT2

	Predicted Grouping								
	CART			FACT1			FACT2		
Actual	G_1	G_2	G_3	G_1	G_2	G_3	G_1	G_2	G_3
G_1	128	37	2	128	38	1	141	23	3
G_2	10	144	19	14	127	32	23	116	34
G_3	2	29	135	2	26	138	5	25	136
Total	140	210	156	144	191	171	169	164	173
	81.3% correct			77.7% correct			77.7% correct		

Source: Adapted from Loh and Vanichsetakul (1988).

Loh and Vanichsetakul (1988) prefer multiway splitting as with FACT than binary splitting as with CART. In particular, they have argued that interpretation may be easier if the root node is able to be decomposed into as many subnodes as the number of groups. The node decompositions of the FACT1 and FACT2 trees in Figures 9.7 and 9.8 show how typically the first split produces subtrees (here three), each with one group dominant. Subsequent splits on a subtree merely try to separate minority cases. Later nodes do not have three splits each, because the F ratios are weighted by the estimated group-priors for the nodes. To understand where most of the cases in a particular group go, the tree may be read top–down. For example, in Figure 9.8, the training cases are split into three subsets according to the value of the variable LSTAT: (a) housing values are high (group G_3) in more affluent tracts (left branch of tree); (b) values in less affluent tracts (right branch) are either low (G_1) or medium (G_2), depending on age; (c) tracts with average values of LSTAT (middle branch) are mostly white. In the latter case, housing values are determined largely by the number of rooms.

On multiway versus binary splits, Breiman and Friedman (1988) have argued that multiway splitting does not make an effective use of the conditional information potentially present in the tree as does binary splitting. Concerning interpretation, they have pointed out how it is difficult to beat the simple nonparametric binary recursive partitioning used by CART in its production of parsimonious trees.

9.9 SOME OTHER NONPARAMETRIC DISCRIMINANT PROCEDURES

9.9.1 Partial Discrimination via Rank Methods

In Section 5.7.2, consideration was given to the use of a rank cutoff point as proposed by Randles et al. (1978a) in order to control the balance between the error rates of the sample NLDR or NQDR when applied with respect to $g = 2$ groups. This rank method had been used previously by Broffitt et al. (1976) to bound simultaneously both unconditional error rates of a partial allocation rule, which has the option of not making an assignment for a highly questionable value of $\mathbf{x}$ on an unclassified entity. We now consider their formulation of partial discriminant analysis in the case of $g = 2$ groups.

As mentioned in Section 1.10.2, Quesenberry and Gessaman (1968) had proposed a nonparametric approach to partial discrimination, using tolerance regions. The idea of using tolerance regions for allocation was first suggested by T. W. Anderson (1966), although it is implicit in the work of Fix and Hodges (1951). Let A_i denote a region in the feature space that is highly suggestive of an entity belonging to group G_i $(i = 1,2)$. Then a partial allocation rule can be formed that assigns an unclassified entity with feature vector $\mathbf{x}$ to G_1 if

$$\mathbf{x} \in A_1 \cap \overline{A}_2,$$

and to G_2 if

$$\mathbf{x} \in A_2 \cap \overline{A}_1.$$

Thus, no allocation is made if $\mathbf{x}$ falls in the region

$$(A_1 \cap A_2) \cup (\overline{A}_1 \cap \overline{A}_2).$$

Quesenbery and Gessaman (1968) suggested using tolerance regions to construct the regions A_1 and A_2.

This formulation, however, does not take into account the direction of group G_{3-i} when defining A_i $(i = 1,2)$. As a consequence, such a rule will often be conservative and will fail to allocate many realizations of $\mathbf{X}$. To reduce the conservative nature of this partial allocation rule, Broffitt et al. (1976) introduced a rank method that takes into account the direction of the groups. That is, the rank approach creates $\overline{A}_1$ in the direction of G_2 and $\overline{A}_2$ in the direction of G_1, and so increases the probability that an allocation will be made.

If this rank method is applied with the sample NLDF $\xi(\mathbf{x};\hat{\boldsymbol{\theta}}_E)$ as in Section 5.7.2, then A_i is defined to be

$$A_i = \{\mathbf{x} : R_i(\mathbf{x};\mathbf{t}) > k_i\} \qquad (i = 1,2),$$

where the rank $R_i(\mathbf{x};\mathbf{t})$ is as defined in Section 5.7.2. The constant k_i ($k_i \leq n_i + 1$) can be so specified to impose a bound α_i on the unconditional error rate $eu_i(F)$ specific to group G_i, regardless of the underlying group-conditional distributions. This can be seen from

$$\begin{aligned} eu_i(F) &= \text{pr}\{\mathbf{X} \in A_{3-i} \cap \overline{A}_i\} \\ &\leq \text{pr}\{\mathbf{X} \in \overline{A}_i\} \\ &= [k_i]/(n_i + 1) \qquad (i = 1,2) \end{aligned} \tag{9.9.1}$$

because, as noted in Section 5.7.2, $R_i(\mathbf{X},\mathbf{T})$ has a uniform distribution over $1,\ldots,n_i + 1$, conditional on $\mathbf{X}$ belonging to G_i. In (9.9.1), $[k_i]$ denotes the greatest integer less than or equal to k_i. By setting

$$\alpha_i = [k_i]/(n_i + 1) \qquad (i = 1,2),$$

a partial allocation rule is obtained with both unconditional error rates bounded simultaneously as desired.

In this rank method of partial allocation as presented above, the sample NLDF $\xi(\mathbf{x};\hat{\boldsymbol{\theta}}_E)$ can be replaced by essentially any discriminant function. The basic requirement is that the chosen discriminant function depends on the training data $\mathbf{t}$ only through statistics that are symmetric functions of $\mathbf{x}_{11},\ldots,\mathbf{x}_{1n_1}$ and of $\mathbf{x}_{21},\ldots,\mathbf{x}_{2n_2}$. In practice, Broffitt et al. (1976) suggest trying a variety of different types of discriminant functions.

Beckman and Johnson (1981) proposed a modification of the rank method of Broffitt et al. (1976), which imposes constraints on the unconditional error rates given that an allocation has been adopted. Broffitt (1982) and Ng and Randles (1983) have considered an extension of the rank method to $g > 2$ groups. For further information, the reader is referred to the review by Ng and Randles (1986) on partial discrimination by nonparametric procedures.

Conover and Iman (1980) proposed a nonparametric allocation rule by first ranking the feature data and then computing the sample NLDF or NQDF on the basis of these ranks. The motivation behind this approach is that hopefully the rank vectors will appear more like normal data than the original feature observations $\mathbf{x}_j$. This approach has been considered also by Koffler and Penfield (1982).

9.9.2 Nonparametric Discrimination with Repeated Measurements

The construction of a nonparametric allocation is facilitated if there are independent repeated measurements available on the entity to be allocated. Various nonparametric approaches in the situation of repeated measurements have been considered by Kanazawa (1974), Das Gupta (1964), Hudimoto (1964), Gupta and Kim (1978), Lin (1979), and Haack and Govindarajulu (1986), among others.

One approach in this situation has been to use standard nonparametric rank tests to devise allocation rules, following the idea of Das Gupta (1964). He proposed a rule that allocates an entity to G_i if $|W_i|$ is the smaller of W_1 and W_2, where W_i is the Wilcoxon statistic based on $\mathbf{x}_{i1},\ldots,\mathbf{x}_{in_i}$ and the repeated measurements $x_{m_1},\ldots,x_{m_k}$ on the unclassified entity ($i = 1,2$). Hudimoto (1964) modified this rule by using W_i instead of $|W_i|$ when $F_1(x) \geq F_2(x)$ for all x. Govindarajulu and Gupta (1977) considered allocation rules based on linear rank statistics for the g-group problem. For a more detailed account of the work above and for additional referees, the reader is referred to Das Gupta (1973) and the aforementioned survey article on nonparametric discrimination by Broffitt (1982).

9.9.3 A Nonmetric Approach to Linear Discriminant Analysis

Recently, Raveh (1989) proposed a metric approach to linear discrimination between $g = 2$ groups. Allocation is performed on the basis of the linear combination $\mathbf{a}'\mathbf{x}$, where $\mathbf{a}$ is chosen so that as many of the discriminant scores $\mathbf{a}'\mathbf{x}_{1j}$ $(j = 1,\ldots,n_1)$ as possible will be greater than $\mathbf{a}'\mathbf{x}_{2k}$ $(k = 1,\ldots,n_2)$. The procedure chooses that $\mathbf{a}$ that maximizes an index of separation between the groups. The method is nonmetric, since the index to be maximized is based on the set of inequalities

$$\mathbf{a}'\mathbf{x}_{1j} \geq \mathbf{a}'\mathbf{x}_{2k} \qquad (j = 1,\ldots,n_1;\ k = 1,\ldots,n_2). \tag{9.9.2}$$

The inequalities (9.9.2) are equivalent to

$$\mathbf{a}'\mathbf{x}_{1j} - \mathbf{a}'\mathbf{x}_{2k} = |\mathbf{a}'\mathbf{x}_{1j} - \mathbf{a}'\mathbf{x}_{2k}| \qquad (j = 1,\ldots,n_1;\ k = 1,\ldots,n_2). \tag{9.9.3}$$

The index of separation is

$$\begin{aligned}\delta(\mathbf{a}) &= \sum_{j=1}^{n_1}\sum_{k=1}^{n_2}(\mathbf{a}'\mathbf{x}_{1j} - \mathbf{a}'\mathbf{x}_{2k}) \Big/ \sum_{j=1}^{n_1}\sum_{k=1}^{n_2}|\mathbf{a}'\mathbf{x}_{1j} - \mathbf{a}'\mathbf{x}_{2k}| \\ &= (\mathbf{a}'\bar{\mathbf{x}}_1 - \mathbf{a}\bar{\mathbf{x}}_2)/(n_1n_2)^{-1}\sum_{j=1}^{n_1}\sum_{k=1}^{n_2}|\mathbf{a}'\mathbf{x}_{1j} - \mathbf{a}'\mathbf{x}_{2k}|.\end{aligned}$$

It varies between -1 and 1 and is equal to 1 if (9.9.2) or (9.9.3) holds. The extreme values ± 1 occur when there is no overlap between the discriminant scores $\mathbf{a}'\mathbf{x}_{1j}$ $(j = 1,\ldots,n_1)$ and the $\mathbf{a}'\mathbf{x}_{2k}$ $(k = 1,\ldots,n_2)$. It was shown in Section 3.3.3 that Fisher's (1936) linear discriminant function is given by $\mathbf{a}'\mathbf{x}$, where $\mathbf{a}$ maximizes the quantity (3.3.8).

As suggested by Raveh (1989), the maximization of $\delta(\mathbf{a})$ can be undertaken by an algorithm such as Powell's conjugate direction algorithm or its modification; see Zangwill (1967). An initial choice of $\mathbf{a}$ in the computation of local maximum by this algorithm is the vector

$$\mathbf{S}^{-1}(\bar{\mathbf{x}}_1 - \bar{\mathbf{x}}_2),$$

defining the coefficients of $\mathbf{x}$ in the sample NLDF $\xi(\mathbf{x};\hat{\boldsymbol{\theta}}_E)$, or equivalently Fisher's LDF, as given by (3.3.9). Another initial choice suggested by Raveh (1989) is $\mathbf{a}_0$, where $(\mathbf{a}_0)_v$ is ± 1, being $+1$ if the $(\mathbf{x}_{1j})_v$ tend to be larger than the $(\mathbf{x}_{2k})_v$ and -1 if the reverse behavior is observed $(v = 1,\ldots,p)$.

The advantage of this proposed nonmetric approach is that there is no need to specify group-conditional distributions and that it produces perfect discrimination when there is no overlap between the group-feature data. In comparisons with the sample NLDF on empirical data and simulations, Raveh (1989) found that the nonmetric LDF yielded fewer misallocations for skewed (log-normal and chi-squared) group-conditional distributions with very different covariance matrices. For multivariate normal group-conditional distributions with the same or comparable covariance matrices, the nonmetric LDF was only slightly inferior to the sample NLDF.

The procedure of Raveh (1989) can be generalized so that each difference of discriminant scores $\mathbf{a}'\mathbf{x}_{1j} - \mathbf{a}'\mathbf{x}_{2k}$ is replaced by a function of the difference. If the sign function is adopted in this role, then Raveh (1989) noted that it leads to the linear rule that minimizes the overall apparent error rate, as considered by Greer (1979); see Section 3.10.1.

CHAPTER 10

Estimation of Error Rates

10.1 INTRODUCTION

In this chapter, we consider the estimation of the error rates associated with a discriminant rule. The relevance of error rates in assessing the performance of a discriminant rule was discussed in Chapter 1, where the various types of error rates were introduced and their properties discussed in some depth. For the allocation of an entity with feature vector $\mathbf{x}$ to one of the g groups $G_1,\ldots,G_g$, let $r(\mathbf{x};\mathbf{t})$ denote a discriminant rule formed from the realized training data,

$$\mathbf{t}' = (\mathbf{y}_1,\ldots,\mathbf{y}_n),$$

where $\mathbf{y}_j = (\mathbf{x}_j', \mathbf{z}_j')'$ and $(\mathbf{z}_j)_i = z_{ij}$ equals one if $\mathbf{x}_j$ belongs to G_i and is zero otherwise ($i = 1,\ldots,g$; $j = 1,\ldots,n$). This is consistent with our previous notation.

For a given realization $\mathbf{t}$ of the training data $\mathbf{T}$, it is the conditional or actual allocation rates of $r(\mathbf{x};\mathbf{t})$, $ec_{ij}(F_i;\mathbf{t})$, that are of central interest, where

$$ec_{ij}(F_i,\mathbf{t}) = \operatorname{pr}\{r(\mathbf{X};\mathbf{t}) = j \mid \mathbf{X} \in G_i, \mathbf{t}\} \qquad (i,j = 1,\ldots,g).$$

That is, $ec_{ij}(F_i;\mathbf{t})$ is the probability, conditional on $\mathbf{t}$, that a randomly chosen entity from G_i is assigned to G_j by $r(\mathbf{x};\mathbf{t})$. As before, F_i denotes the distribution function of $\mathbf{X}$ in group G_i ($i = 1,\ldots,g$), and F denotes the distribution function of $\mathbf{Y} = (\mathbf{X}', \mathbf{Z}')'$ when $\mathbf{X}$ is drawn from a mixture of these g groups and $\mathbf{Z}$ specifies its group of origin.

The unconditional or expected allocation rates of $r(\mathbf{x};\mathbf{t})$ are given by

$$\begin{aligned} eu_{ij}(F) &= \operatorname{pr}\{r(\mathbf{X};\mathbf{T}) = j \mid \mathbf{X} \in G_i\} \\ &= E\{ec_{ij}(F_i;\mathbf{T})\} \qquad (i,j = 1,\ldots,g). \end{aligned}$$

The unconditional rates are useful in providing a guide to the performance of the rule before it is actually formed from the training data.

Concerning the error rates specific to a group, the conditional probability of misallocating a randomly chosen member from G_i is

$$ec_i(F_i;\mathbf{t}) = \sum_{j \neq i}^{g} ec_{ij}(F_i;\mathbf{t}) \qquad (i = 1,\ldots,g).$$

The overall conditional error rate for an entity drawn randomly from a mixture G of $G_1,\ldots,G_g$ in proportions $\pi_1,\ldots,\pi_g$, respectively, is

$$ec(F;\mathbf{t}) = \sum_{i=1}^{g} \pi_i ec_i(F_i;\mathbf{t}).$$

Similarly, the individual group and overall unconditional error rates, $eu_i(F)$ and $eu(F)$, are defined.

If $r(\mathbf{x};\mathbf{t})$ is constructed from $\mathbf{t}$ in a consistent manner with respect to the Bayes rule $r_o(\mathbf{x};F)$, then

$$\text{Lt}_{n \to \infty} eu(F) = eo(F),$$

where $eo(F)$ denotes the optimal error rate. Interest in the optimal error rate in practice is limited to the extent that it represents the error of the best obtainable version of the sample-based rule $r(\mathbf{x};\mathbf{t})$.

Unless F, or the group-conditional distribution functions F_i in the case of separate sampling, is known, the conditional allocation rates, as well as the unconditional and optimal rates, are unknown and therefore must be estimated. This task is far from straightforward and poses a number of formidable problems. Indeed, as described by Glick (1978), "the task of estimating probabilities of correct classification confronts the statistician simultaneously with difficult distribution theory, questions intertwining sample size and dimension, problems of bias, variance, robustness, and computation costs. But coping with such conflicting concerns (at least in my experience) enhances understanding of many aspects of statistical classification—and stimulates insight into general methodology of estimation."

Over the years, there have been numerous investigations on this topic; see, for example, Hills (1966), Lachenbruch and Mickey (1968), and McLachlan (1973a, 1974b, 1974c, 1974d), and the references therein. Toussaint (1974a) has compiled an extensive bibliography, which has been updated by Hand (1986a). An overview of error-rate estimation has been given by McLachlan (1986), and recent work on robust error-rate estimation has been summarized by Knoke (1986). Further advances are covered in McLachlan (1987b). More recently, Fukunaga and Hayes (1989b) have applied the expression derived by Fukunaga and Hayes (1989a) for the assessed error rate of a sample-based discriminant rule to analyze theoretically various methods of error-rate estimation. There is also the empirical study of Ganeshanandam and Krzanowski (1990), which is to be considered further in Section 10.6.

Following the major studies done in the 1960s on error-rate estimation, there has always been a high level of interest in the topic, promoted more recently by the appearance of Efron's (1979) paper on the bootstrap, with its important applications to all aspects of error-rate estimation. In the last few years, there has been a considerable number of papers produced in this area. Much of the recent work has arisen from consideration of the novel ideas presented on the subject in the seminal paper of Efron (1983). The main thrust of the recent studies has been to highlight the usefulness of resampling techniques in producing improved estimators of the error rates through appropriate bias correction of the apparent error rate. Attention has also been given to ways of smoothing the apparent error rate in an attempt to reduce its variance.

McLachlan (1974d) has investigated the relationship between the separate problems of estimating each of the three types of error rate (conditional, unconditional, and optimal). In the subsequent exposition, we concentrate on the estimation of the conditional allocation rates since they are of primary concern once the training data have been obtained. We will only give the definitions of the estimators in their estimation of the group-specific errors $ec_i(F_i;\mathbf{t})$, as they can be extended in an obvious manner to the estimation of the allocation rates $ec_{ij}(F_i;\mathbf{t})$, where the assigned group is specified in addition to the group of origin. In some situations, as with the estimation of the mixing proportions as considered in Section 2.3, the full confusion matrix, which has $ec_{ij}(F_i;\mathbf{t})$ as its (i,j)th element, has to be estimated. Of course, for $g = 2$ groups, the confusion matrix is specified by the $ec_i(F_i;\mathbf{t})$.

10.2 SOME NONPARAMETRIC ERROR-RATE ESTIMATORS

10.2.1 Apparent Error Rate

As the groups can be relabeled, we will consider without loss of generality the estimation of the error rate for the first group. An obvious and easily computed nonparametric estimator of the conditional error rate $ec_1(F_1;\mathbf{t})$ is the apparent error rate $A_1(\mathbf{t})$ of $r(\mathbf{x};\mathbf{t})$ in its application to the observations in $\mathbf{t}$ from G_1. That is, $A_1(\mathbf{t})$ is the proportion of the observations from G_1 in $\mathbf{t}$ misallocated by $r(\mathbf{x};\mathbf{t})$, and so we can write

$$A_1(\mathbf{t}) = \frac{1}{n_1}\sum_{j=1}^{n} z_{1j} Q[1, r(\mathbf{x}_j;\mathbf{t})], \tag{10.2.1}$$

where

$$n_1 = \sum_{j=1}^{n} z_{1j}$$

is the number of observations from G_1 in $\mathbf{t}$, and where, for any u and v, $Q[u,v] = 0$ for $u = v$ and 1 for $u \neq v$. The apparent error rate, or resubstitution estimator as it is often called, was first suggested by Smith (1947) in connection with the sample NQDR $r_o(\mathbf{x};\hat{\boldsymbol{\Psi}}_U)$. As the apparent rate is obtained by applying the rule to the same data from which it has been formed, it provides an optimistic assessment of the true conditional error rates. In particular, for complicated discriminant rules, overfitting is a real danger, resulting in a grossly optimistic apparent error. Although the optimism of the apparent error rate declines as n increases, it usually is of practical concern.

As discussed in Section 8.4, Efron (1986) has derived the bias of the apparent error rate for the logistic model in a regression context, where the aim is to estimate the average of the overall conditional error rate associated with the application of the rule to n future entities with observations fixed at $\mathbf{x}_1,\ldots,\mathbf{x}_n$, respectively. In the present discriminant analysis context, under the homoscedastic normal model for $g = 2$ groups,

$$\mathbf{X} \sim N(\boldsymbol{\mu}_i, \boldsymbol{\Sigma}) \quad \text{in} \quad G_i \qquad (i = 1,2), \tag{10.2.2}$$

McLachlan (1976a) has derived the asymptotic bias of A_1 for the plug-in sample version $r_o(\mathbf{x};\hat{\boldsymbol{\Psi}}_E)$ of the NLDR. The cutoff point was taken to be zero, or equivalently, with the group-prior probabilities taken to be the same. McLachlan (1976a) showed that up to terms of the first order that

$$\begin{aligned}\text{bias}(A_1) &= E\{A_1(\mathbf{T}) - ec_1(F_1;\mathbf{T})\}\\ &= E\{A_1(\mathbf{T})\} - eu_1(\Delta)\\ &\approx -\phi(\tfrac{1}{2}\Delta)[\{\tfrac{1}{4}\Delta + (p-1)\Delta^{-1}\}/n_1 + \tfrac{1}{2}(p-1)(\Delta/N),\end{aligned} \tag{10.2.3}$$

where $N = n_1 + n_2 - 2$, Δ is the Mahalanobis distance between G_1 and G_2, and $\phi(\cdot)$ denotes the standard normal density function. The asymptotic bias of A_2 is obtained by interchanging n_1 with n_2 in (10.2.3). The unconditional error $eu_1(F)$ is written here as $eu_1(\Delta)$, since it depends on only Δ under (10.2.2) for $g = 2$ and $\pi_1 = \pi_2$.

For the use of the sample NLDR with zero cutoff point under (10.2.2), Hills (1966) and Das Gupta (1974) have considered some inequalities between the unconditional error rate $eu_1(F)$, the average apparent error rate $E\{A_1(\mathbf{T})\}$, the optimal error rate $\Phi(-\frac{1}{2}\Delta)$, and the so-called plug-in error rate $\Phi(-\frac{1}{2}D)$, where

$$D = \{(\bar{\mathbf{x}}_1 - \bar{\mathbf{x}}_2)'\mathbf{S}^{-1}(\bar{\mathbf{x}}_1 - \bar{\mathbf{x}}_2)\}^{1/2}$$

is the sample version of the Mahalanobis distance between G_1 and G_2. The latter rate is discussed in more detail in Section 10.6 on parametric estimation of error rates. The questions not answered were whether

$$E\{A_1(\mathbf{T})\} < \Phi(-\tfrac{1}{2}\Delta) \tag{10.2.4}$$

and

$$E\{A_1(\mathbf{T})\} < eu_1(\Delta), \tag{10.2.5}$$

where $n_1 \neq n_2$ and $p \neq 1$.

It can be seen from (10.2.3) that the asymptotic bias of A_1 is always negative, implying that (10.2.5) holds asymptotically. Considering (10.2.4), McLachlan (1976a) showed that

$$E\{A_1(\mathbf{T})\} \approx \Phi(-\tfrac{1}{2}\Delta) + \{\phi(\tfrac{1}{2}\Delta)/(16\Delta)\}[-\{3\Delta^2 + 4(p-1)\}/n_1 + \{\Delta^2 - 4(p-1)\}/n_2 - 4(p-1)(\Delta^2/N)]. \tag{10.2.6}$$

From (10.2.6), it can be seen that as $n_1 \to \infty$, the difference

$$E\{A_1(\mathbf{T})\} - \Phi(-\tfrac{1}{2}\Delta)$$

becomes positive if $\Delta^2 > 4(p-1)$ and n_2 is sufficiently large. Hence, (10.2.5) does not hold for some $n_1 \neq n_2$ and $p \neq 1$.

Schervish (1981b) extended the asymptotic expansions of McLachlan (1976a) to $g > 2$ groups within which $\mathbf{X}$ is normally distributed with a common covariance matrix. If the latter is assumed to hold in practice, then (10.2.3) and its multiple-group analogue provide a way of correcting parametrically the apparent error rate for bias. If $B_1(\Psi_E)$ denotes the first-order asymptotic bias of A_1, then the bias of

$$A_1^{(P)} = A_1(\mathbf{T}) - B_1(\hat{\Psi}_E) \tag{10.2.7}$$

is of the second order. Here $\hat{\Psi}_E$ need only be a consistent estimator of Ψ_E, but usually it would be taken to be the same fully efficient estimator used to form the sample NLDR $r_o(\mathbf{x}; \hat{\Psi}_E)$ in the first instance. Of course, if the present normal model were assumed to hold, then consideration would be given to basing the final estimate of the true error rate on a fully parametric method of estimation, as discussed in Section 10.6.

It is worth noting that although one may be willing to use a particular parametric rule due to its known robustness to mild departures from the adopted model, parametric estimators of the error rates of the rule may not be robust. For example, the sample NLDR is known to be fairly robust, but the normality-based estimators of its error rates are not; see Konishi and Honda (1990). It explains why much attention has been given to the development of nonparametric estimators of the error rate, in particular, to nonparametric methods of correcting the apparent error rate for bias. We now proceed to consider some of these methods.

10.2.2 Cross-Validation

One way of avoiding the bias in the apparent error rate as a consequence of the rule being tested on the same data from which it has been formed (trained) is to use a holdout method as considered by Highleyman (1962), among others. The available data are split into disjoint training and test subsets. The discriminant rule is formed from the training subset and then assessed on the test subset. Clearly, this method is inefficient in its use of the data. There are, however, methods of estimation, such as cross-validation, the

Quenouille–Tukey jackknife, and the recent bootstrap of Efron (1979), that obviate the need for a separate test sample. An excellent account of these three methods has been given by Efron (1982), who has exhibited the close theoretical relationship between them.

One way of almost eliminating the bias in the apparent error rate is through the leave-one-out (LOO) technique as described by Lachenbruch and Mickey (1968) or cross-validation (CV) as discussed in a wider context by M. Stone (1974) and Geisser (1975). For the estimation of $ec_1(F_1;\mathbf{t})$ by the apparent error rate $A_1(\mathbf{t})$, the leave-one-out cross-validated estimate is given by

$$A_1^{(CV)} = \frac{1}{n_1}\sum_{j=1}^{n} z_{1j}Q[1, r(\mathbf{x}_j;\mathbf{t}_{(j)})], \tag{10.2.8}$$

where $\mathbf{t}_{(j)}$ denotes $\mathbf{t}$ with the point $\mathbf{y}_j$ deleted ($j = 1,\ldots,n$). Hence, before the sample rule is applied at $\mathbf{x}_j$, it is deleted from the training set and the rule recalculated on the basis of $\mathbf{t}_{(j)}$. This procedure at each stage can be viewed as the extreme version of the holdout method where the size of the test set is reduced to a single entity.

According to M. Stone (1974), the refinement of this type of assessment appears to have been developed by Lachenbruch (1965) following a suggestion in Mosteller and Wallace (1963). Toussaint (1974a) in his bibliography on error-rate estimation traces the idea back to at least 1964 in the Russian literature on pattern recognition.

It can be seen that, in principle at least, cross-validation requires a considerable amount of computing as the sample rule has to be formed n_1 times in the computation of $A_1^{(CV)}$. For some rules, however, it is possible to calculate $A_1^{(CV)}$ with little additional effort. This is so with the normal-based linear or quadratic discriminant rules. To see this, consider firstly the calculation of $r(\mathbf{x}_j;\mathbf{t}_{(j)})$ in (10.2.8) for the sample NLDR. In order to calculate $r(\mathbf{x}_j;\mathbf{t}_{(j)})$ in this case, we require the value of

$$\begin{aligned}\hat{f}_{i(j)}(\mathbf{x}_j) &= \phi(\mathbf{x}_j;\bar{\mathbf{x}}_{i(j)},\mathbf{S}_{(j)})\\ &\propto \exp\{-\tfrac{1}{2}\delta(\mathbf{x}_j,\bar{\mathbf{x}}_{i(j)};\mathbf{S}_{(j)}\} \qquad (i = 1,\ldots,g),\end{aligned}$$

where

$$\delta(\mathbf{x}_j,\bar{\mathbf{x}}_{i(j)};\mathbf{S}_{(j)}) = (\mathbf{x}_j - \bar{\mathbf{x}}_{i(j)})'\mathbf{S}_{i(j)}^{-1}(\mathbf{x}_j - \bar{\mathbf{x}}_{i(j)}),$$

and $\bar{\mathbf{x}}_{i(j)}$ and $\mathbf{S}_{(j)}$ denote the ith group-sample mean and the bias-corrected pooled (within-group) sample covariance matrix, respectively, based on $\mathbf{t}_{(j)}$ for $j = 1,\ldots,n$. Concerning the computation of $\delta(\mathbf{x}_j,\mathbf{x}_{i(j)};\mathbf{S}_{(j)})$, Lachenbruch and Mickey (1968) noted how $\mathbf{S}_{(j)}^{-1}$ can be expressed in terms of $\mathbf{S}^{-1}$ using a result of Bartlett (1951a). Recently, Hjort (1986a) considered this problem using this now common device. He showed that

$$\delta(\mathbf{x}_j,\bar{\mathbf{x}}_{i(j)};\mathbf{S}_{(j)}) = \left(\sum_{u=1}^{g} z_{uj}c_{iu,j}\right)\delta(\mathbf{x}_j,\bar{\mathbf{x}}_i;\mathbf{S})$$

for $i = 1,\ldots,g$, where

$$c_{ii,j} = \{(n-g-1)n_i^2/(n_i-1)\}/\{(n-g)(n_i-1)-n_i d_{ii,j}\},$$

and where for $u \neq i$,

$$c_{iu,j} = \frac{n-g+1}{n-g}\left[1+\frac{n_i d_{iu,j}^2}{d_{ii,j}\{(n-g)(n_i-1)-n_i d_{uu,j}\}}\right]$$

and

$$d_{iu,j} = (\mathbf{x}_j - \bar{\mathbf{x}}_i)'\mathbf{S}^{-1}(\mathbf{x}_j - \bar{\mathbf{x}}_u) \qquad (i,u = 1,\ldots,g).$$

For the sample NQDR formed under the assumption of unequal group-conditional covariance matrices, we have from Hjort (1986a) that

$$\delta(\mathbf{x}_j,\bar{\mathbf{x}}_{i(j)};\mathbf{S}_{i(j)}) = \{1+z_{ij}(c_{i,j}-1)\}\delta(\mathbf{x}_j,\bar{\mathbf{x}}_i;\mathbf{S}_i)$$

for $i = 1,\ldots,g$, where

$$c_{i,j} = \{n_i^2(n_i-2)/(n_i-1)\}/\{(n_i-1)^2 - n_i\delta(\mathbf{x}_j,\bar{\mathbf{x}}_i;\mathbf{S}_i)\}.$$

This result, which is also available from (6.2.4), suffices for the calculation of the rule $r(\mathbf{x}_j;\mathbf{t}_{(j)})$ formed on the basis of minimum (Mahalanobis) distance. For the sample version $r_o(\mathbf{x};\hat{\boldsymbol{\Psi}}_U)$ of the Bayes rule, we need to calculate $\hat{f}_{i(j)}(\mathbf{x}_j)$, and not just its exponent. Hjort (1986a) showed that

$$\begin{aligned}\hat{f}_{i(j)}(\mathbf{x}_j) &= \phi(\mathbf{x}_j;\bar{\mathbf{x}}_{i(j)},\mathbf{S}_{i(j)})\\ &= \{1+z_{ij}(k_{1i,j}-1)\}\phi(\mathbf{x}_j;\bar{\mathbf{x}}_i,\mathbf{S}_i),\end{aligned}$$

for $i = 1,\ldots,g$, where

$$k_{1i,j} = k_{2i,j}\exp\{-\tfrac{1}{2}(c_{i,j}-1)\delta(\mathbf{x}_j,\bar{\mathbf{x}}_i;\mathbf{S}_i)\},$$

and

$$k_{2i,j} = (n_i-1)\{(n_i-2)/(n_i-1)\}^{(1/2)p}\{(n_i-1)^2 - n_i\delta(\mathbf{x}_j,\bar{\mathbf{x}}_i;\mathbf{S}_i)\}^{-1/2}.$$

Hjort (1986a) and Fukunaga and Hummels (1989) also have considered the computation of $r(\mathbf{x}_j;\mathbf{t}_{(j)})$, where $r(\mathbf{x};\mathbf{t})$ is based nonparametrically on k-NN and kernel density estimates of the group-conditional densities. Lesaffre, Willems, and Albert (1989) have considered the computation of the cross-validated estimate of the error rate for the logistic discriminant rule. As discussed in Section 8.2.4, the maximum likelihood estimates of the parameters in this model have to be computed iteratively. Hence, this can be quite time consuming even with small sample sizes. As a consequence, Lesaffre and Albert (1989b) have proposed a useful way of approximating the cross-validated estimate by one-step approximations to the maximum likelihood estimates.

It can be seen from the definition (10.2.3) of the apparent error rate A_1 that a modest perturbation in an $\mathbf{x}_j$ can switch the indicator function $Q[1,r(\mathbf{x}_j;\mathbf{t})]$

from zero to one and vice versa. As explained by Glick (1978), the leave-one-out modification of A_1 to give $A_1^{(CV)}$ exacerbates rather than smoothes such fluctuations. Thus, although cross-validation circumvents the bias problem, there is a consequent higher level of variability in the estimate.

For $g = 2$ groups, Hand (1986b) considered shrinking the cross-validated version $A^{(CV)}$ of the overall apparent error rate in an attempt to reduce its variance at the expense of increasing its bias in its estimation of the overall conditional rate $ec(F;\mathbf{t})$. For shrinking toward the origin in the case where there is no prior knowledge of the true error rate, Hand (1986b) studied two shrunken estimators, $A^{(SCV)} = \omega A^{(CV)}$, where $\omega = n/(n+1)$ or $n/(n+3)$. These two choices of ω correspond to taking $\pi_1 = 0.5$ and 0.25, respectively, in

$$\omega_o = n\pi_1/(n\pi_1 + \pi_2). \qquad (10.2.9)$$

The shrinking factor (10.2.9) is a conservative estimate of the optimal value of ω under the assumption that the optimal overall error rate $eo(F)$ is less than π_1, where, without loss of generality, π_1 is the minimum of π_1 and π_2.

In terms of mean-squared error,

$$\text{MSE}\{A^{(SCV)}\} = E\{A^{(SCV)} - ec(F;\mathbf{T})\}^2,$$

Hand (1986b) compared these two estimators with $A^{(CV)}$ and some other estimators, using simulations for the sample NLDR formed under (10.2.2) with the common covariance matrix taken to be known. For the combinations of the parameters considered, these two shrunken estimators were found to be superior to $A^{(CV)}$ although, as anticipated, they were more biased. Sometimes, however, they were more biased than A. The use of ω_o in shrinking $A^{(CV)}$ has been considered further by Hand (1987b). The relatively large variability in $A^{(CV)}$ is discussed further in Section 10.3 on a comparison of its performance relative to the bias-corrected version $A^{(B)}$ using the bootstrap.

With the main statistical packages, the error rates in the programs for discriminant analysis are usually estimated by the apparent error rate. The discriminant analysis program BMDP7M in the BMDP (1988) suite of programs has the provision for the calculation of not only the apparent error rate, but also its cross-validated version for the sample NLDR. The latter version is also available in the SAS (1990) package.

10.2.3 The Jackknife

There has been confusion in the literature over the roles of cross-validation and the jackknife in correcting the apparent error rate for bias. This is understandable as both methods delete one or more observations at a time in forming the bias-corrected estimates. According to M. Stone (1974), "Gray and Schucany (1972, pp. 125–136) appear to initiate the confusion in their description of Mosteller and Tukey's sophisticated, simultaneous juggling act with the two concepts." Consider the jackknife version of the apparent error rate given by

$$A_1^{(J_o)} = A_1 + (n_1 - 1)(A_1 - A_{1(\cdot)}), \tag{10.2.10}$$

where

$$A_{1(\cdot)} = \sum_{j=1}^{n} A_{1(j)}/n_1,$$

and $A_{1(j)}$ denotes the apparent error rate of $r(\mathbf{x}, \mathbf{t}_{(j)})$ when applied to the observations in $\mathbf{t}_{(j)}$ from G_1, that is,

$$A_{1(j)} = \sum_{k \neq j}^{n} z_{1k} Q[1, r(\mathbf{x}_k; \mathbf{t}_{(j)})]/(n_1 - 1).$$

This jackknifed form of A_1 is appropriate for the estimation of the limiting value of the unconditional error rate $eu_1(F)$, as $n_1, n_2 \to \infty$, that is, the optimal error $eo_1(F)$ assuming $r(\mathbf{x};\mathbf{t})$ is a Bayes consistent rule. As an estimator then of $eo_1(F)$, the bias of $A_1^{(J_o)}$ is of order $O(n^{-2})$. But $A_1^{(J_o)}$ is frequently used or suggested as an estimate of the conditional error $ec_1(F;\mathbf{t})$, as in Crask and Perreault (1977). However, as an estimator of $ec_1(F_1;\mathbf{t})$, the bias of $A_1^{(J_o)}$ is still of order $O(n^{-1})$.

It follows from Efron (1982, Chapter 7) that the jackknifed version of A_1, which reduces its bias as an estimator of $ec_1(F_1;\mathbf{t})$ to the second order, can be written as

$$A_1^{(J_c)} = A_1 + (n_1 - 1)(\tilde{A}_1 - A_{1(\cdot)}), \tag{10.2.11}$$

where

$$\tilde{A}_1 = \frac{1}{n_1^2} \sum_{j=1}^{n} \sum_{k=1}^{n} z_{1j} z_{1k} Q[1, r(\mathbf{x}_k; \mathbf{t}_{(j)})]; \tag{10.2.12}$$

see also Efron and Gong (1983). Efron (1982) noted that the last term on the right-hand side of (10.2.11) can be rearranged to give

$$A_1^{(J_c)} = A^{(CV)} + A_1 - \tilde{A}, \tag{10.2.13}$$

demonstrating the close relationship between the jackknife and the cross-validation methods of bias correction of the apparent error rate in estimating the conditional error rate of a sample-based rule. Also, he showed how the jackknife estimate of bias, $A_1 - A_1^{(J_c)}$ in this instance, can be considered as a quadratic approximation to the nonparametric bootstrap estimate of bias defined in the next section. The underlying assumption here that $r(\mathbf{x};\mathbf{t})$ is symmetrically defined in $\mathbf{y}_1, \ldots, \mathbf{y}_n$ has to be strengthened to $r(\mathbf{x};\mathbf{t})$ depending on $\mathbf{y}_1, \ldots, \mathbf{y}_n$ through a functional statistic in order to establish the above connection between the bootstrap, cross-validation, and the jackknife.

The misunderstanding over the use of $A_1^{(J_o)}$ to estimate $ec_1(F_1;\mathbf{t})$ would appear to still occur in the literature. For example, for $g = 2$ groups, Rao and Dorvlo (1985) used the formula corresponding to (10.2.10) to jackknife

$\Phi(-\frac{1}{2}D)$, $\Phi(-\frac{1}{2}DS)$, and D for use in $\Phi(-\frac{1}{2}D)$, for the purpose of estimating the optimal error rate $\Phi(-\frac{1}{2}\Delta)$, but then considered these jackknifed versions as estimators also of the unconditional error rate. The version $\Phi(-\frac{1}{2}DS)$ of the plug-in error rate, where

$$DS = \{(n-p-3)/(n-2)\}^{1/2}D,$$

partially reduces its optimistic bias; see Section 10.6 to follow.

Wang (1986b) used the formula (10.2.10) to jackknife the apparent error rate for the estimation of the unconditional error rate in the case of $g = 2$ multinomial distributions. The inappropriate choice of the jackknifed version of the apparent error rate was reflected in his simulation results, where this estimate was generally well below that of the cross-validated estimate.

10.3 THE BOOTSTRAP

10.3.1 Introduction

The bootstrap was introduced by Efron (1979), who has investigated it further in a series of articles; see Efron (1981a, 1981b, 1982, 1983, 1985, 1987, 1990) and the references therein. A useful introductory account of bootstrap methods has been given by Efron and Tibshirani (1986). Over the past 10 years, the bootstrap has become one of the most popular recent developments in statistics. Hence, there now exists an extensive literature on it, as evident in the survey articles of Hinkley (1988) and DiCiccio and Romano (1988); see also Hall (1988). It is a powerful technique that permits the variability in a random quantity to be assessed using just the data at hand. An estimate $\hat{F}$ of the underlying distribution is formed from the observed sample. Conditional on the latter, the sampling distribution of the random quantity of interest with F replaced by $\hat{F}$, defines its so-called bootstrap distribution, which provides an approximation to its true distribution. It is assumed that $\hat{F}$ has been so formed that the stochastic structure of the model has been preserved. Usually, it is impossible to express the bootstrap distribution in simple form, and it must be approximated by Monte Carlo methods whereby pseudo-random samples (bootstrap samples) are drawn from $\hat{F}$. The bootstrap can be implemented nonparametrically by using the empirical distribution function constructed from the original data. In recent times, there has been a number of papers written on improving the efficiency of the bootstrap computations with the latter approach; see, for example, Davison, Hinkley, and Schechtman (1986); Hall (1989a, 1989b); Hinkley and Shi (1989); Efron (1990); and Graham et al. (1990).

10.3.2 Bias Correction of Apparent Error Rate

We now consider the application of the bootstrap in the present context of correcting nonparametrically the apparent error rate of a sample-based rule.

The bias correction of A_1 in its estimation of the conditional error $ec_1(F_1;\mathbf{t})$ may be implemented according to the bootstrap as follows.

Step 1. In the case of mixture sampling, a new set of data,

$$\mathbf{t}^* = \{\mathbf{y}_1^* = (\mathbf{x}_1^{*\prime},\mathbf{z}_1^{*\prime})',\ldots,\mathbf{y}_n^* = (\mathbf{x}_n^{*\prime},\mathbf{z}_n^{*\prime})'\},$$

called the bootstrap sample, is generated according to $\hat{F}$, an estimate of the distribution function of $\mathbf{Y}$ formed from the original training data $\mathbf{t}$. That is, $\mathbf{t}^*$ consists of the observed values of the random sample

$$\mathbf{Y}_1^*,\ldots,\mathbf{Y}_n^* \overset{iid}{\sim} \hat{F}, \tag{10.3.1}$$

where $\hat{F}$ is held fixed at its observed value.

With separate sampling, the bootstrap group-indicator vectors are fixed at their original values. The bootstrap observations $\mathbf{x}_1^*,\ldots,\mathbf{x}_n^*$ are then generated according to this group specification, which ensures that n_i of them are from the ith group G_i $(i = 1,\ldots,g)$. Hence, the bootstrap training set $\mathbf{t}^*$ is given by

$$\mathbf{t}^* = \{\mathbf{y}_1^* = (\mathbf{x}_1^{*\prime},\mathbf{z}_1^{*\prime})',\ldots,\mathbf{y}_n^* = (\mathbf{x}_n^{*\prime},\mathbf{z}_n^{*\prime})'\},$$

where if $(\mathbf{z}_j^*)_i = 1$ for $i = u$, then $\mathbf{x}_j^*$ is the realized value of $\mathbf{X}_j^*$ distributed according to $\hat{F}_u$. Thus, for a given i $(i = 1,\ldots,g)$, there are n_i of the $\mathbf{x}_j^*$ with $(\mathbf{z}_j^*)_i = 1$. If they are relabeled as $\mathbf{x}_{i1}^*,\ldots,\mathbf{x}_{in_i}^*$, then they may be viewed as the observed values of the random sample

$$\mathbf{X}_{i1}^*,\ldots,\mathbf{X}_{in_i}^* \overset{iid}{\sim} \hat{F}_i, \qquad (i = 1,\ldots,g) \tag{10.3.2}$$

where each $\hat{F}_i$ is held fixed at its observed value.

Step 2. As the notation implies, the rule $r(\mathbf{x};\mathbf{t}^*)$ is formed from the training data $\mathbf{t}^*$ in precisely the same manner as $r(\mathbf{x};\mathbf{t})$ was from the original set.

Step 3. The apparent error rate of $r(\mathbf{x};\mathbf{t}^*)$ for the first group, $A_1(\mathbf{t}^*)$, is computed by noting the proportion of the observations in $\mathbf{t}^*$ from G_1 misallocated by $r(\mathbf{x};\mathbf{t}^*)$. That is,

$$A_1(\mathbf{t}^*) = \frac{1}{n_1^*}\sum_{j=1}^{n} z_{1j}^* Q[1, r(\mathbf{x}_j^*;\mathbf{t}^*)],$$

where

$$n_1^* = \sum_{j=1}^{n} z_{1j}^*,$$

and where $n_i^* = n_i$ $(i = 1,\ldots,g)$ with separate sampling. The difference

$$\hat{d}_1^* = A_1(\mathbf{t}^*) - ec_1(\hat{F}_1;\mathbf{t}^*) \tag{10.3.3}$$

is computed, too, where $ec_1(\hat{F}_1;\mathbf{t}^*)$ is the bootstrap analogue of the conditional error rate for the first group.

Step 4. The bootstrap bias of the apparent error for the first group is given by

$$\hat{b}^{(B)} = E^*(\hat{d}_1^*)$$
$$= E^*\{A_1(\mathbf{T}^*) - ec_1(\hat{F}_1; \mathbf{T}^*)\},$$

where E^* refers to expectation over the distribution of $\mathbf{T}^*$ defined according to (10.3.1) for mixture sampling and (10.3.2) for separate sampling. It can be approximated by $\hat{b}_1^{(B)} \approx \overline{\hat{d}_1^*}$, obtained by averaging $\hat{d}_1^*$ over K independently repeated realizations of $\mathbf{T}^*$, that is,

$$\overline{\hat{d}_1^*} = \sum_{k=1}^{K} \hat{d}_{1,k}^* / K, \tag{10.3.4}$$

where

$$\hat{d}_{1,k}^* = A_1(\mathbf{t}_k^*) - ec_1(\hat{F}_1; \mathbf{t}_k^*) \tag{10.3.5}$$

denotes the value of $A_1(\mathbf{T}^*) - ec_1(\hat{F}_1; \mathbf{T}^*)$ for the kth bootstrap replication $\mathbf{t}_k^*$ of $\mathbf{T}^*$. The standard error of the Monte Carlo approximation $\overline{\hat{d}_1^*}$ to the bootstrap bias $E^*(\hat{d}_1^*)$ is calculated as the positive square root of

$$\sum_{k=1}^{K} (\hat{d}_{1,k}^* - \overline{\hat{d}_1^*})^2 / \{K(K-1)\}.$$

The bias-corrected version of A_1 according to the bootstrap, therefore, is given by

$$A_1^{(B)} = A_1 - \hat{b}_1^{(B)},$$

where the Monte Carlo approximation $\overline{\hat{d}_1^*}$ can be used for the bootstrap bias $\hat{b}_1^{(B)}$. Efron (1983) subsequently gave a more efficient way of approximating $\hat{b}_1^{(B)}$, which is described in Section 10.4. In this and other applications of the bootstrap, Efron (1979, 1981a, 1981b) noted that the choice of replication number usually does not seem to be critical past 50 or 100. Of course, as $K \to \infty$, the standard error of $\overline{\hat{d}_1^*}$ tends to zero, but, as explained by Efron (1979), there is no point in taking K to be any larger than necessary to ensure that the standard error of $\overline{\hat{d}_1^*}$ is small relative to the standard deviation of $E^*(\hat{d}^*{}_1)$. More recently, Efron (1985), Tibshirani (1985), Efron and Tibshirani (1986), and Gong (1986) have shown that whereas 50 to 100 bootstrap replications may be sufficient for standard error and bias estimation, a large number, say, 350, is needed to give a useful estimate of a percentile or P-value, and many more for a highly accurate assessment. This problem of the number of bootstrap replications has been studied in some depth recently by Hall (1986b).

In Step 1 of the above algorithm, the nonparametric version of the bootstrap would under mixture sampling take $\hat{F}$ to be the empirical distribution function with mass $1/n$ at each original data point $\mathbf{y}_j$ in $\mathbf{t}$ $(j = 1,\dots,n)$. Under separate sampling, $\hat{F}_i$ would be the empirical distribution function with mass $1/n_i$ at each of the n_i original observations from $\mathbf{x}_1,\dots,\mathbf{x}_n$ in G_i $(i = 1,\dots,g)$. Under either sampling scheme with the nonparametric bootstrap, the rate $ec_1(\hat{F}_1;\mathbf{t}^*)$ in (10.3.3) is given by

$$ec_1(\hat{F}_1;\mathbf{t}^*) = \frac{1}{n_1}\sum_{j=1}^{n} z_{1j}Q[1, r(\mathbf{x}_j;t^*)].$$

10.3.3 Some Other Uses of the Bootstrap

The bootstrap is a very powerful technique and it can be used to assess other sampling properties of the apparent error rate besides its bias. For instance, an estimate of the mean-squared error (MSE) of A_1 in estimating $ec_1(F_1;\mathbf{t})$ is provided by

$$\text{MSE}^{(B)}(A_1) = \sum_{k=1}^{K}\{A_1(t_k^*) - ec_1(\hat{F}_1;\mathbf{t}_k^*)\}^2/K, \tag{10.3.6}$$

where the right-hand side of (10.3.6) is the Monte Carlo approximation to the bootstrap MSE of $ec_1(\hat{F}_1;\mathbf{T}^*)$. As noted by Efron and Gong (1983), the sample variance of the $\hat{d}_{1,k}^*$,

$$\sum_{k=1}^{K}(\hat{d}_{1,k}^* - \overline{\hat{d}_1^*})^2/(K-1),$$

suggests a lower bound for the MSE of $A_1^{(B)}$ in estimating $ec_1(F_1;\mathbf{t})$. This is because the variance of $A_1(\mathbf{T}) - ec_1(F_1;\mathbf{T})$ can be viewed as the mean-squared error of the "ideal constant" estimator,

$$A_1^{(IC)} = A_1(\mathbf{T}) - b_1,$$

which would be used if we knew b_1, the bias of $A_1(\mathbf{T})$. To see this, note that

$$\begin{aligned}\text{var}\{A_1(\mathbf{T}) - ec_1(F_1;\mathbf{T})\} &= E\{A_1(\mathbf{T}) - ec_1(F_1;\mathbf{T}) - b_1\}^2\\ &= E[\{A_1(\mathbf{T}) - b_1\} - ec_1(F_1;\mathbf{T})]^2\\ &= \text{MSE}(A_1^{(IC)}).\end{aligned}$$

It would be anticipated that $A_1^{(B)}$ would have mean-squared error at least as large as $A_1^{(IC)}$.

The bootstrap can be used to assess the performance of the apparent error rate in its estimation of the other types of error rates. Replacing $ec_1(\hat{F}_1;\mathbf{t}_k^*)$ by $eo_1(\hat{F})$ in (10.3.5) and (10.3.6) yields the bootstrap estimates of the bias and mean-squared error, respectively, of A_1 in estimating the optimal error rate $eo_1(F)$. For the nonparametric version of the bootstrap, where $\hat{F}$ is the empirical distribution function, $eo_1(\hat{F}) = A_1$ at least if $r(\mathbf{x};\mathbf{t})$ depends on $\mathbf{t}$ through $\hat{F}$. Similarly, an assessment of the mean-squared error of A_1 in its estimation of the unconditional rate $eu_1(F)$ is obtained by replacing $ec_1(\hat{F}_1;\mathbf{t}_k^*)$ with

$$\sum_{k=1}^{K} ec_1(\hat{F}_1;\mathbf{t}_k^*)/K \tag{10.3.7}$$

in the Monte Carlo approximation (10.3.6). The quantity (10.3.7) is the Monte Carlo approximation to the bootstrap estimate of the unconditional error rate for the first group.

The assessment of the distribution of $A_1(\mathbf{T}) - ec_1(F_1;\mathbf{T})$ provided by the bootstrap can be used also to form approximate confidence intervals for the unobservable error rates, as discussed in Section 10.7.

10.3.4 Parametric Version of the Bootstrap

With the parametric version of the bootstrap, the pseudo-data are generated according to (10.3.1) or (10.3.2) with the vector $\boldsymbol{\Psi}$ of unknown parameters in the parametric form adopted for F or F_i $(i = 1,\ldots,g)$, replaced by an appropriate estimate $\hat{\boldsymbol{\Psi}}$ formed from the original training data $\mathbf{t}$. Usually, $\hat{\boldsymbol{\Psi}}$ is the maximum likelihood estimate, perhaps corrected for bias as typically done with the estimates of the group-covariance matrices under normality. Suppose, for example, that normality is adopted for the group-conditional distribution of $\mathbf{X}$, where

$$\mathbf{X} \sim N(\boldsymbol{\mu}_i, \boldsymbol{\Sigma}_i) \text{ with prob. } \pi_i \text{ in } G_i \; (i = 1,\ldots,g). \tag{10.3.8}$$

Then the generation of the bootstrap data $\mathbf{z}_1^*,\ldots,\mathbf{x}_n^*$ according to (10.3.1) for mixture sampling can be achieved as follows. For each j $(j = 1,\ldots,n)$, a random variable that takes the values $1,\ldots,g$ with probabilities $\hat{\pi}_1,\ldots,\hat{\pi}_g$ is generated where $\hat{\pi}_i = n_i/n$ for $i = 1,\ldots,g$. If the generated value is equal to, say, u, then z_{ij}^* is set equal to 1 for $i = u$ and to zero otherwise. The observation $\mathbf{x}_j^*$ is then generated from the density $\phi(\mathbf{x};\bar{\mathbf{x}}_u,\mathbf{S}_u)$, the multivariate normal density with mean $\bar{\mathbf{x}}_u$ and covariance matrix $\mathbf{S}_u$. This implies that $\mathbf{x}_1^*,\ldots,\mathbf{x}_n^*$ are the observed values of a random sample from the normal mixture density

$$f_X(\mathbf{x};\hat{\boldsymbol{\Psi}}) = \sum_{i=1}^{g} \hat{\pi}_i \phi(\mathbf{x};\bar{\mathbf{x}}_i,\mathbf{S}_i),$$

and that $\mathbf{z}_1^*,\ldots,\mathbf{z}_n^*$ are the realized values of the random sample $\mathbf{Z}_1^*,\ldots,\mathbf{Z}_n^*$, where

$$\mathbf{Z}_1^*,\ldots,\mathbf{Z}_n^* \overset{iid}{\sim} \text{Mult}_g(1,\hat{\boldsymbol{\pi}}),$$

a multinomial distribution consisting of one draw on g categories with probabilities $\hat{\pi}_1,\ldots,\hat{\pi}_g$. For separate sampling according to (10.3.2), $\mathbf{x}_{i1}^*,\ldots,\mathbf{x}_{in_i}^*$ are the realized values of a random sample generated from the multivariate normal density $\phi(\mathbf{x};\bar{\mathbf{x}}_i, \mathbf{S}_i)$ for $i = 1,\dots,g$.

In this parametric framework with mixture sampling for $g = 2$ groups having a common covariance matrix, $ec_1(\hat{F}_1;\mathbf{t}^*)$ is given by

$$ec_1(\hat{F}_1;\mathbf{t}^*) = \Phi\{-(\hat{\beta}_{0E}^* + \hat{\boldsymbol{\beta}}_E^{*\prime}\bar{\mathbf{x}}_1^*)/(\hat{\boldsymbol{\beta}}_E^{*\prime}\mathbf{S}\hat{\boldsymbol{\beta}}_E^*)^{1/2}\}, \tag{10.3.9}$$

where

$$\hat{\boldsymbol{\beta}}_E^* = \mathbf{S}^{*-1}(\bar{\mathbf{x}}_1^* - \bar{\mathbf{x}}_2^*)$$

and

$$\hat{\beta}_{0E}^* = \log(n_1^*/n_2^*) - \tfrac{1}{2}(\bar{\mathbf{x}}_1^* + \bar{\mathbf{x}}_2^*)'\hat{\boldsymbol{\beta}}_E^* \tag{10.3.10}$$

are the bootstrap analogues of the coefficients $\hat{\boldsymbol{\beta}}_E$ and $\hat{\beta}_{0E}$ in the sample NLDR, defined from (3.3.5). For separate sampling, the term $\log(n_1^*/n_2^*)$ on the right-hand side of (10.3.10) is replaced by minus the cutoff point k specified initially with the application of the sample NLDR.

Under the normal model (10.3.8), there is really no need to resort to Monte Carlo approximation of the (parametric) bootstrap bias of $A_1(\mathbf{T}^*)$, as asymptotic approximations are available from McLachlan (1976a) and Schervish (1981b). For example, for $g = 2$ and $\pi_1 = \pi_2 = 0.5$, we have that

$$E^*\{A_1(\mathbf{T}^*) - ec_1(\hat{F}_1;\mathbf{T}^*)\} \approx B_1(D),$$

where $B_1(\Delta)$ is the n^{-1}-order asymptotic bias of $A_1(\mathbf{T})$ given by (10.2.3), and D is the sample counterpart of Δ. McLachlan (1980b) carried out some simulations in which he compared the mean-squared error of $\hat{b}_1^{(B)}$, as obtained with the nonparametric version of the bootstrap, with that of $B_1(D)$. This comparison demonstrated the high efficiency of the nonparametric version of the bootstrap estimator of the bias of the apparent error rate. Similar results were reported by Schervish (1981b) for $g > 2$ groups.

Returning to the nonparametric bootstrap, where the pseudo-data are generated using the empirical distribution function, one might wish to attribute some smoothness to $\hat{F}$ without going all the way to the normal model (10.3.8). The reader is referred to Efron (1982, page 30) and Silverman and Young (1987) for various ways of smoothing the empirical distribution for use in generating the data.

10.3.5 Relationship of Bootstrap with Some Other Methods of Bias Correction

Efron (1983) has reported some simulation results on the performance of the bootstrap relative to other methods such as cross-validation in their bias correction of the overall apparent error rate,

$$A(\mathbf{t}) = \frac{1}{n}\sum_{i=1}^{g}\sum_{j=1}^{n} z_{ij}Q[i, r(\mathbf{x}_j;\mathbf{t})].$$

The simulations were performed in the context of estimating the overall conditional error rate $ec(F,\mathbf{t})$ of $r(\mathbf{x};\mathbf{t})$ formed under the homoscedastic normal model (10.2.2) for $g = 2$ groups in equal proportions. Preliminary simulations, which compared the bootstrap estimate $\hat{b}^{(B)}$ of the bias of A with the jackknife and cross-validation estimates, $\hat{b}^{(J_c)} = A - A^{(J_c)}$ and $\hat{b}^{(CV)} = A - A^{(CV)}$, respectively, showed that $\hat{b}^{(J_c)}$ and $\hat{b}^{(CV)}$ are very close in value. This is not surprising, given the close relationship in the forms of their estimates for the bias of A, as demonstrated by (10.2.12) and (10.2.13) for the bias correction of A_1. Also, under smoothness conditions on the indicator function Q, Gong (1982) has shown that $\hat{b}^{(J_c)}$ and $\hat{b}^{(CV)}$ have asymptotic correlation one. As summarized by Efron (1983), $\hat{b}^{(B)}$ is the obvious nonparametric maximum likelihood estimator for b, the bias of A; $\hat{b}^{(J_c)}$ is a quadratic approximation to $\hat{b}^{(B)}$ and $\hat{b}^{(CV)}$ is similar in form and value to $\hat{b}^{(J_c)}$. Efron (1983) carried out further simulations to compare cross-validation with the bootstrap and some variants, which are described shortly. It was concluded that the cross-validation estimator of the overall error rate, $A^{(CV)}$, is nearly unbiased, but that it has often an unacceptably high variability if n is small. The bootstrap bias-corrected version of $A, A^{(B)}$, has much less variability, but, unfortunately, $\hat{b}^{(B)}$ is negatively correlated with $A(\mathbf{T}) - ec(F;\mathbf{T})$, the actual difference between the apparent error rate and the overall conditional error rate. In contrast, $\hat{b}^{(CV)}$ has a correlation near to zero with this difference.

The mean-squared error of $A^{(B)} = A - \hat{b}^{(B)}$ is given by

$$\begin{aligned} \text{MSE}(A^{(B)}) &= E\{A^{(B)} - ec(F;\mathbf{T})\}^2 \\ &= E\{A - \hat{b}^{(B)} - ec(F;\mathbf{T})\}^2, \end{aligned}$$

which can be expressed as

$$\begin{aligned} \text{MSE}(A^{(B)}) = \text{var}(A - ec(F;\mathbf{T})) + \text{var}(\hat{b}^{(B)}) + \{E(\hat{b}^{(B)}) - b\}^2 \\ - 2\,\text{cov}(A(\mathbf{T}) - ec(F;\mathbf{T}), \hat{b}^{(B)}). \end{aligned}$$

It can be seen that a negative value for the correlation between $A(\mathbf{T})-ec(F;\mathbf{T})$ and $\hat{b}^{(B)}$ inflates the mean-squared error of $A^{(B)}$. Nevertheless, the simulated mean-squared error of $A^{(B)}$ was still less than that of $A^{(CV)}$.

As explained further by Efron (1986), $\hat{b}^{(B)}$ is really intended to estimate the bias $b = E\{A(\mathbf{T}) - ec(F;\mathbf{T})\}$, rather than the actual difference $A(\mathbf{t}) - ec(F;\mathbf{t})$. However, in using $A^{(B)} = A(\mathbf{t}) - \hat{b}^{(B)}$ to estimate $ec(F;\mathbf{t})$, we are employing $\hat{b}^{(B)}$ in the latter role. This point is demonstrated by expressing the mean-squared error of $A^{(B)}$ as

$$\begin{aligned} \text{MSE}(A^{(B)}) &= E\{A(\mathbf{T}) - \hat{b}^{(B)} - ec(F;\mathbf{T})\}^2 \\ &= E[\{A(\mathbf{T}) - ec(F;\mathbf{T})\} - \hat{b}^{(B)}]^2. \end{aligned} \qquad (10.3.11)$$

It can be seen from (10.3.11) that, as an estimator of $ec(F;\mathbf{t})$, $A^{(B)}$ is judged by how well $\hat{b}^{(B)}$ estimates the difference $A(\mathbf{t}) - ec(F;\mathbf{t})$.

TABLE 10.1 The First 10 Trials and Summary Statistics for 100 Trials of a Simulation Experiment for the Bias of the Overall Apparent Error Rate and Some Bias-Corrected Versions with the Bootstrap Based on $K = 200$ Replications

Trial	$A(\mathbf{t})$	$A(\mathbf{t}) - ec(F;\mathbf{t})$	$\hat{b}^{(B)}$	$\hat{b}^{(CV)}$	$\hat{b}^{(J_c)}$
1	0.286	−0.172	−0.083	−0.214	−0.214
2	0.357	0.045	−0.098	0.000	−0.066
3	0.357	0.044	−0.110	−0.071	−0.066
4	0.429	0.078	−0.107	−0.071	−0.066
5	0.357	0.027	−0.102	−1.43	−0.148
6	0.143	−0.175	−0.073	−0.214	−0.194
7	0.071	−0.239	−0.047	−0.071	−0.066
8	0.286	−0.094	−0.097	−0.071	−0.056
9	0.429	0.069	−0.127	−0.071	−0.087
10	0.143	−0.192	−0.048	0.000	−0.010
100 trials					
Average	0.264	−0.096	−0.080	−0.091	−0.093
(S.D.)	(0.123)	(0.113)	(0.028)	(0.073)	(0.068)

Source: Adapted from Efron (1983).

To illustrate the aforementioned simulation experiments of Efron (1983), we have listed in Table 10.1 the values of the overall error rate, the apparent error rate, and the various estimates of its bias for one of the experiments performed for $n = 14$ observations drawn from a mixture in equal proportions of two bivariate normal distributions under the canonical form (4.2.3) with $\Delta = 1$.

In another comparison of the bootstrap with cross-validation and the jackknife, Gong (1986) reported some results for simulations and real data for a moderately complicated rule formed using forward logistic regression. The bootstrap was found to offer a significant improvement over cross-validation and the jackknife for estimation of the bias of the apparent error rate and for providing a bias-corrected estimate. For the same purposes, Wang (1986b) also found the bootstrap to be the best of these three methods for some simulations performed for a discriminant rule in the case of two groups with multinomial distributions.

10.4 VARIANTS OF THE BOOTSTRAP

10.4.1 Double Bootstrap

For $g = 2$ groups, Efron (1983) has developed more sophisticated versions of his ordinary bootstrap, including the double bootstrap, which corrects its downward bias without an increase in its mean-squared error, and the randomized bootstrap and 0.632 estimates which appreciably lower its mean-squared error. The double bootstrap corrects the bias of the ordinary bootstrap apparently without increasing its mean-squared error (Efron, 1983). The bias-corrected version of A so obtained for estimation of the overall conditional error is

$$A^{(DB)} = A^{(B)} - \text{bias}^{(B)}(A^{(B)}),$$

where $\text{bias}^{(B)}(A^{(B)})$ is the bootstrap estimate of the bias of the ordinary bootstrap estimator $A^{(B)}$. Although it appears that the computation of the estimate $\text{bias}^{(B)}(A^{(B)})$ requires two layers of bootstrapping with a total of K^2 bootstrap replications, Efron (1983) has shown that, by using a Monte Carlo "swindle," it can be implemented with just $2K$ replications.

10.4.2 The Randomized Bootstrap

The randomized bootstrap in the case of mixture sampling with respect to $g = 2$ groups generates the bootstrap data with $\hat{F}$ taken to be the distribution function that assigns nonzero mass over the $2n$ points, $\mathbf{y}_j$ and $\tilde{\mathbf{y}}_j$ $(j = 1,\ldots,n)$, with mass $\nu(\mathbf{y}_j)/n$ and $\nu(\tilde{\mathbf{y}}_j)/n$ at $\mathbf{y}_j$ and $\tilde{\mathbf{y}}_j$, respectively, and $\nu(\mathbf{y}_j) + \nu(\tilde{\mathbf{y}}_j) = 1$. Here $\tilde{\mathbf{y}}_j$ is taken to be $\mathbf{y}_j$, but with the group of origin of $\mathbf{x}_j$ switched to the other group, that is, for $j = 1,\ldots,n$,

$$\tilde{\mathbf{y}}_j = (\mathbf{x}_j', \tilde{\mathbf{z}}_j')',$$

where

$$\tilde{\mathbf{z}}_j = (z_{2j}, z_{1j})'.$$

Efron (1983) studied the use of

$$\nu(\mathbf{y}_j) = 0.9, \qquad \nu(\tilde{\mathbf{y}}_j) = 0.1, \tag{10.4.1}$$

and a more complicated version, equivalent here to taking

$$\nu(\mathbf{y}_j) = \hat{\tau}_1(\mathbf{x}_j;\mathbf{t}), \qquad \nu(\tilde{\mathbf{y}}_j) = \hat{\tau}_2(\mathbf{x}_j;\mathbf{t}), \tag{10.4.2}$$

with the restriction that $\nu(\mathbf{y}_j)$, and hence $\nu(\tilde{\mathbf{y}}_j)$, lie in the range 0.1–0.9. In (10.4.2), $\hat{\tau}_i(\mathbf{x}_j;\mathbf{t})$ denotes the estimated posterior probability of membership of G_i $(i = 1,2)$. It can be seen that the randomized bootstrap is an attempt to smooth $\hat{F}$ in the $\mathbf{z}$ direction. The use of either (10.4.1) or (10.4.2) in Step 1 of the bootstrap algorithm was found to substantially lower the mean-squared error of the ordinary bootstrap estimator of the overall conditional error, with (10.4.1) giving almost as much improvement as the more complicated version (10.4.2).

10.4.3 The 0.632 Estimator

We let $A^{(0.632)}$ be the estimator of the overall conditional error rate, termed the "0.632" estimator by Efron (1983), who reported that it was clearly the best in his simulation experiments. It is a weighted sum of the apparent error rate and $\hat{\epsilon}^{(B)}$, the bootstrap error rate at an original data point not in the training set $\mathbf{t}$. It is defined by

$$A^{(0.632)} = 0.368A + 0.632\hat{\epsilon}^{(B)},$$

where $\hat{\epsilon}^{(B)}$ is approximated by

$$\hat{\epsilon}^{(B)} = \sum_{i=1}^{g}\sum_{j=1}^{n}\sum_{k=1}^{K} z_{ij}\tilde{\delta}_{jk}Q[i, r(\mathbf{x}_j;\mathbf{t}_k^*)] \Big/ \sum_{j=1}^{n}\sum_{k=1}^{K}\tilde{\delta}_{jk}, \qquad (10.4.3)$$

and $\tilde{\delta}_{jk} = 1$ if $\mathbf{y}_j$ is not present in the bootstrap training set $\mathbf{t}_k^*$ and is zero otherwise. Efron (1983) developed the 0.632 estimator by consideration of the distribution of the distance $\delta(\mathbf{x},\mathbf{t})$ between the point $\mathbf{x}$ at which the rule is applied and the nearest feature observation in the training set $\mathbf{t}$. The distribution of this distance is quite different in the nonparametric bootstrap context than in the actual situation, with the bootstrap distance $\delta(\mathbf{X}^*,\mathbf{T}^*)$ having a high probability of being zero. This probability is equal to the probability that the point at which the rule is applied is included in the bootstrap sample, which is $1-(1-1/n)^n$ and which tends to 0.632, as $n \to \infty$. In his ingenious argument, Efron (1983) showed that the points that contribute to $\hat{\epsilon}^{(B)}$ (i.e., those with $\delta(\mathbf{X}^*,\mathbf{T}^*) > 0$ in the bootstrap context) are about 1/0.632 too far away from the training set than in the actual situation. This leads to

$$\hat{b}^{(0.632)} = 0.632(A - \hat{\epsilon}^{(B)}) \qquad (10.4.4)$$

as the estimate of the bias of A, and hence

$$\begin{aligned} A^{(0.632)} &= A - \hat{b}^{(0.632)} \\ &= 0.368A + 0.632\hat{\epsilon}^{(B)}, \end{aligned}$$

as the bias-corrected version of A.

10.4.4 Outline of Derivation of $A^{(0.632)}$

We now give a brief outline of the result (10.4.4) as derived by Efron (1983) in the context of sampling from a mixture of $g = 2$ groups. He introduced a notion of distance between the point $\mathbf{x}$ at which the rule $r(\mathbf{x};\mathbf{t})$ is applied and the training data $\mathbf{t}$ by defining $S(\mathbf{x}, a)$ to be a set around the point $\mathbf{x}$ in the feature space having probability content a under the true distribution F for $\mathbf{Y}$. Thus,

$$\mathrm{pr}_F\{\mathbf{X} \in S(\mathbf{x}, a)\} = a.$$

As $a \to 0$, it is assumed that $S(\mathbf{x}, a)$ tends to the single point $\mathbf{x}$. The distance between $\mathbf{x}$ and the observed training data $\mathbf{t}$ is defined then as

$$\delta(\mathbf{x},\mathbf{t}) = \inf\left\{ a : \mathbf{x} \in \bigcup_{j=1}^{n} S(\mathbf{x}_j, a)\right\}.$$

As noted above, in the bootstrap framework $\delta(\mathbf{X}^*, \mathbf{T}^*)$ has a high probability of being zero, with

$$\begin{aligned} \mathrm{pr}_{\hat{F}}\{\delta(\mathbf{X}^*, \mathbf{T}^*) = 0\} &= 1 - (1 - 1/n)^n \\ &\approx 1 - e^{-1} \\ &= 0.632. \end{aligned}$$

Efron (1983) was able to establish that $\delta(\mathbf{X}^*, \mathbf{T}^*)$ is roughly distributed as $\delta(\mathbf{X}, \mathbf{T})/0.632$, so that

$$\begin{aligned} \mathrm{pr}_{\hat{F}}\left\{\delta(\mathbf{X}^*, \mathbf{T}^*) > \frac{\Delta}{0.632} \mid \delta(\mathbf{X}^*, \mathbf{T}^*) > 0\right\} &\approx \mathrm{pr}_F\{\delta(\mathbf{X}, \mathbf{T}) > \Delta\} \\ &= 1 - F_\delta(\Delta), \end{aligned} \tag{10.4.5}$$

where F_δ denotes the distribution function of $\delta(\mathbf{X}, \mathbf{T})$. We will use $\hat{F}_\delta^{(B)}$ to denote its bootstrap analogue. Also, let

$$Q_1(\Delta) = E\{Q_2(\mathbf{X}; \mathbf{T}) \mid \delta(\mathbf{X}; \mathbf{T}) = \Delta\},$$

where

$$Q_2(\mathbf{x}; \mathbf{t}) = \sum_{i=1}^{g} z_i Q[i, r(\mathbf{x}; \mathbf{t})].$$

Then $Q_1(\Delta)$ is the overall unconditional error rate of $r(\mathbf{X}; \mathbf{T})$ given that $\mathbf{X}$ is distributed independently of $\mathbf{T}$ at a distance Δ away. Notice that $Q_1(0) = E\{A(\mathbf{T})\}$. We let $\hat{Q}_1^{(B)}(\Delta)$ denote the bootstrap analogue of $Q_1(\Delta)$. For economy of notation, we are using Δ here to denote a realization of $\delta(\mathbf{X}, \mathbf{T})$ or its bootstrap analogue. Elsewhere, Δ is used always to denote the Mahalanobis distance between two distributions with a common covariance matrix.

With the above notation, it follows that

$$\begin{aligned} &E^*\{Q_2(\mathbf{X}^*, \mathbf{T}^*) \mid \delta(\mathbf{X}^*, \mathbf{T}^*) > 0\} - E\{A(\mathbf{T})\} \\ &\quad = \int_{\Delta > 0} \hat{Q}_1^{(B)}(\Delta) d\hat{F}_\delta^{(B)}(\Delta) - Q_1(0) \\ &\quad \approx \int_{\Delta > 0} \{\hat{Q}_1^{(B)}(\Delta) - \hat{Q}_1^{(B)}(0)\} d\hat{F}_\delta^{(B)}(\Delta) \\ &\quad \approx \int_{\Delta > 0} \{Q_1(\Delta) - Q_1(0)\} dF_\delta(.632\Delta) && (10.4.6) \\ &\quad = \int_{\Delta > 0} \{Q_1(\Delta/0.632) - Q_1(0)\} dF_\delta(\Delta) \\ &\quad \approx (1/0.632) \int_{\Delta > 0} \{Q_1(\Delta) - Q_1(0)\} dF_\delta(\Delta) && (10.4.7) \\ &\quad = -(1/0.632) b. && (10.4.8) \end{aligned}$$

The approximation (10.4.6) follows from (10.4.5), and the final approximation (10.4.7) assumes that $Q_1(\Delta)$ is reasonably linear for small Δ. From (10.4.8), the bias of A is given approximately by

$$b \approx 0.632[E\{A(\mathbf{T})\} - E^*\{ec(\hat{F};\mathbf{T}^*) \mid \delta(\mathbf{X}^*,\mathbf{T}^*) > 0\}], \qquad (10.4.9)$$

which suggests estimating b as

$$\hat{b}^{0.632} = 0.632(A - \hat{\epsilon}^{(B)}),$$

where $\hat{\epsilon}^{(B)}$ is the Monte Carlo approximation (10.4.3) to

$$E^*\{ec(\hat{F};\mathbf{T}^*) \mid \delta(\mathbf{X}^*,\mathbf{T}^*) > 0\}.$$

Hjort (1986a) has since generalized this result to the estimation of the group-conditional error rates for an arbitrary number of groups. For either mixture or separate sampling, the 0.632 estimator of the error rate for the first group is

$$A_1^{(0.632)} = 0.328A_1 + 0.632\hat{\epsilon}_1^{(B)},$$

where $\hat{\epsilon}_1^{(B)}$ is approximated as

$$\hat{\epsilon}_1^{(B)} = \sum_{j=1}^{n}\sum_{k=1}^{K} z_{1j}\tilde{\delta}_{jk} Q[1, r(\mathbf{x}_j;\mathbf{t}_k^*)] \Big/ \sum_{j=1}^{n}\sum_{k=1}^{K} z_{1j}\tilde{\delta}_{jk}.$$

10.4.5 Linear Combinations of the Apparent and Cross-Validated Error Rates

Efron (1983) showed that $\hat{\epsilon}^{(B)}$ is almost the same as $A^{(HCV)}$, the estimated overall error rate obtained after a cross-validation that leaves out half of the observations at a time. For even n, we can define $A^{(HCV)}$ as

$$A^{(HCV)} = \sum_{i=1}^{g}\sum_{j=1}^{n}\sum_{s_j} z_{ij} Q[i, r(\mathbf{x}_j;\mathbf{t}_{(s_j)})]/n \binom{n-1}{\frac{1}{2}n},$$

where for a given j $(j = 1,\ldots,n)$, the third sum is taken over all subsamples $\mathbf{t}_{(s_j)}$ of $\mathbf{t}_j$, containing $\frac{1}{2}n$ observations. As remarked by Efron (1983), cross-validation is often carried out, removing large blocks of observations at a time. Suppose, for example, that the training set is divided into, say, q blocks, each consisting of m data points, where, thus, $n = qm$ $(m \geq 1)$. Let now

$$\mathbf{t}_{(k)} = (\mathbf{y}_1',\ldots,\mathbf{y}_{(k-1)m}',\mathbf{y}_{km+1}',\ldots,\mathbf{y}_n')',$$

that is, the training set after the deletion of the kth block of m observations. Then

$$A^{(CVq)} = \sum_{i=1}^{g}\sum_{j=1}^{m}\sum_{k=1}^{q} z_{ij} Q[i, r(\mathbf{x}_{(k-1)m+j};\mathbf{t}_{(k)})]/n$$

requires only q recomputations of the rule.

As $\hat{\epsilon}^{(B)}$ is almost the same as $A^{(HCV)}$, $A^{(0.632)}$ is almost the same as

$$0.368A + 0.632A^{(HCV)}. \qquad (10.4.10)$$

Estimators of this type were considered previously by Toussaint and Sharpe (1975) and McLachlan (1977c) in the context of choosing the weight ω so that

$$A^{(\omega)} = (1-\omega)A + \omega A^{(CVq)},$$

where, as above, $A^{(CVq)}$ denotes the estimated rate after cross-validation removing $m = n/q$ observations at a time.

For the rule based on the sample NLDF with zero cutoff point, McLachlan (1977c) calculated the value ω_o of ω for which $A^{(\omega)}$ has zero first-order bias. The training data were taken to have been obtained by separate sampling from each of $g = 2$ groups under the homoscedastic normal model (10.2.2) with $\pi_1 = \pi_2 = 0.5$. The value of ω_o was computed as a function of q, Δ, p, and the relative size of n_1 and n_2. For $q = 2$, McLachlan (1977c) showed that ω_o ranged from 0.6 to 0.7 for the combinations of the other parameters considered ($\Delta = 1,2$; $p = 4,8,16$; $n_1/n_2 = 1/3,1,3$). Hence, at least under the homoscedastic normal model, the estimator $A^{(\omega_o)}$ is about the same as Efron's 0.632 estimator. The latter, therefore, should have almost zero first-order bias under (10.2.2), at least for the sample NLDR. Efron (1983, Table 4) did calculate the first-order bias of $A^{(0.632)}$ for this rule in the various cases of (10.2.2) under which it was applied in his simulations, and it was small. With one exception, the asymptotic bias was in a downward direction, and in the simulations, $A^{(0.632)}$ exhibited a moderate downward bias. The reason for the remarkably low mean-squared error of $A^{(0.632)}$ in the simulations was the lack of negative correlation between $\hat{b}^{(0.632)}$ and $A(\mathbf{T}) - ec(F;\mathbf{T})$.

Estimators of the type $A^{(\omega)}$ have been considered, too, by Wernecker, Kalb, and Stürzebecher (1980) and Wernecke and Kalb (1983, 1987). In particular, Wernecke and Kalb (1983) focused on the choice of the weighting function ω for $g > 2$ groups. The recent simulations of Wernecke and Kalb (1987) provide further support of the superiority of this type of estimator over the apparent error rate corrected for bias according to the ordinary bootstrap.

Concerning the use of $\hat{\epsilon}^{(B)}$ as an estimator in its own right, it has been suggested in the literature (Chatterjee and Chatterjee, 1983) that $\hat{\epsilon}^{(B)}$ will be an unbiased estimator, since it is computed by applying the rule to those original data points not in the bootstrap sample. Actually, Chatterjee and Chatterjee (1983) proposed a slightly different version where $\hat{\epsilon}^{(B)}$ was computed for each single bootstrap training set and then averaged over the K replications. However, it follows from the work of Efron (1983) briefly outlined above, that $\hat{\epsilon}^{(B)}$ is biased upward, which has been confirmed empirically in the simulation experiments of Chernick, Murthy, and Nealy (1985, 1986a). These experiments did suggest, though, that in some instances where the error rate is high, $\hat{\epsilon}^{(B)}$ is superior to $A^{(0.632)}$.

Efron (1983) proposed a more efficient way of estimating the bootstrap bias than via (10.3.4), by expressing it in terms of the bootstrap repetition error rates. The hth repetition error rate $\hat{\epsilon}^{(h)}$ is defined to be the bootstrap error rate at a point replicated h times in the bootstrap training data. It was established that

$$E^*\{A(\mathbf{T}^*) - ec(\hat{F};\mathbf{T}^*)\} = \sum_{h=0}^{n} p_n^{(h)}(1-h)\hat{\epsilon}^{(h)}, \tag{10.4.11}$$

where

$$p_n^{(h)} = \binom{n}{h}(n-1)^{n-h}/n^n.$$

As explained by Efron (1983), the method of estimation based on (10.4.11) can be quite more efficient than the obvious Monte Carlo algorithm giving (10.3.4) if the number of bootstrap replications K is small. The improvement arises from not having to estimate by Monte Carlo the theoretical constants $p_n^{(h)}$. Usually, the $\hat{\epsilon}^{(h)}$ must be estimated by Monte Carlo as

$$\hat{\epsilon}^{(h)} \approx \sum_{i=1}^{g}\sum_{j=1}^{n}\sum_{k=1}^{K} z_{ij}\delta_{jk}^{(h)} Q[i, r(\mathbf{x}_j;\mathbf{t}_k^*)] \Big/ \sum_{j=1}^{n}\sum_{k=1}^{K}\delta_{jk}^{(h)},$$

where $\delta_{jk}^{(h)} = 1$ if $\mathbf{y}_j$ occurs h times in $\mathbf{t}_k^*$, the kth bootstrap replication of $\mathbf{t}$, and is zero otherwise. Note that $\hat{\epsilon}^{(0)}$ is the same as $\hat{\epsilon}^{(B)}$ defined earlier by (10.4.3).

10.4.6 Comparative Performance of $A^{(0.632)}$

Much of the work in comparing error-rate estimators has concentrated on multivariate normal group-conditional distributions, and then mostly for the sample NLDR applied under the appropriate homoscedastic normal model (10.2.2) for $g = 2$ groups. For the latter case, the simulation results of Chernick, Murthy, and Nealy (1985, 1986a) on nonparametric error rate estimators confirm the aforementioned findings of Efron (1983) that, in terms of mean-squared error, $A^{(0.632)}$ is superior to its competitors in those situations where bias correction of the apparent error rate is warranted. In the former study, which also considered $g = 3$ groups, some new variants of the bootstrap were proposed, including the convex bootstrap, which takes convex combinations of neighboring training observations in the resampling. Also, some further variants of the bootstrap approach to error-rate estimation are considered in Sánchez and Cepeda (1989), including the Bayesian bootstrap based on the idea of Rubin (1981).

Further support of the superiority of $A^{(0.632)}$ in estimating the overall error rate of the sample NLDR has been provided by the extensive simulation study conducted by Fitzmaurice, Krzanowski, and Hand (1990). Also, previous simulations by Jain, Dubes, and Chen (1987) confirm the superiority of $A^{(0.632)}$ for the $k = 1$ nearest neighbor (1-NN) and sample quadratic discriminant rules applied to data arising under the normal model (10.2.2).

Further simulations by Chernick, Murthy, and Nealy (1986b) for $g = 2$ and 3 groups with nonnormal distributions led to similar conclusions as in the case of normality. However, they caution about the use of $A^{(0.632)}$ for heavy-tailed distributions such as the Cauchy, as their simulations suggest that $\hat{\epsilon}^{(B)}$ does not have its usual positive bias then to compensate for the downward bias of the apparent error rate.

10.5 SMOOTHING OF THE APPARENT ERROR RATE

10.5.1 Introduction

Various attempts have been made to smooth the apparent error rate with a view to reducing its variance in estimating the conditional error rate. Let $h(u)$ denote a function ranging from zero to one, which is continuous, monotonic decreasing, and satisfies

$$h(u) = 1 - h(-u)$$

over the real line. For $g = 2$ groups, suppose that a sample-based rule $r(\mathbf{x};\mathbf{t})$ is 1 or 2 according as some statistic $u(\mathbf{x};\mathbf{t})$ is greater or less than zero. Then a smoothed version of the overall apparent error rate A is given by

$$A^{(S)} = \frac{1}{n}\sum_{j=1}^{n}[z_{1j}h(u(\mathbf{x}_j;\mathbf{t})) + z_{2j}\{1 - h(u(\mathbf{x}_j;\mathbf{t}))\}]. \qquad (10.5.1)$$

Hence, $A^{(S)}$ is formed by replacing the zero-one function $Q[i, r(\mathbf{x}_j;\mathbf{t})]$ in the definition of the apparent error rate by a smoothing function that can take on values between zero and one. It can be seen that a modest perturbation of an $\mathbf{x}_j$ can switch the indicator function Q from zero to one or vice versa, but will cause only a small perturbation for a smooth function.

For univariate normal and uniform distributions, Glick (1978) performed some simulations to study $A^{(S)}$ with $h(u)$ having a straight-line slope. Tutz (1985) proposed the logistic form for $h(u)$,

$$h(u,\zeta) = 1/\{1 + \exp(\zeta u)\},$$

where $\zeta(\zeta > 0)$ determines the slope of the function, and, hence, the properties of the smoothed estimator $A^{(\zeta)}$. For example, for $\zeta = 0$, $A^{(\zeta)}$ is 0.5, and in the other direction, it approaches the apparent error rate as $\zeta \to \infty$. Tutz (1985) suggested that ζ be based on the observed training data by letting $\zeta = \kappa D^2$, where κ is chosen to depend on the sample sizes. He subsequently showed how, using the single function $h(u;\zeta)$, a smoothed version of the apparent error rate for $g > 2$ can be defined recursively. Some simulations were performed to demonstrate the usefulness of this smoothing.

Snapinn and Knoke (1985) extended the ideas of Glick (1978) to the multivariate case. In particular, for the sample NLDF $\xi(\mathbf{x};\hat{\boldsymbol{\theta}}_E)$ applied with a cutoff point k with respect to $g = 2$ groups, they considered a smoothed estimator by taking

$$h(u(\mathbf{x};\mathbf{t});\zeta) = \Phi\{-u(\mathbf{x};\mathbf{t})/(\zeta D)\}, \qquad (10.5.2)$$

where

$$u(\mathbf{x};\mathbf{t}) = \xi(\mathbf{x};\hat{\boldsymbol{\theta}}_E) - k.$$

The smoothing constant,

$$\zeta = [\{(p+2)(n_1-1)+(n_2-1)\}/\{n(n-p-3)\}]^{1/2}, \qquad (10.5.3)$$

was chosen so that the bias of the smoothed estimator of $ec_1(F_1;\mathbf{t})$ is approximately equal to that of the modified plug-in estimator $\Phi(-\frac{1}{2}DS)$ for separate sampling under the homoscedastic normal model (10.2.2). For nonnormal as well as normal group-conditional distributions, Snapinn and Knoke (1985) compared their normally smoothed estimator (10.5.2) with the apparent error rate A_1 and its cross-validated version $A_1^{(CV)}$, and the ideal constant estimator $A_1^{(IC)} = A_1 - b_1$, where b_1 is the bias of A_1. They concluded that the normally smoothed estimator has a smaller mean-squared error than $A_1^{(IC)}$ if the sample size n is sufficiently large relative to p. In a later study, Snapinn and Knoke (1988) demonstrated how the performance of their smoothed estimator can be improved by using the bootstrap to reduce its bias. They also suggested a new smoothed estimator with reduced bias. It corresponds to using (10.5.2) with

$$\zeta = \{D^2/(c_1D^2 - c_2) - (n_1-1)/n_1\}^{1/2}$$

if $D^2 > c_2/c_1$ and $n > p+3$; otherwise ζ is taken to be infinitely large. Here

$$c_1 = (n-p-3)/(n-2)$$

and

$$c_2 = pn/n_1n_2.$$

Note that with $D^2 \leq c_2/c_1$ or $n \leq p+3$, this method returns an estimate of 1/2.

In later work, Snapinn and Knoke (1989) found on the basis of a simulation study performed under both normality and nonnormality that their new smoothed error rate estimated performed generally at least as well as the ordinary bootstrap. In their study, the problem was to estimate the error rate of the final version of the sample NLDR formed from a subset of the available feature variables, using a forward stepwise algorithm.

10.5.2 Posterior Probability-Based Error-Rate Estimator

In many situations, the sample-based rule $r(\mathbf{x};\mathbf{t})$ is defined from (1.4.3) in terms of the estimated posterior probabilities $\hat{\tau}_i(\mathbf{x}_j;\mathbf{t})$. In which case a natural way of smoothing the apparent error rate is to define the smoothing function in terms of the available estimates of the posterior probabilities. The smoothed estimator so obtained is given by

$$A^{(PP)} = \sum_{j=1}^{n} \min_i \hat{\tau}_i(\mathbf{x}_j;\mathbf{t})/n.$$

By virtue of its definition in terms of the estimated posterior probabilities, $A^{(PP)}$ is sometimes referred to as the posterior probability error rate estimator. It is also referred to as the average conditional error-rate estimator since the minimum of the known posterior probabilities of group membership is the error, conditional on $\mathbf{x}$, of misallocating a randomly chosen entity from a mixture of the groups on the basis of the Bayes rule $r_o(\mathbf{x};F)$.

Before proceeding to consider the use of $A^{(PP)}$ as an estimator of the overall conditional error rate $ec(F;\mathbf{t})$, we consider first the case of known posterior probabilities of group membership. The use of the posterior probability-based error-rate estimator originally arose in the context of estimating the overall error rate $eo(F)$ of the Bayes rule $r_o(\mathbf{x};F)$ formed on the basis of the known posterior probabilities $\tau_i(\mathbf{x})$. In their extension of the work of Chow (1970), Fukunaga and Kessell (1972, 1973) showed for $g = 2$ groups that

$$AO^{(PP)} = \frac{1}{n}\sum_{j=1}^{n} \min_i \tau_i(\mathbf{x}_j)$$

is an unbiased estimator of $eo(F)$ with smaller variance than the apparent error rate of $r_o(\mathbf{x};F)$ applied to $\mathbf{t}$,

$$AO = \frac{1}{n}\sum_{i=1}^{g}\sum_{j=1}^{n} z_{ij}Q[i, r_o(\mathbf{x}_j;F)].$$

More specifically, they showed that the variance of $AO^{(PP)}$ can be expressed as

$$\begin{aligned}\operatorname{var}(AO^{(PP)}) &= \operatorname{var}(AO) - n^{-1}E[\tau(\mathbf{X})\{1-\tau(\mathbf{X})\}]\\ &= \operatorname{var}(AO) - n^{-1}eo(F) + n^{-1}E\{\tau(\mathbf{X})\}^2, \qquad (10.5.4)\end{aligned}$$

where

$$\tau(\mathbf{x}) = \min_i \tau_i(\mathbf{x})$$

and

$$\operatorname{var}(AO) = n^{-1}eo(F)\{1-eo(F)\}.$$

The result (10.5.2) provides a lower bound on the reduction in the variance by using $AO^{(PP)}$ instead of AO. For since

$$E\{\tau(\mathbf{X})\}^2 \leq \tfrac{1}{2}E\{\tau(\mathbf{X})\} = \tfrac{1}{2}eo(F), \qquad (10.5.5)$$

it follows from (10.5.4) that

$$\operatorname{var}(AO) - \operatorname{var}(AO^{(PP)}) \geq \tfrac{1}{2}n^{-1}eo(F).$$

More precisely, the bound 1/2 on the right-hand side of (10.5.5) may be replaced by the maximum of $\tau(\mathbf{x})$ over $\mathbf{x}$ to give

$$\operatorname{var}(AO) - \operatorname{var}(AO^{(PP)}) \geq n^{-1}eo(F)\left\{1 - \max_{\mathbf{x}} \tau(\mathbf{x})\right\}, \qquad (10.5.6)$$

as established by Kittler and Devijver (1981) for $g \geq 2$ groups.

Fukunaga and Kessell (1973) also considered the case where the posterior probabilities of group membership were estimated nonparametrically from the training data using k-nearest neighbor (k-NN) and kernel density estimates of the group-conditional densities.

As remarked by Fukunaga and Kessell (1973) and reiterated by Kanal (1974), the result (10.5.6) that $AO^{(PP)}$ has smaller variance than AO may seem paradoxical, since $A^{(PP)}$ does not make use of the known labels of group origin of the feature observations in the training data $\mathbf{t}$, as does AO. This paradox was considered further by Kittler and Devijver (1981) in a wider context. Suppose that $r(\mathbf{x})$ is a rule formed independently of $\mathbf{t}$ for the allocation of an entity with $\mathbf{X} = \mathbf{x}$ to one of g groups. Suppose further that the training data $\mathbf{t}$ is split into two subsets, say, $\mathbf{t}_1' = (\mathbf{y}_1, \ldots, \mathbf{y}_{m_1})$ and $\mathbf{t}_2' = (\mathbf{y}_{m_1+1}, \ldots, \mathbf{y}_n)$. The $m_2 = n - m_1$ observations in $\mathbf{t}_2$ are used to estimate $\tau(\mathbf{x})$, where now $\tau(\mathbf{x})$ denotes the error of misallocation of the rule $r(\mathbf{X})$ given $\mathbf{X} = \mathbf{x}$. The m_1 observations in $\mathbf{t}_1$ are solely for the testing of $r(\mathbf{x})$. We let $\mathbf{t}_{1,x}' = (\mathbf{x}_1, \ldots, \mathbf{x}_{m_1})$ be the matrix containing the feature observations in $\mathbf{t}_1$. Kittler and Devijver (1981) considered the posterior probability-based estimator,

$$A^{(PP)}(\mathbf{t}_{1,x}, \mathbf{t}_2) = \frac{1}{m_1} \sum_{j=1}^{m_1} \hat{\tau}(\mathbf{x}_j; \mathbf{t}_2),$$

where $\hat{\tau}(\mathbf{x}_j; \mathbf{t}_2)$ is the k-NN estimate of $\tau(\mathbf{x}_j)$ computed from $\mathbf{t}_2$, that is, $\hat{\tau}(\mathbf{x}_j; \mathbf{t}_2) = k_j/k$, where k_j is the number of the k nearest neighbors of $\mathbf{x}_j$ in $\mathbf{t}_2$ belonging to groups other than the group to which $\mathbf{x}_j$ is allocated according to $r(\mathbf{x}_j)$. They showed that, although $A^{(PP)}(\mathbf{t}_{1,x}, \mathbf{t}_2)$ does not make use of the known group labels $\mathbf{z}_j$ in $\mathbf{t}_1$, it still has smaller variance than

$$A(\mathbf{t}_1) = \frac{1}{m_1} \sum_{i=1}^{g} \sum_{j=1}^{m_1} z_{ij} Q[i, r(\mathbf{x}_j)].$$

Kittler and Devijver (1981) then went on to show how the known group labels in $\mathbf{t}_1$ can be utilized to produce a more efficient estimator. They showed that

$$A^{(PP)}(\mathbf{t}_1, \mathbf{t}_2) = \frac{1}{m_1} \sum_{j=1}^{m_1} \left[\left\{ k\hat{\tau}(\mathbf{x}_j; \mathbf{t}_2) + \sum_{i=1}^{g} z_{ij} Q[i, r(\mathbf{x}_j)] \right\} \Big/ (k+1) \right] \tag{10.5.7}$$

has smaller variance than $A^{(PP)}(\mathbf{t}_{1,x}, \mathbf{t}_2)$. Notice that in (10.5.7), the known group label of $\mathbf{x}_j$ is used to create an additional neighbor of $\mathbf{x}_j$ ($j = 1, \ldots, m_1$). More recently, Fitzmaurice and Hand (1987) have compared $A^{(PP)}(\mathbf{t}_1, \mathbf{t}_2)$ for two different choices of the metric in specifying what is meant by "nearest" in the formation of the k-NN estimates.

The posterior probability-based error-rate estimator has been considered also by Fukunaga and Hostetler (1973); Lissack and Fu (1976); and Moore, Whitsitt, and Landgrebe (1976), among others. More recent references include Kittler and Devijver (1982) and Hand (1986d). A good deal of work on error-rate estimation in pattern recognition, including that on the posterior probability-based error-rate estimator, has concentrated on the provision of point estimates and bounds for the optimal overall error rate. For references additional to those cited above, the reader is referred to Hand (1986a), who has given an informative account of the pattern recognition literature on this topic.

From here on, we focus on results for $A^{(PP)}$ in its role as an estimator of the overall conditional error rate $ec(F;\mathbf{t})$ of the sample-based rule $r(\mathbf{x};\mathbf{t})$ defined by (1.4.3) in terms of the estimated posterior probabilities $\hat{\tau}_i(\mathbf{x};\mathbf{t})$. Matloff and Pruitt (1984) have given sufficient conditions on the form of the posterior probabilities of group membership $\tau_i(\mathbf{x})$ for $\sqrt{n}\{A^{(PP)} - ec(F;\mathbf{T})\}$ to have a distribution converging to the normal with mean zero and finite variance. Empirical evidence (for example, Glick, 1978; Hora and Wilcox, 1982) suggests that $A^{(PP)}$ has a smaller mean-squared error than A, but is biased in the same direction as A. This is not surprising, as $A^{(PP)}$ is closely related to the optimistic plug-in estimator (Ganesalingam and McLachlan, 1980). The component of the bias of $A^{(PP)}$ due to using the estimated posterior probabilities of the training feature observations rather than new (independent) data in its formation can be reduced to the second order by cross-validation. Let

$$A^{(CVPP)} = \frac{1}{n}\sum_{j=1}^{n}\min_i \hat{\tau}_i(\mathbf{x}_j;\mathbf{t}_{(j)}).$$

Ignoring terms of the second order, $A^{(CVPP)}$ will have the same bias as $A^{(PP)}$ formed using the estimated posterior probabilities of n feature observations independent of the training data. The coefficient of $1/n$ in the leading term of the asymptotic bias of the latter estimator was derived by Ganesalingam and McLachlan (1980) for the $\hat{\tau}_i(\mathbf{x};\mathbf{t})$ and the consequent sample rule $r(\mathbf{x};\mathbf{t})$ formed parametrically for $g = 2$ groups under the homoscedastic normal model (10.2.2). Their tabulated results for a variety of the combinations of π_1, p, and Δ suggest that the bias of $A^{(CVPP)}$ is still of practical concern for small Δ or large p relative to n. Further, their results suggest that $A^{(CVPP)}$ will provide a more optimistic assessment if a logistic model were adopted rather than directly using the appropriate normal model in obtaining estimates of the posterior probabilities. One way of correcting $A^{(PP)}$ for bias is weighting it with an estimator biased in the opposite direction. Hand (1986b) reported some encouraging results for $A^{(PP)}$ weighted with $A^{(CV)}$ with arbitrarily chosen weights 0.1 and 0.9, respectively.

An advantage of the posterior probability-based estimator $A^{(PP)}$ is that it does not require knowledge of the origin of the data used in its formation, and so is particularly useful when data of known origin are in short supply. Basford and McLachlan (1985) used this to propose an estimator of the same form as $A^{(PP)}$ for assessing the performance of a rule formed from training data that are unclassified with respect to the underlying groups. The bootstrap was found to be quite effective in reducing the optimistic bias of the assessment so obtained; see McLachlan and Basford (1988, Chapter 5) for further details.

There is not an obvious extension of the posterior probability-based error-rate estimator for a specified group or for a specified group and a specified assignment. We consider this extension now, firstly for the allocation rate $eo_{ij}(F)$, $i,j = 1,\ldots,g$, of the Bayes rule $r_o(\mathbf{x};F)$. Proceeding similarly to Schwemer and Dunn (1980), we can write

$$\begin{aligned} e_{ij}(F) &= \operatorname{pr}\{r_o(\mathbf{X};F) = j \mid \mathbf{X} \in G_i\} \\ &= (1/\pi_i)E[\tau_i(\mathbf{X})\{1 - Q[j, r_o(\mathbf{X};F)]\}] \\ &= E[\tau_i(\mathbf{X})\{1 - Q[j, r_o(\mathbf{X};F)]\} \mid \mathbf{X} \in G_i] \\ &\quad + \sum_{u \neq i}^{g} (\pi_u/\pi_i)E[\tau_i(\mathbf{X})\{1 - Q[j, r_o(\mathbf{X};F)]\} \mid \mathbf{X} \in G_u]. \end{aligned} \tag{10.5.8}$$

From (10.5.8), it can be seen that

$$AO_{ij}^{(PP)} = \sum_{k=1}^{n} \left[(1/n_i)\tau_i(\mathbf{x}_k)z_{ik}\hat{z}_{jk} + (1/\pi_i)\sum_{u \neq i}(\pi_u/n_u)\tau_i(\mathbf{x}_k)z_{uk}\hat{z}_{jk} \right] \tag{10.5.9}$$

is an unbiased estimator of $e_{ij}(F)$, where $\hat{z}_{jk}$ is taken to be one or zero according as $Q[j, r_o(\mathbf{x}_k;F)]$ is zero or one, that is, according as

$$\tau_j(\mathbf{x}_k) \geq \tau_u(\mathbf{x}_k) \qquad (u = 1,\ldots,g;\ u \neq j) \tag{10.5.10}$$

holds or not. It would be unusual for the prior probabilities π_i to be unknown when the posterior probabilities $\tau_i(\mathbf{x})$ are known. However, if the former were unknown and $\mathbf{t}$ were obtained by mixture sampling, then n_i/n provides an estimate of π_i $(i = 1,\ldots,g)$. On substitution of this estimate into the right-hand side of (10.5.9), the expression for $AO_{ij}^{(PP)}$ reduces to

$$AO_{ij}^{(PP)} = (1/n_i)\sum_{k=1}^{n} \tau_i(\mathbf{x}_k)\hat{z}_{jk}. \tag{10.5.11}$$

The form (10.5.11) for $AO_{ij}^{(PP)}$ suggests using

$$A_{ij}^{(PP)} = (1/n_i)\sum_{k=1}^{n} \hat{\tau}_i(\mathbf{x}_k;\mathbf{t})\hat{z}_{jk} \tag{10.5.12}$$

for the estimation of the conditional allocation rate $ec_{ij}(F_i;\mathbf{t})$ of the rule $r(\mathbf{x};\mathbf{t})$ defined on the basis of the maximum of the estimated posterior probabilities $\hat{\tau}_u(\mathbf{x};\mathbf{t})$ over $u = 1,\ldots,g$. Accordingly, in (10.5.12), $\hat{z}_{jk}$ is defined on the basis of (10.5.10) now, using the estimated posterior probabilities of group membership.

From (10.5.12),

$$\sum_{j=1}^{g} A_{ij}^{(PP)} = (1/n_i)\sum_{k=1}^{n} \hat{\tau}_i(\mathbf{x}_k;\mathbf{t}). \tag{10.5.13}$$

Although (10.5.13) converges in probability to one as $n \to \infty$, it is desirable that it equals one for finite n, since

$$\sum_{j=1}^{g} ec_{ij}(F_i;\mathbf{t}) = 1.$$

This can be achieved by replacing n_i in (10.5.12) by $\sum_{k=1}^{n} \hat{\tau}_i(\mathbf{x}_k;\mathbf{t})$ and so defining $A_{ij}^{(PP)}$ as

$$A_{ij}^{(PP)} = \sum_{k=1}^{n} \hat{\tau}_i(\mathbf{x}_k;\mathbf{t})\hat{z}_{jk} \bigg/ \sum_{k=1}^{n} \hat{\tau}_i(\mathbf{x}_k;\mathbf{t}). \tag{10.5.14}$$

This approach was adopted by Basford and McLachlan (1985) in their estimation of the conditional error rate for the ith group on the basis of a completely unclassified sample. In the latter situation, n_i is unknown and so there is no option but to use (10.5.14) in place of (10.5.12). It can be seen from (10.5.14) that if the entity with $\mathbf{x}_k$ has a high estimated posterior probability of belonging to G_i, then $\hat{z}_{jk}$ gets almost full weight, but if this posterior probability is estimated to be low, then $\hat{z}_{jk}$ is downweighted accordingly.

In the case of separate sampling, n_i/n cannot be taken as an estimate of π_i $(i = 1,\ldots,g)$. However, $A_{ij}^{(PP)}$ as given by (10.5.14) is still applicable for the estimation of $e_{ij}(F_i;\mathbf{t})$, provided we continue to set $\hat{\pi}_i = n_i/n$ in the formation of $\hat{\tau}_i(\mathbf{x}_k;\mathbf{t})$.

10.6 PARAMETRIC ERROR-RATE ESTIMATORS

With any application of a sample-based rule, it would be usual to compute, at least in the first instance, the apparent error rate with respect to each group and overall to provide an initial guide to the performance of the rule based on the training data at hand. In the previous sections, we have considered how the apparent error rate can be modified to give an improved estimate of the conditional error rate, concentrating on nonparametric methods of bias correction.

In the case of a sample-based rule formed parametrically from the training data $\mathbf{t}$, we may wish to adopt a parametric method of estimation of its associated error rates, not withstanding the cautionary note in Section 10.2 on this approach. We denote the Bayes rule here as $r_o(\mathbf{x};\boldsymbol{\Psi})$ to emphasize its dependence on the vector of unknown parameters in the parametric form adopted for the distribution of $\mathbf{Y} = (\mathbf{X}',\mathbf{Z}')'$. The corresponding sample-based rule is given by $r(\mathbf{x};\mathbf{t}) \equiv r_o(\mathbf{x};\hat{\boldsymbol{\Psi}})$, where $\hat{\boldsymbol{\Psi}}$ is an appropriate estimate of $\boldsymbol{\Psi}$ formed from $\mathbf{t}$ under the adopted parametric model. The associated conditional error rate for the ith group is denoted by $ec_i(\boldsymbol{\Psi};\hat{\boldsymbol{\Psi}})$ and overall by $ec(\boldsymbol{\Psi};\hat{\boldsymbol{\Psi}})$. Note that $ec_i(\boldsymbol{\Psi};\boldsymbol{\Psi})$ and $ec(\boldsymbol{\Psi};\boldsymbol{\Psi})$ are the optimal error rates for the ith group and overall, respectively, that is, $eo_i(\boldsymbol{\Psi})$ and $eo(\boldsymbol{\Psi})$. As in the previous sections, we focus without loss of generality on the estimation of the conditional error rate for the first group.

A number of parametric error-rate estimators have been proposed and evaluated over the years, mainly under the assumption of a homoscedastic normal model for the group-conditional distributions; see Dunn and Varady (1966), Lachenbruch (1968), Lachenbruch and Mickey (1968), Dunn (1971), Sedransk and Okamoto (1971), Sorum (1971, 1972, 1973), Broffitt and Williams (1973), McLachlan (1974b, 1974c, 1974d), and the references therein. More recent studies include those by Snapinn and Knoke (1984), Page (1985), and Huberty, Wisenbaker, and Smith (1987).

A common approach is to use plug-in estimates whereby unknown parameters are estimated by the same estimates used in the formation of the rule. Thus, the plug-in estimator $\hat{e}^{(PI)}$ of $ec_1(\Psi;\hat{\Psi})$ is $ec_1(\hat{\Psi};\hat{\Psi})$. Hence, we are really estimating the conditional error rate by the optimal error rate of the Bayes rule for $\Psi = \hat{\Psi}$.

For the homoscedastic normal model (10.2.2) for $g = 2$ groups with equal prior probabilities, the plug-in estimator is given by (Fisher, 1936),

$$ec_1(\hat{\Psi};\hat{\Psi}) = \Phi(-\tfrac{1}{2}D).$$

The sample Mahalanobis distance D tends to overestimate the true value Δ of this distance, as can be seen from

$$E(D^2) = \{\Delta^2 + p(1/n_1 + 1/n_2)\}\{N/(N-p-1)\}, \qquad (10.6.1)$$

where $N = n_1 + n_2 - 2$. Hence, as $\Phi(-\frac{1}{2}D)$ is the optimal error rate under the normal model (10.2.2) with a true Mahalanobis distance D between the group-conditional distributions, and D overestimates Δ, the plug-in estimator $\Phi(-\frac{1}{2}D)$ provides an optimistic estimate not only of the conditional error rate, but also mostly of the optimal error rate $\Phi(-\frac{1}{2}\Delta)$. McLachlan (1973b) expanded the expectation of $\Phi(-\frac{1}{2}D)$ to give

$$E\{\Phi(-\tfrac{1}{2}D)\} \approx \Phi(-\tfrac{1}{2}\Delta) + B_o^{(PI)}(\Delta),$$

where the asymptotic bias $B_o^{(PI)}(\Delta)$ of the plug-in estimator of $\Phi(-\frac{1}{2}\Delta)$ was given up to terms of order $O(n^{-2})$. Up to terms of the first order,

$$\begin{aligned} B_o^{(PI)}(\Delta) = \{\phi(\tfrac{1}{2}\Delta)/32\}[2\{\Delta - (p-1)(4/\Delta)\}(1/n_1 + 1/n_2) \\ + \Delta\{\Delta^2 - 4(2p+1)\}(1/N)]. \end{aligned} \qquad (10.6.2)$$

For the sample NLDF $\xi(\mathbf{x};\hat{\boldsymbol{\theta}}_E)$ applied with a zero cutoff point, McLachlan (1973b) used the result (10.6.2) to derive under (10.2.2) the N^{-2}-order bias of $\Phi(-\frac{1}{2}D)$ in its estimation of the conditional error rate $ec_1(\Psi;\hat{\Psi})$, which is given by (4.3.5) on putting $i = 1$ and $k = 0$. This asymptotic bias is given up to terms of the first order by

$$\begin{aligned} B_1^{(PI)}(\Delta) = -\phi(\tfrac{1}{2}\Delta)[(p-1)(\Delta/n_1) \\ + \{4(4p-1) - \Delta^2\}\{\Delta/(32N)\}]. \end{aligned} \qquad (10.6.3)$$

As established in McLachlan (1976a), the asymptotic expectations of the plug-in and apparent error rates under (10.2.2) are related as

$$E\{\Phi(-\tfrac{1}{2}D)\} \approx E\{A_1(\mathbf{T})\} + \phi(\tfrac{1}{2}\Delta)(\Delta/32)\{8/n_1 + (\Delta^2 - 12)/N\}.$$

Using (10.6.3), the bias-corrected version of $\Phi(-\frac{1}{2}D)$,

$$\Phi(-\tfrac{1}{2}D) - B_1^{(PI)}(D), \tag{10.6.4}$$

has bias of the second order. The estimator (10.6.4) corresponds to applying the bootstrap parametrically to correct $\Phi(-\frac{1}{2}D)$ for bias, where the bootstrap bias is calculated analytically up to terms of the first order. As with the asymptotic result (10.2.3) of McLachlan (1976a) for the bias of the apparent error rate, Schervish (1981b) extended (10.6.4) to the case of $g > 2$ groups. He performed also some simulations in the case of $g = 3$ groups with bivariate normal distributions having a common covariance matrix. They demonstrated the superiority in terms of mean-squared error of the parametric bias correction analogous to (10.6.4) over nonparametric methods in the situation where the adopted parametric model is valid.

From (10.6.1), an unbiased estimator of Δ^2 for separate sampling under (10.2.2) is given by

$$\hat{\Delta}^2 = \{(N-p-1)/N\}D^2 - p(1/n_1 + 1/n_2), \tag{10.6.5}$$

but, if n_1 and n_2 are small relative to p, it can be negative. As a consequence, Lachenbruch and Mickey (1968) suggested using $\Phi(-\frac{1}{2}DS)$ as an estimate, both of the optimal and conditional error rates, where

$$DS = \{(N-p-1)/N\}^{1/2}D.$$

Lachenbruch and Mickey (1968) also proposed $eua_1(\hat{\Delta})$ as an error-rate estimator, where $eua_1(\Delta)$ is the N^{-2}-order asymptotic expansion of the unconditional error rate $eu_1(\Delta)$ under the homoscedastic normal model (10.2.2). They labeled $eua_1(\Delta)$ as the O or OS estimator, corresponding to the use of $\hat{\Delta} = D$ or DS, respectively. The bias of $\hat{e}_1^{(O)} = eua_1(D)$ or of $\hat{e}_1^{(OS)} = eua_1(DS)$ is of the second order. The estimator $\hat{e}_1^{(O)}$ can be viewed as the N^{-2}-order expansion of the bootstrap unconditional error rate for the bootstrap applied parametrically; see Chen and Tu (1987).

For the estimation of the conditional error rate $ec_1(\Psi;\hat{\Psi})$, McLachlan (1974b, 1975b) showed that the bias can be reduced to the third order only by using

$$\begin{aligned}\hat{e}_1^{(M)} = \Phi(-\tfrac{1}{2}D) + \phi(\tfrac{1}{2}D)[&(p-1)/(Dn_1)\\ &+ D\{4(4p-1) - D^2\}/(32N) + \{(p-1)(p-2)\}/(4Dn_1^2)\\ &+ (p-1)\{-D^3 + 8(2p+1)D + (16/D)\}/(64n_1N)\\ &+ (D/12288)\{3D^6 - 4(24p+7)D^4 + 16(48p^2 - 48p - 53)D^2\\ &\qquad + 192(-8p+15)\}/N^2].\end{aligned} \tag{10.6.6}$$

Lachenbruch (1968) proposed another parametric estimator obtained by replacing the numerator and denominator of the argument in (4.3.5) with their expectations in which Δ^2 is replaced by its unbiased estimate (10.6.5). This so-called L estimator is given by

$$\hat{e}_1^{(L)} = \Phi[-M^{-1/2}\{\tfrac{1}{2}D - (pND^{-1})/(n_1(N-p-1))\}],$$

where

$$M = N(N-1)/(N-p)(N-p-3).$$

Note that by interchanging n_1 and n_2 in the expressions above for the various parametric error estimators for the first group, the corresponding error-rate estimators for the second group are obtained.

Lachenbruch and Mickey (1968) proposed their so-called $\overline{U}$ estimator by using cross-validation in conjunction with a normal-based approach for $g = 2$ groups. The estimator of the conditional error rate $ec_i(\Psi;\hat{\Psi})$ so produced is given by

$$\hat{e}_i^{(\overline{U})} = \Phi\{(-1)^i\overline{\xi}_i/s_{\xi_i}\} \qquad (i = 1,2),$$

where $\overline{\xi}_i$ and $s^2_{\xi_i}$ denote the sample mean and variance of those $\xi(\mathbf{x}_j;\hat{\boldsymbol{\theta}}_{(j)E})$, n_i in number, from G_i $(i = 1,2)$, and where $\hat{\boldsymbol{\theta}}_{(j)E}$ denotes the estimate of $\boldsymbol{\theta}_E$ formed from $\mathbf{t}_{(j)}$ for $j = 1,\ldots,n$.

Results on the relative superiority of parametric estimators of the various types of error rates of the sample NLDR are available from the comparative studies reported in the references cited near the beginning of this section. For example, Lachenbruch and Mickey (1968) put forward recommendations on the basis of an empirical study conducted under (10.2.2), and McLachlan (1974b) provided theoretical confirmation using the criterion of asymptotic mean-squared error. Among more recent studies, Page (1985) performed a series of simulation experiments devoted exclusively to parametric error-rate estimators, including all of those defined above. The results reported by Page (1985) are generally consistent with the results of the earlier studies of Lachenbruch and Mickey (1968) and McLachlan (1974b, 1974c). Page (1985) concluded that, so far as the estimation of the conditional error rate is concerned, the estimators $\hat{e}_1^{(L)}$, $\hat{e}_1^{(M)}$, and $\hat{e}_1^{(OS)}$ are to be preferred. The latter estimator appears to be best for small and medium p, whereas the former two are best for large p.

Recently, Ganeshanandam and Krzanowski (1990) have compared several parametric estimators of the overall conditional error rate of the sample NLDF $\xi(\mathbf{x};\hat{\boldsymbol{\theta}}_E)$ applied with a zero cutoff point with respect to $g = 2$ groups. The group-conditional distributions were taken to be multivariate normal with a common covariance matrix (ideal conditions) and multivariate binary (nonideal conditions). Their study also included nonparametric estimators. Altogether, they compared 11 estimators: $\hat{e}^{(L)}$, $\hat{e}^{(M)}$, $\hat{e}^{(OS)}$, $\hat{e}^{(PI)}$, $\hat{e}^{(\overline{U})}$, A, $A^{(J_c)}$, $A^{(0.632)}$, $A^{(S)}$ with the smoothing constant defined by (10.5.3), $A^{(CV)}$, and also $A^{(CV)}$ as computed for the sample NQDF $\xi(\mathbf{x};\hat{\boldsymbol{\theta}}_U)$. For multivariate normal group-conditional distributions, Ganeshanandam and Krzanowski (1990) found that in terms of mean-squared error, the best estimators are, in order of decreasing superiority, $\hat{e}^{(M)}$, $\hat{e}^{(L)}$, $\hat{e}^{(\overline{U})}$, $\hat{e}^{(OS)}$, $A^{(CV)}$, and $A^{(CV)}$ as for the sample NQDR. These estimators performed well above the remaining ones. In terms of bias, $A^{(CV)}$ was found to be the least optimistic, followed by $\hat{e}^{(M)}$, $A^{(J_c)}$, $\hat{e}^{(L)}$, and $\hat{e}^{(\overline{U})}$. They noted that although $\hat{e}^{(OS)}$ is in the class of best estimators on the basis of mean-squared error, it has an unacceptably large optimistic bias.

For multivariate binary group-conditional distributions representing nonideal conditions for the parametric estimators, Ganeshanandam and Krzanowski (1990) concluded that $A^{(CV)}$, $\hat{e}^{(\overline{U})}$, $A^{(J_c)}$, $\hat{e}^{(OS)}$, $\hat{e}^{(L)}$, and $\hat{e}^{(M)}$ are the best. Again $\hat{e}^{(OS)}$ was found to be overly optimistic in its assessment.

Of course, there is little absolute difference between the various parametric estimators when the sample size n is large relative to the dimension p of the feature vector. The formation of parametric error-rate estimators under models more general than (10.2.2) is limited by the difficulty in calculating manageable analytical expressions for the unconditional or even optimal error rates; see Section 4.6.

10.7 CONFIDENCE INTERVALS

McLachlan (1975b) considered interval estimation of the conditional error rate $ec_1(F_1;\mathbf{t})$ of the sample NLDR under the homoscedastic normal model (10.2.2) with equal group-prior probabilities. With his proposal, an approximate $100(1-\alpha)\%$ confidence interval for $ec_1(F_1;\mathbf{t})$ has the form

$$\hat{e}_1^{(M)} \pm \{v_1(DS)\}^{1/2}\Phi^{-1}(1-\tfrac{1}{2}\alpha), \tag{10.7.1}$$

where $\hat{e}_1^{(M)}$ is the asymptotically unbiased estimator of $ec_1(F_1;\mathbf{t})$ defined by (10.6.6). The interval (10.7.1) is based on the result that for sufficiently large n_1 and n_2 under separate sampling, $\hat{e}_1^{(M)} - ec_1(F_1;\mathbf{T})$ is approximately normal with mean zero and variance $v_1(\Delta)$, where

$$\begin{aligned} v_1(\Delta) = \{\phi(\tfrac{1}{2}\Delta)\}^2[n_1^{-1} &+ (\Delta^2/8)N^{-1} + \{\Delta^2 + 4(3p-4) \\ &\qquad + (p^2-4p+5)(16/\Delta^2)\}(4n_1)^{-2} \\ &+ \{(\Delta^2-2p)/8\}(n_1n_2)^{-1} + \{\Delta^4 + 2(11p-16)\Delta^2 \\ &\qquad + 8(5p-4)\}(64n_1N)^{-1} \\ &+ \{2\Delta^6 + 16(2p-5)\Delta^4 - 32(4p-13)\Delta^2\}(32N)^{-2}]. \end{aligned}$$

Previously, Lachenbruch (1967) had considered interval estimation of the unconditional error rate $eu_1(F)$ for $g=2$ groups, using his proposed estimator $A_1^{(CV)}$. The lower and upper limits of an approximate $100(1-\alpha)\%$ confidence interval for $eu_1(F)$ can be obtained as roots of the equation

$$n_1\{A_1^{(CV)} - eu_1(F)\}^2/[eu_1(F)\{1 - eu_1(F)\}] = \{\Phi^{-1}(1 - \tfrac{1}{2}\alpha)\}^2.$$

This interval is formed on the basis that $A_1^{(CV)}$ has a binomial distribution with parameters n_1 and $eu_1(F)$. As noted by Lachenbruch (1967), the expectation of $A_1^{(CV)}$ actually equals $eu_1(F; n_1 - 1, n_2)$, the unconditional error rate for a discriminant rule based on $n_1 - 1$ observations from G_1 and n_2 from G_2. Also, the correlation between the summands in the definition (10.2.8) of $A_1^{(CV)}$ has to be ignored for the variance of $A_1^{(CV)}$ to be

$$eu_1(F; n_1 - 1, n_2)\{1 - eu_1(F; n_1 - 1, n_2)\}/n_1.$$

However, this correlation decreases as n_1 becomes large, and Lachenbruch (1967) performed some simulations to demonstrate that it is small for the normal model (10.2.2).

For an arbitrary sample-based discriminant rule $r(\mathbf{x}; \mathbf{t})$, let $\hat{e}_1(\mathbf{t})$ denote an estimate of its unconditional error rate $eu_1(F)$. Typically, $\sqrt{n}\{\hat{e}_1(\mathbf{T}) - eu_1(F)\}$ is asymptotically normal with mean zero and finite variance, say, v_1. In which case, a standard approximate confidence interval for $eu_1(F)$ is

$$\hat{e}_1(\mathbf{t}) \pm (\hat{v}_1/n)^{1/2}\Phi^{-1}(1 - \tfrac{1}{2}\alpha),$$

where $\hat{v}_1$ denotes an estimate of the asymptotic variance formed from the training data $\mathbf{t}$. For special cases such as multivariate normal group-conditional distributions, expressions are available for v_1; see McLachlan (1972b) and Schervish (1981a). In those instances where there is no result available for v_1, it can be assessed using the bootstrap.

However, although the standard approximation may be asymptotically correct, it can be quite misleading in small samples. In a series of papers, Efron (1981b, 1982, 1985, 1987) has developed methods for constructing approximate confidence intervals for a real-valued parameter using the bootstrap. Let $\mathbf{t}_1^*, \ldots, \mathbf{t}_K^*$ denote K bootstrap replications of $\mathbf{t}$ generated according to the bootstrap algorithm described in Section 10.3. For this generation, the distribution function F of $\mathbf{Y} = (\mathbf{X}', \mathbf{Z}')'$ is replaced by an appropriate estimate $\hat{F}$ formed from $\mathbf{t}$. The nonparametric bootstrap uses $\hat{F} = \hat{F}_n$, the empirical distribution function of $\mathbf{Y}$ formed from $\mathbf{t}$. If a parametric form is adopted for the distribution function of $\mathbf{Y}$, where $\boldsymbol{\Psi}$ denotes the vector of unknown parameters, then the parametric bootstrap uses an estimate $\hat{\boldsymbol{\Psi}}$ formed from $\mathbf{t}$ in place of $\boldsymbol{\Psi}$. That is, if we write F as $F_{\boldsymbol{\Psi}}$ to signify its dependence on $\boldsymbol{\Psi}$, then the bootstrap data are generated from $\hat{F} = F_{\hat{\boldsymbol{\Psi}}}$.

In the subsequent discussion, we will assume a mixture sampling scheme for $\mathbf{t}$, but the results apply similarly under separate sampling. As above, $\hat{e}_1(\mathbf{t})$ denotes an arbitrary error-rate estimator formed from $\mathbf{t}$. The distribution function of $\hat{e}_1(\mathbf{T})$ can be approximated by

$$\widehat{CDF}(u) = \#\{\hat{e}_1(\mathbf{t}_k^*) \le u\}/K, \tag{10.7.2}$$

the empirical distribution function formed from the K bootstrap replications $\hat{e}_1(\mathbf{t}_1^*),\ldots,\hat{e}_1(\mathbf{t}_K^*)$. With the percentile method of Efron (1981b, 1982), a nominal $100(1-\alpha)\%$ confidence interval for $eu_1(F)$ is given by

$$[\widehat{CDF}^{-1}(\tfrac{1}{2}\alpha), \widehat{CDF}^{-1}(1-\tfrac{1}{2}\alpha)], \qquad (10.7.3)$$

where

$$\widehat{CDF}^{-1}(\alpha) = \sup\{u : \widehat{CDF}(u) \leq \alpha\}. \qquad (10.7.4)$$

That is, the percentile method interval consists of the central $1-\alpha$ proportion of the (simulated) bootstrap distribution of $\hat{e}(\mathbf{T}^*)$.

Efron (1982) showed that (10.7.3) will have the desired coverage probability $1-\alpha$ if $\hat{e}_1(\mathbf{T}) - eu_1(F)$ has the same distribution, symmetric about the origin, as $\hat{e}_1(\mathbf{T}^*) - eu_1(\hat{F})$ does with $\hat{F}$ held fixed at its observed value. This will be so for the parametric bootstrap if $\hat{e}_1(\mathbf{T}) - eu_1(F_{\boldsymbol{\Psi}})$ is a pivotal quantity, that is, if its distribution does not depend on $\boldsymbol{\Psi}$. Also, if the latter holds, it should be approximately so for the nonparametric bootstrap. The accuracy of the approximation depends on how close the empirical distribution function $\hat{F}_n$ is to $F_{\hat{\boldsymbol{\Psi}}}$. It can be much different for small n; see Schenker (1985).

As explained by Efron (1982), a bias correction to the percentile method is called for if $\hat{e}_1(\mathbf{T}^*)$ is median-biased, that is, if

$$\text{pr}_*\{\hat{e}_1(\mathbf{T}^*) \leq eu_1(\hat{F})\} \neq 0.50,$$

where pr_* refers to probability with respect to $\hat{F}$ held fixed at its observed value. The bias-corrected percentile interval has the form

$$[\widehat{CDF}^{-1}(\Phi(2z_o + z_{(1/2)\alpha})), \widehat{CDF}^{-1}(\Phi(2z_o + z_{1-\alpha/2}))], \qquad (10.7.5)$$

where $z_\alpha = \Phi^{-1}(\alpha)$, and where z_o is estimated by

$$\Phi^{-1}\{\widehat{CDF}(eu_1(\hat{F}))\}.$$

The assumption underlying the bias-corrected percentile method is that there exists a monotonic transformation h such that $h(\hat{e}_1(\mathbf{T})) - h(eu_1(F))$ has the same normal distribution as $h(\hat{e}_1(\mathbf{T}^*)) - h(eu_1(\hat{F}))$ does with $\hat{F}$ held fixed at its observed value. Efron (1987) subsequently proposed the bias-corrected percentile acceleration method where the assumption underlying the bias-corrected percentile method is relaxed by allowing the scaling of the distribution of $h(\hat{e}_1(\mathbf{T}) - h(eu_1(F))$ to vary linearly with $h(eu_1(F))$. Note that although the validity of these bootstrap intervals rely on the existence of such a transformation h, it is not necessary to know h to form them.

In recent times, there has been much attention devoted to the bootstrap approach to the formation of confidence intervals. In particular, Efron (1985) showed that there is a wide class of problems for which the bootstrap intervals are an order of magnitude more accurate than the standard intervals. However, there are only a few results available on bootstrap confidence intervals in a discriminant analysis context. Hence, these intervals should be used with caution. If the assumptions underlying them are inapplicable, then the coverage probability may be substantially different from the nominal level. These assumptions are difficult or impossible to check in complicated situations such as the present one of error-rate estimation. Indeed, it is unlikely that the underlying assumptions are satisfied for this problem.

There is also the percentile-t method of Efron (1982) for forming bootstrap confidence intervals. Additional references on this method include Abramovitch and Singh (1985), Hall (1986a, 1986b, 1988), and Efron (1987), who consider applications where it has the capacity to provide improved confidence intervals over the percentile method. In the present context of error-rate estimation, suppose that there is available an estimate $\hat{v}_1(\mathbf{t})$ of the variance of $\hat{e}_1(\mathbf{T})$. Then with the percentile-t method, a nominal $100(1-\alpha)\%$ confidence interval for $eu_1(F)$ is given by

$$[\hat{e}_1(\mathbf{t}) - \{\hat{v}_1(\mathbf{t})\}^{1/2}\widehat{CDF}^{-1}(\tfrac{1}{2}\alpha),\ \hat{e}_1(\mathbf{t}) + \{\hat{v}_1(\mathbf{t})\}^{1/2}\widehat{CDF}^{-1}(1-\tfrac{1}{2}\alpha)],$$

where $\widehat{CDF}(u)$ denotes the empirical distribution function defined by (10.7.2), but with $\hat{e}_1(\mathbf{t}_k^*)$ now replaced by its Studentized version

$$\{\hat{e}_1(\mathbf{t}_k^*) - eu_1(\hat{F})\}/\{\hat{v}_1(\mathbf{t}_k^*)\}^{1/2}$$

for $k = 1,\ldots,K$.

10.8 SOME OTHER TOPICS IN ERROR-RATE ESTIMATION

10.8.1 Rule Selection via Apparent Error Rate

Recently, Devroye (1988) has considered the use of the overall apparent error rate as the basis for selecting the best rule from a class of discriminant rules based on some given training data $\mathbf{t}$. For this purpose, it is proposed that in addition to $\mathbf{t}$, there are test data $\mathbf{t}_m$, where

$$\mathbf{t}_m' = (\mathbf{y}_{n+1},\ldots,\mathbf{y}_{n+m}),$$

and $\mathbf{y}_j = (\mathbf{x}_j', \mathbf{z}_j')'$ for $j = n+1,\ldots,n+m$. With this provision for test data $\mathbf{t}_m$, the overall apparent error rate of a sample rule $r(\mathbf{x};\mathbf{t})$ formed from $\mathbf{t}$ is given by

$$A^{(r)}(\mathbf{t},\mathbf{t}_m) = \sum_{i=1}^{g}\sum_{j=n+1}^{n+m} z_{ij}Q[i, r(\mathbf{x}_j;\mathbf{t})]/m.$$

It is an unbiased estimator of the overall conditional error rate of $r(\mathbf{x};\mathbf{t})$, which is given by

$$ec^{(r)}(F;\mathbf{t}) = \sum_{i=1}^{g}\pi_i E\{Q[i, r(\mathbf{X};\mathbf{t})] \mid \mathbf{X} \in G_i, \mathbf{t}\}.$$

As a class of rules is under consideration here, the notation now for the various error rates explicitly shows the particular rule to which reference is made.

On the drawback of splitting up the data into training and test subsets, Devroye (1988) argued that the size m of the test subset can often be taken much smaller than the size n of the training subset ($m = o(n)$). Also, he surmised that more sophisticated methods such as cross-validation would do equally well or better than the split-data method.

For a specified class R of sample rules based on the training data $\mathbf{t}$, the automatic selection procedure considered by Devroye (1988) chooses the rule that minimizes the overall apparent error rate $A^{(r)}(\mathbf{t},\mathbf{t}_m)$ over all rules $r(\mathbf{x};\mathbf{t})$ in the class R. We let $r_{\mathbf{t}_m}(\mathbf{x};\mathbf{t})$ denote the rule so chosen, where the subscript $\mathbf{t}_m$ is to emphasize that this rule also depends on $\mathbf{t}_m$, as well as $\mathbf{t}$. For brevity, we henceforth abbreviate $r_{\mathbf{t}_m}$ to r_m. By the definition of $r_m(\mathbf{x};\mathbf{t})$,

$$A^{(r_m)}(\mathbf{t},\mathbf{t}_m) = \min_{r \in R} A^{(r)}(\mathbf{t},\mathbf{t}_m).$$

The overall conditional error rate of $r_m(\mathbf{x};\mathbf{t})$ is given by

$$ec^{(r_m)}(F;\mathbf{t}) = \sum_{i=1}^{g} \pi_i E\{Q[i, r_m(\mathbf{X};\mathbf{t})] \mid \mathbf{X} \in G_i, \mathbf{t}, \mathbf{t}_m\}.$$

For a finite class R with cardinality bounded by N_n, Devroye (1988) established that, for all $\epsilon > 0$,

$$\operatorname{pr}\left\{\sup_{r \in R} \mid A^{(r)}(\mathbf{T},\mathbf{T}_m) - ec^{(r)}(F;\mathbf{T}) \mid > \epsilon\right\} \leq 2N_n \exp(-\tfrac{1}{2}\epsilon^2) \tag{10.8.1}$$

and that

$$E\left\{\sup_{r \in R} \mid A^{(r)}(\mathbf{T},\mathbf{T}_m) - ec^{(r)}(F;\mathbf{T})\mid\right\} \leq \sqrt{\frac{\log(2N_n)}{2m}} + \frac{1}{\sqrt{8m\log(2N_n)}}. \tag{10.8.2}$$

Since

$$\left| ec^{(r_m)}(F;\mathbf{t}) - \inf_{r \in R} ec^{(r)}(F;\mathbf{t}) \right| \leq 2\sup_{r \in R} \mid A^{(r)}(\mathbf{t},\mathbf{t}_m) - ec^{(r)}(F;\mathbf{t})\mid,$$

(10.8.2) provides an upper bound for the suboptimality of $r_m(\mathbf{x};\mathbf{t})$ in R. For instance, if we take $m = n$ and assume that N_n is large, then (10.8.2) shows that the chosen rule $r_m(\mathbf{x};\mathbf{t})$ has on the average an error rate within $2\{\log(N_n)/(2n)\}^{1/2}$ of the best possible rate, whatever it is.

As explained by Devroye (1988), the apparent error rate is employed in the present context as a tool for the selection of a rule and not for the provision of estimates of the error rates per se. However, it turns out that the apparent error rate $A^{(r_m)}(\mathbf{t},\mathbf{t}_m)$ of the selected rule can serve as an estimate of its error rate. This follows from (10.8.2) on noting the inequality

$$|A^{(r_m)}(\mathbf{t},\mathbf{t}_m) - ec^{(r_m)}(F;\mathbf{t})| \leq \sup_{r \in R} \mid A^{(r)}(\mathbf{t},\mathbf{t}_m) - ec^{(r)}(F;\mathbf{t})\mid, \tag{10.8.3}$$

which is trivially true. Thus, although $A^{(r_m)}(\mathbf{t}, \mathbf{t}_m)$ will usually provide an optimistic estimate of $ec^{(r_m)}(F; \mathbf{t})$, (10.8.2) and (10.8.3) show that it is within given bounds of the latter rate.

On the specification of the class R, Devroye (1988) advises that it is a compromise between having R rich enough so that every Bayes rule can be approached asymptotically by a sequence of rules selected from a sequence of R's, but not having R so rich that trivial selection ensues. In practice, R can be specified by a variety of parametric and nonparametric rules, including, say, a few linear discriminant rules, nearest neighbor rules, a couple of tree classifiers, and perhaps a kernel-type rule.

Devroye (1988) also has derived bounds corresponding to (10.8.1) and (10.8.2) for infinite classes. He showed how these bounds for finite and infinite classes can be used to prove the consistency and asymptotic optimality for several popular classes, including linear, nearest neighbor, kernel, histogram, binary tree, and Fourier series-type discriminant rules.

10.8.2 Estimation of Realized Signal-to-Noise Ratio

Reed, Mallet, and Brennan (1974); Khatri, Rao, and Sun (1986); and Khatri and Rao (1987) have approached the estimation of the conditional error rates of the sample NLDR in terms of the realized signal-to-noise ratio in the context of signal detection. The feature vector represents an incoming message that is taken to have a multivariate normal distribution with unknown covariance matrix Σ and mean vector either $\boldsymbol{\mu}_1 = \boldsymbol{\mu}$, corresponding to a known signal $\boldsymbol{\mu}$, or $\boldsymbol{\mu}_2 = 0$, corresponding to pure white noise. In this situation, since $\boldsymbol{\mu}_1$ and $\boldsymbol{\mu}_2$ are both known, only Σ needs to be estimated. We let $\hat{\Sigma}$ denote an estimator of Σ for which

$$M\hat{\Sigma} \sim W(M, \Sigma), \tag{10.8.4}$$

a Wishart distribution with M degrees of freedom and expectation matrix $M\Sigma$. Khatri and Rao (1987) suggest that the estimation of Σ is best done by generating n_2 independent observations $\mathbf{x}_{2j}$ $(j = 1, \ldots, n_2)$ from the pure white noise process and taking $\hat{\Sigma}$ to be the maximum likelihood estimate of Σ,

$$\hat{\Sigma} = \sum_{j=1}^{n_2} \mathbf{x}_{2j}\mathbf{x}_{2j}'/n_2.$$

In which case, (10.8.4) holds with $M = n_2$.

With Σ estimated by $\hat{\Sigma}$, a message can be filtered using the simplified normal-based linear discriminant function

$$(\mathbf{x} - \tfrac{1}{2}\boldsymbol{\mu})'\hat{\Sigma}^{-1}\boldsymbol{\mu}$$

applied with a zero cutoff point. Its conditional error rates have the common value of

$$\Phi\left\{-\sqrt{\rho(\hat{\Sigma}, \Sigma)}\right\},$$

where

$$\rho(\hat{\Sigma}, \Sigma) = (\mu'\hat{\Sigma}^{-1}\mu)^2/(\mu'\hat{\Sigma}^{-1}\Sigma\hat{\Sigma}^{-1}\mu)$$

is the realized signal-to-noise ratio. As noted by Reed et al. (1974), $\rho(\hat{\Sigma}, \Sigma) \leq \mu'\Sigma^{-1}\mu = \Delta^2$. Thus, $\Delta^2 - \rho(\hat{\Sigma}, \Sigma)$ represents the loss in information in using $\hat{\Sigma}$ to estimate Σ.

An obvious point estimate of $\rho(\hat{\Sigma}, \Sigma)$ is

$$\begin{aligned}\rho(\hat{\Sigma}, \hat{\Sigma}) &= \mu'\hat{\Sigma}^{-1}\mu \\ &= D^2,\end{aligned}$$

which overestimates unless M is large compared to p. Khatri and Rao (1987) consequently suggested estimating $\rho(\hat{\Sigma}, \Sigma)$ by cD^2, where c is chosen so that

$$E\{\rho(\hat{\Sigma}, \Sigma) - cD^2\}^2$$

is a minimum. They showed that this minimum is achieved at

$$c = (M - p + 2)(M - p - 3)/M(M + 1).$$

In order to provide a lower confidence bound on $\rho(\hat{\Sigma}, \Sigma)$, Khatri and Rao (1985) considered the distributions of the statistics $\rho(\hat{\Sigma}, \Sigma)/\Delta^2$ and $\frac{1}{2}M\Delta^2/D^2$. They proved that $\rho(\hat{\Sigma}, \Sigma)/\Delta^2$ is distributed according to a beta distribution with parameters $\frac{1}{2}(M - p + 2)$ and $\frac{1}{2}(p - 1)$, independently of $\frac{1}{2}M\Delta^2/D^2$, which is distributed as a gamma distribution with parameter $\frac{1}{2}(M - p + 1)$. Also, they have derived the corresponding results for a complex feature vector. Previously, Reed et al. (1974) had obtained the distribution of $\rho(\hat{\Sigma}, \Sigma)/\Delta^2$ for the complex case.

An exact confidence bound for $\rho(\hat{\Sigma}, \Sigma)$ can be obtained from the distribution of the product of these two statistics

$$U = \tfrac{1}{2}(M/D^2)\rho(\hat{\Sigma}, \Sigma). \tag{10.8.5}$$

Khatri and Rao (1987) showed that this pivotal statistic U has the confluent hypergeometric density

$$\{\Gamma(a_3)\}^{-1}e^{-u}u^{a_3-1}\{\Gamma(a_1 + a_3 - a_2 + 1)/\Gamma(a_3 - a_2 + 1)\}h(u; a_1, a_2),$$

where $a_1 = \frac{1}{2}(p - 1)$, $a_2 = \frac{1}{2}$, $a_3 = \frac{1}{2}(M - p + 1)$, and where

$$h(u; a_1, a_2) = \{\Gamma(a_1)\}^{-1}\int_0^\infty v^{a_1-1}(1 + v)^{a_2-a_1-1}\exp(-uv)\,dv$$

is the confluent hypergeometric function of the second kind. If u_α denotes the quantile of order α of this distribution, then

$$\rho(\hat{\Sigma}, \Sigma) \geq (2D^2/M)u_\alpha$$

provides a lower confidence bound on the realized signal-to-noise ratio with a confidence coefficient of $1 - \alpha$. Khatri, Rao, and Sun (1986) have tabulated

u_α for various combinations of α, p, and M. They also have given several approximations to the distribution of U.

Khattree and Gill (1987) have derived some more estimators of $\rho(\hat{\Sigma}, \Sigma)$ under various loss functions that are dimension-free. Recently, Khattree and Gupta (1990) have considered the bivariate extension of this problem, where the incoming message is either pure white noise or one of a pair of known multivariate signals.

CHAPTER 11

Assessing the Reliability of the Estimated Posterior Probabilities of Group Membership

11.1 INTRODUCTION

In this chapter, we focus on the assessment of the reliability of the posterior probabilities of group membership as estimated from the training data **t**. As before, the posterior probability that an entity with $\mathbf{X} = \mathbf{x}$ belongs to G_i is denoted by $\tau_i(\mathbf{x})$, where

$$\tau_i(\mathbf{x}) = \pi_i f_i(\mathbf{x})/f_X(\mathbf{x}) \qquad (i = 1,\ldots,g),$$

and

$$f_X(\mathbf{x}) = \sum_{h=1}^{g} \pi_h f_h(\mathbf{x})$$

is the mixture density of **X**.

The point estimation of the posterior probabilities of group membership $\tau_i(\mathbf{x})$ has been considered in the previous chapters. It was seen there that the Bayes or optimal discriminant rule $r_o(\mathbf{x};F)$ is defined on the basis of the relative size of the $\tau_i(\mathbf{x})$. A sample version of $r_o(\mathbf{x})$ can be formed by replacing the $\tau_i(\mathbf{x})$ with some point estimates $\hat{\tau}_i(\mathbf{x};\mathbf{t})$. A variety of ways of procuring suitable point estimates $\hat{\tau}_i(\mathbf{x};\mathbf{t})$ has been discussed. For example, in Chapter 8, the logistic approach was described, whereby the posterior probabilities of group membership $\tau_i(\mathbf{x})$ are estimated by adopting a log-linear model for the ratios of the group-conditional densities. The assessment of the reliability of the estimates of the posterior probabilities so formed was considered there for this approach.

Here we consider the problem of assessing the reliability of the estimated posterior probability of group membership for the approach where the $\tau_i(\mathbf{x})$ are estimated indirectly through estimation of the group-conditional densities $f_i(\mathbf{x})$, so that

$$\hat{\tau}_i(\mathbf{x};\mathbf{t}) = \hat{\pi}_i \hat{f}_i(\mathbf{x})/\hat{f}_X(\mathbf{x}) \qquad (i = 1,\ldots,g).$$

This is called the sampling approach by Dawid (1976); see Section 1.7.

As discussed in Section 1.11, if the posterior probabilities have been estimated for the express purpose of forming a discriminant rule, then their overall reliability can be assessed through the global performance of this rule as measured by estimates of the associated error rates. However, often it is more appropriate to proceed conditionally on the observed value $\mathbf{x}$ and to provide standard errors for the estimated posterior probabilities $\hat{\tau}_i(\mathbf{x};\mathbf{t})$ and also interval estimates for the $\tau_i(\mathbf{x})$ if possible. For example, one of the objectives of the Glasgow dyspepsia study (Spiegelhalter and Knill-Jones, 1984) was to obtain an accurate assessment of the probability of a peptic ulcer being present, based on symptoms alone, in order to provide an objective basis for further selection of patients for investigation, treatment, or clinical trials.

We will concentrate here on interval estimation of the posterior probabilities of group membership $\tau_i(\mathbf{x})$, firstly in the special case of multivariate normal group-conditional distributions for which there are some analytical results available. We will then describe methods of forming interval estimates, in particular the bootstrap, in the general case of arbitrary group-conditional distributions. An excellent account of interval estimation of the posterior probabilities $\tau_i(\mathbf{x})$ was given recently by Critchley et al. (1987). An empirical comparison of several methods of interval estimation was provided recently by Hirst, Ford and Critchley (1990).

11.2 DISTRIBUTION OF SAMPLE POSTERIOR PROBABILITIES

11.2.1 Distribution of Sample Log Odds ($g = 2$ Homoscedastic Groups)

We adopt the same notation as introduced in Section 3.1 in the case of multivariate normal group-conditional densities. The vector $\boldsymbol{\theta}_U$ contains the elements of the group means $\boldsymbol{\mu}_1,\ldots,\boldsymbol{\mu}_g$ and the distinct elements of the group-covariance matrices $\boldsymbol{\Sigma}_1,\ldots,\boldsymbol{\Sigma}_g$, and $\boldsymbol{\Psi}_U = (\boldsymbol{\pi}',\boldsymbol{\theta}_U')'$. Also, $\boldsymbol{\theta}_E$ and $\boldsymbol{\Psi}_E$ denote $\boldsymbol{\theta}_U$ and $\boldsymbol{\Psi}_U$, respectively, under the constraint that $\boldsymbol{\Sigma}_1 = \cdots = \boldsymbol{\Sigma}_g = \boldsymbol{\Sigma}$.

As seen in previous sections, for $g = 2$ groups, it is more convenient to work with the posterior log odds rather than directly with the posterior probabilities of group membership. In the case of a mixture of two homoscedastic normal groups, so that

$$\mathbf{X} \sim N(\boldsymbol{\mu}_i,\boldsymbol{\Sigma}) \quad \text{with prob. } \pi_i \text{ in } G_i \qquad (i = 1,2), \tag{11.2.1}$$

the posterior log odds can be expressed as

$$\begin{aligned}\eta(\mathbf{x};\Psi_E) &= \log\{\tau_1(\mathbf{x};\Psi_E)/\tau_2(\mathbf{x};\Psi_E)\}\\ &= \log(\pi_1/\pi_2)+\xi(\mathbf{x};\boldsymbol{\theta}_E),\end{aligned}$$

where

$$\xi(\mathbf{x};\boldsymbol{\theta}_E) = -\tfrac{1}{2}\{\delta_{1,E}(\mathbf{x})-\delta_{2,E}(\mathbf{x})\},$$

and where

$$\begin{aligned}\delta_{i,E}(\mathbf{x}) &= \delta(\mathbf{x},\boldsymbol{\mu}_i;\boldsymbol{\Sigma})\\ &= (\mathbf{x}-\boldsymbol{\mu}_i)'\boldsymbol{\Sigma}^{-1}(\mathbf{x}-\boldsymbol{\mu}_i) \qquad (i=1,2).\end{aligned}$$

As discussed in Section 3.3, $\xi(\mathbf{x};\boldsymbol{\theta}_E)$ is commonly estimated by

$$\xi(\mathbf{x};\hat{\boldsymbol{\theta}}_E) = -\tfrac{1}{2}\{\hat{\delta}_{1,E}(\mathbf{x})-\hat{\delta}_{2,E}(\mathbf{x})\}, \tag{11.2.2}$$

where $\hat{\delta}_{i,E}(\mathbf{x}) = \delta(\mathbf{x},\bar{\mathbf{x}}_i;\mathbf{S})$. In their work leading to the provision of an interval estimate for $\xi(\mathbf{x};\boldsymbol{\theta}_E)$, Critchley et al. (1987) chose to work with its uniform minimum variance unbiased estimator $\hat{\xi}_E(\mathbf{x})$, given by (3.3.16). Schaafsma (1984) noted that the choice of estimate of $\xi(\mathbf{x};\boldsymbol{\theta}_E)$ is not crucial in the context of interval estimation. This is because differences between estimators such as $\xi(\mathbf{x};\hat{\boldsymbol{\theta}}_E)$ and $\hat{\xi}_E(\mathbf{x})$ are of order n^{-1}, whereas their standard errors are of order $n^{-1/2}$. For brevity, $\xi(\mathbf{x};\boldsymbol{\theta}_E)$ is written henceforth as $\xi_E(\mathbf{x})$.

The exact sampling distribution of $\hat{\xi}_E(\mathbf{x})$ appears to be intractable, although Critchley and Ford (1985) have shown that it depends upon only three parameters, the Mahalanobis distance Δ, $\xi_E(\mathbf{x})$, and

$$\zeta(\mathbf{x}) = \tfrac{1}{2}\{\delta_{1,E}(\mathbf{x})+\delta_{2,E}(\mathbf{x})\},$$

which may be viewed as the average atypicality of $\mathbf{x}$ with respect to G_1 and G_2. The exact variance of $\hat{\xi}_E(\mathbf{x})$, however, has been obtained by Schaafsma (1982) and Critchley and Ford (1984). They showed that

$$\operatorname{var}\{\hat{\xi}_E(\mathbf{x})\} = v(\xi_E(\mathbf{x}),\zeta(\mathbf{x}),\Delta),$$

where

$$\begin{aligned}(n-p-2)&(n-p-5)v(\xi_E(\mathbf{x}),\zeta(\mathbf{x}),\Delta)\\ &= (n-p-1)\{\xi_E(\mathbf{x})-\tfrac{1}{2}(n-3)(n_1^{-1}-n_2^{-1})\}^2\\ &\quad+(n-p-3)[\zeta(\mathbf{x})\{(n-3)(n_1^{-1}+n_2^{-1})+\Delta^2\}-\tfrac{1}{4}\Delta^4]\\ &\quad+\tfrac{1}{4}(n-3)(n-p-3)\{2p(n_1^{-2}+n_2^{-2})-(n-1)(n_1^{-1}-n_2^{-1})^2\}.\end{aligned} \tag{11.2.3}$$

Schaafsma (1982) reported that this exact variance is a substantial improvement over the asymptotic variance given in Schaafsma and van Vark (1979).

By using the asymptotic normality of $\hat{\xi}_E(\mathbf{x})$, an approximate $100(1-\alpha)\%$ confidence interval for $\xi_E(\mathbf{x})$ is

$$\hat{\xi}_E(\mathbf{x}) \pm \Phi^{-1}(1-\tfrac{1}{2}\alpha)\{v(\hat{\xi}_E(\mathbf{x}), \hat{\zeta}(\mathbf{x}), \hat{\Delta})\}^{1/2}.$$

Hirst et al. (1990) have shown by a practical example that this method can be unstable in high-dimensional, small sample size situations.

Critchley and Ford (1985) also considered another method, which substitutes estimates for $\zeta(\mathbf{x})$ and Δ only in $v(\xi_E(\mathbf{x}), \zeta(\mathbf{x}), \Delta)$. An approximate $100(1-\alpha)\%$ confidence interval is given then by the set of $\xi_E(\mathbf{x})$ values satisfying

$$\{\hat{\xi}_E(\mathbf{x}) - \xi_E(\mathbf{x})\}^2 < \{\Phi^{-1}(1-\tfrac{1}{2}\alpha)\}^2 v(\xi_E(\mathbf{x}), \hat{\zeta}(\mathbf{x}), \hat{\Delta}), \tag{11.2.4}$$

which, for $n > p + q$, is an interval whose end points are found by solving the associated quadratic equations. They concluded from some simulations performed that the latter method, which explicitly allows for the quadratic dependence of the variance on $\xi_E(\mathbf{x})$, is to be preferred. They suggested that their method would be improved if information on the skewness and kurtosis of $\hat{\xi}_E(\mathbf{x})$ was used.

A. W. Davis (1987) subsequently extended the approach of Critchley and Ford (1985) by incorporating the cumulants of $\hat{\xi}_E(\mathbf{x})$ up to the fourth order. His initial attempt using only plug-in estimates of $\zeta(\mathbf{x})$ and Δ^2 proved to be unsuccessful. It was concluded that the joint distribution of $\hat{\xi}_E(\mathbf{x})$, $\hat{\zeta}(\mathbf{x})$, and $\hat{\Delta}^2$ needed to be taken into account. A. W. Davis (1987) showed how this could be effected using the procedure of Peers and Iqbal (1985) for constructing confidence intervals in the presence of nuisance parameters. From a simulation study carried out following the basic scheme of Critchley and Ford (1985), A. W. Davis (1987) found that a considerable improvement is achieved with this procedure of Peers and Iqbal (1985). The asymptotic expressions for the joint cumulants of $\hat{\xi}_E(\mathbf{x})$, $\hat{\zeta}(\mathbf{x})$, and $\hat{\Delta}^2$ are not given here, as they are quite lengthy; see A. W. Davis (1987) for details.

11.2.2 Distribution of Sample Log Odds ($g = 2$ Heteroscedastic Groups)

We consider now the case of a mixture of $g = 2$ heteroscedastic normal components, so that

$$\mathbf{X} \sim N(\boldsymbol{\mu}_i, \boldsymbol{\Sigma}_i) \quad \text{with prob. } \pi_i \text{ in } G_i \qquad (i = 1, 2). \tag{11.2.5}$$

Under (11.2.5), we have from (3.2.2) that the posterior log odds are given by

$$\eta(\mathbf{x}; \boldsymbol{\Psi}_U) = \log(\pi_1/\pi_2) + \xi(\mathbf{x}; \boldsymbol{\theta}_U),$$

where

$$\xi(\mathbf{x}; \boldsymbol{\theta}_U) = \frac{1}{2}\sum_{i=1}^{2}(-1)^i\{\delta_i(\mathbf{x}) + \log|\boldsymbol{\Sigma}_i|\} \tag{11.2.6}$$

and $\delta_i(\mathbf{x}) = \delta(\mathbf{x}, \boldsymbol{\mu}_i; \boldsymbol{\Sigma}_i)$.

The uniform minimum variance unbiased estimator of $\xi(\mathbf{x};\boldsymbol{\theta}_U)$, denoted by $\hat{\xi}_U(\mathbf{x})$, is given by (3.2.10). The exact variance of $\hat{\xi}_U(\mathbf{x})$ appears intractable. Critchley et al. (1987) established asymptotically for $n_i > p+4$ that

$$\operatorname{var}\{\hat{\xi}_U(\mathbf{x})\} \approx v_U\{\delta_1(\mathbf{x}),\delta_2(\mathbf{x})\},$$

where

$$v_U\{\delta_1(\mathbf{x}),\delta_2(\mathbf{x})\} = \sum_{i=1}^{2}\left[\frac{\{\delta_i(\mathbf{x})\}^2}{2(n_i-p-4)} + \left\{\frac{1}{n_i} - \frac{(n_i-2)}{(n_i-p-1)(n_i-p-4)}\right\}\delta_i(\mathbf{x}) + \frac{p(n_i-2)}{2(n_i-p-1)(n_i-p-4)}\right]. \tag{11.2.7}$$

By using the asymptotic normality of $\hat{\xi}_U(\mathbf{x})$, an approximate $100(1-\alpha)\%$ confidence interval for $\xi(\mathbf{x};\boldsymbol{\theta}_U)$ is given by

$$\hat{\xi}_U(\mathbf{x}) \pm \Phi^{-1}(1-\tfrac{1}{2}\alpha)[v_U\{\hat{\delta}_1(\mathbf{x}),\hat{\delta}_2(\mathbf{x})\}]^{1/2}. \tag{11.2.8}$$

Critchley et al. (1987) explored by simulation the performance of (11.2.8) with $\alpha = 0.05$ as a nominal 95% confidence interval for $\xi(\mathbf{x};\boldsymbol{\theta}_U)$. Various combinations of the parameters were taken with $n_i = 20$, 40, or 400 ($i = 1,2$), and $p = 2$ or 5. In summary, they found that (11.2.8) gave a very reliable approximation for very large sample sizes ($n_1 = n_2 = 400$), a good approximation for moderate sample sizes ($n_1 = n_2 = 40$), and rather uncertain results for smaller sizes ($n_1 = n_2 = 20$).

Ambergen and Schaafsma (1984) have considered approximate interval estimates for the estimated posterior probabilities rather than for the log posterior odds. Also, Machin et al. (1983) have given the asymptotic standard error of

$$\tau_1(\mathbf{x};\boldsymbol{\pi},\hat{\boldsymbol{\theta}}_U) = (\pi_1/\pi_2)\exp\{\xi(\mathbf{x};\hat{\boldsymbol{\theta}}_U\}/\{1+(\pi_1/\pi_2)\exp(\xi(\mathbf{x};\hat{\boldsymbol{\theta}}_U)\}$$

in the univariate case of (11.2.5), where the prior probabilities π_1 and π_2 are specified.

11.2.3 Joint Distribution of Sample Posterior Probabilities ($g > 2$ Groups)

In the case of $g > 2$ groups under normality, the relative sizes of the posterior probabilities of group membership can be considered in terms of the quantities $\eta_{ig}(\mathbf{x};\boldsymbol{\Psi}_U)$ for $i = 1,\ldots,g-1$, where, from (3.3.2),

$$\eta_{ig}(\mathbf{x};\boldsymbol{\Psi}_U) = \log(\pi_i/\pi_g) + \xi_{ig}(\mathbf{x};\boldsymbol{\theta}_U)$$

and

$$\xi_{ig}(\boldsymbol{\theta}_U) = -\tfrac{1}{2}\{\delta_i(\mathbf{x}) - \delta_g(\mathbf{x})\} - \tfrac{1}{2}\log\{|\boldsymbol{\Sigma}_i|/|\boldsymbol{\Sigma}_g|\}$$

for $i = 1,\ldots,g-1$. For pairwise group comparisons, an unbiased estimator of $\xi_{ig}(\mathbf{x};\boldsymbol{\theta}_U)$ and an expression for its asymptotic variance are available from the results in the previous sections for $g = 2$ groups. Also, in the case of homoscedasticity, an exact expression exists for the variance of the unbiased estimator of $\xi_{ig}(\mathbf{x};\boldsymbol{\theta}_E)$.

Concerning the joint distribution of the estimated posterior probabilities of group membership, Ambergen and Schaafsma (1985) have considered this problem under the homoscedastic normal (11.2.1). We let

$$\boldsymbol{\tau}(\mathbf{x};\boldsymbol{\Psi}_E) = (\tau_1(\mathbf{x};\boldsymbol{\Psi}_E),\ldots,\tau_g(\mathbf{x};\boldsymbol{\Psi}_E))'$$

be the vector containing the g posterior probabilities of group membership. Further, for specified $\boldsymbol{\pi}$, we let $\boldsymbol{\tau}(\mathbf{x};\boldsymbol{\pi},\hat{\boldsymbol{\theta}}_E)$ denote the plug-in estimate of $\boldsymbol{\tau}(\mathbf{x};\boldsymbol{\Psi}_E)$.

For separate sampling under (11.2.1), Ambergen and Schaafsma (1985) showed that the asymptotic distribution of $\sqrt{n}\{\boldsymbol{\tau}(\mathbf{x};\boldsymbol{\pi},\hat{\boldsymbol{\theta}}_E) - \boldsymbol{\tau}(\mathbf{x};\boldsymbol{\Psi}_E)\}$ is multivariate normal with mean zero and singular covariance matrix $\boldsymbol{\Upsilon}\boldsymbol{\Gamma}\boldsymbol{\Upsilon}$,

$$\sqrt{n}\{\boldsymbol{\tau}(\mathbf{x};\boldsymbol{\pi},\hat{\boldsymbol{\theta}}_E) - \boldsymbol{\tau}(\mathbf{x};\boldsymbol{\Psi}_E)\} \sim N(\mathbf{0},\boldsymbol{\Upsilon}\boldsymbol{\Gamma}\boldsymbol{\Upsilon}), \tag{11.2.9}$$

where

$$(\boldsymbol{\Gamma})_{ii} = (4n/n_i)\delta_{i,E}(\mathbf{x}) + 2\{\delta_{i,E}(\mathbf{x})\}^2$$

and

$$(\boldsymbol{\Gamma})_{ij} = 2\{(\mathbf{x}-\boldsymbol{\mu}_i)'\boldsymbol{\Sigma}^{-1}(\mathbf{x}-\boldsymbol{\mu}_j)\}^2 \qquad (i \neq j),$$

and where

$$(\boldsymbol{\Upsilon})_{ii} = \tfrac{1}{2}\tau_i(\mathbf{x};\boldsymbol{\Psi}_E)\{-1+\tau_i(\mathbf{x};\boldsymbol{\Psi}_E)\}$$

and

$$(\boldsymbol{\Upsilon})_{ij} = \tfrac{1}{2}\{\tau_i(\mathbf{x};\boldsymbol{\Psi}_E) + \tau_j(\mathbf{x};\boldsymbol{\Psi}_E)\} \qquad (i \neq j).$$

By plugging in estimates for the unknown parameters in the covariance matrix in the asymptotic distribution (11.2.9), simultaneous confidence intervals, for example, Scheffé-type intervals, can be constructed for the $\tau_i(\mathbf{x};\boldsymbol{\Psi}_E)$. Only estimates of the diagonal elements of the covariance matrix are needed for interval estimation of each $\tau_i(\mathbf{x};\boldsymbol{\Psi}_E)$ separately. For example, an approximate $100(1-\alpha)\%$ confidence interval for $\tau_i(\mathbf{x};\boldsymbol{\Psi}_E)$ is provided by

$$\tau_i(\mathbf{x};\boldsymbol{\pi},\hat{\boldsymbol{\theta}}_E) \pm \Phi^{-1}(1-\tfrac{1}{2}\alpha)\{(\hat{\boldsymbol{\Upsilon}}\hat{\boldsymbol{\Gamma}}\hat{\boldsymbol{\Upsilon}})_{ii}/n\}^{1/2} \qquad (i = 1,\ldots,g).$$

Ambergen and Schaafsma (1985) concluded from some simulations performed under (11.2.1) that, for a nominal 95% confidence interval, each n_i should not be smaller than 50(25) if the actual confidence probability is not to fall below 0.90(0.85). Ambergen and Schaafsma (1984) have considered the extension of the result to the case of arbitrary group-covariance matrices with no assumption on their equality.

11.3 FURTHER APPROACHES TO INTERVAL ESTIMATION OF POSTERIOR PROBABILITIES OF GROUP MEMBERSHIP

11.3.1 Profile Likelihood Approach

Critchley, Ford, and Rijal (1988) have considered the profile likelihood approach to the provision of an interval estimate for the posterior log odds $\xi(\mathbf{x};\boldsymbol{\theta}_U)$ in the case of $g = 2$ normal groups with unequal covariance matrices. For brevity, we will write in this section $\boldsymbol{\theta}_U$ as $\boldsymbol{\theta}$ and $\xi(\mathbf{x};\boldsymbol{\theta}_U)$ as ξ, where, for the given feature vector $\mathbf{x}$, $\xi = h(\boldsymbol{\theta})$. With ξ the parameter of interest, the profile log likelihood function $p(\xi)$ is defined to be the supremum of the log likelihood function $L(\boldsymbol{\theta})$ over all admissible values of $\boldsymbol{\theta}$ subject to $h(\boldsymbol{\theta}) = \xi$, that is,

$$p(\xi) = \sup_{h(\boldsymbol{\theta})=\xi} L(\boldsymbol{\theta}). \tag{11.3.1}$$

A confidence region for ξ based on $p(\xi)$ has the form

$$\{\xi : 2[p(\hat{\xi}) - p(\xi)] < \epsilon\} \tag{11.3.2}$$

for some constant ϵ and where $\hat{\xi}$ is the maximum likelihood estimate of ξ. Large-sample theory suggests taking $\epsilon = \chi^2_{1,1-\alpha}$ for an approximate $100(1-\alpha)\%$ confidence region.

The major obstacle in constructing the interval defined by (11.3.2) is the computation of the profile likelihood $p(\xi)$. However, Critchley et al. (1988) showed that the computations are simplified by using strong Lagrangian theory. It follows from this theory, which was presented in a general context, that, provided certain conditions hold, the profile likelihood can be obtained through the maximization of

$$L(\boldsymbol{\theta}) - \lambda h(\boldsymbol{\theta}) \tag{11.3.3}$$

with respect to $\boldsymbol{\theta}$, for a given $\lambda \in \Lambda$, where Λ denotes the set of values of λ for which a solution $\hat{\boldsymbol{\theta}}_\lambda$ to the maximization of (11.3.3) exists. The entire function $p(\xi)$ is obtained by letting λ vary throughout Λ since, if $h(\hat{\boldsymbol{\theta}}_\lambda) = \hat{\xi}_\lambda$, say, then

$$p(\hat{\xi}_\lambda) = L(\hat{\boldsymbol{\theta}}_\lambda).$$

For the present problem of estimating the posterior log odds under separate sampling with respect to $g = 2$ groups under normality, (11.3.3) has exactly the same form as $L(\boldsymbol{\theta})$, and so $\hat{\boldsymbol{\theta}}_\lambda$ can be computed exactly. As noted by Critchley et al. (1988), the Lagrange multiplier λ has an interpretation as a weight with which the feature vector $\mathbf{x}$ is subtracted from the training set from the first group and added to the second. For each $\lambda \in \Lambda$, the unique $\hat{\boldsymbol{\theta}}_\lambda$ consists of the elements of $\hat{\boldsymbol{\mu}}_i(\lambda)$ and the distinct elements of $\hat{\boldsymbol{\Sigma}}_i(\lambda)$ for $i = 1,2$, where

$$\hat{\boldsymbol{\mu}}_i(\lambda) = \{n_i\bar{\mathbf{x}}_i + (-1)^i\lambda\mathbf{x}\}/n_i(\lambda),$$

$$\hat{\boldsymbol{\Sigma}}_i(\lambda) = [(n_i - 1)\mathbf{S}_i + \{(-1)^i\lambda n_i/n_i(\lambda)\}(\mathbf{x} - \bar{\mathbf{x}}_i)(\mathbf{x} - \bar{\mathbf{x}}_i)']/n_i(\lambda),$$

and

$$n_i(\lambda) = n_i + (-1)^i \lambda$$

for $i = 1,2$. Here Λ is the open interval

$$(-n_2(1 + D_2^2)^{-1}, n_1(1 + D_1^2)^{-1}),$$

where

$$D_i^2 = \{n_i/(n_i - 1)\}(\mathbf{x} - \bar{\mathbf{x}}_i)'\mathbf{S}_i^{-1}(\mathbf{x} - \bar{\mathbf{x}}_i) \qquad (i = 1,2).$$

Also, for each $\lambda \in \Lambda$,

$$\hat{\xi}_\lambda = \frac{1}{2}\sum_{i=1}^{2}(-1)^i[\{n_i D_i^2/\tilde{n}_i(\lambda)\} - m_i(\lambda)]$$

and, apart from an additive constant depending only upon n_1, n_2, and p,

$$p(\hat{\xi}_\lambda) = \lambda\hat{\xi}_\lambda + \frac{1}{2}\sum_{i=1}^{2} n_i(\lambda)m_i(\lambda),$$

where

$$\tilde{n}_i(\lambda) = n_i(\lambda) + (-1)^i \lambda D_i^2,$$

and

$$m_i(\lambda) = (p + 1)\log n_i(\lambda) - \log \tilde{n}_i(\lambda) - \log|(n_i - 1)\mathbf{S}_i|$$

for $i = 1,2$.

Critchley et al. (1987) have considered the profile likelihood approach for multivariate normal linear discrimination, corresponding to the assumption of equal covariance matrices. Hirst et al. (1990) have noted that the use of a Bartlett-type correction could improve this approach and the approach that yields (11.2.4).

11.3.2 Illustration of the Profile Likelihood Approach

Critchley et al. (1988) illustrated the profile likelihood approach, using data from the example on the diagnosis of Conn's syndrome, as reported in Aitchison and Dunsmore (1975, Chapter 1) and broadly referred to in Section 1.1. More specifically, the data set consisted of observations on $n_1 = 20$ patients from group G_1, having a benign tumor (i.e., confirmed adenoma), and on $n_2 = 11$ patients from group G_2, having a more diffuse condition, bilateral hyperplasia. For this illustration, the feature vectors on the patients were taken to contain the measurements of their age (in years) and the concentrations (in meq/L) of potassium, carbon dioxide, and renin in their blood plasma. As recommended by Aitchison and Dunsmore (1975, Section 11.6), Critchley et al. (1988) worked with the logarithms of the raw data on these four variables, which were taken to have a multivariate normal distribution within each group; see also Aitchison et al. (1977) and Hawkins (1981) on the feasibility of this assumption.

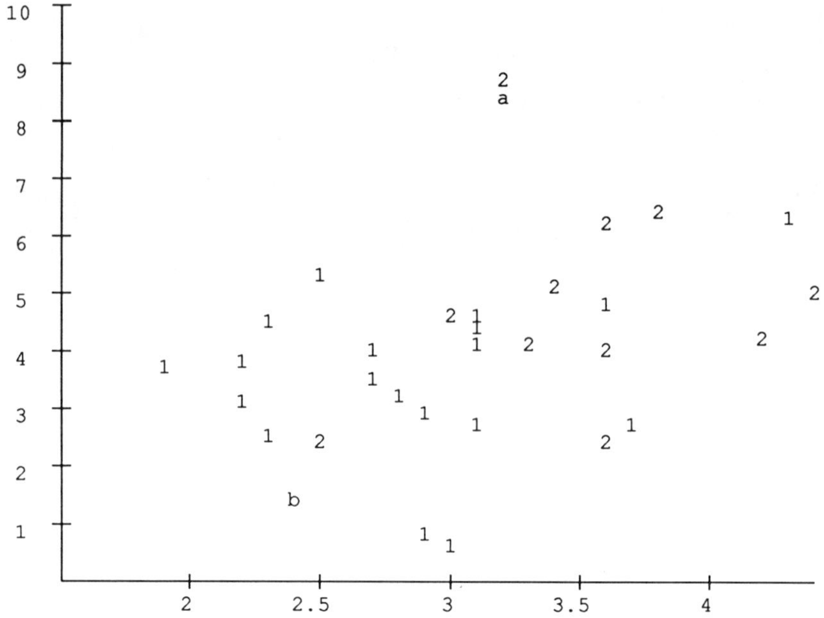

FIGURE 11.1. Plot of potassium versus renin concentrations for the unclassified case (a) and the classified case (b) from G_1, along with the other 30 classified points, labeled as 1 or 2 according as they belong to G_1 or G_2, respectively.

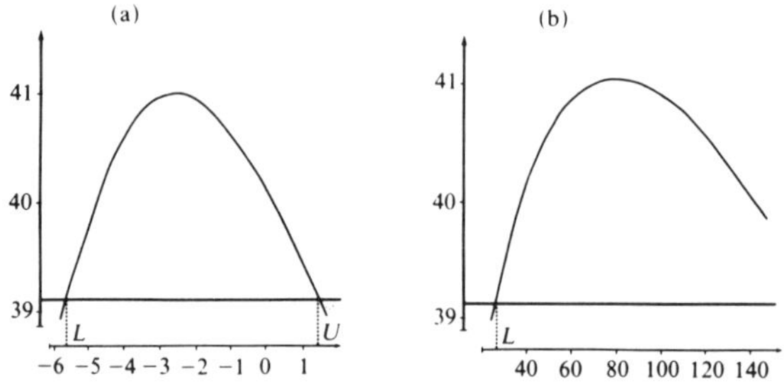

FIGURE 11.2. Profile log likelihood $p(\xi)$ for two cases (a) and (b). (From Critchley et al., 1988.)

Critchley et al. (1988) focused on two cases, (a) and (b). Case (a), which is labeled u1 in Table 11.4 in Aitchison and Dunsmore (1975), is an unclassified patient with feature vector $\mathbf{x}_a = (50, 3.2, 27.0, 8.5)'$. Case (b), which is labeled A8 in Table 1.6 in Aitchison and Dunsmore (1975), is a member of the 30 patients classified from group G_1, and has feature vector $\mathbf{x}_b = (18, 2.5, 30.0, 2.5)'$. The raw values of the potassium and renin variables for these two cases are plotted in Figure 11.1, along with the corresponding values of these two variables for the other 19 classified patients from G_1 and the 11 from G_2.

The profile log likelihood $p(\xi)$ is illustrated in Figure 11.2 for cases (a) and (b) under the assumption of the heteroscedastic normal model (11.2.1) with $\pi_1 = \pi_2 = 0.5$. The horizontal line in Figure 11.2 corresponds to taking $\epsilon = \chi^2_{1;.95}$ in the definition (11.3.2) of the interval estimate. The upper and lower bounds, L and U, are indicated on the figures. Note that, due to the extreme nature of Figure 11.2(b), U is not indicated on the plot.

In Figure 11.2(a), the point estimate for $\xi(\mathbf{x}_a;\boldsymbol{\theta}_U)$ would suggest that case (a) is a bilateral hyperplasma. However, the interval estimate includes zero, indicating uncertainty over whether $\xi(\mathbf{x}_a;\boldsymbol{\theta}_U)$ is negative or positive. It can be seen in Figure 11.2(b) that, although we are uncertain about the exact value of $\xi(\mathbf{x}_b;\boldsymbol{\theta}_U)$, we could be certain that $\xi(\mathbf{x}_b;\boldsymbol{\theta}_U)$ is positive. Thus, on a probability scale, the interval for the posterior probability of an adenoma is very close to 1.

11.3.3 Bayesian Approach

As discussed in Section 2.2, the semi-Bayesian approach to the estimation of the group-posterior probabilities is carried out in terms of their posterior distribution computed on the basis of the information in $\mathbf{t}$ but not $\mathbf{x}$. Rigby (1982) has investigated this approach applied to the estimation of $\tau_1(\mathbf{x};\boldsymbol{\Psi}_U)$. His proposal for interval estimation of $\tau_1(\mathbf{x};\boldsymbol{\Psi}_U)$ is to calculate the moment-generating function of the posterior distribution of $\tau_1(\mathbf{x};\boldsymbol{\Psi}_U)$. The first four moments of the posterior density are then computed and Pearson curves are used to approximate this density. From this approximated density, a Bayesian posterior or credibility interval can be formed for $\tau_1(\mathbf{x};\boldsymbol{\Psi}_U)$.

Although this approach of Rigby (1982) is not designed to produce a classical confidence interval, Hirst et al. (1990) have found it to have extremely good properties as such. Indeed, in their simulation study, it performed the best out of the methods investigated in achieving the target confidence coefficient.

11.3.4 Bootstrap

The use of the bootstrap to assess the distribution of the estimated error rates of a discriminant rule was described in the previous chapter. It can be applied in a similar manner to assess the sampling distribution of the estimated posterior probabilities of group membership or the log odds conditional on the observed value $\mathbf{x}$ of the feature vector. Let

$$\hat{\boldsymbol{\tau}}(\mathbf{x};\mathbf{t}) = (\hat{\tau}_1(\mathbf{x};\mathbf{t}),\ldots,\hat{\tau}_g(\mathbf{x};\mathbf{t}))'$$

be the vector of estimated posterior probabilities. Also, let $\mathbf{t}_1^*,\ldots,\mathbf{t}_K^*$ denote K bootstrap applications of $\mathbf{t}$ generated with the distribution function F of $\mathbf{Y} = (\mathbf{X}',\mathbf{Z}')'$ replaced by an estimate $\hat{F}$ formed from $\mathbf{t}$. Then, conditional on $\mathbf{x}$, the bootstrap distribution of $\hat{\boldsymbol{\tau}}(\mathbf{x};\mathbf{T}^*)$, approximated on the basis of $\hat{\boldsymbol{\tau}}(\mathbf{x};\mathbf{t}_1^*),\ldots,\hat{\boldsymbol{\tau}}(\mathbf{x};\mathbf{t}_K^*)$, can be used to assess the distribution of $\hat{\boldsymbol{\tau}}(\mathbf{x};\mathbf{T})$.

hand, the aim may be to identify those feature variables that are most useful in describing differences among the possible groups. In this context, it is appropriate to assess the adequacy of a subset of feature variables in terms of the separation they provide among the groups. Thus, the choice of the selection criterion should depend on the aim of the discriminant analysis. It was already shown in Section 3.9 that allocatory and separatory criteria can lead to quite different decisions being made in the reduction of the dimension of the feature vector for use with the sample NLDR for $g > 2$ groups.

In many situations, attention is focused not on just one subset of the available feature variables. Rather the intention is to find the "best subset" in some sense. Thus, consideration of a number of subsets has to be undertaken. Ideally, the performance of the discriminant rule should be assessed on the basis of the specified criterion for each possible subset. But unless the total number p of variates is small, an exhaustive search is computationally prohibitive; see McCabe (1975). As a consequence, stepwise selection procedures, either forward or backward, are commonly employed; see, for example, Farmer and Freund (1975) and Jennrich (1977). Their use raises a number of important issues, such as the overall significance level at which the final subset of variables is concluded to be "best." Another problem with variable selection is how to provide an unbiased and not an overly optimistic assessment of the final version of the discriminant rule in its application to subsequent feature data of unknown origin. These issues are discussed further in the following sections.

The selection problem for $g > 2$ groups is much more difficult than for $g = 2$ groups, where there is a close tie with multiple regression. Variable selection in the latter context is a well-studied problem in the literature; see Miller (1984, 1990) for a recent account. Fowlkes, Gnanadesikan, and Kettenring (1987) have considered the variable-selection problem in three different contexts, namely, multiple regression, discriminant analysis, and cluster analysis. They noted that the strong theoretical underpinnings of regression and discriminant analysis make possible a variety of formal and informal approaches to variable selection in these two contexts, although the state of development in the latter context is not at the same stage as in the former.

Initial studies of subset selection methods in discriminant analysis include those by Dixon and Massey (1957); Kendall (1957); Cochran (1964a); Weiner and Dunn (1966); Eisenbeis, Gilbert, and Avery (1973); McCabe (1975); McLachlan (1976b); Habbema and Hermans (1977); Costanza and Afifi (1979); and Farver and Dunn (1979). A two-stage approach was considered by Zielezny and Dunn (1975), and Zielezny (1976). The focus in these studies is on applications of the sample NLDF for $g = 2$ groups under the assumption of normality and homoscedasticity. It should be noted that although the sample NLDF may be robust to moderate departures from this assumption as considered in Section 5.6, little is known about the robustness of subset selection procedures per se.

Studies that address subset selection for discrete feature variables include those by Elashoff, Elashoff, and Goldman (1967); Hills (1967); Goldstein and

Rabinowitz (1975); Goldstein and Dillon (1977, 1978, Chapter 4); Haerting (1983); and Krusińska and Liebhart (1988, 1989). Krzanowski (1983b) and Daudin (1986) have considered the problem for mixed feature variables.

In other work on variable selection, Fu, Min, and Li (1970) applied the Kullback–Leibler criterion, and Chen (1975) proposed the use of the Bhattacharyya distance as a criterion. Fu (1968), C. Y. Chang (1973), and Hand (1981b) devised mathematical programming approaches. Lam and Cox (1981) presented an algorithm that selects variables in pairs. More recently, Ridout (1988) has provided an improved version of the branch and bound algorithm as described by Fukunaga and Narendra (1977) for selecting a subset of the feature variables that maximizes a criterion function over all possible subsets of a specified size; see also Roberts (1984).

McKay and Campbell (1982a, 1982b) have given an excellent review of variable-selection methods in discriminant analysis. Their first article focuses on methods that are appropriate to selection where the primary aim is to obtain a more parsimonious description of the groups, and the second addresses the problem where the aim is to allocate future entities of unknown origin to the groups. In the first of their two papers, they have provided a table that concisely summarizes the main issues and their recommendations on variable-selection methods. Some references to more recent theoretical results in the special case of the homoscedastic normal model for $g = 2$ groups can be found in Steerneman (1987).

Another way of variable selection is to adapt the approach of Rissanen (1989) and work in terms of the stochastic complexity of the data. Variables are then selected according to the minimum description length (MDL) principle. In his monograph, Rissanen (1989, Chapter 7) has sketched two applications of stochastic complexity and the MDL principle to variable selection in discriminant analysis.

12.1.2 Peaking Phenomenon

For training samples of finite size, the performance of a given discriminant rule in a frequentist framework does not keep on improving as the number p of feature variables is increased. Rather, its overall unconditional error rate will stop decreasing and start to increase as p is increased beyond a certain threshold, depending on the particular situation. Consider, for example, the case of $g = 2$ groups under the homoscedastic normal model (3.3.1). Although the deletion of some variables from the feature vector $\mathbf{x}$ can never increase the Mahalanobis distance between the groups G_1 and G_2, the overall unconditional error rate of the sample NLDR can be reduced if the subsequent reduction in the distance is sufficiently small relative to the number of variables deleted. This was first illustrated by Rao (1949). A more recent example was given by Boullion, Odell, and Duran (1975) in the context of remote sensing; see also Kanal and Chandrasekaran (1971), Foley (1972), and Kanal (1974).

An appreciation of the peaking phenomenon as the number of feature variables is increased can be obtained from the expansion (4.3.2), which gives the asymptotic overall error rate of the sample NLDR in terms of Δ, p, n_1, and n_2, under the homoscedastic normal model (3.3.1). For further details, the reader is referred to Jain and Waller (1978), who have studied the peaking phenomenon under (3.3.1) in order to relate the optimal number of feature variables to the total sample size n and Δ. This situation was considered also by Chandrasekaran and Jain (1975, 1977), Van Ness and Simpson (1976), Raudys (1979), and Van Ness (1979). The latter study extended the model to allow for unequal covariance matrices, as well as considering nonparametric discriminant rules based on kernel density estimation of the group-conditional densities; see also Van Ness (1977, 1980).

Jain and Chandrasekaran (1982) have given a comprehensive review of this problem of dimensionality versus sample size. A good deal of the work has been on the sample NLDR as considered above. However, the problem has been studied in a wider context, including in a Bayesian framework where the overall unconditional error rate is averaged over the prior distribution adopted for the vector of unknown parameters in the group-conditional distributions. Hughes (1968) showed that for multinomial group-conditional distributions, that by increasing the number of cells, corresponding to the increasing of the dimensionality of the feature vector, the overall error rate decreases to a minimum and then deteriorates. Lindley (1978) attributed this apparently paradoxical behavior of the error rate to the incoherence of the prior distributions used in Hughes' (1968) model. Lindley (1978) furthermore derived an explicit expression for the error rate with an infinite training sample size and coherent priors, showing that it tends monotonically to zero. Although coherence guarantees monotonicity on the overall error rate, it does not imply the latter will be zero in the limit, even with infinite training sample sizes. An example demonstrating this was provided by Brown (1980).

The abovementioned apparent paradox was considered also by Van Campenhout (1978) and Waller and Jain (1978). More recent references on the effect of dimensionality from the Bayesian point of view include Kokolakis (1981, 1985), who has derived exact and asymptotic results for two particular linear discrimination problems, demonstrating the positive effect of increasing the number of features.

12.2 TEST FOR NO ADDITIONAL INFORMATION

12.2.1 Multiple Groups

Let

$$S = \{s;\ s \text{ is a nonempty subset of } 1,\ldots,p\}$$

and, for each s in S, let p_s denote the number of elements in s. If

$$s = \{k_1,\ldots,k_{p_s}\},$$

then $\mathbf{x}^{(s)}$ denotes the subvector of the full feature vector $\mathbf{x}$, defined by

$$\mathbf{x}^{(s)} = (x_{k_1}, \ldots, x_{k_{p_s}})'.$$

The complement of s with respect to S is denoted by $\bar{s}$. Thus, the subvector $\mathbf{x}^{(\bar{s})}$ contains those $p - p_s$ variables in $\mathbf{x}$ that are not in $\mathbf{x}^{(s)}$. Note that without loss of generality, we can relabel the elements of $\mathbf{x}$ so that

$$\mathbf{x} = (\mathbf{x}^{(s)\prime}, \mathbf{x}^{(\bar{s})\prime})'. \tag{12.2.1}$$

We let

$$\boldsymbol{\mu}_i = (\boldsymbol{\mu}_{i1}', \boldsymbol{\mu}_{i2}')' \qquad (i = 1, \ldots, g)$$

and

$$\boldsymbol{\Sigma} = \begin{pmatrix} \boldsymbol{\Sigma}_{11} & \boldsymbol{\Sigma}_{12} \\ \boldsymbol{\Sigma}_{21} & \boldsymbol{\Sigma}_{22} \end{pmatrix}$$

be the corresponding partitions of $\boldsymbol{\mu}_i$ and $\boldsymbol{\Sigma}$, respectively, under the usual assumption of the homoscedastic normal model,

$$\mathbf{X} \sim N(\boldsymbol{\mu}_i, \boldsymbol{\Sigma}) \quad \text{in} \quad G_i \qquad (i = 1, \ldots, g). \tag{12.2.2}$$

In terms of the conditional distribution of $\mathbf{X}^{(\bar{s})}$ given $\mathbf{x}^{(s)}$, the notion of the additional information supplied by the subvector $\mathbf{x}^{(\bar{s})}$ in the presence of $\mathbf{x}^{(s)}$ plays an important role in many variable-selection methods; see Rao (1973). Additional references on this notion can be found in McKay and Campbell (1982a), Krishnaiah (1982) and Fujikoshi, Krishnaiah, and Schmidhammer (1987). Under (12.2.2), the hypothesis H_0 that $\mathbf{x}^{(\bar{s})}$ provides no additional information over that supplied by $\mathbf{x}^{(s)}$ can be formulated as

$$H_0^{(s)} : \boldsymbol{\mu}_{i2} - \boldsymbol{\mu}_{h2} - \boldsymbol{\Sigma}_{21}\boldsymbol{\Sigma}_{11}^{-1}(\boldsymbol{\mu}_{i1} - \boldsymbol{\mu}_{h1}) = \mathbf{0} \qquad (i \neq h = 1, \ldots, g).$$

Let $\tilde{\mathbf{B}} = (g-1)\mathbf{B}$ be the between-group matrix of sums of squares and products and let $\mathbf{W} = (n-g)\mathbf{S}$ be the pooled within-group sums of squares and products matrix, as defined by (6.5.4) and (6.5.5), respectively. Corresponding to the partition (12.2.1) of $\mathbf{x}$ into the subvectors $\mathbf{x}^{(s)}$ and $\mathbf{x}^{(\bar{s})}$, we partition $\tilde{\mathbf{B}}$ and $\mathbf{W}$ as

$$\tilde{\mathbf{B}} = \begin{pmatrix} \tilde{\mathbf{B}}_{11} & \tilde{\mathbf{B}}_{12} \\ \tilde{\mathbf{B}}_{21} & \tilde{\mathbf{B}}_{22} \end{pmatrix}$$

and

$$\mathbf{W} = \begin{pmatrix} \mathbf{W}_{11} & \mathbf{W}_{12} \\ \mathbf{W}_{21} & \mathbf{W}_{22} \end{pmatrix}.$$

Rao (1973, page 551) has described a procedure for testing the hypothesis $H_0^{(s)}$ based on the matrices $\mathbf{W}$ and $\tilde{\mathbf{B}}$, adjusted for $\mathbf{x}^{(s)}$, namely,

$$\mathbf{W}_{2.1} = \mathbf{W}_{22} - \mathbf{W}_{21}\mathbf{W}_{11}^{-1}\mathbf{W}_{12}$$

and

$$\tilde{\mathbf{B}}_{2.1} = (\tilde{\mathbf{B}}_{22} + \mathbf{W}_{22}) - (\tilde{\mathbf{B}}_{21} + \mathbf{W}_{21})(\tilde{\mathbf{B}}_{11} + \mathbf{W}_{11})^{-1}(\tilde{\mathbf{B}}_{12} + \mathbf{W}_{12}) - \mathbf{W}_{2.1}$$

with degrees of freedom $n-g-p_s$ and $g-1$, respectively. Test criteria parallel those used in MANOVA, being functions of the nonzero eigenvalues of the matrix $\mathbf{W}_{2.1}^{-1}\tilde{\mathbf{B}}_{2.1}$, such as the largest root or the likelihood ratio test statistic

$$\Lambda_{\bar{s}\cdot s} = |\mathbf{W}_{2.1}|/|\tilde{\mathbf{B}}_{2.1} + \mathbf{W}_{2.1}|$$

as obtained under (12.2.2).

It was seen in Section 3.9 on separatory measures that a class of measures of sample spread of the g groups on the basis of the full feature vector $\mathbf{x}$ is provided by any scalar function that is increasing in the nonzero eigenvalues of $\mathbf{W}^{-1}\tilde{\mathbf{B}}$. Likewise for $\mathbf{x}^{(\bar{s})}$ adjusted for the subvector $\mathbf{x}^{(s)}$, a class of measures of sample spread is provided by any scalar function that is increasing in the eigenvalues of $\mathbf{W}_{2.1}^{-1}\tilde{\mathbf{B}}_{2.1}$. The hypothesis of no additional information can be interpreted as one concerning the adequacy of $\mathbf{x}^{(s)}$. For as noted by McKay (1977), the hypothesis $H_0^{(s)}$ is true if and only if the variables in $\mathbf{x}^{(s)}$ provide the same overall separation as the full set of p variables in $\mathbf{x}$, that is, if and only if the eigenvalues of $\mathbf{\Sigma}_{11}^{-1}\mathbf{B}_{o,11}$ are the same in number and magnitude as those of $\mathbf{\Sigma}^{-1}\mathbf{B}_o$, where $\mathbf{B}_o$ is the population analogue of $\mathbf{B}$.

The statistic $\Lambda_{\bar{s}\cdot s}$ provides further insight into the test for no additional information. As noted by McKay and Campbell (1982a), it satisfies the relation

$$1 + \Lambda_{\bar{s}\cdot s} = (1+\Lambda)/(1+\Lambda_s),$$

where

$$\Lambda_s = |\mathbf{W}_{11}|/|\tilde{\mathbf{B}}_{11} + \mathbf{W}_{11}| \tag{12.2.3}$$

and

$$\Lambda = |\mathbf{W}|/|\tilde{\mathbf{B}} + \mathbf{W}|$$

are the statistics arising from the likelihood ratio test that there are no differences between the means of the groups on the basis of $\mathbf{x}^{(s)}$ and $\mathbf{x}$, respectively.

In testing for the adequacy of $p_s = p-1$ of the p feature variables, $\tilde{\mathbf{B}}_{2.1}$ and $\mathbf{W}_{2.1}$ are scalars, being the adjusted between- and within-group sums of squares, respectively. It follows in this special case of $p - p_s = 1$ that

$$(n-g-p+1)(1-\Lambda_{\bar{s}\cdot s})/(g-1)\Lambda_{\bar{s}\cdot s} \tag{12.2.4}$$

is distributed under $H_0^{(s)}$ according to the F-distribution with $g-1$ and $n-g-p+1$ degrees of freedom; see Rao (1973, page 553).

12.2.2 Two-Group Case

As previously in the case of $g=2$ groups, Δ denotes the Mahalanobis distance (and D its sample counterpart) between G_1 and G_2 on the basis of the group-conditional means and common covariance matrix of the full feature vector $\mathbf{X}$. Corresponding to the partition (12.2.1) of $\mathbf{x}$, we let

$$\Delta_s = \{(\boldsymbol{\mu}_{11} - \boldsymbol{\mu}_{21})'\mathbf{\Sigma}_{11}^{-1}(\boldsymbol{\mu}_{11} - \boldsymbol{\mu}_{21})'\}^{1/2}$$

be the Mahalanobis distance between G_1 and G_2 on the basis of the group-conditional means and common covariance matrix of $\mathbf{X}^{(s)}$. It follows that

$$\Delta_{\bar{s}} = (\Delta^2 - \Delta_s^2)^{1/2}$$

can be expressed as

$$\Delta_{\bar{s}} = \{(\boldsymbol{\mu}_{1,2.1} - \boldsymbol{\mu}_{2,2.1})'\boldsymbol{\Sigma}_{2.1}^{-1}(\boldsymbol{\mu}_{1,2.1} - \boldsymbol{\mu}_{2,2.1})\}^{1/2},$$

where, conditional on $\mathbf{x}^{(s)}$,

$$\boldsymbol{\mu}_{i,2.1} = \boldsymbol{\mu}_{i2} - \boldsymbol{\Sigma}_{21}\boldsymbol{\Sigma}_{11}^{-1}\boldsymbol{\mu}_{i1},$$

and

$$\boldsymbol{\Sigma}_{2.1} = \boldsymbol{\Sigma}_{22} - \boldsymbol{\Sigma}_{21}\boldsymbol{\Sigma}_{11}^{-1}\boldsymbol{\Sigma}_{12}$$

are the expectation and covariance matrix, respectively, of $\mathbf{X}^{(\bar{s})}$ in group G_i $(i = 1,2)$.

For $g = 2$ groups, the hypothesis of no additional information is equivalent to the hypothesis

$$H_0^{(s)} : \Delta_{\bar{s}} = 0.$$

It follows (Rao, 1973) that $H_0^{(s)}$ can be tested using the statistic

$$\mathcal{F}_s = (N - p + 1)\{(n_1 n_2)/(p - p_s)\}(D^2 - D_s^2)/(n_1 n_2 D_s^2 + nN), \qquad (12.2.5)$$

which, under the homoscedastic normal model (12.2.2), is distributed according to the F-distribution with $p - p_s$ and $N - p + 1$ degrees of freedom. In (12.2.5), D_s is the sample analogue of Δ_s and $N = n_1 + n_2 - 2$. The hypothesis $H_0^{(s)}$ is rejected for large $\mathcal{F}_s$, so that the significance probability P_s is given by the area to the right of the observed value of $\mathcal{F}_s$ under the $F_{p-p_s,N-p+1}$ density. Under $H_0^{(s)}$, P_s is thus uniformly distributed on (0,1); otherwise P_s will tend to be small.

The test statistic (12.2.5) was obtained originally by Rao (1946). Subrahmaniam and Subrahmaniam (1976) have since shown that the test based on

$$\tilde{\mathcal{F}}_s = \{(n_1 n_2)/(nN)\}(D^2 - D_s^2) \qquad (12.2.6)$$

has superior power to that based on $\mathcal{F}_s$ if D_s^2 is not small. However, unless $\Delta_s = 0$, its null distribution is quite complicated and percentage points are difficult to obtain.

The test statistic $\tilde{\mathcal{F}}_s$ was originally suggested by Rao (1949) as an alternative to $\mathcal{F}_s$ in the case where $\mathbf{x}^{(s)}$ is a covariate vector, that is, where $\Delta_s = 0$. However, as noted by Rao (1966b), it is customary to still use $\mathcal{F}_s$ to test $H_0^{(s)}$ in the case where there is prior knowledge that $\Delta_s = 0$. This test statistic is formed and applied regardless of the value of Δ_s. But as a result, it is not as efficient for $\Delta_s = 0$ as the test statistic $\tilde{\mathcal{F}}_s$, which has been formed with the knowledge that $\Delta_s = 0$. Subrahmaniam and Subrahmaniam (1973) have concluded in the case of $\Delta_s = 0$ that the test based on $\tilde{\mathcal{F}}_s$ is superior to $\mathcal{F}_s$. They have provided tables of percentage points of the null distribution of $\tilde{\mathcal{F}}_s$. Also, they found that the chi-squared approximation to the null distribution of $\tilde{\mathcal{F}}_s$ is better than the approximation given by the F-distribution. Both of these approximations were suggested by Rao (1950). They are conservative in that the actual level of significance is less than the prescribed level of significance. The F-distribution-based approximation to the null distribution of $\tilde{\mathcal{F}}_s$ is

$$\frac{N-p+1}{p-p_s} \cdot \frac{N-p_s+1}{N+1} \tilde{\mathcal{F}}_s \sim F_{p-p_s, N-p+1},$$

and the chi-squared approximation is

$$\{N(N-p+1)\} \frac{N-p_s+1}{N+1} \tilde{\mathcal{F}}_s \sim \chi^2_{p-p_s}.$$

Further discussion on tests of H_0 in the case of $\Delta_s = 0$ can be found in Rao (1966b) and Kshirsagar (1972, Chapter 6).

A number of procedures for reducing the number of variables in discriminant analysis is based on the use of $\mathcal{F}_s$. Das Gupta and Perlman (1974) have suggested a modified version of the F-test for variable selection. They showed that the power of Hotelling's T^2-test may be increased by elimination of a subset of the variables represented as $\mathbf{x}^{(\bar{s})}$, provided the reduction in the noncentrality parameter is small. This led them to consider the problem of testing for reduced power due to the exclusion of $\mathbf{x}^{(\bar{s})}$ and they presented a test on the basis of a preliminary sample. In the present context, their procedure would test $H_0^* : \Delta_{\bar{s}} \leq h$, rather than $H_0^{(s)} : \Delta_{\bar{s}} = 0$, where h is the bound on $\Delta_{\bar{s}}$ in order for the power to be reduced on deleting $\mathbf{x}^{(\bar{s})}$; it depends on the unknown population parameters and so must be estimated. Their test rejects H_0^* at level α if $\mathcal{F}_s$ exceeds the $100(1-\alpha)$ percentile of the noncentral F-distribution with degrees of freedom $p - p_s$ and $N - p + 1$ and noncentrality parameter based on estimated h. A similar problem has been studied by Sinha, Clement, and Giri (1978).

12.3 SOME SELECTION PROCEDURES

12.3.1 Selection via Canonical Variate Analysis

One way of proceeding with variable selection is through a sample canonical variate analysis, which was described in Section 6.5. For variable selection, the aim is to eliminate those variables that do not appear to contribute to the canonical variates. This procedure in the two-group situation was used by Weiner and Dunn (1966) and Eisenbeis et al. (1973). The more complicated multiple-group case was considered by Hawkins and Rasmussen (1973) and McKay and Campbell (1982a).

12.3.2 All-Subsets Approach

McCabe (1975) has advocated that the selection of variables in discriminant analysis should begin with an examination of all possible subsets wherever feasible. He proposed an algorithm DISCRIM, using Furnival's (1971) algorithm for evaluating the Wilks' statistic Λ_s, defined by (12.2.3), for all possible subsets s. As a guide for finding important variables, subsets of the same size can then be ordered with respect to their values for Λ_s. McCabe (1975) proposed the use of Λ_s simply as a descriptive statistic that measures the separatory potential of the subset selected. However, he noted that it can be used to construct hypothesis tests about any subsets under the imposition of the normal distributional assumption (12.2.2).

When all subsets are to be considered simultaneously, there is the problem of how to choose critical values for individual tests on its subsets so as to specify a significance level for the entire tests that are performed. A formal treatment of the variable-selection problem in discriminant analysis was provided by McKay (1976, 1977). For $g = 2$ groups, he showed that a simultaneous test procedure (STP) of the type described and illustrated by Gabriel (1968, 1969) can be used to isolate those subsets of the feature variables that provide adequate discrimination. His procedure for ensuring that the family type I probability error rate does not exceed a specified value α considers the variables contained in $\mathbf{x}^{(s)}$ to be adequate for discrimination purposes if

$$(p - p_s)\mathcal{F}_s < pF_{p,N-p+1;1-\alpha}, \tag{12.3.1}$$

where $F_{p,N-p+1;1-\alpha}$ denotes the $100(1-\alpha)$ percentile of the F-distribution with p and $N-p+1$ degrees of freedom. That is, the subset of variables defined by s is adequate if

$$P_s \geq \alpha_s,$$

where α_s satisfies

$$p - p_s F_{p-p_s,N-p+1;1-\alpha_s} = pF_{p,N-p+1;1-\alpha},$$

and where, as before, P_s is the area to the right of $\mathcal{F}_s$ under the $F_{p-p_s,N-p+1}$ density. Hence, α_s is the significance level of the test concerning the adequacy of the particular subvector $\mathbf{x}^{(s)}$ in the STP. The set of subsets of the feature variables for which this condition is satisfied is denoted here by B_α; it includes all adequate subsets with probability at least $1-\alpha$.

In order to isolate all adequate subsets, it is in general not necessary to examine all subsets because of a certain coherence property. If the subset s (really the subset of the feature variables defined by s) belongs to B_α, then so must any subset containing s. Alternatively, if s does not belong to B_α, then neither does any subset of s. As illustrated in the example presented in Section 12.4, if p_s is near to p, then α_s is very small and so the test of adequacy of the particular subset s is quite conservative. As explained by McKay (1978), this is the price paid for the protection of the family type I probability error rate α. To overcome this problem, McKay (1978) has suggested either using a large α or reducing B_α to the set A_α, in which each subset s must satisfy either $P_s \geq \alpha$ (i.e., s is considered adequate at significance level α) or it must contain a subset s satisfying $P_s \geq \alpha$.

In a similar manner to the two-group case, an STP can be provided to handle variable selection for $g > 2$ groups. Of course, now the choice of test criterion on which the STP is based arises (e.g., the likelihood ratio or largest root statistics), since they are no longer identical. McKay and Campbell (1982a) prefer the likelihood ratio test statistic because it is easier to handle computationally (McHenry, 1978; Hintze, 1980). Also, it can be expected, in general, to have greater power. The latter compensates for the greater resolution provided by the STP based on the largest root statistic.

In multiple regression, the C_p plot (Mallows, 1973) is a well-known graphical aid to variable selection. Alternatives to C_p have been suggested by Spjøtvoll (1977). One of Spjøtvoll's proposals was applied by McKay (1978) in the two-group discriminant analysis case. A graphical representation is obtained by plotting P_s against p_s. On such a graph, points above the curve $P_s = \alpha_s$ correspond to members of the set B_α. The idea can also be utilized in the multiple-group case; see McKay and Campbell (1982a) and Fowlkes et al. (1987).

12.3.3 Stepwise Procedures

As mentioned in the previous sections of this chapter, it is not always computationally feasible to examine every possible subset of the variables in the full feature vector $\mathbf{x}$ in the selection of a best subset with respect to some specified criterion. Hence, many selection methods proceed in a stepwise fashion, assessing the variables one at a time for their contribution. They often involve both forward selection and backward elimination. Algorithms for stepwise selection are widely available in the major statistical packages, for example, the programs BMDP7M, DISCRIMINANT, and STEPDISC in the packages BMDP (1988), SPSS (1986), and SAS (1990), respectively. There is also the well-known ALLOC program discussed in the next section.

We now describe a stepwise forward selection procedure, based on the criterion of no additional information. For each k considered in turn ($k = 1,\ldots,p$), compute the statistic

$$(n-g)(1-\Lambda_s)/(g-1)\Lambda_s \tag{12.3.2}$$

for $s = \{k\}$, where Λ_s is given by (12.2.3). The statistic (12.3.2) is the usual univariate analysis of variance F-statistic with $g-1$ and $n-g$ degrees of freedom. The variable corresponding to the largest of the p values of this statistic is selected, provided it exceeds a specified value, say, $F_{g-1,n-g;1-\alpha}$.

Let x_{k_o} denote the variable selected. Then for each k in turn ($k = 1,\ldots,p$; $k \neq k_o$), compute the (partial) F-statistic given by (12.2.4) for $s = \{k_o\}$ and $\bar{s} = \{k\}$ to assess the individual contribution of each of the remaining $p-1$ variables in the presence of x_{k_o}. The variable with the largest value of the partial F-statistic is added to the current subset consisting of x_{k_o}, provided it exceeds the specified threshold. Otherwise, the process is terminated. If the process is not terminated, then consideration is given to the selection of an additional variable in the same manner as on the previous step. The selection continues along these lines and is terminated at the current subset of p_s variables if none of the $p - p_s$ values of the partial F-statistic for an additional variable exceeds the specified threshold. Otherwise, the variable with the largest value of the F-statistic is added to constitute a subset of $p_s + 1$ variables, and consideration is given to the selection of an additional variable in the same manner as on the previous steps.

The sequence of partial F-statistics so produced occur in a factorization of the likelihood ratio statistic; see, for example, Kshirsagar (1972, Chapter 8) and McKay and Campbell (1982a). More recently, Fatti and Hawkins (1986) have considered the decomposition of the likelihood ratio statistic in the heteroscedastic case, where the group-conditional covariance matrices are allowed to be unequal. Their decomposition gives rise to three components, testing the residual homoscedasticity of each variable, the parallelism of its regression on its predecessor, and the identity of location. They have proposed a variety of uses of this decomposition in selecting variables.

With a backward elimination procedure, consideration is given first to the deletion of a single variable from the full set of p variables. For each variable in turn, the partial F-statistic is computed for the test of whether it provides additional information over the remaining $p - 1$ variables. The variable with the smallest value of this partial F-statistic is eliminated provided that the statistic does not exceed a specified initial value. If this variable can be eliminated, the process is repeated on the remaining $p - 1$ variables.

Stepwise procedures employed in practice often use a combination of forward selection and backward elimination. One such way of proceeding is to use backward elimination following each forward step. For example, suppose that the best subset of size $p_s + 1$ variables has been selected on the basis of the criterion of no additional information. Then before consideration is given to the selection of the best subset of $p_s + 2$ variables, the partial F-statistic is computed in turn for each of the $p_s + 1$ variables in the current subset to examine what information it supplies in addition to that supplied by the remaining p_s variables. If the smallest value of this partial F-statistic does not exceed a specified critical value, it is deleted and so the process is shifted back one stage with the current subset containing p_s variables. The whole procedure terminates when none of the subset variables can be excluded, and no further variables can be included. Note that although forward selection may stop at a subset of size, say, p_{s_o}, the best subset of size $p_{s_o} - 1$ should be formed to allow for the possibility of the removal of a variable from it through backward elimination.

As noted in Section 12.1, the use of stepwise procedures raises a number of issues. For example, although with the stepwise selection procedure described above, the F-distribution is usually adopted to specify the threshold value, it is not really appropriate given that it is the smallest of a number of F-statistics that is under consideration. Thus, at each stage the test is not carried out at the nominal significance level. Further, as cautioned by McKay and Campbell (1982a), the tests are not independent and it is extremely difficult to judge the magnitude of the simultaneous significance level for the sequence of tests.

Hawkins (1976), however, has given some guidelines to adopt if the desire is to ensure that the overall probability of including an irrelevant variable will be less than a predetermined value α. In the forward selection phase, if p_s variables have been included, and $p - p_s$ remain for possible inclusion, the suggested critical value for the partial F-statistic is $F_{g-1,n-g-p_s;1-\alpha^*}$, where $\alpha^* = \alpha/(p - p_s)$. In the elimination phase, if p_s variables have been included, the suggested critical value is $F_{g-1,n-g-p_s;1-\alpha^*}$, where $\alpha^* = \tilde{\alpha}/(p - p_s + 1)$ and $\tilde{\alpha} > \alpha$. The exclusion level $\tilde{\alpha}$ is (slightly) larger than α to ensure that the procedure does not continually delete and include variables, never terminating. Hawkins (1976) indicated that his rules for inclusion and deletion may be a little conservative at each step.

For the homoscedastic normal model (12.4.3) with $g = 2$ groups, Costanza and Afifi (1979, 1986) have performed some Monte Carlo experiments to investigate the choice of the critical value of the partial F-statistic for stepwise forward selection based on this statistic.

Although up to now we have described stepwise selection procedures in terms of the no additional information criterion, other criteria may be used. For example, another criterion considers the contribution of an additional variable on the basis of the increase in Rao's (1952, page 257) generalized distance measure, which is equivalent to using the increase in Hotelling's (1951) trace statistic. Another criterion adds the variable, which, together with the variables already selected, maximizes the likelihood ratio statistic, as in the example in Horton, Russell, and Moore (1968). This leads to the same decision as choosing the variable corresponding to the largest partial F-statistic. J. A. Anderson (1982) has suggested a forward selection procedure, using the likelihood ratio statistic in the context of logistic discrimination. There is also the use of the error rate as a criterion, which is considered now.

12.4 ERROR-RATE-BASED PROCEDURES

12.4.1 Introduction

As explained in the previous section, the use of the error rate as a criterion for selection of suitable feature variables in discriminant analysis is particularly appropriate where the aim is to form a discriminant rule for the subsequent allocation of unclassified entities to the groups. The use of this criterion in selection problems has been advocated by McLachlan (1976b) and Habbema and Hermans (1977), among others. Menzefricke (1981) used this criterion in a Bayesian framework. More recently, Ganeshanandam and Krzanowski (1989) considered a stepwise procedure with combined forward selection and backward elimination, as described in the previous section, but with the criterion of no additional information replaced by the overall error rate. A variable is added or deleted if there is a consequent decrease or increase in the estimated overall error rate. For the location model in mixed-variable discrimination, Krzanowski (1983b) has proposed a stepwise backward elimination method to reduce the number of discrete variables.

In keeping with our previous notation, we let $ec(F;\mathbf{t})$ and $eu(F)$ denote the conditional and unconditional overall error rate, respectively, of the sample-based rule $r(\mathbf{x};\mathbf{t})$. Pertaining to the use of the subvector $\mathbf{x}^{(s)}$, we let $F^{(s)}$ denote the distribution function of $\mathbf{Y}^{(s)} = (\mathbf{X}^{(s)\prime}, \mathbf{Z}')'$ and $\mathbf{t}^{(s)}$ denote the reduced form of $\mathbf{t}$ obtained when $\mathbf{x}_j$ is replaced by $\mathbf{x}_j^{(s)}$ for $j = 1,\ldots,n$ in (1.6.1). Then the conditional and unconditional overall error rates of the sample rule $r(\mathbf{x}^{(s)};\mathbf{t}^{(s)})$ based on the subvector $\mathbf{x}^{(s)}$ are given by $ec(F^{(s)};\mathbf{t}^{(s)})$ and $eu(F^{(s)})$, respectively.

The change in the overall conditional error rate consequent to the deletion of the variables in $\mathbf{x}^{(\bar{s})}$ from $\mathbf{x}$ is therefore given by

$$ec(F;\mathbf{t}) - ec(F^{(s)};\mathbf{t}^{(s)}). \qquad (12.4.1)$$

In practice, F, and hence also $F^{(s)}$, is unknown, and so the change in the conditional error rate must be estimated from the training data $\mathbf{t}$, employing one of the methods of error-rate estimation, as described in Chapter 10. Let

$$\hat{e}(\mathbf{t}) - \hat{e}(\mathbf{t}^{(s)}) \qquad (12.4.2)$$

denote an estimate of (12.4.1). An assessment of the usefulness of the variables $\mathbf{x}^{(\bar{s})}$ can be made on the basis of $\hat{e}(\mathbf{t}) - \hat{e}(\mathbf{t}^{(s)})$. Obviously, if this difference is positive, or is only slightly less than zero relative to $\hat{e}(\mathbf{t})$, then it would appear that the variables in $\mathbf{x}^{(\bar{s})}$ can be deleted without any detrimental effect on the error rate. The greater the magnitude of this difference, the more confident one can be that $\mathbf{x}^{(\bar{s})}$ should be either deleted or retained, depending on whether (12.4.2) is positive or negative.

12.4.2 Variable Selection Based on the Assessed Error Rate of the Sample NLDR

To demonstrate the use of the error rate as a criterion in the selection of suitable feature variables, McLachlan (1976b) considered the special case where

$$\mathbf{X} \sim N(\boldsymbol{\mu}_i, \boldsymbol{\Sigma}) \quad \text{with prob. } \pi_i \text{ in } G_i \qquad (i = 1,2), \qquad (12.4.3)$$

and where π_1 and π_2 are set equal to 0.5. It follows from (10.6.6), that under (12.4.3), an N^{-2}-order unbiased estimator of the difference (12.4.1) between the conditional error rates is given by

$$\hat{e}^{(M)}(D,p) - \hat{e}^{(M)}(D_s,p_s). \qquad (12.4.4)$$

Let

$$\{v(\Delta,\Delta_s)\}^2 = c(\Delta) + c(\Delta_s) - 2c(\Delta_s)\{\Delta_s\phi(\tfrac{1}{2}\Delta)\}/\{\Delta\phi(\tfrac{1}{2}\Delta_s)\},$$

where

$$c(\Delta) = \{\tfrac{1}{2}\phi(\tfrac{1}{2}\Delta)\}^2\{n_1^{-1} + n_2^{-1} + \tfrac{1}{2}\Delta^2 N^{-1}\}.$$

McLachlan (1976b) established that the limiting distribution of

$$\{\hat{e}^{(M)}(D,p) - \hat{e}^{(M)}(D_s,p_s)\} - \{ec(F;\mathbf{T}) - ec(F^{(s)};\mathbf{T}^{(s)})\}$$

scaled by $v(\Delta,\Delta_s)$ is a standard normal. It follows that

$$\text{pr}\{ec(F;\mathbf{T}) - ec(F^{(s)};\mathbf{T}^{(s)}) \geq \hat{e}^{(M)}(D,p) - \hat{e}^{(M)}(D_s,p_s) - \Phi^{-1}(1-\alpha_s)v(\Delta,\Delta_s)\}$$
$$\simeq \alpha_s$$

By finding the value of α_s for which

$$\hat{e}^{(M)}(D,p) - \hat{e}^{(M)}(D_s,p_s) - \Phi^{-1}(1-\alpha_s)v(D,D_s) = 0, \tag{12.4.5}$$

an approximate confidence level α_s is obtained that corresponds to no increase in the overall conditional error rate of the sample NLDR on deleting $\mathbf{x}^{(s)}$ from $\mathbf{x}$. By replacing zero with ϵ on the right-hand side of (12.4.4), we can test whether

$$ec(F;\mathbf{t}) - ec(F^{(s)};\mathbf{t}^{(s)}) \geq \epsilon$$

for $\epsilon < 0$, rather than $\epsilon = 0$.

The magnitude and sign of the difference (12.4.2) is reflected in α_s, which indicates the additional discrimination by $\mathbf{x}^{(\bar{s})}$. If the elimination of a subset of variables is to be considered by examining the separate contribution of more than one subset of variables, then the various subsets can be ranked according to their associated values of α_s.

In practice, we do not wish to delete $\mathbf{x}^{(\bar{s})}$ if there is a fair chance that the overall error rate can be increased; on the other hand, we do not wish to retain $\mathbf{x}^{(\bar{s})}$ if the error rate is not likely to be decreased. If α_s falls in the interval, say, [0.9,1] or, say, [0,0.1], then we can either delete or retain $\mathbf{x}^{(\bar{s})}$, respectively, with a high degree of confidence that the error rate is not increased as a consequence of the selection process. For α_s in (0.1, 0.9), there will be reasonable doubt as to whether the error rate is reduced, no matter what decision is taken. If considerable expense or inconvenience is involved with using the feature variables in $\mathbf{x}^{(\bar{s})}$, then we might retain $\mathbf{x}^{(\bar{s})}$ only if α_s is in [0.0.1] or perhaps in a slightly longer interval. If the inclusion of $\mathbf{x}^{(\bar{s})}$ does not involve any appreciable effort with subsequent applications of $r(\mathbf{x};\mathbf{t})$, then it would be reasonable to delete or retain $\mathbf{x}^{(\bar{s})}$, according as α_s is greater or less than 0.5, that is, according as the estimated change in the error rate is greater or less than zero. Wolde-Tsadik and Yu (1979) considered an error-rate-based criterion. They proposed that the adequacy of the subvector $\mathbf{x}^{(s)}$ be measured by the probability of concordance, equal to the probability that the entity is allocated to the same group on the basis of $\mathbf{x}^{(s)}$ as with the full vector $\mathbf{x}$; see McLachlan (1980e).

12.4.3 Akaike's Information Criterion ($g = 2$ Groups)

For the special case (12.4.3) considered by McLachlan (1976b), Fujikoshi (1985) has compared the use of the estimated error rate (12.4.4) with Akaike's information criterion (AIC) for the selection of feature variables. More recently, Daudin (1986) has used this criterion for variable selection in a MANOVA log-linear formulation of the location model for mixed feature variables. Akaike's (1974) information criterion is given by

$$\mathrm{AIC}(\mathbf{x}^{(s)}) = 2d_s - 2\log L(\hat{\boldsymbol{\theta}}_E^{(s)}),$$

where $\hat{\boldsymbol{\theta}}_E^{(s)}$ denotes the maximum likelihood estimator of $\boldsymbol{\theta}_E^{(s)}$ computed under the hypothesis $H_0^{(s)}$ of no additional information in $\mathbf{x}^{(\bar{s})}$, and d_s denotes the dimension of the parameter space under $H_0^{(s)}$. Here $\boldsymbol{\theta}_E$ contains the elements of $\boldsymbol{\mu}_1$, $\boldsymbol{\mu}_2$, and the distinct elements of $\boldsymbol{\Sigma}$. Suppose, as before, the variables of $\mathbf{x}$ are relabeled so that $\mathbf{x} = (\mathbf{x}^{(s)\prime}, \mathbf{x}^{(\bar{s})\prime})'$. The hypothesis $H_0^{(s)}$ of no additional information is equivalent then to

$$H_0^{(s)} : \beta_{kE} = 0 \qquad (k = p_s + 1, \ldots, p), \tag{12.4.6}$$

where

$$\boldsymbol{\beta}_E = (\beta_{1E}, \ldots, \beta_{pE})' = \boldsymbol{\Sigma}^{-1}(\boldsymbol{\mu}_1 - \boldsymbol{\mu}_2) \tag{12.4.7}$$

is the vector of the coefficients in the NLDF $\xi(\mathbf{x}; \boldsymbol{\theta}_E)$ defined by (3.3.4). It can be seen from (12.4.3) and (12.4.6) that d_s is given by

$$\begin{aligned} d_s &= 2p + \tfrac{1}{2}p(p+1) - (p - p_s) \\ &= p + p_s + \tfrac{1}{2}p(p+1). \end{aligned}$$

Fujikoshi (1985) established that

$$\mathrm{AIC}(\mathbf{x}^{(s)}) - \mathrm{AIC}(\mathbf{x}) = N \log\{1 + (p - p_s)\mathcal{F}_s/(N - p - 1)\} + 2(p_s - p),$$

where $\mathcal{F}_s$ is the F-statistic for the test of $H_0^{(s)}$ as defined by (12.2.5).

With a view to selecting the "best" subset of feature variables in $\mathbf{x}$ for discrimination between G_1 and G_2, Fujikoshi (1985) adopted the model where, in addition to (12.4.6),

$$\beta_{kE} \neq 0 \qquad (k = 1, \ldots, p_s). \tag{12.4.8}$$

An equivalent condition in terms of the Mahalanobis distance Δ_s for the subset of the feature variables defined by s is that

$$\Delta_{s_1} < \Delta_s$$

for any proper subset s_1 of s. Thus, $\mathbf{x}^{(s)}$ contains the smallest subset of the original set of variables in $\mathbf{x}$ that provides the same separation between G_1 and G_2 as $\mathbf{x}$.

Fujikoshi (1985) addressed the problem of finding the subvector $\mathbf{x}^{(s)}$, or equivalently the subset s, by selecting the subset s that minimizes AIC($\mathbf{x}^{(s)}$). Let $\hat{s}_A$ be the subset so obtained and $\hat{s}_M$ denote the subset obtained by minimizing the estimated difference (12.4.4) between the error rates. Fujikoshi (1985) showed that these two criteria are asymptotically equivalent in that $\hat{s}_A$ and $\hat{s}_M$ have the same asymptotic null distributions. Also, under (12.4.6) and (12.4.8), the sample discriminant rules using the subvectors of $\mathbf{x}$ corresponding to $\hat{s}_A$ and $\hat{s}_M$ have asymptotically the same overall error rates.

More recently, Zhang (1988) showed that $\hat{s}_A$ is a strongly consistent estimator of s_0, where (12.4.6) and (12.4.8) both hold for the subset s_0. Zhang (1988) proposed also a modified version of this criterion that requires much less computation. For example, he noted that if $p = 20$, then consideration of all possible subsets would require the calculation of the sample Mahalanobis distance D_s for 1,048,575 subvectors $\mathbf{x}^{(s)}$.

With his proposed criterion, this quantity has to calculated only 21 times. The selected subset $\hat{s}_Z$ is taken to be the subset for which

$$\hat{s}_Z = \{k : 1 \leq k \leq p \text{ and } B(k, C_m) > 0\},$$

where

$$\begin{aligned} B(k, C_m) &= m(D - D_{s_k}) - C_m, \\ s_k &= \{j : 1 \leq j \leq p;\ j \neq k\}, \\ m &= \min(n_1, n_2), \end{aligned}$$

and where C_m is chosen to satisfy

$$\lim_{m \to \infty} C_m / m = 0$$

and

$$\lim_{m \to \infty} C_m / \log\log m = \infty.$$

Zhang (1988) showed that $\hat{s}_Z$ is a strongly consistent estimator of s_0.

12.4.4 Variable Selection Based on the Assessed Error Rates of the Sample NQDR

Young and Odell (1986) have considered the variable-selection problem in the situation in which there may be more than $g = 2$ groups and in which the covariance matrices are allowed to be unequal in the specification of the normal model for the group-conditional distributions. Their approach is now described.

For a specified value of p_s ($1 \leq p_s < p$), let $\mathbf{U}_{p_s}$ be the sample analogue of the matrix defined from (3.10.3), which is the approximate solution proposed by Young, Marco, and Odell (1987) to the problem of finding the linear projection of rank p_s that minimizes the overall error rate of the quadratic Bayes rule. Corresponding to the choice of the subvector $\mathbf{x}^{(s)}$ of the feature variables, let $\mathbf{C}_{p_s}$ denote the $p_s \times p$ permutation matrix defined by

$$\mathbf{x}^{(s)} = \mathbf{C}_{p_s}\mathbf{x}.$$

Then Young and Odell (1986) have suggested that the subvector containing the optimal subset of size p_s of the feature variables be determined by the matrix $\mathbf{C}_{p_s}$ that is closest in Euclidean norm to $\mathbf{U}'_{p_s}$. This requires the evaluation of

$$\|\mathbf{U}'_{p_s} - \mathbf{C}_{p_s}\| \tag{12.4.9}$$

over all $\binom{p}{p_s}$ distinct choices of $\mathbf{C}_{p_s}$.

In situations where it is not possible to consider all such choices of $\mathbf{C}_{p_s}$, Young and Odell (1986) have proposed computationally efficient methods of considering (12.4.9) that use either forward or backward selection procedures. They concluded from some simulation studies performed that their proposed methods of variable selection are superior to several methods currently used in popular statistical packages.

12.4.5 ALLOC Algorithm

Habbema et al. (1974a) developed an algorithm ALLOC1 for selection of variables in terms of the overall error rate. Hermans et al. (1982) have given an extension of this algorithm ALLOC80. In order to handle the cases of multiple groups and nonnormality, Habbema et al. (1974a) use the plug-in sample version of the Bayes rule, where the group-conditional densities are estimated nonparametrically by the kernel method. Their algorithm uses the multivariate normal density with a diagonal covariance matrix as the kernel. The smoothing parameter is estimated by the program. A subsequent modification allows the program to use variable kernels to provide better estimates of the group-conditional densities.

For the subvector $\mathbf{x}^{(s)}$ of p_s feature variables, let $\hat{e}(\mathbf{t}^{(s)})$ denote an estimate of the overall conditional error rate of the sample-based rule $r(\mathbf{x}^{(s)}; \mathbf{t}^{(s)})$. Habbema et al. (1974a) proposed that $\hat{e}(\mathbf{t}^{(s)})$ be formed nonparametrically, using the method of cross-validation, which was described in Section 10.2. Actually, they presented their algorithm in terms of $1 - \hat{e}(\mathbf{t}^{(s)})$, that is, the estimate of the overall correct allocation rate. With their algorithm, consideration of all possible subvectors $\mathbf{x}^{(s)}$ is not possible if p is large, particularly since the group-conditional densities and the associated error rate of the subsequent discriminant rule are estimated nonparametrically, which requires extensive computation.

As a consequence, Habbema et al. (1974a) proposed a forward stepwise selection procedure to find a suboptimal subset of the p feature variables. Let $s_{o,k}$ define the subset of variables having the smallest estimated error rate for subsets of size k. Suppose that the selection process is at the qth step. Then it is terminated at this step if

$$\hat{e}(\mathbf{t}^{(s_{o,q+1})}) > \hat{e}(\mathbf{t}^{(s_{o,q})}).$$

12.5 THE *F*-TEST AND ERROR-RATE-BASED VARIABLE SELECTIONS

12.5.1 Introduction

For the sample NLDR applied to three data sets, McLachlan (1980c) has examined the relationship between the significance probability P_s of the F-test for no additional information in the subvector $\mathbf{x}^{(\bar{s})}$ and the approximate confidence coefficient α_s corresponding to no increase in the overall conditional error rate, as determined from (12.4.5).

If P_s is very large, then it suggests that the variables in $\mathbf{x}^{(\bar{s})}$ make little contribution to the Mahalanobis distance between G_1 and G_2 (i.e., $\Delta_{\bar{s}}$ is small), and so it is likely that the overall conditional error rate will not be increased on using $\mathbf{x}^{(s)}$ in place of $\mathbf{x}$. On the other hand, a small P_s would suggest that the overall conditional error rate would be increased.

For the three data sets examined, McLachlan (1980c) concluded that provided the significance level of the F-test is not set at too conservative a level, there should be a fairly high degree of confidence that the overall conditional error rate is not increased by the selection decision based on the F-test of no additional information. For instance, if the feature variables in $\mathbf{x}^{(\bar{s})}$ are deleted according to the F-test with a significance level α greater than, say, 0.10, then the associated confidence α_s of no increase in the overall error rate should be quite high. If we are willing to allow a small increase in the error rate relative to its value for the entire set, then the F-test with a more conservative α, say, around 0.05, should also give a selection decision in accordance with α_s, now corresponding to the confidence that the error rate is not increased beyond the specified bound.

We report here two of the three examples considered in McLachlan (1980c).

12.5.2 Example 1

This example concerns some data of Lubischew (1962) on the characteristics of flea-beetles from the genus *Chaetocnema*. They were analyzed by McKay (1977, 1978), whose aim was to find which subsets of $p = 6$ variables provide adequate discrimination between the two species *Ch. concinna* and *Ch. heikertingeri*. The value of D^2, based on the $p = 6$ variables, was 42.971, and the group-sample sizes were $n_1 = 21$ and $n_2 = 31$. However, N, the degrees of freedom for the estimate of Σ, was 71 (and not $n_1 + n_2 - 2 = 50$) since the estimate was computed by pooling over an additional sample of $n_3 = 22$ observations.

Consistent with our previous notation, we let s refer to the subset of the available feature variables to be retained and hence its complement $\bar{s}$ to the subset of variables to be deleted. The values of $\mathcal{F}_s$ and P_s for the various subsets s, or equivalently $\bar{s}$, considered by McKay (1978) are listed in Table 12.1, along with the corresponding values of D_s^2, $\hat{e}^{(M)}(D,p) - \hat{e}^{(M)}(D_s,p_s)$, which is written henceforth as $\hat{e} - \hat{e}_s$, and α_s. The subsets $\bar{s}$ are listed in order of

TABLE 12.1 Significance Probabilities for F-Test of No Additional Information in $\bar{s}$ versus Asymptotic Confidence of No Increase in the Overall Error Rate: Example 1

Subset Deleted $\bar{s}$	$\mathcal{F}_s$	P_s	α_s	D_s^2	$\hat{e} - \hat{e}_s$
35*	0.66	0.51	0.77	42.02	0.0001
5*	0.53	0.48	0.76	42.59	0.0001
3*	0.73	0.41	0.66	42.44	0.0000
356*	2.53	0.064	0.30	37.95	−0.0003
36*	3.30	0.042	0.233	38.55	−0.0004
56*	3.37	0.039	0.229	38.46	−0.0004
235*	3.50	0.019	0.227	36.29	−0.0006
6*	5.87	0.017	0.1597	39.00	−0.0004
23*	5.08	0.0089	0.1604	36.48	−0.0008
25*	5.25	0.0078	0.1559	36.29	−0.0008
2*	10.10	0.0026	0.1046	36.52	−0.0010
26	10.07	0.00033	0.0869	31.60	−0.0023
1	26.27	0.00003	0.04444	29.12	−0.0041
4	45.21	0.00000	0.0211	23.20	−0.0104

Source: From McLachlan (1980c), with permission from the Biometric Society.

decreasing P_s and, for example, the entry 35 under $\bar{s}$ refers to the situation in which the subset consisting of variables 3 and 5 is deleted from the entire system of $p = 6$ variables. An asterisk denotes a subset whose complement satisfies the condition (12.3.1) of McKay (1976) for adequacy at simultaneous level $\alpha = 0.05$.

McLachlan (1980c) ranked the various subsets according to their associated P_s values so as to contrast this ranking with that obtained on the basis of α_s. This is because he wished to illustrate how large P_s must be for one to be reasonably confident that the error rate is not increased on deleting the particular subset $\bar{s}$. This confidence is approximately measured by the corresponding value of α_s. Note that it was not the objective of McKay (1976) to rank the subsets s. He was concerned solely with isolating the set B_α of adequate subsets s and considered that the final choice among the subsets in B_α must be a subjective one.

It can be seen from Table 12.1 that the ranking of the subsets $\bar{s}$ according to P_s is almost the same as using α_s and, as explained in the previous section, this is to be expected. The subsets whose rankings are different when using α_s are $\bar{s} = \{6\}$ and $\bar{s} = \{23\}$; their ranks according to P_s are interchanged on the basis of α_s.

As D^2 is very large for this problem, the overall conditional error rate, at least for the entire $p = 6$ variables, is small; indeed, it is estimated by $\hat{e}$ to be only 0.0008.

Consequently, unless D^2 is drastically reduced by the deletion of a subset $\bar{s}$, any change in the error rate should only be small in the absolute sense, although it can be appreciable in relation to the size of the error rate for the entire system.

For the first three subsets, their P_s values are quite large and the subsets would certainly be deleted at any conventional significance level specified with the F-test in practice. The α_s values associated with these three subsets range from 0.66 to 0.77, indicating a fair degree of confidence that the error rate is not increased on the separate elimination of each of these three subsets.

The fourth subset $\bar{s} = \{356\}$ is a more interesting case to discuss. According to its P_s value of 0.064, it would be retained using the F-test at $\alpha = 0.10$, but deleted at $\alpha = 0.05$. The corresponding α_s value of 0.3 suggests that there is over twice the chance that the error rate is increased rather than decreased on deleting $\bar{s}$. The increase in the error is estimated by $\hat{e} - \hat{e}_s$ to be 0.0003, suggesting that the error rate has increased by over a third of its estimated value for $p = 6$ variables. If we are willing to accept an increase of this size [i.e., $\epsilon = -0.0003$ replaces zero on the right-hand side of (12.4.5)], then α_s, now corresponding to the confidence that any increase in the error rate does not exceed 0.0003, is equal to 0.5.

As regards the fifth and sixth subsets, their P_s values show that each would be deleted only if the significance level α of the F-test were smaller than 0.042 and 0.039, respectively; otherwise, they would be retained. It can be seen that the associated α_s values suggest that the chance of the error rate being reduced is only a little over 20%.

McLachlan (1980c) noted that the simultaneous test procedure of McKay (1978) leads to the complement of each subset in Table 12.1, except for those of the last three subsets, being considered adequate at simultaneous level $\alpha =$ 0.05. This is because the α_s value that P_s must exceed for s to be considered adequate is very small for p_s near p, as explained earlier.

12.5.3 Example 2

As another example, we consider the third example in McLachlan (1980c) on the data of Zaslavski and Sycheva (1965), concerning the diagnosis of stenocardia on the basis of 13 ballisto-cardiographical variables and described by Urbakh (1971). The latter author suggested that a variable be deleted if $\hat{\Delta}_{\bar{s}}^2 - \hat{\Delta}^2 \geq 0$, where $\hat{\Delta}^2$ and $\hat{\Delta}_{\bar{s}}^2$ are possible estimates of Δ^2 and $\Delta_{\bar{s}}^2$, respectively; for example, the unbiased estimates defined from (10.6.5). For this data set,

$$D^2 = 10.90, \qquad \hat{e} = 0.0639, \qquad n_1 = 100, \qquad n_2 = 93, \qquad \text{and} \qquad p = 13.$$

According to Urbakh's criterion described above, any one of the variables 3, 4, or 10 can be omitted; McLachlan (1976b) showed that these selections were in agreement with those based on α_s. Subsequently, Schaafsma and van Vark (1979) showed that the same selections are obtained from their crite-

TABLE 12.2 Significance Probabilities for F-Test of No Additional Information in $\bar{s}$ versus Approximate Confidence of No Increase in the Overall Error Rate: Example 2

Variable Deleted $\bar{s}$	$\mathcal{F}_s$	P_s	α_s	D_s^2	$\hat{e} - \hat{e}_s$
8	0.1205	0.73	0.99	10.89	0.0011
4	0.3620	0.55	0.89	10.87	0.0009
3	0.4830	0.49	0.83	10.86	0.0008
10	0.6042	0.44	0.78	10.85	0.0007
9	2.1943	0.14	0.40	10.72	−0.0005
7	6.1021	0.014	0.14	10.41	−0.0035
5	6.4892	0.012	0.12	10.38	−0.0038
1	12.9114	0.00042	0.031	9.90	−0.0087
13	13.6060	0.00030	0.027	9.85	−0.0093
2	14.1653	0.00022	0.024	9.81	−0.0097
11	23.7325	0.00000	0.0043	9.16	−0.0172
12	24.6636	0.00000	0.0036	9.10	−0.0180
6	46.2130	0.00000	0.0001	7.85	−0.0352

Source: From McLachlan (1980c), with permission from the Biometric Society.

rion based on the mean difference of $\mathbf{x}'\mathbf{S}^{-1}(\bar{\mathbf{x}}_1 - \bar{\mathbf{x}}_2)$ between the groups relative to its standard deviation.

The values of P_s and α_s corresponding to the deletion of a single variable from the system of $p = 13$ variables are displayed in Table 12.2. The deletion of subsets of more than one variable was not considered by McLachlan (1980c). This is because the corresponding values of D_s^2 could not be computed with sufficient precision from the tabulated data in Urbakh (1971), which were given to two decimal places only.

We see from Table 12.2 that the ranking of each variable according to P_s is the same as for α_s. For the first four variables in the table, P_s is quite large, and so according to the F-test, each can be deleted separately with no loss of discrimination. The associated values of α_s are correspondingly large, indicating that there is a high degree of confidence that the overall conditional error rate is not increased if either of these four variables is deleted separately. The case of variable 9 is of interest. With its P_s value of 0.14, it would be deleted according to the F-test at any conventional significance level α. However, the value of α_s suggests that the error rate is slightly more likely to be increased than decreased on its deletion. Regarding the next two variables (7 and 5) listed in Table 12.2, each would be deleted according to the F-test if α was specified at the conservative level of 0.01. The associated α_s values indicate only a small chance that the error rate is not increased on deleting either variable. If we are willing to accept a relatively small increase in the error rate, provided it is less than, say, 0.01 then α_s, now corresponding to the confidence that any increase in the error rate does not exceed 0.01, is equal to 0.98 and 0.97 for variables 7 and 5, respectively.

TABLE 12.3 Apparent Error Rate of Sample NLDR Formed from Best Subset s_o of Specified Size p_{s_o}

p_{s_o}	1	2	3	4	5	6	7	8	9	10
$A(t^{(s_o)})$	25.2	18.0	13.9	11.5	10.6	10.0	10.8	12.2	15.3	21.4
$eo(F^{(s_o)})$	40.1	36.2	33.3	30.8	28.8	27.0	25.4	24.0	22.7	21.5

Source: Murray (1977b).

12.6 ASSESSMENT OF THE ALLOCATORY CAPACITY OF THE SELECTED FEATURE VARIABLES

12.6.1 Selection Bias

Caution has to be exercised in selecting a small number of variables from a large set, as there will be a selection bias associated with choosing the optimal of a large number of possible subsets, regardless of the criterion used. This problem in discriminant analysis has been considered by Murray (1977b), Hecker and Wagner (1978) and, more recently, by Gong (1986), Queiros and Gelsema (1989), Snapinn and Knoke (1989), and Ganeshanandam and Krzanowski (1989), among others. Rencher and Larson (1980) considered this bias problem with the use of Wilks' ratio Λ_s, which is biased downward, in particular if p exceeds $n-g$. In the latter case, a value of Λ_s cannot be obtained for the entire set of p variables. Miller (1984, 1990) makes a number of allied points regarding the selection of variables in multiple regression.

Consider the selection of the subvector $\mathbf{x}^{(s)}$ of the full feature vector $\mathbf{x}$. Suppose that s_o defines the subset of feature variables of some specified size p_{s_o} that minimizes the adopted estimate $\hat{e}(\mathbf{t}^{(s)})$ over all possible $\binom{p}{p_{s_o}}$ distinct subsets s of size p_{s_o}. Although $\hat{e}(\mathbf{t}^{(s)})$ may be an unbiased estimator of the overall conditional error rate, $\hat{e}(\mathbf{t}^{(s_o)})$ is obviously not as it is obtained by taking the smallest of the estimated error rates after they have been ordered. As noted by Murray (1977b), the situation is exacerbated if an optimistic estimator of the error rate, such as the apparent error rate, is used. There is then a double layer of overoptimistic bias inherent in the assessment. For a given selection method, the magnitude of the bias obviously depends on the sample size n, being greatest for small values of n.

To illustrate this bias, Murray (1977b) performed some simulation experiments for the sample NLDR under the homoscedastic normal model (12.2.2) with $g=2$ and independent feature variables having unit variances in each group and means 1/4 and $-1/4$ in G_1 and G_2, respectively. For the case of $p=10$ and $n_1=n_2=25$, we list in Table 12.3 as a percentage his simulated value of the apparent error rate $A(\mathbf{t}^{(s_o)})$, corresponding to the best subset s_o of size p_{s_o}, for $p_{s_o}=1,\ldots,10$. For comparative purposes, Murray (1977b) also tabulated the optimal overall error rate given by

$$eo(F^{(s_o)}) = \Phi(-\tfrac{1}{4}\sqrt{p_{s_o}})$$

in the case where s_o contains p_{s_o} elements.

It can be seen that as p_{s_o} increases, the apparent error rate corresponding to the selected subset s_o falls initially, but then increases again. This behavior can be explained by the fact that the number of subsets $\binom{p}{p_{s_o}}$ increases rapidly as p_o increases to $\frac{1}{2}p$, and then falls as p_{s_o} increases to p. Hence, the selection bias is greatest for p_{s_o} near $\frac{1}{2}p$ and decreases to zero as p_{s_o} tends to zero or p.

The empirical results of Murray (1977b) emphasize the point that one should not use the apparent error rate of the sample rule based on the best subset to provide a reliable assessment of its performance in allocating new entities of unknown origin. Murray (1977b) also noted that they raise a more difficult point of whether it is in fact sensible to choose the best subset of feature variables that appears to be optimal for a given set of data.

12.6.2 Reduction of Selection Bias

Suppose that $\mathbf{x}^{(s_o)}$ contains the subset of feature variables selected as being the best of size p_{s_o}, according to some criterion. Let $r(\mathbf{x}^{(s_o)};\mathbf{t}^{(s_o)})$ denote some arbitrary sample discriminant rule formed from the classified training data $\mathbf{t}^{(s_o)}$ on the feature vector $\mathbf{x}^{(s_o)}$.

By using the same notation as adopted in Chapter 10, the overall apparent error rate of this rule can be expressed as

$$A(\mathbf{t}^{(s_o)}) = \frac{1}{n}\sum_{i=1}^{g}\sum_{j=1}^{n} z_{ij}Q[i, r(\mathbf{x}_j^{(s_o)};\mathbf{t}^{(s_o)})]. \tag{12.6.1}$$

The optimism arising from the use of the apparent error rate can be almost eliminated using cross-validation. The cross-validated estimate is

$$A^{(CV)}(\mathbf{t}^{(s_o)}) = \frac{1}{n}\sum_{i=1}^{g}\sum_{j=1}^{n} z_{ij}Q[i, r(\mathbf{x}_j^{(s_o)};\mathbf{t}_{(j)}^{(s_o)})], \tag{12.6.1}$$

where $\mathbf{t}_{(j)}^{(s_o)}$ denotes the training data $\mathbf{t}^{(s_o)}$ with $\mathbf{y}_j^{(s_o)} = (\mathbf{x}_j^{(s_o)\prime}, \mathbf{z}_j)'$ deleted. In order to reduce the selection bias that is still present in the estimate (12.6.1), Ganeshanandam and Krzanowski (1989) proposed that cross-validation should precede the variable selection itself. Their proposed estimate of the overall error rate of $r(\mathbf{x}^{(s_o)};\mathbf{t}^{(s_o)})$ is given by

$$\tilde{A}^{(CV)}(\mathbf{t}) = \frac{1}{n}\sum_{i=1}^{g}\sum_{j=1}^{n} z_{ij}Q[i, r(\mathbf{x}_j^{(s_{oj})};\mathbf{t}_{(j)}^{(s_{oj})})],$$

where s_{oj} denotes the optimal subset, according to the adopted selection criterion applied to the training data $\mathbf{t}_{(j)}$ without $\mathbf{y}_j = (\mathbf{x}_j', \mathbf{z}_j')'$. As the notation implies, the selected subset s_{oj} for the allocation of the jth entity can be different for each j $(j = 1, \ldots, n)$.

As conceded by Ganeshanandam and Krzanowski (1989), this way of overcoming the selection bias involves a high computing penalty. But they note there are many situations in practice where its implementation is computationally feasible. To demonstrate the usefulness of their approach in overcoming this problem of selection bias, Ganashanandam and Krzanowski (1989) performed a number of simulation experiments under ideal conditions as represented by the normal model (12.2.2) and also under nonideal conditions as represented by multivariate binary data. They concluded that their approach leads to much greater accuracy relative to that of other available methods of assessment of the selected subset. In their simulation experiments, they employed a fully stepwise procedure that included the possibility of deleting as well as entering variables at each stage.

CHAPTER 13

Statistical Image Analysis

13.1 INTRODUCTION

13.1.1 Image Processing

In this chapter, we consider applications and extensions of discriminant analysis motivated by problems in statistical image analysis, in particular in the area of remote sensing. Image processing is required in a very wide range of practical problems. In addition to the analysis of remotely sensed images from satellites, examples include problems in medicine (e.g., automatic analysis and classification of photomicrographs of tissue cells in blood and cancer tests, and recognition of chromosome properties for genetic studies), nuclear medicine (e.g., photon emission tomography and scans obtained by nuclear magnetic resonance or gamma camera), ultrasound, computer vision (e.g., automatic object recognition), and astronomy, where it is now common to collect data using two-dimensional arrays of detectors. In the past, contributions by statisticians in the latter area were limited possibly, as conjectured by Ripley (1986), because of the barrier imposed by a different jargon. Much of image analysis presents challenges that have been traditionally statistical, such as data reduction and detecting patterns amongst noise. Statisticians have begun to work on these challenges, in particular, on those in remote sensing with the segmentation of satellite images, in which pioneering work using statistical theory was undertaken by Switzer (1969, 1980, 1983). There is now an extensive literature on statistical image analysis. A detailed and systematic overview of the field was given recently by Ripley (1986, 1988, Chapter 5) and Switzer (1987). Venetoulias (1988) provides a useful annotated bibliography, containing 85 papers of a statistical nature on image processing. In a further report, Venetoulias (1989) considered problems of parameter estimation in the context of image-processing applications. A concise but informative account of statistical image analysis was given by Besag (1986) in the presentation of his ICM algorithm, which is considered in Section 13.6. Further references on the topic can be found in the work by Fu and Yu (1980); Yu and Fu (1983); Geman and Geman (1984); Hjort and Mohn (1984, 1987); Kittler and Föglein (1984); Mardia (1984); Haslett (1985); Saebo et al. (1985); Kay and Titterington (1986); Mohn, Hjort, and Storvik (1986); Glaz (1988); and Owen (1984, 1989). There is also the special issue of the *Journal of Applied Statistics* on Statistical Methods in Image Analysis (1989, **16**, 125–290), which is a source of additional references.

13.1.2 Remote Sensing

Remote sensing is a generic term that includes aerial surveys and sonar and radar mappings, but which is principally applied to digital imaging from satellites. Integrated studies on earth resources have assumed a lot of importance recently, with the availability of data from a variety of satellites such as LANDSAT Multispectral Scanner (MSS), Thematic Mapper (TM), the French satellite SPOT, and the Indian Remote Sensing Satellite (IRS).

Every point on the earth's surface is constantly reflecting and emitting electromagnetic radiation. The intensities vary for the different wavelengths of the electromagnetic spectrum. This spectral distribution, or spectral signature, depends upon several factors, of which the most important ones are surface conditions, type of land cover, temperature, biological activity, and the angle of incoming radiation. Satellites equipped with a multispectral scanner are able to measure the intensity in several various bands of the spectrum. A portion of the surface of the earth is partitioned into a grid of small squares called pixels. For LANDSAT 1–3, these pixels, or picture elements, are nominally 80 m square but overlap. For each pixel, a satellite records the intensity of reflected electromagnetic radiation at each of several wavelengths, typically four (green, red, and two infrared bands). More recent satellites have resolutions down to 10 m $\times$ 10 m and/or more colors. For example, the French SPOT satellite has 20 m $\times$ 20 m surface elements.

The traditional analysis of remotely sensed images has been done by photointerpreters. However, human interpretation is slow, expensive, and open to dispute. Thus, much attention has been directed in recent times on the automatic classification of pixels in remotely sensed scenes from satellites such as LANDSAT and SPOT. The output is an image segmentation, partitioning the scene into blocks, each belonging to one of a small number of different classes or groups representing the type of vegetation, rock type, etc.

As explained by Ripley (1986), the spatial structure of pixels is relevant in two distinct ways. It has already been mentioned that the pixels overlap. The true situation is much worse, with 50% of the light received by the scanner when pointing at one nominal pixel coming from nearby pixels. The noise that corrupts the signal is also spatially autocorrelated. Hence, the whole observation process has a spatial component. Of perhaps even greater importance is the need to use contextual rules in allocating the pixels to the specified groups. By a contextual allocation rule is meant one formed under a model that attempts to incorporate the *a priori* knowledge that spatially neighboring pixels tend to belong to the same group. Thus, with a contextual allocation rule, a pixel is allocated not only on the basis of its observed feature vector, but also on the feature data of neighboring pixels. The use of a noncontextual rule that allocates a pixel on the basis of its feature vector only and thereby ignores the information on neighboring pixels leads to a "patchwork quilt" of colors representing the different groups. However, in the situation where the colors represent different categories of land use, it is known *a priori* that the land-use categories tend to occur in blocks of moderate size (a few to tens of pixels across).

13.1.3 Notation

It is supposed that there is a two-dimensional scene S of pixels, forming a $M_1 \times M_2$ rectangular array. For the (u,v)th pixel, a p-dimensional vector $\mathbf{X}_{uv}$ of intensities is observed with $\mathbf{x}_{uv}$ denoting its observed value. With some applications, each $\mathbf{X}_{uv}$ is augmented with further variables, perhaps related to topography and texture. For the present, it is computationally convenient to relabel the pixels in some manner by the integers $j = 1,\ldots,m$, where $m = M_1M_2$. The associated feature vectors containing the intensities are denoted correspondingly by $\mathbf{X}_1,\ldots,\mathbf{X}_m$. It is assumed that each pixel has a true color, lying in a prescribed set. If the colors have a natural ordering (usually ranging from black to white), they are referred to as grey levels. In this case, the colors represent the value per pixel of some underlying variable, such as intensity. We will focus here on the situation where the colors are unordered. In this case, the colors are usually tokens for other characteristics of the scene, such as the types of vegetation or the predominant rock type. The assumption that each pixel has a true color is unlikely to hold for pixels on the boundary between different surface categories. They would be expected to contain at least two different categories. Such pixels are said to be mixed; see Kent and Mardia (1986) for a recent treatment of mixed pixels.

We let the groups representing the g possible (unordered) colors be denoted by $G_1,\ldots,G_g$. That is, a pixel belongs to group G_i if it is of the ith color ($i = 1,\ldots,g$). The possibility that none of the pixels belong to these groups can be handled by the introduction of another group if so desired; see Hjort and Mohn (1984). Also consistent with our previous notation, $\mathbf{Z}_j$ is the group-indicator vector defining the color of the jth pixel with feature vector $\mathbf{X}_j$, where $Z_{ij} = (\mathbf{Z}_j)_i = 1$ if the jth pixel belongs to G_i and is zero otherwise ($i = 1,\ldots,g$). We let

$$\mathbf{X} = (\mathbf{X}_1',\ldots,\mathbf{X}_m')'.$$

Note that we are now using $\mathbf{X}$ to denote an m-tuple of points in $\mathbb{R}^p$, whereas in the previous chapters, it was used to denote a single random vector of feature variables in $\mathbb{R}^p$, having $\mathbf{X}_j$ as its jth replication ($j = 1,\ldots,m$). Similarly, we now let

$$\mathbf{Z} = (\mathbf{Z}_1',\ldots,\mathbf{Z}_m')'$$

and

$$\mathbf{Y} = (\mathbf{Y}_1',\ldots,\mathbf{Y}_m')',$$

where $\mathbf{Y}_j = (\mathbf{X}_j',\mathbf{Z}_j')'$ for $j = 1,\ldots,m$. The observed values of $\mathbf{X}$, $\mathbf{Y}$, and $\mathbf{Z}$ are denoted by $\mathbf{x}$, $\mathbf{y}$, and $\mathbf{z}$, respectively.

13.1.4 Image Segmentation or Restoration

In the framework above, the problem is to produce an estimate $\hat{\mathbf{z}}$ of the unknown vector $\mathbf{z}$ that defines the colors of the m pixels in the given scene S, on the basis of the observed feature data $\mathbf{x}$. This process is referred to as segmentation. It can also be referred to as image restoration.

An estimate $\hat{\mathbf{z}}$ can be produced either by simultaneous estimation of its m subvectors $\mathbf{z}_j$ ($j = 1,\ldots,m$) or by estimation of each $\mathbf{z}_j$ considered individually. An example of the former approach is taking $\hat{\mathbf{z}}$ to be the value of $\mathbf{z}$ that maximizes

$$\text{pr}\{\mathbf{Z} = \mathbf{z} \mid \mathbf{x}\}. \tag{13.1.1}$$

That is, $\hat{\mathbf{z}}$ is taken to be the mode of the posterior distribution of $\mathbf{Z}$. It is therefore referred to as the MAP (maximum *a posteriori*) estimate. From a decision-theoretic viewpoint, $\hat{\mathbf{z}}$ corresponds to the adoption of a zero-one loss function according as to whether the reconstructed image is perfect or not. The maximization of (13.1.1) would appear at first sight to be an ambitious task given there are g^m possible values of $\mathbf{z}$. Nevertheless, this is the approach of Geman and Geman (1984) described in Section 13.7.

The pixels can be individually allocated by choosing $\hat{\mathbf{z}}_j$ to be the value of $\mathbf{z}_j$ that has maximum posterior probability given $\mathbf{X} = \mathbf{x}$, that is, $\hat{\mathbf{z}}_j$ maximizes

$$\text{pr}\{\mathbf{Z}_j = \mathbf{z}_j \mid \mathbf{x}\}. \tag{13.1.2}$$

This approach of maximizing the posterior marginal probability for each pixel corresponds to maximizing the expected number of correctly assigned pixels in the scene. It is thus biased toward a low rate of misallocated pixels rather than overall appearance. A low misallocation rate does not necessarily produce a good-looking segmentation; see Figure 2 of Ripley (1986). Individual allocation of the pixels according to (13.1.2) may be more pertinent than a simultaneous allocation procedure in some circumstances. An example where this would be the case is the construction of a crop inventory from satellite data, where the reconstruction of the scene itself is of secondary importance; see the special issue of *Communications in Statistics–Theory and Methods* (1984, **13**, 2857–2996) on Crop Surveys Using Satellite Data.

A key aspect in the development of an estimate $\hat{\mathbf{z}}$ of the true colors of the pixels, whether undertaken simultaneously or individually, is the prior distribution specified for $\mathbf{Z}$. In modeling this distribution, the intent is to probabilistically reflect the extent to which spatially neighboring pixels are of the same color. In this endeavor, models known as Markov random fields can play a very useful role.

13.2 MARKOV RANDOM FIELDS

13.2.1 Definitions

Markov random fields are commonly employed in image-processing problems to model departures from independence in the prior distribution of the colors of the pixels within a given scene. We therefore provide here a brief account of Markov random fields, focusing on two broad classes of such models: symmetric (noncausal) models and mesh (causal) models.

Markov random fields are models that extend the concept of one-dimensional Markov processes to the two-dimensional plane. Let $P(\mathbf{z})$ define a probability distribution for $\mathbf{Z}$ that assigns colors to the pixels in the scene S. Let $\mathbf{z}_{(j)}$ denote $\mathbf{z}$ with $\mathbf{z}_j$ deleted and consider the conditional probability

$$\text{pr}\{\mathbf{Z}_j = \mathbf{z}_j \mid \mathbf{z}_{(j)}\}, \tag{13.2.1}$$

where, as noted by Besag (1986), $\mathbf{z}_{(j)}$ is the only natural conditioning set for spatial distributions. Viewed through the conditional probability (13.2.1) for each pixel j, the probability distribution is termed a Markov random field (MRF).

A Markov random field is said to be locally dependent if the conditional distribution (13.2.1) depends only on the colors of the pixels in the immediate vicinity of pixel j. That is, if $N_j = \{j_1, \ldots, j_s\}$ is a subset of $\{1, \ldots, m\}$, containing the labels of the s pixels in some so-called neighborhood of pixel j, the Markov random field is locally dependent if and only if

$$\text{pr}\{\mathbf{Z}_j = \mathbf{z}_j \mid \mathbf{z}_{(j)}\} = P_j(\mathbf{z}_j; \mathbf{z}_{N_j}) \qquad (j = 1, \ldots, m), \tag{13.2.2}$$

where the function $P_j(\cdot;\cdot)$ is specific to pixel j and

$$\mathbf{z}_{N_j} = (\mathbf{z}'_{j_1}, \ldots, \mathbf{z}'_{j_s})'$$

specifies the colors of the pixels in the neighborhood of pixel j.

In practice, the construction of a Markov random field on S is usually carried out by first imposing a neighborhood structure on S and then choosing a probability distribution that satisfies the condition (13.2.2). The alternative approach of trying to fit a model to a particular image by choosing, for example, the most "appropriate" neighborhood structure has been used occasionally (Kashyap and Chellappa, 1983), but is not very common. The first approach where the neighborhood structure is specified *a priori* may at first sight seem unrealistic, but is less restrictive than it appears. As discussed in the following sections, locally dependent Markov random fields have formed the basis for a number of relatively successful image-segmentation algorithms.

The simplest departure from independence is a first-order Markov random field, in which the neighbors of each pixel j comprise its available N, S, E, and W adjacencies. On the boundary, where pixels have less neighbors, assumptions are made for convenience. A first-order assumption is viewed by Besag (1986) to be unrealistic for most practical purposes. For a second-order field, the available pixels that are diagonally adjacent to pixel j are included also. Thus, each interior pixel j has eight neighbors.

The first- and second-order neighborhoods are the first two members of a whole hierarchy of symmetric neighborhoods. It should be noted, however, that the probability distribution (13.2.2) is generally subject to restrictive consistency conditions, which are not all obvious. These conditions are identified by the Hammersley–Clifford theorem; see Besag (1974). For instance, this theorem implies that there is a symmetry in designating neighbors, that is, if pixel k is a neighborhood of j, then j must be a neighbor of k.

A Markovian random field should not be viewed as a definitive representation of an image. Rather, locally dependent Markov random fields serve only to provide a way of formalizing the notion that nearby pixels are likely to have the same color. As explained by Besag (1986), the effect of choosing a nondegenerate field to describe the local properties of the scene is to induce large-scale characteristics of the model that are somewhat undesirable. Even relatively simple Markov random fields can exhibit positive correlations over arbitrarily large distances when adjacent pixels are very likely to be the same color. Indeed, on the infinite lattice, there is a strong tendency to form infinite single-color patches. Thus, the aim has been to devise methods of image segmentation that, besides being relatively straightforward to implement, are unaffected by the large-scale characteristics of the chosen Markov random field.

The conditional probability distribution (13.2.2) is not the only possible definition of a Markov random field. There exists an equivalent definition in terms of energy functions (or Gibbs potentials), but it is not needed here. For further details on Markov random fields, the reader is referred to Besag (1974), Kinderman and Snell (1980), Dubes and Jain (1989), Qian and Titterington (1989), and Venetoulias (1989).

13.2.2 Spatially Symmetric Models

Besag (1986) has suggested the use of the following Markov random field:

$$P(\mathbf{z}) \propto \exp\left\{ \sum_{j} H_j(\mathbf{z}_j) + \sum_{j \neq k} H_{jk}(\mathbf{z}_j, \mathbf{z}_k) \right\}, \tag{13.2.3}$$

where the functions H_j and H_{jk} are arbitrary, subject to the restriction that H_{jk} must be zero if pixels j and k are not neighbors. This is a pairwise interaction Markov random field. It is possible to have interactions among three or more pixels; see Kinderman and Snell (1980) and Venetoulias (1989) for further details. The normalizing constant for (13.2.3) is known as the partition function and is usually intractable as it is a sum of g^m terms. Fortunately, it is not required in the implementation of the image-segmentation algorithms that use (13.2.3) as the prior distribution for **Z**. Besag (1986) discussed some interesting versions of (13.2.3) that can be applied to discrete or continuous grey levels.

The present situation of g groups representing g unordered colors of the image can be modeled (Besag, 1976; Strauss, 1977) by a special case of (13.2.3), namely,

$$P(\mathbf{z}) \propto \exp\left(\sum_{i=1}^{g} \beta_i m_i - \sum_{1 \leq h < i \leq g}^{g} \beta_{hi} m_{hi}\right), \tag{13.2.4}$$

where m_i is the number of pixels that belong to G_i, and m_{hi} is the number of distinct pairs of neighboring pixels that belong to groups G_h and G_i, respectively. The constants β_i and β_{hi} are arbitrary parameters. In the subsequent work, β_{hi} is taken to be postive, which discourages pixels from G_h and G_i being neighbors. Note that there is a redundancy among the β_i's, since $m_1 + \cdots + m_g = m$.

By considering any two realizations of **Z** that differ only at pixel j, one can compute the conditional probability that pixel j belongs to G_i (i.e., $Z_{ij} = (\mathbf{Z}_j)_i = 1$), given the group membership of all other pixels (i.e., given $\mathbf{z}_{(j)}$). It follows that

$$\text{pr}\{Z_{ij} = 1 \mid \mathbf{z}_{(j)}\} \propto \exp\left(\beta_i - \sum_{h \neq i} \beta_{hi} u_{hj}\right), \tag{13.2.5}$$

where u_{hj} denotes the number of neighbors of pixel j belonging to G_h ($h = 1,\ldots,g$). This defines the probability distribution for $P(\mathbf{z})$ through the conditional probabilities (13.2.5), which confirms the Markov property of (13.2.4). The interaction parameters β_{hi} are required to satisfy the condition of symmetry,

$$\beta_{hi} = \beta_{ih} \qquad (h, i = 1,\ldots,g).$$

A further simplication of (13.2.5) is to take the β_{hi} to have a common value (β), so that

$$\text{pr}\{Z_{ij} = 1 \mid \mathbf{z}_{(j)}\} \propto \exp(\beta_i + \beta u_{ij}), \tag{13.2.6}$$

on noting that $u_{1j} + \cdots + u_{gj}$ is the total number of neighbors of pixel j. The result (13.2.6) implies that the conditional odds in favor of pixel j belonging to G_i depend only on the number of its neighbors belonging to the same group G_i. A final simplification occurs if all the groups $G_1,\ldots,G_g$ are equally likely, in which case the β_i must all be equal and so can be set equal to zero. This yields

$$\text{pr}\{Z_{ij} = 1 \mid \mathbf{z}_{(j)}\} \propto \exp(\beta u_{ij}) \qquad (i = 1,\ldots,g;\ j = 1,\ldots,m). \tag{13.2.7}$$

This simple model for binary scenes ($g = 2$) is the autologistic model due to Besag (1972). It is known as the Ising model in statistical physics, where it has been used to describe the behavior of magnetic particles in which the two groups represent the two polarities (Kinderman and Snell, 1980).

It can be seen from (13.2.7) that for this model, the geometry of the neighborhood does not affect the conditional probability that pixel j has a certain color, given the colors of all the other pixels. Rather, it depends only on the number of neighbors of pixel j that are of the same color.

13.2.3 Markov Mesh Models

For image-segmentation methods using a Markov random field to model the prior distribution of the scene, a drawback is that

$$\text{pr}\{\mathbf{Z}_j = \mathbf{z}_j, \mathbf{Z}_{N_j} = \mathbf{z}_{N_j}\}$$

does not exist in closed form, even for N_j corresponding to simple choices of a neighborhood. The exceptions to this fall within the special class of Markov random field models known as Markov mesh models (Abend, Harley, and Kanal, 1965), which have been designed specifically with this aim. The Markov mesh model is defined by the conditional distribution of $\mathbf{Z}_j$ given the value of $\mathbf{z}_k$ for all pixels k that precede pixel j. To define the concept of precedence, suppose that pixel j_k corresponds to the (u_k, v_k)th pixel in the rectangular array comprising the scene S. Then pixel j_1 precedes pixel j_2 if and only if $u_1 < u_2$ or $u_1 = u_2$ and $v_1 < v_2$.

Markov mesh models are also called causal Markov random fields because their unilateral nature resembles the time causality present in time series. In accordance with the Markovian condition (13.2.2), the conditional distribution of $\mathbf{Z}_j$ given the group membership of the pixels that precede it is taken to depend on only a few of the predecessors of pixel j. The unilateral nature of these models implies that the unobvious consistency conditions do not arise.

The specifications underlying Markov mesh models are unnatural in a spatial context and can be very restrictive; see Besag (1986) for further discussion on this. The one exception to the asymmetry that pervades the family of Markov mesh models is the model of Pickard (1977, 1980), adopted in image processing, implicitly by Kittler and Föglein (1984) and explicitly by Haslett (1985); see also Derin et al. (1984). It corresponds to the Markov mesh model in which the predecessors of the (u, v)th pixel are taken to be the pixels with coordinates

$$\{(u-1, v), (u, v-1), (u-1, v-1)\}. \tag{13.2.8}$$

As such, the Pickard (1977, 1980) model lays emphasis inappropriately on the orientation of a pixel with respect to the scene. Computation, however, is simplified considerably. This model has the curious property that, given the color of any pixel, the colors of immediately adjacent pixels are conditionally independent. Thus,

$$\mathrm{pr}\{\mathbf{Z}_{N_j} = \mathbf{z}_{N_j} \mid \mathbf{z}_j\} = \prod_{k \in N_j} \mathrm{pr}\{\mathbf{Z}_k = \mathbf{z}_k \mid \mathbf{z}_j\}, \tag{13.2.9}$$

where N_j denotes the labels of the predecessors of pixel j, corresponding to (13.2.8). It will be seen in Section 13.5 that the model (13.2.9) for the transition probabilities for the colors of neighboring pixels provides the basis for several approaches to contextual segmentation performed pixel by pixel.

13.3 NONCONTEXTUAL METHODS OF SEGMENTATION

The simplest way of proceeding with the allocation of the pixels is to ignore all their spatial characteristics and to take their associated feature vectors $\mathbf{X}_j$ to be independently distributed. Segmentation is then performed by assigning each pixel j on the basis of its observed feature vector $\mathbf{x}_j$ only. This is a standard problem in discriminant analysis, as surveyed in the preceding chapters. In very low noise settings, noncontextual methods will work well. However, they will perform poorly if the group-conditional distributions of the feature data overlap substantially. In any event, a noncontextual allocation rule may be of use in that it can be applied to provide an initial specification $\mathbf{z}^{(0)}$ of the colors of the pixels from which an estimate of $\mathbf{z}$ can be computed iteratively via a contextual method of segmentation.

If the group-conditional densities for the feature vector of intensities are not completely specified, then they have to be assessed from the available data. For this purpose with typical applications of discriminant analysis, there are available training data, consisting of observations on independent feature vectors whose classification with respect to the underlying groups is known.

However, given the spatial structure in a scene of pixels, the assumption of independence may not be valid if the training feature observations correspond to several rows of pixels rather than to pixels selected at intervals sufficiently far apart. This selection aspect is discussed further in Section 13.9, where the effect of correlated training data on a sample discriminant rule is considered.

In the absence of any training data of known origin, the group-conditional densities can be estimated parametrically by maximum likelihood under the assumption that $\mathbf{X}_1, \ldots, \mathbf{X}_m$ constitute a random sample from a mixture of the g possible groups. The fitting of this mixture model was discussed in a general context in Section 2.10 and for the special case of multivariate normal group-conditional distributions in Section 3.8.

A common assumption in practice is to take the group-conditional distribution of the feature vector of intensities to be multivariate normal. Although the intensity measurements are not really continuous (the spectral reflectance is usually recorded as a number between 0 and 255), the scale is sufficiently fine to make the conceptual leap to continuous data acceptable. Moreover, the histograms of the feature data from the same group are well modeled by multivariate normal densities; see Hjort and Mohn (1987).

13.4 SMOOTHING METHODS

13.4.1 Spatial Assumptions

Switzer (1980, 1983) and Switzer, Kowalik, and Lyon (1982) have incorporated contextual information in ways they describe as smoothing. Before defining the allocation rule of Switzer (1980), we will set up this smoothing problem in a wider context, along the lines of the formulation in Mardia (1984).

Consider the scene S of m pixels with associated feature vectors $\mathbf{X}_1,\ldots,\mathbf{X}_m$, where

$$\mathbf{X}_j = \sum_{i=1}^{g} z_{ij}\boldsymbol{\mu}_i + \boldsymbol{\epsilon}_j, \tag{13.4.1}$$

where z_{ij} equals one if pixel j is in G_i and is zero otherwise ($i = 1,\ldots,g;\ j = 1,\ldots,m$), and where the noise $\boldsymbol{\epsilon}_1,\ldots,\boldsymbol{\epsilon}_m$ are the realizations of a zero-mean stationary spatially correlated random process. The first assumption is that this noise process is Gaussian with a locally spatial isotropic covariance. Hence, the (common) group-conditional covariance between any two feature vectors $\mathbf{X}_j$ and $\mathbf{X}_k$ can be factored as

$$\operatorname{cov}(\mathbf{X}_j, \mathbf{X}_k) = \rho(d)\boldsymbol{\Sigma} \qquad (j \neq k), \tag{13.4.2}$$

where $\rho(\cdot)$ is the isotropic correlation function, $\rho(0) = 1$, and $d = |j - k|$ refers to the Euclidean distance between pixels j and k in the two-dimensional representation of the scene S. For $p = 1$, (13.4.2) is the usual assumption in kriging; see Matheron (1971).

We let N_j be the subset of $\{1,2,\ldots,m\}$ containing the labels of the pixels that belong to the prescribed neighborhood of pixel j. If the neighborhood contains s pixels labeled $j_1,\ldots,j_s$, then we let $\mathbf{X}_{N_j}$ contain the feature vectors on those pixels in the prescribed neighbor of pixel j, that is,

$$\mathbf{X}_{N_j} = (\mathbf{X}'_{j_1},\ldots,\mathbf{X}'_{j_s})'.$$

For example, $s = 4$ for the first-order neighborhood of adjacent pixels, while $s = 8$ for the second-order neighborhood including also the diagonally adjacent pixels.

The second assumption about the joint distribution of $\mathbf{X}_1,\ldots,\mathbf{X}_m$ assumes local spatial continuity of the neighborhood in that if pixel j belongs to G_i, then so does every neighbor. Thus, on letting

$$\mathbf{X}_j^+ = (\mathbf{X}'_j, \mathbf{X}'_{N_j})',$$

we have that

$$\begin{aligned}\boldsymbol{\mu}_i^+ &= E\{\mathbf{X}_j^+ \mid \mathbf{X}_j \in G_i\} \\ &= \mathbf{1}_{s+1} \otimes \boldsymbol{\mu}_i \qquad (i = 1,\ldots,g),\end{aligned} \tag{13.4.3}$$

where $\otimes$ is the Kronecker delta, and $\mathbf{1}_{s+1}$ is the $(s+1)$-dimensional vector of ones. The covariance matrix of $\mathbf{X}_j^+$, given that $\mathbf{X}_j$ belongs to G_i, is

$$\boldsymbol{\Sigma}^+ = \mathbf{C} \otimes \boldsymbol{\Sigma}, \tag{13.4.4}$$

where $\mathbf{C}$ is the spatial correlation matrix of order $(s+1) \times (s+1)$. It can be written down once the neighborhood of pixel j has been specified.

Local spatial continuity of the neighborhood was assumed also by Sclove (1981). However, he adopted a two-dimensional Markov process to model the feature observations $\mathbf{X}_{j_1},\ldots,\mathbf{X}_{j_s}$, which were taken to be univariate.

13.4.2 Presmoothing of Data

Presmoothing of the data is accomplished by basing the assignment of pixel j on the basis of the observed value $\mathbf{x}_j^+$ of the augmented feature vector $\mathbf{X}_j^+$. Under the assumptions above, which imply that the ith group-conditional distribution of $\mathbf{X}_j^+$ is multivariate normal with mean (13.4.3) and covariance matrix (13.4.4), the optimal rule of allocation is given by the NLDR formed for the augmented feature $\mathbf{x}_j^+$, that is, by $r_o(\mathbf{x}_j^+;\boldsymbol{\Psi}_E^+)$, where

$$\boldsymbol{\Psi}_E^+ = (\boldsymbol{\pi}', \boldsymbol{\theta}_E^{+\prime})',$$

and $\boldsymbol{\theta}_E^+$ denotes $\boldsymbol{\theta}_E$ with $\boldsymbol{\mu}_i$ and $\boldsymbol{\Sigma}$ replaced by $\boldsymbol{\mu}_i^+$ and $\boldsymbol{\Sigma}^+$, respectively ($i = 1,\ldots,g$). It can be seen from equations (3.3.2) and (3.3.3) defining the NLDR $r_o(\mathbf{x}_j;\boldsymbol{\theta}_E)$ that in order to apply $r_o(\mathbf{x}_j^+;\boldsymbol{\Psi}_E^+)$ at the augmented feature vector $\mathbf{x}_j^+$, we have to compute

$$\xi_{ig}(\mathbf{x}_j^+;\boldsymbol{\theta}_E^+) = \{\mathbf{x}_j^+ - \tfrac{1}{2}(\boldsymbol{\mu}_i^+ + \boldsymbol{\mu}_g^+)\}'(\boldsymbol{\Sigma}^+)^{-1}(\boldsymbol{\mu}_i^+ - \boldsymbol{\mu}_g^+) \qquad (i = 1,\ldots,g-1).$$

Mardia (1984) showed that $\xi_{ig}(\mathbf{x}_j^+;\boldsymbol{\theta}_E^+)$ can be reduced to the form

$$\xi_{ig}(\mathbf{x}_j^+;\boldsymbol{\theta}_E^+) = \left\{\gamma_0\mathbf{x}_j + \sum_{k=1}^{s}\gamma_k\mathbf{x}_{j_k} - \frac{1}{2}c^2(\boldsymbol{\mu}_i + \boldsymbol{\mu}_g)\right\}'\boldsymbol{\Sigma}^{-1}(\boldsymbol{\mu}_i - \boldsymbol{\mu}_g) \qquad (i = 1,\ldots,g-1), \tag{13.4.5}$$

where

$$\begin{aligned}c^2 &= \mathbf{1}_{s+1}'\mathbf{C}^{-1}\mathbf{1}_{s+1} \\ &= \sum_{k=0}^{s}\gamma_k,\end{aligned}$$

and where

$$(\gamma_0,\ldots,\gamma_s)' = \mathbf{C}^{-1}\mathbf{1}_{s+1}.$$

The result (13.4.5) implies that allocation is effectively based on the observed value of the linear combination of the feature vectors given by

$$\gamma_0\mathbf{X}_j + \sum_{k=1}^{s}\gamma_k\mathbf{X}_{j_k}, \tag{13.4.6}$$

which has mean $c^2\boldsymbol{\mu}_i$ and covariance matrix $c^2\boldsymbol{\Sigma}$ in G_i $(i = 1,\ldots,g)$.

13.4.3 Reduction in Error Rate

From this last result, it follows on noting the expression in Section 3.3.2 for the optimal error rate of the NLDR that the overall optimal error rate of $r_o(\mathbf{x}_j^+;\boldsymbol{\theta}_E^+)$ in the case of equal group-prior probabilities is equal to

$$e_o(\boldsymbol{\Psi}_E^+) = \Phi(-\tfrac{1}{2}c\Delta), \tag{13.4.7}$$

where

$$\Delta = \{(\boldsymbol{\mu}_1 - \boldsymbol{\mu}_2)'\boldsymbol{\Sigma}^{-1}(\boldsymbol{\mu}_1 - \boldsymbol{\mu}_2)\}^{1/2}.$$

As the optimal error rate of $r_o(\mathbf{x};\boldsymbol{\Psi}_E)$ in the same situation is $\Phi(-\frac{1}{2}\Delta)$, the error rate is reduced by augmenting the feature vector with the intensities of neighboring pixels, since it is easy to establish that $c > 1$; see Mardia (1984). The maximum value of c is $s+1$ in the case in which $\mathbf{X}_j,\mathbf{X}_{j_1},\ldots,\mathbf{X}_{j_s}$ are independently distributed.

For the first-order neighborhood containing the $s = 4$ adjacencies of pixel j,

$$\gamma_0 = (\omega_1 - \omega_2)/(\omega_1 - \omega_2^2)$$

and

$$\gamma_k = (1-\omega_2)/\{4(\omega_1 - \omega_2^2)\} \qquad (k = 1,2,3,4),$$

where

$$\omega_1 = \tfrac{1}{4}\{1 + 2\rho(\sqrt{2}) + \rho(2)\}$$

and

$$\omega_2 = \rho(1) \qquad (\omega_1 > \omega_2^2).$$

Thus, it can be seen from (13.4.6) that for this prescribed neighborhood, allocation is performed on the basis of the linear combination

$$\{(\omega_1 - \omega_2)\mathbf{x}_j + (1-\omega_2)\bar{\mathbf{x}}_{N_j}\}/(\omega_1 - \omega_2^2), \tag{13.4.8}$$

where

$$\bar{\mathbf{x}}_{N_j} = \frac{1}{4}\sum_{k=1}^{4}\mathbf{x}_{j_k}.$$

The use of (13.4.8) was proposed initially by Switzer (1980).

Concerning the reduction in error rate in this case, the value of c is

$$c = \{(1+\omega_1 - 2\omega_2)/(\omega_1 - \omega_2^2)\}^{1/2}.$$

Mardia (1984) also gives the corresponding result for the second-order neighborhood in which the diagonally adjacent pixels are included, too.

13.4.4 Estimation of Parameters

In practice, the parameters of the model proposed above for smoothing the data will be unknown. Switzer (1980) and Mardia (1984) have considered their estimation from available training data; see also Lawoko (1990). In particular, regarding the estimation of ω_1 and ω_2, Mardia (1984) suggests that they be estimated by first estimating $\rho(\cdot)$, as in kriging procedures (Journel and Huijbregts, 1978). The prior probabilities of the groups can be estimated by the method proposed by Switzer, Kowalik, and Lyon (1982), which introduces some spatial smoothing in the prior distribution for the colors of the pixels. This method is based on the so-called discriminant analysis approach to the estimation of mixing proportions as described in Section 2.3; see equation (2.3.4).

13.4.5 Another Method of Presmoothing

Switzer (1985) has proposed the use of min/max autocorrelation factors (MAF) as a noise separation procedure for general-purpose processing of multivariate spatial imagery. The feature vector $\mathbf{X}_j$ is transformed to a vector $\tilde{\mathbf{X}}_j = \mathbf{A}\mathbf{X}_j$, which has the following property. The first few variates in $\tilde{\mathbf{X}}_j$ have minimal spatial autocorrelation, identified as mainly noise, and the latter variates have maximal spatial autocorrelation, identified as mainly signal. The procedure operates pointwise to avoid the signal blurring introduced through smoothing or spatial averaging procedures. However, the pointwise operator itself is defined using primitive global spatial characteristics of the data.

The MAF procedure has other formulations that permit the use of standard multivariate routines to extract the factors. Specifically, the factors are obtained as eigenvectors of the matrix $\boldsymbol{\Sigma}_d\boldsymbol{\Sigma}^{-1}$, where

$$\boldsymbol{\Sigma} = \mathrm{cov}(\mathbf{X}_j),$$

$$\boldsymbol{\Sigma}_d = \mathrm{cov}(\mathbf{X}_j - \mathbf{X}_k),$$

and where $d = |j-k|$ is the specified spatial log.

13.5 INDIVIDUAL CONTEXTUAL ALLOCATION OF PIXELS

One way of providing a contextual method of segmentation is to consider the allocation of each pixel individually on the basis of its posterior probabilities of group membership given the recorded feature vectors on all the m pixels in the scene S. The jth pixel is allocated then on the basis of the maximum of the posterior probability

$$\mathrm{pr}\{\mathbf{Z}_j = \mathbf{z}_j \mid \mathbf{x}\}$$

with respect to $\mathbf{z}_j$, where $\mathbf{z}_j$ defines the group of origin (color) of pixel j ($j = 1,\ldots,m$).

A common assumption is to form this posterior probability under the assumption of white noise, that is, the feature vectors are conditionally independent given their group of origin (color). In many image problems, this assumption of white noise is reasonable. But in remote sensing applications with high-resolution satellite data, this is not so; see Hjort and Mohn (1985, 1987) and Mohn, Hjort, and Storvik (1987) for further discussion on this. As noted in these papers, contextual rules that assume white noise offer less improvement in terms of error rate over noncontextual rules in situations where the feature data are autocorrelated.

Under the assumption that the feature vectors $\mathbf{X}_j$ are group-conditionally independent with the same density $f_i(\cdot;\boldsymbol{\theta}_i)$ in group G_i, we have up to an additive term not involving $\mathbf{z}_j$ that

$$\mathrm{pr}\{\mathbf{Z}_j = \mathbf{z}_j \mid \mathbf{x}\} \propto \sum_{\mathbf{z}_{(j)}} P(\mathbf{z}) \exp\left\{\sum_{i=1}^{g}\sum_{k=1}^{m} z_{ik} \log f_i(\mathbf{x}_k;\boldsymbol{\theta}_i)\right\}, \qquad (13.5.1)$$

where the assumption is over all admissible values of $\mathbf{z}_{(j)}$. It can be seen that (13.5.1) depends on all the observed feature vectors $\mathbf{x}_j$ for (almost) any probability function $P(\mathbf{z})$ for $\mathbf{Z}$. Therefore, in order to reduce the complexity of the problem, $\mathbf{z}_j$ might be chosen to maximize

$$\mathrm{pr}\{\mathbf{Z}_j = \mathbf{z}_j \mid \mathbf{x}_j, \mathbf{x}_{N_j}\},$$

where N_j is the subset of $\{1,\ldots,m\}$, containing the labels of the prescribed neighbors of the jth pixel. We have that

$$\mathrm{pr}\{\mathbf{Z}_j = \mathbf{z}_j \mid \mathbf{x}_j, \mathbf{x}_{N_j}\} \propto \sum_{\mathbf{z}_{N_j}} \mathrm{pr}\{\mathbf{Z}_j = \mathbf{z}_j, \mathbf{Z}_{N_j} = \mathbf{z}_{N_j}\}$$
$$\times \exp\left\{\sum_{i=1}^{g}\sum_{k\in N_j^+} z_{ik} \log f_i(\mathbf{x}_k;\boldsymbol{\theta}_i)\right\}, \qquad (13.5.2)$$

where the summation is over all admissible values of $\mathbf{z}_{N_j}$ defining the group membership (colors) of the pixels in the prescribed neighborhood of pixel j, and where N_j^+ is the union of N_j and $\{j\}$.

Even with this simplification, analytical progress is barred, in general, because $\mathrm{pr}\{\mathbf{Z}_j = \mathbf{z}_j, \mathbf{Z}_{N_j} = \mathbf{z}_{N_j}\}$ is unavailable in closed form. The exceptions to the general rule fall within the Markov mesh family as defined in Section

13.2.3. With the Pickard (1977, 1980) model defined there,

$$\mathrm{pr}\{\mathbf{Z}_j = \mathbf{z}_j, \mathbf{Z}_{N_j} = \mathbf{z}_{N_j}\} = \mathrm{pr}\{\mathbf{Z}_j = \mathbf{z}_j\} \prod_{k \in N_j} \mathrm{pr}\{\mathbf{Z}_k = \mathbf{z}_k \mid \mathbf{z}_j\}, \qquad (13.5.3)$$

since given $\mathbf{z}_j$, the $\mathbf{Z}_k$ for the neighbors of pixel j are independently distributed. The use of (13.5.3) or generalizations of it in (13.5.2) have provided the basis for the contextual approaches to allocation as considered by Welch and Salter (1971), Owen (1984), Hjort and Mohn (1984), and Haslett (1985), among others.

Haslett (1985) has extended the calculation of the left-hand side of (13.5.3), where the prescribed neighborhood consists of all pixels in the same row and column as pixel j. The transition probabilities $\mathrm{pr}\{\mathbf{Z}_k = \mathbf{z}_k \mid \mathbf{z}_j\}$ were assumed to be stationary and were taken to be known along with the group-conditional densities $f_i(\mathbf{x}_k; \boldsymbol{\theta}_i)$ and the common prior distribution for the $\mathbf{Z}_j$. Using (13.5.3), Haslett (1985) was able to compute the right-hand side of (13.5.2) exactly. Haslett (1985) also has given an approximation to include second-order neighbors.

The procedures of Hjort and Mohn (1984) and Owen (1984) also maximize the posterior probability (13.5.2) for N_j corresponding to first-order neighbors of pixel j. Whereas Haslett (1985) assumes the $\mathbf{Z}_k$ with k belonging to N_j to be conditionally independent given $\mathbf{z}_j$, Owen (1984) and Hjort and Mohn (1984) allow only certain configurations of the prescribed neighbors. Suppose that pixel j belongs to group G_i, that is, it is of color i. Then its neighbors must either all be of this color or have at most two members of some other color (say, $h \neq i$), with only configurations of the form

$$\begin{array}{ccc} & i & \\ i & i & i \\ & i & \end{array} \qquad \begin{array}{ccc} & i & \\ i & i & h \\ & i & \end{array} \qquad \begin{array}{ccc} & i & \\ i & i & h \\ & h & \end{array}$$

allowed. Hence, the number of configurations for pixel j and its neighbors has been reduced from g^5 to

$$g + 4g(g-1) + 4g(g-1) = g(8g-7).$$

Hence, there are only $g(8g-7)$ possible values for $\mathbf{z}_{N_j}$. Owen (1984) specifies the transition probabilities from a geometrical probability model with one parameter, whereas Hjort and Mohn (1984) allow an arbitrary (exchangeable) distribution. Hjort (1985) has generalized these procedures to include diagonal neighbors. Also, Hjort and Mohn (1985) have obtained a natural generalization of this to the case where spatial autocorrelation between the feature vectors is allowed for. More recently, Devijver (1988) has discussed approximation techniques for parameter estimation in Markov mesh models.

13.6 ICM ALGORITHM

13.6.1 Underlying Assumptions

This section is devoted to the so-called ICM (iterated conditional modes) algorithm, which is one of the standard algorithms for the segmentation of images. The ICM algorithm was introduced by Besag (1983, 1986). The image it produces was proposed as an approximation to the maximum *a posteriori* (MAP) image in his first paper and as an improvement in his second paper. The ICM algorithm is based on the imposition of a locally dependent Markov random field for the prior distribution for the image as represented by **Z**. The algorithm makes use of the desirable local properties of the Markov model, but is not influenced by some of their undesirable properties, notably strong long-term dependence that favors near constant colors.

More specifically, the ICM algorithm was developed under two main assumptions. The first is that given $\mathbf{z}$ specifying the colors of the pixels, the feature vectors $\mathbf{X}_1,\ldots,\mathbf{X}_m$ containing the intensities measured on pixels 1 to m are conditionally independent, with each $\mathbf{X}_j$ having the same density $f_i(\cdot;\boldsymbol{\theta}_i)$ in G_i $(i = 1,\ldots,g)$. In the notation of the previous sections, it implies that the joint density of $\mathbf{X}_1,\ldots,\mathbf{X}_m$ given $\mathbf{z}$ is equal to

$$\exp\left\{\sum_{i=1}^{g}\sum_{j=1}^{m} z_{ij}\log f_i(\mathbf{x}_j;\boldsymbol{\theta}_i)\right\}. \qquad (13.6.1)$$

The applicability of this assumption was discussed in the previous section.

The second assumption is that the true image $\mathbf{z}$ is a realization of a locally dependent Markov random field with probability distribution $P(\mathbf{z};\boldsymbol{\beta})$, specified up to a vector $\boldsymbol{\beta}$ of unknown parameters. As discussed in Section 13.2, where Markov random fields were defined, the intended role of this model is that it should be consistent with the true scene as regards its *local* rather than *global* characteristics. We have from (13.2.2) that for a locally dependent Markov field,

$$\operatorname{pr}\{\mathbf{Z}_j = \mathbf{z}_j \mid \mathbf{z}_{(j)}\} = P_j(\mathbf{z}_j;\mathbf{z}_{N_j},\boldsymbol{\beta}) \qquad (j = 1,\ldots,m), \qquad (13.6.2)$$

where the function P_j is specific to pixel j, and N_j is the subset of $\{1,\ldots,m\}$ containing the labels of the pixels in the immediate vicinity of pixel j. This neighborhood of pixels is prescribed before implementation of the algorithm; see the discussion on this in Section 13.2.1.

13.6.2 Definition

Besag (1986) proposed that segmentation be carried out by allocating each pixel individually on the basis of

$$\operatorname{pr}\{\mathbf{Z}_j = \mathbf{z}_j \mid \mathbf{x},\mathbf{z}_{(j)}\},$$

where $\mathbf{z}_{(j)}$ specifies the colors of all the pixels in the scene apart from pixel j $(j = 1,\ldots,m)$. Under assumptions (13.6.1) and (13.6.2), it follows from Bayes' theorem that

$$\log \mathrm{pr}\{\mathbf{Z}_j = \mathbf{z}_j \mid \mathbf{x}, \mathbf{z}_{(j)}\} = \sum_{i=1}^{g} z_{ij} \log f_i(\mathbf{x}_j; \boldsymbol{\theta}_i) + \log P_j(\mathbf{z}_j; \mathbf{z}_{N_j}, \boldsymbol{\beta}) \qquad (13.6.3)$$

up to an additive term not involving $\mathbf{z}_j$.

An estimate $\hat{\mathbf{z}}_j$ is computed iteratively, using the following algorithm, which is described firstly in the case where the $\boldsymbol{\theta}_i$ and $\boldsymbol{\beta}$ are treated as known. Let $\mathbf{z}^{(0)}$ denote an initial estimate of $\mathbf{z}$ obtained, for example, by using a non-contextual rule, as discussed in Section 13.3. Then starting with pixel j, $\hat{\mathbf{z}}_j$ is chosen to maximize (13.6.3) at $\mathbf{z}_{N_j} = \mathbf{z}_{N_j}^{(0)}$. The initial estimate $\mathbf{z}^{(0)}$ of $\mathbf{z}$ is then updated so that the current estimate $\hat{\mathbf{z}}$ becomes $\mathbf{z}^{(0)}$ but with $\mathbf{z}_j^{(0)}$ replaced by $\hat{\mathbf{z}}_j$. The procedure is then applied to the next pixel, say $j+1$, by choosing $\hat{\mathbf{z}}_{j+1}$ to maximize (13.6.3) at $\mathbf{z}_{N_{j+1}} = \hat{\mathbf{z}}_{N_{j+1}}$. The estimate of $\mathbf{z}$ is then updated and the procedure applied to the next pixel. After application to each pixel in turn, this procedure defines a single cycle of the ICM algorithm for the iterative computation of the estimate of $\mathbf{z}$. The algorithm is applied for a prespecified number of cycles, or until convergence. Since

$$\mathrm{pr}\{\mathbf{Z} = \mathbf{z} \mid \mathbf{x}\} = \mathrm{pr}\{\mathbf{Z}_j = \mathbf{z}_j \mid \mathbf{x}, \mathbf{z}_{(j)}\} \mathrm{pr}\{\mathbf{Z}_{(j)} = \mathbf{z}_{(j)} \mid \mathbf{x}\},$$

it can be seen that $\mathrm{pr}\{\mathbf{Z} = \hat{\mathbf{z}} \mid \mathbf{x}\}$ is never decreased at any stage and eventual convergence is assured.

Besag (1986) has found in practice that convergence, to what must be a local maximum of $\mathrm{pr}\{\mathbf{Z} = \mathbf{z} \mid \mathbf{x}\}$, is extremely rapid with few if any changes occurring after the sixth cycle. As remarked earlier, it was as an approximation to the maximization of $\mathrm{pr}\{\mathbf{Z} = \mathbf{z} \mid \mathbf{x}\}$ that this algorithm was originally proposed by Besag (1983). This iterative computation of the estimate $\hat{\mathbf{z}}_j$ was suggested independently by Kittler and Föglein (1984), who applied it to LANDSAT data, as did Kiiveri and Campbell (1986). As stressed by Besag (1986), its dependence only on the local characteristics of the scene is ensured by the rapid convergences. The guarantee of convergence holds no matter in what order the pixels are scanned. However, there is no guarantee of convergence if the image is updated synchronously, whereby each new estimate of $\mathbf{z}_j$ is calculated in turn according to (13.6.3), but the current estimate of $\mathbf{z}$ is not updated until each cycle is completed.

13.6.3 Estimation of Parameters by the ICM Algorithm

We now consider the estimation of the vector $\boldsymbol{\theta}_i$ of parameters in the ith group-conditional density of the feature vector $(i = 1,\ldots,g)$ and the vector $\boldsymbol{\beta}$ of parameters in the locally dependent Markov random field specified for the prior distribution of $\mathbf{Z}$. It is supposed that there are no training data of known origin to provide suitable estimates of these parameters for adoption in the application of the ICM algorithm as outlined above. Besag (1986) suggested that the ICM algorithm be modified as follows.

Let $\mathbf{z}^{(k)}$ be the estimate of $\mathbf{z}$ after completion of the kth cycle of the algorithm. Then an estimate $\boldsymbol{\theta}^{(k)}$ of the vector $\boldsymbol{\theta}$ of the distinct elements of $\boldsymbol{\theta}_1,\ldots,\boldsymbol{\theta}_g$ is obtained by maximization of

$$\sum_{i=1}^{g}\sum_{j=1}^{m} z_{ij}^{(k)} \log f_i(\mathbf{x}_j;\boldsymbol{\theta}_i) \tag{13.6.4}$$

with respect to $\boldsymbol{\theta}$. An estimate $\boldsymbol{\beta}^{(k)}$ of $\boldsymbol{\beta}$ can be obtained by maximum pseudo likelihood, whereby $\boldsymbol{\beta}^{(k)}$ is chosen to maximize

$$\sum_{j=1}^{m} \log P_j(\mathbf{z}_j^{(k)};\mathbf{z}_{N_j}^{(k)},\boldsymbol{\beta}), \tag{13.6.5}$$

where terms in this sum corresponding to boundary pixels might be excluded because of the added artificiality of the model there. The sum (13.6.5) is not in general a genuine log likelihood, since it is formed under the assumption of independence. Exact maximum likelihood estimation of $\boldsymbol{\beta}$ is generally not computationally tractable. This is because the normalizing constant in (13.2.3) cannot be evaluated in general. Exceptions under (13.2.4) are the Markov mesh models and first-order Ising models; see Besag (1975, 1986) and Venetoulias (1989) for further discussion of the estimation of $\boldsymbol{\beta}$.

After computation of $\boldsymbol{\theta}^{(k)}$ and $\boldsymbol{\beta}^{(k)}$, a new estimate of $\mathbf{z}$ is obtained by carrying out another cycle of the algorithm with $\boldsymbol{\theta}$ and $\boldsymbol{\beta}$ replaced by $\boldsymbol{\theta}^{(k)}$ and $\boldsymbol{\beta}^{(k)}$, respectively, in (13.6.3). As reported by Besag (1986), little is known of the convergence properties of this process, but limited results so far are encouraging.

13.6.4 *A Priori* Specification of Parameter in Prior Distribution of Image

Besag (1986) found that his ICM algorithm performed well for the (second-order) locally dependent Markov random field model for which

$$\text{pr}\{Z_{ij} = 1 \mid \mathbf{z}_{(j)}\} \propto \exp(\beta u_{ij}), \tag{13.6.6}$$

where u_{ij} is the number of the prescribed neighbors of pixel j, belonging to G_i ($i = 1,\ldots,g$; $j = 1,\ldots,m$), and β is a single unknown parameter to be estimated or specified. The development of this spatially symmetric model (13.6.6) was described in Section 13.2.2. Here the prescribed neighborhood of pixel j consists of the four adjacent pixels and the four diagonally adjacent pixels, where available. Thus, segmentation is local to the extent that it depends only on data up to eight pixels distant from the pixel undergoing allocation.

The model (13.6.6) for the prior distribution of the image has the advantage of only one parameter. As an alternative to its estimation during the implementation of the ICM algorithm, it can be specified beforehand by *a priori* reasons, as demonstrated by Ripley (1986) in the case of $g = 2$ colors. If all of the eight neighbors of a pixel are of the same color, then the probability that the central pixel has the same color as its eight like neighbors should be very high. Ripley (1986) suggests that this probability, which is given from (13.6.6) by

$$\exp(8\beta)/\{1 + \exp(8\beta)\},$$

should be at least 0.999, requiring $\beta \geq 0.86$. On the other hand, if the colors of the neighbors of a pixel are split five versus three, then the probability that the central pixel has the same color as its three like neighbors, which is

$$\exp(3\beta)/\{\exp(3\beta) + \exp(5\beta)\},$$

should be at least 0.01, requiring $\beta \leq 2.3$. Previously, Besag (1986) recommended empirically that β be set at 1.5, or better still slowly increased to 1.5 as the iterations continue. To some extent, the higher values of β compensate for the fast convergence and very local behavior of the ICM algorithm. Experiments of Ripley (1986) suggest that smaller values of β may be preferable. Owen (1986) finds $\beta = 0.7$ more appealing. This value in the two aforementioned examples considered by Ripley (1986) gives probability 0.996 that a pixel is of the same color as its eight like neighbors and a probability of 0.2 that it is of the same color as its three like neighbors.

13.6.5 Examples of ICM Segmentation

We consider firstly a binary example taken from Ripley (1986). There are $g = 2$ groups representing two colors, white (G_1) and black (G_2). In G_i, each feature observation X_j is univariate normal with mean μ_i and variance σ^2 ($i = 1,2$), where $\mu_1 = 0$ and $\mu_2 = 1$. Corresponding to (13.6.6), the locally dependent Markov random field for the prior distribution of the image is the Ising model, for which

$$\text{pr}\{Z_{1j} = 1 \mid \mathbf{z}_{(j)}\} = e^{\beta u_{1j}}/(e^{\beta u_{1j}} + e^{\beta u_{2j}})$$

except at the edges, where u_{1j} is the number of white neighbors of pixel j, and $u_{2j} = 8 - u_{1j}$ is the number of black neighbors. For known parameters μ_1, μ_2, σ^2, and β, we have from (13.6.3) that

$$\begin{aligned}\log[\text{pr}\{Z_{1j} = 1 \mid \mathbf{x}, \mathbf{z}_{(j)}\}/\text{pr}\{Z_{2j} = 1 \mid \mathbf{x}, \mathbf{z}_{(j)}\}] \\ = -(x_j - \tfrac{1}{2})/\sigma^2 + \beta(u_{1j} - u_{2j}).\end{aligned} \tag{13.6.7}$$

Hence, with the ICM algorithm, pixel j is allocated on the basis of (13.6.7), where u_{ij} is replaced by its current estimate $\hat{u}_{ij}$ ($i = 1,2$). Thus, $\hat{z}_{1j} = 1$ (allocates pixel j to white) if

$$x_j < \tfrac{1}{2} + \beta\sigma^2(\hat{u}_{1j} - \hat{u}_{2j}),$$

which is both simple and intuitive. The noncontextual version of this rule, corresponding to $\beta = 0$, would take $\hat{z}_{1j} = 1$ if $x_j < \frac{1}{2}$, thereby ignoring the information $\hat{u}_{1j}$ and $\hat{u}_{2j}$ on the colors of the neighboring pixels.

We now generalize the model in the example above to multivariate features and multiple groups (colors), where the group means $\boldsymbol{\mu}_i$, common group-covariance matrix $\boldsymbol{\Sigma}$, and the parameter β in the Markov random field model (13.6.6) are all unknown. Then it follows from (13.6.3) that the $(k+1)$th cycle of the EM algorithm requires the minimization of

$$\sum_{i=1}^{g} z_{ij}(\mathbf{x}_j - \boldsymbol{\mu}_i^{(k)})'\boldsymbol{\Sigma}^{(k)^{-1}}(\mathbf{x}_j - \boldsymbol{\mu}_i^{(k)}) - \beta^{(k)} \sum_{i=1}^{g} z_{ij}\hat{u}_{ij} \tag{13.6.8}$$

with respect to $\mathbf{z}_j$, for each pixel j in turn. In (13.6.8), $\hat{u}_{ij}$ denotes the current number of neighbors of pixel j that belong to G_i $(i = 1,\ldots,g;\ j = 1,\ldots,m)$. Also,

$$\boldsymbol{\mu}_i^{(k)} = \sum_{j=1}^{m} z_{ij}^{(k)}\mathbf{x}_j \bigg/ \sum_{j=1}^{m} z_{ij}^{(k)} \qquad (i = 1,\ldots,g)$$

and

$$\boldsymbol{\Sigma}^{(k)} = \sum_{i=1}^{g}\sum_{j=1}^{m} z_{ij}^{(k)}(\mathbf{x}_j - \boldsymbol{\mu}_i^{(k)})(\mathbf{x}_j - \boldsymbol{\mu}_i^{(k)})'/m,$$

where $\mathbf{z}^{(k)}$ denotes the estimate of $\mathbf{z}$ at the completion of the kth cycle. That is, they are the ith group-sample means and the pooled within-group sample covariance matrix of the feature data $\mathbf{x}_1,\ldots,\mathbf{x}_m$ partitioned according to $\mathbf{z}^{(k)}$.

In the present context, we give an artificial example taken from Besag (1986). The true scene contains $g = 6$ colors, on a 120×120 array. It was originally hand-drawn and chosen to display a wide variety of characteristics. The univariate feature observations were generated from the color labels by superimposing Gaussian noise with $\sigma^2 = 0.36$. The first 64 rows and the last 64 columns are displayed in Figure 13.1(a), in which the adjacencies are less contrived than in the scene as a whole. A color key, which is part of the pattern, is shown at the top right-hand corner of Figure 13.1(a), where "minus"= 1, "cross"= 2, and so on. The initial reconstruction using the noncontextual NLDR (i.e., $\beta = 0$ in the univariate version of (13.6.8)), produced an overall misallocation rate of 32%, as in Figure 13.1(b). With the correct value of σ^2 and with $\beta = 1.5$ throughout, the ICM algorithm gave an overall error rate of 2.1% on the eighth cycle. The ICM algorithm applied with β increased by equal increments from 0.5 to 1.5 over the first six cycles, reduced the overall error rate to 1.2% on the eighth cycle, as in Figure 13.1(c). For σ^2 known but β estimated as in Section 13.6.3, the eventual error rate was 1.1% after eight iterations, with $\hat{\beta} = 1.8$, as in Figure 13.1(d). With both β and σ^2 estimated during reconstruction, the error rate again settled at 1.2%, now with

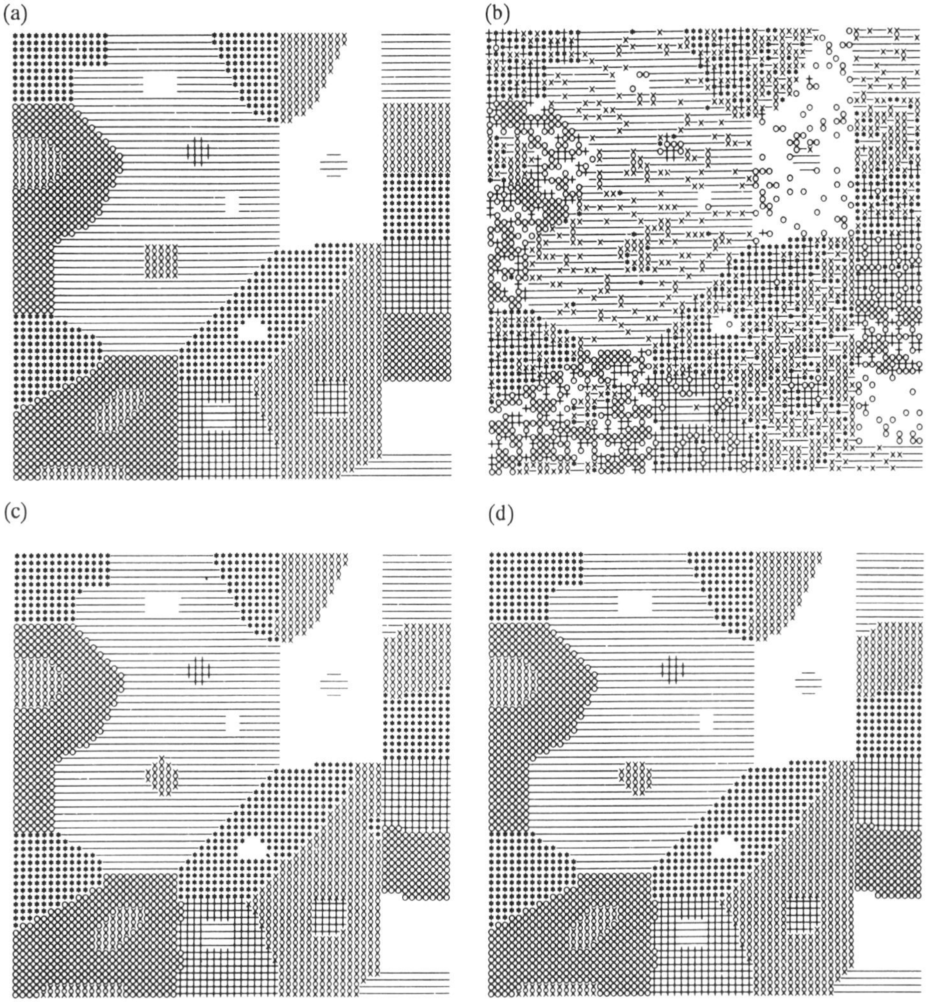

FIGURE 13.1. Image restoration: (a) true color scene: 64×64; (b) noncontextual reconstruction ($\beta = 0$): 32% error rate; (c) ICM reconstruction with $\beta \uparrow 1.5$: 1.2%; and (d) ICM reconstruction with β estimated ($\hat{\beta} = 1.80$): 1.1%. (From Besag, 1986.)

$\hat{\beta} = 1.83$ and $\hat{\sigma}^2 = 0.366$. As commented by Switzer (1986), this simulated example with spatially uncorrelated feature data in conjunction with single color patches that consist of ten or more pixels does not imply that very low error rates will be achievable in more complicated situations with real images.

13.6.6 Modifications of the ICM Algorithm

Besag (1986) has considered various modifications of his ICM algorithm to handle, for example, grey-level scenes and overlap of the pixels, among other departures from the basic model considered here. Further modifications of the ICM algorithm have been proposed by other authors.

As discussed in McLachlan and Basford (1988, Chapter 1) on the fitting of finite mixture models, if the fractional weights given by the current estimates of the posterior probabilities of group membership were replaced by zero-one weights, then inconsistent estimators of the group-conditional parameters are obtained. This led Hjort and Mohn (1987) to suggest using the posterior probabilities of group membership $\tau_{ij}^{(k)}$ in place of the zero-one weights $z_{ij}^{(k)}$ in the estimation process on the $(k+1)$th cycle of the ICM algorithm. For the Markov random field defined by (13.6.6), the posterior probability of membership of G_i for pixel j is assessed on the $(k+1)$th cycle as

$$\tau_{ij}^{(k)} = \pi_{i,j}^{(k)} f_i(\mathbf{x}_j;\boldsymbol{\theta}_i^{(k)}) \Big/ \sum_{h=1}^{g} \pi_{h,j}^{(k)} f_h(\mathbf{x}_j;\boldsymbol{\theta}_h^{(k)}), \tag{13.6.9}$$

where

$$\pi_{i,j}^{(k)} = \exp(\beta^{(k)}\hat{u}_{ij}) \Big/ \sum_{h=1}^{g} \exp(\beta^{(k)}\hat{u}_{hj}),$$

and where $\hat{u}_{ij}$ denotes the current number of neighbors of pixel j that belong to G_i $(i = 1,\ldots,g;\ j = 1,\ldots,m)$.

In order to make the assumption of group-conditional independence for the feature vectors $\mathbf{X}_1,\ldots,\mathbf{X}_m$ more tenable, Hjort and Mohn (1987) suggest a further modification to the estimation process of the ICM algorithm. They suggest that the log likelihood (13.6.4) be formed for only a subset of the pixels by choosing distant grids. They also demonstrate an alternative approach by proposing a simple model that allows for the presence of spatial autocorrelation between the feature vectors. The ith group-conditional density of $\mathbf{X}_j$ given $\mathbf{z}$ and $\mathbf{x}_{(j)}$ is taken to be multivariate normal with mean

$$\boldsymbol{\mu}_i + \frac{1}{4}\gamma\left\{\sum_{v\in N_j}\left(\mathbf{x}_v - \sum_{h=1}^{g} z_{hv}\boldsymbol{\mu}_h\right)\right\} \tag{13.6.10}$$

and covariance matrix $\boldsymbol{\Sigma}$, where N_j contains the labels of the four first-order neighbors of pixel j. As a consequence of (13.6.10), the log likelihood (13.6.4) has the additional term

$$-\frac{1}{2}\sum_{h=1}^{g}\sum_{i=1}^{g}\sum_{v<w=1}^{m} z_{hv}z_{iw}\gamma_{vw}(\mathbf{x}_v - \boldsymbol{\mu}_h)'\boldsymbol{\Sigma}^{-1}(\mathbf{x}_w - \boldsymbol{\mu}_i),$$

where $\gamma_{vw} = \gamma$ if pixels v and w are first-order neighbors and equals zero otherwise. The relaxation of the assumption of group-conditional independence of the feature data was considered in a similar manner by Kittler and Föglein (1984) and Kiiveri and Campbell (1986), among others.

Modifications to ICM have been suggested also by Owen (1986, 1989) and by Jubb and Jennison (1988). As noted by Owen (1989), the number of iterations with the ICM algorithm does not seem to increase rapidly with the noise level. Thus, ICM is like a smoothing algorithm with a small bandwidth built in.

With the simple modification of the ICM algorithm proposed by Owen (1986), the number u_{ij} of neighbors of pixel j that belong to G_i on the $(k+1)$th cycle is replaced by its assessed conditional expectation

$$\sum_{v \in N_j} \tau_{iv}^{(k)}, \tag{13.6.11}$$

where $\tau_{ij}^{(k)}$ is the estimate (13.6.9) of the posterior probability that pixel j belongs to G_i $(i = 1,\dots,g;\ j = 1,\dots,m)$. This avoids the discretization in counting the neighbors. For most pixels, the fractional neighbor count (13.6.11) is near 8 or 0, but near boundaries it can differ appreciably from u_{ij}.

Owen (1986) termed this modified version of the ICM algorithm the ICE (iterated conditional expectations) algorithm. He found in a low-noise situation that ICE produced smoother image estimates than did ICM when both used the same value of β. An intuitive explanation put forward by him is that information from neighboring pixels can flow through without their having to change colors. Owen (1986) found no special advantage to ICE over ICM in accuracy. More recently, however, Owen (1989) has demonstrated that in very noisy images, the ICE algorithm dramatically improves image segmentation.

Jubb and Jennison (1988) suggest another modification that extends the range of the ICM algorithm to very noisy images and greatly reduces the computational costs through aggregation. Owen (1989) has since shown that there is a natural combination of the methods of Owen (1986) and Jubb and Jennison (1988) that retains the speed of the latter method, while improving its accuracy.

13.7 GLOBAL MAXIMIZATION OF THE POSTERIOR DISTRIBUTION OF THE IMAGE

13.7.1 Maximum *A Posteriori* (MAP) Estimate

It was seen in the previous section that the ICM algorithm finds a local maximum of the posterior distribution $\mathrm{pr}\{\mathbf{Z} = \mathbf{z} \mid \mathbf{x}\}$ near the starting point. The method of Geman and Geman (1984) to be described now is directed at finding the global maximum. Their procedure, therefore, estimates simultaneously the group-indicator vectors $\mathbf{z}_j$ defining the colors of the pixels by taking $\hat{\mathbf{z}}$ to be the mode of the posterior distribution

$$\mathrm{pr}\{\mathbf{Z} = \mathbf{z} \mid \mathbf{x}\}. \tag{13.7.1}$$

Under the assumptions (13.6.1) and (13.6.2), it follows that

$$\log \mathrm{pr}\{\mathbf{Z} = \mathbf{z} \mid \mathbf{x}\} = \sum_{i=1}^{g} \sum_{j=1}^{m} z_{ij} \log f_i(\mathbf{x}_j; \boldsymbol{\theta}_i) + \log P(\mathbf{z}), \qquad (13.7.2)$$

where $P(\mathbf{z})$ denotes the probability distribution for the specified Markov random field, for example, as given by (13.2.4). The last term on the right-hand side of (13.7.2) can be viewed as a smoothness penalty; see Titterington (1985) for an account of smoothing techniques in a general framework.

13.7.2 Computation of the MAP Estimate

The maximization of (13.7.1) with respect to $\mathbf{z}$ is clearly formidable, as there are g^m different colorings of the scene. Geman and Geman (1984) approached it, using a procedure from combinatorial optimization known as simulated annealing. This technique was popularized by Kirkpatrick, Gelatt, and Vecchi (1983), but dates back to Pincus (1968, 1970).

Let

$$P_T(\mathbf{z}) = [\mathrm{pr}\{\mathbf{Z} = \mathbf{z} \mid \mathbf{x}\}]^{1/T}, \qquad (13.7.3)$$

where $T > 0$ is a parameter, and, in physical terms, represents the "absolute temperature" of the system. Then, as $T \to 0$, P_T concentrates on the mode of the posterior distribution (13.7.1), the required solution. If multiple global maxima exist, it is uniformly distributed over them. Further, $P_T(\mathbf{z})$ also specifies the probability distribution of a Markov random field. Thus, if a series of samples is taken from P_T with $T \to 0$, these samples should converge to the required solution. Unfortunately, it is not possible to simulate discrete Markov random fields directly, except in special cases; see Besag (1986) for further discussion of this point.

In simulating (13.7.3), Geman and Geman (1984) used an approach they termed the "Gibbs sampler." In the basic version, the pixels are scanned in turn in some prescribed order. At pixel j, $\mathbf{Z}_j$ is chosen to have value $\mathbf{z}_j$ with probability

$$[\mathrm{pr}\{\mathbf{Z}_j = \mathbf{z}_j \mid \mathbf{x}, \mathbf{z}_{(j)}\}]^{1/T},$$

where the current estimate is used for $\mathbf{z}_{(j)}$. Under the assumption (13.6.1) and (13.6.2), $\mathrm{pr}\{\mathbf{Z}_j = \mathbf{z}_j \mid \mathbf{x}, \mathbf{z}_{(j)}\}$ can be evaluated from (13.6.3). It follows that this process has (13.7.3) as its limit distribution.

The random choice of $\mathbf{z}_j$ enables the algorithm to escape from any local maxima. However, to ensure such escapes, the process must be simulated long enough and T must eventually be decreased extremely slowly. On the schedule for progressively decreasing T to zero, Geman and Geman (1984) suggest decreasing T at each scan, according to

$$T = 3/\log(1 + \text{scan no.}),$$

for several hundred scans. To ensure convergence of the Gibbs sampler and of simulated annealing, Geman and Geman (1984) make only weak demands on the manner in which the pixels are scanned; see also Besag (1986) and Lakshmanan and Derin (1989).

13.7.3 Comparison with the ICM Algorithm

Suppose that the homoscedastic normal model holds for the group-conditional distributions of $\mathbf{X}_j$ and that the locally dependent Markov random field corresponding to (13.6.6) is adopted, where all parameters are specified. Then it follows from (13.2.4) and (13.7.2) that the simultaneous estimation approach of Geman and Geman (1984) requires the minimization of

$$\sum_{i=1}^{g}\sum_{j=1}^{m} z_{ij}(\mathbf{x}_j - \boldsymbol{\mu}_i)'\boldsymbol{\Sigma}^{-1}(\mathbf{x}_j - \boldsymbol{\mu}_i) - \beta u(\mathbf{z}) \tag{13.7.4}$$

with respect to $\mathbf{z}$, where $u(\mathbf{z})$ is the number of neighbor pairs having like colors. From the previous section, it can be seen that the ICM algorithm of Besag (1983) requires the minimization of

$$\sum_{i=1}^{g} z_{ij}(\mathbf{x}_j - \boldsymbol{\mu}_i)'\boldsymbol{\Sigma}^{-1}(\mathbf{x}_j - \boldsymbol{\mu}_i) - \beta\sum_{i=1}^{g} z_{ij}\hat{u}_{ij} \tag{13.7.5}$$

with respect to $\mathbf{z}_j$, for each pixel j considered in turn. The coefficient of β in (13.7.5) is the current number of neighbors of the same color as pixel j.

It is of interest to examine the extremes of β in (13.7.4) and (13.7.5). For $\beta = 0$, both methods reduce simply to noncontextual allocation, where pixel j is assigned on the basis of the sample NLDR applied to its observed feature vector $\mathbf{x}_j$. However, the two methods are radically different, as $\beta \to \infty$. With the ICM algorithm, $\mathbf{z}_j$ is estimated recursively on the basis of a majority vote among the neighbors of pixel j, using only $\mathbf{x}_j$ to break ties. On the other hand, (13.7.4) produces an estimate $\hat{\mathbf{z}}$ for which

$$\hat{\mathbf{z}}_1 = \hat{\mathbf{z}}_2 = \cdots = \hat{\mathbf{z}}_m.$$

That is, it produces a single-color solution, corresponding to the color that the majority of the pixels would be assigned on the basis of the minimum-distance rule, using

$$(\mathbf{x}_j - \boldsymbol{\mu}_i)'\boldsymbol{\Sigma}^{-1}(\mathbf{x}_j - \boldsymbol{\mu}_i).$$

13.7.4 Exact Computation of the MAP Estimate for Binary Images

Recently, Greig, Porteous, and Seheult (1989) have shown how in the special case of $g = 2$ colors (binary images), the MAP estimate can be computed exactly using efficient variants of the Ford–Fulkerson algorithm for finding the maximum flow in a certain capacitated network. Their results indicate that simulated annealing, applied according to *practical* "temperature" schedules, can produce a poor approximation to the MAP estimate. However, their experimental results suggest that good approximations are more likely for smaller values of the parameter β in the prior distribution (13.6.6) for the image, and in such cases, geometric schedules outperform logarithmic schedules. An informal explanation of this behavior offered by Greig et al. (1989) is that the amount of smoothing required to obtain an MAP estimate increases with β and, as conjectured by Besag (1986, page 298), the simulated annealing algorithm then appears to become increasingly bogged down by local maxima, resulting in undersmooth approximations. Concerning the use of the MAP estimate for image restoration, Greig et al. (1989) demonstrate that as β increases, the global properties of the prior distribution of the image very rapidly dominate the likelihood contribution to the posterior distribution. They conjecture that corresponding multicolor estimates will behave similarly.

13.8 INCOMPLETE-DATA FORMULATION OF IMAGE SEGMENTATION

13.8.1 EM Framework for Incomplete Data

The problem of estimating the vector $\mathbf{z}$ of group-indicator variables defining the colors of the m pixels can be viewed as an incomplete-data problem. This suggests that the estimation of $\mathbf{z}$ can be approached by attempting to apply the EM algorithm of Dempster et al. (1977), as proposed for maximum likelihood estimation from incomplete data. This line of approach was undertaken by Kay and Titterington (1986), who related some of the relaxation algorithms for image analysis (Rosenfeld et al., 1976; Hummel and Zucker, 1983) with methods in the literature on the statistical analysis of incomplete data. The incomplete data are represented by $\mathbf{X}$, consisting of the observable feature vectors associated with the m pixels. The complete-data vector is given by $\mathbf{Y} = (\mathbf{X}', \mathbf{Z}')'$, where $\mathbf{Z}$ is unobservable. We let $\log L(\Psi)$ denote the incomplete-data log likelihood for Ψ, as formed from $\mathbf{x}$, where Ψ denotes the vector of unknown parameters in the parametric model adopted for the density of $\mathbf{X}$, that is, for the joint density of the feature vectors $\mathbf{X}_1, \ldots, \mathbf{X}_m$.

13.8.2 Application of the EM Algorithm

An estimate of Ψ can be attained by solving the likelihood equation

$$\partial \log L(\Psi)/\partial \Psi = \mathbf{0}.$$

This can be attempted iteratively, using the EM algorithm of Dempster et al. (1977), which was considered in Section 2.10 in the context of independent unclassified data. At the $(k+1)$th cycle of this algorithm, the E-step requires the calculation of

$$H(\Psi, \Psi^{(k)}) = E\{\log L_C(\Psi) \mid \mathbf{X} = \mathbf{x}; \Psi^{(k)}\}, \tag{13.8.1}$$

which is the conditional expectation of the log likelihood for the complete data $\mathbf{Y}$, given $\mathbf{X} = \mathbf{x}$ and the current fit $\Psi^{(k)}$ for Ψ. On the M-step, the intent is to find the value of Ψ that maximizes $H(\Psi, \Psi^{(k)})$, which gives $\Psi^{(k+1)}$. Recently, Silverman et al. (1990) have introduced a simple smoothing step at each EM iteration.

If we make the common assumption that the feature vectors $\mathbf{X}_j$ are independently distributed given $\mathbf{z}$, with a common density $f_i(\cdot;\boldsymbol{\theta}_i)$ in G_i $(i = 1,\ldots,g)$, the log likelihood formed from the complete-data vector $\mathbf{Y}$ is given by

$$\log L_C(\Psi) = \sum_{i=1}^{g}\sum_{j=1}^{m} z_{ij}\log f_i(\mathbf{x}_j;\boldsymbol{\theta}_i) + \sum_{i_1,\ldots,i_m=1}^{g} z_{i_1 1}\cdots z_{i_m m}\log \pi_{i_1\cdots i_m}, \tag{13.8.2}$$

where $z_{ij} = (\mathbf{z}_j)_i$ $(i = 1,\ldots,g;\ j = 1,\ldots,m)$, $f_i(\mathbf{x}_j;\boldsymbol{\theta}_i)$ denotes the density of $\mathbf{X}_j$ in G_i $(i = 1,\ldots,g)$, and where

$$\pi_{i_1\cdots i_m} = \text{pr}\{Z_{i_j j} = 1\ (j = 1,\ldots,m)\}.$$

Here Ψ consists of the elements of $\boldsymbol{\theta}_1,\ldots,\boldsymbol{\theta}_g$ known *a priori* to be distinct and the joint prior probabilities $\pi_{i_1\cdots i_m}$ $(i_j = 1,\ldots,g;\ j = 1,\ldots,m)$.

13.8.3 Noncontextual Case: Exact E-Step

We consider first the case in which contextual information is ignored, by taking the $\mathbf{Z}_j$ to be independently distributed. Under this independence,

$$\pi_{i_1,\ldots,i_m} = \prod_{j=1}^{m}\pi_{i_j,j},$$

where

$$\pi_{i_j,j} = \text{pr}\{Z_{i_j j} = 1\} \qquad (i_j = 1,\ldots,g;\ j = 1,\ldots,m).$$

The complete-data log likelihood then reduces to

$$\log L_C(\Psi) = \sum_{i=1}^{g}\sum_{j=1}^{m} z_{ij}\{\log f_i(\mathbf{x}_j;\boldsymbol{\theta}_i) + \log \pi_{i,j}\}. \tag{13.8.3}$$

Corresponding to (13.8.1), the E-step on the (k+1)th cycle gives

$$H(\Psi, \Psi^{(k)}) = \sum_{i=1}^{g}\sum_{j=1}^{m} \tau_{ij}(\mathbf{x}_j;\Psi^{(k)})\{\log f_i(\mathbf{x}_j;\boldsymbol{\theta}_i) + \log \pi_{i,j}\},$$

where

$$\tau_{ij}(\mathbf{x}_j;\Psi) = \pi_{i,j} f_i(\mathbf{x}_j;\boldsymbol{\theta}_i) \bigg/ \sum_{h=1}^{g}\pi_{h,j} f_h(\mathbf{x}_j;\boldsymbol{\theta}_h).$$

where

$$\tau_{ij}(\mathbf{x}_j;\mathbf{\Psi}^{(k)}) = q_{i,j}^{(k)} f_i(\mathbf{x}_j;\boldsymbol{\theta}_i^{(k)}) \Big/ \sum_{h=1}^{g} q_{h,j}^{(k)} f_h(\mathbf{x}_j;\boldsymbol{\theta}_h^{(k)}),$$

and where

$$q_{i,j}^{(k)} = \sum \mathrm{pr}^{(k)}\{Z_{ij} = 1, \mathbf{Z}_{N_j} = \mathbf{z}_{N_j}\}$$

$$= \sum \mathrm{pr}^{(k)}\{\mathbf{Z}_{N_j} = \mathbf{z}_{N_j}\}\mathrm{pr}^{(k)}\{Z_{ij} = 1 \mid \mathbf{z}_{N_j}\}, \qquad (13.8.11)$$

where again the summation is over all admissible values of $\mathbf{z}_{N_j}$, specifying the colors of the neighbors of pixel j.

13.8.5 Link with the ICM Algorithm

The ICM algorithm can be viewed in the incomplete-data scheme as developed above (Kay, 1986; Kay and Titterington, 1986). On the $(k+1)$th cycle of the iterative computation of $\hat{\mathbf{z}}$ with the ICM algorithm, the estimate of the ith component of $\mathbf{z}_j$ is taken to be one or zero, according as

$$\tau_{ij}(\mathbf{x}_j;\mathbf{\Psi}^{(k)}) \geq \tau_{hj}(\mathbf{x}_j;\mathbf{\Psi}^{(k)}) \qquad (h = 1,\ldots,g;\ h \neq i)$$

holds or not, where $q_{i,j}^{(k)}$ is computed as follows. In forming $q_{i,j}^{(k)}$ from (13.8.11), $\mathrm{pr}^{(k)}\{\mathbf{Z}_{N_j} = \mathbf{z}_{N_j}\}$ is taken to be one for $\mathbf{z}_{N_j} = \hat{\mathbf{z}}_{N_j}$ and zero otherwise, where $\hat{\mathbf{z}}_{N_j}$ is the current estimate of $\mathbf{z}_{N_j}$. Thus, $q_{i,j}^{(k)}$ is computed as

$$q_{i,j}^{(k)} = \mathrm{pr}^{(k)}\{Z_{ij} = 1 \mid \mathbf{z}_{N_j} = \hat{\mathbf{z}}_{N_j}\}.$$

Under the model (13.6.6) specifying a locally dependent second-order Markov random field for the prior distribution of $\mathbf{Z}$,

$$q_{i,j}^{(k)} = \exp(\beta^{(k)}\hat{u}_{ij}) \Big/ \sum_{h=1}^{g} \exp(\beta^{(k)}\hat{u}_{hj}),$$

where $\hat{u}_{hj}$ is the current number of the prescribed (second-order) neighbors of pixel j that belong to G_h $(h = 1,\ldots,g;\ j = 1,\ldots,m)$.

Qian and Titterington (1989) have outlined the link between the ICM algorithm and the EM algorithm for parameter estimation in the case of autonormal models for continuous intensities.

13.8.6 Links with Some Other Methods of Segmentation

If the prescribed neighborhood of pixel j consists of a single pixel, say, pixel s, so that $N_j = \{s\}$, then from (13.8.11),

$$q_{i,j}^{(k)} = \sum_{h=1}^{g} \mathrm{pr}^{(k)}\{Z_{hs} = 1\}\mathrm{pr}^{(k)}\{Z_{ij} = 1 \mid Z_{hs} = 1\}. \qquad (13.8.12)$$

Kay and Titterington (1986) noted that the use of (13.8.12) in conjunction with (13.8.10) is equivalent to the relaxation algorithm of Hjort and Mohn (1985). In the case where the prescribed neighborhood contains more than a single pixel, they suggested evaluating (13.8.12) for each pixel in the neighborhood and then taking a weighted average in order to create a value for $q_{i,j}^{(k)}$.

Kay and Titterington (1986) also noted that the expression (13.8.12) summed over those s that are labels of the prescribed neighbors of pixel j gives a version of the first-order compatibility function of Rosenfeld et al. (1976) and Hummel and Zucker (1983).

On other links of (13.8.12) with existing methods of segmentation, Kay and Titterington (1986) noted how the method of Haslett (1985), which was outlined in Section 13.5, fits into this incomplete-data framework under the additional assumption of a common prior distribution for the color of each pixel.

13.9 CORRELATED TRAINING DATA

13.9.1 Introduction

It was seen in the previous sections that some of the approaches to image segmentation or restoration require the group-conditional densities to be completely specified beforehand. One way of achieving this is to adopt some parametric family, in particular the multivariate normal, for each group-conditional density, where the parameters taken to be unknown are estimated from training data of known origin. In a typical discriminant analysis, the training data are taken to be independent for the purposes of forming the sample discriminant rule of interest. However, it was seen in the previous sections with applications in image analysis, such as in remote sensing, that there is a positive correlation between the feature vectors of neighboring pixels, which attenuates as the distance between the pixels increases.

Craig (1979) experimentally examined LANDSAT pictures covering several areas at different times. He concluded that the data must be sampled no closer than every tenth pixel in order to treat the observations as independent. However, in practice, one might not have the luxury of sampling at intervals of a sufficient number of pixels apart for the training data to be taken to be independent. Hence, consideration needs to be given to the effect of correlated training data on the performance of the sample discriminant rule formed from such data. We now report some results on this effect for noncontextual applications of the sample NLDR.

13.9.2 Effect of Equicorrelated Training Data on Sample NLDR

The effect of correlated training data on the performance of a sample discriminant rule has been investigated mainly for the sample NLDR, given the widespread use of the latter in practice; see, for example, Weber and Baldessari (1988). One of the initial studies of the performance of the sample NLDR formed from dependent data was by Basu and Odell (1974). They were led to relaxing the usual assumption of independent training data, as it had been frequently observed with applications of the sample NLDF to remotely sensed data that its actual error rates were higher than their theoretically anticipated values under the assumption of independence. Basu and Odell (1974) investigated the effect of equicorrelated training data classified with respect to $g = 2$ groups. The replicated observations $\mathbf{X}_{i1},\ldots,\mathbf{X}_{in_i}$ from G_i are said to be equicorrelated if

$$\operatorname{cov}(\mathbf{X}_{ij},\mathbf{X}_{ik}) = \boldsymbol{\Upsilon}_i \qquad (j \neq k = 1,\ldots,n_i),$$

where $\boldsymbol{\Upsilon}_i$ is a symmetric matrix ($i = 1,2$). The same observations are said to be simply equicorrelated if

$$\boldsymbol{\Upsilon}_i = \rho_i \boldsymbol{\Sigma}_i,$$

where $0 < |\rho_i| < 1$, and $\boldsymbol{\Sigma}_i$ is the common covariance matrix of the observations from G_i ($i = 1,2$). In their study of the sample NLDR, Basu and Odell (1974) assumed the same covariance structure within each group, where $\rho_1 = \rho_2 = \rho$, and $-(n_i - 1)^{-1} < \rho < 1$ ($i = 1,2$). Basu and Odell (1974) distinguished between simply equicorrelated and equicorrelated training data because they felt that the error rates of the sample NLDR were unaffected by simple equicorrelation. However, McLachlan (1976c) showed that the error rates do change in the presence of simple equicorrelation; indeed, the error rates asymptotically increase in size for positive correlation.

13.9.3 Effect of Autocorrelated Training Data on Sample NLDR

As explained by Föglein and Kittler (1983), physical characteristics of the process for obtaining the LANDSAT imagery allow one-dimensional models to be used. For example, the spatial location of a pixel along a single scan line can be considered. These one-dimensional models can be adopted when there exists a trade-off between the precision and the complexity of two-dimensional spatial models. Of course, LANDSAT multispectral scanner data are neither simply equicorrelated nor equicorrelated, but are instead serially correlated. Tubbs and Coberly (1978) suggested that the correlation structure is similar to that of a stationary autoregressive process of order one, AR(1), with the correlation ρ between successive observations typically in the 0.65–0.90 range. Craig (1979) suggested a moving average model of order one-one, ARMA(1,1), would be a more appropriate model for the correlation structure of the LANDSAT data. Föglein and Kittler (1983) have considered the effect of correlated data on the separation between the groups when pixel dependencies are modeled as an ARMA process.

Previously, Tubbs (1980) had investigated the effect of serially correlated training data on the unconditional error rates of the sample NLDR under the homoscedastic normal model (3.3.1). He supposed further that the n_i training observations $\mathbf{X}_{ij}$ ($j = 1,\ldots,n_i$) from group G_i ($i = 1,2$), follow a simple multivariate stationary AR(1) model, where

$$\text{cov}(\mathbf{X}_{ij}, \mathbf{X}_{ik}) = \rho_d \mathbf{\Sigma} \qquad (j \neq k = 1, \ldots, n_i)$$

for $d = |j-k|$ and $0 < |\rho_d| < 1$ $(i = 1, 2)$.

The effect of correlated training data on the sample NLDR under the homoscedastic normal model (3.3.1) has been explored further in a series of papers by Lawoko and McLachlan (1983, 1985, 1986, 1988, 1989). For the same model considered by Tubbs (1980), Lawoko and McLachlan (1983) have derived the asymptotic expansions of the group-specific unconditional error rates of the sample NLDR. To illustrate the magnitude of the increase in the unconditional error rates for positively correlated training data, they evaluated the asymptotic expansions for a univariate stationary autoregressive process of order one, for which $\rho_d = \rho^d$, where $0 < |\rho| < 1$. These numerical results demonstrated how the group-specific unconditional error rates increase with positive correlation ρ and with the dimension p of the feature vector.

For univariate training data following a stationary autoregressive process of order k, AR(k), Lawoko and McLachlan (1985) compared the performance of the sample NLDR with the plug-in sample version formed using the maximum likelihood estimates of the group means and variances appropriate for an AR(k) model. They showed that there was no first-order difference between the overall unconditional error rates, but that there are first-order changes of varying degrees between the group-specific error rates. For $k = 1$, they demonstrated that the use of the appropriate maximum likelihood estimates improves the balance between the group-specific error rates.

As considered in Section 10.6, the plug-in method of estimation of the error rates of the sample NLDR provides an optimistic assessment of its conditional error rates. Lawoko and McLachlan (1988) showed that the optimism of this method is magnified by positively correlated training data following a stationary AR(1) model as above. This is because the effect of positive correlation is to increase the expectation of each conditional error rate, but to decrease the expectation of the corresponding plug-in rate. To demonstrate the magnitude of the increase in the bias, Lawoko and McLachlan (1988) tabulated the asymptotic bias of the plug-in estimator $\Phi(-\frac{1}{2}D)$ for equal group-sample sizes and a zero cutoff point. In this special case, the plug-in estimator has the same bias in estimating both group-specific error rates of the sample NLDR. Their results are reported in Table 13.1, in which we have listed the coefficient of n^{-1} in the n^{-1}-order expansion of the bias of $\Phi(-\frac{1}{2}D)$ for $n_1 = n_2 = \frac{1}{2}n$.

In related work on dependent training data, Lawoko and McLachlan (1986) have shown asymptotically that the Z-statistic defined by (3.4.7) is preferable to the sample NLDF for positively correlated univariate training data following a stationary AR(1) process. Also, as mentioned in Section 2.3, Lawoko and McLachlan (1989) have studied the effect of correlated training data on the estimates of the mixing proportions obtained by an application of the sample NLDR, as in area estimation via remote sensing. The effect of serially correlated training data on the sample NQDR has been investigated by Young,

TABLE 13.1 Coefficient of n^{-1} in the n^{-1}-Order Expansion of the Bias of $\Phi(-\frac{1}{2}D)$ Under AR(1) Model for $n_1 = n_2$

ρ	$\Delta = 1$	$\Delta = 2$	$\Delta = 3$
		$p = 1$	
0	−0.121	−0.121	−0.036
0.1	−0.163	−0.177	−0.080
0.2	−0.219	−0.252	−0.137
0.4	−0.402	−0.490	−0.309
0.6	−0.785	−0.983	−0.660
		$p = 4$	
0	−2.761	−1.573	−0.878
0.1	−3.283	−1.805	−0.991
0.2	−3.960	−2.127	−1.156
0.4	−6.060	−3.186	−1.719
0.6	−10.357	−5.429	−2.935

Source: From Lawoko and McLachlan (1988), with permission from the Pattern Recognition Society.

Turner, and Marco (1988) in the special case where in the multivariate normal group-conditional densities, the means are taken be equal, and where the uniform covariance structure (3.2.11) is adopted for the covariance matrices.

REFERENCES

Abend, K., Harley, T. J., and Kanal, L. N. (1965). Classification of binary random patterns. *IEEE Trans. Inform. Theory* **11**, 538–544.

Abramovitch, L., and Singh, K. (1985). Edgeworth corrected pivotal statistics and the bootstrap. *Ann. Statist.* **13**, 116–132.

Abramowitz, M., and Stegun, I. A. (Eds.). (1965). *Handbook of Mathematical Functions*. New York: Dover.

Adegboye, O. S. (1987). A classification rule whose probability of misclassification is independent of the variance. *Austral. J. Statist.* **29**, 208–213.

Adhikari, B. P., and Joshi, D. D. (1956). Distance, discrimination et résumé exhaustif. *Publ. Inst. Statist. University Paris* **5**, 57–74.

Afifi, A. A., and Elashoff, R. M. (1969). Multivariate two sample tests with dichotomous and continuous variables. I. The location model. *Ann. Math. Statist.* **40**, 290–298.

Agrawala, A. K. (Ed.). (1977). *Machine Recognition of Patterns*. New York: IEEE Press.

Ahmad, I. A. (1982). Matusita's distance. In *Encyclopedia of Statistical Sciences* (Vol. 5), S. Kotz and N. L. Johnson (Eds.). New York: Wiley, pp. 334–336.

Ahmed, S. W., and Lachenbruch, P. A. (1975). Discriminant analysis when one or both of the initial samples is contaminated: large sample results. *EDV Med. Biol.* **6**, 35–42.

Ahmed, S. W., and Lachenbruch, P. A. (1977). Discriminant analysis when scale contamination is present in the initial sample. In *Classification and Clustering*, J. Van Ryzin (Ed.). New York: Academic Press, pp. 331–353.

Aitchison, J. (1975). Goodness of prediction fit. *Biometrika* **62**, 547–554.

Aitchison, J. (1986). *The Statistical Analysis of Compositional Data*. London: Chapman and Hall.

Aitchison, J., and Aitken, C. G. G. (1976). Multivariate binary discrimination by the kernel method. *Biometrika* **63**, 413–420.

Aitchison, J., and Begg, C. B. (1976). Statistical diagnosis when basic cases are not classified with certainty. *Biometrika* **63**, 1–12.

Aitchison, J., and Dunsmore, I. R. (1975). *Statistical Predication Analysis*. Cambridge: Cambridge University Press.

Aitchison, J., and Kay, J. W. (1975). Principles, practice and performance in decision making in clinical medicine. *Proc. 1973 NATO Conf. on the Role and Effectiveness of Theories of Decision in Practice*. London: Hodder and Stoughton, pp. 252–272.

Aitchison, J., Habbema, J. D. F., and Kay, J. W. (1977). A critical comparison of two methods of statistical discrimination. *Appl. Statist.* **26**, 15–25.

Aitken, C. G. G. (1978). Methods of discrimination in multivariate binary data. *Compstat 1978, Proc. Computational Statistics*. Vienna: Physica-Verlag, pp. 155–161.

Aitken, C. G. G. (1983). Kernel methods for the estimation of discrete distribution. *J. Statist. Comput. Simul.* **16**, 189–200.

Aitkin, M., and Tunnicliffe Wilson, G. (1980). Mixture models, outliers, and the EM algorithm. *Technometrics* **22**, 325–331.

Aitkin, M., Anderson, D., and Hinde, J. (1981). Statistical modelling of data on teaching styles (with discussion). *J. R. Statist. Soc. A* **144**, 414–461.

Akaike, H. (1974). A new look at the statistical model identification. *IEEE Trans. Automat. Control* **AC-19**, 716–723.

Albert, A., and Anderson, J. A. (1981). Probit and logistic discriminant functions. *Commun. Statist.–Theory Meth.* **A10**, 641–657.

Albert, A., and Anderson, J. A. (1984). On the existence of maximum likelihood estimates in logistic regression models. *Biometrika* **71**, 1–10.

Albert, A., and Harris, E. K. (1987). *Multivariate Interpretation of Clinical Laboratory Data*. New York: Marcel Dekker.

Albert, A., and Lesaffre, E. (1986). Multiple group logistic discrimination. *Comput. Math. Applic.* **12A**, 209–224.

Albert, A., Chapelle, J. P., and Smeets, J. P. (1981). Stepwise probit discrimination with specific application to short term prognosis in acute myocardial infarction. *Comput. Biomed. Research* **14**, 391–398.

Ali, S. M., and Silvey, S. D. (1966). A general class of coefficients of divergence of one distribution from another. *J. R. Statist. Soc. B* **28**, 131–142.

Ambergen, A. W., and Schaafsma, W. (1984). Interval estimates for posterior probabilities, applications to Border Cave. In *Multivariate Statistical Methods in Physical Anthropology*, G. N. van Vark and W. W. Howells (Eds.). Dordrecht: Reidel, pp. 115–134.

Ambergen, A. W., and Schaafsma, W. (1985). Interval estimates for posterior probabilities in a multivariate normal classification model. *J. Multivar. Anal.* **16**, 432–439.

Amemiya, T. (1980). The n^{-2}-order mean-squared errors of the maximum likelihood and the minimum logit chi-square estimator. *Ann. Statist.* **8**, 488–505. Correction (1984). *Ann. Statist.* **12**, 783.

Amemiya, T. (1985). *Advanced Econometrics*. Cambridge, Massachusetts: Harvard University Press.

Amemiya, T., and Powell, J. L. (1983). A comparison of the logit model and normal discriminant analysis when the independent variables are binary. In *Studies in*

Econometrics, Time Series and Multivariate Statistics, S. Karlin, T. Amemiya, and L. A. Goodman (Eds.). New York: Academic Press, pp. 3–30.

Amoh, R. K. (1985). Estimation of a discriminant function from a mixture of two inverse Gaussian distributions when sample size is small. *J. Statist. Comput. Simul.* **20**, 275–286.

Amoh, R. K., and Kocherlakota, K. (1986). Errors of misclassification associated with the inverse Gaussian distribution. *Commun. Statist.–Theory Meth.* **15**, 589–612.

Anderson, E. (1935). The irises of the Gaspé Peninsula. *Bull. Amer. Iris Soc.* **59**, 2–5.

Anderson, J. A. (1969). Constrained discrimination between *k* populations. *J. R. Statist. Soc. B* **31**, 123–139.

Anderson, J. A. (1972). Separate sample logistic discrimination. *Biometrika* **59**, 19–35.

Anderson, J. A. (1974). Diagnosis by logistic discriminant function: further practical problems and results. *Appl. Statist.* **23**, 397–404.

Anderson, J. A. (1975). Quadratic logistic discrimination. *Biometrika* **62**, 149–154.

Anderson, J. A. (1979). Multivariate logistic compounds. *Biometrika* **66**, 7–16.

Anderson, J. A. (1982). Logistic discrimination. In *Handbook of Statistics* (Vol. 2), P. R. Krishnaiah and L. Kanal (Eds.). Amsterdam: North-Holland, pp. 169–191.

Anderson, J. A. (1984). Regression and ordered categorical variables (with discussion). *J. R. Statist. Soc. B* **46**, 1–30.

Anderson, J. A., and Blair, V. (1982). Penalized maximum likelihood estimation in logistic regression and discrimination. *Biometrika* **69**, 123–136.

Anderson, J. A., and Philips, P. R. (1981). Regression, discrimination and measurement models for ordered categorical variables. *Appl. Statist.* **30**, 22–31.

Anderson, J. A., and Richardson, S. C. (1979). Logistic discrimination and bias correction in maximum likelihood estimation. *Technometrics* **21**, 71–78.

Anderson, T. W. (1951). Classification by multivariate analysis. *Psychometrika* **16**, 31–50.

Anderson, T. W. (1958). *An Introduction to Multivariate Statistical Analysis*. First Edition. New York: Wiley.

Anderson, T. W. (1966). Some nonparametric multivariate procedures based on statistically equivalent blocks. In *Multivariate Analysis*, P. R. Krishnaiah (Ed.). New York: Academic Press, pp. 5–27.

Anderson, T. W. (1973a). Asymptotic evaluation of the probabilities of misclassification by linear discriminant functions. In *Discriminant Analysis and Applications*, T. Cacoullos (Ed.). New York: Academic Press, pp. 17–35.

Anderson, T. W. (1973b). An asymptotic expansion of the distribution of the Studentized classification statistic *W*. *Ann. Statist.* **1**, 964–972.

Anderson, T. W. (1984). *An Introduction to Multivariate Statistical Analysis*. Second Edition. New York: Wiley.

Anderson, T. W., and Bahadur, R. R. (1962). Classification into two multivariate normal distributions with different covariance matrices. *Ann. Math. Statist.* **33**, 420–431.

Andrews, D. F., and Herzberg, A. M. (1985). *Data: A Collection of Problems from Many Fields for the Student and Research Worker*. New York: Springer-Verlag.

Andrews, D. F., Gnanadesikan, R., and Warner, J. L. (1971). Transformations of multivariate data. *Biometrics* **27**, 825–840.

Andrews, D. F., Gnanadesikan, R., and Warner, J. L. (1973). Methods for assessing multivariate normality. In *Multivariate Analysis* (Vol. III), P. R. Krishnaiah (Ed.). New York: Academic Press, pp. 95–116.

Andrews, D. F., Brant, R., amd Percy, M. E. (1986). Bayesian incorporation of repeated measurements in logistic discrimination. *Canad. J. Statist.* **14**, 263–266.

Aoyama, H. (1950). A note on the classification of data. *Ann. Inst. Statist. Math.* **2**, 17–20.

Armitage, P. (1950). Sequential analysis with more than two alternative hypotheses and its relation to discriminant function analysis. *J. R. Statist. Soc. B* **12**, 137–144.

Ashikaga, T., and Chang, P. C. (1981). Robustness of Fisher's linear discriminant function under two-component mixed normal models. *J. Amer. Statist. Assoc.* **76**, 676–680.

Ashton, E. H., Healy, M. J. R., and Lipton, S. (1957). The descriptive use of discriminant functions in physical anthropology. *Proc. R. Soc. B* **146**, 552–572.

Atkinson, C., and Mitchell, A. F. S. (1981). Rao's distance measure. *Sankhyā A* **43**, 345–365.

Bacon-Shone, J., and Fung, W. K. (1987). A new graphical method for detecting single and multiple outliers in univariate and multivariate data. *Appl. Statist.* **36**, 153–162.

Bagnold, R. A. (1941). *The Physics of Blown Sand and Desert Dunes*. London: Methuen.

Bagnold, R. A., and Barndorff-Nielsen, O. (1980). The pattern of natural size distributions. *Sedimentology* **27**, 199–207.

Bahadur, R. R. (1961). A representation of the joint distribution of responses to n dichotomous items. In *Studies in Item Analysis and Prediction*, H. Solomon (Ed.). Stanford: Stanford University Press, pp. 158–168.

Bailey, T., and Jain, A. K. (1978). A note on distance-weighted k-nearest neighbor rules. *IEEE Trans. Syst. Man Cybern.* **SMC-8**, 311–313.

Balakrishnan, N., and Kocherlakota, S. (1985). Robustness to nonnormality of the linear discriminant function: mixtures of normal distributions. *Commun. Statist.–Theory Meth.* **14**, 465–478.

Balakrishnan, N., and Tiku, M. L. (1988). Robust classification procedures. In *Classification and Related Methods of Data Analysis*, H. H. Bock (Ed.). Amsterdam: North-Holland, pp. 269–276.

Balakrishnan, N., Tiku, M. L., and El Shaarawi, A. H. (1985). Robust univariate two-way classification. *Biom. J.* **27**, 123–138.

Balakrishnan, N., Kocherlakota, S., and Kocherlakota, K. (1986). On the errors of misclassification based on dichotomous and normal variables. *Ann. Inst. Statist. Math.* **38**, 529–538.

Balakrishnan, N., Kocherlakota, S., and Kocherlakota, K. (1988). Robustness of the double discriminant function in nonnormal situations. *South African Statist. J.* **22**, 15–43.

Balakrishnan, V., and Sanghvi, L. D. (1968). Distance between populations on the basis of attribute data. *Biometrics* **24**, 859–865.

Bandyopadhyay, S. (1977). Probability inequalities involving estimates of probability of correct classification using dependent samples. *Sankhyā* **39**, 145–150.

Bandyopadhyay, S. (1978). Classification, distributions and admissibility in autoregressive process. *Austral. J. Statist.* **20**, 229–239.

Bandyopadhyay, S. (1979). Two population classification in Gaussian process. *J. Statist. Planning Inference* **3**, 225–233.

Bandyopadhyay, S. (1982). Covariate classification using dependent samples. *Austral. J. Statist.* **24**, 309–317.

Bandyopadhyay, S. (1983). Admissibility of likelihood ratio rules in covariate classification using dependent samples. *Austral. J. Statist.* **25**, 482–486.

Banjeree, K. S., and Marcus, L. F. (1965). Bounds in a minimax classification procedure. *Biometrika* **53**, 653–654.

Baringhaus, L., Danschke, R., and Henze, N. (1989). Recent and classical tests for normality—a comparative study. *Commun. Statist.–Simula.* **18**, 363–379.

Barnard, G. A. (1963). Contribution to the discussion of paper by M. S. Bartlett. *J. R. Statist. Soc. B* **25**, 294.

Barnard, M. M. (1935). The secular variations of skull characters in four series of Egyptian skulls. *Ann. Eugenics* **6**, 352–371.

Barndorff-Nielsen, O. (1977). Exponentially decreasing distributions for the logarithm of particle size. *Proc. R. Soc. Lond. A* **353**, 401–419.

Barnett, V. (1983). Contribution to the discussion of paper by R. J. Beckman and R. D. Cook. *Technometrics* **25**, 150–152.

Barnett, V., and Lewis, T. (1978). *Outliers in Statistical Data*. New York: Wiley.

Bartlett, M. S. (1938). Further aspects of multiple regression. *Proc. Camb. Phil. Soc.* **34**, 33–40.

Bartlett, M. S. (1939). A note on tests of significance in multivariate analysis. *Proc. Camb. Phil. Soc.* **35**, 180–185.

Bartlett, M. S. (1947). Multivariate analysis. *J. R. Statist. Soc. Suppl.* **9**, 176–190.

Bartlett, M. S. (1951a). An inverse matrix adjustment arising in discriminant analysis. *Ann. Math. Statist.* **22**, 107–111.

Bartlett, M. S. (1951b). The goodness-of-fit of a single hypothetical discriminant function in the case of several groups. *Ann. Eugen.* **16**, 199–214.

Bartlett, M. S., and Please, N. W. (1963). Discrimination in the case of zero mean differences. *Biometrika* **50**, 17–21.

Basford, K. E., and McLachlan, G. J. (1985). Estimation of allocation rates in a cluster analysis context. *J. Amer. Statist. Assoc.* **80**, 286–293.

Basu, A. P., and Gupta, A. K. (1974). Classification rules for exponential populations. In *Reliability and Biometry: Statistical Analysis of Lifelength,* F. Proschan and J. Serfling (Eds.). Philadelphia: SIAM, pp. 637–650.

Basu, A. P., and Odell, P. L. (1974). Effects of intraclass correlation among training samples on the misclassification probabilities of Bayes' procedure. *Pattern Recognition* **6**, 13–16.

Bayne, C. K., and Tan, W. Y. (1981). QDF misclassification probabilities for known population parameters. *Commun. Statist.–Theory Meth.* **A10**, 2315–2326.

Bayne, C. K., Beauchamp, J. J., Kane, V. E., and McCabe, G. P. (1983). Assessment of Fisher and logistic linear and quadratic discrimination models. *Comput. Statist. Data Anal.* **1**, 257–273.

Broemeling, L. D., and Son, M. S. (1987). The classification problem with autoregressive processes. *Commun. Statist.–Theory Meth.* **16**, 927–936.

Broffitt, B., Clarke, W. R., and Lachenbruch, P. A. (1980). The effect of Huberizing and trimming on the quadratic discriminant function. *Commun. Statist.–Theory Meth.* **A9**, 13–25.

Broffitt, B., Clarke, W. R., and Lachenbruch, P. A. (1981). Measurement errors—a location contamination problem in discriminant analysis. *Commun. Statist.–Simula.* **B10**, 129–141.

Broffitt, J. D. (1982). Nonparametric classification. In *Handbook of Statistics* (Vol. 2), P. R. Krishnaiah and L. N. Kanal (Eds.). Amsterdam: North-Holland, pp. 139–168.

Broffitt, J. D., and Williams, J. S. (1973). Minimum variance estimators for misclassification probabilities in discriminant analysis. *J. Multivar. Anal.* **3**, 311–327.

Broffitt, J. D., Randles, R. H., and Hogg, R. V. (1976). Distribution-free partial discriminant analysis. *J. Amer. Statist. Assoc.* **71**, 934–939.

Brooks, C. A., Clark, R. R., Hodgu, A., and Jones, A. M. (1988). The robustness of the logistic risk function. *Commun. Statist.–Simula.* **17**, 1–24.

Browdy, B. L., and Chang, P. C. (1982). Bayes procedures for the classification of multiple polynomial trends with dependent residuals. *J. Amer. Statist. Assoc.* **77**, 483–487.

Brown, P. J. (1980). Coherence and complexity in classification problems. *Scand. J. Statist.* **7**, 95–98.

Brunk, H. D., and Pierce, D. A. (1974). Estimation of discrete multivariate densities for computer-aided differential diagnosis of disease. *Biometrika* **61**, 493–499.

Bryant, J. (1989). A fast classifier for image data. *Pattern Recognition* **22**, 45–48.

Bryant, J., and Guseman, L. F. (1979). Distance preserving linear feature selection. *Pattern Recognition* **11**, 345–352.

Bull, S. B., and Donner, A. (1987). The efficiency of multinomial logistic regression compared with multiple group discriminant analysis. *J. Amer. Statist. Assoc.* **82**, 1118–1122.

Burbea, J. (1984). The convexity with respect to Gaussian distributions of divergences of order α. *Utilitas Math.* **26**, 171–192.

Burbea, J., and Rao, C. R. (1982). On the convexity of some divergence measures based on entropy functions. *IEEE Trans. Inform. Theory* **IT-28**, 489–495.

Burnaby, T. P. (1966). Growth invariant discriminant functions and generalized distances. *Biometrics* **22**, 96–110.

Butler, R. W. (1986). Predictive likelihood inference with applications (with discussion). *J. R. Statist. Soc. B* **48**, 1–38.

Byth, K., and McLachlan, G. J. (1978). The biases associated with maximum likelihood methods of estimation of the multivariate logistic risk function. *Commun. Statist.–Theory Meth.* **A7**, 877–890.

Byth, K., and McLachlan, G. J. (1980). Logistic regression compared to normal discrimination for non-normal populations. *Austral. J. Statist.* **22**, 188–196.

Cacoullos, T. (1965a). Comparing Mahalanobis distances, I: comparing distances between k normal populations and another unknown. *Sankhyā A* **27**, 1–22.

Cacoullos, T. (1965b). Comparing Mahalanobis distances, II: Bayes procedures when the mean vectors are unknown. *Sankhyā A* **27**, 23–32.

Cacoullos, T. (1966). Estimation of a multivariate density. *Ann. Inst. Statist. Math.* **18**, 179–189.

Cacoullos, T. (Ed.). (1973). *Discriminant Analysis and Applications*. New York: Academic Press.

Campbell, M. N., and Donner, A. (1989). Classification efficiency of multinomial logistic regression relative to ordinal logistic regression. *J. Amer. Statist. Assoc.* **84**, 587–591.

Campbell, N. A. (1978). The influence function as an aid in outlier detection in discriminant analysis. *Appl. Statist.* **27**, 251–258.

Campbell, N. A. (1980a). Robust procedures in multivariate analysis. I: Robust covariance estimation. *Appl. Statist.* **29**, 231–237.

Campbell, N. A. (1980b). Shrunken estimators in discriminant and canonical variate analysis. *Appl. Statist.* **29**, 5–14.

Campbell, N. A. (1982). Robust procedures in multivariate analysis II. Robust canonical variate analysis. *Appl. Statist.* **31**, 1–8.

Campbell, N. A. (1984a). Canonical variate analysis—a general model formulation. *Austral. J. Statist.* **26**, 86–96.

Campbell, N. A. (1984b). Canonical variate analysis with unequal covariance matrices: generalizations of the usual solution. *Math. Geol.* **16**, 109–124.

Campbell, N. A. (1984c). Mixture models and atypical values. *Math. Geol.* **16**, 465–477.

Campbell, N. A. (1985). Mixture models—some extensions. Unpublished manuscript.

Campbell, N. A., and Atchley, W. R. (1981). The geometry of canonical variate analysis. *Syst. Zool.* **30**, 268–280.

CART (1984). *CART Version 1.1.* Lafayette, California: California Statistical Software Inc.

Castagliola, P., and Dubuisson, B. (1989). Two classes linear discrimination. A min-max approach. *Pattern Recognition Letters* **10**, 281–287.

Čencov, N. N. (1962). Evaluation of an unknown distribution density from observations. *Soviet Math.* **3**, 1559–1562.

Chaddha, R. L., and Marcus, L. F. (1968). An empirical comparison of distance statistics for populations with unequal covariance matrices. *Biometrics* **24**, 683–694.

Chan, L. S., and Dunn, O. J. (1972). The treatment of missing values in discriminant analysis–I. The sampling experiment. *J. Amer. Statist. Assoc.* **67**, 473–477.

Chan, L. S., and Dunn, O. J. (1974). A note on the asymptotic aspect of the treatment of missing values in discriminant analysis. *J. Amer. Statist. Assoc.* **69**, 672–673.

Chan, L. S., Gilman, J. A., and Dunn, O. J. (1976). Alternative approaches to missing values in discriminant analysis. *J. Amer. Statist. Assoc.* **71**, 842–844.

Chanda, K. C. (1990). Asymptotic expansions of the probabilities of misclassification for k-NN discriminant rules. *Statist. Prob. Letters* **10**, 341–349.

Chanda, K. C., and Ruymgaart, F. H. (1989). Asymptotic estimate of probability of misclassification for discriminant rules based on density estimates. *Statist. Prob. Letters* **8**, 81–88.

Chandrasekaran, B., and Jain, A. K. (1975). Independence, measurement complexity, and classification performance. *IEEE Trans. Syst. Man. Cybern.* **SMC-5**, 240–244.

Chandrasekaran, B., and Jain, A. K. (1977). Independence, measurement complexity, and classification performance: an emendation. *IEEE Trans. Syst. Man. Cybern.* **SMC-7**, 564–566.

Chang, C. Y. (1973). Dynamic programming as applied to feature subset selection in a pattern recognition system. *IEEE Trans. Syst. Man. Cybern.* **SMC-3**, 166–171.

Chang, P. C., and Afifi, A. A. (1974). Classification based on dichotomous and continuous variables. *J. Amer. Statist. Assoc.* **69**, 336–339.

Chang, W. C. (1983). On using principal components before separating a mixture of two multivariate normal distributions. *Appl. Statist.* **32**, 267–275.

Chang, W. C. (1987). A graph for two training samples in a discriminant analysis. *Appl. Statist.* **36**, 82–91.

Chatterjee, S., and Chatterjee, S. (1983). Estimation of misclassification probabilities by bootstrap methods. *Commun. Statist.–Simula.* **12**, 645–656.

Chen, C. H. (1975). On a class of computationally efficient feature selection criteria. *Pattern Recognition* **7**, 87–94.

Chen, Y., and Tu, D. S. (1987). Estimating the error rate in discriminant analysis: by the delta, jackknife and bootstrap method. *Chinese J. Appl. Prob. Statist.* **3**, 203–210.

Chernick, M. R., Murthy, V. K., and Nealy, C. D. (1985). Application of bootstrap and other resampling techniques: evaluation of classifier performance. *Pattern Recognition Letters* **3**, 167–178.

Chernick, M. R., Murthy, V. K., and Nealy, C. D. (1986a). Correction note to Application of bootstrap and other resampling techniques: evaluation of classifier performance. *Pattern Recognition Letters* **4**, 133–142.

Chernick, M. R., Murthy, V. K., and Nealy, C. D. (1986b). Estimation of error rate for linear discriminant functions by resampling: non-Gaussian populations. *Comput. Math. Applic.* **15**, 29–37.

Chernoff, H. (1952). A measure of asymptotic efficiency for tests of a hypothesis based on the sum of observations. *Ann. Math. Statist.* **23**, 493–507.

Chernoff, H. (1972). The selection of effective attributes for deciding between hypotheses using linear discriminant functions. In *Frontiers of Pattern Recognition*, S. Watanbe (Ed.). New York: Academic Press, pp. 55–60.

Chernoff, H. (1973). Some measures for discriminating between normal multivariate distributions with unequal covariance matrices. In *Multivariate Analysis* (Vol. III), P. R. Krishnaiah (Ed.). New York: Academic Press, pp. 337–344.

Chhikara, R. S. (1986). Error analysis of crop acreage estimation using satellite data. *Technometrics* **28**, 73–80.

Chhikara, R. S., and Folks, J. L. (1989). *The Inverse Gaussian Distribution: Theory, Methodology, and Applications*. New York: Marcel Dekker.

Chhikara, R. S., and McKeon, J. (1984). Linear discriminant analysis with misallocation in training samples. *J. Amer. Statist. Assoc.* **79**, 899–906.

Chhikara, R. S., and Odell, P. L. (1973). Discriminant analysis using certain normed exponential densities with emphasis on remote sensing application. *Pattern Recognition* **5**, 259–272.

Chinganda, E. F., and Subrahmaniam, K. (1979). Robustness of the linear discriminant function to nonnormality: Johnson's system. *J. Statist. Planning Inference* **3**, 69–77.

Chittineni, C. B. (1980). Learning with imperfectly labeled patterns. *Pattern Recognition* **12**, 281–291.

Chitteneni, C. B. (1981). Estimation of probabilities of label imperfections and correction of mislabels. *Pattern Recognition* **13**, 257–268.

Choi, K. (1969). Empirical Bayes procedure for (pattern) classification with stochastic learning. *Ann. Inst. Statist. Math.* **21**, 117–125.

Choi, S. C. (1972). Classification of multiply observed data. *Biom. J.* **14**, 8–11.

Choi, S. C. (Ed.). (1986). *Statistical Methods of Discrimination and Classification—Advances in Theory and Applications*. New York: Pergamon Press.

Chow, C. K. (1970). On optimum recognition error and reject tradeoff. *IEEE Trans. Inform. Theory* **IT-16**, 41–46.

Chow, Y. S., Geman, S., and Wu, L. D. (1983). Consistent cross-validated density estimation. *Ann. Statist.* **11**, 25–38.

Clarke, W. R., Lachenbruch, P. A., and Broffitt, B. (1979). How nonnormality affects the quadratic discriminant function. *Commun. Statist.–Theory Meth.* **A8**, 1285–1301.

Clarkson, D. B. (1988). Algorithm AS R74. A least-squares version of algorithlm AS 211: the $F - G$ diagonalization algorithm. *Appl. Statist.* **37**, 317–321.

Cléroux, R., Helbling, J.-M., and Ranger, N. (1986). Some methods of detecting multivariate outliers. *Comput. Statist. Quart.* **3**, 177–195.

Cleveland, W. S., and Lachenbruch, P. A. (1974). A measure of divergence among several populations. *Commun. Statist.* **3**, 201–211.

Clunies-Ross, C. W., and Riffenburgh, R. H. (1960). Geometry and linear discrimination. *Biometrika* **47**, 185–189.

Cochran, W. G. (1964a). On the performance of the linear discriminant function. *Technometrics* **6**, 179–190.

Cochran, W. G. (1964b). Comparison of two methods of handling covariates in discriminant analysis. *Ann. Inst. Statist. Math.* **16**, 43–53.

Cochran, W. G. (1966). Contribution to the discussion of paper by M. Hills. *J. R. Statist. Soc. B* **28**, 28–29.

Cochran, W. G., and Bliss, C. I. (1948). Discriminant functions with covariance. *Ann. Math. Statist.* **19**, 151–176.

Cochran, W. G., and Hopkins, C. E. (1961). Some classification problems with multivariate qualitative data. *Biometrics* **17**, 10–32.

Collins, J. R. (1982). Robust M-estimators of location vectors. *J. Multivar. Anal.* **12**, 480–492.

Collins, J. R., and Wiens, D. P. (1985). Minimax variance M-estimators in ϵ-contamination models. *Ann. Statist.* **13**, 1078–1096.

Conover, W. J., and Iman, R. L. (1980). The rank transformation as a method of discrimination with some examples. *Commun. Statist.–Theory Meth.* **A9**, 465–487.

Coomans, D., and Broeckaert, I. (1986). *Potential Pattern Recognition in Chemical and Medical Decision Making*. Letchworth: Research Studies Press.

Cooper, P. W. (1962a). The hyperplane in pattern recognition. *Cybern.* **5**, 215–238.

Cooper, P. W. (1962b). The hypersphere in pattern recognition. *Inform. Control* **5**, 324–346.

Cooper, P. W. (1963). Statistical classification with quadratic forms. *Biometrika* **50**, 439–448.

Cooper, P. W. (1965). Quadratic discriminant functions in pattern recognition. *IEEE Trans. Inform. Theory* **IT-11**, 313–315.

Copas, J. B. (1981). Contribution to the discussion of paper by D. M. Titterington et al. *J. R. Statist. Soc. A* **144**, 165–166.

Copas, J. B. (1983). Plotting p against x. *Appl. Statist.* **32**, 25–31.

Copas, J. B. (1988). Binary regression models for contaminated data (with discussion). *J. R. Statist. Soc. B* **50**, 225–265.

Cornfield, J. (1962). Joint dependence of risk of coronary heart disease on serum cholesterol and systolic blood pressure: a discriminant function approach. *Fed. Amer. Socs. Exper. Biol. Proc. Suppl.* **11**, 58–61.

Cosslett, S. (1981a). Maximum likelihood estimators for choice-based samples. *Econometrica* **49**, 1289–1316.

Cosslett, S. (1981b). Efficient estimation of discrete-choice models. In *Structural Analysis of Discrete Data with Econometric Applications*, C. F. Manski and D. McFadden (Eds.). Cambridge, Massachusetts. MIT Press, pp. 51–111.

Costanza, M. C., and Afifi, A. A. (1979). Comparison of stoppinig rules in forward stepwise discriminant analysis. *J. Amer. Statist. Assoc.* **74**, 777–785.

Costanza, M. C., and Ashikaga, T. (1986). Monte Carlo study of forward stepwise discrimination based on small samples. *Comput. Math. Applic.* **12A**, 245–252.

Cover, T. M. (1968). Rates of convergence for nearest neighbor procedures. *Proc. Hawaii Inter. Conf. on System Sciences*. Honolulu: Western Periodicals, pp. 413–415.

Cover, T. M., and Hart, P. (1967). Nearest neighbor pattern classification. *IEEE Trans. Inform. Theory* **IT-11**, 21–27.

Cox, D. R. (1966). Some procedures associated with the logistic qualitative response curve. In *Research Papers on Statistics: Festschrift for J. Neyman,* F. N. David (Ed.). New York: Wiley, pp. 55–71.

Cox, D. R. (1970). *The Analysis of Binary Data*. First Edition. London: Methuen.

Cox, D. R. (1972a). Regression models and life tables (with discussion). *J. R. Statist. Soc. B* **34**, 187–220.

Cox, D. R. (1972b). The analysis of binary multivariate data. *Appl. Statist.* **21**, 113–120.

Cox, D. R., and Small, N. J. H. (1978). Testing multivariate normality. *Biometrika* **65**, 263–272.

Cox, D. R., and Snell, E. J. (1968). A general definition of residuals. *J. R. Statist. Soc. B* **30**, 248–275.

Cox, D. R., and Snell, E. J. (1989). *Analysis of Binary Data*. Second Edition. London: Chapman and Hall.

Craig, R. G. (1979). Autocorrelation in LANDSAT data. *Proc. 13th Inter. Symp. Remote Sensing of Environment*. Ann Arbor: Environmental Research Institute of Michigan, pp. 1517–1524.

Cramér, H. (1946). *Mathematical Methods of Statistics*. Princeton: Princeton University Press.

Crask, M. R., and Perreault, W. D. (1977). Validation of discriminant analysis in marketing research. *J. Marketing Res.* **14**, 60–68.

Crawley, D. R. (1979). Logistic discrimination as an alternative to Fisher's linear discriminant function. *N. Z. Statist.* **14**, 21–25.

Cressie, N. A. C., and Read, T. R. C. (1984). Multinomial goodness-of-fit tests. *J. R. Statist. Soc. B* **46**, 440–464.

Critchley, F., and Ford, I. (1984). On the covariance of two noncentral F random variables and the variance of the estimated linear discriminant function. *Biometrika* **71**, 637–638.

Critchley, F., and Ford, I. (1985). Interval estimation in discrimination: the multivariate normal equal covariance case. *Biometrika* **72**, 109–116.

Critchley, F., Ford, I., and Rijal, O. (1987). Uncertainty in discrimination. *Proc. Conf. DIANA II*. Prague: Math. Inst. of the Czechoslovak Academy of Sciences, pp. 83–106.

Critchley, F., Ford, I., and Rijal, O. (1988). Interval estimation based on the profile likelihood: strong Lagrangian theory with applications to discrimination. *Biometrika* **75**, 21–28.

Csörgö, S. (1986). Testing for normality in arbitrary dimensions. *Ann. Statist.* **14**, 708–723.

Ćwik, J., and Mielniczuk, J. (1989). Estimating density ratio with application to discriminant ratio. *Commun. Statist.–Theory Meth.* **18**, 3057–3069.

Dargahi-Noubary, G. R. (1981). An application of discrimination when covariance matrices are proportional. *Austral. J. Statist.* **23**, 38–44.

Das Gupta, S. (1964). Non-parametric classification rules. *Sankhyā A* **26**, 25–30.

Das Gupta, S. (1965). Optimum classification rules for classification into two multivariate normal populations. *Ann. Math. Statist.* **36**, 1174–1184.

Das Gupta, S. (1968). Some aspects of discrimination function coefficients. *Sankhyā A* **30**, 387–400.

Das Gupta, S. (1973). Theories and methods in classification: a review. In *Discriminant Analysis and Applications*, T. Cacoullos (Ed.). New York: Academic Press, pp. 77–137.

Das Gupta, S. (1974). Probability of inequalities and errors in classification. *Ann. Statist.* **2**, 751–762.

Das Gupta, S. (1980). Discriminant analysis. In *R. A. Fisher: An Appreciation,* S. E. Fienberg and D. V. Hinkley (Eds.). New York: Springer-Verlag, pp. 161–170.

Das Gupta, S. (1982). Optimum rules for classification into two multivariate normal populations with the same covariance matrix. In *Handbook of Statistics* (Vol. 2), P. R. Krishnaiah and L. Kanal (Eds.). Amsterdam: North-Holland, pp. 47–60.

Das Gupta, S., and Bandyopadhyay, S. (1977). Asymptotic expansions of the distributions of some classification statistics and the probabilities of misclassification when the training samples are dependent. *Sankhyā* **39**, 12–25.

Das Gupta, S., and Perlman, M. D. (1974). Power of the noncentral F-test: effect of additional variates on Hotelling's T^2 test. *J. Amer. Statist. Assoc.* **69**, 174–180.

Daudin, J. J. (1986). Selection of variables in mixed-variable discriminant analysis. *Biometrics* **42**, 473–481.

Dröge, J. B. M., Rinsma, W. J., van't Klooster, H. A., Tas, A. C., and Van Der Greef, J. (1987). An evaluation of SIMCA. Part 2—classification of pyrolysis mass spectra of pseudomonas and serratia bacteria by pattern recognition using the SIMCA classifier. *J. Chemometrics* **1**, 231–241.

Dubes, R. C., and Jain, A. K. (1989). Random field models in image analysis. *J. Appl. Statist.* **16**, 131–164.

Duchene, J., and Leclercq, S. (1988). An optimal transformation for discriminant and principal component analysis. *IEEE Trans. Pattern Anal. Machine Intell.* **10**, 978–983.

Duda, R. O., and Hart, P. E. (1973). *Pattern Classification and Scene Analysis.* New York: Wiley.

Dudewicz, E. J., and Bishop, T. A. (1979). The heteroscedastic method. In *Optimizing Methods in Statistics*, J. S. Rustagi (Ed.). New York: Academic Press, pp. 183–203.

Dudewicz, E. J., and Taneja, V. S. (1989). Heteroscedastic discriminant analysis with expert systems and NMR applications. *Amer. J. Math. Management Sciences* **9**, 229–242.

Dudewicz, E. J., Levy, G. C., Lienhart, J., and Wehrli, F. (1989). Statistical analysis of magnetic resonance imaging data in the normal brain (data, screening, normality, discrimination, variability), & implications for expert statistical programming for ESS^{TM}. *Amer. J. Math. Management Sciences* **9**, 299–359.

Duffy, D. E., and Santner, T. J. (1989). On the small sample properties of norm-restricted maximum likelihood estimators for logistic regression models. *Commun. Statist.–Theory Meth.* **18**, 959–980.

Duin, R. P. W. (1976). On the choice of smoothing parameters for Parzen estimators of probablities density functions. *IEEE Trans. Comput.* **C-25**, 1175–1179.

Dunn, C. L. (1982). Comparison of combinatoric and likelihood ratio procedures for classifying samples. *Commun. Statist.–Theory Meth.* **11**, 2361–2377.

Dunn, C. L., and Smith, W. B. (1980). Combinatoric classification of multivariate normal variates. *Commun. Statist.–Theory Meth.* **A9**, 1317–1340.

Dunn, C. L., and Smith, W. B. (1982). Normal combinatoric classification: the sampling case. *Commun. Statist.–Theory Meth.* **11**, 271–289.

Dunn, J. E., and Tubbs, J. D. (1980). VARSTAB: a procedure for determining homoscedastic transformations of multivariate normal populations. *Commun. Statist.–Simula.* **B9**, 589–598.

Dunn, O. J. (1971). Some expected values for probabilities of correct classification in discriminant analysis. *Technometrics* **13**, 345–353.

Dunn, O. J., and Varady, P. D. (1966). Probabilities of correct classification in discriminant analysis. *Biometrics* **22**, 908–924.

Dunsmore, I. R. (1966). A Bayesian approach to classification. *J. R. Statist. Soc. B* **28**, 568–577.

Durbin, J. (1987). Statistics and Statistical Sciences. *J. R. Statist. Soc. A* **150**, 177–191.

Eastment, H. T., and Krzanowski, W. J. (1972). Cross-validatory choice of the number of components from a principal component analysis. *Technometrics* **24**, 73–77.

Edwards, A. W. F. (1971). Distances between populations on the basis of gene frequencies. *Biometrics* **27**, 873–882.

Efron, B. (1975). The efficiency of logistic regression compared to normal discriminant analysis. *J. Amer. Statist. Assoc.* **70**, 892–898.

Efron, B. (1977). The efficiency of Cox's likelihood function for censored data. *J. Amer. Statist. Assoc.* **72**, 557–565.

Efron, B. (1979). Bootstrap methods: another look at the jackknife. *Ann. Statist.* **7**, 1–26.

Efron, B. (1981a). Nonparametric estimates of standard error: the jackknife, the bootstrap and other methods. *Biometrika* **68**, 589–599.

Efron, B. (1981b). Nonparametric standard errors and confidence intervals (with discussion). *Canad. J. Statist.* **9**, 139–172.

Efron, B. (1982). *The Jackknife, the Bootstrap and Other Resampling Plans*. Philadelphia: SIAM.

Efron, B. (1983). Estimating the error rate of a prediction rule: improvement on cross-validation. *J. Amer. Statist. Assoc.* **78**, 316–331.

Efron, B. (1985). Bootstrap confidence intervals for a class of parametric problems. *Biometrika* **72**, 45–58.

Efron, B. (1986). How biased is the apparent error rate of a prediction rule? *J. Amer. Statist. Assoc.* **81**, 461–470.

Efron, B. (1987). Better bootstrap confidence intervals (with discussion). *J. Amer. Statist. Assoc.* **82**, 171–200.

Efron, B. (1990). More efficient bootstrap computations. *J. Amer. Statist. Assoc.* **85**, 79–89.

Efron, B., and Gong, G. (1983). A leisurely look at the bootstrap, the jackknife, and cross-validation. *Amer. Statistician* **37**, 36–48.

Efron, B., and Hinkley, D. V. (1978). Assessing the accuracy of the maximum likelihood estimator: observed versus expected Fisher information (with discussion). *Biometrika* **65**, 457–487.

Efron, B., and Morris, C. (1976). Multivariate empirical Bayes and estimation of covariance matrives. *Ann. Statist.* **4**, 22–32.

Efron, B., and Tibshirani, R. (1986). Bootstrap methods for standard errors, confidence intervals, and other measures of statistical accuracy (with discussion). *Statist. Science* **1**, 54–77.

Eisenbeis, R. A., Gilbert, G. G., and Avery, R. B. (1973). Investigating the relative importance of individual variables and variable subsets in discriminant analysis. *Commun. Statist.* **2**, 205–219.

Elashoff, J. D., Elashoff, R. M., and Goldman, G. E. (1967). On the choice of variables in classification problems with dichotomous variables. *Biometrika* **54**, 668–670.

Elfving, G. (1961). An expansion principle for distribution functions with applications to Student's statistic and the one-dimensional classification statistic. In *Studies in Item Analysis and Prediction*, H. Solomon (Ed.). Stanford: Stanford University Press, pp. 276–284.

El Khattabi, S., and Streit, F. (1989). Further results on identification when the parameters are partially unknown. In *Statistical Data Analysis and Inference*, Y. Dodge (Ed.). Amsterdam: North-Holland, pp. 347–352.

Ellison, B. E. (1962). A classification problem in which information about alternative distributions is based on samples. *Ann. Math. Statist.* **33**, 213–223.

Ellison, B. E. (1965). Multivariate-normal classification with covariances known. *Ann. Math. Statist.* **36**, 1787–1793.

El-Sayyad, G. M., Samiuddin, M., and Al-Harbey, A. A. (1989). On parametric density estimation. *Biometrika* **76**, 343–348.

Elvers, E. (1977). Statistical discrimination in seismology. In *Recent Developments in Statistics*, J. R. Barra et al. (Eds.). Amsterdam: North-Holland, pp. 663–672.

Emery, J. L., and Carpenter, R. G. (1974). Pulmonary mast cells in infants and their relation to unexpected death in infancy. *SIDS 1974, Proc. Francis E. Camps Inter. Symp. on Sudden and Unexpected Deaths in Infancy*. Toronto: Canada Foundation for the Study of Infant Deaths, pp. 7–19.

Enas, G. G., and Choi, S. C. (1986). Choice of the smoothing parameter and efficiency of k-nearest neighbour classification. *Comput. Math. Applic.* **12A**, 235–244.

Enis, P., and Geisser, S. (1970). Sample discriminants which minimize posterior squared error loss. *South African Statist. J.* **4**, 85–93.

Enis, P., and Geisser, S. (1974). Optimal predictive linear discriminants. *Ann. Statist.* **2**, 403–410.

Epanechnikov, V. A. (1969). Non-parametric estimation of a multivariate probability density. *Theor. Prob. Appl.* **14**, 153–158.

Eriksen, P. S. (1987). Proportionality of covariance matrices. *Ann. Statist.* **15**, 732–748.

Estes, S. E. (1965). *Measurement Selection for Linear Discriminant Used in Pattern Classification*. Unpublished Ph. D. thesis, Stanford University.

Everitt, B. S. (1984). *An Introduction to Latent Variable Models*. London: Chapman and Hall.

Farewell, V. T. (1979). Some results on the estimation of logistic models based on retrospective data. *Biometrika* **66**, 27–32.

Farmer, J. H., and Freund, R. J. (1975). Variable selection in the multivariate analysis of variance (MANOVA). *Commun. Statist.* **4**, 87–98.

Farver, T. B., and Dunn, O. J. (1979). Stepwise variable selection in classification problems. *Biom. J.* **21**, 145–153.

Fatti, L. P. (1983). The random-effects model in discriminant analysis. *J. Amer. Statist. Assoc.* **78**, 679–687.

Fatti, L. P., and Hawkins, D. M. (1986). Variable selection in heteroscedastic discriminant analysis. *J. Amer. Statist. Assoc.* **81**, 494–500.

Fatti, L. P., Hawkins, D. M., and Raath, E. L. (1982). Discriminant analysis. In *Topics in Applied Multivariate Analysis*, D. M. Hawkins (Ed.). Cambridge: Cambridge University Press, pp. 1–71.

Fieller, N. R. J., Gilbertson, D. D., and Olbright, W. (1984). A new method for the environmental analysis of particle size data from shoreline environments. *Nature* **311**, 648–651.

Fieller, N. R. J., Flenley, E. C., and Olbright, W. (1990). Statistics of particle size data. *Technical Report No. 358/90*. Sheffield: Dept. of Probability and Statistics, University of Sheffield.

Fienberg, S. E. (1980). *The Analysis of Cross-Classified Categorical Data*. Second Edition. Cambridge, Massachusetts. MIT Press.

Fienberg, S. E., and Holland, P. W. (1973). Simultaneous estimation of multinomial probabilities. *J. Amer. Statist. Assoc.* **68**, 683–691.

Fisher, R. A. (1936). The use of multiple measurements in taxonomic problems. *Ann. Eugen.* **7**, 179–188.

Fisher, R. A. (1938). The statistical utilization of multiple measurements. *Ann. Eugen.* **8**, 376–386.

Fitzmaurice, G. M., and Hand, D. J. (1987). A comparison of two average conditional error rate estimators. *Pattern Recognition Letters* **6**, 221–224.

Fitzmaurice, G. M., Krzanowski, W. J., and Hand, D. J. (1990). A Monte Carlo study of the 632 bootstrap estimator of error rate. *J. Classification*. To appear.

Fix, E., and Hodges, J. L. (1951). Discriminatory analysis-nonparametric discrimination: consistency properties. *Report No. 4*. Randolph Field, Texas: U. S. Air Force School of Aviation Medicine. (Reprinted as pp. 261–279 of Agrawala, 1977.)

Fix, E., and Hodges, J. L. (1952). Discriminatory analysis. Nonparametric: small sample performance. *Report No. 11*. Randolph Field, Texas: U. S. Air Force School of Aviation Medicine. (Reprinted as pp. 280–322 of Agrawala, 1977.)

Flick, T. E., Jones, L. K., Priest, R. G., and Herman, C. (1990). Pattern classification using projection pursuit. *Pattern Recognition* **23**, 1367–1376.

Flury, B. (1984). Common principal components in k groups. *J. Amer. Statist. Assoc.* **79**, 892–898.

Flury, B. (1986). Proportionality of k covariance matrices. *Statist. Prob. Letters* **4**, 29–33.

Flury, B. (1987a). Two generalizations of the common principal component model. *Biometrika* **74**, 59–69.

Flury, B. (1987b). A hierarchy of relationships between covariance matrices. In *Advances in Multivariate Statistical Analysis*, A. K. Gupta (Ed.). Dordrecht: Reidel, pp. 31–43.

Flury, B. (1988). *Common Principal Components and Related Multivariate Models*. New York: Wiley.

Flury, B., and Gautschi, W. (1986). An algorithm for simultaneous orthogonal transformation of several positive definite symmetric matrices to nearly diagonal form. *SIAM J. Scientific Statist. Comput.* **7**, 169–184.

Föglein, J., and Kittler, J. (1983). The effect of pixel correlations on class separability. *Pattern Recognition Letters* **1**, 401–407.

Foley, D. H. (1972). Consideration of sample and feature size. *IEEE Trans. Inform. Theory* **IT-18**, 618–626.

Foley, D. H., and Sammon, J. W. (1975). An optimal set of discrimination vectors. *IEEE Trans. Comput.* **C-24**, 281–289.

Folks, J. R., and Chhikara, R. S. (1978). The inverse Gaussian distribution and its statistical applications—a review. *J. R. Statist. Soc. B* **40**, 263–289.

Fowlkes, E. B. (1979). Some methods for studying the mixture of two normal (lognormal) distributions. *J. Amer. Statist. Assoc.* **74**, 561–575.

Fowlkes, E. B. (1987). Some diagnostics for binary logistic regression via smoothing. *Biometrika* **74**, 503–515.

Fowlkes, E. B., Gnanadesikan, R., and Kettenring, J. R. (1987). Variable selection in clustering and other contexts. In *Design, Data, and Analysis*, C. L. Mallows (Ed.). New York: Wiley, pp. 13–34.

Frank, I. E., and Friedman, J. H. (1989). Classification: oldtimers and newcomers. *J. Chemometrics* **3**, 463–475.

Frank, I. E., and Lanteri, S. (1989). *Chemometrics Intell. Lab. Syst.* **5**, 247–256.

Freed, N., and Glover, F. (1981). A linear programming approach to the discriminant problem. *Decision Sciences* **12**, 68–74.

Freedman, D. A. (1967). A remark on sequential discrimination. *Ann. Math. Statist.* **38**, 1666–1676.

Friedman, J. H. (1977). A recursive partitioning decision rule for nonparametric classification. *IEEE Trans. Comput.* **C-26**, 404–408.

Friedman, J. H. (1987). Exploratory projection pursuit. *J. Amer. Statist. Assoc.* **82**, 249–266.

Friedman, J. H. (1989). Regularized discriminant analysis. *J. Amer. Statist. Assoc.* **84**, 165–175.

Friedman, J. H., and Tukey, J. W. (1974). A projection pursuit algorithm for exploratory data analysis. *IEEE Trans. Comput.* **C-23**, 881–890.

Friedman, J. H., Baskett, F., and Shustek, J. (1975). An algorithm for finding nearest neighbors. *IEEE Trans. Comput.* **C-24**, 1000–1006.

Friedman, J. H., Stuetzle, W., and Schroeder, A. (1984). Projection pursuit density estimation. *J. Amer. Statist. Assoc.* **79**, 599–608.

Fritz, J. (1975). Distribution-free exponential error bound for nearest neighbour pattern classification. *IEEE Trans. Inform. Theory* **21**, 552–557.

Fryer, M. J. (1977). A review of some non-parametric methods of density estimation. *J. Inst. Math. Appl.* **20**, 335–354.

Fu, K.-S. (1968). *Sequential Methods in Pattern Recognition and Machine Learning.* New York: Academic Press.

Fu, K.-S. (1986). A step towards unification of syntactic and statistical pattern recognition. *IEEE Trans. Pattern Anal. Machine Intell.* **PAMI-8**, 398–404.

Fu, K.-S., and Yu, T. S. (1980). *Statistical Pattern Classification Using Contextual Information.* Chichester: Research Studies Press.

Fu, K.-S., Min, P. J., and Li, T. J. (1970). Feature selection in pattern recognition. *IEEE Trans. Syst. Science Cybern.* **SSC-6**, 33–39.

Fujikoshi, Y. (1985). Selection of variables in two-group discriminant analysis by error rate and Akaike's information criteria. *J. Multivar. Anal.* **17**, 27–37.

Fujikoshi, Y., and Kanazawa, M. (1976). The ML classification statistic in covariate discriminant analysis and its asymptotic expansions. In *Essays in Probability and Statistics*, S. Ikeda et al. (Eds.). Tokyo: Shinko-Tsusho, pp. 305–320.

Fujikoshi, Y., Krishnaiah, P. R., and Schmidhammer, J. (1987). Effect of additional variables in principal component analysis, discriminant analysis and canonical correlation analysis. In *Advances in Multivariate Statistical Analysis*, A. K. Das Gupta (Ed.). Dordrecht: Reidel, pp. 45–61.

Fukunaga, K. (1972). *Introduction to Statistical Pattern Recognition.* First Edition. New York: Academic Press.

Fukunaga, K. (1990). *Introduction to Statistical Pattern Recognition.* Second Edition. New York. Academic Press.

Fukunaga, K., and Ando, S. (1977). The optimum nonlinear features for a scatter criterion in discriminant analysis. *IEEE Trans. Inform. Theory* **IT-23**, 453–459.

Fukunaga, K., and Flick, T. E. (1984). An optimal global nearest neighbor metric. *IEEE Trans. Pattern Anal. Machine Intell.* **PAMI-6**, 313–318.

Fukunaga, K., and Flick, T. E. (1985). The 2-NN rule for more accurate NN risk estimation. *IEEE Trans. Pattern Anal. Machine Intell.* **PAMI-7**, 107–112.

Fukunaga, K., and Hayes, R. R. (1989a). Effects of sample size in classifier design. *IEEE Trans. Pattern Anal. Machine Intell.* **PAMI-11**, 873–885.

Fukunaga, K., and Hayes, R. R. (1989b). Estimation of classifier performance. *IEEE Trans. Pattern Anal. Machine Intell.* **PAMI-11**, 1087–1101.

Fukunaga, K., and Hostetler, L. D. (1973). Optimization of k-nearest neighbor density estimates. *IEEE Trans. Inform. Theory* **IT-19**, 320–326.

Fukunaga, K., and Hummels, D. M. (1987a). Bias of nearest neighbor error estimates. *IEEE Trans. Pattern Anal. Machine Intell.* **PAMI-9**, 103–112.

Fukunaga, K., and Hummels, D. M. (1987b). Bayes error estimation using Parzen and k-NN procedures. *IEEE Trans. Pattern Anal. Machine Intell.* **PAMI-9**, 634–643.

Fukunaga, K., and Hummels, D. M. (1989). Leave-one-out procedures for nonparametric error estimates. *IEEE Trans. Pattern Anal. Machine Intell.* **PAMI-11**, 421–423.

Fukunaga, K., and Kessell, D. L. (1972). Application of optimum error-reject function. *IEEE Trans. Inform. Theory* **IT–18**, 814–817.

Fukunaga, K., and Kessell, D. L. (1973). Nonparametric Bayes error estimation using unclassified samples. *IEEE Trans. Inform. Theory* **IT-18**, 814–817.

Fukunaga, K., and Narendra, P. M. (1975). A branch and bound algorithm for computing k-nearest neighbours. *IEEE Trans. Comput.* **C-24**, 750–753.

Fukunaga, K., and Narendra, P. M. (1977). A branch and bound algorithm for feature subset selection. *IEEE Trans. Comput.* **C-26**, 917–922.

Fung, K. Y., and Wrobel, B. A. (1989). The treatment of missing values in logistic regression. *Biom. J.* **31**, 35–47.

Furnival, G. M. (1971). All possible regression with less computation. *Technometrics* **13**, 403–408.

Gabriel, K. R. (1968). Simultaneous test procedures in multivariate analysis of variance. *Biometrika* **55**, 489–504.

Gabriel, K. R. (1969). Simultaneous test procedures—some theory of multiple comparisons. *Ann. Math. Statist.* **40**, 224–250.

Ganesalingam, S., and McLachlan, G. J. (1978). The efficiency of a linear discriminant function based on unclassified initial samples. *Biometrika* **65**, 658–662.

Ganesalingam, S., and McLachlan, G. J. (1979). Small sample results for a linear discriminant function estimated from a mixture of normal populations. *J. Statist. Comput. Simul.* **9**, 151–158.

Ganesalingam, S., and McLachlan, G. J. (1980). Error rate estimation on the basis of posterior probabilities. *Pattern Recognition* **12**, 405–413.

Ganesalingam, S., and McLachlan, G. J. (1981). Some efficiency results for the estimation of the mixing proportion in a mixture of two normal distributions. *Biometrika* **37**, 23–33.

Gordon, L., and Olshen, R. A. (1978). Asymptotically efficient solutions to the classification problem. *Ann. Statist.* **6**, 515–533.

Gordon, L., and Olshen, R. A. (1980). Consistent nonparametric regression from recursive partitioning schemes. *J. Multivar. Anal.* **10**, 611–627.

Govindarajulu, Z., and Gupta, A. K. (1977). Certain nonparametric classification rules and their asymptotic efficiencies. *Canad. J. Statist.* **5**, 167–178.

Gower, J. C. (1966). Some distance properties of latent root and vector methods used in multivariate analysis. *Biometrika* **53**, 325–338.

Graham, R. L., Hinkley, D. V., John, P. W. M., and Shi, S. (1990). Balanced design of bootstrap simulations. *J. R. Statist. Soc. B* **52**, 185–202.

Gray, H. L., and Schucany, W. R. (1972). *The Generalized Jackknife Statistic*. New York: Marcel Dekker.

Grayson, D. A. (1987). Statistical diagnosis and the influence of diagnostic error. *Biometrics* **43**, 975–984.

Green, M. S. (1988). Evaluating the discriminatory power of a multiple logistic regression model. *Statist. Med.* **7**, 519–524.

Greene, T., and Rayens, W. S. (1989). Partially pooled covariance matrix estimation in discriminant analysis. *Commun. Statist.–Theory Meth.* **18**, 3679–3702.

Greenland, S. (1985). An application of logistic models to the analysis of ordinal responses. *Biom. J.* **27**, 189–197.

Greenstreet, R. L., and Connor, R. J. (1974). Power of tests for equality of covariance matrices. *Technometrics* **16**, 27–30.

Greer, R. L. (1979). Consistent nonparametric estimation of best linear classification rule/solving inconsistent systems of linear inequalities. *Technical Report No. 129*. Stanford: Department of Statistics, Stanford University.

Greer, R. L. (1984). *Trees and Hills: Methodology for Maximizing Functions of Systems of Linear Relations*. Amsterdam: North-Holland.

Greig, D. M., Porteous, B. T., and Seheult, A. H. (1989). Exact maximum *a posteriori* estimation for binary images. *J. R. Statist. Soc. B* **51**, 271–279.

Gupta, A. K. (1980). On a multivariate statistical classification model. In *Multivariate Statistical Analysis*, R. P. Gupta (Ed.). Amsterdam: North-Holland, pp. 83–93.

Gupta, A. K. (1986). On a classification rule for multiple measurements. *Comput. Math. Applic.* **12A**, 301–308.

Gupta, A. K., and Govindarajulu, Z. (1973). Some new classification rules for *c* univariate normal populations. *Canad. J. Statist.* **1**, 139–157.

Gupta, A. K., and Kim, B. K. (1978). On a distribution-free discriminant analysis. *Biom. J.* **20**, 729–736.

Gupta, A. K., and Logan, T. P. (1990). On a multiple observations model in discriminant analysis. *J. Statist. Comput. Simul.* **34**, 119–132.

Guseman, L. F., Peters, B. C., and Walker, H. F. (1975). On minimizing the probability of misclassification for linear feature selection. *Ann. Statist.* **3**, 661–668.

Haack, D., and Govindarajulu, Z. (1986). A decision-theoretic approach to a distribution-free compound classification problem. *Commun. Statist.–Theory Meth.* **15**, 2877–2897.

Habbema, J. D. F., and Hermans, J. (1977). Selection of variables in discriminant analysis by F-statistic and error rate. *Technometrics* **19**, 487–493.

Habbema, J. D. F., and Hilden, J. (1981). The measurement of performance in probabilistic diagnosis IV. Utility considerations in therapeutics and prognostics. *Meth. Inform. Med.* **20**, 80–96.

Habbema, J. D. F., Hermans, J., and van den Broek, K. (1974a). A stepwise discriminant analysis program using density estimation. *Compstat 1974, Proc. Computational Statistics.* Vienna: Physica-Verlag, pp. 101–110.

Habbema, J. D. F., Hermans, J., and van der Burgt, A. T. (1974b). Cases of doubt in allocation problems. *Biometrika* **61**, 313–324.

Habbema, J. D. F., Hermans, J., and Remme, J. (1978a). Variable kernel density estimation in discriminant analysis. *Compstat 1978, Proc. Computational Statistics.* Vienna: Physica-Verlag, pp. 178–185.

Habbema, J. D. F., Hilden, J., and Bjerregaard, B. (1978b). The measurement of performance in probabilistic diagnosis I. The problem, descriptive tools, and measures based on classification matrices. *Meth. Inform. Med.* **17**, 217–226.

Habbema, J. D. F., Hilden, J., and Bjerregaard, B. (1981). The measurement of performance in probabilistic diagnosis V. General recommendations. *Meth. Inform. Med.* **20**, 97–100.

Haberman, S. J. (1974). *The Analysis of Frequency Data.* Chicago: University of Chicago Press.

Haberman, S. J. (1978). *Analysis of Qualitative Data* (Vol. 1). New York: Academic Press.

Haberman, S. J. (1979). *Analysis of Qualitative Data* (Vol. 2). New York: Academic Press.

Haerting, J. (1983). Special properties in selection performance of qualitative variables in discriminant analysis. *Biom. J.* **25**, 215–222.

Haff, L. R. (1980). Empirical Bayes estimation of the multivariate normal covariance matrix. *Ann. Statist.* **8**, 586–597.

Haff, L. R. (1986). On linear log-odds and estimation of discriminant coefficients. *Commun. Statist.–Theory Meth.* **15**, 2131–2144.

Haldane, J. B. S. (1956). The estimation and significance of the logarithm of a ratio of frequencies. *Ann. Hum. Gen.* **20**, 309–311.

Hall, P. (1981a). On nonparametric multivariate binary discrimination. *Biometrika* **68**, 287–294.

Hall, P. (1981b). Optimal near neighbour estimator for use in discriminant analysis. *Biometrika* **68**, 572–575.

Hall, P. (1983a). Large-sample optimality of least-squares cross-validation in density estimation. *Ann. Statist.* **11**, 1156–1174.

Hall, P. (1983b). Orthogonal series methods for both qualitative and quantitative data. *Ann. Statist.* **11**, 1004–1007.

Hall, P. (1986a). On the bootstrap and confidence intervals. *Ann. Statist.* **14**, 1431–1452.

Hall, P. (1986b). On the number of bootstrap simulations required to construct a confidence interval. *Ann. Statist.* **14**, 1453–1462.

Hall, P. (1986c). Cross-validation in nonparametric density estimation. *Proc. XIIIth Inter. Biometric Conf.* Alexandria, Virginia: Biometric Society, 15 pp.

Hall, P. (1987a). On the use of compactly supported density estimates in problems of discrimination. *J. Multivar. Anal.* **23**, 131–158.

Hall, P. (1987b). On Kullback–Leibler loss and density estimation. *Ann. Statist.* **15**, 1491–1519.

Hall, P. (1988). Theoretical comparison of bootstrap confidence intervals (with discussion). *Ann. Statist.* **16**, 927–985.

Hall, P. (1989a). On efficient bootstrap simulation. *Biometrika* **76**, 613–617.

Hall, P. (1989b). Antithetic resampling for the bootstrap. *Biometrika* **76**, 713–724.

Hall, P., and Marron, J. S. (1987a). Extent to which least-squares cross-validation minimises integrated squared error in nonparametric density estimation. *Prob. Theory Related Fields* **74**, 567–581.

Hall, P., and Marron, J. S. (1987b). Estimation of integrated squared density derivatives. *Statist. Prob. Letters* **6**, 109–115.

Hall, P., and Titterington, D. M. (1987). On smoothing sparse multinomial data. *Austral. J. Statist.* **29**, 19–37.

Hall, P., and Titterington, D. M. (1989). The effect of simulation order on level accuracy and power of Monte Carlo tests. *J. R. Statist. Soc. B* **51**, 459–467.

Hall, P., and Wand, M. P. (1988). On nonparametric discrimination using density differences. *Biometrika* **75**, 541–547.

Halperin, M., Blackwelder, W. C., and Verter, J. I. (1971). Estimation of the multivariate logistic risk function: a comparison of the discriminant and maximum likelihood approaches. *J. Chron. Dis.* **24**, 125–158.

Halperin, M., Abbott, R. D., Blackwelder, W. C., Jacobowitz, R., Lan, K. K. G., Verter, J. I., and Wedel, H. (1985). On the use of the logistic model in prospective studies. *Statist. Med.* **4**, 227–235.

Hampel, F. R. (1973). Robust estimation: a condensed partial survey. *Z. Wahrscheinlickeitstheorie verw. Gebiete* **27**, 87–104.

Hampel, F. R., Ronchetti, E. M., Rousseeuw, P. J., and Stahel, W. A. (1986). *Robust Statistics.* New York: Wiley.

Han, C. P. (1968). A note on discrimination in the case of unequal covariance matrices. *Biometrika* **55**, 586–587.

Han, C. P. (1969). Distribution of discriminant function when covariance matrices are proportional. *Ann. Math. Statist.* **40**, 970–985.

Han, C. P. (1974). Asymptotic distribution of discriminant function when covariance matrices are proportional and unknown. *Ann. Inst. Statist. Math.* **26**, 127–133.

Han, C. P. (1979). Alternative methods of estimating the likelihood ratio in classification of multivariate normal observations. *Amer. Statistician.* **33**, 204–206.

Hand, D. J. (1981a). *Discrimination and Classification*. New York: Wiley.

Hand, D. J. (1981b). Branch and bound in statistical data analysis. *Statistician* **30**, 1–13.

Hand, D. J. (1982). *Kernel Discriminant Analysis*. New York: Research Studies Press.

Hand, D. J. (1983). A comparison of two methods of discriminant analysis applied to binary data. *Biometrics* **39**, 683–694.

Hand, D. J. (1986a). Recent advances in error rate estimation. *Pattern Recognition Letters* **4**, 335–346.

Hand, D. J. (1986b). Cross-validation in error rate estimation. *Proc. XIIIth Inter. Biometric Conference*. Alexandria, Virginia: Biometric Society, 13 pp.

Hand, D. J. (1986c). Estimating class sizes by adjusting fallible classifier results. *Comput. Math. Applic.* **12A**, 261–272.

Hand, D. J. (1986d). An optimal error rate estimator based on average conditional error rate: asymptotic results. *Pattern Recognition Letters* **4**, 347–350.

Hand, D. J. (1987a). Screening vs prevalence estimation. *Appl. Statist.* **36**, 1–7.

Hand, D. J. (1987b). A shrunken leaving-one-out estimator of error rate. *Comput. Math. Applic.* **14**, 161–167.

Harrison, D., and Rubinfeld, D. L. (1978). Hedonic prices and the demand for clean air. *J. Environ. Econ. Management.* **5**, 81–102.

Harter, H. L. (1951). On the distribution of Wald's classification statistic. *Ann. Math. Statist.* **22**, 58–67.

Haslett, J. (1985). Maximum likelihood discriminant analysis on the plane using a Markovian model of spatial context. *Pattern Recognition* **18**, 287–296.

Hastie, T., and Tibshirani, R. J. (1986). Generalized additive models (with discussion). *Statist. Science* **1**, 297–318.

Hastie, T., and Tibshirani, R. J. (1990). *Generalized Additive Models*. London: Chapman and Hall.

Hawkins, D. M. (1976). The subset problem in multivariate analysis of variance. *J. R. Statist. Soc. B* **38**, 132–139.

Hawkins, D. M. (1980). *Identification of Outliers*. London: Chapman and Hall.

Hawkins, D. M. (1981). A new test for multivariate normality and homoscedasticity. *Technometrics* **23**, 105–110.

Hawkins, D. M., and Raath, E. L. (1982). An extension of Geisser's discrimination model to proportional covariance matrices. *Canad. J. Statist.* **10**, 261–270.

Hawkins, D. M., and Rasmussen, S. E. (1973). Use of discriminant analysis for classification of strata in sedimentary succession. *Math. Geol.* **5**, 163–177.

Hawkins, D. M., and Wicksley, R. A. J. (1986). A note on the transformation of chi-squared variables to normality. *Amer. Statistician* **40**, 296–298.

Hawkins, D. M., Muller, M. W., and ten Krooden, J. A. (1982). Cluster analysis. In *Topics in Applied Multivariate Analysis*, D. M. Hawkins (Ed.). Cambridge: Cambridge University Press, pp. 303–356.

Hecker, R., and Wagner, H. (1978). The valuation of classification rates in stepwise discriminant analyses. *Biom. J.* **20**, 713–727.

Hellman, M. E. (1970). The nearest neighbor classification rule with a reject option. *IEEE Trans. Syst. Science Cybern.* **SSC-6**, 179–185.

Hermans, J., and Habbema, J. D. F. (1975). Comparison of five methods to estimate posterior probabilities. *EDV Med. Biol.* **6**, 14–19.

Hermans, J., van Zomeren, B., Raatgever, J. W., Sterk, P. J., and Habbema, J. D. F. (1981). Use of posterior probabilities to evaluate methods of discriminant analysis. *Meth. Inform. Med.* **20**, 207–212.

Hermans, J., Habbema, J. D. F., and Schaefer, J. R. (1982). The ALLOC80 package for discriminant analysis. *Statist. Software Newsletter* **8**, 15–20.

Highleyman, W. H. (1962). The design and analysis of pattern recognition experiments. *Bell Syst. Tech. J.* **41**, 723–744.

Hildebrandt, B., Michaelis, J., and Koller, S. (1973). Die Häufigkeit der Fehlklassifikation bei der quadratische Diskriminanzanalyse. *Biom. J.* **15**, 3–12.

Hilden, J., and Bjerregaard, B. (1976). Computer-aided diagnosis and the atypical case. In *Decision Making and Medical Care: Can Information Science Help?*, F. T. de Dombal and F. Gremy (Eds.). Amsterdam: North-Holland, pp. 365–378.

Hilden, J., Habbema, J. D. F., and Bjerregaard, B. (1978a). The measurement of performance in probabilistic diagnosis II. Trustworthiness of the exact values of the diagnostic probabilities. *Meth. Inform. Med.* **17**, 227–237.

Hilden, J., Habbema, J. D. F., and Bjerregaard, B. (1978b). The measurement of performance in probabilistic diagnosis III. Methods based on continuous functions of the diagnosis probabilities. *Meth. Inform. Med.* **17**, 238–246.

Hills, M. (1966). Allocation rules and their error rates (with discussion). *J. R. Statist. Soc. B* **28**, 1–31.

Hills, M. (1967). Discrimination and allocation with discrete data. *Appl. Statist.* **16**, 237–250.

Hinkley, D. V. (1988). Bootstrap methods (with discussion). *J. R. Statist. Soc. B* **50**, 321–370.

Hinkley, D. V., and Shi, S. (1989). Importance sampling and the nested bootstrap. *Biometrika* **76**, 435–446.

Hintze, J. L. (1980). On the use of *elemental analysis* in multivariate variable selection. *Technometrics* **22**, 609–612.

Hirst, D., Ford, I., and Critchley, F. (1988). Applications of measures of uncertainity in discriminant analysis. In *Lecture Notes in Computer Science*, J. Kittler (Ed.). Berlin: Springer-Verlag, pp. 487–496.

Hirst, D. J., Ford, I., and Critchley, F. (1990). An empirical investigation of methods for interval estimation of the log odds ratio in discriminant analysis. *Biometrika* **77**, 609–615.

Hjort, N. (1985). Neighbourhood based classification of remotely sensed data based on geometric probability models. *Technical Report No. 10*. Stanford: Department of Statistics, Stanford University.

Hjort, N. (1986a). Notes on the theory of statistical symbol recognition. *Report No. 778*. Oslo: Norwegian Computing Centre.

Hjort, N. (1986b). Contribution to the discussion of paper by P. Dianconis and D. Freedman. *Ann. Statist.* **14**, 49–55.

Hjort, N. L., and Mohn, E. (1984). A comparison of some contextual methods in remote sensing. *Proc. 18th Symp. Remote Sensing of the Environment*. Paris: CNES, pp. 1693–1702.

Hjort, N. L., and Mohn, E. (1985). On the contextual classification of data from high resolution satellites. *Proc. 4th Scandinavian Conf. on Image Analysis*. Trondheim: Norwegian Institute of Technology, 10 pp.

Hjort, N. L., and Mohn, E. (1987). Topics in the statistical analysis of remotely sensed data. *Proc. 46th Session of the ISI.*. Voorburg: International Statistical Institute, 18 pp.

Hoaglin, D. C. (1985). Using quantiles to study shape. In *Exploring Data Tables, Trends, and Shapes*, D. C. Hoaglin, F. Mosteller, and J. W. Tukey (Eds.). New York: Wiley, pp. 417–460.

Hoel, P. G., and Peterson, R. P. (1949). A solution to the problem of optimum classification. *Ann. Math. Statist.* **20**, 433–438.

Hope, A. C. A. (1968). A simplified Monte Carlo significance test procedure. *J. R. Statist. Soc. B* **30**, 582–598.

Hora, S. C., and Wilcox, J. B. (1982). Estimation of error rates in several-population discriminant analysis. *J. Marketing Res.* **19**, 57–61.

Horton, I. R., Russell, J. S., and Moore, A. W. (1968). Multivariate covariance and canonical analysis: a method for selecting the most effective discriminators in a multivariate situation. *Biometrics* **24**, 845–858.

Hosmer, D. W., and Lemeshow, S. (1980). Goodness-of-fit tests for the multiple logistic regression model. *Commun. Statist.–Theory Meth.* **A9**, 1043–1069.

Hosmer, D. W., and Lemeshow, S. (1985). Goodness-of-fit tests for the logistic regression model for matched case-control studies. *Biom. J.* **27**, 511–520.

Hosmer, D. W., and Lemeshow, S. (1989). *Applied Logistic Regression*. New York: Wiley.

Hosmer, T. A., Hosmer, D. W., and Fisher, L. (1983a). A comparison of the maximum likelihood and discriminant function estimators of the coefficients of the logistic regression model for mixed continuous and discrete variables. *Commun. Statist.–Simula.* **12**, 23–43.

Hosmer, T. A., Hosmer, D. W., and Fisher, L. (1983b). A comparison of three methods of estimating the logistic regression coefficients. *Commun. Statist.–Simula.* **12**, 577–593.

Hosmer, D. W., Lemeshow, S., and Klar, J. (1988). Goodness-of-fit testing for the logistic regression model when the estimated probabilities are small. *Biom. J.* **30**, 911–924.

Hotelling, H. (1931). The generalization of Student's ratio. *Ann. Math. Statist.* **2**, 360–378.

Hotelling, H. (1935). The most predictable criterion. *J. Educat. Psychol.* **26**, 139–142.

Hotelling, H. (1936). Relations between two sets of variates. *Biometrika* **28**, 321–377.

Hotelling, H. (1951). A generalized T test and measure of multivariate dispersion. *Proc. 2nd Berkeley Symp.* Berkeley: University of California Press, pp. 23–41.

Huber, P. J. (1964). Robust estimation of a local parameter. *Ann. Math. Statist.* **35**, 73–101.

Huber, P. J. (1977). Robust covariances. In *Statistical Decision Theory and Related Topics* (Vol. II), S. S. Gupta and D. S. Moore (Eds.). New York: Academic Press, pp. 165–191.

Huber, P. J. (1981). *Robust Statistics*. New York: Wiley.

Huber, P. J. (1985). Projection pursuit (with discussion). *Ann. Statist.* **13**, 435–475.

Huber, P. J. (1987). Experience with three-dimensional scatterplots. *J. Amer. Statist. Assoc.* **82**, 448–453.

Huberty, C. J., Wisenbaker, J. M., and Smith, J. C. (1987). Assessing predictive accuracy in discriminant analysis. *Multivar. Behav. Res.* **22**, 307–329.

Hudimoto, H. (1956–1957). On the distribution-free classification of an individual into one of two groups. *Ann. Inst. Statist. Math.* **8**, 105–112.

Hudimoto, H. (1957–1958). A note on the probability of the correct classification when the distributions are not specified. *Ann. Inst. Statist. Math.* **9**, 31–36.

Hudimoto, H. (1964). On a distribution-free two-way classification. *Ann. Inst. Statist. Math.* **16**, 247–253.

Hudimoto, H. (1968). On the empirical Bayes procedure (1). *Ann. Inst. Statist. Math.* **20**, 169–185.

Hudlet, R., and Johnson, R. (1977). Linear discrimination and some further results on best lower dimensional representations. In *Classification and Applications*, J. Van Ryzin (Ed.). New York: Academic Press, pp. 371–394.

Hufnagel, G. (1988). On estimating missing values in linear discriminant analysis: Part I. *Biom. J.* 30, 69–75.

Hughes, G. F. (1968). On the mean accuracy of statistical pattern recognizers. *IEEE Trans. Inform. Theory* **IT-14**, 55–63.

Hummel, R. A., and Zucker, S. W. (1983). On the foundation of relaxation labeling processes. *IEEE Trans. Pattern Anal. Machine Intell.* **PAMI-5**, 267–287.

Hung, Y.-T., and Kshirsagar, A. M. (1984). A note on optimum grouping and the relative discriminating power of qualitative to continuous normal variates. *Statist. Prob. Letters* **2**, 19–21.

IMSL (1984). *IMSL Library: FORTRAN Subroutines for Mathematics and Statistics.* Edition 9. 2. Houston: Author.

Jain, A. K., and Chandrasekaran, B. (1982). Dimensionality and sample size considerations in pattern recognition practice. In *Handbook of Statistics* (Vol. 2), P. R. Krishnaiah and L. Kanal (Eds.). Amsterdam: North-Holland, pp. 835–855.

Jain, A. K., and Waller, W. G. (1978). On the optimal number of features in the classification of multivariate Gaussian data. *Pattern Recognition* **10**, 365–374.

Jain, A. K., Dubes, R. C., and Chen, C.-C. (1987). Bootstrap techniques for error estimation. *IEEE Trans. Pattern Anal. Machine Intell.* **PAMI-9**, 628–633.

James, M. (1985). *Classification Algorithms*. New York: Wiley.

James, W., and Stein, C. (1961). Estimation with quadratic loss. *Proc. 4th Berkeley Symp.* (Vol. 1). Berkeley: University of California Press, pp. 361–379.

Jeffreys, H. (1948). *Theory of Probability*. Second Edition. Oxford: Clarendon Press.

Jennrich, R. I. (1962). Linear discrimination in the case of unequal covariance matrices. Unpublished manuscript.

Jennrich, R. I. (1977). Stepwise discriminant analysis. In *Statistical Methods for Digital Computers*, K. Enslein, A. Ralston, and H. S. Wilf (Eds.). New York: Wiley, pp. 76–95.

Jensen, S. T., and Johansen, S. (1987). Estimation of proportional covariances. *Statist. Prob. Letters* **6**, 83–85.

Joachimsthaler, E. A., and Stam, A. (1988). Four approaches to the classification problem in discriminant analysis: an experimental study. *Decision Sciences* **19**, 322–333.

John, S. (1959). The distribution of Wald's classification statistic when the dispersion matrix is known. *Sankhyā* **21**, 371–375.

John, S. (1960a). On some classification problems. *Sankhyā* **22**, 301–308.

John, S. (1960b). On some classification statistics. *Sankhyā* **22**, 309–316.

John, S. (1961). Errors in discrimination. *Ann. Math. Statist.* **32**, 1125–1144.

John, S. (1963). On classification by the statistics R and Z. *Ann. Inst. Statist Math.* **14**, 237–246.

John, S. (1964). Further results on classification by W. *Sankhyā* **26**, 39–46.

John, S. (1973). On inferring the probability of misclassification by the linear discriminant function. *Ann. Inst. Math. Statist.* **25**, 363–371.

Johns, M. V. (1961). An empirical Bayes approach to non-parametric two-way classification. In *Studies in Item Analysis and Prediction*, H. Solomon (Ed.). Stanford: Stanford University Press, pp. 221–232.

Johnson N. L., and Kotz, S. (1972). *Distributions in Statistics: Continuous Multivariate Distributions* (Vol. 4). New York: Wiley.

Johnson, R. A., and Wichern, D. W. (1982). *Applied Multivariate Statistical Analysis.* Englewood Cliffs, New Jersey: Prentice-Hall.

Johnson, W. O. (1985). Influence measures for logistic regression: another point of view. *Biometrika* **72**, 59–65.

Jolliffe, I. T. (1986). *Principal Component Analysis.* New York: Springer-Verlag.

Jones, M. C., and Sibson, R. (1987). What is projection pursuit? (with discussion). *J. R. Statist. Soc. A* **150**, 1–36.

Jones, P. N., and McLachlan, G. J. (1989). Modelling mass-size particle data by finite mixtures. *Commun. Statist.–Theory Meth.* **18**, 2629–2646.

Jones, P. N., and McLachlan, G. J. (1990a). Algorithm AS 254. Maximum likelihood estimation from grouped and truncated data with finite normal mixture models. *Appl. Statist.* **39**, 273–282.

Jones, P. N., and McLachlan, G. J. (1990b). Improving the convergence rate of the EM algorithm for a mixture model fitted to grouped truncated data. Unpublished manuscript.

Journel, A. G., and Huijbregts, Ch. J. (1978). *Mining Geostatistics*. London: Academic Press.

Jubb, M., and Jennison, C. (1988). Aggregation and refinement in binary image restoration. *Technical Report*. Bath: School of Mathematical Sciences, University of Bath.

Kailath, T. (1967). The divergence and Bhattacharyya distance measures in signal selection. *IEEE Trans. Commun. Tech.* **COM-15**, 52–60.

Kanal, L. (1974). Patterns in pattern recognition: 1968–1974. *IEEE Trans. Inform. Theory* **IT-20**, 697–722.

Kanal, L., and Chandrasekaran, B. (1971). On dimensionality and sample size in statistical pattern classification. *Pattern Recognition* **3**, 225–234.

Kanazawa, M. (1974). A non-parametric classification rule for several multivariate populations. *Canad. J. Statist.* **2**, 145–156.

Kanazawa, M. (1979). The asymptotic cut-off point and comparison of error probabilities in covariance discriminant analysis. *J. Japan Statist. Soc.* **9**, 7–17.

Kanazawa, M., and Fujikoshi, Y. (1977). The distribution of the Studentized classification statistic W^* in covariate discriminant analysis. *J. Japan Statist. Soc.* **7**, 81–88.

Kappenman, R. F. (1987). A nonparametric data based univariate function estimate. *Comput. Statist. Data Anal.* **5**, 1–7.

Kashyap, R. L., and Chellappa, R. (1983). Estimation and choice of neighbors in spatial-interaction models of images. *IEEE Trans. Inform. Theory* **IT-29**, 60–72.

Katre, U. A., and Krishnan, T. (1989). Pattern recognition with an imperfect supervisor. *Pattern Recognition* **22**, 423–431.

Kay, J. (1986). Contribution to the discussion of paper by J. Besag. *J. R. Statist. Soc. B* **48**, 293.

Kay, J., and Titterington, D. M. (1986). Image labelling and the statistical analysis of incomplete data. *Proc. 2nd Inter. Conf. Image Processing and Applications*. London: Institute of Electrical Engineers, pp. 44–48.

Kay, R., and Little, S. (1986). Assessing the fit of the logistic model: a case study of children with the haemolytic uraemic syndrome. *Appl. Statist.* **35**, 16–30.

Kay, R., and Little, S. (1987). Transformations for the explanatory variables in the logistic regression model for binary data. *Biometrika* **74**, 495–501.

Kendall, M. G. (1957). *A Course in Multivariate Analysis*. First Edition. London: Griffin.

Kendall, M. G. (1966). Discrimination and classification. In *Multivariate Analysis*, P. R. Krishnaiah (Ed.). New York: Academic Press, pp. 165–185.

Kent, J. T., and Mardia, K. V. (1986). Spatial classification using fuzzy membership models. *IEEE Trans. Pattern Anal. Machine Intell.* **PAMI-10**, 659–671.

Kharin, Y. S. (1984). The investigation of risk for statistical classifiers using minimum estimators. *Theory Prob. Appl.* **28**, 623–630.

Khatri, C. G., and Rao, C. R. (1985). Test for a specified signal when the noise covariance matrix is unknown. *Technical Report 85-47*. Pittsburgh: Centre for Multivariate Analysis, University of Pittsburgh.

Khatri, C. G., and Rao, C. R. (1987). Effects of estimated noise covariance matrix in optimal signal detection. *IEEE Trans. Acoustic Speech Signal Processing.* **ASSP-35**, 671–679.

Khatri, C. G., Rao, C. R., and Sun, Y. N. (1986). Tables for obtaining confidence bounds for realized signal to noise ratio with an estimated discriminant function. *Commun. Statist.–Simula.* **15**, 1–14.

Khattree, R., and Gill, D. S. (1987). Estimation of signal to noise ratio using Mahalanobis distance. *Commun. Statist.–Theory Meth.* **16**, 897–907.

Khattree, R., and Gupta, R. D. (1990). Estimation of realized signal to noise ratio for a pair of multivariate signals. *Austral. J. Statist.* **32**, 239–246.

Kiiveri, H. T., and Campbell, N. A. (1986). Allocation of remotely sensed data using Markov models for spectral variables and pixel labels. *Technical Report*. Perth: Division of Mathematics & Statistics, CSIRO.

Kimura, F., Takashina, K., Tsuruoka, S., and Miyake, Y. (1987). Modified quadratic discriminant functions and the application to Chinese character recognition. *IEEE Trans. Pattern Anal. Machine Intell.* **PAMI-9**, 149–153.

Kinderman, R., and Snell, J. L. (1980). *Markov Random Fields and Their Applications*. Providence: American Mathematical Society.

Kirkpatrick, S., Gelatt, C. D., and Vecchi, M. P. (1983). Optimization by simulated annealing. *Science* **220**, 671–680.

Kittler, J., and Devijver, P. A. (1981). An efficient estimator of pattern recognition system error probability. *Pattern Recognition* **13**, 245–249.

Kittler, J., and Devijver, P. A. (1982). Statistical properties of error estimators in performance assessment of recognition systems. *IEEE Trans. Pattern Anal. Machine Intell.* **PAMI-4**, 215–220.

Kittler, J., and Föglein, J. (1984). Contextual classification of multispectral pixel data. *Image Vision Comput.* **2**, 13–29.

Klecka, W. R. (1980). *Discriminant Analysis*. Beverly Hills, California: Sage Publications.

Kleinbaum, D. G., Kupper, L. L., and Chambless, L. E. (1982). Logistic regression analysis of epidemiologic data: theory and practice. *Commun. Statist.–Theory Meth.* **11**, 485–547.

Knoke, J. D. (1982). Discriminant analysis with discrete and continuous variables. *Biometrics* **38**, 191–200.

Knoke, J. D. (1986). The robust estimation of classification error rates. *Comput. Math. Applic.* **12A**, 253–260.

Kocherlakota, S., Balakrishnan, N., and Kocherlakota, K. (1987). The linear discriminant function: sampling from the truncated normal distribution. *Biom. J.* **29**, 131–139.

Kocherlakota, S., Kocherlakota, K., and Balakrishnan, N. (1987). Asymptotic expansions for errors of misclassification: nonnormal situations. In *Advances in Multivariate Statistical Analysis*, A. K. Gupta (Ed.). Dordrecht: Reidel, pp. 191–211.

Koffler, S. L., and Penfield, D. A. (1979). Nonparametric discrimination procedures for non-normal distributions. *J. Statist. Comput. Simul.* **8**, 281–299.

Koffler, S. L., and Penfield, D. A. (1982). Nonparametric classification based upon inverse normal scores and rank transformations. *J. Statist. Comput. Simul.* **15**, 51–68.

Kokolakis, G. E. (1981). On the expected probability of correct classification. *Biometrika* **68**, 477–483.

Kokolakis, G. E. (1985). A martingale approach to the problem of dimensionality and performance of Bayesian classifiers: asymptotic results. *Commun. Statist.–Theory Meth.* **14**, 927–935.

Konishi, S., and Honda, M. (1990). Comparison of procedures for estimation of error rates in discriminant analysis under nonnormal populations. *J. Statist. Comput. Simul.* **36**, 105–115.

Kowalski, B., and Wold, S. (1982). Pattern recognition in chemistry. In *Handbook of Statistics* (Vol. 2), P. R. Krishnaiah and L. Kanal (Eds.). Amsterdam: North-Holland, pp. 673–697.

Koziol, J. A. (1986). Assessing multivariate normality: a compendium. *Commun. Statist.–Theory Meth.* **15**, 2763–2783.

Krishnaiah, P. R. (1982). Selection of variables in discriminant analysis. In *Handbook of Statistics* (Vol. 2), P. R. Krishnaiah and L. Kanal (Eds.). Amsterdam: North-Holland, pp. 883–892.

Krishnaiah. P. R., and Kanal, L. (Eds.). (1982). *Handbook of Statistics* (Vol. 2). Amsterdam: North-Holland.

Krishnan, T. (1988). Efficiency of learning with imperfect supervision. *Pattern Recognition* **21**, 183–188.

Krishnan, T., and Nandy, S. C. (1987). Discriminant analysis with a stochastic supervisor. *Pattern Recognition* **20**, 379–384.

Krishnan, T., and Nandy, S. C. (1990a). Efficiency of discriminant analysis when initial samples are classified stochastically. *Pattern Recognition* **23**, 529–537.

Krishnan, T., and Nandy, S. C. (1990b). Efficiency of logistic-normal stochastic supervision. *Pattern Recognition* **23**, 1275–1279.

Kronmal, R. A., and Tarter, M. (1968). The estimation of probability densities and cumulatives by Fourier series methods. *J. Amer. Statist. Assoc.* **63**, 925–952.

Krusińska, E. (1987). A valuation of state of object based on weighted Mahalanobis distance. *Pattern Recognition* **20**, 413–418.

Krusińska, E., and Liebhart, J. (1988). Robust selection of the most discriminative variables in the dichotomous problem with application to some respiratory disease data. *Biom. J.* **30**, 295–303.

Krusińska, E., and Liebhart, J. (1989). Some further remarks on robust selection of variables in discriminant analysis. *Biom. J.* **31**, 227–233.

Krzanowski, W. J. (1975). Discrimination and classification using both binary and continuous variables. *J. Amer. Statist. Assoc.* **70**, 782–790.

Krzanowski, W. J. (1976). Canonical representation of the location model for discrimination or classification. *J. Amer. Statist. Assoc.* **71**, 845–848.

Krzanowski, W. J. (1977). The performance of Fisher's linear discriminant function under non-optimal conditions. *Technometrics* **19**, 191–200.

Krzanowski, W. J. (1979a). Some linear transformations for mixtures of binary and continuous variables, with particular reference to linear discriminant analysis. *Biometrika* **66**, 33–39.

Krzanowski, W. J. (1979b). Between-groups comparison of principal components. *J. Amer. Statist. Assoc.* **76**, 1022.

Krzanowski, W. J. (1980). Mixtures of continuous and categorical variables in discriminant analysis. *Biometrics* **36**, 493–499.

Krzanowski, W. J. (1982a). Mixtures of continuous and categorical variables in discriminant analysis: a hypothesis-testing approach. *Biometrics* **38**, 991–1002.

Krzanowski, W. J. (1982b). Between-group comparison of principal components—some sampling results. *J. Statist. Comput. Simul.* **15**, 141–154.

Krzanowski, W. J. (1983a). Distance between populations using mixed continuous and categorical variables. *Biometrika* **70**, 235–243.

Krzanowski, W. J. (1983b). Stepwise location model choice in mixed-variable discrimination. *Appl. Statist.* **32**, 260–266.

Krzanowski, W. J. (1984a). On the null distribution of distance between two groups, using mixed continuous and categorical variables. *J. Classification* **1**, 243–253.

Krzanwoski, W. J. (1984b). Principal component analysis in the presence of group structure. *Appl. Statist.* **33**, 164–168.

Krzanowski, W. J. (1986). Multiple discriminant analysis in the presence of mixed continuous and categorical data. *Comput. Math. Applic.* **12A**, 179–185.

Krzanowski, W. J. (1987a). A comparison between two distance-based discriminant principles. *J. Classification* **4**, 73–84.

Krzanowski, W. J. (1987b). Selection of variables to preserve multivariate data structure, using principal components. *Appl. Statist.* **36**, 22–33.

Krzanowski, W. J. (1988). *Principles of Multivariate Analysis*. Oxford: Clarendon Press.

Krzanowski, W. J. (1989). On confidence regions in canonical variate analysis. *Biometrika* **76**, 107–116.

Krzanowski, W. J. (1990). Between-group analysis with heterogeneous covariance matrices: the common principal component model. *J. Classification* **7**, 81–98.

Krzanowski, W. J., and Radley, D. (1989). Nonparametric confidence and tolerance regions in canonical variate analysis. *Biometrics* **45**, 1163–1173.

Krzyśko, M. (1983). Asymptotic distribution of the discriminant function. *Statist. Prob. Letters* **1**, 243–250.

Kshirsagar, A. M. (1964). Distributions of the direction and collinearity factors in discriminant analysis. *Proc. Camb. Phil. Soc.* **60**, 217–225.

Kshirsagar, A. M. (1969). Distributions associated with Wilks' Λ. *J. Austral. Math. Soc.* **10**, 269–277.

Kshirsagar, A. M. (1970). An alternative derivation of the direction and collinearity statistics in discriminant analysis. *Calcutta Statist. Assoc. Bull.* **19**, 123–134.

Kshirsagar, A. M. (1971). Goodness-of-fit of a discriminant function from the vector space of dummy variables. *J. R. Statist. Soc.* **B 33**, 111–116.

Kshirsagar, A. M. (1972). *Multivariate Analysis*. New York: Marcel Dekker.

Kshirsagar, A. M., and Arvensen, E. (1975). A note on the equivalency of two discrimination procedures. *Amer. Statistician* **29**, 38–39.

Kshirsagar, A. M., and Musket, S. F. (1972). A note on discrimination in the case of zero mean differences. *J. Indian Statist. Assoc.* **10**, 42–51.

Kudo, A. (1959). The classificatory problem viewed as a two decision problem. *Mem. Fac. Science Kyushu University A* **13**, 96–125.

Kudo, A. (1960). The classificatory problem viewed as a two decision problem–II. *Mem. Fac. Science Kyushu University A* **14**, 63–83.

Kullback, S., and Leibler, A. (1951). On information and sufficiency. *Ann. Math. Statist.* **22**, 79–86.

Kurezynski, T. W. (1970). Generalized distance and discrete variables. *Biometrics* **26**, 525–534.

Lachenbruch, P. A. (1965). *Estimation of Error Rates in Discriminant Analysis*. Unpublished Ph. D. thesis, University of Los Angeles.

Lachenbruch, P. A. (1966). Discriminant analysis when the initial samples are misclassified. *Technometrics* **8**, 657–662.

Lachenbruch, P. A. (1967). An almost unbiased method of obtaining confidence intervals for the probability of misclassification in discriminant analysis. *Biometrics* **23**, 639–645.

Lachenbruch, P. A. (1968). On expected probabilities of misclassification in discriminant analysis, necessary sample size, and a relation with the multiple correlation coefficient. *Biometrics* **24**, 823–834.

Lachenbruch, P. A. (1974). Discriminant analysis when the initial samples are misclassified II: non-random misclassification models. *Technometrics* **16**, 419–424.

Lachenbruch, P. A. (1975a). *Discriminant Analysis*. New York: Hafner Press.

Lachenbruch, P. A. (1975b). Zero-mean difference discrimination and the absolute linear discriminant function. *Biometrika* **62**, 397–401.

Lachenbruch, P. A. (1977). Covariance adjusted discriminant functions. *Ann. Inst. Statist. Math.* **29**, 247–257.

Lachenbruch, P. A. (1979). Note on initial misclassification effects on the quadratic discriminant function. *Technometrics* **21**, 129–132.

Lachenbruch, P. A. (1980). Note on combining risks using a logistic discriminant function approach. *Biom. J.* **22**, 759–762.

Lachenbruch, P. A. (1982). Discriminant analysis. In *Encyclopedia of Statistical Sciences* (Vol. 2), S. Kotz and N. L. Johnson (Eds.). New York: Wiley, pp. 389–397.

Lachenbruch, P. A., and Broffit, B. (1980). On classifying observations when one population is a mixture of normals. *Biom. J.* **22**, 295–301.

Lachenbruch, P. A., and Goldstein, M. (1979). Discriminant analysis. *Biometrics* **35**, 69–85.

Lachenbruch, P. A., and Kupper, L. L. (1973). Discriminant analysis when one population is a mixture of normals. *Biom. J.* **15**, 191–197.

Lachenbruch, P. A., and Mickey, M. R. (1968). Estimation of error rates in discriminant analysis. *Technometrics* **10**, 1–11.

Lachenbruch, P. A., Sneeringer, C., and Revo, L. T. (1973). Robustness of the linear and quadratic discriminant function to certain types of non-normality. *Commun. Statist.* **1**, 39–57.

Lakshmanan, S., and Derin, H. (1989). Simultaneous parameter estimation and segmentation of Gibbs random fields using simulated annealing. *IEEE Trans. Pattern Anal. Machine Intell.* **PAMI-11**, 799–813.

Lam, C. F., and Cox, M. (1981). A discriminant analysis procedure for paired variables. *Technometrics* **23**, 185–187.

Lancaster, H. O. (1969). *The Chi-Squared Distribution.* New York: Wiley.

Landwehr, J. M., Pregibon, D., and Shoemaker, A. C. (1984). Graphical methods for assessing logistic regression models (with discussion). *J. Amer. Statist. Assoc.* **79**, 61–83.

Lauder, I. J. (1978). Computational problems in predictive diagnosis. *Compstat 1978, Proc. Computational Statistics*. Vienna: Physica-Verlag, pp. 186–192.

Lauder, I. J. (1983). Direct kernel assessment of diagnostic probabilities. *Biometrika* **70**, 251–256.

Lawoko, C. R. O. (1990). On the maximum likelihood estimation of parameters of a spatial discrimination model. *Commun. Statist–Theory Meth.* **19**, 4627–4641.

Lawoko, C. R. O., and McLachlan, G. J. (1983). Some asymptotic results on the effect of autocorrelation on the error rates of the sample linear discriminant function. *Pattern Recognition* **16**, 119–121.

Lawoko, C. R. O., and McLachlan, G. J. (1985). Discrimination with autocorrelated observations. *Pattern Recognition* **18**, 145–149.

Lawoko, C. R. O., and McLachlan, G. J. (1986). Asymptotic error rates of the W and Z statistics when the training observations are dependent. *Pattern Recognition* **19**, 467–471.

Lawoko, C. R. O., and McLachlan, G. J. (1988). Further results on discrimination with autocorrelated observations. *Pattern Recognition* **21**, 69–72.

Lawoko, C. R. O., and McLachlan, G. J. (1989). Bias associated with the discriminant analysis approach to the estimation of mixing proportions. *Pattern Recognition* **22**, 763–766.

Layard, M. W. J. (1974). A Monte Carlo comparison of tests for equality of covariance matrices. *Biometrika* **16**, 461–465.

Lazarsfeld, P. F. (1956). Some observations on dichotomous systems. *Duplicated Report*. New York: Department of Sociology, Columbia University.

Lazarsfeld, P. F. (1961). The algebra of dichotomous systems. In *Studies in Item Analysis and Prediction*, H. Solomon (Ed.). Stanford: Stanford University Press, pp. 111–157.

Lazarsfeld, P. F., and Henry, N. W. (1968). *Latent Structure Analysis*. New York: Houghton Mifflin.

Le Cam, L. (1970). On the assumptions used to prove asymptotic normality of maximum likelihood estimates. *Ann. Math. Statist.* **41**, 802–828.

Lee, J. C. (1975). A note on equal-mean discrimination. *Commun. Statist.–Theory Meth.* **A4**, 251–254.

Lee, J. C. (1977). Bayesian classification of data from growth curves. *South African Statist. J.* **11**, 155–166.

Lee, J. C. (1982). Classification of growth curves. In *Handbook of Statistics* (Vol. 2), P. R. Krishnaiah and L. Kanal (Eds.). Amsterdam: North-Holland, pp. 121–137.

Lehmann, E. L. (1980). Efficient likelihood estimators. *Amer. Statistician* **34**, 233–235.

Lehmann, E. L. (1983). *Theory of Point Estimation*. New York: Wiley.

Lemeshow, S., and Hosmer, D. W. (1982). A review of goodness-of-fit statistics for use in the development of logistic regression models. *Amer. J. Epidem.* **115**, 92–106.

Leonard T. (1977). A Bayesian approach to some multinomial estimation and pretesting problems. *J. Amer. Statist. Assoc.* **72**, 869–874.

Lesaffre, E. (1983). Normality tests and transformations. *Pattern Recognition Letters* **1**, 187–199.

Lesaffre, E., and Albert, A. (1988). An uncertainty measure in logistic discrimination. *Statistics in Medicine* **7**, 525–533.

Lesaffre, E., and Albert, A. (1989a). Partial separation in logistic discrimination. *J. R. Statist. Soc. B* **51**, 109–116.

Lesaffre, E., and Albert, A. (1989b). Multiple-group logistic regression diagnostics. *Appl. Statist.* **38**, 425–440.

Lesaffre, E., Willems, J. L., and Albert, A. (1989). Estimation of error rate in multiple group logistic discrimination: the approximate leaving-one-out method. *Commun. Statist.–Theory Meth.* **18**, 2989–3007.

Leung, C. Y. (1988a)., anderson's classification statistic based on a post-stratified training sample. *Commun. Statist.–Theory Meth.* **17**, 1659–1667.

Leung, C. Y. (1988b). Covariate classification by post-stratification. *Commun. Statist.–Theory Meth.* **17**, 3869–3880.

Leung, C. Y. (1989). The Studentized location linear discriminant function. *Commun. Statist.–Theory Meth.* **18**, 3977–3990.

Leung, C. Y., and Srivastava, M. S. (1983a). Covariate classification for two correlated populations. *Commun. Statist.–Theory Meth.* **12**, 223–241.

Leung, C. Y., and Srivastava, M. S. (1983b). Asymptotic comparison of two discriminants used in normal covariate classification. *Commun. Statist.–Theory Meth.* **12**, 1637–1646.

Lin, H. E. (1979). Classification rules based on U-statistics. *Sankhyā B* **41**, 41–52.

Lin, S. P., and Perlman, M. D. (1984). A Monte Carlo comparison of four estimators for a covariance matrix. In *Multivariate Analysis* (Vol. 6), P. R. Krishnaiah (Ed.). Amsterdam: North-Holland, pp. 415–429.

Lindley, D. (1966). Contribution to the discussion of paper by M. Hills. *J. R. Statist. Soc. B* **28**, 24–26.

Lindley, D. (1978). The Bayesian approach. *Scand. J. Statist.* **5**, 1–26.

Linnett, K. (1988). Testing normality of transformed data. *Appl. Statist.* **37**, 180–186.

Lissack, T., and Fu, K. S. (1976). Error estimation in pattern recognition via L^α-distance between posterior density functions. *IEEE Trans. Inform. Theory* **IT-22**, 34–45.

Little, R. J. A. (1978). Consistent regression methods for discriminant analysis with incomplete data. *J. Amer. Statist. Assoc.* **73**, 319–322.

Little, R. J. A. (1988). Robust estimation of the mean and covariance matrix from data with mixing values. *Appl. Statist.* **37**, 23–38.

Little, R. J. A., and Rubin, D. B. (1987). *Statistical Analysis with Missing Data*. New York: Wiley.

Little, R. J. A., and Schluchter, M. D. (1985). Maximum likelihood estimation for mixed continuous and categorical data with missing values. *Biometrika* **72**, 497–512.

Loftsgaarden, D. O., and Quesenberry, C. P. (1965). A nonparametric estimate of a multivariate density function. *Ann. Math. Statist* **36**, 1049–1051.

Logan, T. P. (1990). Bayesian discrimination using multiple observations. Unpublished manuscript.

Loh, W.-Y., and Vanichsetakul, N. (1988). Tree-structured classification via generalized discriminant analysis (with discussion). *J. Amer. Statist. Assoc.* **83**, 715–728

Loizou, G., and Maybank, S. J. (1987). The nearest neighbor and the Bayes error rates. *IEEE Trans. Pattern Anal. Machine Intell.* **PAMI-9**, 254–262.

Looney, S. W. (1986). A review of techniques for assessing multivariate normality. *Proc. Statistical Computing Section of the Annual Meeting of the American Statistical Association.* Alexandria, Virginia: American Statistical Association, pp. 280–285.

Louis, T. A. (1982). Finding the observed information matrix when using the EM algorithm. *J. R. Statist. Soc. B* **44**, 226–233.

Lubischew, A. A. (1962). On the use of discriminant functions in taxonomy. *Biometrics* **18**, 455–477.

Luk, A., and MacLeod, J. E. S. (1986). An alternative nearest neighbour classification scheme. *Pattern Recogniti n Letters* **4**, 375–381.

Macdonald, P. D. M. (1975). Estimation of finite distribution mixtures. In *Applied Statistics*, R. P. Gupta (Ed.). Amsterdam: North-Holland, pp. 231–245.

Machin, D., Dennis, N. R., Tippett, P. A., and Andrews, V. (1983). On the standard error of the probability of a particular diagnosis. *Statist. Med.* **2**, 87–93.

Macleod, J. E. S., Luk, A., and Titterington, D. M. (1987). A re-examination of the distance-weighted k-nearest neighbor classification rule. *IEEE Trans. Syst. Man Cybern.* **SMC-17**, 689–696.

Mahalanobis, P. C. (1927). Analysis of race mixture in Bengal. *J. Proc. Asiatic Soc. Bengal* **23**, 301–333.

Mahalanobis, P. C. (1928). A statistical study of the Chinese head. *Man in India* **8**, 107–122.

Mahalanobis, P. C. (1936). On the generalized distance in statistics. *Proc. Nat. Inst. Sciences India* **2**, 49–55.

Malina, W. (1987). Some multiclass Fisher feature selection algorithms and their comparison with Karhunen–Loève algorithms. *Pattern Recognition Letters* **6**, 279–285.

Mallows, C. L. (1953). Sequential discrimination. *Sankhyā* **12**, 321–338.

Mallows, C. L. (1973). Some comments on C_p. *Technometrics* **15**, 661–675.

Manly, B. F. J., and Rayner, J. C. W. (1987). The comparison of sample covariance matrices using likelihood ratio tests. *Biometrika* **74**, 841–847.

Mantas, J. (1987). Methodologies in pattern recognition and image analysis —a brief survey. *Pattern Recognition* **20**, 1–6.

Marco, V. R., Young, D. M., and Turner, D. W. (1987a). The Euclidean distance classifier: an alternative to the linear discriminant function. *Commun. Statist.–Simula.* **16**, 485–505.

Marco, V. R., Young, D. M., and Turner, D. W. (1987b). Asymptotic expansions and estimation of the expected error rate for equal-mean discrimination with uniform covariance structure. *Biom. J.* **29**, 103–113.

Marco, V. R., Young, D. M., and Turner, D. W. (1987c). A note on the effect of simple equicorrelation in detecting a spurious multivariate observation. *Commun. Statist.–Theory Meth.* **16**, 1027–1036.

Marco, V. R., Young, D. M., and Turner, D. W. (1988). Predictive discrimination for autoregressive processes. *Pattern Recognition Letters* **7**, 145–149.

Mardia, K. V. (1970). Measures of multivariate skewness and kurtosis with applications. *Biometrika* **57**, 519–520.

Mardia, K. V. (1974). Applications of some measures of multivariate skewness and kurtosis in testing normality and robustness studies. *Sankhyā B* **36**, 115–128.

Mardia, K. V. (1980). Tests of univariate and multivariate normality. In *Handbook of Statistics* (Vol. 1), P. R. Krishnaiah (Ed.). Amsterdam: North–Holland, pp. 279–320.

Mardia, K. V. (1984). Spatial discrimination and classification maps. *Commun. Statist.–Theory Meth.* **13**, 2181–2197. Correction (1987). *Commun. Statist.–Theory Meth.* **16**, 3749.

Marks, S., and Dunn, O. J. (1974). Discriminant functions when covariance matrices are unequal. *J. Amer. Statist. Assoc.* **69**, 555–559.

Maronna, R. A. (1976). Robust M-estimators of multivariate location and scatter. *Ann. Statist.* **4**, 51–67.

Marron, J. S. (1983). Optimal rates of convergence to Bayes risk in nonparametric discrimination. *Ann. Statist.* **11**, 1142–1155.

Marron, J. S. (1989). Comments on a data based bandwidth selector. *Comput. Statist. Data Anal.* **8**, 155–170.

Marshall, A. W., and Olkin, I. (1968). A general approach to some screening and classification problem (with discussion). *J. R. Statist. Soc. B* **30**, 407–443.

Marshall, R. J. (1986). Partitioning methods for classification and decision making in medicine. *Statist. Med.* **5**, 517–526.

Martin, D. C., and Bradley, R. A. (1972). Probability models, estimation, and classification for multivariate dichotomous populations. *Biometrics* **28**, 203–221.

Martin, G. G., Lienhart, J., Anand, R., Dudewicz, E. J., and Levy, G. C. (1989). From statistical-expert-system to expert-statistical-system. *Amer. J. Math. Management Sciences.* **9**, 273–297.

Matheron, G. (1971). *The Theory of Regionalized Variables and Its Applications.* Paris: Les Cahiers du Centre de Morphologie Mathématique de Fontainebleau, Ecole Nationale Supérieure des Mines de Paris.

Matloff, N., and Pruitt, R. (1984). The asymptotic distribution of an estimator of the Bayes error rate. *Pattern Recognition Letters* **2**, 271–274.

Matusita, K. (1956). Decision rule, based on distance, for the classification problem. *Ann. Inst. Statist. Math.* **8**, 67–77.

Matusita, K. (1964). Distance and decision rule. *Ann. Inst. Statist. Math.* **16**, 305–315.

Matusita, K. (1967a). On the notion of affinity of several distributions and some of its applications. *Ann. Inst. Statist. Math.* **19**, 181–192.

Matusita, K. (1967b). Classification based on distance in multivariate Gaussian cases. *Proc. 5th Berkeley Symp.* (Vol. 1). Berkeley: University of California Press, pp. 299–304.

Matusita, K. (1971). Some properties of affinity and applications. *Ann. Inst. Statist. Math.* **23**, 137–155.

Matusita, K. (1973). Discrimination and the affinity of distributions. In *Discriminant Analysis and Applications*, T. Cacoullos (Ed.). New York: Academic Press, pp. 213–223.

McCabe, G. P., Jr. (1975). Computations for variable selection in discriminant analysis. *Technometrics* **17**, 103–109.

McCulloch, R. E. (1986). Some remarks on allocatory and separatory linear discrimination. *J. Statist. Planning Inference* **14**, 323–330.

McDonald, L. L., Lowe, V. W., Smidt, R. K., and Meister, K. A. (1976). A preliminary test for discriminant analysis based on small samples. *Biometrics* **32**, 417–422.

McGee, R. I. (1976). *Comparison of the W^* and Z^* Procedures in Covariate Discriminant Analysis.* Unpublished Ph. D. thesis, Kansas State University.

McHenry, C. E. (1978). An efficient computational procedure for selecting a best subset. *Appl. Statist.* **27**, 291–296.

McKay, R. J. (1976). Simultaneous procedures in discriminant analysis involving two groups. *Technometrics* **18**, 47–53.

McKay, R. J. (1977). Simultaneous procedures for variable selection in multiple discriminant analysis. *Biometrika* **64**, 283–290.

McKay, R. J. (1978). A graphical aid to selection of variables in two-group discriminant analysis. *Appl. Statist.* **27**, 259–263.

McKay, R. J. (1989). Simultaneous procedures for covariance matrices. *Commun. Statist.–Theory Meth.* **18**, 429–443.

McKay, R. J., and Campbell, N. A. (1982a). Variable selection techniques in discriminant analysis I. Description. *Br. J. Math. Statist. Psychol.* **35**, 1–29.

McKay, R. J., and Campbell, N. A. (1982b). Variable selection techniques in discriminant analysis II. Allocation. *Br. J. Math. Statist. Psychol.* **35**, 30–41.

McLachlan, G. J. (1972a). Asymptotic results for discriminant analysis when the initial samples are misclassified. *Technometrics* **14**, 415–422.

McLachlan, G. J. (1972b). An asymptotic expansion for the variance of the errors of misclassification of the linear discriminant function. *Austral. J. Statist.* **14**, 68–72.

McLachlan, G. J. (1973a). *The Errors of Allocation and their Estimators in the Two-Population Discrimination Problem*. Abstract of unpublished Ph. D. thesis, University of Queensland. *Bull. Austral. Math. Soc.* **9**, 149–150.

McLachlan, G. J. (1973b). An asymptotic expansion of the expectation of the estimated error rate in discriminant analysis. *Austral. J. Statist.* **15**, 210–214.

McLachlan, G. J. (1974a). The asymptotic distributions of the conditional error rate and risk in discriminant analysis. *Biometrika* **61**, 131–135.

McLachlan, G. J. (1974b). An asymptotic unbiased technique for estimating the error rates in discriminant analysis. *Biometrics* **30**, 239–249.

McLachlan, G. J. (1974c). Estimator of the errors of misclassification on the criterion of asymptotic mean square error. *Technometrics* **16**, 255–260.

McLachlan, G. J. (1974d). The relationship in terms of asymptotic mean square error between the separate problems of estimating each of the three types of error rate of the linear discriminant function. *Technometrics* **16**, 569–575.

McLachlan, G. J. (1975a). Iterative reclassification procedure for constructing an asymptotically optimal rule of allocation in discriminant analysis. *J. Amer. Statist. Assoc.* **70**, 365–369.

McLachlan, G. J. (1975b). Confidence intervals for the conditional probability of misallocation in discriminant analysis. *Biometrics* **32**, 161–167.

McLachlan, G. J. (1975c). Some expected values for the error rates of the sample quadratic discriminant function. *Austral. J. Statist.* **17**, 161–165.

McLachlan, G. J. (1976a). The bias of the apparent error rate in discriminant analysis. *Biometrika* **63**, 239–244.

McLachlan, G. J. (1976b). A criterion for selecting variables for the linear discriminant function. *Biometrics* **32**, 529–515.

McLachlan, G. J. (1976c). Further results on the effect of intraclass correlation among training samples in discriminant analysis. *Pattern Recognition* **7**, 273–275.

McLachlan, G. J. (1977a). Estimating the linear discriminant function from initial samples containing a small number of unclassified observations. *J. Amer. Statist. Assoc.* **72**, 403–406.

McLachlan, G. J. (1977b). Constrained sample discrimination with the Studentized classification statistic *W*. *Commun. Statist.–Theory Meth.* **A6**, 575–583.

McLachlan, G. J. (1977c). A note on the choice of a weighting function to give an efficient method for estimating the probability of misclassification. *Pattern Recognition* **9**, 147–149.

McLachlan, G. J. (1977d). The bias of sample-based posterior probabilities. *Biom. J.* **19**, 421–426.

Murray, G. D., and Titterington, D. M. (1978). Estimation problems with data from a mixture. *Appl. Statist.* **27**, 325–334.

Murray, G. D., Murray, L. S., Barlow, P., Teasdale, G. M., and Jennett, W. B. (1986). Assessing the performance and clinical impact of a computerized prognostic system in severe head injury. *Statist. Med.* **5**, 403–410.

Myles, J. P., and Hand, D. J. (1990). The multi-class metric problem in nearest neighbour discrimination rules. *Pattern Recognition* **23**, 1291–1297.

Nagel, P. J. A., and de Waal, D. J. (1979). Bayesian classification, estimation and prediction of growth curves. *South. African Statist. J.* **13**, 127–137.

Ng, T.-H., and Randles, R. H. (1983). Rank procedures in many population forced discrimination problems. *Commun. Statist.–Theory Meth.* **12**, 1943–1959.

Ng, T-H., and Randles, R. H. (1986). Distribution-free partial discrimination procedures. *Comput. Math. Applic.* **12A**, 225–234.

Niemann, H., and Goppert, R. (1988). An efficient branch-and-bound nearest neighbour classifier. *Pattern Recognition Letters* **7**, 67–72.

Novotny, T. J., and McDonald, L. L. (1986). Model selection using discriminant analysis. *J. Appl. Statist.* **13**, 159–165.

Odell, P. L. (1979). A model for dimension reduction in pattern recognition using continuous data. *Pattern Recognition* **11**, 51–54.

O'Hara, T. F., Hosmer, D. W., Lemeshow, S., and Hartz, S. C. (1982). A comparison of discriminant function and maximum likelihood estimates of logistic coefficients for categorical-scaled data. *J. Statist. Comput. Simul.* **14**, 169–178.

Okamoto, M. (1961). Discrimination for variance matrices. *Osaka Math. J.* **13**, 1–39.

Okamoto, M. (1963). An asymptotic expansion for the distribution of the linear discriminant function. *Ann. Math. Statist.* **34**, 1286–1301. Correction (1968). *Ann. Math. Statist.* **39**, 1358–1359.

Oliver, L. H., Poulsen, R. S., Toussaint, G. T., and Louis, C. (1979). Classification of atypical cells in the automatic cytoscreening for cervical cancer. *Pattern Recognition* **11**, 205–212.

Olkin, I., and Sellian, J. B. (1977). Estimating covariances in a multivariate normal distribution. In *Statistical Decision Theory and Related Topics II*, S. S. Gupta and D. Moore (Eds.). New York: Academic Press, pp. 313–326.

Olkin, I., and Tate, R. F. (1961). Multivariate correlation models with mixed discrete and continuous variables. *Ann. Math. Statist.* **22**, 92–96.

O'Neill, T. J. (1976). *Efficiency Calculations in Discriminant Analysis*. Unpublished Ph. D. thesis, Stanford University.

O'Neill, T. J. (1978). Normal discrimination with unclassified observations. *J. Amer. Statist. Assoc.* **73**, 821–826.

O'Neill, T. J. (1980). The general distribution of the error rate of a classification procedure with application to logistic regression discrimination. *J. Amer. Statist. Assoc.* **75**, 154–160.

O'Neill, T. J. (1984a). Fisher's versus logistic discrimination: the unequal covariance matrices case. *Technical Report No. 99.* Stanford: Department of Statistics, Stanford University.

O'Neill, T. J. (1984b). A theoretical method of comparing classification rules under non-optimal conditions with applications to the estimates of Fisher's linear and the quadratic discriminant rules under unequal covariance matrices. *Technical Report No. 217.* Stanford: Department of Statistics, Stanford University.

O'Neill, T. J. (1986). The variance of the error rates of classification rules. *Comput. Math. Applic.* **12A**, 273–287.

O'Sullivan, F. (1986). A statistical perspective on ill-posed inverse problems. *Statist. Science* **1**, 502–527.

Ott, J., and Kronmal, R. A. (1976). Some classification procedures for multivariate binary data using orthogonal functions. *J. Amer. Statist. Assoc.* **71**, 391–399.

Owen, A. (1984). A neighbourhood-based classifier for LANDSAT data. *Canad. J. Statist.* **12**, 191–200.

Owen, A. (1986). Contribution to the discussion of paper by B. D. Ripley. *Canad. J. Statist.* **14**, 106–110.

Owen, A. (1989). Image segmentation via iterated conditional expectations. *Technical Report No. 18*. Stanford: Department of Statistics, Stanford University.

Page, E. (1977). Approximations to the cumulative normal function and its inverse for use on a pocket calculator. *Appl. Statist.* **26**, 75–76.

Page, J. T. (1985). Error-rate estimation in discriminant analysis. *Technometrics* **27**, 189–198.

Panel on Discriminant Analysis, Classification, and Clustering (1989). Discriminant analysis and clustering. *Statist. Science* **4**, 34–69.

Parzen, E. (1962). On estimation of a probability density function and mode. *Ann. Math. Statist.* **33**, 1065–1076.

Patrick, E. A. (1972). *Fundamentals of Pattern Recognition*. Englewood Cliffs, New Jersey: Prentice-Hall.

Patrick, E. A. (1990). The Outcome Adviser. *Pattern Recognition* **23**, 1427–1439.

Patrick, E. A., and Fischer, F. P. (1970). Generalized k nearest neighbour decision rule. *J. Inform. Control* **16**, 128–152.

Pearson, K. (1916). On the probability that two independent distributions of frequency are really from the same population. *Biometrika* **8**, 250–254.

Peck, R., and Van Ness, J. (1982). The use of shrinkage estimators in linear discriminant analysis. *IEEE Trans. Pattern Anal. Machine Intell.* **PAMI-4**, 530–537.

Peck, R., Jennings, L. W., and Young, D. M. (1988). A comparison of several biased estimators for improving the expected error rate of the sample quadratic discriminant function. *J. Statist. Comput. Simul.* **29**, 143–156.

Peers, H. W., and Iqbal, M. (1985). Asymptotic expansions for confidence limits in the presence of nuisance parameters, with applications. *J. R. Statist. Soc. B* **47**, 547–554.

Pelto, C. R. (1969). Adaptive nonparametric classification. *Technometrics* **11**, 775–792.

Penrose, L. S. (1946–1947). Some notes on discrimination. *Ann. Eugen.* **13**, 228–237.

Perlman, M. D. (1980). Unbiasedness of the likelihood ratio test for equality of several covariance matrices and equality of several multivariate normal populations. *Ann. Statist.* **8**, 247–263.

Peters, C. (1979). Feature selection for best mean square approximation of class densities. *Pattern Recognition* **11**, 361–364.

Pickard, D. K. (1977). A curious binary lattice process. *J. Appl. Prob.* **14**, 717–731.

Pickard, D. K. (1980). Unilateral Markov fields. *Adv. Appl. Prob.* **12**, 655–671.

Pincus, M. (1968). A closed form solution of certain programming problems. *Oper. Res.* **16**, 690–694.

Pincus, M. (1970). A Monte-Carlo method for the approximate solution of certain types of constrained optimization problems. *Oper. Res.* **18**, 1225–1228.

Plomteux, G. (1980). Multivariate analysis of an enzyme profile for the differential diagnosis of viral hepatitis. *Clin. Chem.* **26**, 1897–1899.

Potthoff, R. F., and Roy, S. N. (1964). A generalized multivariate analysis of variance model useful especially for growth curve problems. *Biometrika* **51**, 313–326.

Prakasa Rao, B. L. S. (1983). *Nonparametric Functional Estimation*. Orlando, Florida: Academic Press.

Pregibon, D. (1981). Logistic regression diagnostics. *Ann. Statist.* **9**, 705–724.

Pregibon, D. (1982). Resistant fits for some commonly used logistic models with medical applications. *Biometrics* **38**, 485–498.

Prentice, R. L., and Breslow, N. E. (1978). Retrospective studies and failure time models. *Biometrika* **65**, 153–158.

Prentice, R. L., and Pyke, R. (1979). Logistic disease incidence models and case-control studies. *Biometrika* **66**, 403–411.

Press, S. J., and Wilson, S. (1978). Choosing between logistic regression and discriminant analysis. *J. Amer. Statist. Assoc.* **73**, 699–705.

P-STAT, Inc. (1984). *P-STAT: Data Management and Statistical Software Package*. Princeton: Author.

Qian, W., and Titterington, D. M. (1989). On the use of Gibbs Markov chain models in the analysis of images based on second-order pairwise interactive distributions. *J. Appl. Statist.* **16**, 267–281.

Queiros, C. E., and Gelsema, E. S. (1989). A note on some feature selection criteria. *Pattern Recognition Letters* **10**, 155–158.

Quesenberry, C. P., and Gessaman, M. P. (1968). Nonparametric discrimination using tolerance regions. *Ann. Math. Statist.* **39**, 664–673.

Quinn, B. G., McLachlan, G. J., and Hjort, N. L. (1987). A note on the Aitkin–Rubin approach to hypothesis testing in mixture models. *J. R. Statist. Soc. B* **49**, 311–314.

Radcliffe, J. (1966). Factorizations of the residual likelihood criterion in discriminant analysis. *Proc. Camb. Phil. Soc.* **62**, 743–752.

Radcliffe, J. (1968). A note on the construction of confidence intervals for the coefficients of a second canonical variable. *Proc. Camb. Phil. Soc.* **64**, 471–475.

Radhakrishnan, R. (1985). Influence functions for certain parameters in discriminant analysis when a single discriminant function is not adequate. *Commun. Statist.–Theory Meth.* **14**, 535–549.

Raiffa, H. (1961). Statistical decision theory approach to item selection for dichotomous test and criterion variables. In *Studies in Item Analysis and Prediction*, H. Solomon (Ed.). Stanford: Stanford University Press, pp. 221–232.

Ramsey, P. H. (1982). Empirical power of procedures for comparing two groups on p variables. *J. Educat. Statist.* **7**, 139–156.

Randles, R. H., Broffit, J. D., Ramberg, J. S., and Hogg, R. V. (1978a). Discriminant analysis based on ranks. *J. Amer. Statist. Assoc.* **73**, 379–384.

Randles, R. H., Broffit, J. D., Ramberg, J. S., and Hogg, R. V. (1978b). Generalized linear and quadratic discriminant functions using robust estimates. *J. Amer. Statist. Assoc.* **73**, 564–568.

Rao, C. R. (1945). Information and the accuracy attainable in the estimation of statistical parameters. *Bull. Calcutta Math. Soc.* **37**, 81–91.

Rao, C. R. (1946). Tests with discriminant functions in multivariate analysis. *Sankhyā* **7**, 407–414.

Rao, C. R. (1948). The utilization of multiple measurements in problems of biological classification. *J. R. Statist. Soc. B* **10**, 159–203.

Rao, C. R. (1949). On some problems arising out of discrimination with multiple characters. *Sankhyā* **9**, 343–366.

Rao, C. R. (1950). A note on the distance of $D^2_{p+q} - D^2_p$ and some computational aspects of D^2 statistic and discriminant functions. *Sankhyā* **10**, 257–268.

Rao, C. R. (1952). *Advanced Statistical Methods in Biometric Research.* New York: Wiley.

Rao, C. R. (1954). A general theory of discrimination when the information about alternative population distributions is based on samples. *Ann. Math. Statist.* **25**, 651–670.

Rao, C. R. (1966a). Discriminant function between composite hypotheses and related problems. *Biometrika* **53**, 339–345.

Rao, C. R. (1966b). Covariance adjustment and related problems in multivariate analysis. In *Multivariate Analysis,* P. R. Krishnaiah (Ed.). New York: Academic Press, pp. 87–103.

Rao, C. R. (1973). *Linear Statistical Inference and Its Applications.* Second Edition. New York: Wiley.

Rao, C. R. (1977). Cluster analysis applied to a study of race mixture in human populations. In *Classification and Clustering,* J. Van Ryzin (Ed.). New York: Academic Press, pp. 175–197.

Rao, P. S. R. S., and Dorvlo, A. S. (1985). The jackknife procedure for the probabilities of misclassification. *Commun. Statist.–Simula.* **14**, 779–790.

Raudys, S. J. (1972). On the amount of *a priori* information in designing the classification algorithm. *Tech. Cybern.* **4**, 168–174 (in Russian).

Raudys, S. J. (1979). Determination of optimal dimensionality in statistical pattern classification. *Pattern Recognition* **11**, 263–270.

Raudys, S. J., and Pikelis, V. (1980). On dimensionality, sample size, classification error, and complexity of classification algorithm in pattern recognition. *IEEE Trans. Pattern Anal. Machine Intell.* **PAMI-2**, 242–252.

Raveh, A. (1989). A nonmetric approach to discriminant analysis. *J. Amer. Statist. Assoc.* **84**, 176–183.

Rawlings, R. R., and Faden, V. B. (1986). A study on discriminant analysis techniques applied to multivariate lognormal data. *J. Statist. Comput. Simul.* **26**, 79–100.

Rawlings, R. R., Graubard, B. K., Rae, D. S., Eckardt, M. J., and Ryback, R. S. (1984). Two-group classification when both groups are mixtures of normals. *Biom. J.* **26**, 923–930.

Rawlings, R. R., Faden, V. B., Graubard, B. I., and Eckardt, M. J. (1986). A study on discriminant analysis techniques applied to multivariate lognormal data. *J. Statist. Comput. Simul.* **26**, 79–100.

Ray, S. (1989a). On a theoretical property of the Bhattacharyya coefficient as a feature evaluation criterion. *Pattern Recognition Letters* **9**, 315–319.

Ray, S. (1989b). On looseness of error bounds provided by the generalized separability measures of Lissack and Fu. *Pattern Recognition Letters* **9**, 321–325.

Rayens, W., and Greene, T. (1991). Covariance pooling and stabilization for classification. *Comput. Statist. Data Anal.* **11**, 17–42.

Reed, T. R. C., and Cressie, N. A. C. (1988). *Goodness-of-Fit Statistics for Discrete Multivariate Data*. New York: Springer-Verlag.

Reaven, G. M., and Miller, R. G. (1979). An attempt to define the nature of chemical diabetes using a multidimensional analysis. *Diabetologia* **16**, 17–24.

Reed, I. S., Mallet, J. D., and Brennan, L. E. (1974). Rapid convergence rate in adoptive rays. *IEEE Trans. Aerosp. Electron. Syst.* **AES-10**, 853–863.

Remme, J., Habbema, J. D. F., and Hermans, J. (1980). A simulative comparison of linear, quadratic and kernel discrimination. *J. Statist. Comput. Statist.* **11**, 87–106.

Rencher, A. C., and Larson, S. F. (1980). Bias in Wilks' Λ in stepwise discriminant analysis. *Technometrics* **22**, 349–356.

Rényi, A. (1961). On measures of entropy and information. *Proc. 4th Berkeley Symp.* (Vol. 1). Berkeley: University of California Press, pp. 547–561.

Richardson, S. C. (1985). Bias correction in maximum likelihood logistic regression (Letter to the Editor). *Statist. in Med.* **4**, 243.

Ridout, M. S. (1988). Algorithm AS 233. An improved branch and bound algorithm for feature subset selection. *Appl. Statist.* **37**, 139–147.

Riffenburgh, R. H., and Clunies-Ross, C. W. (1960). Linear discriminant analysis. *Pacific Science* **14**, 251–256.

Rigby, R. A. (1982). A credibility interval for the probability that a new observation belongs to one of two multivariate normal populations. *J. R. Statist. Soc. B* **44**, 212–220.

Ripley, B. D. (1986). Statistics, images, and pattern recognition (with discussion). *Canad. J. Statist.* **14**, 83–111.

Ripley, B. D. (1988). *Statistical Inference for Spatial Processes*. Cambridge: Cambridge University Press.

Rissanen, J. (1989). *Stochastic Complexity in Statistical Inquiry*. Singapore: World Scientific Publishing Company.

Robbins, H. (1951). Asymptotically subminimax solutions of compound statistical decision functions. *Proc. 2nd Berkeley Symp.* Berkeley: University of California Press, pp. 131-148.

Robbins, H. (1964). The empirical Bayes approach to statistical decision functions. *Ann. Math. Statist.* **35**, 1–20.

Roberts, R. A., and Mullis, C. T. (1970). A Bayes sequential test of *m* hypotheses. *IEEE Trans. Inform. Theory* **IT-16**, 91–94.

Roberts, S. J. (1984). Algorithm AS 199. A branch and bound algorithm for determining the optimal feature subset of given size. *Appl. Statist.* **33**, 236–241.

Rodriguez, A. F. (1988). Admissibility and unbiasedness of the ridge classification rules for two normal populations with equal covariance matrices. *Statistics* **19**, 383–388.

Rogan, W. J., and Gladen, B. (1978). Estimating prevalence from the results of a screening test. *Amer. J. Epidem.* **107**, 71–76.

Rogers, W. H., and Wagner, T. J. (1978). A finite sample distribution—free performance bound for local discrimination rules. *Ann. Statist.* **6**, 506–514.

Rosenbaum, P. R., and Rubin, D. B. (1983). The central role of the propensity score in observational studies. *Biometrika* **70**, 41–55.

Rosenblatt, M. (1956). Remarks on some nonparametric estimates of a density function. *Ann. Math. Statist.* **27**, 832–837.

Rosenfeld, A., Hummel, R. A., and Zucker, S. W. (1976). Scene labeling by relaxation operations. *IEEE Trans. Syst. Man Cybern.* **SMC-6**, 420–433.

Rubin, D. B. (1981). The Bayesian bootstrap. *Ann. Statist.* **9**, 130–134.

Rubin, D. B. (1984). Comment on *Graphical Methods of Assessing Logistic Regression Models. J. Amer. Statist. Assoc.* **79**, 79–80.

Rudemo, M. (1982). Empirical choice of histograms and kernel density estimators. *Scand. J. Statist.* **9**, 65–78.

Ryan, T., Joiner, B., and Ryan, B. (1982). *Minitab Reference Manual.* Boston: Duxbury Press.

Saebo, H. V., Braten, K., Hjort, N. L., Llewellyn, B., and Mohn, E. (1985). Contextual classification of remotely sensed data: statistical methods and development of a system. *Technical Report No. 768*. Oslo: Norwegian Computer Centre.

Sammon, J. W., (1970). An optimal discriminant plane. *IEEE Trans. Comput.* **C-19**, 826–829.

Samuel, E. (1963a). Asymptotic solutions of the sequential compound decision problem. *Ann. Math. Statist.* **34**, 1079–1094.

Samuel, E. (1963b). Notes on a sequential classification problem. *Ann. Math. Statist.* **34**, 1095–1097.

Sánchez, J. M. P., and Cepeda, X. L. O. (1989). The use of smooth bootstrap techniques for estimating the error rate of a prediction rule. *Commun. Statist.–Simula.* **18**, 1169–1186.

Sanghvi, L. D. (1953). Comparison of genetical and morphological methods for a study of biological differences. *Amer. J. Phys. Anthrop. n. s.* **11**, 385–404.

Santner, T. J., and Duffy, D. E. (1986). A note on A. Albert and J. A., anderson's conditions for the existence of maximum likelihood estimates in logistic regression models. *Biometrika* **73**, 755–758.

SAS Institute, Inc. (1990). *SAS User's Guide: Statistics*. Edition 6.04. Cary, North Carolina: Author.

Sayre, J. W. (1980). The distribution of the actual error rates in linear discriminant analysis. *J. Amer. Statist. Assoc.* **75**, 201–205.

Schaafsma W. (1982). Selecting variables in discriminant analysis for improving upon classical procedures. In *Handbook of Statistics* (Vol. 2), P. R. Krishnaiah and L. Kanal (Eds.). Amsterdam: North-Holland, pp. 857–881.

Siotani, M., and Wang, R.-H. (1977). Asymptotic expansions for error rates and comparison of the W-procedure and the Z-procedure in discriminant analysis. In *Multivariate Analysis* (Vol. IV), P. R. Krishnaiah (Ed.). Amsterdam: North-Holland, pp. 523–545.

Siotani, M., Hayakawa, T., and Fujikoshi, Y. (1985). *Modern Multivariate Statistical Analysis: A Graduate Course and Handbook.* Columbus: American Sciences Press.

Sitgreaves, R. (1952). On the distribution of two random matrices used in classification procedures. *Ann. Math. Statist.* **23**, 263–270.

Sitgreaves, R. (1961). Some results on the W-classification statistic. In *Studies in Item Analysis and Prediction*, H. Solomon (Ed.). Stanford: Stanford University Press, pp. 241–251.

Skene, A. M. (1978). Discrimination using latent structure models. *Compstat 1978, Proc. Computational Statistics.* Vienna: Physica–Verlag, pp. 199–204.

Smith, A. F. M., and Makov, U. E. (1978). A quasi-Bayes sequential procedure for mixtures. *J. R. Statist. Soc. B* **40**, 106–112.

Smith, A. F. M., and Spiegelhalter, D. J. (1982). Bayesian approach to multivariate structure. In *Interpreting Multivariate Data*, V. Barnett (Ed.). Chichester: Wiley, pp. 335–348.

Smith, C. A. B. (1947). Some examples of discrimination. *Ann. Eugen.* **13**, 272–282.

Snapinn, S. M., and Knoke, J. D. (1984). Classification error rate estimators evaluated by unconditional mean squared error. *Technometrics* **26**, 371–378.

Snapinn, S. M., and Knoke, J. D. (1985). An evaluation of smoothed classification error-rate estimators. *Technometrics* **27**, 199–206.

Snapinn, S. M., and Knoke, J. D. (1988). Bootstrapped and smoothed classification error rate estimators. *Commun. Statist.–Simula.* **17**, 1135–1153.

Snapinn, S. M., and Knoke, J. D. (1989). Estimation of error rates in discriminant analysis with selection of variables. *Biometrics* **45**, 289–299.

Sorum, M. J. (1971). Estimating the conditional probability of misclassification. *Technometrics* **13**, 333–343.

Sorum, M. J. (1972). Estimating the expected and the optimal probabilities of misclassification. *Technometrics* **14**, 935–943.

Sorum, M. J. (1973). Estimating the expected probability of misclassification for a rule based on the linear discrimination function: univariate normal case. *Technometrics* **15**, 329–339.

Specht, D. F. (1967). Generation of polynomial discriminant functions for pattern recognition. *IEEE Trans. Electron. Comput.* **EC-16**, 308–319.

Spiegelhalter, D. J. (1986). Probabilistic prediction in patient management and clinical trials. *Statist. Med.* **5**, 421–433.

Spiegelhalter, D. J., and Knill-Jones, R. P. (1984). Statistical and knowledge-based approaches to clinical decision-support systems, with an application in gastroenterology (with discussion). *J. R. Statist. Soc. A* **147**, 35–76.

Spjøtvoll, E. (1977). Alternatives to C_p in multiple regression. *Biometrika* **64**, 1–8.

SPSS, Inc. (1986). *SPSS-X User's Guide*. Second Edition. Chicago: Author.

Srivastava, J. N., and Zaatar, M. K. (1972). On the maximum likelihood classification rule for incomplete multivariate samples and its admissibility. *J. Multivar. Anal.* **2**, 115–126.

Srivastava, M. S. (1967a). Classification into multivariate normal populations when the population means are linearly restricted. *Ann. Inst. Statist. Math.* **40**, 473–478.

Srivastava, M. S. (1967b). Comparing distances between multivariate populations—the problem of minimum distance. *Ann. Math. Statist.* **38**, 550–556.

Srivastava, M. S. (1973a). A sequential approach to classification: cost of not knowing the covariance matrix. *J. Multivar. Anal.* **3**, 173–183.

Srivastava, M. S. (1973b). Evaluation of misclassification errors. *Canad. J. Statist.* **1**, 35–50.

Stablein, D. M., Miller, J. D., Choi, S. C., and Becker, D. P. (1980). Statistical methods for determining prognosis in severe head injury. *Neurosurgery* **6**, 213–248.

Steerneman, A. G. M. (1987). *On the Choice of Variables in Discriminant Analysis and Regression Analysis.* Unpublished Ph. D. thesis, University of Groningen.

Stefanski, L. A., Carroll, R. J., and Ruppert, D. (1986). Bounded score functions for generalized linear models. *Biometrika* **73**, 413–424.

Stein, C. (1945). A two-sample test of a linear hypothesis whose power is independent of the variance. *Ann. Math. Statist.* **16**, 243–258.

Stein, C. (1973). Estimation of the mean of a multivariate normal distribution . *Proc. Prague Symp. on Asymptotic Statistics.* Karlova, Czechoslovakia: Prague University Press, pp. 345–381.

Stein, C., Efron, B., and Morris, C. (1972). Improving the usual estimator of a normal covariance matrix. *Technical Report No. 37*. Stanford: Department of Statistics, Stanford University.

Stirling, W. C. and Swindlehurst, A. L. (1987). Decision-directed multivariate empirical Bayes classification with nonstationary priors. *IEEE Trans. Pattern Anal. Machine Intell.* **PAMI-9**, 644–660.

Stocks, P. (1933). A biometric investigation of twins. Part II. *Ann. Eugen.* **5**, 1–55.

Stoller, D. S. (1954). Univariate two-population distribution-free discrimination. *J. Amer. Statist. Assoc.* **49**, 770–777.

Stone, C. J. (1977). Consistent nonparametric regression (with discussion). *Ann. Statist.* **5**, 595–645.

Stone, C. J. (1984). An asymptotically optimal window selection rule for kernel density estimates. *Ann. Statist.* **12**, 1285–1297.

Stone, M. (1974). Cross-validatory choice and assessment of statistical predictions. *J. R. Statist. Soc. B* **36**, 111–147.

Strauss, D. J. (1977). Clustering on coloured lattices. *J. Appl. Prob.* **14**, 135–143.

Streit, F. (1979). Multivariate linear discrimination when the covariance matrix is unknown. *South. African Statist. J.* **14**, 76.

Subrahmaniam, K. (1971). Discrimination in the presence of covariables. *South African Statist. J.* **5**, 5–14.

Subrahmaniam, K., and Chinganda, E. F. (1978). Robustness of the linear discriminant function to nonnormality: Edgeworth series distribution. *J. Statist. Planning Inference* **2**, 79–91.

Subrahmaniam, Kathleen and Subrahmaniam, Kocherlakota (1973). On the distribution of $(D^2_{p+q} - D^2_q)$ statistic: percentage points and the power of the test. *Sankhyā B* **35**, 51–78.

Subrahmaniam, Kocherlakota and Subrahmaniam, Kathleen (1976). On the performance of some statistics in discriminant analysis based on covariates. *J. Multivar. Anal.* **6**, 330–337.

Sutherland, M., Fienberg, S. E., and Holland, P. W. (1974). Combining Bayes and frequency approaches to estimate a multinomial parameter. In *Studies in Bayesian Econometrics and Statistics*, S. E. Fienberg and A. Zellner (Eds.). Amsterdam: North-Holland, pp. 585–617.

Sutradhar, B. C. (1990). Discrimination of observations into one of two t populations. *Biometrics* **46**, 827–835.

Switzer, P. (1969). Mapping a geographically correlated environment (with discussion). *Proc. Inter. Symp. Statistical Ecology: Spatial Patterns and Statistical Distribution* (Vol. 1). pp. 235–269.

Switzer, P. (1980). Extensions of linear discriminant analysis for statistical classification of remotely sensed satellite imagery. *Math. Geol.* **12**, 367–376.

Switzer, P. (1983). Some spatial statistics for the interpretation of satellite data (with discussion). *Bull. Inter. Statist. Inst.* **50**, 962–972.

Switzer, P. (1985). Min/max autocorrelation factors for multivariate spatial imagery. In *Computer Science and Statistics: The Interface*, L. Billard (Ed.). Amsterdam: North-Holland, pp. 13–16.

Switzer, P. (1986). Contribution to the discussion of paper by J. Besag. *J. R. Statist. Soc. B* **48**, 295.

Switzer, P. (1987). Statistical image processing. *Technical Report No. 15.* Stanford: Department of Statistics, Stanford University.

Switzer, P., Kowalik, W. S., and Lyon, R. J. F. (1982). A prior probability method for smoothing discriminant classification maps. *Math. Geol.* **14**, 433–444.

Symons, M. J. (1981). Clustering criteria and multivariate normal mixtures. *Biometrics* **37**, 35–43.

SYSTAT, Inc. (1988). *SYSTAT: The System for Statistics*. Evanston, Illinois: Author.

Taneja, I. J. (1983). On characterization of J–divergence and its generalizations. *J. Combin. Inform. Syst. Science* **8**, 206–212.

Taneja, I. J. (1987). Statistical aspects of divergence measures. *J. Statist. Planning Inference* **16**, 137–145.

Tapia, R. A., and Thompson, J. R. (1978). *Nonparametric Probability Density Estimation*. Baltimore: Johns Hopkins University Press.

Tarter, M., and Kronmal, R. (1970). On multivariate density estimates based on orthogonal expansions. *Ann. Math. Statist.* **4**, 718–722.

Teichroew, D., and Sitgreaves, R. (1961). Computation of an empirical sampling distribution for the W-classification statistic. In *Studies in Item Analysis and Prediction*, H. Solomon (Ed.). Stanford: Stanford University Press, pp. 252–275.

Tiao, G. C., and Zellner, A. (1964). On the Bayesian estimation of multivariate regression. *J. R. Statist. Soc. B* **26**, 277–285.

Tibshirani, R. (1985). How many bootstraps? *Technical Report No. 362.* Stanford: Department of Statistics, Stanford University.

Tiku, M. L. (1967). Estimating the mean and standard deviation from censored normal samples. *Biometrika* **54**, 155–165.

Tiku, M. L. (1980). Robustness of MML estimators based on censored samples and robust test statistics. *J. Statist. Planning Inference* **4**, 123–143.

Tiku, M. L. (1983). Robust location-tests and classification procedures. In *Robustness of Statistical Methods and Nonparametric Statistics*, D. Rasch and M. L. Tiku (Eds.). East Berlin: VEB deutscher Verlag der Wissenschaften, pp. 152–155.

Tiku, M. L., and Balakrishnan, N. (1984). Robust multivariate classification procedures based on the MML estimators. *Commun. Statist.–Theory Meth.* **13**, 967–986.

Tiku, M. L., and Balakrishnan, N. (1989). Robust classification procedures based on the MML estimators. *Commun. Statist.–Theory Meth.* **18**, 1047–1066.

Tiku, M. L., and Singh, M. (1982). Robust statistics for testing mean vectors of multivariate distributions. *Commun. Statist.–Theory Meth.* **11**, 985–1001.

Tiku, M. L., Tan W. Y., and Balakrishnan, N. (1986). *Robust Inference*. New York: Marcel Dekker.

Tiku, M. L., Balakrishnan, N., and Amagaspitiya, R. S. (1989). Error rates of a robust classification procedure based on dichotomous and continuous random variables. *Commun. Statist.–Simula.* **18**, 571–588.

Titterington, D. M. (1976). Updating a diagnostic system using unconfirmed cases. *Appl. Statist.* **24**, 238–247.

Titterington, D. M. (1977). Analysis of incomplete multivariate binary data by the kernel method. *Biometrika* **64**, 455–460.

Titterington D. M. (1980). A comparative study of kernel-based density estimates for categorical data. *Technometrics* **22**, 259–268.

Titterington, D. M. (1984). Comments on *Application of the Conditional Population-Mixture Model to Image Segmentation. IEEE Trans. Pattern Anal. Machine Intell.* **PAMI-6**, 656–658.

Titterington, D. M. (1985). Common structure of smoothing techniques in statistics. *Inter. Statist. Rev.* **53**, 141–170.

Titterington, D. M. (1989). An alternative stochastic supervisor in discriminant analysis. *Pattern Recognition* **22**, 91–95.

Titterington, D. M., and Bowman, A. W. (1985). A comparative study of smoothing procedures for ordered categorical data. *J. Statist. Comput. Simul.* **21**, 291–312.

Titterington, D. M., and Mill, G. M. (1983). Kernel-based density estimates from incomplete data. *J. R. Statist. Soc. B* **45**, 258–266.

Titterington, D. M., Murray, G. D., Murray, L. S., Spiegelhalter, D. J., Skene, A. M., Habbema, J. D. F., and Gelpke, G. J. (1981). Comparison of discrimination techniques applied to a complex data set of head injured patients (with discussion). *J. R. Statist. Soc. A* **144**, 145–175.

Titterington, D. M., Smith, A. F. M., and Makov, U. E. (1985). *Statistical Analysis of Finite Mixture Distributions*. New York: Wiley.

Todeschini, R. (1989). k-Nearest neighbour method: the influence of data transformations and metrics. *Chemometrics Intell. Labor. Syst.* **6**, 213–220.

Toussaint, G. T. (1974a). Bibliography on estimation of misclassification. *IEEE Trans. Inform. Theory* **IT-20**, 472–479.

Toussaint, G. T. (1974b). On some measures of information and their application in pattern recognition. *Proc. Measures of Information and their Applications*. Bombay: Indian Institute of Technology, pp. 21–28.

Toussaint, G. T., and Sharpe, P. M. (1975). An efficient method for estimating the probability of misclassification applied to a problem in medical diagnosis. *Comput. Biol. Med.* **4**, 269–278.

Trampisch, H. J. (1976). A discriminant analysis for qualitative data with interactions. *Comput. Programs Biomed.* **6**, 50–60.

Trampisch, H. J. (1983). On the performance of some classification rules for qualitative data for simulated underlying distributions. *Biom. J.* **25**, 689–698.

Truett, J., Cornfield, J., and Kannel, W. B. (1967). A multivariate analysis of the risk of coronary heart disease in Framingham. *J. Chron. Dis.* **20**, 511–524.

Tsiatis, A. A. (1980). A note on a goodness-of-fit test for the logistic regression model. *Biometrika* **67**, 250–251.

Tu, C., and Han, C. (1982). Discriminant analysis based on binary and continuous variables. *J. Amer. Statist. Assoc.* **77**, 447–454.

Tubbs, J. D. (1980). Effect of autocorrelated training samples on Bayes' probabilities of misclassification. *Pattern Recognition* **12**, 351–354.

Tubbs, J. D., and Coberly, W. (1978). Spatial correlation and its effect upon classification results in LANDSAT. *Proc. 12th Inter. Symp. Remote Sensing of Environment*. Ann Arbor: Environmental Research Institute of Michigan, pp. 775–781.

Tubbs, J. D., Coberly, W. A., and Young, D. M. (1982). Linear dimension reduction and Bayes classification with unknown population parameters. *Pattern Recognition* **15**, 167–172.

Tutz, G. E. (1985). Smoothed additive estimators for non-error rates in multiple discriminant analysis. *Pattern Recognition* **18**, 151–159.

Tutz, G. E. (1986). An alternative choice of smoothing for kernel-based density estimates in discrete discriminant analysis. *Biometrika* **73**, 405–411.

Tutz, G. E. (1988). Smoothing for discrete kernels in discrimination. *Biom. J.* **6**, 729–739.

Tutz, G. E. (1989). On cross-validation for discrete kernel estimates in discrimination. *Commun. Statist.–Theory Meth.* **18**, 4145–4162.

Urbakh, V.Yu. (1971). Linear discriminant analysis: loss of discriminating power when a variate is omitted. *Biometrics* **27**, 531–534.

Van Campenhout, J. M. (1978). On the peaking of the Hughes mean recognition accuracy: the resolution of an apparent paradox. *IEEE Trans. Syst. Man. Cybern.* **SMC-8**, 390–395.

Van Houwelingen, J. C., and le Cessie, S. (1988). Logistic regression, a review. *Statist. Neerlandica* **42**, 215–232.

Van Ness, J. W. (1977). Dimensionality and classification performance with independent coordinates. *IEEE Trans. Syst. Cybern.* **SC-7**, 560–564.

Van Ness, J. W. (1979). On the effects of dimension in discriminant analysis for unequal covariance populations. *Technometrics* **21**, 119–127.

Van Ness, J. W. (1980). On the dominance of non-parametric Bayes rule discriminant algorithm in high dimensions. *Pattern Recognition* **12**, 355–368.

Van Ness, J. W., and Simpson, C. (1976). On the effects of dimension in discriminant analysis. *Technometrics* **18**, 175–187.

Van Ryzin, J. (1966). Bayes risk consistency of classification procedures using density estimation. *Sankhyā* **A28**, 261–270.

Van Ryzin, J. (Ed.). (1977). *Classification and Clustering*. New York: Academic Press.

Van Ryzin, J., and Wang, M.-C. (1978). Two methods for smooth estimation of discrete distributions. *Commun. Statist.–Theory Meth.* **A7**, 211–228.

Venetoulias, A. (1988). Statistical image processing: an annotated bibliography. *Technical Report No. 119*. Stanford: Department of Statistics, Stanford University.

Venetoulias, A. (1989). Parameter estimation in image processing. *Technical Report No. 19*. Stanford: Department of Statistics, Stanford University.

Victor, N., Trampisch, H. J., and Zentgraf, R. (1974). Diagnostic rules for qualitative variables with interactions. *Meth. Inform. Med.* **13**, 184–186.

Villalobos, M. A., and Wahba, G. (1983). Multivariate thin plate spline estimates for the posterior probabilities in the classification problem. *Commun. Statist.–Theory Meth.* **12**, 1449–1479.

Vlachonikolis, I. G. (1985). On the asymptotic distribution of the location linear discriminant function. *J. R. Statist. Soc. B* **47**, 498–509.

Vlachonikolis, I. G. (1986). On the estimation of the expected probability of misclassification in discriminant analysis with mixed binary and continuous variables. *Comput. Math. Applic.* **12A**, 187–195.

Vlachonikolis, I. G. (1990). Predictive discrimination and classification with mixed binary and continuous variables. *Biometrika* **77**, 657–662.

Vlachonikolis, I. G., and Marriott, F. H. C. (1982). Discrimination with mixed binary and continuous data. *Appl. Statist.* **31**, 23–31.

von Mises, R. (1945). On the classification of observation data into distinct groups. *Ann. Math. Statist.* **16**, 68–73.

Wagner, G., Tautu, P., and Wolbler, U. (1978). Problems of medical diagnosis —a bibliography. *Meth. Inform. Med.* **17**, 55–74.

Wagner, T. J. (1971). Convergence of the nearest neighbor rule. *IEEE Trans. Inform. Theory* **IT-17**, 566–571.

Wahl, P. W., and Kronmal, R. A. (1977). Discriminant functions when covariances are unequal and sample sizes are moderate. *Biometrics* **33**, 479–484.

Wakaki, H. (1990). Comparison of linear and quadratic discriminant functions. *Biometrika* **77**, 227–229.

Wald, A. (1939). Contributions to the theory of statistical estimation and testing hypothesis. *Ann. Math. Statist.* **10**, 299–326.

Wald, A. (1944). On a statistical problem arising in the classification of an individual into one of two groups. *Ann. Math. Statist.* **15**, 145–162.

Wald, A. (1947). *Sequential Analysis*. New York: Wiley.

Wald, A. (1949). Statistical decision functions. *Ann. Math. Statist.* **20**, 165–205.

Wald, A. (1950). *Statistical Decision Functions*. New York: Wiley.

Walker, P. H., and Chittleborough, D. J. (1986). Development of particle-size distributions in some alfisols of south-eastern Australia. *Soil Science Soc. Amer. J.* **50**, 394–400.

Walker, S. H., and Duncan, D. B. (1967). Estimation of the probability of an event as a function of several independent variables. *Biometrika* **54**, 167–169.

Waller, W. G., and Jain, A. K. (1978). On the monotonicity of the performance of Bayesian classifiers. *IEEE Trans. Inform. Theory* **IT-24**, 392–394.

Walter, S. D. (1985). Small sample estimation of log odds ratios from logistic regression and fourfold tables. *Statist. Med.* **4**, 437–444.

Wang, M.-C. (1986a). A prior-valued estimator applied to multinomial classification. *Commun. Statist.–Theory Meth.* **15**, 405–427.

Wang, M.-C. (1986b). Re-sampling procedures for reducing bias of error rate estimation in multinomial classification. *Comput. Statist. Data Anal.* **4**, 15–39.

Wang, M.-C., and Van Ryzin, J. (1981). A class of smooth estimators for discrete distributions. *Biometrika* **68**, 301–309.

Warner, H. R., Toronto, A. F., Veasey, L. G., and Stephenson, R. (1961). A mathematical approach to medical diagnosis. *J. Amer. Med. Assoc.* **177**, 177–183.

Weber, J., and Baldessari, B. (1988). Optimal classification regions for normal random vectors with separable and partially exchangeable dependence. In *Classification and Related Methods of Data Analysis*, H. H. Boch (Ed.). Amsterdam: North–Holland, pp. 285–291.

Wegman, E. J. (1972a). Non-parametric probability density estimation. I: *Technometrics* **14**, 533–546.

Wegman, E. J. (1972b). Non-parametric probability density estimation. II: *J. Statist. Comput. Simul.* **1**, 225–246.

Weiner, J. M., and Dunn, O. J. (1966). Elimination of variates in linear discrimination problems. *Biometrics* **22**, 268–275.

Weiss, S. H. et al. (1985). Screening test for HTLV–III (AIDS agent) antibodies. *J. Amer. Med. Assoc.* **253**, 221–225.

Welch, B. L. (1939). Note on discriminant functions. *Biometrika* **31**, 218–220.

Welch, J. R., and Salter, K. G. (1971). A context algorithm for pattern recognition and image segmentation. *IEEE Trans. Syst. Man. Cybern.* **SMC-1**, 24–30.

Wernecke, K. D., and Kalb, G. (1983). Further results in estimating the classification error in discriminance analysis. *Biom. J.* **25**, 247–258.

Wernecke, K. D., and Kalb, G. (1987). Estimation of error rates by means of simulated bootstrap distributions. *Biom. J.* **29**, 287–292.

Wernecke, K. D., Kalb, G., and Stürzebecher, E. (1980). Comparison of various procedures for estimation of the classification error in discriminance analysis. *Biom. J.* **22**, 639–649.

Wernecke, K.-D., Haerting, J., Kalb, G., and Stürzebecher, E. (1989). On model-choice in discrimination with categorical variables. *Biom. J.* **31**, 289–296.

Wertz, W., and Schneider, B. (1979). Statistical density estimation: a bibliography. *Inter. Statist. Rev.* **47**, 155–175.

Whittle, P. (1958). On the smoothing of probability density functions. *J. R. Statist. Soc. B* **20**, 334–343.

Wiens, D. P., and Zheng, Z. (1986). Robust M-estimation of multivariate location and scatter in the presence of asymmetry. *Canad. J. Statist.* **14**, 161–176.

Wilks, S. S. (1932). Certain generalizations of the analysis of variance. *Biometrika* **39**, 471–494.

Williams, E. J. (1952). Some exact tests in multivariate analysis. *Biometrika* **39**, 17–31.

Williams, E. J. (1955). Significance tests for discriminant functions and linear functional relationship. *Biometrika* **42**, 360–381.

Williams, E. J. (1961). Tests for discriminant functions. *J. Austral. Math. Soc.* **2**, 243–252.

Williams, E. J. (1967). The analysis of association among many variates (with discussion). *J. R. Statist. Soc. B* **29**, 199–242.

Wilson, E. B., and Hilferty, M. M. (1931). The distribution of chi-square. *Proc. Nat. Acad. Science U. S. A.* **28**, 94–100.

Wilson, S. R. (1982). Sound and exploratory data analysis. *Compstat 1982, Proc. Computational Statistics*. Vienna: Physica–Verlag, pp. 447–450.

Wojciechowski, T. J. (1985). The empirical Bayes classification rules for mixtures of discrete and continuous variables. *Biom. J.* **27**, 521–532.

Wojciechowski, T. J. (1987). Nearest neighbor classification rule for mixtures of discrete and continuous random variables. *Biom. J.* **29**, 953–959.

Wojciechowski, T. J. (1988). An application of the orthogonal series estimators in the classification of mixed variables. *Biom. J.* **30**, 931–938.

Wold, S. (1976). Pattern recognition by means of disjoint principal component models. *Pattern Recognition* **8**, 127–139.

Wold, S. (1978). Cross-validatory estimation of the number of components in factor and principal components models. *Technometrics* **20**, 397–405.

Wolde-Tsadik, G., and Yu, M. C. (1979). Concordance in variable-subset discriminant analysis. *Biometrics* **35**, 641–644.

Wolfe, J. H. (1971). A Monte Carlo study of the sampling distribution of the likelihood ratio for mixtures of multinormal distributions. *Technical Bulletin STB 72–2*. San Diego: U. S. Naval Personnel and Training Research Laboratory.

Worlund, D. D., and Fredin, R. A. (1962). Differentiation of stocks. *Symp. on Pink Salmon*, N. J. Wilimovsky (Ed.). Vancouver: Institute of Fisheries, University of British Columbia, pp. 143–153.

Worton, B. J. (1989). Optimal smoothing parameters for multivariate fixed and adaptive kernel methods. *J. Statist. Comput. Simul.* **32**, 45–57.

Wright, R. M., and Switzer, P. (1971). Numerical classification applied to certain Jamaican Eocena Nummulitids. *Math. Geol.* **3**, 297–311.

Wu, C. F. J. (1983). On the convergence properties of the EM algorithm. *Ann. Statist.* **11**, 95–103.

Wyman, F. J., Young, D. M., and Turner, D. W. (1990). A comparison of asymptotic error rate expansions for the sample linear discriminant function. *Pattern Recognition* **23**, 775–783.

Wyrwoll, K.-H., and Smyth, G. K. (1985). On using the log-hyperbolic distribution to describe the textural characteristics of eolian sediments. *J. Sed. Petrology* **55**, 471–478.

Yarborough, D. A. (1971). Sequential discrimination with likelihood-ratios. *Ann. Math. Statist.* **42**, 1339–1347.

Yerushalmy, J., et al. (1965). Birth weight and gestation as indices of immaturity. *Amer. J. Dis. Child.* **109**, 43–57.

Young, D. M., and Odell, P. L. (1984). A formulation and comparison of two linear feature selection techniques applicable to statistical classification. *Pattern Recognition* **17**, 331–337.

Young, D. M., and Odell, P. L. (1986). Feature-subset selection for statistical classification problems involving unequal covariance matrices. *Commun. Statist.–Theory Meth.* **15**, 137–157.

Young, D. M., Odell, P. L., and Marco, V. R. (1985). Optimal linear feature selection for a general class of statistical pattern recognition models. *Pattern Recognition Letters* **3**, 161–165.

Young, D. M., Marco, V. R., and Odell, P. L. (1986). Dimension reduction for predictive discrimination. *Comput. Statist. Data Anal.* **4**, 243–255.

Young, D. M., Marco, V. R., and Odell, P. L. (1987). Quadratic discrimination: some results on optimal low-dimensional representation. *J. Statist. Planning Inference* **17**, 307–319.

Young, D. M., Turner, D. W., and Marco, V. R. (1987). Some results on error rates for quadratic discrimination with known population parameters. *Biom. J.* **6**, 721–731.

Young, D. M., Turner, D. W., and Marco, V. R. (1988). On the robustness of the equal-mean discrimination rule with uniform covariance structure against serially correlated training data. *Pattern Recognition* **21**, 189–194.

Young, T. Y., and Calvert, T. W. (1974). *Classification, Estimation and Pattern Recognition*. New York: American Elsevier.

Yu, T. S., and Fu, K.-S. (1983). Recursive contextual classification using a spatial stochastic model. *Pattern Recognition* **16**, 89–108.

Zangwill, W. (1967). Minimizing a function without calculation derivatives. *Comput. J.* **10**, 293–296.

Zaslavski, A. E., and Sycheva, N. M. (1965). On the problem of optimal pattern recognition. *Comput. Syst.* **19**, 35–65 (in Russian).

Zentgraf, R. (1975). A note on Lancaster's definition of higher-order interactions. *Biometrika* **62**, 375–378.

Zhang, L. (1988). Selection of variables in two group discriminant analysis using information theoretical criteria. Unpublished manuscript.

Zielezny, M. (1976). Maximum gain in the two-stage classification procedure. *Biometrics* **32**, 481–484.

Zielezny, M., and Dunn, O. J. (1975). Cost evaluation of a two-stage classification procedure. *Biometrics* **31**, 37–47.

Author Index

Abbott, R. D., 472
Abend, K., 420, 447
Abramovitch, L., 373, 447
Abramowitz, M., 57, 447
Adegboye, O. S., 67, 447
Adhikari, B. P., 24, 447
Afifi, A. A., 221, 222, 390, 400, 456, 458, 477
Agrawala, A. K., 447
Ahmad, I. A., 26, 447
Ahmed, S. W., 154, 447
Aitchison, J., 2, 22, 30, 49, 50, 68, 70, 74, 133, 185, 254, 265, 266, 288, 296, 297, 298, 304, 305, 306, 309, 316, 317, 385, 386, 447, 448
Aitken, C. G. G., 266, 288, 296, 297, 298, 304, 305, 306, 309, 311, 316, 317, 447, 448
Aitkin, M., 48, 177, 182, 448
Akaike, H., 403, 448
Al-Harbey, A. A., 70, 464
Albert, A., 256, 257, 258, 263, 264, 265, 271, 275, 276, 279, 280, 281, 282, 343, 448, 483
Ali, S. M., 24, 448
Amagaspitiya, R. S., 232, 501
Ambergen, A. W., 382, 383, 448
Amemiya, T., 278, 279, 448
Amoh, R. K., 15, 240, 449
Anand, R., 486
Anderson, D., 48, 177, 448
Anderson, E., 192, 211, 449
Anderson, J. A., 3, 19, 35, 38, 256, 257, 258, 259, 262, 263, 265, 268, 269, 270, 400, 448, 449
Anderson, T. W., 10, 14, 20, 64, 65, 66, 96, 97, 102, 118, 119, 175, 188, 191, 283, 324, 333, 449
Ando, S., 196, 467
Andrews, D. F., 80, 172, 178, 180, 211, 214, 449
Andrews, V., 484
Aoyama, H., 96, 450
Armitage, P., 86, 450
Arvensen, E., 90, 481
Ashikaga, T., 154, 450, 458
Ashton, E. H., 191, 450
Atchley, W. R., 194, 455
Atkinson, C., 25, 450
Avery, R. B., 390, 463

Bacon-Shone, J., 181, 450
Bagnold, R. A., 251, 450
Bahadur, R. R., 96, 97, 156, 449, 450
Bailey, T., 322, 450
Balakrishnan, N., 104, 154, 165, 166, 167, 223, 232, 450, 479, 501
Balakrishnan, V., 25, 450
Baldessari, B., 444, 504
Bandyopadhyay, S., 83, 84, 450, 459
Banjeree, K. S., 97, 451
Baringhaus, L., 170, 451
Barker, D. J. P., 468
Barlow, P., 490
Barnard, G. A., 48, 451
Barnard, M. M., 12, 451
Barndorff-Nielsen, O., 250, 251, 450, 451
Barnett, V., 181, 182, 451
Bartlett, M. S., 58, 148, 177, 188, 189, 190, 342, 451
Basford, K. E., 28, 31, 32, 33, 39, 41, 42, 45, 46, 47, 48, 87, 170, 174, 210, 213, 220, 364, 366, 434, 451, 488
Baskett, F., 323, 466
Basu, A. P., 239, 444, 451
Bayne, C. K., 122, 133, 154, 451, 452
Beauchamp, J. J., 122, 181, 451, 452
Becker, D. P., 62, 499
Becker, R. A., 199, 200, 294, 452
Beckman, R. J., 181, 334, 425, 452

Begg, C. B., 49, 50, 266, 448
Bellman, R. E., 15, 452
Belsley, D. A., 329, 452
Ben-Bassat, M., 23, 452
Beran, R., 23, 452
Bernardo, J. M., 57, 452
Bertolino, F., 82, 452
Besag, J., 414, 417, 418, 419, 420, 428, 429, 430, 431, 432, 433, 436, 437, 438, 452
Bhattacharya, P. K., 239, 453
Bhattacharyya, A., 22, 23, 25, 391, 452
Bidasaria, H. B., 100, 453
Biscay, R., 199, 453
Bishop, T., 67, 462
Bishop, Y. M. M., 218, 453
Bjerregaard, B., 21, 217, 471, 474
Blackhurst, D. W., 43, 453
Blackwelder, W. C., 278, 472
Blair, V., 262, 449
Bliss, C. I., 75, 457
Bloomfield, P., 233, 453
Boullion, T. L., 391, 453
Bowker, A. H., 102, 103, 453
Bowman, A. W., 288, 297, 302, 303, 310, 453, 501
Box, G. E. P., 178, 453
Boys, R. J., 35, 453
Bozdogan, H., 180, 181, 453
Bradley, R. A., 290, 486
Brant, R., 80, 450
Braten, K., 495
Breiman, L., 20, 311, 324, 325, 326, 327, 328, 329, 330, 332, 453
Brennan, L. E., 375, 494
Breslow, N. E., 257, 262, 453, 492
Broeckaert, I., 291, 457
Broemeling, L. D., 83, 454
Broffitt, B., 50, 154, 163, 454, 457, 482
Broffitt, J. D., 20, 164, 165, 333, 334, 367, 454, 493
Brooks, C. A., 278, 454
Browdy, B. L., 82, 454
Brown, P. J., 392, 454
Brunk, H. D., 219, 454
Bryant, J., 100, 323, 454
Buhler, F. R., 497
Bull, S. B., 277, 454
Burbea, J., 22, 454
Burnaby, T. P., 67, 454
Butler, R. W., 182, 454
Byth, K., 268, 270, 277, 278, 279, 454, 488

Cacoullos, T., 4, 183, 294, 295, 454
Calvert, T. W., 4, 506
Campbell, M. N., 258, 455
Campbell, N. A., 145, 163, 164, 183, 187, 194, 195, 391, 393, 394, 396, 398, 399, 400, 429, 434, 455, 478, 487
Carpenter, R. G., 50, 464
Carroll, R. J., 271, 499
Castagliola, P., 97, 455
Čencov, N. N., 289, 455
Cepeda, X. L. O., 359, 495
Chaddha, R. L., 97, 455
Chambers, J. M., 62, 294, 452
Chambless, L. E., 257, 479
Chan, L. S., 43, 455
Chanda, K. C., 309, 312, 455
Chandrasekaran, B., 391, 392, 455, 456, 476, 477
Chang, C. Y., 391, 456
Chang, P. C., 82, 154, 221, 450, 454, 456
Chang, W. C., 196, 197, 198, 199, 200, 456
Chapelle, J. P., 258, 448
Chatterjee, S., 358, 456
Chellappa, R., 417, 478
Chen, C.-C., 359, 476
Chen, C. H., 391, 456
Chen, Y., 368, 456
Chernick, M. R., 358, 359, 456
Chernoff, H., 23, 97, 456
Chhikara, R. S., 31, 36, 238, 240, 456, 465
Chinganda, E. F., 153, 154, 456
Chittineni, C. B., 36, 457
Chittleborough, D. J., 251, 503
Choi, K., 4, 13, 457
Choi, S. C., 78, 79, 82, 323, 457, 464, 499
Chow, C. K., 362, 457
Chow, Y. S., 302, 457
Christi, R., 460
Clark, R. R., 454
Clarke, W. R., 154, 454, 457
Clarkson, D. B., 141, 457
Clement, B., 396, 497
Cléroux, R., 181, 457
Cleveland, W. S., 26, 199, 452, 457
Clunies-Ross, C. W., 96, 457, 494
Coberly, W. A., 98, 99, 444, 460, 502
Cochran, W. G., 18, 75, 78, 284, 390, 457
Collins, J. R., 161, 457
Connor, R. J., 177, 470
Conover, W. J., 334, 457
Cook, R. D., 181, 452
Coomans, D., 291, 457
Cooper, P. W., 238, 457, 458
Copas, J. B., 152, 271, 272, 458
Cornfield, J., 235, 256, 458, 502
Cosslett, S., 262, 458

Costanza, M. C., 390, 400, 458
Cover, T. M., 321, 322, 458
Cox, D. R., 170, 178, 217, 256, 262, 267, 453, 458
Cox, M., 391, 482
Craig, R. G., 443, 444, 458
Cramér, H., 28, 109, 458
Crask, M. R., 345, 459
Crawley, D. R., 278, 459
Cressie, N. A. C., 22, 459, 494
Critchley, F., 21, 30, 57, 134, 135, 379, 380, 381, 382, 384, 385, 386, 388, 459, 474
Csörgö, S., 170, 459
Ćwik, J., 307, 459

Danschke, R., 170, 451
Dargahi-Noubary, G. R., 138, 459
Das Gupta, S., 8, 12, 56, 67, 80, 83, 84, 127, 239, 334, 340, 396, 453, 459
Datta, L., 99, 489
Daudin, J. J., 391, 403, 459
Davies, E. R., 320, 460
Davis, A. W., 381, 460
Davis, L., 279, 460
Davison, A. C., 346, 460
Dawid, A. P., 13, 460
Day, N. E., 238, 256, 257, 263, 264, 265, 278, 453, 460
Decell, H. P., 98, 100, 460
Deev, A. D., 108, 460
Deheuvels, P., 295, 460
Dempster, A. P., 39, 41, 129, 220, 438, 460
Dennis, N. R., 484
Derin, H., 420, 437, 460, 482
Desu, M. M., 58, 59, 71, 460
Devijver, P. A., 4, 322, 363, 427, 460, 478, 479
Devlin, S. J., 460
Devroye, L. P., 133, 291, 292, 301, 303, 321, 322, 373, 374, 375, 461
de Waal, D. J., 84, 490
Dey, D. K., 131, 461
Diaconis, P., 34, 461
DiCiccio, T. J., 346, 461
Dillon, W. R., 4, 157, 226, 227, 286, 287, 290, 390, 461, 469
Di Pillo, P. J., 145, 461
Dixon, W. J., 390, 461
Do, K., 31, 461
Donner, A., 258, 277, 454, 455
Donoho, A. W., 199, 461
Donoho, D. L., 199, 461
Dorvlo, A. S., 345, 493
Dröge, J. B. M., 141, 461, 462
Dubes, R. C., 359, 418, 462, 476
Dubuisson, B., 97, 455
Duchene, J., 199, 462
Duda, R. O., 4, 462
Dudewicz, E. J., 67, 462, 486
Duffy, D. E., 264, 265, 462, 495
Duin, R. P. W., 302, 462
Duncan, D. B., 256, 503
Dunn, C. L., 67, 462
Dunn, J. E., 180, 462
Dunn, O. J., 43, 97, 133, 365, 367, 390, 396, 455, 462, 464, 485, 497, 504, 506
Dunsmore, I. R., 2, 30, 35, 68, 74, 185, 265, 385, 386, 448, 453, 462
Duran, B. S., 391, 453
Durbin, J., 74, 462

Eastment, H. T., 462
Eckardt, M. J., 494
Edwards, A. W. F., 462
Efron, B., 41, 48, 113, 115, 116, 131, 152, 263, 266, 273, 274, 275, 276, 277, 339, 340, 342, 345, 346, 348, 349, 352, 353, 354, 355, 356, 357, 358, 359, 371, 372, 373, 463, 499
Eisenbeis, R. A., 390, 396, 463
Elashoff, J. D., 390, 463
Elashoff, R. M., 222, 390, 447, 463
Elfving, G., 463
El Khattabi, S., 240, 463
Elliott, H., 460, 463
Ellison, B. E., 67, 82, 464
El-Sayyad, G. M., 70, 464
El Shaarawi, A. H., 165, 450
Elvers, E., 2, 464
Emery, J. L., 50, 464
Enas, G. G., 323, 464
Enis, P., 30, 59, 69, 464
Epanechnikov, V. A., 294, 295, 311, 464
Eriksen, P. S., 139, 464
Estes, S. E., 103, 464
Everitt, B. S., 220, 464

Faden, V. B., 155, 464, 493, 494
Farewell, V. T., 262, 464
Farmer, J. H., 464
Farver, T. B., 390, 464
Fatti, L. P., 108, 155, 170, 172, 173, 175, 187, 399, 464
Fieller, N. R. J., 252, 254, 464
Fienberg, S. E., 219, 288, 453, 464, 500
Fischer, F. P., 309, 491
Fisher, L., 160, 475
Fisher, R. A., 8, 12, 62, 63, 89, 164, 180, 192, 197, 211, 335, 367, 465
Fitzmaurice, G. M., 359, 363, 465

Fix, E., 14, 291, 292, 309, 319, 321, 333, 465, 497
Flenley, E. C., 254, 464
Flick, T. E., 309, 322, 465, 467
Flury, B., 138, 139, 140, 141, 177, 465
Föglein, J., 414, 420, 429, 434, 444, 464, 465, 479
Foley, D. H., 196, 391, 465
Folkert, J. E., 181, 452
Folks, J. L., 456
Folks, J. R., 240, 465
Ford, I., 21, 30, 57, 379, 380, 381, 384, 459, 474
Fowlkes, E. B., 200, 271, 390, 398, 465
Frank, I. E., 141, 142, 143, 144, 466
Fredin, R. A., 32, 505
Freed, N., 97, 466
Freedman, D. A., 34, 86, 461, 466
Freund, R. J., 390, 464
Friedman, J. H., 131, 133, 138, 141, 142, 143, 144, 145, 146, 147, 148, 149, 150, 151, 152, 199, 201, 211, 309, 319, 323, 324, 328, 329, 331, 332, 389, 453, 466
Fritz, J., 321, 322, 466
Fryer, M. J., 291, 466
Fu, K.-S., 4, 24, 86, 363, 391, 414, 466, 484, 506
Fujikoshi, Y., 64, 78, 127, 393, 403, 404, 466, 498
Fukunaga, K., 4, 123, 199, 322, 323, 338, 343, 362, 363, 391, 466, 467, 497
Fung, K. Y., 43, 467
Fung, W. K., 181, 450
Furnival, G. M., 397, 467

Gabriel, K. R., 397, 467
Ganesalingam, S., 32, 38, 46, 364, 467, 488
Ganeshanandam, S., 338, 369, 370, 400, 410, 411, 412, 467, 468
Gardner, M. J., 468
Gaskins, R. A., 308, 469
Gasko, M., 199, 461
Gastwirth, J. L., 3, 34, 468
Gautschi, W., 141, 465
Geisser, S., 29, 30, 58, 59, 69, 71, 86, 87, 91, 92, 94, 183, 342, 460, 464, 468
Gelatt, C. D., 436, 478
Gelpke, G. J., 501
Gelsema, E. S., 410, 492
Geman, D., 414, 416, 435, 436, 437, 460
Geman, S., 302, 414, 416, 435, 436, 437, 468
Gessaman, M. P., 19, 20, 313, 333, 468, 492
Gessaman, P. H., 20, 313, 468
Ghurye, S. G., 56, 468
Gilbert, E. S., 122, 133, 156, 468
Gilbert, G. G., 390, 463
Gilbertson, D. D., 252, 464
Gill, D. S., 377, 478
Gill, P. E., 263, 468
Gilman, J. A., 43, 455
Giri, N. C., 396, 497
Gladen, B., 33, 495
Glaz, J., 414, 469
Glick, N., 14, 24, 26, 97, 238, 285, 338, 344, 360, 364, 469
Glover, F., 97, 466
Gnanadesikan, R., 4, 163, 170, 172, 173, 178, 180, 390, 450, 460, 465, 469
Goin, J. E., 323, 469
Goldman, G. E., 390, 463
Goldstein, M., 4, 153, 157, 226, 227, 285, 286, 287, 290, 313, 390, 391, 461, 469, 482
Gong, G., 345, 348, 349, 352, 353, 410, 463, 469
Good, I. J., 288, 308, 469
Goodall, C., 164, 469
Goodman, I. R., 238, 469
Goppert, R., 323, 490
Gordon, L., 323, 324, 470
Gordon, R. D., 43, 50, 243, 244, 247, 488
Govindarajulu, Z., 20, 334, 470
Gower, J. C., 194, 470
Graham, R. L., 346, 470
Graubard, B. I., 494
Gray, H. L., 344, 470
Grayson, D. A., 36, 37, 470
Green, M., 213, 488
Green, M. S., 271, 470
Greene, T., 145, 149, 151, 470, 494
Greenland, S., 258, 470
Greenstreet, R. L., 177, 470
Greer, R. L., 97, 265, 336, 470
Gregory, G. G., 302, 496
Greig, D. M., 437, 438, 470
Gupta, A. K., 20, 79, 334, 451, 470
Gupta, R. D., 377, 478
Guseman, L., 88, 98, 100, 454, 460, 470
Györfi, L., 291, 292, 461

Haack, D., 334, 470
Habbema, J. D. F., 19, 21, 70, 93, 158, 201, 294, 298, 302, 313, 317, 318, 390, 400, 405, 448, 471, 473, 494, 496, 501
Haberman, S. J., 219, 471
Haerting, J., 391, 471, 504
Haff, L. R., 131, 471
Haldane, J. B. S., 268, 471
Hall, P., 48, 288, 289, 290, 301, 302, 303, 304, 305, 306, 307, 308, 317, 346, 348, 373, 453, 471, 472

Halperin, M., 278, 472
Hampel, F. R., 161, 163, 472
Han, C. P., 59, 70, 122, 123, 232, 472, 502
Hand, D. J., 4, 35, 290, 291, 313, 314, 316, 317, 318, 320, 338, 344, 359, 363, 364, 391, 465, 472, 473, 490
Harley, T. J., 420, 447
Harris, E. K., 280, 448
Harrison, D., 329, 473
Hart, P. E., 4, 321, 458, 462
Harter, H. L., 102, 473
Hartz, S. C., 490
Haslett, J., 414, 420, 427, 443, 473
Hastie, T., 258, 271, 473
Hawkins, D. M., 122, 138, 155, 163, 170, 171, 172, 174, 181, 182, 201, 209, 247, 385, 396, 399, 400, 464, 473
Hayakawa, T., 64, 498
Hayes, R. R., 123, 338, 467
Healy, M. J. R., 191, 450
Hecker, R., 410, 473
Helbling, J.-M., 181, 457
Hellman, M. E., 320, 473
Henry, N. W., 220, 483
Henze, N., 170, 451
Herman, C., 70, 465
Hermans, J., 19, 93, 158, 180, 201, 294, 298, 302, 313, 317, 390, 400, 405, 471, 473, 494, 496
Herzberg, A. M., 211, 214, 449
Highleyman, W. H., 341, 474
Hildebrandt, B., 122, 474
Hilden, J., 21, 217, 471, 474
Hilferty, M. M., 163, 505
Hills, M., 18, 285, 305, 310, 316, 338, 340, 390, 474
Hinde, J., 48, 177, 448
Hinkley, D. V., 41, 346, 460, 463, 470, 474
Hintze, J. L., 398, 474
Hirst, D., 379, 381, 385, 387, 474
Hjort, N. L., 4, 29, 48, 172, 178, 184, 342, 343, 357, 414, 415, 422, 426, 427, 434, 474, 489, 492, 495
Hoaglin, D. C., 48, 475
Hodges, J. L., 14, 291, 292, 309, 319, 321, 333, 465, 497
Hodgu, A., 454
Hoel, P. G., 12, 14, 475
Hogg, R. V., 20, 454, 493
Holland, P. W., 219, 288, 453, 464, 500
Honda, M., 341, 479
Hope, A. C. A., 48, 475
Hopkins, C. E., 284, 457
Hora, S. C., 364, 475
Horton, I. R., 400, 475
Hosmer, D. W., 160, 256, 271, 475, 483, 490
Hosmer, T. A., 160, 278, 475
Hostetler, L. D., 323, 363, 467
Hotelling, H., 12, 89, 400, 475
Huber, P. J., 161, 162, 163, 164, 199, 201, 207, 475
Huberty, C. J., 367, 475
Hudimoto, H., 13, 23, 334, 476
Hudlet, R., 88, 476
Hufnagel, G., 43, 476
Hughes, G. F., 392, 476
Huijbregts, Ch. J., 425, 477
Hummel, R. A., 438, 443, 476, 495
Hummels, D. M., 322, 343, 467
Hung, Y.-T., 284, 476

Iman, R. L., 334, 457
Iqbal, M., 381, 491

Jacobowitz, R., 472
Jain, A. K., 322, 359, 392, 418, 450, 455, 456, 462, 476, 503
James, M., 62, 476
James W., 131, 476
Jeffreys, H., 22, 476
Jennett, W. B., 490
Jennings, L. W., 145, 491
Jennison, C., 435, 477
Jennrich, R. I., 390, 476
Jensen, S. T., 476
Joachimsthaler, E. A., 155, 476
Johansen, S., 139, 476
John, P. W. M., 470
John, S., 60, 102, 108, 112, 476, 477
Johns, M. V., 13, 477
Johnson, M. E., 334, 452
Johnson, N. L., 68, 477
Johnson, R., 88, 476
Johnson, R. A., 180, 477
Johnson, W. O., 271, 477
Joiner, B., 495
Jolliffe, I. T., 194, 198, 477
Jones, A. M., 454
Jones, L. K., 465
Jones, M. C., 14, 199, 201, 291, 319, 497
Jones, P. N., 41, 56, 200, 250, 251, 252, 477, 488
Joshi, D. D., 24, 447
Journel, A. G., 425, 477
Jubb, M., 435, 477

Kailath, T., 23, 477
Kalb, G., 358, 504

Kanal, L., 4, 363, 391, 420, 447, 477, 479
Kanazawa, M., 78, 127, 128, 334, 466, 477
Kane, V. E., 122, 451, 452
Kannel, W. B., 235, 256, 502
Kappenman, R. F., 303, 304, 477
Kashyap, R. L., 417, 478
Katre, U. A., 50, 478
Kay, J. W., 22, 30, 414, 438, 440, 441, 442, 443, 448, 478
Kay, R., 271, 272, 280, 281, 478
Kendall, M. G., 86, 390, 478
Kent, J. T., 415, 478
Kerridge, D. F., 256, 263, 264, 265, 279, 460
Kessell, D. L., 362, 363, 467
Kettenring, J. R., 163, 390, 460, 465
Kharin, Y. S., 108, 478
Khatri, C. G., 375, 376, 478
Khattree, R., 377, 478
Kiiveri, H. T., 429, 434, 478
Kim, B. K., 334, 470
Kimura, F., 145, 478
Kinderman, R., 418, 419, 420, 478
Kirkpatrick, S., 436, 478
Kittler, J., 4, 322, 363, 414, 420, 429, 434, 444, 460, 465, 478, 479
Klar, J., 271, 475
Klecka, W. R., 4, 479
Kleinbaum, D. G., 257, 479
Knill-Jones, R. P., 379, 498
Knoke, J. D., 158, 235, 317, 338, 360, 361, 367, 410, 479, 498
Kocherlakota, K., 104, 232, 240, 449, 450, 479
Kocherlakota, S., 104, 232, 450, 479
Koffler, S. L., 313, 334, 479
Kokolakis, G. E., 329, 479
Koller, S., 122, 474
Konishi, S., 341, 479
Kotz, S., 68, 238, 469, 477
Kowalik, W. S., 422, 425, 500
Kowalski, B., 141, 479
Koziol, J. A., 170, 479
Krishnaiah, P. R., 4, 393, 466, 479
Krishnan, T., 49, 50, 478, 479, 480
Kronmal, R. A., 133, 289, 290, 480, 491, 500, 503
Krooden, J. A., 473
Krusińska, E., 227, 391, 480
Krzanowski, W. J., 22, 25, 26, 141, 144, 155, 158, 191, 192, 195, 196, 199, 219, 221, 222, 224, 225, 226, 227, 231, 233, 234, 285, 287, 338, 359, 369, 370, 391, 400, 401, 410, 411, 412, 465, 468, 480, 481, 489
Krzyśko, M., 82, 481
Kshirsagar, A. M., 58, 90, 188, 190, 284, 396, 399, 476, 481
Kudo, A., 67, 481
Kuh, E., 329, 452
Kullback, S., 22, 391, 481
Kupper, L. L., 50, 257, 479, 482
Kurezynski, T. W., 25, 481

Lachenbruch, P. A., 4, 26, 36, 50, 58, 62, 76, 81, 153, 154, 338, 342, 367, 368, 369, 370, 371, 447, 454, 457, 481, 482
Laird, N. M., 39, 220, 460
Lakshmanan, S., 437, 482
Lam, C. F., 391, 482
Lan, K. K. G., 472
Lancaster, H. O., 219, 482
Landgrebe, D. A., 363, 489
Landwehr, J. M., 271, 482
Lanteri, S., 141, 466
Laragh, J. H., 497
Larson, S. F., 410, 494
Lauder, I. J., 266, 312, 482
Lawoko, C. R. O., 32, 425, 445, 446, 482, 483
Layard, M. W. J., 175, 483
Lazarsfeld, P. F., 156, 220, 483
Le Cam, L., 23, 483
Le Cessie, S., 25, 502
Leclercq, S., 199, 462
Lee, J. C., 84, 85, 483
Lehmann, E. L., 28, 115, 483
Leibler, A., 22, 391, 481
Lemeshow, S., 256, 271, 475, 483, 490
Leonard, T., 288, 483
Lesaffre, E., 180, 256, 263, 264, 265, 271, 275, 276, 279, 280, 281, 282, 343, 448, 483
Leung, C. Y., 78, 84, 104, 483, 484
Levy, G. C., 462, 486
Lewis, T., 181, 451
Li, T. J., 391, 466
Liebhart, J., 391, 480
Lienhart, J., 462, 486
Lin, H. E., 334, 484
Lin, S. P., 131, 484
Lindley, D., 18, 392, 484
Linnett, K., 172, 484
Lipton, S., 191, 450
Lissack, T., 24, 363, 484
Little, R. J. A., 43, 225, 484
Little, S., 271, 272, 280, 281, 478
Llewellyn, B., 495
Loftsgaarden, D. O., 309, 484
Logan, T. P., 80, 82, 470, 484
Loh, W.-Y., 324, 328, 329, 330, 331, 332, 484

Loizou, G., 320, 484
Looney, S. W., 170, 484
Louis, C., 490
Louis, T. A., 41, 484
Lowe, V. W., 486
Lubischew, A. A., 242, 406, 484
Luk, A., 320, 323, 484, 485
Lyon, R. J. F., 422, 425, 500

Macdonald, P. D. M., 32, 484
Machin, D., 382, 484
MacLeod, J. E. S., 320, 323, 484, 485
Mahalanobis, P. C., 12, 19, 485
Makov, U. E., 38, 86, 498, 501
Malina, W., 199, 485
Mallet, J. D., 375, 494
Mallows, C. L., 86, 398, 485
Manly, B. F. J., 139, 140, 175, 177, 485
Mantas, J., 4, 485
Marani, S. K., 100, 460
Marco, V. R., 58, 83, 88, 98, 99, 100, 122, 123, 137, 181, 404, 446, 485, 506
Marcus, L. F., 97, 451, 455
Mardia, K. V., 170, 172, 173, 181, 414, 415, 422, 423, 424, 425, 478, 485
Marks, S., 97, 133, 485
Maronna, R. A., 161, 162, 485
Marriott, F. H. C., 158, 235, 256, 317, 503
Marron, J. S., 301, 303, 304, 312, 472, 485
Marshall, A. W., 19, 486
Marshall, R. J., 326, 486
Martin, D. C., 290, 486
Martin, G. G., 67, 486
Massey, F. J., 390, 461
Matheron, G., 422, 486
Matloff, N., 364, 486
Matusita, K., 23, 25, 26, 196, 225, 286, 486
Maybank, S. J., 320, 484
Mayekar, S. M., 100, 460
McCabe, G. P., Jr., 390, 397, 451, 486
McCulloch, R. E., 88, 91, 92, 95, 96, 486
McDonald, L. L., 184, 249, 486, 490
McGee, R. I., 78, 486
McHenry, C. E., 398, 486
McKay, R. J., 177, 391, 393, 394, 396, 397, 398, 399, 400, 406, 407, 408, 486, 487
McKeon, J., 36, 456
McLachlan, G. J., 20, 28, 31, 32, 33, 36, 38, 39, 41, 42, 43, 45, 46, 47, 48, 50, 56, 71, 87, 97, 109, 110, 112, 120, 121, 122, 123, 170, 174, 200, 210, 213, 220, 243, 244, 247, 250, 251, 252, 268, 269, 270, 277, 278, 279, 338, 339, 340, 341, 351, 358, 364, 366, 367, 368, 369, 370, 390, 400, 401, 402, 403, 406, 407, 408, 409, 434, 444, 445, 446, 451, 454, 461, 467, 477, 482, 483, 487, 488, 492
McLeish, D. L., 44, 488
Meilijson, I., 41, 86, 488, 489
Meisel, W., 311, 453
Meister, K. A., 486
Memon, A. Z., 77, 78, 124, 126, 127, 489
Menzefricke, U., 400, 489
Messenger, R. C., 324, 489
Michaelis, J., 122, 474
Michalek, J. E., 36, 489
Mickey, M. R., 58, 338, 342, 368, 369, 482, 497
Mielniczuk, J., 307, 459
Mill, G. M., 300, 501
Miller, A. J., 390, 410, 489
Miller, J. D., 499
Miller, R. G., 47, 199, 206, 207, 208, 209, 494
Min, P. J., 391, 466
Mitchell, A. F. S., 25, 450, 489
Miyake, Y., 478
Mohn, E., 414, 415, 422, 426, 427, 434, 474, 489, 495
Moore, A. W., 400, 475
Moore, D. H., 155, 156, 220, 489
Moore, D. S., 363, 489
Moran, M. A., 30, 56, 67, 73, 74, 102, 108, 185, 314, 489
Morgan, J. N., 324, 489
Morgera, S. D., 99, 489
Morris, C., 131, 463, 499
Mosteller, F., 342, 489
Muller, M. W., 174, 473
Mullis, C. T., 86, 494
Murphy, B. J., 30, 56, 73, 74, 185, 314, 489
Murray, G. D., 3, 38, 70, 299, 300, 410, 411, 489, 490, 501
Murray, L. S., 490, 501
Murray, W., 263, 468
Murthy, V. K., 358, 359, 456
Musket, S. F., 58, 481
Myles, J. P., 320, 490

Nagel, P. J. A., 84, 490
Nandy, S. C., 49, 50, 480
Narendra, P. M., 323, 391, 467
Nealy, C. D., 358, 359, 456
Ng, T.-H., 334, 490
Niemann, H., 323, 490
Novotny, T. J., 249, 490
Nychka, D. W., 497

Odell, P. L., 88, 98, 99, 100, 238, 391, 404, 405, 444, 451, 453, 456, 460, 490, 505, 506
O'Hara, T. F., 160, 278, 490
Okamoto, M., 19, 58, 77, 78, 103, 104, 105, 106, 107, 108, 110, 112, 116, 122, 124, 126, 127, 231, 367, 489, 490, 497
Olbright, W., 252, 254, 464
Oliver, L. H., 51, 490
Olkin, I., 19, 56, 131, 221, 486, 490
Olshen, R. A., 323, 324, 453, 470
O'Neill, T. J., 38, 46, 113, 114, 115, 116, 117, 123, 134, 277, 490, 491
O'Sullivan, F., 131, 491
Ott, J., 289, 290, 491
Owen, A., 138, 139, 414, 427, 431, 435, 491

Page, E., 258, 491
Page, J. T., 367, 369, 491
Parzen, E., 294, 491
Pascual, R., 199, 453
Patrick, E. A., 4, 309, 491
Pearson, K., 12, 491
Peck, R., 145, 491
Peers, H. W., 381, 491
Pelto, C. R., 309, 491
Penfield, D. A., 313, 334, 479
Penrose, L. S., 58, 491
Percy, M. E., 80, 450
Perlman, M. D., 131, 177, 396, 459, 484, 491
Perreault, W. D., 345, 459
Peters, B. C., 88, 470
Peters, C., 100, 492
Peterson, R. P., 12, 14, 475
Philips, P. R., 258, 449
Pickard, D. K., 420, 492
Pierce, D. A., 219, 454
Pikelis, V., 102, 137, 493
Pincus, M., 436, 492
Please, N. W., 58, 451
Plomteux, G., 279, 492
Porteous, B. T., 437, 470
Potthoff, R. F., 85, 492
Poulsen, R. S., 490
Powell, J. L., 278, 448
Prakasa Rao, B. L. S., 291, 292, 492
Pregibon, D., 271, 482, 492
Prentice, R. L., 262, 263, 492
Press, S. J., 278, 492
Priest, R. G., 465
Pruitt, R., 364, 486
Purcell, E., 311, 453
Pyke, R., 262, 263, 492

Qian, W., 418, 442, 492
Queiros, C. E., 410, 492
Quesenberry, C. P., 19, 309, 333, 484, 492
Quinn, B. G., 48, 492

Raatgever, J. W., 496
Raath, E. L., 122, 138, 155, 170, 464, 473
Rabinowitz, M., 391, 469
Radcliffe, J., 190, 492
Radhakrishnan, R., 183, 492
Radley, D., 192, 481
Rae, D. S., 494
Raiffa, H., 10, 492
Ramberg, J. S., 493
Ramirez, D. E., 180, 181, 453
Ramsey, P. H., 493
Randles, R. H., 20, 145, 164, 332, 334, 454, 490, 493
Ranger, N., 181, 457
Rao, C. R., 8, 12, 14, 22, 24, 25, 66, 67, 74, 75, 89, 375, 376, 391, 393, 394, 395, 396, 400, 454, 478, 493
Rao, P. S. R. S., 345, 493
Rasmussen, S. E., 396, 473
Raudys, S. J., 102, 108, 137, 392, 493
Raveh, A., 335, 336, 493
Rawlings, R. R., 50, 76, 155, 493, 494
Ray, S., 23, 24, 494
Rayens, W. S., 145, 149, 151, 470, 494
Rayner, J. C. W., 139, 140, 175, 177, 485
Read, T. R. C., 22, 459, 494
Reaven, G. M., 47, 199, 206, 207, 208, 209, 494
Reed, I. S., 375, 494
Remme, J., 298, 302, 310, 313, 471, 494
Rencher, A. C., 410, 494
Rényi, A., 22, 494
Revo, L. T., 153, 482
Richardson, S. C., 268, 269, 270, 449, 494
Ridout, M. S., 391, 494
Riffenburgh, R. H., 96, 457, 494
Rigby, R. A., 30, 387, 494
Rijal, O., 30, 384, 459
Rinsma, W. J., 462
Ripley, B. D., 413, 414, 416, 431, 494
Rissanen, J., 391, 494
Robbins, H., 13, 495
Roberts, R. A., 86, 494
Roberts, S. J., 391, 494
Robson, D. S., 181, 452
Rodriguez, A. F., 145, 495
Rogan, W. J., 33, 495
Rogers, W. H., 322, 495
Romano, J. P., 346, 461

Ronchetti, E. M., 472
Rosenbaum, P. R., 495
Rosenblatt, M., 292, 294, 495
Rosenfeld, A., 272, 438, 495
Rousseeuw, P. J., 472
Roy, S. N., 85, 492
Rubin, D. B., 39, 43, 220, 272, 359, 460, 484, 495
Rubinfeld, D. L., 329, 473
Rudemo, M., 303, 495
Ruppert, D., 271, 499
Russell, J. S., 400, 475
Ruymgaart, F. H., 312, 455
Ryan, B., 495
Ryan, T., 62, 495
Ryback, R. S., 494

Saebo, H. V., 414, 495
Salter, K. G., 427, 504
Samiuddin, M., 70, 464
Sammon, J. W., 196, 465, 495
Samuel, E., 13, 495
Sánchez, J. M. P., 359, 495
Sanghvi, L. D., 25, 450, 495
Santner, T. J., 264, 265, 462, 495
Sayre, J. W., 108, 111, 113, 118, 495
Schaafsma, W., 102, 380, 382, 383, 389, 408, 448, 495, 496
Schaefer, J. R., 294, 473
Schaefer, R. L., 264, 268, 496
Schechtman, E., 460
Schenker, N., 372, 496
Schervish, M. J., 88, 94, 95, 96, 111, 112, 341, 351, 368, 496
Schluchter, M. D., 43, 225, 453, 484
Schmid, M. J., 141, 496
Schmidhammer, J., 393, 466
Schmitz, P. I. M., 158, 217, 236, 298, 317, 318, 496
Schneider, B., 291, 504
Schork, M. A., 174, 496
Schork, N. J., 174, 496
Schott, J. R., 141, 496
Schroeder, A., 309, 466
Schucany, W. R., 344, 470
Schumway, R. H., 82, 83, 496
Schuster, E. F., 302, 496
Schwartz, S. C., 289, 496
Schwemer, G. T., 58, 365, 497
Sclove, S. L., 423, 440, 497
Scott, A. J., 262, 497
Sealey, J. E., 243, 497
Seber, G. A. F., 89, 189, 191, 497
Sedransk, N., 367, 497
Seheult, A. H., 437, 470
Sellian, J. B., 131, 490
Shannon, C. E., 22, 497
Sharpe, P. M., 358, 502
Shi, S., 346, 470, 474
Shoemaker, A. C., 271, 482
Short, R. D., 322, 497
Shustek, J., 323, 466
Sibson, R., 22, 199, 201, 477, 497
Silverman, B. W., 14, 291, 292, 300, 301, 308, 310, 311, 319, 351, 439, 497
Silvey, S. D., 24, 448
Simons, G., 86, 497
Simpson, C., 133, 305, 313, 315, 392, 502
Singh, K., 373, 447
Singh, M., 167, 501
Sinha, B. K., 396, 497
Siotani, M., 64, 78, 102, 103, 111, 113, 122, 124, 125, 127, 497, 498
Sitgreaves, R., 102, 103, 453, 498, 500
Skene, A. M., 220, 498, 501
Small, C. G., 45, 488
Small, N. J. H., 170, 458
Smeets, J. P., 258, 448
Smidt, R. K., 486
Smith, A. F. M., 38, 86, 133, 498, 501
Smith, C. A. B., 55, 339, 498
Smith, J. C., 367, 475
Smith, W. B., 67, 462
Smyth, G. K., 254, 505
Snapinn, S. M., 360, 361, 367, 410, 498
Sneeringer, C., 153, 482
Snell, E. J., 256, 267, 418, 419, 420, 458, 478
Son, M. S., 83, 454
Sonquist, J. A., 324, 489
Sorum, M. J., 367, 498
Specht, D. F., 315, 498
Spiegelhalter, D. J., 21, 133, 379, 498, 501
Spjøtvoll, E., 398, 498
Srinivasan, C., 131, 461, 484
Srivastava, J. N., 43, 498
Srivastava, M. S., 67, 78, 84, 86, 108, 183, 483, 499
Stablein, D. M., 275, 499
Stahel, W. A., 472
Stam, A., 155, 476
Steerneman, A. G. M., 391, 499
Stefanski, L. A., 271, 499
Stegun, I. A., 57, 447
Stein, C., 67, 131, 476, 499
Stephenson, R., 504
Sterk, P. J., 473

Stirling, W. C., 13, 499
Stocks, P., 58, 499
Stoller, D. S., 96, 499
Stone, C. J., 303, 319, 320, 321, 453, 499
Stone, M., 288, 302, 342, 344, 499
Storvik, G., 414, 426, 489
Strauss, D. J., 419, 499
Streit, F., 108, 240, 463, 499
Stuetzle, W., 309, 466
Stürzebecher, E., 358, 504
Subrahmaniam, Kathleen, 75, 395, 499, 500
Subrahmaniam, Kocherlakota, 75, 153, 154, 395, 456, 499, 500
Sun, Y. N., 375, 376, 478
Sutherland, M., 288, 500
Sutradhar, B. C., 241, 242, 500
Swindlehurst, A. L., 13, 499
Switzer, P., 138, 201, 413, 422, 424, 425, 433, 500, 505
Sycheva, N. M., 408, 506
Symons, M. J., 208, 500

Takashina, K., 478
Tan, W. Y., 122, 166, 451, 501
Taneja, I. J., 22, 67, 500
Taneja, V. S., 462
Tapia, R. A., 291, 500
Tarter, M., 289, 480, 500
Tate, R. F., 221, 490
Tautu, P., 2, 503
Teasdale, G. M., 490
Teichroew, D., 103, 500
ten Krooden, J. A., 174, 473
Thompson, J. R., 291, 500
Tiao, G. C., 228, 500
Tibshirani, R. J., 258, 271, 346, 348, 463, 473, 500
Tiku, M. L., 165, 166, 167, 223, 232, 450, 500, 501
Tippett, P. A., 484
Titterington, D. M., 3, 38, 48, 49, 50, 86, 131, 157, 158, 174, 217, 220, 288, 290, 297, 299, 300, 302, 316, 317, 323, 414, 418, 436, 438, 440, 441, 442, 443, 453, 472, 478, 485, 490, 492, 501
Todeschini, R., 320, 501
Toronto, A. F., 504
Toussaint, G. T., 26, 338, 342, 358, 490, 501, 502
Trampisch, H. J., 220, 502, 503
Tripathi, R. C., 36, 489
Truett, J., 235, 256, 502
Tsiatis, A. A., 271, 502
Tsuruoka, S., 478
Tu, C., 232, 502
Tu, D. S., 368, 456
Tubbs, J. D., 98, 99, 180, 444, 445, 462, 502
Tukey, J. W., 201, 466
Tunnicliffe Wilson, G., 182, 448
Turner, D. W., 58, 83, 108, 122, 123, 137, 181, 446, 485, 505, 506
Tutz, G. E., 305, 360, 502

Urbakh, V. Yu., 408, 409, 502

Valdes, P., 199, 453
Van Campenhout, J. M., 392, 502
van den Broeck, K., 294, 471
van der Burgt, A. T., 19, 471
van Houwelingen, J. C., 256, 502
Vanichsetakul, N., 324, 328, 329, 330, 331, 332, 484
Van Ness, J. W., 133, 145, 305, 313, 315, 392, 491, 502
Van Ryzin, J., 4, 14, 288, 502, 503, 504
van't Klooster, H. A., 141, 461, 462
van Vark, G. N., 102, 380, 408, 496
van Zomeren, B., 473
Varady, P. D., 367, 462
Vaughan, E. D., 497
Veasey, L. G., 504
Vecchi, M. P., 436, 478
Venetoulias, A., 413, 418, 419, 430, 503
Verter, J. I., 278, 472
Victor, N., 503
Villalobos, M. A., 309, 503
Vlachonikolis, I. G., 158, 227, 228, 229, 231, 235, 236, 317, 503
von Mises, R., 8, 10, 503

Wagner, G., 2, 503
Wagner, H., 410, 473
Wagner, T. J., 321, 322, 461, 495, 503
Wahba, G., 309, 503
Wahl, P. W., 133, 503
Wakaki, H., 123, 135, 503
Wald, A., 8, 12, 14, 62, 86, 102, 503
Walker, H. F., 88, 470
Walker, P. H., 251, 503
Walker, S. H., 256, 503
Wallace, D. L., 342, 489
Waller, W. G., 392, 476, 503
Walter, S. D., 268, 504
Wand, M. P., 305, 306, 307, 308, 472
Wang, M.-C., 288, 346, 353, 503, 504
Wang, R.-H., 103, 124, 127, 497, 498
Warner, H. R., 217, 504
Warner, J. L., 172, 178, 450

Weber, J., 443, 504
Wedel, H., 472
Wegman, E. J., 291, 504
Wehrli, F., 462
Weiner, J. M., 390, 396, 504
Weiss, S. H., 35, 504
Welch, B. L., 8, 504
Welch, J. R., 427, 504
Welsch, R. E., 329, 452
Wernecke, K.-D., 219, 358, 504
Wertz, W., 291, 504
Whitsitt, S. J., 363, 489
Whittle, P., 289, 504
Wichern, D. W., 180, 477
Wicksley, R. A. J., 163, 473
Wiens, D. P., 161, 457, 504
Wilcox, J. B., 364, 475
Wild, C. J., 262, 497
Wilks, A. R., 62, 199, 294, 452
Wilks, S. S., 504
Willems, J. L., 343, 483
Williams, E. J., 189, 190, 504, 505
Williams, J. S., 367, 454
Wilson, E. B., 163, 505
Wilson, J. D., 497
Wilson, S., 278, 492
Wilson, S. R., 211, 214, 505
Wisenbaker, J. M., 367, 475
Wojciechowski, T., 13, 291, 320, 505
Wolbler, U., 2, 503
Wold, S., 141, 144, 479, 505
Wolde-Tsadik, G., 402, 505
Wolf, E., 285, 469
Wolfe, J. H., 212, 505
Worlund, D. D., 32, 505
Worton, B. J., 310, 311, 312, 313, 314, 505
Wright, R. M., 201, 505
Wrobel, B. A., 43, 467
Wu, C. F. J., 41, 505
Wu, L. D., 302, 457
Wyman, F. J., 108, 505
Wyrwoll, K.-H., 254, 505

Yarborough, D. A., 86, 505
Yerushalmy, J., 156, 505
Young, D. M., 58, 83, 88, 98, 99, 100, 108, 122, 123, 137, 145, 181, 196, 404, 405, 446, 485, 491, 502, 505, 506,
Young, G. A., 351, 497
Young, T. Y., 4, 506
Yu, M. C., 402, 505
Yu, T. S., 414, 466, 506

Zaatar, M. K., 43, 498
Zangwill, W., 335, 506
Zaslavski, A. E., 408, 506
Zellner, A., 228, 500
Zentgraf, R., 219, 220, 503, 506
Zhang, L., 404, 506
Zheng, Z., 161, 504
Zielezny, M., 390, 506
Zucker, S. W., 438, 443, 476, 495

Subject Index

Affinity, measures of, 23
Akaike's information criterion, *see* Selection of feature variables, criteria for selection, Akaike's information criterion
Allocation rates, *see* Error rates
Allocation rules, 6–15
  admissibility of, 10
  Bayes rule, 7
  decision-theoretic approach, 7–9
  estimation of, 12. *See also* Likelihood-based approaches to discrimination
    Bayes risk consistency, 14
    diagnostic paradigm, 13
    empirical Bayes approach, 13
    estimative method, 14
    minimum-distance approach, 15, 137, 225–227, 285–287
    nonparametric approach, 13, 283–284
    parametric approach, 13–15
      nonnormal models, 216–254
      normal models, 52–87
    plug-in method, *see* Allocation rules, estimation of, estimative method
    predictive method, 29–31
    regularization approach, 129–152
    sampling approach, 13
    semiparametric approach, 255
  formulation of, 6
  likelihood ratio rule, 65–67
  minimax criterion, 10
  optimal rule, *see* Allocation rules, Bayes rule
Atypicality index, *see* Typicality index

BASIC, algorithms in, 62
Bayesian approach to discrimination, *see* Likelihood-based approaches to discrimination, Bayesian approach
Bayes risk consistency, *see* Allocation rules, estimation of, Bayes risk consistency
Bias correction of allocation rules, *see* Discrimination via normal models, heteroscedastic model, plug-in sample NQDR, bias correction of; Discrimination via normal models, homoscedastic model, plug-in sample NLDR, bias correction of; Logistic discrimination, maximum likelihood estimates of logistic regression coefficients, bias correction of
BMDP statistical package, 62, 398
Bootstrap, *see* Error-rate estimation, apparent error rate, bias correction of, bootstrap method; Error-rate estimation, bootstrap-based estimators; Number of groups, testing for, resampling approach, bootstrap method; Posterior probabilities of group membership, interval estimation of, bootstrap approach

Canonical variate analysis:
  known group parameters, 88–90
    discrimination in terms of canonical variates, 90–91
  unknown group parameters, 185–187
    alternative formulation, 193–194
    confidence regions, 190–193
      bootstrap approach, 192–193
      jackknife approach, 192
    example, 192–193
    heteroscedastic data, 194–196
    tests for number of canonical variates, 187–190
CART method of discrimination, *see* Nonparametric discrimination, tree-structured allocation rules, CART method

Classifiers, *see* Allocation rules
Cluster analysis, 169. *See also* Likelihood-based approaches to discrimination, maximum likelihood estimation, absence of classified training data
Common principal-component model, *see* Variants of normal theory-based discrimination, common principal-component model
Confusion matrix, 32
Consistency of allocation rules, *see* Allocation rules, estimation of, Bayes risk consistency
Constrained allocation, *see* Discrimination via normal models, constrained allocation
Correct allocation rates, *see* Error rates
Costs of misallocation, 7–9
Covariance-adjusted discrimination, 74–78
  asymptotic error rates, 77–78
  covariance-adjusted sample NLDF, 77
  formulation of, 74–76
Cross-validation, *see* Error-rate estimation, apparent entry rate, bias correction of, cross-validation method

DASCO method of discrimination, *see* Variants of normal theory-based discrimination, DASCO method of discrimination
Data:
  analyte data on liver diseases, 279–282
  ballisto-cardiographical data for diagnosis of stenocardia, 408–409
  Boston housing data, 329–332
  on Conn's syndrome, 385–387
  on diabetes, 206–208
  Fisher's *Iris* data, 192–193, 211–215
  on fleabeetles, 406–408
  on hemophilia A carriers, 201
  on keratoconjunctivitis sica, 307–308
  mass-size data on soil profiles, 251–254
  on pixels, 432–433
  psychological data on normals and psychotics, 55
  on renal venous renin ratios, 243–244
  on sensory characteristics of wines, 151–152
  simulated data for bias of apparent error rate of a logistic regression and its bootstrap correction, 274
  simulated data for bias of overall apparent error rate and of some bias-corrected versions, 353
  simulated data for bivariate kernel density estimates, 311–312
  simulated data for regularized discriminant analysis, 149–151
  simulated data for univariate kernel density estimates, 292–294
  simulated data for variable selection via apparent rate, 410
Diagnostic tests:
  sensitivity of, 6
  specificity of, 6
Dimension reduction, *See also* Canonical variate analysis; Principal components; Projections, linear
  mean-variance plot, 196–197
  projection pursuit, unclassified feature data, 200–201
Discriminant analysis:
  applications of, 2–4
  definition of, 1–2
Discriminant rules, *see* Allocation rules
Discrimination via normal models:
  constrained allocation, 19–20, 118
    confidence bounds on one conditional error, 120–121
    confidence bounds on both conditional errors, 121–122
    constraint on one unconditional error, 118–120
  heteroscedastic model, 52
    comparison of estimative and predictive methods, 70
    equal-mean rule, 57–58
    estimative method, 54
    likelihood ratio rule, 65
    normal-based quadratic discriminant rule (NQDR), 53
    plug-in sample NQDR, 54
      bias correction of, 56–57
    predictive method, 67–68
    semi-Bayesian approach, 69
  homoscedastic model, 59
    comparison of estimative and predictive methods, 71–74
      biases of posterior log odds, 72–73
    estimative method, 61
    likelihood ratio rule, 66
    normal-based linear discriminant rule (NLDR), 60
    optimal error rates, 60–61
    plug-in sample NLDR, 61
      bias correction of, 64
    predictive method, 69
    sample NLDF, 62
    semi-Bayesian approach, 69
    W-statistic, *see* Discrimination via normal

models, homoscedastic model, predictive method, sample NLDF
linear *vs*. quadratic normal-based discriminant rules:
comparison of plug-in sample versions of NLDR and NQDR, 132–133
loss of efficiency, 135–137
theoretical results, 134–135
partially classified data, examples, 201, 204–206, 208–215
Distances between groups, measures of:
Bhattacharyya, 23
Chernoff, 23
Hellinger, *see* Distances between groups, measures of, Matusita
Jeffreys, 22
Kullback–Leibler, 22
Mahalanobis, 19, 25
Matusita, 23
Shannon entropy, 22
Distributional results for normal models:
asymptotic results for distribution of:
group-specific conditional error rates of sample NLDR, 112–114
overall conditional error rate of a plug-in sample rule, 114–116
overall conditional error rate of sample NLDR, 116–118
sample NLDF (W-statistic), 101–107
Studentized version, 119
sample NQDF, 122–123
Z-statistic, 123–125
Studentized version, 127–128
comparison of rules based on Z- and W-statistics, 126–127
moments of conditional error rates of sample NLDR:
means, 107–108, 111–112
variances, 108–111

EM algorithm, *see* Likelihood-based approaches to discrimination, maximum likelihood estimation, partially classified training data, EM algorithm
Equal group-covariance matrices, *see* Discrimination via normal models, homoscedastic model
Error-rate estimation, 337–339
apparent error rate, 339–340
bias of, 340
bias correction of:
bootstrap method, 346–348
nonparametric version, 348–350
other uses of, 349–350
parametric version, 350–351
relationship with some other methods of bias correction, 351–353
cross-validation method, 341–344
jackknife method, 344–346
rule selection via, 370–373
smoothing of, 360–361
bootstrap-based estimators:
double bootstrap, 353–354
randomized bootstrap, 354
the 0.623 estimator, 354–360
confidence intervals for error rates, 370–373
linear combinations of apparent and cross-validated rates, 357–359
nonparametric error rate estimators, 339–346
parametric error rate estimators, 366–370
posterior probability-based estimators, 361–366
signal to noise ratio, estimation of, 375–377
Error rates, 6
estimation of, *see* Error-rate estimation
relevance of, 18
types of:
actual, 17
conditional, 17
expected, 17
optimal, 17–18, 60
unconditional, 17–18
Estimative method of discrimination, *see* Allocation rules, estimation of, estimative method
Expert statistical systems, 67

FACT method of discrimination, *see* Nonparametric discrimination, tree-structured allocation rules, FACT method
Feature variables:
definition of, 1
selection of, 389–412
Fisher's linear regression approach, 63

GENSTAT statistical package, 62
GLIM statistical package, 62
Group-prior probabilities, 15
estimation of, 31–35
discriminant analysis approach, 31–32
maximum likelihood method, 32, 39–40
method of moments, 33
unavailability of, 9–10
Group proportions, *see* Group-prior probabilities
Growth curves, allocation of, 84–85

Image analysis, *see* Statistical image analysis
IMSL statistical package, 62, 398
Independence models, *see* Parametric discrimination via nonnormal models, models for discrete feature data, independence

Jackknife, *see* Canonical variate analysis, unknown group parameters, confidence regions, jackknife approach; Error-rate estimation, apparent error rate, bias correction of, jackknife method

Kernel discriminant analysis, 284
  kernel method of density estimation, 291–308
    alternatives to fixed kernel density estimates, 308–309
      adaptive method, 310–312
      nearest neighbor method, 309–310
    choice of smoothing parameters, criteria for:
      joint selection approach, 305–307
        example, 307–308
      separate selection for each group, 300
        criteria:
        least squares, 308
        minimization of MISE, 301
          multivariate binary data, 304–305
        moment cross-validation, 303–304
        pseudo-likelihood cross-validation, 302–303
  comparative performance of kernel-based rules, 312–313
  definition of, 291–294
    continuous data, 313–314
    discrete data, 316–317
    mixed data, 317–319
  kernels for:
    categorical data with an infinite number of cells, 298
    continuous data, 291–292
      product kernels, 295–296
        loss of generality under canonical transformations, 314–316
    incomplete data, 299–300
    mixed data, 298–299
    multivariate binary data, 296
    ordered categorical data, 297–298
    unordered categorical data, 297
  large sample properties, 294–295

Lancaster models, *see* Parametric discrimination via nonnormal models, models for discrete feature data, Lancaster
Latent class models, *see* Parametric discrimination via nonnormal models, models for discrete feature data, latent class
Likelihood-based approaches to discrimination:
  Bayesian approach, 29–31
  classification approach, 45–46
  maximum likelihood estimation:
    absence of classified training data, 46–48
    of group parameters, 27–29
    of group proportions, 31–33, 40
    misclassified training data, 49–50
    mixture group-conditional densities, 50–51
    nonrandom partially classified training data, 43–45
    partially classified training data:
      EM algorithm, 39, 86–87
      information matrix, 41–42
        approximations, 41–42
        expected, 41
        observed, 41
  updating a discriminant rule, 37–38
    example, 204–206
Linear discriminant function, *see* Discrimination via normal models, homoscedastic model
Location model, *see* Parametric discrimination via nonnormal models, models for mixed feature data, location model
Logistic discrimination:
  apparent error rate of a logistic regression, 272–275
  applicability of logistic model, 256–258
  β-confidence discrimination, 275–276
  example, 279–282
  fit of model, assessment of, 255–256
  formulation of, 255–256
  logistic *vs.* normal-based linear discriminant analysis, 276–279
    relative efficiency:
      under nonnormality, 278–279
      under normality, 276–278
  maximum likelihood estimates of logistic regression coefficients:
    asymptotic bias of:
      for mixture sampling, 267–269
      for separate sampling, 261–263
    bias correction of, 266–267

computation of, 263–264
for mixture sampling, 260–261
for separate sampling, 261–263
for x-conditional sampling, 259–260
existence of, 264–265
minimum chi-squared estimate of logistic regression coefficients, 279
predictive version, 265–266
quadratic version, 258–259
Log-linear models, *see* Parametric discrimination via nonnormal models, models for discrete feature data, log-linear

Mahalanobis distance, *see* Distances between groups, measures of, Mahalanobis
Minimax rule, *see* Allocation rules, minimax criterion
Minimum-distance rules, 15, 137, 225–227, 285–287. *See also* Allocation rules, estimation of, minimum-distance approach
Minitab statistical package, 62
Mixing proportions, *see* Group-prior probabilities
Model fit, assessment of, 16
Models for discrete feature data, *see* Parametric discrimination via nonnormal models, models for discrete feature data
Models for mixed feature data, *see* Parametric discrimination via nonnormal models, models for mixed feature data
Multinomial models, *see* Parametric discrimination via nonnormal models, models for discrete feature data, multinomial
Multivariate normality, assessment of, 169
absence of classified data, 174
Anderson–Darling statistic, 171
data-based transformations, 178
Box–Cox transformation, 178–181
Hawkins' method, 170–172
examples:
diabetes data, 209
hemophilia data, 201–203
likelihood ratio test, 174–175
hierarchical partitions of, 175–178
other tests of multivariate normality, 172–174

Nonmetric approach, *see* Nonparametric discrimination, nonmetric approach
Nonnormal discrimination, *see* Parametric discrimination via nonnormal models; Nonparametric discrimination
Nonnormal models for continuous feature data, *see* Parametric discrimination via nonnormal models, models for continuous feature data
Nonparametric discrimination, *see also* Kernel discriminant analysis
multinomial-based discrimination, 216–217, 283–284
error rates, 285
minimum-distance rule, 285–287
smoothing of multinomial probabilities, 217–218
convex method, 287–288
orthogonal series method, 288–291
nearest neighbor rules, 319
k-NN rules:
asymptotic results, 320–322
choice of k, 323
definition of, 319–320
finite sample size behavior, 322–323
nonmetric approach, 335–336
rank methods, 332–334
for repeated measurements, 334
tree-structured allocation rules, 323–332
CART method, 324–332
FACT method, 324, 328–332
Nonparametric estimation of group-conditional distributions, *see* Kernel discriminant analysis; Nonparametric discrimination
Normal discrimination, *see* Discrimination via normal models
Normality, *see* Multivariate normality, assessment of
Notation, basic, 4–6
statistical image analysis, 415–416
Number of groups, testing for, 47
likelihood ratio statistic, 48
resampling approach:
bootstrap method, 48
examples, 212–215

Odds ratio, posterior log odds, 255, 257, 266, 275. *See also* Posterior probabilities of group membership, distribution of, sample log odds
Outliers, detection of, *see* Robust estimation of group parameters; Typicality index

Parametric discrimination via nonnormal models:
case study, 243–247
example, 249–254

Parametric discrimination (*Continued*)
models for continuous feature data, 238–254
exponential, 239
inverse normal, 240–241
multivariate-t, 68, 241–242
normal mixtures, 50–51, 238, 244–248
θ-generalized normal, 238–239
models for discrete feature data:
independence, 217–218
Lancaster, 219–220
latent class, 220
log-linear, 218–219
multinomial, 216–217, 283–284
models for mixed feature data:
adjustments to sample NLDR:
augmenting the sample NLDF, 234–237
linear transformations, 232–234
location model, 220–232
error rates of location model-based rules:
conditional, 229–230
optimal, 229–230
unconditional, 230–232
maximum likelihood estimation:
full saturated version, 223–224
reduced version, 224–225
minimum-distance rules, 225–227
optimal rule, 222–223
predictive approach, 227–229
Pattern recognition, *see* Statistical pattern recognition, definition of
Plug-in sample rules, *see* Allocation rules, estimation of, estimative method; Discrimination via normal models, heteroscedastic model, plug-in sample NQDR; Discrimination via normal models, homoscedastic model, plug-in sample NLDR
Posterior probabilities of group membership:
definition of, 5
distribution of:
sample log odds ($g = 2$ heteroscedastic groups), 381–382
sample log odds ($g = 2$ homoscedastic groups), 379–381
sample posterior probabilities ($g > 2$ groups), 382–383
interval estimation of:
Bayesian approach, 387
bootstrap approach, 387–388
profile likelihood approach, 384–385
illustration of, 385–387
point estimation of, 20, 378–379
Bayesian approach, 29–31
semi-Bayesian approach, 30–31
reliability of, 379
Predictive method of discrimination, *see* Allocation rules, estimation of, predictive method
Prevalence rates, *see* Group-prior probabilities
Principal components:
classified feature data, 197–199
unclassified feature data, 199–200
Prior probabilities, *see* Group-prior probabilities
Projections, linear, *see also* Canonical variate analysis; Dimension reduction; Principal components
heteroscedastic data:
best linear rule ($g = 2$ groups), 96–97
best quadratic rule, 97–99
homoscedastic data:
allocatory aspects, 92–93
allocatory *vs.* separatory solution, 93–95
best linear rule in terms of error rate, 92–93
computation of, 95–96
separatory measures, 91
Proportional group-covariance matrices, *see* Variants of normal theory-based discrimination, proportional group-covariance matrices
P-STAT statistical package, 62

Quadratic discriminant function, *see* Discrimination via normal models, heteroscedastic model, normal-based quadratic discriminant rule (NQDR)

Rank methods, *see* Nonparametric discrimination, rank methods
Regression approach, *see* Fisher's linear regression approach
Regularized discriminant analysis:
assessment of regularization parameters, 146–149
effectiveness of, 149–151
examples:
real, 151–152
simulated, 150–151
formulation, 144–146
Repeated measurements, discrimination with, 80–81
Risk of misallocation, 7–8. *See also* Error rates
Robust estimation of group parameters:
M-estimates, 161–164
MML estimates, 165–167
use of rank-cutoff point, 164–165

Robustness of sample NLDR and NQDR:
continuous feature data, 152–155
discrete feature data, 155–158
mixed feature data, 158–161

Sampling schemes:
mixture sampling, 11
prospective sampling, 11
retrospective sampling, 12
separate sampling, 11
SAS statistical package, 62, 398
Screening tests, *see* Diagnostic tests
Selection of feature variables, 389–391
ALLOC algorithm, 294, 398, 405
criteria for selection:
additional information criterion:
multiple groups, 392–394
two-group case, 394–396
Akaike's information criterion, 403–404
all-subsets approach, 397–398
canonical variate analysis, 396
error rate-based criteria, 400–403, 404–405
F-test, 395
vs. error rate-based selections, 406
examples, 406–409
stepwise procedures, 398–400
peaking phenomenon, 391–392
selection bias, 410–411
reduction of, 411–412
Sensitivity, *see* Diagnostic tests
Separatory measures, *see* Distances between groups, measures of; Projections, linear, homoscedastic data, separatory measures
Sequential discrimination, 85–86
SIMCA method of discrimination, *see* Variants of normal theory-based discrimination, SIMCA method of discrimination
Specificity, *see* Diagnostic tests
SPSS-X statistical package, 62, 398
S statistical package, 62, 398
Statistical image analysis, 413–414
correlated training data, 443
effect of autocorrelation on sample NLDR, 444–446
effect of equicorrelation on sample NLDR, 443–444
ICM algorithm:
definition of, 428–429
estimation of parameters, 429–431
examples, 431–433
underlying assumptions, 428
modifications, 433–435
image processing, 413–414
image restoration, 416
image segmentation, 416–417
contextual methods, 415, 422–438
incomplete-data formulation, 438
EM framework, 438
E-step, contextual case, 440–442
E-step, noncontextual case, 439–440
link with ICM algorithm, 442
link with some other methods of segmentation, 442–443
noncontextual methods, 441–442
individual allocation of pixels, 425–428
Markov mesh models, 420–421
Markov random fields:
definitions, 417–418
spatially symmetric models, 418–420
MAP (Maximum *A Posteriori*) estimate, 435–436
comparison with ICM algorithm, 437–438
computation of, 436–437
binary images, 437–438
posterior distribution of image, global maximization of, *see* Statistical image analysis, MAP (Maximum *A Posteriori*) estimate
remote sensing, 414–415
smoothing methods, 422–425
estimation of parameters, 425
presmoothing approaches, 423–425
reduction in error rate, 424–425
spatial assumptions, 422–423
Statistical pattern recognition, definition of, 4
SYSTAT statistical package, 62

Time series:
discrimination between, 82–83
discrimination within, 83–84
Training data:
correlated data, 443. *See also* Statistical image analysis
misclassified data:
nonrandom, 36
random, 36, 49–50
missing data 43,
partially classified data, 37–39
Transformations, *see* Multivariate normality, assessment of, data-based transformations
Typicality index, 181, 185
assessment of:
classified feature data, 181–183
predictive approach, 185
unclassified feature data, 183

Typicality index (*Continued*)
examples:
diabetes data, 209–210
hemophilia data, 203–204
viewed as a *P*-value, 183–184

Unequal (unrestricted) group-covariance matrices, *see* Discrimination via normal models, heteroscedastic model
Updating a discriminant rule, *see* Likelihood-based approaches to discrimination, updating a discriminant rule

Variants of normal theory-based discrimination, 129. *See also* Regularized discriminant analysis
common principal-component model, 140–141
DASCO method of discrimination, 143–144
equal group-correlation matrices, 139–140
equal spherical group-covariance matrices, 137–138
proportional group-covariance matrices, 138–139
regularization in quadratic discrimination, 130–131
SIMCA method of discrimination, 141–144

W-statistic, *see* Discrimination via normal models, homoscedastic model, sample NLDF

Z-statistic, 66. *See also* Distributional results for normal models, asymptotic results for distribution of, *Z*-statistic